PILOT'S HANDBOOK OF AERONAUTICAL KNOWLEDGE

2023

FAA-H-8083-25C

U.S. Department of Transportation
FEDERAL AVIATION ADMINISTRATION
Flight Standards Service

Skyhorse Publishing

First published in 2023
First Skyhorse Publishing edition 2023

All rights to any and all materials in copyright owned by the publisher are strictly reserved by the publisher. All inquiries should be addressed to Skyhorse Publishing, 307 West 36th Street, 11th Floor, New York, NY 10018.

Skyhorse Publishing books may be purchased in bulk at special discounts for sales promotion, corporate gifts, fund-raising, or educational purposes. Special editions can also be created to specifications. For details, contact the Special Sales Department, Skyhorse Publishing, 307 West 36th Street, 11th Floor, New York, NY 10018 or info@skyhorsepublishing.com.

Skyhorse® and Skyhorse Publishing® are registered trademarks of Skyhorse Publishing, Inc.®, a Delaware corporation.

Visit our website at www.skyhorsepublishing.com.
Please follow our publisher Tony Lyons on Instagram @tonylyonsisuncertain

10 9 8 7 6 5 4

Library of Congress Cataloging-in-Publication Data is available on file.

Cover design by Federal Aviation Administration

ISBN: 978-1-5107-7987-7
eBook ISBN: 978-1-5107-7988-4

Printed in China

Preface

This handbook provides the basic knowledge that is essential for pilots. It introduces pilots to the broad spectrum of knowledge that will be needed as they progress in their pilot training. Except for the Code of Federal Regulations pertinent to civil aviation, most of the knowledge areas applicable to pilot certification are presented. This handbook is useful to beginning pilots, as well as those pursuing more advanced pilot certificates.

Occasionally the word "must" or similar language is used where the desired action is deemed critical. The use of such language is not intended to add to, interpret, or relieve a duty imposed by Title 14 of the Code of Federal Regulations (14 CFR).

It is essential for persons using this handbook to become familiar with and apply the pertinent parts of 14 CFR and the Aeronautical Information Manual (AIM). The AIM is available online at www.faa.gov. The current Flight Standards Service airman training and testing material and learning statements for all airman certificates and ratings can be obtained from www.faa.gov.

This handbook supersedes FAA-H-8083-25B, Pilot's Handbook of Aeronautical Knowledge, dated 2016; the Pilot's Handbook of Aeronautical Knowledge Addendum A, dated February 2021; the Pilot's Handbook of Aeronautical Knowledge Addendum B, dated January 2022; and the Pilot's Handbook of Aeronautical Knowledge Addendum C, dated March 2023.

This handbook is available for download, in PDF format, from www.faa.gov.

This handbook is published by the United States Department of Transportation, Federal Aviation Administration, Airman Testing Standards Branch, AFS-630, P.O. Box 25082, Oklahoma City, OK 73125.

Comments regarding this publication should be emailed to AFS630comments@faa.gov.

Major Enhancements

This revision of the handbook has only been updated to include the following:

- Pilot's Handbook of Aeronautical Knowledge Addendum A, dated February 2021

- Pilot's Handbook of Aeronautical Knowledge Addendum B, dated January 2022

- Pilot's Handbook of Aeronautical Knowledge Addendum C, dated March 2023

- Included language to answer National Transportation Safety Board (NTSB) Safety Recommendation (SR) A-21-020

This revision is considered a minor revision. A major revision is underway and is planned for release June 2024.

Acknowledgments

The Pilot's Handbook of Aeronautical Knowledge was produced by the Federal Aviation Administration (FAA) with the assistance of Safety Research Corporation of America. The FAA wishes to acknowledge the following contributors:

Mrs. Nancy A. Wright for providing imagery of a de Haviland DH-4 inaugural air mail flight (Chapter 1)

The Raab Collection, Philadelphia, Pennsylvania, for images of the first pilot license (Chapter 1)

Sandy Kenyon and Rod Magner (magicair.com) for photo of 1929 TravelAir 4000 (Chapter 1)

Dr. Pat Veillette for information used on decision-making (Chapter 2)

Adventure Seaplanes for photos of a ski and float training plane (Chapter 3)

Jack Davis, Stearman Restorers Asociation, for photo of a 1941 PT-17 Army Air Corps trainer (Chapter 3)

Michael J. Hoke, Abaris Training Resources, Inc., for images and information about composite aircraft (Chapter 3)

Colin Cutler, Boldmethod, for images and content on the topic of ground effect (Chapter 5)

Mark R. Korin, Alpha Systems, for images of AOA disaplys (Chapter 5)

M. van Leeuwen (www.zap16.com) for image of Piaggio P-180 (Chapter 6)

Greg Richter, Blue Mountain Avionics, for autopilot information and imagery (Chapter 6)

Mountain High E&S Company for various images provided regarding oxygen systems (Chapter 7)

Jeff Callahan, Aerox, for image of MSK-AS Silicone Mask without Microphone (Chapter 7)

Nonin Medical, Inc. for image of Onyx pulse oximeter (Chapter 7)

Pilotfriend.com for photo of a TKS Weeping Wing (Chapter 7)

Chelton Flight Systems for image of FlightLogic (Chapter 8)

Avidyne Corporation for image of the Entegra (Chapter 8)

Teledyne Controls for image of an air data computer (Chapter 8)

Watson Industries, Inc. (www.watson-gyro.com) for image of Attitude and Heading Reference system (Chapter 8)

Engineering Arresting Systems Corporation (www.esco.zodiacaerospace.com) for EMAS imagery and EMASMAX technical digrams (Chapter 14)

Caasey Rose and Jose Roggeveen (burningholesinthesky.wordpress.com) for flight checklist image (Chapter 14)

Tim Murnahan for images of EMAS at Yeager Airport, Charleston, West Virginia, and EMAS arrested aircraft (Chapter 14)

Cessna Aircraft Company, Columbia Aircraft Manufacturing Corporation, Eclipse Aviation Corporation, Garmin Ltd., The Boeing Company for images provided and used throughout the Handbook.

Additional appreciation is extended to the Aircraft Owners and Pilots Association (AOPA), the AOPA Air Safety Foundation, the General Aviation Manufacturers Association (GAMA), and the National Business Aviation Association (NBAA) for their technical support and input.

Disclaimer: Information in Chapter 14 pertaining to Runway Incursion Avoidance was created using FAA orders, documents, and Advisory Circulars that were current at the date of publication. Users should not assume that all references are current and should check often for reference updates.

Table of Contents

Chapter 1

Introduction To Flying

Introduction

The Pilot's Handbook of Aeronautical Knowledge provides basic knowledge for the student pilot learning to fly, as well as pilots seeking advanced pilot certification. For detailed information on a variety of specialized flight topics, see specific Federal Aviation Administration (FAA) handbooks and Advisory Circulars (ACs).

This chapter offers a brief history of flight, introduces the history and role of the FAA in civil aviation, FAA regulations and standards, government references and publications, eligibility for pilot certificates, available routes to flight instruction, the role of the Certificated Flight Instructor (CFI) and Designated Pilot Examiner (DPE) in flight training, Practical Test Standards (PTS), and new, industry-developed Airman Certification Standards (ACS) framework that will eventually replace the PTS.

History of Flight

From prehistoric times, humans have watched the flight of birds, and longed to imitate them, but lacked the power to do so. Logic dictated that if the small muscles of birds can lift them into the air and sustain them, then the larger muscles of humans should be able to duplicate the feat. No one knew about the intricate mesh of muscles, sinew, heart, breathing system, and devices not unlike wing flaps, variable-camber and spoilers of the modern airplane that enabled a bird to fly. Still, thousands of years and countless lives were lost in attempts to fly like birds.

The identity of the first "bird-men" who fitted themselves with wings and leapt off of cliffs in an effort to fly are lost in time, but each failure gave those who wished to fly questions that needed to be answered. Where had the wing flappers gone wrong? Philosophers, scientists, and inventors offered solutions, but no one could add wings to the human body and soar like a bird. During the 1500s, Leonardo da Vinci filled pages of his notebooks with sketches of proposed flying machines, but most of his ideas were flawed because he clung to the idea of birdlike wings. *[Figure 1-1]* By 1655, mathematician, physicist, and inventor Robert Hooke concluded that the human body does not possess the strength to power artificial wings. He believed human flight would require some form of artificial propulsion.

The quest for human flight led some practitioners in another direction. In 1783, the first manned hot air balloon, crafted by Joseph and Etienne Montgolfier, flew for 23 minutes. Ten days later, Professor Jacques Charles flew the first gas balloon. A madness for balloon flight captivated the public's imagination and for a time flying enthusiasts turned their expertise to the promise of lighter-than-air flight. But for all its majesty in the air, the balloon was little more than a billowing heap of cloth capable of no more than a one-way, downwind journey.

Balloons solved the problem of lift, but that was only one of the problems of human flight. The ability to control speed and direction eluded balloonists. The solution to that problem lay in a child's toy familiar to the East for 2,000 years, but not introduced to the West until the 13th century—the kite. The kites used by the Chinese for aerial observation, to test winds for sailing, as a signaling device, and as a toy, held many of the answers to lifting a heavier-than-air device into the air.

One of the men who believed the study of kites unlocked the secrets of winged flight was Sir George Cayley. Born in England 10 years before the Mongolfier balloon flight, Cayley spent his 84 years seeking to develop a heavier-than-air vehicle supported by kite-shaped wings. *[Figure 1-2]* The "Father of Aerial Navigation," Cayley discovered the basic principles on which the modern science of aeronautics is founded; built what is recognized as the first successful flying model; and tested the first full-size man-carrying airplane.

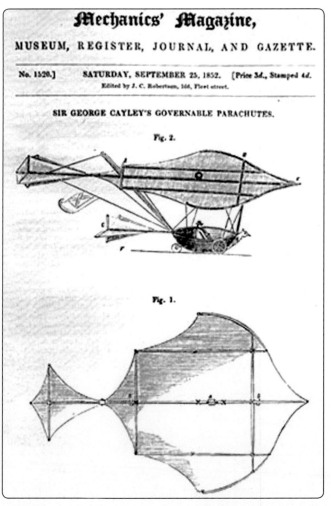

Figure 1-2. *Glider from 1852 by Sir George Cayley, British aviator (1773–1857).*

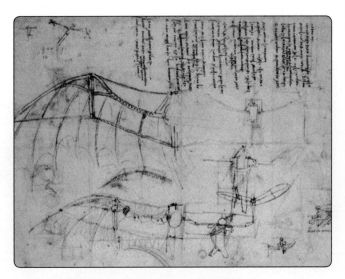

Figure 1-1. *Leonardo da Vinci's ornithopter wings.*

For the half-century after Cayley's death, countless scientists, flying enthusiasts, and inventors worked toward building a powered flying machine. Men, such as William Samuel Henson, who designed a huge monoplane that was propelled by a steam engine housed inside the fuselage, and Otto Lilienthal, who proved human flight in aircraft heavier than air was practical, worked toward the dream of powered flight. A dream turned into reality by Wilbur and Orville Wright at Kitty Hawk, North Carolina, on December 17, 1903.

The bicycle-building Wright brothers of Dayton, Ohio, had experimented for 4 years with kites, their own homemade wind tunnel, and different engines to power their biplane. One of their great achievements in flight was proving the value of the scientific, rather than a build-it-and-see approach. Their biplane, The Flyer, combined inspired design and engineering with superior craftsmanship. *[Figure 1-3]* By the afternoon of December 17th, the Wright brothers had flown a total of 98 seconds on four flights. The age of flight had arrived.

History of the Federal Aviation Administration (FAA)

During the early years of manned flight, aviation was a free for all because no government body was in place to establish policies or regulate and enforce safety standards. Individuals were free to conduct flights and operate aircraft with no government oversight. Most of the early flights were conducted for sport. Aviation was expensive and became the playground of the wealthy. Since these early airplanes were small, many people doubted their commercial value. One group of individuals believed otherwise and they became the genesis for modern airline travel.

P. E. Fansler, a Florida businessman living in St. Petersburg, approached Tom Benoist of the Benoist Aircraft Company in St. Louis, Missouri, about starting a flight route from St. Petersburg across the waterway to Tampa. Benoist suggested using his "Safety First" airboat and the two men signed an agreement for what would become the first scheduled airline in the United States. The first aircraft was delivered to St. Petersburg and made the first test flight on December 31, 1913. *[Figure 1-4]*

A public auction decided who would win the honor of becoming the first paying airline customer. The former mayor of St. Petersburg, A. C. Pheil, made the winning bid of $400.00, which secured his place in history as the first paying airline passenger.

On January 1, 1914, the first scheduled airline flight was conducted. The flight length was 21 miles and lasted 23 minutes due to a headwind. The return trip took 20 minutes. The line, which was subsidized by Florida businessmen, continued for 4 months and offered regular passage for $5.00 per person or $5.00 per 100 pounds of cargo. Shortly after the opening of the line, Benoist added a new airboat that afforded more protection from spray during takeoff and landing. The routes were also extended to Manatee, Bradenton, and Sarasota giving further credence to the idea of a profitable commercial airline.

The St. Petersburg-Tampa Airboat Line continued throughout the winter months with flights finally being suspended when the winter tourist industry began to dry up. The airline operated for only 4 months, but 1,205 passengers were carried without injury. This experiment proved commercial passenger airline travel was viable.

The advent of World War I offered the airplane a chance to demonstrate its varied capabilities. It began the war as a reconnaissance platform, but by 1918, airplanes were being

Figure 1-3. *First flight by the Wright brothers.*

Figure 1-4. *Benoist airboat.*

mass produced to serve as fighters, bombers, trainers, as well as reconnaissance platforms.

Aviation advocates continued to look for ways to use airplanes. Airmail service was a popular idea, but the war prevented the Postal Service from having access to airplanes. The War Department and Postal Service reached an agreement in 1918. The Army would use the mail service to train its pilots in flying cross-country. The first airmail flight was conducted on May 15, 1918, between New York and Washington, DC. The flight was not considered spectacular; the pilot became lost and landed at the wrong airfield. In August of 1918, the United States Postal Service took control of the airmail routes and brought the existing Army airmail pilots and their planes into the program as postal employees.

Transcontinental Air Mail Route

Airmail routes continued to expand until the Transcontinental Mail Route was inaugurated. *[Figure 1-5]* This route spanned from San Francisco to New York for a total distance of 2,612 miles with 13 intermediate stops along the way. *[Figure 1-6]* On May 20, 1926, Congress passed the Air Commerce Act, which served as the cornerstone for aviation within the United States. This legislation was supported by leaders in the aviation industry who felt that the airplane could not reach its full potential without assistance from the Federal Government in improving safety.

The Air Commerce Act charged the Secretary of Commerce with fostering air commerce, issuing and enforcing air traffic rules, licensing pilots, certificating aircraft, establishing airways, and operating and maintaining aids to air navigation. The Department of Commerce created a new Aeronautics Branch whose primary mission was to provide oversight for the aviation industry. In addition, the Aeronautics Branch took over the construction and operation of the nation's system of lighted airways. The Postal Service, as part of the Transcontinental Air Mail Route system, had initiated this system. The

Figure 1-6. *The transcontinental airmail route ran from New York to San Francisco.*

Department of Commerce made significant advances in aviation communications, including the introduction of radio beacons as an effective means of navigation.

Built at intervals of approximately 10 miles apart, the standard beacon tower was 51 feet high, and was topped with a powerful rotating light. Below the rotating light, two course lights pointed forward and back along the airway. The course lights flashed a code to identify the beacon's number. The tower usually stood in the center of a concrete arrow 70 feet long. A generator shed, where required, stood at the "feather" end of the arrow. *[Figure 1-7]*

Federal Certification of Pilots and Mechanics

The Aeronautics Branch of the Department of Commerce began pilot certification with the first license issued on April 6, 1927. The recipient was the Chief of the Aeronautics Branch, William P. MacCracken, Jr. *[Figure 1-8]* (Orville Wright, who was no longer an active flier, had declined the honor.) MacCracken's license was the first issued to a pilot by a civilian agency of the Federal Government. Some 3 months later, the Aeronautics Branch issued the first Federal aircraft mechanic license.

Equally important for safety was the establishment of a system of certification for aircraft. On March 29, 1927, the Aeronautics Branch issued the first airworthiness type certificate to the Buhl Airster CA-3, a three-place open biplane.

In 1934, to recognize the tremendous strides made in aviation and to display the enhanced status within the department, the Aeronautics Branch was renamed the Bureau of Air Commerce. *[Figure 1-9]* Within this time frame, the Bureau of Air Commerce brought together a group of airlines

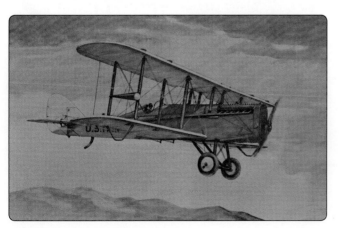

Figure 1-5. *The de Haviland DH-4 on the New York to San Francisco inaugural route in 1921.*

Figure 1-7. *A standard airway beacon tower.*

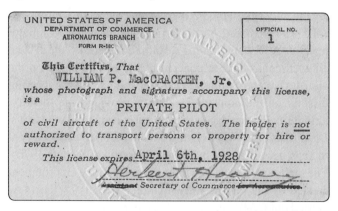

Figure 1-8. *The first pilot license was issued to William P. MacCracken, Jr.*

Figure 1-9. *The third head of the Aeronautics Branch, Eugene L. Vidal, is flanked by President Franklin D. Roosevelt (left) and Secretary of Agriculture Henry A. Wallace (right). The photograph was taken in 1933. During Vidal's tenure, the Aeronautics Branch was renamed the Bureau of Air Commerce on July 1, 1934. The new name more accurately reflected the status of the organization within the Department of Commerce.*

and encouraged them to form the first three Air Traffic Control (ATC) facilities along the established air routes. Then in 1936, the Bureau of Air Commerce took over the responsibilities of operating the centers and continued to advance the ATC facilities. ATC has come a long way from the early controllers using maps, chalkboards, and performing mental math calculations in order to separate aircraft along flight routes.

The Civil Aeronautics Act of 1938

In 1938, the Civil Aeronautics Act transferred the civil aviation responsibilities to a newly created, independent body, named the Civil Aeronautics Authority (CAA). This Act empowered the CAA to regulate airfares and establish new routes for the airlines to service.

President Franklin Roosevelt split the CAA into two agencies—the Civil Aeronautics Administration (CAA) and the Civil Aeronautics Board (CAB). Both agencies were still part of the Department of Commerce but the CAB functioned independently of the Secretary of Commerce. The role of the CAA was to facilitate ATC, certification of airmen and aircraft, rule enforcement, and the development of new airways. The CAB was charged with rule making to enhance safety, accident investigation, and the economic regulation of the airlines. Then in 1946, Congress gave the CAA the responsibility of administering the Federal Aid

Airport Program. This program was designed to promote the establishment of civil airports throughout the country.

The Federal Aviation Act of 1958

By mid-century, air traffic had increased and jet aircraft had been introduced into the civil aviation arena. A series of mid-air collisions underlined the need for more regulation of the aviation industry. Aircraft were not only increasing in numbers, but were now streaking across the skies at much higher speeds. The Federal Aviation Act of 1958 established a new independent body that assumed the roles of the CAA and transferred the rule making authority of the CAB to the newly created Federal Aviation Agency (FAA). In addition, the FAA was given complete control of the common civil-military system of air navigation and ATC. The man who was given the honor of being the first Administrator of the FAA was former Air Force General Elwood Richard "Pete" Quesada. He served as the administrator from 1959–1961. [Figure 1-10]

Department of Transportation (DOT)

On October 15, 1966, Congress established the Department of Transportation (DOT), which was given oversight of the transportation industry within the United States. The result was a combination of both air and surface transportation. Its mission was and is to serve the United States by ensuring a fast, safe, efficient, accessible, and convenient transportation system meeting vital national interests and enhancing the quality of life of the American people, then, now, and into

Figure 1-10. *First Administrator of the FAA was General Elwood Richard "Pete" Quesada, 1959–1961.*

the future. The DOT began operation on April 1, 1967. At this same time, the Federal Aviation Agency was renamed to the Federal Aviation Administration (FAA).

The role of the CAB was assumed by the newly created National Transportation Safety Board (NTSB), which was charged with the investigation of all transportation accidents within the United States.

As aviation continued to grow, the FAA took on additional duties and responsibilities. With the highjacking epidemic of the 1960s, the FAA was responsible for increasing the security duties of aviation both on the ground and in the air. After September 11, 2001, the duties were transferred to a newly created body called the Department of Homeland Security (DHS).

With numerous aircraft flying in and out of larger cities, the FAA began to concentrate on the environmental aspect of aviation by establishing and regulating the noise standards of aircraft. Additionally, in the 1960s and 1970s, the FAA began to regulate high altitude (over 500 feet) kite and balloon flying. In 1970, more duties were assumed by the FAA in the addition of a new federal airport aid program and increased responsibility for airport safety.

ATC Automation

By the mid-1970s, the FAA had achieved a semi-automated ATC system based on a marriage of radar and computer technology. By automating certain routine tasks, the system allowed controllers to concentrate more efficiently on the vital task of providing aircraft separation. Data appearing directly on the controllers' scopes provided the identity, altitude, and groundspeed of aircraft carrying radar beacons. Despite its effectiveness, this system required enhancement to keep pace with the increased air traffic of the late 1970s. The increase was due in part to the competitive environment created by the Airline Deregulation Act of 1978. This law phased out CAB's economic regulation of the airlines, and CAB ceased to exist at the end of 1984.

To meet the challenge of traffic growth, the FAA unveiled the National Airspace System (NAS) Plan in January 1982. The new plan called for more advanced systems for en route and terminal ATC, modernized flight service stations, and improvements in ground-to-air surveillance and communication.

The Professional Air Traffic Controllers Organization (PATCO) Strike

While preparing the NAS Plan, the FAA faced a strike by key members of its workforce. An earlier period of discord between management and the Professional Air

Traffic Controllers Organization (PATCO) culminated in a 1970 "sickout" by 3,000 controllers. Although controllers subsequently gained additional wage and retirement benefits, another period of tension led to an illegal strike in August 1981. The government dismissed over 11,000 strike participants and decertified PATCO. By the spring of 1984, the FAA ended the last of the special restrictions imposed to keep the airspace system operating safely during the strike.

The Airline Deregulation Act of 1978

Until 1978, the CAB regulated many areas of commercial aviation such as fares, routes, and schedules. The Airline Deregulation Act of 1978, however, removed many of these controls, thus changing the face of civil aviation in the United States. After deregulation, unfettered free competition ushered in a new era in passenger air travel.

The CAB had three main functions: to award routes to airlines, to limit the entry of air carriers into new markets, and to regulate fares for passengers. Much of the established practices of commercial passenger travel within the United States went back to the policies of Walter Folger Brown, the United States Postmaster General during the administration of President Herbert Hoover. Brown had changed the mail payments system to encourage the manufacture of passenger aircraft instead of mail-carrying aircraft. His influence was crucial in awarding contracts and helped create four major domestic airlines: United, American, Eastern, and Transcontinental and Western Air (TWA). Similarly, Brown had also helped give Pan American a monopoly on international routes.

The push to deregulate, or at least to reform the existing laws governing passenger carriers, was accelerated by President Jimmy Carter, who appointed economist and former professor Alfred Kahn, a vocal supporter of deregulation, to head the CAB. A second force to deregulate emerged from abroad. In 1977, Freddie Laker, a British entrepreneur who owned Laker Airways, created the Skytrain service, which offered extraordinarily cheap fares for transatlantic flights. Laker's offerings coincided with a boom in low-cost domestic flights as the CAB eased some limitations on charter flights (i.e., flights offered by companies that do not actually own planes but leased them from the major airlines). The big air carriers responded by proposing their own lower fares. For example, American Airlines, the country's second largest airline, obtained CAB approval for "SuperSaver" tickets.

All of these events proved to be favorable for large-scale deregulation. In November 1977, Congress formally deregulated air cargo. In late 1978, Congress passed the Airline Deregulation Act of 1978, legislation that had been principally authored by Senators Edward Kennedy and Howard Cannon. *[Figure 1-11]* There was stiff opposition to the bill—from the major airlines who feared free competition, from labor unions who feared non-union employees, and from safety advocates who feared that safety would be sacrificed. Public support was, however, strong enough to pass the act. The act appeased the major airlines by offering generous subsidies and pleased workers by offering high unemployment benefits if they lost their jobs as a result. The most important effect of the act, whose laws were slowly phased in, was on the passenger market. For the first time in 40 years, airlines could enter the market or (from 1981) expand their routes as they saw fit. Airlines (from 1982) also had full freedom to set their fares. In 1984, the CAB was finally abolished since its primary duty of regulating the airline industry was no longer necessary.

The Role of the FAA

The Code of Federal Regulations (CFR)

The FAA is empowered by regulations to promote aviation safety and establish safety standards for civil aviation. The FAA achieves these objectives under the Code of Federal Regulations (CFR), which is the codification of the general and permanent rules published by the executive departments and agencies of the United States Government. The regulations are divided into 50 different codes, called Titles, that represent broad areas subject to Federal regulation. FAA regulations are listed under Title 14, "Aeronautics and Space," which encompasses all aspects of civil aviation from how to earn a pilot's certificate to maintenance of an aircraft. Title 14 CFR Chapter 1, Federal Aviation Administration, is broken down into subchapters A through N as illustrated in *Figure 1-12*.

For the pilot, certain parts of 14 CFR are more relevant than others. During flight training, it is helpful for the pilot to become familiar with the parts and subparts that relate

Figure 1-11. *President Jimmy Carter signs the Airline Deregulation Act in late 1978.*

Code of Federal Regulations				
Title	**Volume**	**Chapter**	**Subchapters**	
Title 14 Aeronautics and Space	1	I	A	Definitions and Abbreviations
			B	Procedural Rules
			C	Aircraft
	2		D	Airmen
			E	Airspace
			F	Air Traffic and General Rules
	3		G	Air Carriers and Operators for Compensation or Hire: Certification and Operations
			H	Schools and Other Certified Agencies
			I	Airports
			J	Navigational Facilities
			K	Administrative Regulations
			L–M	Reserved
			N	War Risk Insurance
	4	II	A	Economic Regulations
			B	Procedural Regulations
			C	Reserved
			D	Special Regulations
			E	Organization
			F	Policy Statements
		III	A	General
			B	Procedure
			C	Licensing
	5	V		
		VI	A	Office of Management and Budget
			B	Air Transportation Stabilization Board

Figure 1-12. *Overview of 14 CFR, available online free from the FAA and for purchase through commercial sources.*

to flight training and pilot certification. For instance, 14 CFR part 61 pertains to the certification of pilots, flight instructors, and ground instructors. It also defines the eligibility, aeronautical knowledge, and flight proficiency, as well as training and testing requirements for each type of pilot certificate issued. 14 CFR part 91 provides guidance in the areas of general flight rules, visual flight rules (VFR), and instrument flight rules (IFR), while 14 CFR part 43 covers aircraft maintenance, preventive maintenance, rebuilding, and alterations.

Primary Locations of the FAA

The FAA headquarters are in Washington, DC, and there are nine regional offices strategically located across the United States. The agency's two largest field facilities are the Mike Monroney Aeronautical Center (MMAC) in Oklahoma City, Oklahoma, and the William J. Hughes Technical Center (WJHTC) in Atlantic City, New Jersey. Home to FAA training and logistics services, the MMAC provides a number of aviation safety-related and business support services. The WJHTC is the premier aviation research and development and test and evaluation facility in the country. The center's programs include testing and evaluation in ATC, communication, navigation, airports, aircraft safety, and security. Furthermore, the WJHTC is active in long-range

development of innovative aviation systems and concepts, development of new ATC equipment and software, and modification of existing systems and procedures.

Field Offices

Flight Standards Service

Within the FAA, the Flight Standards Service promotes safe air transportation by setting the standards for certification and oversight of airmen, air operators, air agencies, and designees. It also promotes safety of flight of civil aircraft and air commerce by:

- Accomplishing certification, inspection, surveillance, investigation, and enforcement.

- Setting regulations and standards.

- Managing the system for registration of civil aircraft and all airmen records.

The focus of interaction between Flight Standards Service and the aviation community/general public is the Flight Standards District Office (FSDO).

Flight Standards District Office (FSDO)

The FAA has approximately 80 FSDOs. *[Figure 1-13]* These offices provide information and services for the aviation community. FSDO phone numbers are listed in the telephone directory under Government Offices, DOT, FAA. Another convenient method of finding a local office is to use the FSDO locator available at: www.faa.gov/about/office_org/field_offices/fsdo.

In addition to accident investigation and the enforcement of aviation regulations, the FSDO is also responsible for the certification and surveillance of air carriers, air operators, flight schools/training centers, and airmen including pilots and flight instructors. Each FSDO is staffed by Aviation Safety Inspectors (ASIs) who play a key role in making the nation's aviation system safe.

Aviation Safety Inspector (ASI)

The ASIs administer and enforce safety regulations and standards for the production, operation, maintenance, and/or modification of aircraft used in civil aviation. They also specialize in conducting inspections of various aspects of the aviation system, such as aircraft and parts manufacturing, aircraft operation, aircraft airworthiness, and cabin safety. ASIs must complete a training program at the FAA Academy in Oklahoma City, Oklahoma, which includes airman evaluation and pilot testing techniques and procedures. ASIs also receive extensive on-the-job training and recurrent training on a regular basis. The FAA has approximately 3,700 inspectors located in its FSDO offices. All questions concerning pilot certification (and/or requests for other aviation information or services) should be directed to the local FSDO.

FAA Safety Team (FAASTeam)

The FAA is dedicated to improving the safety of United States civilian aviation by conveying safety principles and practices through training, outreach, and education. The FAA

Figure 1-13. *Atlanta Flight Standards District Office (FSDO).*

Safety Team (FAASTeam) exemplifies this commitment. The FAASTeam has replaced the Aviation Safety Program (ASP), whose education of airmen on all types of safety subjects successfully reduced accidents. Its success led to its demise because the easy-to-fix accident causes have been addressed. To take aviation safety one step further, Flight Standards Service created the FAASTeam, which is devoted to reducing aircraft accidents by using a coordinated effort to focus resources on elusive accident causes.

Each of the FAA's nine regions has a Regional FAASTeam Office dedicated to this new safety program and managed by the Regional FAASTeam Manager (RFM). The FAASTeam is "teaming" up with individuals and the aviation industry to create a unified effort against accidents and tip the safety culture in the right direction. To learn more about this effort to improve aviation safety, to take a course at their online learning center, or to join the FAASTeam, visit their website at www.faasafety.gov.

Obtaining Assistance from the FAA

Information can be obtained from the FAA by phone, Internet/e-mail, or mail. To talk to the FAA toll-free 24 hours a day, call 1-866-TELL-FAA (1-866-835-5322). To visit the FAA's website, go to www.faa.gov. Individuals can also e-mail an FAA representative at a local FSDO office by accessing the staff e-mail address available via the "Contact FAA" link at the bottom of the FAA home page. Letters can be sent to:

> Federal Aviation Administration
> 800 Independence Ave, SW
> Washington, DC 20591

FAA Reference Material

The FAA provides a variety of important reference material for the student, as well as the advanced civil aviation pilot. In addition to the regulations provided online by the FAA, several other publications are available to the user. Almost all reference material is available online at www.faa.gov in downloadable format. Commercial aviation publishers also provide published and online reference material to further aid the aviation pilot.

Aeronautical Information Manual (AIM)

The Aeronautical Information Manual (AIM) is the official guide to basic flight information and ATC procedures for the aviation community flying in the NAS of the United States. *[Figure 1-14]* An international version, containing parallel information as well as specific information on international airports, is also available. The AIM also contains information of interest to pilots, such as health and medical facts, flight

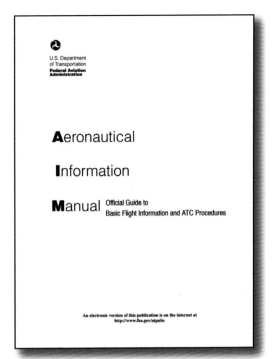

Figure 1-14. *Aeronautical Information Manual.*

Aeronautical Information Manual (AIM)

The Aeronautical Information Manual is designed to provide the aviation community with basic flight information and ATC procedures for use in the NAS of the United States. It also contains the fundamentals required in order to fly in the United States NAS, including items of interest to pilots concerning health/medical facts, factors affecting flight safety, etc.

Aircraft Flying Handbooks (by category)

The Aircraft Flying Handbooks are designed as technical manuals to introduce basic pilot skills and knowledge that are essential for piloting aircraft. They provide information on transition to other aircraft and the operation of various aircraft systems.

Aviation Instructor's Handbook

The Aviation Instructor's Handbook provides the foundation for beginning instructors to understand and apply the fundamentals of instructing. This handbook also provides aviation instructors with up-to-date information on learning and teaching, and how to relate this information to the task of conveying aeronautical knowledge and skills to students. Experienced aviation instructors also find the new and updated information useful for improving their effectiveness in training activities.

Instrument Flying Handbook

The Instrument Flying Handbook is designed for use by instrument flight instructors and pilots preparing for instrument rating tests. Instructors find this handbook a valuable training aid as it includes basic reference material for knowledge testing and instrument flight training.

Instrument Procedures Handbook

The Instrument Procedures Handbook is designed as a technical reference for professional pilots who operate under IFR in the NAS and expands on information contained in the Instrument Flying Handbook.

Figure 1-15. *A sample of handbooks available to the public. Most can be downloaded free of charge from the FAA website.*

---

safety, a pilot/controller glossary of terms used in the system, and information on safety, accidents, and reporting of hazards. This manual is offered for sale on a subscription basis or is available online at: http://bookstore.gpo.gov.

Order forms are provided at the beginning of the manual or online and should be sent to the Superintendent of Documents, United States Government Printing Office (GPO). The AIM is complemented by other operational publications that are available via separate subscriptions or online.

Handbooks

Handbooks are developed to provide specific information about a particular topic that enhances training or understanding. The FAA publishes a variety of handbooks that generally fall into three categories: aircraft, aviation, and examiners and inspectors. *[Figure 1-15]* These handbooks can be purchased from the Superintendent of Documents or downloaded at www.faa.gov/regulations_policies. Aviation handbooks are also published by various commercial aviation companies. Aircraft flight manuals commonly called Pilot Operating Handbooks (POH) are documents developed by the airplane manufacturer, approved by the FAA, and are specific to a particular make and model aircraft by serial number. This subject is covered in greater detail in Chapter 8, "Flight Manuals and Other Documents," of this handbook. *[Figure 1-16]*

Advisory Circulars (ACs)

An AC is an informational document that the FAA wants to distribute to the aviation community. This can be in the form of a text book used in a classroom or a one page document. Some ACs are free while others cost money. They are to be used for information only and are not regulations. The FAA website www.faa.gov/regulations_policies/advisory_circulars/ provides a database that is a searchable repository of all aviation safety ACs. All ACs, current and historical, are provided and can be viewed as a portable document format (PDF) copy.

ACs provide a single, uniform, agency-wide system that the FAA uses to deliver advisory material to FAA customers, industry, the aviation community, and the public. An AC may be needed to:

- Provide an acceptable, clearly understood method for complying with a regulation

Figure 1-16. *Pilot Operating Handbooks from manufacturers.*

- Standardize implementation of a regulation or harmonize implementation for the international aviation community

- Resolve a general misunderstanding of a regulation

- Respond to a request from some government entity, such as General Accounting Office, NTSB, or the Office of the Inspector General

- Help the industry and FAA effectively implement a regulation

- Explain requirements and limits of an FAA grant program

- Expand on standards needed to promote aviation safety, including the safe operation of airports

There are three parts to an AC number, as in 25-42C. The first part of the number identifies the subject matter area of the AC and corresponds to the appropriate 14 CFR part. For example, an AC on "Certification: Pilots and Flight and Ground Instructors" is numbered as AC 61-65E. Since ACs are numbered sequentially within each subject area, the second part of the number beginning with the dash identifies this sequence. The third part of the number is a letter assigned by the originating office and shows the revision sequence if an AC is revised. The first version of an AC does not have a revision letter. In *Figure 1-17*, this is the fifth revision, as designated by the "E."

Flight Publications

The FAA, in concert with other government agencies, orchestrates the publication and changes to publications that are key to safe flight. *Figure 1-18* illustrates some publications a pilot may use.

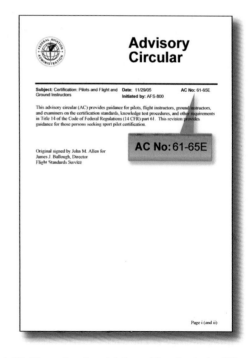

Figure 1-17. *Example of an Advisory Circular in its fifth revision.*

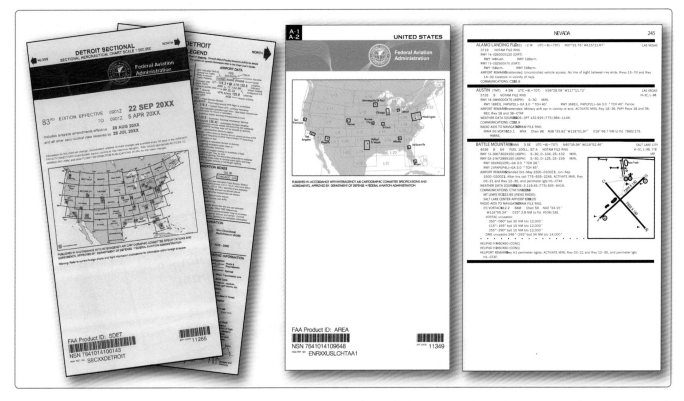

Figure 1-18. *From left to right, a sectional VFR chart, IFR chart, and chart supplement U.S. (formerly Airport/Facility Directory) with a sample of a page from the supplement.*

Pilot and Aeronautical Information
Notices to Airmen (NOTAMs)

Notices to Airmen, or NOTAMs, are time-critical aeronautical information either temporary in nature or not sufficiently known in advance to permit publication on aeronautical charts or in other operational publications. The information receives immediate dissemination via the National Notice to Airmen (NOTAM) System. NOTAMs contain current notices to airmen that are considered essential to the safety of flight, as well as supplemental data affecting other operational publications. There are many different reasons that NOTAMs are issued. Following are some of those reasons:

- Hazards, such as air shows, parachute jumps, kite flying, and rocket launches

- Flights by important people such as heads of state

- Closed runways

- Inoperable radio navigational aids

- Military exercises with resulting airspace restrictions

- Inoperable lights on tall obstructions

- Temporary erection of obstacles near airfields

- Passage of flocks of birds through airspace (a NOTAM in this category is known as a BIRDTAM)

- Notifications of runway/taxiway/apron status with respect to snow, ice, and standing water (a SNOWTAM)

- Notification of an operationally significant change in volcanic ash or other dust contamination (an ASHTAM)

- Software code risk announcements with associated patches to reduce specific vulnerabilities

NOTAM information is generally classified into four categories: NOTAM (D) or NOTAMs that receive distant dissemination, distant and Flight Data Center (FDC) NOTAMs, Pointer NOTAMs, and Military NOTAMs pertaining to military airports or NAVAIDs that are part of the NAS. NOTAMs are available through Flight Service Station (FSS), Direct User Access Terminal Service (DUATS), private vendors, and many online websites.

NOTAM (D) Information

NOTAM (D) information is disseminated for all navigational facilities that are part of the NAS, and all public use airports, seaplane bases, and heliports listed in the Chart Supplement U.S. (formerly Airport/Facility Directory). NOTAM (D) information now includes such data as taxiway closures, personnel and equipment near or crossing runways, and airport lighting aids that do not affect instrument approach criteria, such as visual approach slope indicator (VASI). All D NOTAMs are required to have one of the following keywords as the first part of the text: RWY, TWY, RAMP, APRON, AD, OBST, NAV, COM, SVC, AIRSPACE, (U), or (O). *[Figure 1-19]*

FDC NOTAMs

FDC NOTAMs are issued by the National Flight Data Center and contain information that is regulatory in nature pertaining to flight including, but not limited to, changes to charts, procedures, and airspace usage. FDC NOTAMs refer to information that is regulatory in nature and includes the following:

- Interim IFR flight procedures:
  1. Airway structure changes
  2. Instrument approach procedure changes (excludes Departure Procedures (DPs) and Standard Terminal Arrivals (STARs)
  3. Airspace changes in general
  4. Special instrument approach procedure changes

- Temporary flight restrictions (discussed in Chapter 15):
  1. Disaster areas
  2. Special events generating a high degree of interest
  3. Hijacking

- Flight restrictions in the proximity of the President and other parties
- 14 CFR part 139 certificated airport condition changes
- Snow conditions affecting glide slope operation
- Air defense emergencies
- Emergency flight rules
- Substitute airway routes
- Special data
- U.S. Government charting corrections
- Laser activity

NOTAM Composition

NOTAMs contain the elements below from left to right in the following order:

- An exclamation point (!)
- Accountability Location (the identifier of the accountability location)

Keyword	Example	Meaning
RWY	RWY 3/21 CLSD	Runways 3 and 21 are closed to aircraft.
TWY	TWY F LGTS OTS	Taxiway F lights are out of service.
RAMP	RAMP TERMINAL EAST SIDE CONSTRUCTION	The ramp in front of the east side of the terminal has ongoing construction.
APRON	APRON SW TWY C NEAR HANGARS CLSD	The apron near the southwest taxiway C in front of the hangars is closed.
AD	AD ABN OTS	Aerodromes: The airport beacon is out of service.
OBST	OBST TOWER 283 (245 AGL) 2.2 S LGTS OTS (ASR 1065881) TIL 0707272300	Obstruction: The lights are out of service on a tower that is 283 feet above mean sea level (MSL) or 245 feet above ground level (AGL) 2.2 statute miles south of the field. The FCC antenna structure registration (ASR) number is 1065881. The lights will be returned to service 2300 UTC (Coordinated Universal Time) on July 27, 2007.
NAV	NAV VOR OTS	Navigation: The VOR located on this airport is out of service.
COM	COM ATIS OTS	Communications: The Automatic Terminal information Service (ATIS) is out of service.
SVC	SVC TWR 1215-0330 MON -FRI/1430-2300 SAT/1600-0100 SUN TIL 0707300100	Service: The control tower has new operating hours, 1215-0330 UTC Monday Thru Friday. 1430-2300 UTC on Saturday and 1600-0100 UTC on Sunday until 0100 on July 30, 2007.
	SVC FUEL UNAVBL TIL 0707291600	Service: All fuel for this airport is unavailable until July 29, 2007, at 1600 UTC.
	SVC CUSTOMS UNAVBL TIL 0708150800	Service: United States Customs service for this airport will not be available until August 15, 2007, at 0800 UTC.
AIRSPACE	AIRSPACE AIRSHOW ACFT 5000/BLW 5 NMR AIRPORT AVOIDANCE ADZD WEF 0707152000-0707152200	Airspace. There is an airshow being held at this airport with aircraft flying 5,000 feet and below within a 5 nautical mile radius. Avoidance is advised from 2000 UTC on July 15, 2007, until 2200 on July 15, 2007.
U	ORT 6K8 (U) RWY ABANDONED VEHICLE	Unverified aeronautical information.
O	LOZ LOZ (O) CONTROLLED BURN OF HOUSE 8 NE APCH END RWY 23 WEF 0710211300-0710211700	Other aeronautical information received from any authorized source that may be beneficial to aircraft operations and does not meet defined NOTAM criteria.

Figure 1-19. *NOTAM (D) Information.*

- Affected Location (the identifier of the affected facility or location)
- KEYWORD (one of the following: RWY, TWY, RAMP, APRON, AD, COM, NAV, SVC, OBST, AIRSPACE, (U) and (O))
- Surface Identification (optional—this shall be the runway identification for runway related NOTAMs, the taxiway identification for taxiway-related NOTAMs, or the ramp/apron identification for ramp/apron-related NOTAMs)
- Condition (the condition being reported)
- Time (identifies the effective time(s) of the NOTAM condition)

Altitude and height are in feet mean sea level (MSL) up to 17,999; e.g., 275, 1225 (feet and MSL is not written), and in flight levels (FL) for 18,000 and above; e.g., FL180, FL550. When MSL is not known, above ground level (AGL) will be written (304 AGL).

When time is expressed in a NOTAM, the day begins at 0000 and ends at 2359. Times used in the NOTAM system are universal time coordinated (UTC) and shall be stated in 10 digits (year, month, day, hour, and minute). The following are two examples of how the time would be presented:

!DCA LDN NAV VOR OTS WEF
0708051600-0708052359

!DCA LDN NAV VOR OTS WEF
0709050000-0709050400

NOTAM Dissemination and Availability

The system for disseminating aeronautical information is made up of two subsystems: the Airmen's Information System (AIS) and the NOTAM System. The AIS consists of charts and publications and is disseminated by the following methods:

Aeronautical charts depicting permanent baseline data:

- IFR Charts—Enroute High Altitude Conterminous U.S., Enroute Low Altitude Conterminous U.S., Alaska Charts, and Pacific Charts
- U.S. Terminal Procedures—Departure Procedures (DPs), Standard Terminal Arrivals (STARs) and Standard Instrument Approach Procedures (SIAPs)
- VFR Charts—Sectional Aeronautical Charts, Terminal Area Charts (TAC), and World Aeronautical Charts (WAC)

Flight information publications outlining baseline data:

- Notices to Airmen (NTAP)—Published by System Operations Services, System Operations and Safety, Publications, every 28 days)
- Chart Supplement U.S. (formerly Airport/Facility Directory)
- Pacific Chart Supplement
- Alaska Supplement
- Alaska Terminal
- Aeronautical Information Manual (AIM)

NOTAMs are available in printed form through subscription from the Superintendent of Documents, from an FSS, or online at PilotWeb (www.pilotweb.nas.faa.gov), which provides access to current NOTAM information. Local airport NOTAMs can be obtained online from various websites. Some examples are www.fltplan.com and www.aopa.org/whatsnew/notams.html. Most sites require a free registration and acceptance of terms but offer pilots updated NOTAMs and TFRs.

Safety Program Airmen Notification System (SPANS)

In 2004, the FAA launched the Safety Program Airmen Notification System (SPANS), an online event notification system that provides timely and easy-to-assess seminar and event information notification for airmen. The SPANS system is taking the place of the current paper-based mail system. This provides better service to airmen while reducing costs for the FAA. Anyone can search the SPANS system and register for events. To read more about SPANS, visit www.faasafety.gov/spans.

Aircraft Classifications and Ultralight Vehicles

The FAA uses various ways to classify or group machines operated or flown in the air. The most general grouping uses the term aircraft. This term is in 14 CFR 1.1 and means a device that is used or intended to be used for flight in the air.

Ultralight vehicle is another general term the FAA uses. This term is defined in 14 CFR 103. As the term implies, powered ultralight vehicles must weigh less than 254 pounds empty weight and unpowered ultralight vehicles must weigh less than 155 pounds. Rules for ultralight vehicles are significantly different from rules for aircraft; ultralight vehicle certification, registration, and operation rules are also contained in 14 CFR 103.

The FAA differentiates aircraft by their characteristics and physical properties. Key groupings defined in 14 CFR 1.1 include:

- Airplane—an engine-driven fixed-wing aircraft heavier than air, that is supported in flight by the dynamic reaction of the air against its wings.

- Glider—a heavier-than-air aircraft, that is supported in flight by the dynamic reaction of the air against its lifting surfaces and whose free flight does not depend principally on an engine.

- Lighter-than-air aircraft—an aircraft that can rise and remain suspended by using contained gas weighing less than the air that is displaced by the gas.

  - Airship—an engine-driven lighter-than-air aircraft that can be steered.

  - Balloon—a lighter-than-air aircraft that is not engine driven, and that sustains flight through the use of either gas buoyancy or an airborne heater.

- Powered-lift—a heavier-than-air aircraft capable of vertical takeoff, vertical landing, and low speed flight that depends principally on engine-driven lift devices or engine thrust for lift during these flight regimes and on nonrotating airfoil(s) for lift during horizontal flight.

- Powered parachute—a powered aircraft comprised of a flexible or semi-rigid wing connected to a fuselage so that the wing is not in position for flight until the aircraft is in motion. The fuselage of a powered parachute contains the aircraft engine, a seat for each occupant and Is attached to the aircraft's landing gear.

- Rocket—an aircraft propelled by ejected expanding gases generated in the engine from self-contained propellants and not dependent on the intake of outside substances. It includes any part which becomes separated during the operation.

- Rotorcraft—a heavier-than-air aircraft that depends principally for its support in flight on the lift generated by one or more rotors.

  - Gyroplane- -a rotorcraft whose rotors are not engine-driven, except for Initial starting, but are made to rotate by action of the air when the rotorcraft Is moving; and whose means of propulsion, consisting usually of conventional propellers, is Independent of the rotor system.

  - Helicopter—a rotorcraft that, for its horizontal motion, depends principally on its engine-driven rotors.

- Weight-shift-control—a powered aircraft with a framed pivoting wing and a fuselage controllable only in pitch and roll by the pilot's ability to change the aircraft's center of gravity with respect to the wing. Flight control of the aircraft depends on the wing's ability to flexibly deform rather than the use of control surfaces.

Size and weight are other methods used in 14 CFR 1.1 to group aircraft:

- Large aircraft—an aircraft of more than 12,500 pounds, maximum certificated takeoff weight.

- Light-sport aircraft (LSA)—an aircraft, other than a helicopter or powered-lift that, since its original certification, has continued to meet the definition in 14 CFR 1.1. (LSA can include airplanes, airships, balloons, gliders, gyro planes, powered parachutes, and weight-shift-control.)

- Small Aircraft—aircraft of 12,500 pounds or less, maximum certificated takeoff weight.

We also use broad classifications of aircraft with respect to the certification of airmen or with respect to the certification of the aircraft themselves. See the next section, Pilot Certifications, and Chapter 3, for further discussion of certification. These definitions are in 14 CFR 1.1:

- Category

  1. As used with respect to the certification, ratings, privileges, and limitations of airmen, means a broad classification of aircraft. Examples include: airplane; rotorcraft; glider; and lighter-than-air; and

  2. As used with respect to the certification of aircraft, means a grouping of aircraft based upon intended use or operating limitations. Examples include: transport, normal, utility, acrobatic, limited, restricted, and provisional.

- Class

  1. As used with respect to the certification, ratings, privileges, and limitations of airmen, means a classification of aircraft within a category having similar operating characteristics. Examples Include: single engine; multiengine; land; water; gyroplane, helicopter, airship, and free balloon; and

  2. As used with respect to the certification of aircraft, means a broad grouping of aircraft having similar characteristics of propulsion, flight, or landing. Examples include: airplane, rotorcraft, gilder, balloon, landplane, and seaplane.

- Type

  1. As used with respect to the certification, ratings, privileges, and limitations of airmen, means

a specific make and basic model of aircraft, Including modifications thereto that do not change its handling or flight characteristics. Examples include: 737-700, G-IV, and 1900; and

2. As used with respect to the certification of aircraft, means those aircraft which are similar in design. Examples include: 737-700 and 737-700C; G-IV and G-IV-X; and 1900 and 1900C.

This system of definitions allows the FAA to group and regulate aircraft to provide for their safe operation.

Pilot Certifications

The type of intended flying influences what type of pilot's certificate is required. Eligibility, training, experience, and testing requirements differ depending on the type of certificates sought. *[Figure 1-20]* Each type of pilot's certificate has privileges and limitations that are inherent within the certificate itself. However, other privileges and limitations may be applicable based on the aircraft type, operation being conducted, and the type of certificate. For example, a certain certificate may have privileges and limitations under 14 CFR part 61 and part 91.

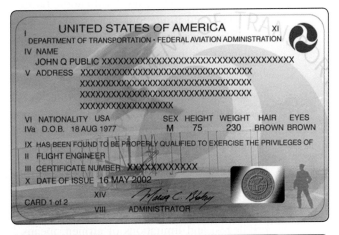

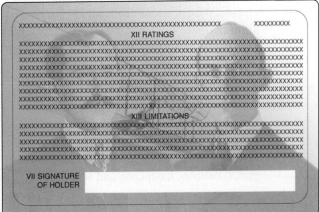

Figure 1-20. *Front side (top) and back side (bottom) of an airman certificate issued by the FAA.*

- Privileges—define where and when the pilot may fly, with whom they may fly, the purpose of the flight, and the type of aircraft they are allowed to fly.

- Limitations—the FAA may impose limitations on a pilot certificate if, during training or the practical test, the pilot does not demonstrate all skills necessary to exercise all privileges of a privilege level, category, class, or type rating.

Endorsements, a form of authorization, are written to establish that the certificate holder has received training in specific skill areas. Endorsements are written and signed by an authorized individual, usually a certificated flight instructor (CFI), and are based on aircraft classification. *[Figure 1-21]*

Sport Pilot

To become a sport pilot, the student pilot is required to have flown, at a minimum, the following hours depending upon the aircraft:

- Airplane: 20 hours

- Powered Parachute: 12 hours

- Weight-Shift Control (Trikes): 20 hours

- Glider: 10 hours

- Rotorcraft (gyroplane only): 20 hours

- Lighter-Than-Air: 20 hours (airship) or 7 hours (balloon)

To earn a Sport Pilot Certificate, one must:

- Be at least 16 years old to become a student sport pilot (14 years old for gliders or balloons)

- Be at least 17 years old to test for a sport pilot certificate (16 years old for gliders or balloons)

- Be able to read, write, and understand the English language

- Hold a current and valid driver's license as evidence of medical eligibility

When operating as a sport pilot, some of the following privileges and limitations may apply.

Privileges:

- Operate as pilot in command (PIC) of a light-sport aircraft

- Carry a passenger and share expenses (fuel, oil, airport expenses, and aircraft rental)

- Fly during the daytime using VFR, a minimum of 3 statute miles visibility and visual contact with the ground are required

Recreational pilot to conduct solo flights for the purpose of obtaining an additional certificate or rating while under the supervision of an authorized flight instructor: section 61.101(i).

I certify that (First name, MI, Last name) has received the required training of section 61.87 in a (make and model aircraft). I have determined he/she is prepared to conduct a solo flight on (date) under the following conditions: (List all conditions which require endorsement, e.g., flight which requires communication with air traffic control, flight in an aircraft for which the pilot does not hold a category/class rating, etc.).

Figure 1-21. *Example endorsement for a recreational pilot to conduct solo flights for the purpose of determining an additional certificate or rating.*

Limitations:

- Prohibited from flying in Class A airspace
- Prohibited from flying in Class B, C, or D airspace until you receive training and a logbook endorsement from an instructor
- No flights outside the United States without prior permission from the foreign aviation authority
- May not tow any object
- No flights while carrying a passenger or property for compensation or hire
- Prohibited from flying in furtherance of a business

The sport pilot certificate does not list aircraft category and class ratings. After successfully passing the practical test for a sport pilot certificate, regardless of the light-sport aircraft privileges you seek, the FAA will issue you a sport pilot certificate without any category and class ratings. The Instructor will provide you with the appropriate logbook endorsement for the category and class of aircraft in which you are authorized to act as pilot in command.

Recreational Pilot

To become a recreational pilot, one must:

- Be at least 17 years old
- Be able to read, write, speak, and understand the English language
- Pass the required knowledge test
- Meet the aeronautical experience requirements in either a single-engine airplane, a helicopter, or a gyroplane.
- Obtain a logbook endorsement from an instructor
- Pass the required practical test
- Obtain a third-class medical certificate issued under 14 CFR part 67

As a recreational pilot, cross-country flight is limited to a 50 NM range from the departure airport but is permitted with additional training per 14 CFR part 61, section 61.101(c). Additionally, recreational pilots are restricted from flying at night and flying in airspace where communications with ATC are required.

The minimum aeronautical experience requirements for a recreational pilot license involve:

- 30 hours of flight time including at least:
  - 15 hours of dual instruction
  - 2 hours of en route training
  - 3 hours in preparation for the practical test
  - 3 hours of solo flight

When operating as a recreational pilot, some of the following privileges and limitations may apply.

Privileges:

- Carry no more than one passenger;
- Not pay less than the pro rata share of the operating expenses of a flight with a passenger, provided the expenses involve only fuel, oil, airport expenses, or aircraft rental fees

Limitations:

- A recreational pilot may not act as PIC of an aircraft that is certificated for more than four occupants or has more than one powerplant.

Private Pilot

A private pilot is one who flies for pleasure or personal business without accepting compensation for flying except in some very limited, specific circumstances. The Private Pilot Certificate is the certificate held by the majority of active pilots. It allows command of any aircraft (subject to appropriate ratings) for any noncommercial purpose and gives almost unlimited authority to fly under VFR. Passengers may be carried and flight in furtherance of a business is permitted; however, a private pilot may not be compensated in any way for services as a pilot, although passengers can pay a pro rata share of flight expenses, such as fuel or rental costs. If training under 14 CFR part 61, experience requirements include at least 40 hours of piloting time, including 20 hours of flight with an instructor and 10 hours of solo flight. *[Figure 1-22]*

Commercial Pilot

A commercial pilot may be compensated for flying. Training for the certificate focuses on a better understanding of aircraft systems and a higher standard of airmanship. The Commercial Pilot Certificate itself does not allow a pilot to fly in instrument meteorological conditions (IMC), and commercial pilots without an instrument rating are restricted to daytime flight within 50 NM when flying for hire.

A commercial airplane pilot must be able to operate a complex airplane, as a specific number of hours of complex (or turbine-powered) aircraft time are among the prerequisites, and at least a portion of the practical examination is performed in a complex aircraft. A complex aircraft must have retractable landing gear, movable flaps, and a controllable-pitch propeller. See 14 CFR part 61, section 61.31(e) for additional information. *[Figure 1-23]*

Airline Transport Pilot

The airline transport pilot (ATP) is tested to the highest level of piloting ability. The ATP certificate is a prerequisite for serving as a PIC and second in command (SIC) of scheduled airline operations. It is also a prerequisite for serving as a PIC in select charter and fractional operations. The minimum pilot experience is 1,500 hours of flight time. In addition, the pilot must be at least 23 years of age, be able to read, write, speak,

Figure 1-23. *A complex aircraft.*

and understand the English language, and be "of good moral standing." A pilot may obtain an ATP certificate with restricted privileges enabling him/her to serve as an SIC in scheduled airline operations. The minimum pilot experience is reduced based upon specific academic and flight training experience. The minimum age to be eligible is 21 years. *[Figure 1-24]*

Selecting a Flight School

Selecting a flight school is an important consideration in the flight training process. FAA-approved training centers, FAA-approved pilot schools, noncertificated flying schools, and independent flight instructors conduct flight training in the United States. All flight training is conducted under the auspices of the FAA following the regulations outlined in 14 CFR parts 142, 141, or 61. Training centers, also referred to as flight academies, operate under 14 CFR part 142 and are certificated by the FAA. Application for certification is voluntary and the training center must meet stringent requirements for personnel, equipment, maintenance, facilities, and must teach a curriculum approved by the FAA. Training centers typically utilize a number of flight simulation training devices as part of its curricula. Flight training conducted at a training center is primarily done under contract to airlines and other commercial operators in transport or turbine aircraft, however many also provide

Figure 1-22. *A typical aircraft a private pilot might fly.*

Figure 1-24. *Type of aircraft flown by an airline transport pilot.*

flight training for the private pilot certificate, commercial pilot certificate, instrument rating, and ATP certificate.

Flight schools operating under 14 CFR part 141 are certificated by the FAA. Application for certification is voluntary and the school must meet stringent requirements for personnel, equipment, maintenance, facilities, and must teach an established curriculum, which includes a training course outline (TCO) approved by the FAA. The certificated schools may qualify for a ground school rating and a flight school rating. In addition, the school may be authorized to give its graduates practical (flight) tests and knowledge (computer administered written) tests. The FAA Pilot School Search database located at http://av-info.faa.gov/PilotSchool.asp, lists certificated ground and flight schools and the pilot training courses each school offers.

Enrollment in a 14 CFR part 141 flight school ensures quality, continuity, and offers a structured approach to flight training because these facilities must document the training curriculum and have their flight courses approved by the FAA. These strictures allow 14 CFR part 141 schools to complete certificates and ratings in fewer flight hours, which can mean a savings on the cost of flight training for the student pilot. For example, the minimum requirement for a Private Pilot Certificate is 35 hours in a part 141-certificated school and 40 hours in a part 61-certificated school. (This difference may be insignificant for a Private Pilot Certificate because the national average indicates most pilots require 60 to 75 hours of flight training.)

Many excellent flight schools find it impractical to qualify for the FAA part 141 certificates and are referred to as part 61 schools. 14 CFR part 61 outlines certificate and rating requirements for pilot certification through noncertificated schools and individual flight instructors. It also states what knowledge-based training must be covered and how much flight experience is required for each certificate and rating. Flight schools and flight instructors who train must adhere to the statutory requirements and train pilots to the standards found in 14 CFR part 61.

One advantage of flight training under 14 CFR part 61 is its flexibility. Flight lessons can be tailored to the individual student, because 14 CFR part 61 dictates the required minimum flight experience and knowledge-based training necessary to gain a specific pilot's license, but it does not stipulate how the training is to be organized. This flexibility can also be a disadvantage because a flight instructor who fails to organize the flight training can cost a student pilot time and expense through repetitive training. One way for a student pilot to avoid this problem is to ensure the flight instructor has a well-documented training syllabus.

How To Find a Reputable Flight Program

To obtain information about pilot training, contact the local FSDO, which maintains a current file on all schools within its district. The choice of a flight school depends on what type of certificate is sought, and whether an individual wishes to fly as a sport pilot or wishes to pursue a career as a professional pilot. Another consideration is the amount of time that can be devoted to training. Ground and flight training should be obtained as regularly and frequently as possible because this assures maximum retention of instruction and the achievement of requisite proficiency.

Do not make the determination based on financial concerns alone, because the quality of training is very important. Prior to making a final decision, visit the schools under consideration and talk with management, instructors, and students. Request a personal tour of the flight school facility.

Be inquisitive and proactive when searching for a flight school, do some homework, and develop a checklist of questions by talking to pilots and reading articles in flight magazines. The checklist should include questions about aircraft reliability and maintenance practices, and questions for current students such as whether or not there is a safe, clean aircraft available when they are scheduled to fly.

Questions for the training facility should be aimed at determining if the instruction fits available personal time. What are the school's operating hours? Does the facility have dedicated classrooms available for ground training required by the FAA? Is there an area available for preflight briefings, postflight debriefings, and critiques? Are these rooms private in nature in order to provide a nonthreatening environment in which the instructor can explain the content and outcome of the flight without making the student feel self-conscious?

Examine the facility before committing to any flight training. Evaluate the answers on the checklist, and then take time to think things over before making a decision. This proactive approach to choosing a flight school will ensure a student pilot contracts with a flight school or flight instructor best suited to their individual needs.

How To Choose a Certificated Flight Instructor (CFI)

Whether an individual chooses to train under 14 CFR part 141 or part 61, the key to an effective flight program is the quality of the ground and flight training received from the CFI. The flight instructor assumes total responsibility for training an individual to meet the standards required for certification within an ever-changing operating environment. A CFI should possess an understanding of the learning process, knowledge of the fundamentals of teaching, and the ability to communicate effectively with the student pilot. During the certification process, a flight instructor applicant is tested on the practical application of these skills in specific teaching situations. The flight instructor is crucial to the scenario-based training program endorsed by the FAA. He or she is trained to function in the learning environment as an advisor and guide for the learner. The duties, responsibilities, and authority of the CFI include the following:

- Orient the student to the scenario-based training system

- Help the student become a confident planner and inflight manager of each flight and a critical evaluator of their own performance

- Help the student understand the knowledge requirements present in real world applications

- Diagnose learning difficulties and help the student overcome them

- Evaluate student progress and maintain appropriate records

- Provide continuous review of student learning

Should a student pilot find the selected CFI is not training in a manner conducive for learning, or the student and CFI do not have compatible schedules, the student pilot should find another CFI. Choosing the right CFI is important because the quality of instruction and the knowledge and skills acquired from their flight instructor affect a student pilot's entire flying career.

The Student Pilot

The first step in becoming a pilot is to select a type of aircraft. FAA rules for obtaining a pilot's certificate differ depending on the type of aircraft flown. Individuals can choose among airplanes, gyroplanes, weight-shift, helicopters, powered parachutes, gliders, balloons, or airships. A pilot does not need a certificate to fly ultralight vehicles.

Basic Requirements

A student pilot is one who is being trained by an instructor pilot for his or her first full certificate, and is permitted to fly alone (solo) under specific, limited circumstances. Before a student pilot may be endorsed to fly solo, that student must have a Student Pilot Certificate. There are multiple ways that an aspiring pilot can obtain their Student Pilot Certificate. The application may be processed by an FAA inspector or technician, an FAA-Designated Pilot Examiner, a Certified Flight Instructor (CFI), or an Airman Certification Representative (ACR). If the application is completed electronically, the authorized person will submit the application to the FAA's Airman Certification Branch (AFS-760) in Oklahoma City, OK, via the Integrated

Airman Certification and Rating Application (IACRA). If the application is completed on paper, it must be sent to the local Flight Standards District Office (FSDO), who will forward it to AFS-760. Once the application is processed, the applicant will receive the Student Pilot Certificate by mail at the address provided on the application.

The aforementioned process will become effective on April 1, 2016. The new certificate will be printed on a plastic card, which will replace the paper certificate that was issued in the past. The plastic card certificate will not have an expiration date. Paper certificates issued prior to the new process will still expire according to the date on the certificate; however, under the new process, paper certificates cannot be renewed. Once the paper certificate expires, the Student Pilot must submit a new application under the new process. Another significant change in the new process is that flight instructors will now make endorsements for solo privileges in the Student Pilot's logbook, instead of endorsing the Student Pilot Certificate.

To be eligible for a Student Pilot Certificate, the applicant must:

- Be at least 16 years of age (14 years of age to pilot a glider or balloon).

- Be able to read, speak, write, and understand the English language.

Medical Certification Requirements

The second step in becoming a pilot is to obtain a medical certificate (if the choice of aircraft is an airplane, helicopter, gyroplane, or an airship). (The FAA suggests the individual get a medical certificate before beginning flight training to avoid the expense of flight training that cannot be continued due to a medical condition.) Balloon or glider pilots do not need a medical certificate, but do need to write a statement certifying that no medical defect exists that would prevent them from piloting a balloon or glider. The new sport pilot category does not require a medical examination; a driver's license can be used as proof of medical competence. Applicants who fail to meet certain requirements or who have physical disabilities which might limit, but not prevent, their acting as pilots, should contact the nearest FAA office. Anyone requesting an FAA Medical Clearance, Medical Certificate, or Student Pilot Medical Certificate can electronically complete an application through the FAA's MedXPress system available at https://medxpress.faa.gov/.

A medical certificate is obtained by passing a physical examination administered by a doctor who is an FAA-authorized AME. There are approximately 6,000 FAA-authorized AMEs in the nation. To find an AME near you, go to the FAA's AME locator at www.faa.gov/pilots/amelocator/. Medical certificates are designated as first class, second class, or third class. Generally, first class is designed for the airline transport pilot; second class for the commercial pilot; and third class for the student, recreational, and private pilot. A Student Pilot Certificate can be processed by an FAA inspector or technician, an FAA Designated pilot examiner (DPE), an Airman Certification Representative (ACR), or a Certified Flight Instructor (CFI). This certificate allows an individual who is being trained by a flight instructor to fly alone (solo) under specific, limited circumstances and must be carried with the student pilot while exercising solo flight privileges. The Student Pilot Certificate is only required when exercising solo flight privileges. The new plastic student certificate does not have an expiration date. For airmen who were issued a paper certificate, that certificate will remain valid until its expiration date. A paper certificate cannot be renewed. When the paper certificate expires, a new application must be completed via the IACRA system, and a new plastic certificate will be issued.

Student Pilot Solo Requirements

Once a student has accrued sufficient training and experience, a CFI can endorse the student's logbook to authorize limited solo flight in a specific type (make and model) of aircraft. A student pilot may not carry passengers, fly in furtherance of a business, or operate an aircraft outside of the various endorsements provided by the flight instructor. There is no minimum aeronautical knowledge or experience requirement for the issuance of a Student Pilot Certificate, however, the applicant must be at least 16 years of age (14 years of age for a pilot for glider or balloon), and they must be able to read, speak, write and understand the English language. There are, however, minimum aeronautical knowledge and experience requirements for student pilots to solo.

Becoming a Pilot

The course of instruction a student pilot follows depends on the type of certificate sought. It should include the ground and flight training necessary to acquire the knowledge and skills required to safely and efficiently function as a certificated pilot in the selected category and class of aircraft. The specific knowledge and skill areas for each category and class of aircraft are outlined in 14 CFR part 61. Eligibility, aeronautical knowledge, proficiency, and aeronautical requirements can be found in 14 CFR part 61.

- Recreational Pilot, see subpart D
- Private Pilot, see subpart E
- Sport Pilot, see subpart J

The knowledge-based portion of training is obtained through FAA handbooks such as this one, textbooks, and other sources of training and testing materials which are available in print form from the Superintendent of Documents, GPO, and online at the Regulatory Support Division: www.faa.gov/about/office_org/headquarters_offices/avs/offices/afs/afs600.

The CFI may also use commercial publications as a source of study materials, especially for aircraft categories where government materials are limited. A student pilot should follow the flight instructor's advice on what and when to study. Planning a definite study program and following it as closely as possible will help in scoring well on the knowledge test. Haphazard or disorganized study habits usually result in an unsatisfactory score.

In addition to learning aeronautical knowledge, such as the principles of flight, a student pilot is also required to gain skill in flight maneuvers. The selected category and class of aircraft determines the type of flight skills and number of flight hours to be obtained. There are four steps involved in learning a flight maneuver:

- The CFI introduces and demonstrates flight maneuver to the student.
- The CFI talks the student pilot through the maneuver.
- The student pilot practices the maneuver under CFI supervision.
- The CFI authorizes the student pilot to practice the maneuver solo.

Once the student pilot has shown proficiency in the required knowledge areas, flight maneuvers, and accrued the required amount of flight hours, the CFI endorses the student pilot logbook, which allows the student pilot to take the written and practical tests for pilot certification.

Knowledge and Skill Tests

Knowledge Tests

The knowledge test is the computer portion of the tests taken to obtain pilot certification. The test contains questions of the objective, multiple-choice type. This testing method conserves the applicant's time, eliminates any element of individual judgment in determining grades, and saves time in scoring.

FAA Airman Knowledge Test Guides for every type of pilot certificate address most questions you may have regarding the knowledge test process. The guides are available on-line (free of charge) at http://www.faa.gov/training_testing/testing/test_guides/.

When To Take the Knowledge Test

The knowledge test is more meaningful to the applicant and more likely to result in a satisfactory grade if it is taken after beginning the flight portion of the training. Therefore, the FAA recommends the knowledge test be taken after the student pilot has completed a solo cross-country flight. The operational knowledge gained by this experience can be used to the student's advantage in the knowledge test. The student pilot's CFI is the best person to determine when the applicant is ready to take the knowledge test.

Practical Test

The FAA has developed PTS for FAA pilot certificates and associated ratings. *[Figure 1-25]* In 2015, the FAA began transitioning to the ACS approach. The ACS is essentially an "enhanced" version of the PTS. It adds task-specific knowledge and risk management elements to each PTS Area of Operation and Task. The result is a holistic, integrated presentation of specific knowledge, skills, and risk management elements and performance metrics for each Area of Operation and Task The ACS evaluation program will eventually replace the PTS program for evaluating and certifying pilots.

The practical tests are administered by FAA ASIs and DPEs. Title 14 CFR part 61 specifies the areas of operation in which knowledge and skill must be demonstrated by the applicant. Since the FAA requires all practical tests be conducted in accordance with the appropriate PTS and the policies set forth in the Introduction section of the PTS book. The pilot applicant should become familiar with this book during training.

The PTS book is a testing document and not intended to be a training syllabus. An appropriately-rated flight instructor is responsible for training the pilot applicant to acceptable standards in all subject matter areas, procedures, and maneuvers. Descriptions of tasks and information on how to perform maneuvers and procedures are contained in reference and teaching documents such as this handbook. A list of reference documents is contained in the Introduction section of each PTS book. Copies may obtained by:

- Downloading from the FAA website at www.faa.gov

- Purchasing print copies from the GPO, Pittsburgh, Pennsylvania, or via their official online bookstore at www.access.gpo.gov

The flight proficiency maneuvers listed in 14 CFR part 61 are the standard skill requirements for certification. They are outlined in the PTS as "areas of operation." These are phases of the practical test arranged in a logical sequence within the standard. They begin with preflight preparation and end with postflight procedures. Each area of operation

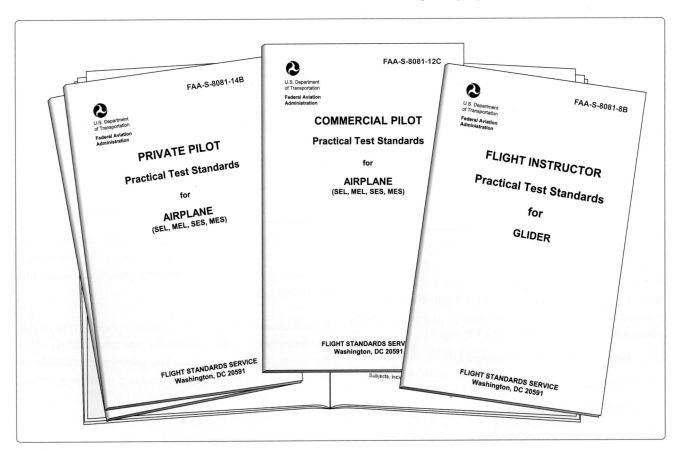

Figure 1-25. *Examples of Practical Test Standards.*

contains "tasks," which are comprised of knowledge areas, flight procedures, and/or flight maneuvers appropriate to the area of operation. The candidate is required to demonstrate knowledge and proficiency in all tasks for the original issuance of all pilot certificates.

When To Take the Practical Test

14 CFR part 61 establishes the ground school and flight experience requirements for the type of certification and aircraft selected. However, the CFI best determines when an applicant is qualified for the practical test. A practice practical test is an important step in the flight training process.

The applicant will be asked to present the following documentation:

- FAA Form 8710-1 (8710.11 for sport pilot applicants), Application for an Airman Certificate and/or Rating, with the flight instructor's recommendation

- An Airman Knowledge Test Report with a satisfactory grade

- A medical certificate (not required for glider or balloon), a Student Pilot Certificate, and a pilot logbook endorsed by a flight instructor for solo, solo cross-country (airplane and rotorcraft), and for the make and model aircraft to be used for the practical test (driver's license or medical certificate for sport pilot applicants)

- The pilot log book records

- A graduation certificate from an FAA-approved school (if applicable)

The applicant must provide an airworthy aircraft with equipment relevant to the areas of operation required for the practical test. He or she will also be asked to produce and explain the:

- Aircraft's registration certificate

- Aircraft's airworthiness certificate

- Aircraft's operating limitations or FAA-approved aircraft flight manual (if required)

- Aircraft equipment list

- Required weight and balance data

- Maintenance records

- Applicable airworthiness directives (ADs)

For a detailed explanation of the required pilot maneuvers and performance standards, refer to the PTS pertaining to the type of certification and aircraft selected. These standards may be downloaded free of charge from the FAA at www.faa.gov. They may also be purchased from the Superintendent of Documents

or GPO bookstores. Most airport fixed-base operators and flight schools carry a variety of government publications and charts, as well as commercially published materials.

Who Administers the FAA Practical Tests?

Due to the varied responsibilities of the FSDOs, practical tests are usually given by DPEs. An applicant should schedule the practical test by appointment to avoid conflicts and wasted time. A list of examiner names can be obtained from the local FSDO. Since a DPE serves without pay from the government for conducting practical tests and processing the necessary reports, the examiner is allowed to charge a reasonable fee. There is no charge for the practical test when conducted by an FAA inspector.

Role of the Certificated Flight Instructor

To become a CFI, a pilot must meet the provisions of 14 CFR part 61. The FAA places full responsibility for student flight training on the shoulders of the CFI, who is the cornerstone of aviation safety. It is the job of the flight instructor to train the student pilot in all the knowledge areas and teach the skills necessary for the student pilot to operate safely and competently as a certificated pilot in the NAS. The training includes airmanship skills, pilot judgment and decision-making, and good operating practices.

A pilot training program depends on the quality of the ground and flight instruction the student pilot receives. The flight instructor must possess a thorough understanding of the learning process, knowledge of the fundamentals of teaching, and the ability to communicate effectively with the student pilot. Use of a structured training program and formal course syllabus is crucial for effective and comprehensive flight training. It should be clear to the student in advance of every lesson what the course of training will involve and the criteria for successful completion. This should include the flight instructor briefing and debriefing the student before and after every lesson. Additionally, scenario-based training has become the preferred method of flight instruction today. This involves presenting the student with realistic flight scenarios and recommended actions for mitigating risk.

Insistence on correct techniques and procedures from the beginning of training by the flight instructor ensures that the student pilot develops proper flying habits. Any deficiencies in the maneuvers or techniques must immediately be emphasized and corrected. A flight instructor serves as a role model for the student pilot who observes the flying habits of his or her flight instructor during flight instruction, as well as when the instructor conducts other pilot operations. Thus, the flight instructor becomes a model of flying proficiency for the student who, consciously or unconsciously, attempts to imitate the instructor. For this reason, a flight instructor should observe recognized safety practices, as well as regulations during all flight operations.

The student pilot who enrolls in a pilot training program commits considerable time, effort, and expense to achieve a pilot certificate. Students often judge the effectiveness of the flight instructor and the success of the pilot training program based on their ability to pass the requisite FAA practical test. A competent flight instructor stresses to the student that practical tests are a sampling of pilot ability compressed into a short period of time. The goal of a flight instructor is to train the "total" pilot.

Role of the Designated Pilot Examiner

The Designated Pilot Examiner (DPE) plays an important role in the FAA's mission of promoting aviation safety by administering FAA practical tests for pilot and Flight Instructor Certificates and associated ratings. Although administering these tests is a responsibility of the ASI, the FAA's highest priority is making air travel safer by inspecting aircraft that fly in the United States. To satisfy the need for pilot testing and certification services, the FAA delegates certain responsibilities to private individuals who are not FAA employees.

Appointed in accordance with 14 CFR part 183, section 183.23, a DPE is an individual who meets the qualification requirements of the Pilot Examiner's Handbook, FAA Order 8710.3, and who:

- Is technically qualified

- Holds all pertinent category, class, and type ratings for each aircraft related to their designation

- Meets requirements of 14 CFR part 61, sections 61.56, 61.57, and 61.58, as appropriate

- Is current and qualified to act as PIC of each aircraft for which he or she is authorized

- Maintains at least a Third-Class Medical Certificate, if required

- Maintains a current Flight Instructor Certificate, if required

Designated to perform specific pilot certification tasks on behalf of the FAA, a DPE may charge a reasonable fee. Generally, a DPE's authority is limited to accepting applications and conducting practical tests leading to the issuance of specific pilot certificates and/or ratings. The majority of FAA practical tests at the private and commercial pilot levels are administered by DPEs.

DPE candidates must have good industry reputations for professionalism, integrity, a demonstrated willingness to serve the public, and must adhere to FAA policies and procedures in certification matters. The FAA expects the DPE to administer practical tests with the same degree of professionalism, using the same methods, procedures, and standards as an FAA ASI.

Chapter Summary

The FAA has entered the second century of civil aviation as a robust government organization and is taking full advantage of technology, such as Global Positioning System (GPS) satellite technology to enhance the safety of civil aviation. The Internet has also become an important tool in promoting aviation safety and providing around-the-clock resources for the aviation community. Handbooks, regulations, standards, references, and online courses are now available at www.faa.gov.

In keeping with the FAA's belief that safety is a learned behavior, the FAA offers many courses and seminars to enhance air safety. The FAA puts the burden of instilling safe flying habits on the flight instructor, who should follow basic flight safety practices and procedures in every flight operation he or she undertakes with a student pilot. Operational safety practices include, but are not limited to, collision avoidance procedures consisting of proper scanning techniques, use of checklists, runway incursion avoidance, positive transfer of controls, and workload management. These safety practices are discussed more fully within this handbook. Safe flight also depends on Scenario-Based Training (SBT) that teaches the student pilot how to respond in different flight situations. The FAA has incorporated these techniques along with decision-making methods, such as aeronautical decision-making (ADM), risk management, and crew resource management (CRM), which are covered more completely in Chapter 2, Aeronautical Decision-Making.

Chapter 2
Aeronautical Decision-Making

Introduction

Aeronautical decision-making (ADM) is decision-making in a unique environment—aviation. It is a systematic approach to the mental process used by pilots to consistently determine the best course of action in response to a given set of circumstances. It is what a pilot intends to do based on the latest information he or she has.

| Flight Skills Basic Maneuvering | Introduction and Use of Autopilot | Introduction and Use of Electronic Flight Systems To Manage Aircraft and Flight |

PHASE I — Piloting Tasks

PHASE II — Use of Autopilot

PHASE III — Use of Electronic Onboard Management systems

A. Analytical

B. Autom...

Situation

Pilot   Aircraft   Enviroment   External factors

Detection

Evaluation of event

- Risk or hazard
- Potential outcomes
- Capabilities of pilot
- Aircraft capabilities
- Outside factors

Outcome desired

Solutions to get you there
Solution 1
Solution 2
Solution 3
Solution 4

What is best action to do

Effect of decision

Problem remains

Done

23.4%   24.1%   15.7%   13%   3.5%   3.3%   2.6%   .7%

The Five Hazardous Attitudes

Anti-authority: "Don't tell me."
This attitude is found in people who do not like anyone telling sense, they are saying, "No one can tell me what to do." They m having someone tell them what to do or may regard rules, regulations as silly or unnecessary. However, it is always your prerogative to que if you feel it is in error.

Impulsivity:
This is the attitude of people who frequently feel the need to do s immediately. They do not stop to think about what they are about select the best alternative, and they do the first thing that come

Invulnerability:
Many people falsely believe that accidents happen to others know accidents can happen, and they know that anyone c they never really feel or believe that they will be personal this way are more likely to take chances and increase ri

Macho:
Pilots who are always trying to prove that they are bet do it—I'll show them." Pilots with this type of attitude taking risks in order to impress others. While this p characteristic, women are equally susceptible.

of Safety

Pilot capabilities

Task requirements

Time

Cruise

Approach & landing

The importance of learning and understanding effective ADM skills cannot be overemphasized. While progress is continually being made in the advancement of pilot training methods, aircraft equipment and systems, and services for pilots, accidents still occur. Despite all the changes in technology to improve flight safety, one factor remains the same: the human factor which leads to errors. It is estimated that approximately 80 percent of all aviation accidents are related to human factors and the vast majority of these accidents occur during landing (24.1 percent) and takeoff (23.4 percent). *[Figure 2-1]*

ADM is a systematic approach to risk assessment and stress management. To understand ADM is to also understand how personal attitudes can influence decision-making and how those attitudes can be modified to enhance safety in the flight deck. It is important to understand the factors that cause humans to make decisions and how the decision-making process not only works, but can be improved.

This chapter focuses on helping the pilot improve his or her ADM skills with the goal of mitigating the risk factors associated with flight. Advisory Circular (AC) 60-22, "Aeronautical Decision-Making," provides background references, definitions, and other pertinent information about ADM training in the general aviation (GA) environment. *[Figure 2-2]*

History of ADM

For over 25 years, the importance of good pilot judgment, or aeronautical decision-making (ADM), has been recognized as critical to the safe operation of aircraft, as well as accident avoidance. The airline industry, motivated by the need to reduce accidents caused by human factors, developed the first training programs based on improving ADM. Crew resource management (CRM) training for flight crews is focused on the effective use of all available resources: human resources, hardware, and information supporting ADM to facilitate crew cooperation and improve decision-making. The goal of all flight crews is good ADM and the use of CRM is one way to make good decisions.

Research in this area prompted the Federal Aviation Administration (FAA) to produce training directed at improving the decision-making of pilots and led to current FAA regulations that require that decision-making be taught as part of the pilot training curriculum. ADM research, development, and testing culminated in 1987 with the publication of six manuals oriented to the decision-making needs of variously rated pilots. These manuals provided multifaceted materials designed to reduce the number of decision-related accidents. The effectiveness of these materials was validated in independent studies where student pilots received such training in conjunction with the standard flying curriculum. When tested, the pilots who had received ADM-training made fewer in-flight errors than those who had

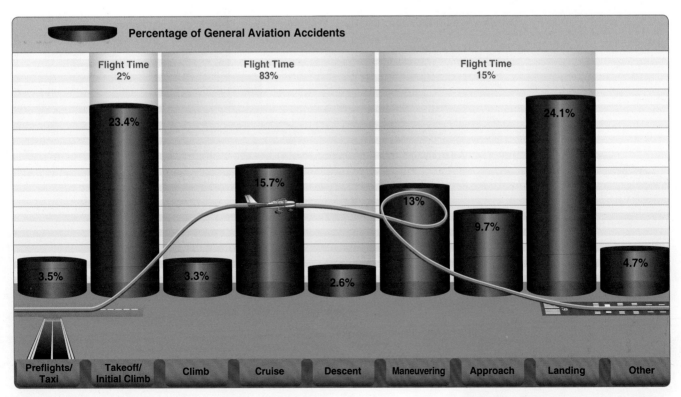

Figure 2-1. *The percentage of aviation accidents as they relate to the different phases of flight. Note that the greatest percentage of accidents take place during a minor percentage of the total flight.*

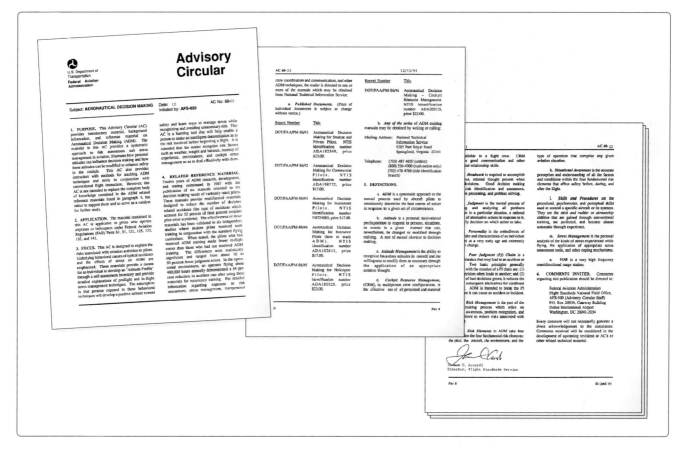

Figure 2-2. *Advisory Circular (AC) 60-22, "Aeronautical Decision Making," carries a wealth of information for the pilot to learn.*

not received ADM training. The differences were statistically significant and ranged from about 10 to 50 percent fewer judgment errors. In the operational environment, an operator flying about 400,000 hours annually demonstrated a 54 percent reduction in accident rate after using these materials for recurrency training.

Contrary to popular opinion, good judgment can be taught. Tradition held that good judgment was a natural by-product of experience, but as pilots continued to log accident-free flight hours, a corresponding increase of good judgment was assumed. Building upon the foundation of conventional decision-making, ADM enhances the process to decrease the probability of human error and increase the probability of a safe flight. ADM provides a structured, systematic approach to analyzing changes that occur during a flight and how these changes might affect the safe outcome of a flight. The ADM process addresses all aspects of decision-making in the flight deck and identifies the steps involved in good decision-making.

Steps for good decision-making are:

1. Identifying personal attitudes hazardous to safe flight

2. Learning behavior modification techniques

3. Learning how to recognize and cope with stress

4. Developing risk assessment skills

5. Using all resources

6. Evaluating the effectiveness of one's ADM skills

Risk Management

The goal of risk management is to proactively identify safety-related hazards and mitigate the associated risks. Risk management is an important component of ADM. When a pilot follows good decision-making practices, the inherent risk in a flight is reduced or even eliminated. The ability to make good decisions is based upon direct or indirect experience and education. The formal risk management decision-making process involves six steps as shown in *Figure 2-3.*

Consider automotive seat belt use. In just two decades, seat belt use has become the norm, placing those who do not wear seat belts outside the norm, but this group may learn to wear a seat belt by either direct or indirect experience. For example, a driver learns through direct experience about the value of wearing a seat belt when he or she is involved in a car accident that leads to a personal injury. An indirect learning experience occurs when a loved one is injured during a car accident because he or she failed to wear a seat belt.

As you work through the ADM cycle, it is important to remember the four fundamental principles of risk management.

Figure 2-3. *Risk management decision-making process.*

1. Accept no unnecessary risk. Flying is not possible without risk, but unnecessary risk comes without a corresponding return. If you are flying a new airplane for the first time, you might determine that the risk of making that flight in low visibility conditions is unnecessary.

2. Make risk decisions at the appropriate level. Risk decisions should be made by the person who can develop and implement risk controls. Remember that you are pilot-in-command, so never let anyone else—not ATC and not your passengers—make risk decisions for you.

3. Accept risk when benefits outweigh dangers (costs). In any flying activity, it is necessary to accept some degree of risk. A day with good weather, for example, is a much better time to fly an unfamiliar airplane for the first time than a day with low IFR conditions.

4. Integrate risk management into planning at all levels. Because risk is an unavoidable part of every flight, safety requires the use of appropriate and effective risk management not just in the preflight planning stage, but in all stages of the flight.

While poor decision-making in everyday life does not always lead to tragedy, the margin for error in aviation is thin. Since ADM enhances management of an aeronautical environment, all pilots should become familiar with and employ ADM.

Crew Resource Management (CRM) and Single-Pilot Resource Management

While CRM focuses on pilots operating in crew environments, many of the concepts apply to single-pilot operations. Many CRM principles have been successfully applied to single-pilot aircraft and led to the development of Single-Pilot Resource Management (SRM). SRM is defined as the art and science of managing all the resources (both on-board the aircraft and from outside sources) available to a single pilot (prior to and during flight) to ensure the successful outcome of the flight. SRM includes the concepts of ADM, risk management (RM), task management (TM), automation management (AM), controlled flight into terrain (CFIT) awareness, and situational awareness (SA). SRM training helps the pilot maintain situational awareness by managing the automation and associated aircraft control and navigation tasks. This enables the pilot to accurately assess and manage risk and make accurate and timely decisions.

SRM is all about helping pilots learn how to gather information, analyze it, and make decisions. Although the flight is coordinated by a single person and not an onboard flight crew, the use of available resources such as auto-pilot and air traffic control (ATC) replicates the principles of CRM.

Hazard and Risk

Two defining elements of ADM are hazard and risk. Hazard is a real or perceived condition, event, or circumstance that a pilot encounters. When faced with a hazard, the pilot makes an assessment of that hazard based upon various factors. The pilot assigns a value to the potential impact of the hazard, which qualifies the pilot's assessment of the hazard—risk.

Therefore, risk is an assessment of the single or cumulative hazard facing a pilot; however, different pilots see hazards differently. For example, the pilot arrives to preflight and discovers a small, blunt type nick in the leading edge at the middle of the aircraft's prop. Since the aircraft is parked on the tarmac, the nick was probably caused by another aircraft's prop wash blowing some type of debris into the propeller. The nick is the hazard (a present condition). The risk is prop fracture if the engine is operated with damage to a prop blade.

The seasoned pilot may see the nick as a low risk. He realizes this type of nick diffuses stress over a large area, is located in the strongest portion of the propeller, and based on experience; he does not expect it to propagate a crack that can lead to high risk problems. He does not cancel his flight.

The inexperienced pilot may see the nick as a high risk factor because he is unsure of the affect the nick will have on the operation of the prop, and he has been told that damage to a prop could cause a catastrophic failure. This assessment leads him to cancel his flight.

Therefore, elements or factors affecting individuals are different and profoundly impact decision-making. These are called human factors and can transcend education, experience, health, physiological aspects, etc.

Another example of risk assessment was the flight of a Beechcraft King Air equipped with deicing and anti-icing. The pilot deliberately flew into moderate to severe icing conditions while ducking under cloud cover. A prudent pilot would assess the risk as high and beyond the capabilities of the aircraft, yet this pilot did the opposite. Why did the pilot take this action?

Past experience prompted the action. The pilot had successfully flown into these conditions repeatedly although the icing conditions were previously forecast 2,000 feet above the surface. This time, the conditions were forecast from the surface. Since the pilot was in a hurry and failed to factor in the difference between the forecast altitudes, he assigned a low risk to the hazard and took a chance. He and the passengers died from a poor risk assessment of the situation.

Hazardous Attitudes and Antidotes

Being fit to fly depends on more than just a pilot's physical condition and recent experience. For example, attitude affects the quality of decisions. Attitude is a motivational predisposition to respond to people, situations, or events in a given manner. Studies have identified five hazardous attitudes that can interfere with the ability to make sound decisions and exercise authority properly: anti-authority, impulsivity, invulnerability, macho, and resignation. *[Figure 2-4]*

Hazardous attitudes contribute to poor pilot judgment but can be effectively counteracted by redirecting the hazardous attitude so that correct action can be taken. Recognition of hazardous thoughts is the first step toward neutralizing them. After recognizing a thought as hazardous, the pilot should label it as hazardous, then state the corresponding antidote. Antidotes should be memorized for each of the hazardous attitudes so they automatically come to mind when needed.

The Five Hazardous Attitudes	Antidote
Anti-authority: "Don't tell me." This attitude is found in people who do not like anyone telling them what to do. In a sense, they are saying, "No one can tell me what to do." They may be resentful of having someone tell them what to do or may regard rules, regulations, and procedures as silly or unnecessary. However, it is always your prerogative to question authority if you feel it is in error.	**Follow the rules. They are usually right.**
Impulsivity: "Do it quickly." This is the attitude of people who frequently feel the need to do something, anything, immediately. They do not stop to think about what they are about to do, they do not select the best alternative, and they do the first thing that comes to mind.	**Not so fast. Think first.**
Invulnerability: "It won't happen to me." Many people falsely believe that accidents happen to others, but never to them. They know accidents can happen, and they know that anyone can be affected. However, they never really feel or believe that they will be personally involved. Pilots who think this way are more likely to take chances and increase risk.	**It could happen to me.**
Macho: "I can do it." Pilots who are always trying to prove that they are better than anyone else think, "I can do it—I'll show them." Pilots with this type of attitude will try to prove themselves by taking risks in order to impress others. While this pattern is thought to be a male characteristic, women are equally susceptible.	**Taking chances is foolish.**
Resignation: "What's the use?" Pilots who think, "What's the use?" do not see themselves as being able to make a great deal of difference in what happens to them. When things go well, the pilot is apt to think that it is good luck. When things go badly, the pilot may feel that someone is out to get them or attribute it to bad luck. The pilot will leave the action to others, for better or worse. Sometimes, such pilots will even go along with unreasonable requests just to be a "nice guy."	**I'm not helpless. I can make a difference.**

Figure 2-4. *The five hazardous attitudes identified through past and contemporary study.*

Risk

During each flight, the single pilot makes many decisions under hazardous conditions. To fly safely, the pilot needs to assess the degree of risk and determine the best course of action to mitigate the risk.

Assessing Risk

For the single pilot, assessing risk is not as simple as it sounds. For example, the pilot acts as his or her own quality control in making decisions. If a fatigued pilot who has flown 16 hours is asked if he or she is too tired to continue flying, the answer may be "no." Most pilots are goal oriented and when asked to accept a flight, there is a tendency to deny personal limitations while adding weight to issues not germane to the mission. For example, pilots of helicopter emergency services (EMS) have been known (more than other groups) to make flight decisions that add significant weight to the patient's welfare. These pilots add weight to intangible factors (the patient in this case) and fail to appropriately quantify actual hazards, such as fatigue or weather, when making flight decisions. The single pilot who has no other crew member for consultation must wrestle with the intangible factors that draw one into a hazardous position. Therefore, he or she has a greater vulnerability than a full crew.

Examining National Transportation Safety Board (NTSB) reports and other accident research can help a pilot learn to assess risk more effectively. For example, the accident rate during night visual flight rules (VFR) decreases by nearly 50 percent once a pilot obtains 100 hours and continues to decrease until the 1,000 hour level. The data suggest that for the first 500 hours, pilots flying VFR at night might want to establish higher personal limitations than are required by the regulations and, if applicable, apply instrument flying skills in this environment.

Several risk assessment models are available to assist in the process of assessing risk. The models, all taking slightly different approaches, seek a common goal of assessing risk in an objective manner. The most basic tool is the risk matrix. *[Figure 2-5]* It assesses two items: the likelihood of an event occurring and the consequence of that event.

Likelihood of an Event

Likelihood is nothing more than taking a situation and determining the probability of its occurrence. It is rated as probable, occasional, remote, or improbable. For example, a pilot is flying from point A to point B (50 miles) in marginal visual flight rules (MVFR) conditions. The likelihood of encountering potential instrument meteorological conditions (IMC) is the first question the pilot needs to answer. The experiences of other pilots, coupled with the forecast, might

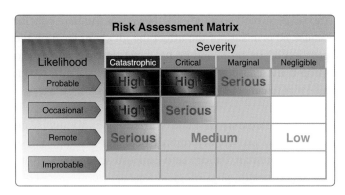

Figure 2-5. *This risk matrix can be used for almost any operation by assigning likelihood and consequence. In the case presented, the pilot assigned a likelihood of occasional and the severity as catastrophic. As one can see, this falls in the high risk area.*

cause the pilot to assign "occasional" to determine the probability of encountering IMC.

The following are guidelines for making assignments.

- Probable—an event will occur several times
- Occasional—an event will probably occur sometime
- Remote—an event is unlikely to occur, but is possible
- Improbable—an event is highly unlikely to occur

Severity of an Event

The next element is the severity or consequence of a pilot's action(s). It can relate to injury and/or damage. If the individual in the example above is not an instrument rated pilot, what are the consequences of him or her encountering inadvertent IMC conditions? In this case, because the pilot is not IFR rated, the consequences are catastrophic. The following are guidelines for this assignment.

- Catastrophic—results in fatalities, total loss
- Critical—severe injury, major damage
- Marginal—minor injury, minor damage
- Negligible—less than minor injury, less than minor system damage

Simply connecting the two factors as shown in *Figure 2-5* indicates the risk is high and the pilot must either not fly or fly only after finding ways to mitigate, eliminate, or control the risk.

Although the matrix in *Figure 2-5* provides a general viewpoint of a generic situation, a more comprehensive program can be made that is tailored to a pilot's flying. *[Figure 2-6]* This program includes a wide array of aviation-related activities specific to the pilot and assesses health, fatigue, weather,

RISK ASSESSMENT

Pilot's Name

Flight From **To**

SLEEP
1. Did not sleep well or less than 8 hours	2
2. Slept well	0

HOW DO YOU FEEL?
1. Have a cold or ill	4
2. Feel great	0
3. Feel a bit off	2

WEATHER AT TERMINATION
1. Greater than 5 miles visibility and 3,000 feet ceilings	1
2. At least 3 miles visibility and 1,000 feet ceilings, but less than 3,000 feet ceilings and 5 miles visibility	3
3. IMC conditions	4

Column total ⬭

HOW IS THE DAY GOING?
1. Seems like one thing after another (late, making errors, out of step)	3
2. Great day	0

IS THE FLIGHT
1. Day?	1
2. Night?	3

PLANNING
1. Rush to get off ground		3
2. No hurry		1
3. Used charts and computer to assist		0
4. Used computer program for all planning	Yes	3
	No	0
5. Did you verify weight and balance?	Yes	0
	No	3
6. Did you evaluate performance?	Yes	0
	No	3
7. Do you brief your passangers on the ground and in flight?	Yes	0
	No	2

Column total ⬭

TOTAL SCORE

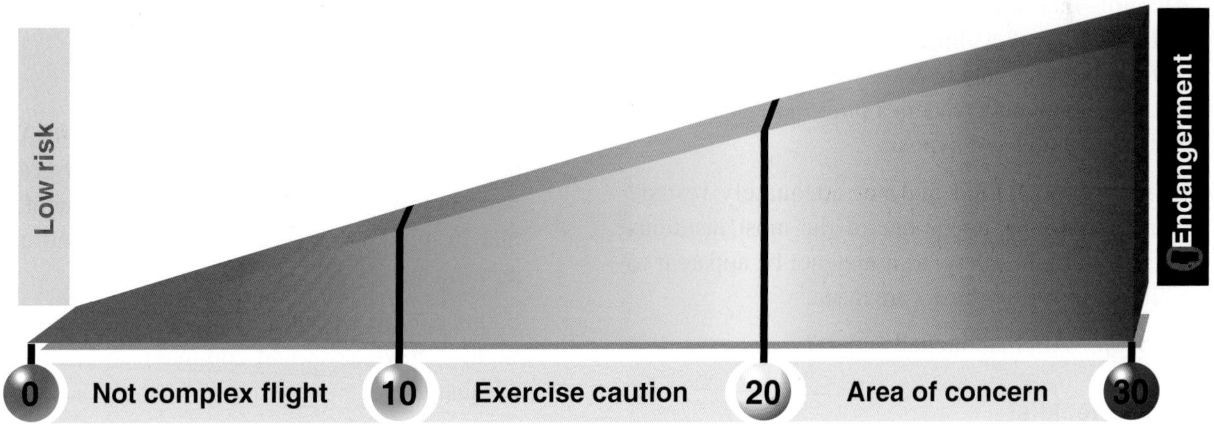

Low risk | 0 Not complex flight | 10 Exercise caution | 20 Area of concern | 30 Endangerment

Figure 2-6. *Example of a more comprehensive risk assessment program.*

capabilities, etc. The scores are added and the overall score falls into various ranges, with the range representative of actions that a pilot imposes upon himself or herself.

Mitigating Risk

Risk assessment is only part of the equation. After determining the level of risk, the pilot needs to mitigate the risk. For example, the pilot flying from point A to point B (50 miles) in MVFR conditions has several ways to reduce risk:

- Wait for the weather to improve to good visual flight rules (VFR) conditions.

- Take an instrument-rated pilot.

- Delay the flight.

- Cancel the flight.

- Drive.

One of the best ways single pilots can mitigate risk is to use the IMSAFE checklist to determine physical and mental readiness for flying:

1. Illness—Am I sick? Illness is an obvious pilot risk.

2. Medication—Am I taking any medicines that might affect my judgment or make me drowsy?

3. Stress—Am I under psychological pressure from the job? Do I have money, health, or family problems? Stress causes concentration and performance problems. While the regulations list medical conditions that require grounding, stress is not among them. The pilot should consider the effects of stress on performance.

4. Alcohol—Have I been drinking within 8 hours? Within 24 hours? As little as one ounce of liquor, one bottle of beer, or four ounces of wine can impair flying skills. Alcohol also renders a pilot more susceptible to disorientation and hypoxia.

5. Fatigue—Am I tired and not adequately rested? Fatigue continues to be one of the most insidious hazards to flight safety, as it may not be apparent to a pilot until serious errors are made.

6. Emotion—Am I emotionally upset?

The PAVE Checklist

Another way to mitigate risk is to perceive hazards. By incorporating the PAVE checklist into preflight planning, the pilot divides the risks of flight into four categories: **P**ilot-in-command (PIC), **A**ircraft, en**V**ironment, and **E**xternal pressures (PAVE) which form part of a pilot's decision-making process.

With the PAVE checklist, pilots have a simple way to remember each category to examine for risk prior to each flight.

Once a pilot identifies the risks of a flight, he or she needs to decide whether the risk, or combination of risks, can be managed safely and successfully. If not, make the decision to cancel the flight. If the pilot decides to continue with the flight, he or she should develop strategies to mitigate the risks. One way a pilot can control the risks is to set personal minimums for items in each risk category. These are limits unique to that individual pilot's current level of experience and proficiency.

For example, the aircraft may have a maximum crosswind component of 15 knots listed in the aircraft flight manual (AFM), and the pilot has experience with 10 knots of direct crosswind. It could be unsafe to exceed a 10 knot crosswind component without additional training. Therefore, the 10 knot crosswind experience level is that pilot's personal limitation until additional training with a certificated flight instructor (CFI) provides the pilot with additional experience for flying in crosswinds that exceed 10 knots.

One of the most important concepts that safe pilots understand is the difference between what is "legal" in terms of the regulations, and what is "smart" or "safe" in terms of pilot experience and proficiency.

P = Pilot in Command (PIC)

The pilot is one of the risk factors in a flight. The pilot must ask, "Am I ready for this trip?" in terms of experience, recency, currency, physical, and emotional condition. The IMSAFE checklist provides the answers.

A = Aircraft

What limitations will the aircraft impose upon the trip? Ask the following questions:

- Is this the right aircraft for the flight?

- Am I familiar with and current in this aircraft? Aircraft performance figures and the AFM are based on a brand new aircraft flown by a professional test pilot. Keep that in mind while assessing personal and aircraft performance.

- Is this aircraft equipped for the flight? Instruments? Lights? Navigation and communication equipment adequate?

- Can this aircraft use the runways available for the trip with an adequate margin of safety under the conditions to be flown?

- Can this aircraft carry the planned load?

- Can this aircraft operate at the altitudes needed for the trip?

- Does this aircraft have sufficient fuel capacity, with reserves, for trip legs planned?

- Does the fuel quantity delivered match the fuel quantity ordered?

V = EnVironment

Weather

Weather is a major environmental consideration. Earlier it was suggested pilots set their own personal minimums, especially when it comes to weather. As pilots evaluate the weather for a particular flight, they should consider the following:

- What is the current ceiling and visibility? In mountainous terrain, consider having higher minimums for ceiling and visibility, particularly if the terrain is unfamiliar.

- Consider the possibility that the weather may be different than forecast. Have alternative plans and be ready and willing to divert, should an unexpected change occur.

- Consider the winds at the airports being used and the strength of the crosswind component.

- If flying in mountainous terrain, consider whether there are strong winds aloft. Strong winds in mountainous terrain can cause severe turbulence and downdrafts and be very hazardous for aircraft even when there is no other significant weather.

- Are there any thunderstorms present or forecast?

- If there are clouds, is there any icing, current or forecast? What is the temperature/dew point spread and the current temperature at altitude? Can descent be made safely all along the route?

- If icing conditions are encountered, is the pilot experienced at operating the aircraft's deicing or anti-icing equipment? Is this equipment in good condition and functional? For what icing conditions is the aircraft rated, if any?

Terrain

Evaluation of terrain is another important component of analyzing the flight environment.

- To avoid terrain and obstacles, especially at night or in low visibility, determine safe altitudes in advance by using the altitudes shown on VFR and IFR charts during preflight planning.

- Use maximum elevation figures (MEFs) and other easily obtainable data to minimize chances of an inflight collision with terrain or obstacles.

Airport

- What lights are available at the destination and alternate airports? VASI/PAPI or ILS glideslope guidance? Is the terminal airport equipped with them? Are they working? Will the pilot need to use the radio to activate the airport lights?

- Check the Notices to Airmen (NOTAM) for closed runways or airports. Look for runway or beacon lights out, nearby towers, etc.

- Choose the flight route wisely. An engine failure gives the nearby airports supreme importance.

- Are there shorter or obstructed fields at the destination and/or alternate airports?

Airspace

- If the trip is over remote areas, is there appropriate clothing, water, and survival gear onboard in the event of a forced landing?

- If the trip includes flying over water or unpopulated areas with the chance of losing visual reference to the horizon, the pilot must be prepared to fly IFR.

- Check the airspace and any temporary flight restriction (TFRs) along the route of flight.

Nighttime

Night flying requires special consideration.

- If the trip includes flying at night over water or unpopulated areas with the chance of losing visual reference to the horizon, the pilot must be prepared to fly IFR.

- Will the flight conditions allow a safe emergency landing at night?

- Perform preflight check of all aircraft lights, interior and exterior, for a night flight. Carry at least two flashlights—one for exterior preflight and a smaller one that can be dimmed and kept nearby.

E = External Pressures

External pressures are influences external to the flight that create a sense of pressure to complete a flight—often at the expense of safety. Factors that can be external pressures include the following:

- Someone waiting at the airport for the flight's arrival

- A passenger the pilot does not want to disappoint

- The desire to demonstrate pilot qualifications

- The desire to impress someone (Probably the two most dangerous words in aviation are "Watch this!")

- The desire to satisfy a specific personal goal ("get-home-itis," "get-there-itis," and "let's-go-itis")

- The pilot's general goal-completion orientation

- Emotional pressure associated with acknowledging that skill and experience levels may be lower than a pilot would like them to be. Pride can be a powerful external factor!

Managing External Pressures

Management of external pressure is the single most important key to risk management because it is the one risk factor category that can cause a pilot to ignore all the other risk factors. External pressures put time-related pressure on the pilot and figure into a majority of accidents.

The use of personal standard operating procedures (SOPs) is one way to manage external pressures. The goal is to supply a release for the external pressures of a flight. These procedures include but are not limited to:

- Allow time on a trip for an extra fuel stop or to make an unexpected landing because of weather.

- Have alternate plans for a late arrival or make backup airline reservations for must-be-there trips.

- For really important trips, plan to leave early enough so that there would still be time to drive to the destination, if necessary.

- Advise those who are waiting at the destination that the arrival may be delayed. Know how to notify them when delays are encountered.

- Manage passengers' expectations. Make sure passengers know that they might not arrive on a firm schedule, and if they must arrive by a certain time, they should make alternative plans.

- Eliminate pressure to return home, even on a casual day flight, by carrying a small overnight kit containing prescriptions, contact lens solutions, toiletries, or other necessities on every flight.

The key to managing external pressure is to be ready for and accept delays. Remember that people get delayed when traveling on airlines, driving a car, or taking a bus. The pilot's goal is to manage risk, not create hazards. *[Figure 2-7]*

Human Factors

Why are human conditions, such as fatigue, complacency and stress, so important in aviation? These conditions, along with many others, are called human factors. Human factors directly cause or contribute to many aviation accidents and

Pilot

A pilot must continually make decisions about competency, condition of health, mental and emotional state, level of fatigue, and many other variables. For example, a pilot may be called early in the morning to make a long flight. If a pilot has had only a few hours of sleep and is concerned that the congestion being experienced could be the onset of a cold, it would be prudent to consider if the flight could be accomplished safely.

A pilot had only 4 hours of sleep the night before being asked by the boss to fly to a meeting in a city 750 miles away. The reported weather was marginal and not expected to improve. After assessing fitness as a pilot, it was decided that it would not be wise to make the flight. The boss was initially unhappy, but later convinced by the pilot that the risks involved were unacceptable.

Environment

This encompasses many elements not pilot or airplane related. It can include such factors as weather, air traffic control, navigational aids (NAVAIDS), terrain, takeoff and landing areas, and surrounding obstacles. Weather is one element that can change drastically over time and distance.

A pilot was landing a small airplane just after a heavy jet had departed a parallel runway. The pilot assumed that wake turbulence would not be a problem since landings had been performed under similar circumstances. Due to a combination of prevailing winds and wake turbulence from the heavy jet drifting across the landing runway, the airplane made a hard landing. The pilot made an error when assessing the flight environment.

Aircraft

A pilot will frequently base decisions on the evaluations of the aircraft, such as performance, equipment, or airworthiness.

During a preflight, a pilot noticed a small amount of oil dripping from the bottom of the cowling. Although the quantity of oil seemed insignificant at the time, the pilot decided to delay the takeoff and have a mechanic check the source of the oil. The pilot's good judgment was confirmed when the mechanic found that one of the oil cooler hose fittings was loose.

External pressures

The interaction between the pilot, airplane, and the environment is greatly influenced by the purpose of each flight operation. The pilot must evaluate the three previous areas to decide on the desirability of undertaking or continuing the flight as planned. It is worth asking why the flight is being made, how critical is it to maintain the schedule, and is the trip worth the risks?

On a ferry flight to deliver an airplane from the factory, in marginal weather conditions, the pilot calculated the groundspeed and determined that the airplane would arrive at the destination with only 10 minutes of fuel remaining. The pilot was determined to keep on schedule by trying to "stretch" the fuel supply instead of landing to refuel. After landing with low fuel state, the pilot realized that this could have easily resulted in an emergency landing in deteriorating weather conditions. This was a chance that was not worth taking to keep the planned schedule.

Figure 2-7. *The PAVE checklist.*

have been documented as a primary contributor to more than 70 percent of aircraft accidents.

Typically, human factor incidents/accidents are associated with flight operations but recently have also become a major concern in aviation maintenance and air traffic management as well. *[Figure 2-8]* Over the past several years, the FAA has made the study and research of human factors a top priority by working closely with engineers, pilots, mechanics, and ATC to apply the latest knowledge about human factors in an effort to help operators and maintainers improve safety and efficiency in their daily operations.

Human factors science, or human factors technologies, is a multidisciplinary field incorporating contributions from psychology, engineering, industrial design, statistics, operations research, and anthropometry. It is a term that covers the science of understanding the properties of human capability, the application of this understanding to the design, development and deployment of systems and services, and the art of ensuring successful application of human factor principles into all aspects of aviation to include pilots, ATC, and aviation maintenance. Human factors is often considered synonymous with CRM or maintenance resource management (MRM) but is really much broader in both its knowledge base and scope. Human factors involves gathering research specific to certain situations (i.e., flight, maintenance, stress levels, knowledge) about human abilities, limitations, and other characteristics and applying it to tool design, machines, systems, tasks, jobs, and environments to produce safe, comfortable, and effective human use. The entire aviation community benefits greatly from human factors research and development as it helps better understand how humans can most safely and efficiently perform their jobs and improve the tools and systems in which they interact.

Human Behavior

Studies of human behavior have tried to determine an individual's predisposition to taking risks and the level of an individual's involvement in accidents. In 1951, a study regarding injury-prone children was published by Elizabeth Mechem Fuller and Helen B. Baune, of the University of Minnesota. The study was comprised of two separate groups of second grade students. Fifty-five students were considered accident repeaters and 48 students had no accidents. Both groups were from the same school of 600 and their family demographics were similar.

The accident-free group showed a superior knowledge of safety, was considered industrious and cooperative

Figure 2-8. *Human factors effects pilots, aviation maintenance technicians (AMTs) and air traffic control (ATC).*

with others, but were not considered physically inclined. The accident-repeater group had better gymnastic skills, was considered aggressive and impulsive, demonstrated rebellious behavior when under stress, were poor losers, and liked to be the center of attention. One interpretation of this data—an adult predisposition to injury stems from childhood behavior and environment—leads to the conclusion that any pilot group should be comprised only of pilots who are safety-conscious, industrious, and cooperative.

Clearly, this is not only an inaccurate inference, it is impossible. Pilots are drawn from the general population and exhibit all types of personality traits. Thus, it is important that good decision-making skills be taught to all pilots.

Historically, the term "pilot error" has been used to describe an accident in which an action or decision made by the pilot was the cause or a contributing factor that led to the accident. This definition also includes the pilot's failure to make a correct decision or take proper action. From a broader perspective, the phrase "human factors related" more aptly describes these accidents. A single decision or event does not lead to an accident, but a series of events and the resultant decisions together form a chain of events leading to an outcome.

In his article "Accident-Prone Pilots," Dr. Patrick R. Veillette uses the history of "Captain Everyman" to demonstrate how aircraft accidents are caused more by a chain of poor choices rather than one single poor choice. In the case of Captain Everyman, after a gear-up landing accident, he became involved in another accident while taxiing a Beech 58P Baron out of the ramp. Interrupted by a radio call from the dispatcher, Everyman neglected to complete the fuel cross-feed check before taking off. Everyman, who was flying solo, left the right-fuel selector in the cross-feed position. Once aloft and cruising, he noticed a right roll tendency and corrected with aileron trim. He did not realize that both engines were feeding off the left wing's tank, making the wing lighter.

After two hours of flight, the right engine quit when Everyman was flying along a deep canyon gorge. While he was trying to troubleshoot the cause of the right engine's failure, the left engine quit. Everyman landed the aircraft on a river sand bar but it sank into ten feet of water.

Several years later Everyman flew a de Havilland Twin Otter to deliver supplies to a remote location. When he returned to home base and landed, the aircraft veered sharply to the left, departed the runway, and ran into a marsh 375 feet from the runway. The airframe and engines sustained considerable damage. Upon inspecting the wreck, accident investigators found the nose wheel steering tiller in the fully deflected position. Both the after takeoff and before landing checklists require the tiller to be placed in the neutral position. Everyman had overlooked this item.

Now, is Everyman accident prone or just unlucky? Skipping details on a checklist appears to be a common theme in the preceding accidents. While most pilots have made similar mistakes, these errors were probably caught prior to a mishap due to extra margin, good warning systems, a sharp copilot, or just good luck. What makes a pilot less prone to accidents?

The successful pilot possesses the ability to concentrate, manage workloads, and monitor and perform several simultaneous tasks. Some of the latest psychological screenings used in aviation test applicants for their ability to multitask, measuring both accuracy, as well as the individual's ability to focus attention on several subjects simultaneously. The FAA oversaw an extensive research study on the similarities and dissimilarities of accident-free pilots and those who were not. The project surveyed over 4,000 pilots, half of whom had "clean" records while the other half had been involved in an accident.

Five traits were discovered in pilots prone to having accidents. These pilots:

- Have disdain toward rules
- Have very high correlation between accidents on their flying records and safety violations on their driving records
- Frequently fall into the "thrill and adventure seeking" personality category
- Are impulsive rather than methodical and disciplined, both in their information gathering and in the speed and selection of actions to be taken
- Have a disregard for or tend to under utilize outside sources of information, including copilots, flight attendants, flight service personnel, flight instructors, and ATC

The Decision-Making Process

An understanding of the decision-making process provides the pilot with a foundation for developing ADM and SRM skills. While some situations, such as engine failure, require an immediate pilot response using established procedures, there is usually time during a flight to analyze any changes that occur, gather information, and assess risks before reaching a decision.

Risk management and risk intervention is much more than the simple definitions of the terms might suggest. Risk management and risk intervention are decision-making processes designed to systematically identify hazards, assess the degree of risk, and determine the best course of action. These processes involve the identification of hazards, followed by assessments of the risks, analysis of the controls, making control decisions, using the controls, and monitoring the results.

The steps leading to this decision constitute a decision-making process. Three models of a structured framework for problem-solving and decision-making are the 5P, the 3P using PAVE, CARE and TEAM, and the DECIDE models. They provide assistance in organizing the decision process. All these models have been identified as helpful to the single pilot in organizing critical decisions.

Single-Pilot Resource Management (SRM)

Single-Pilot Resource Management (SRM) is about how to gather information, analyze it, and make decisions. Learning how to identify problems, analyze the information, and make informed and timely decisions is not as straightforward as the training involved in learning specific maneuvers. Learning how to judge a situation and "how to think" in the endless variety of situations encountered while flying out in the "real world" is more difficult.

There is no one right answer in ADM, rather each pilot is expected to analyze each situation in light of experience level, personal minimums, and current physical and mental readiness level, and make his or her own decision.

The 5 Ps Check

SRM sounds good on paper, but it requires a way for pilots to understand and use it in their daily flights. One practical application is called the "Five Ps (5 Ps)." *[Figure 2-9]* The 5 Ps consist of "the Plan, the Plane, the Pilot, the Passengers, and the Programming." Each of these areas consists of a set of challenges and opportunities that every pilot encounters. Each challenge and opportunity can substantially increase or decrease the risk of successfully completing the flight based on the pilot's ability to make informed and timely decisions. The 5 Ps are used to evaluate the pilot's current situation at key decision points during the flight or when an emergency arises. These decision points include preflight, pretakeoff, hourly or at the midpoint of the flight, pre-descent, and just prior to the final approach fix or for VFR operations, just prior to entering the traffic pattern.

The 5 Ps are based on the idea that pilots have essentially five variables that impact his or her environment and forcing him or her to make a single critical decision, or several less critical decisions, that when added together can create a critical outcome. These variables are the Plan, the Plane, the Pilot, the Passengers, and the Programming. This concept stems from the belief that current decision-making models tended to be reactionary in nature. A change has to occur and be detected to drive a risk management decision by the pilot. For instance, many pilots complete risk management sheets prior to takeoff. These form a catalog of risks that may be encountered that day. Each of these risks is assigned a numerical value. If the total of these numerical values exceeds a predetermined level, the flight is altered or cancelled. Informal research shows that while these are useful documents for teaching risk factors, they are almost never used outside of formal training programs. The 5P concept is an attempt to take the information contained in those sheets and in the other available models and use it.

The 5P concept relies on the pilot to adopt a "scheduled" review of the critical variables at points in the flight where decisions are most likely to be effective. For instance, the easiest point to cancel a flight due to bad weather is before the pilot and passengers walk out the door and load the aircraft. So the first decision point is preflight in the flight planning room, where all the information is readily available to make a sound decision, and where communication and Fixed Base Operator (FBO) services are readily available to make alternate travel plans.

The second easiest point in the flight to make a critical safety decision is just prior to takeoff. Few pilots have ever had to make an "emergency takeoff." While the point of the 5P check is to help the pilot fly, the correct application of the 5 P before takeoff is to assist in making a reasoned go/no-go decision based on all the information available. That decision will usually be to "go," with certain restrictions and changes, but may also be a "no-go." The key idea is that these two points in the process of flying are critical go/no-go points on each and every flight.

The third place to review the 5 Ps is at the midpoint of the flight. Often, pilots may wait until the Automated Terminal information Service (ATIS) is in range to check weather, yet, at this point in the flight, many good options have already passed behind the aircraft and pilot. Additionally, fatigue

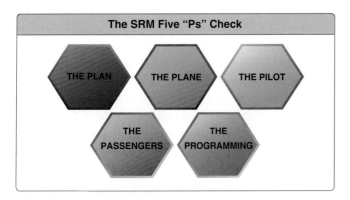

The SRM Five "Ps" Check

THE PLAN   THE PLANE   THE PILOT

THE PASSENGERS   THE PROGRAMMING

Figure 2-9. *The Five Ps checklist.*

and low-altitude hypoxia serve to rob the pilot of much of his or her energy by the end of a long and tiring flight day. This leads to a transition from a decision-making mode to an acceptance mode on the part of the pilot. If the flight is longer than 2 hours, the 5 P check should be conducted hourly.

The last two decision points are just prior to descent into the terminal area and just prior to the final approach fix, or if VFR, just prior to entering the traffic pattern as preparations for landing commence. Most pilots execute approaches with the expectation that they will land out of the approach every time. A healthier approach requires the pilot to assume that changing conditions (the 5 Ps again) will cause the pilot to divert or execute the missed approach on every approach. This keeps the pilot alert to all manner of conditions that may increase risk and threaten the safe conduct of the flight. Diverting from cruise altitude saves fuel, allows unhurried use of the autopilot and is less reactive in nature. Diverting from the final approach fix, while more difficult, still allows the pilot to plan and coordinate better, rather than executing a futile missed approach. Let's look at a detailed discussion of each of the Five Ps.

The Plan

The "Plan" can also be called the mission or the task. It contains the basic elements of cross-country planning, weather, route, fuel, publications currency, etc. The "Plan" should be reviewed and updated several times during the course of the flight. A delayed takeoff due to maintenance, fast moving weather, and a short notice TFR may all radically alter the plan. The "plan" is not only about the flight plan, but also all the events that surround the flight and allow the pilot to accomplish the mission. The plan is always being updated and modified and is especially responsive to changes in the other four remaining Ps. If for no other reason, the 5 P check reminds the pilot that the day's flight plan is real life and subject to change at any time.

Obviously, weather is a huge part of any plan. The addition of datalink weather information gives the advanced avionics pilot a real advantage in inclement weather, but only if the pilot is trained to retrieve and evaluate the weather in real time without sacrificing situational awareness. And of course, weather information should drive a decision, even if that decision is to continue on the current plan. Pilots of aircraft without datalink weather should get updated weather in flight through an FSS and/or Flight Watch.

The Plane

Both the "plan" and the "plane" are fairly familiar to most pilots. The "plane" consists of the usual array of mechanical and cosmetic issues that every aircraft pilot, owner, or operator can identify. With the advent of advanced avionics, the "plane" has expanded to include database currency, automation status, and emergency backup systems that were unknown a few years ago. Much has been written about single pilot IFR flight, both with and without an autopilot. While this is a personal decision, it is just that—a decision. Low IFR in a non-autopilot equipped aircraft may depend on several of the other Ps to be discussed. Pilot proficiency, currency, and fatigue are among them.

The Pilot

Flying, especially when business transportation is involved, can expose a pilot to risks such as high altitudes, long trips requiring significant endurance, and challenging weather. Advanced avionics, when installed, can expose a pilot to high stresses because of the inherent additional capabilities which are available. When dealing with pilot risk, it is always best to consult the "IMSAFE" checklist (see page 2-6).

The combination of late nights, pilot fatigue, and the effects of sustained flight above 5,000 feet may cause pilots to become less discerning, less critical of information, less decisive, and more compliant and accepting. Just as the most critical portion of the flight approaches (for instance a night instrument approach, in the weather, after a 4-hour flight), the pilot's guard is down the most. The 5 P process helps a pilot recognize the physiological challenges that they may face towards the end of the flight prior to takeoff and allows them to update personal conditions as the flight progresses. Once risks are identified, the pilot is in a better place to make alternate plans that lessen the effect of these factors and provide a safer solution.

The Passengers

One of the key differences between CRM and SRM is the way passengers interact with the pilot. The pilot of a highly capable single-engine aircraft maintains a much more personal relationship with the passengers as he/she is positioned within an arm's reach of them throughout the flight.

The necessity of the passengers to make airline connections or important business meetings in a timely manner enters into this pilot's decision-making loop. Consider a flight to Dulles Airport in which and the passengers, both close friends and business partners, need to get to Washington, D.C. for an important meeting. The weather is VFR all the way to southern Virginia, then turns to low IFR as the pilot approaches Dulles. A pilot employing the 5 P approach might consider reserving a rental car at an airport in northern North Carolina or southern Virginia to coincide with a refueling stop. Thus, the passengers have a way to get to Washington, and the pilot has an out to avoid being pressured into continuing the flight if the conditions do not improve.

Passengers can also be pilots. If no one is designated as pilot in command (PIC) and unplanned circumstances arise, the decision-making styles of several self-confident pilots may come into conflict.

Pilots also need to understand that non-pilots may not understand the level of risk involved in flight. There is an element of risk in every flight. That is why SRM calls it risk management, not risk elimination. While a pilot may feel comfortable with the risk present in a night IFR flight, the passengers may not. A pilot employing SRM should ensure the passengers are involved in the decision-making and given tasks and duties to keep them busy and involved. If, upon a factual description of the risks present, the passengers decide to buy an airline ticket or rent a car, then a good decision has generally been made. This discussion also allows the pilot to move past what he or she thinks the passengers want to do and find out what they actually want to do. This removes self-induced pressure from the pilot.

The Programming

The advanced avionics aircraft adds an entirely new dimension to the way GA aircraft are flown. The electronic instrument displays, GPS, and autopilot reduce pilot workload and increase pilot situational awareness. While programming and operation of these devices are fairly simple and straightforward, unlike the analog instruments they replace, they tend to capture the pilot's attention and hold it for long periods of time. To avoid this phenomenon, the pilot should plan in advance when and where the programming for approaches, route changes, and airport information gathering should be accomplished, as well as times it should not. Pilot familiarity with the equipment, the route, the local ATC environment, and personal capabilities vis-à-vis the automation should drive when, where, and how the automation is programmed and used.

The pilot should also consider what his or her capabilities are in response to last minute changes of the approach (and the reprogramming required) and ability to make large-scale changes (a reroute for instance) while hand flying the aircraft. Since formats are not standardized, simply moving from one manufacturer's equipment to another should give the pilot pause and require more conservative planning and decisions.

The SRM process is simple. At least five times before and during the flight, the pilot should review and consider the "Plan, the Plane, the Pilot, the Passengers, and the Programming" and make the appropriate decision required by the current situation. It is often said that failure to make a decision is a decision. Under SRM and the 5 Ps, even the decision to make no changes to the current plan is made through a careful consideration of all the risk factors present.

Perceive, Process, Perform (3P) Model

The Perceive, Process, Perform (3P) model for ADM offers a simple, practical, and systematic approach that can be used during all phases of flight. To use it, the pilot will:

- Perceive the given set of circumstances for a flight

- Process by evaluating their impact on flight safety

- Perform by implementing the best course of action

Use the Perceive, Process, Perform, and Evaluate method as a continuous model for every aeronautical decision that you make. Although human beings will inevitably make mistakes, anything that you can do to recognize and minimize potential threats to your safety will make you a better pilot.

Depending upon the nature of the activity and the time available, risk management processing can take place in any of three timeframes. *[Figure 2-10]* Most flight training activities take place in the "time-critical" timeframe for risk management. The six steps of risk management can be combined into an easy-to-remember 3P model for practical risk management: Perceive, Process, Perform with the PAVE, CARE and TEAM checklists. Pilots can help perceive hazards by using the PAVE checklist of: Pilot, Aircraft, enVironment, and External pressures. They can process hazards by using the CARE checklist of: Consequences, Alternatives, Reality, External factors. Finally, pilots can perform risk management by using the TEAM choice list of: Transfer, Eliminate, Accept, or Mitigate.

PAVE Checklist: Identify Hazards and Personal Minimums

In the first step, the goal is to develop situational awareness by perceiving hazards, which are present events, objects, or circumstances that could contribute to an undesired future event. In this step, the pilot will systematically identify and

	Strategic	Deliberate	Time-Critical
Purpose	Used in a complex operation (e.g., introduction of new equipment); involves research, use of analysis tools, formal testing, or long term tracking of risks.	Uses experience and brainstorming to identify hazards, assess risks, and develop controls for planning operations, review of standard operating or training procedures, etc.	"On the fly" mental or verbal review using the basic risk management process during the execution phase of an activity.

Figure 2-10. *Risk management processing can take place in any of three timeframes.*

list hazards associated with all aspects of the flight: **P**ilot, **A**ircraft, en**V**ironment, and **E**xternal pressures, which makes up the PAVE checklist. *[Figure 2-11]* For each element, ask "what could hurt me, my passengers, or my aircraft?" All four elements combine and interact to create a unique situation for any flight. Pay special attention to the pilot-aircraft combination, and consider whether the combined "pilot-aircraft team" is capable of the mission you want to fly. For example, you may be a very experienced and proficient pilot, but your weather flying ability is still limited if you are flying a 1970s-model aircraft with no weather avoidance gear. On the other hand, you may have a new technically advanced aircraft with moving map GPS, weather datalink, and autopilot—but if you do not have much weather flying experience or practice in using this kind of equipment, you cannot rely on the airplane's capability to compensate for your own lack of experience.

CARE Checklist: Review Hazards and Evaluate Risks

In the second step, the goal is to process this information to determine whether the identified hazards constitute risk, which is defined as the future impact of a hazard that is not controlled or eliminated. The degree of risk posed by a given hazard can be measured in terms of exposure (number of people or resources affected), severity (extent of possible loss), and probability (the likelihood that a hazard will cause a loss). The goal is to evaluate their impact on the safety of your flight, and consider "why must I CARE about these circumstances?"

For each hazard that you perceived in step one, process by using the CARE checklist of: **C**onsequences, **A**lternatives, **R**eality, **E**xternal factors. *[Figure 2-12]* For example, let's evaluate a night flight to attend a business meeting:

> **C**onsequences—departing after a full workday creates fatigue and pressure

> **A**lternatives—delay until morning; reschedule meeting; drive

> **R**eality —dangers and distractions of fatigue could lead to an accident

> **E**xternal pressures—business meeting at destination might influence me

A good rule of thumb for the processing phase: if you find yourself saying that it will "probably" be okay, it is definitely time for a solid reality check. If you are worried about missing a meeting, be realistic about how that pressure will affect not just your initial go/no-go decision, but also your inflight decisions to continue the flight or divert.

TEAM Checklist: Choose and Implement Risk Controls

Once you have perceived a hazard (step one) and processed its impact on flight safety (step two), it is time to move to the third step, perform. Perform risk management by using the TEAM checklist of: **T**ransfer, **E**liminate, **A**ccept, **M**itigate to deal with each factor. *[Figure 2-13]*

> **T**ransfer—Should this risk decision be transferred to someone else (e.g., do you need to consult the chief flight instructor?)

> **E**liminate—Is there a way to eliminate the hazard?

> **A**ccept—Do the benefits of accepting risk outweigh the costs?

> **M**itigate—What can you do to mitigate the risk?

The goal is to perform by taking action to eliminate hazards or mitigate risk, and then continuously evaluate the outcome of this action. With the example of low ceilings at destination, for instance, the pilot can perform good ADM by selecting a suitable alternate, knowing where to find good weather,

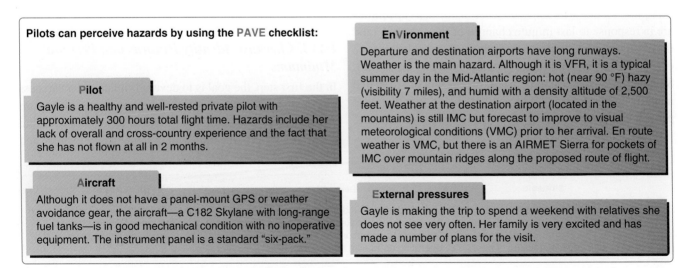

Figure 2-11. *A real-world example of how the 3P model guides decisions on a cross-country trip using the PAVE checklist.*

Pilots can perceive hazards by using the CARE checklist:

Pilot

- **C**onsequences: Gayle's inexperience and lack of recent flight time create some risks for an accident, primarily because she plans to travel over mountains on a hazy day and land at an unfamiliar mountain airport that is still in IMC conditions.
- **A**lternatives: Gayle might mitigate the pilot-related risk by hiring a CFI to accompany her and provide dual cross-country instruction. An added benefit is the opportunity to broaden her flying experience in safe conditions.
- **R**eality: Accepting the reality that limited experience can create additional risks is a key part of sound risk management and mitigation.
- **E**xternal Factors: Like many pilots, Gayle must contend with the emotional pressure associated with acknowledging that her skill and experience levels may be lower than she would like them to be. Pride can be a powerful external factor!

Aircraft

- **C**onsequences: This area presents low risk because the aircraft is in excellent mechanical condition and Gayle is familiar with its avionics.
- **A**lternatives: Had there been a problem with her aircraft, Gayle might have considered renting another plane from her flight school. Bear in mind, however, that alternatives sometimes create new hazards. In this instance, there may be hazards associated with flying an unfamiliar aircraft with different avionics.
- **R**eality: It is important to recognize the reality of an aircraft's mechanical condition. If you find a maintenance discrepancy and then find yourself saying that it is "probably" okay to fly with it anyway, you need to revisit the consequences part of this checklist.
- **E**xternal Factors: Pilot decision-making can sometimes be influenced by the external pressure of needing to return the airplane to the FBO by a certain date and time. Because Gayle owns the airplane, there was no such pressure in this case.

Environment

- **C**onsequences: For a pilot whose experience consists mostly of local flights in good VMC, launching a long cross-country flight over mountainous terrain in hazy conditions could lead to pilot disorientation and increase the risk of an accident.
- **A**lternatives: Options include postponing the trip until the visibility improves, or modifying the route to avoid extended periods of time over the mountains.
- **R**eality: Hazy conditions and mountainous terrain clearly create risks for an inexperienced VFR-only pilot.
- **E**xternal Factors: Few pilots are immune to the pressure of "get-there-itis," which can sometimes induce a decision to launch or continue in less than ideal weather conditions.

External pressures

- **C**onsequences: Any number of factors can create the risk of emotional pressure from a "get-there" mentality. In Gayle's case, the consequences of her strong desire to visit family, her family's expectations, and personal pride could induce her to accept unnecessary risks.
- **A**lternatives: Gayle clearly needs to develop a mitigating strategy for each of the external factors associated with this trip.
- **R**eality: Pilots sometimes tend to discount or ignore the potential impact of these external factors. Gayle's open acknowledgement of these factors (e.g., "I might be pressured into pressing on so my mother won't have to worry about our late arrival.") is a critical element of effective risk management.
- **E**xternal Factors: (see above)

Figure 2-12. *A real-world examples of how the 3P model guides decisions on a cross-country trip using the CARE checklist.*

and carrying sufficient fuel to reach it. This course of action would mitigate the risk. The pilot also has the option to eliminate it entirely by waiting for better weather.

Once the pilot has completed the 3P decision process and selected a course of action, the process begins anew because now the set of circumstances brought about by the course of action requires analysis. The decision-making process is a continuous loop of perceiving, processing, and performing. With practice and consistent use, running through the 3P cycle can become a habit that is as smooth, continuous, and automatic as a well-honed instrument scan. This basic set of practical risk management tools can be used to improve risk management.

Your mental willingness to follow through on safe decisions, especially those that require delay or diversion is critical. You can bulk up your mental muscles by:

- Using personal minimums checklist to make some decisions in advance of the flight. To develop a good personal minimums checklist, you need to assess your abilities and capabilities in a non-flying environment, when there is no pressure to make a specific trip. Once developed, a personal minimums checklist will give you a clear and concise reference point for making your go/no-go or continue/discontinue decisions.

- In addition to having personal minimums, some pilots also like to use a preflight risk assessment checklist to help with the ADM and risk management processes. This kind of form assigns numbers to certain risks and situations, which can make it easier to see when a particular flight involves a higher level of risk

- Develop a list of good alternatives during your processing phase. In marginal weather, for instance, you might mitigate the risk by identifying a reasonable

Pilots can perform risk management by using the TEAM checklist:

Pilot

To manage the risk associated with her inexperience and lack of recent flight time, Gayle can:
- **T**ransfer the risk entirely by having another pilot act as PIC.
- **E**liminate the risk by canceling the trip.
- **A**ccept the risk and fly anyway.
- **M**itigate the risk by flying with another pilot.

Gayle chooses to mitigate the major risk by hiring a CFI to accompany her and provide dual cross-country instruction. An added benefit is the opportunity to broaden her flying experience.

Aircraft

To manage risk associated with any doubts about the aircraft's mechanical condition, Gayle can:
- **T**ransfer the risk by using a different airplane.
- **E**liminate the risk by canceling the trip.
- **A**ccept the risk.
- **M**itigate the remaining (residual) risk through review of aircraft performance and careful preflight inspection.

Since she finds no problems with the aircraft's mechanical condition, Gayle chooses to mitigate any remaining risk through careful preflight inspection of the aircraft.

Environment

To manage the risk associated with hazy conditions and mountainous terrain, Gayle can:
- **T**ransfer the risk of VFR in these conditions by asking an instrument-rated pilot to fly the trip under IFR.
- **E**liminate the risk by canceling the trip.
- **A**ccept the risk.
- **M**itigate the risk by careful preflight planning, filing a VFR flight plan, requesting VFR flight following, and using resources such as Flight Watch.

Detailed preflight planning must be a vital part of Gayle's weather risk mitigation strategy. The most direct route would put her over mountains for most of the trip. Because of the thick haze and pockets of IMC over mountains, Gayle might mitigate the risk by modifying the route to fly over valleys. This change will add 30 minutes to her estimated time of arrival (ETA), but the extra time is a small price to pay for avoiding possible IMC over mountains. Because her destination airport is IMC at the time of departure, Gayle needs to establish that VFR conditions exist at other airports within easy driving distance of her original destination. In addition, Gayle should review basic information (e.g., traffic pattern altitude, runway layout, frequencies) for these alternate airports. To further mitigate risk and practice good cockpit resource management, Gayle should file a VFR flight plan, use VFR flight following, and call Flight Watch to get weather updates en route. Finally, basic functions on her handheld GPS should also be practiced.

External pressures

To mitigate the risk of emotional pressure from family expectations that can drive a "get-there" mentality, Gayle can:
- **T**ransfer the risk by having her co-pilot act as PIC and make the continue/divert decision.
- **E**liminate the risk by canceling the trip.
- **A**ccept the risk.
- **M**itigate the risk by managing family expectations and making alternative arrangements in the event of diversion to another airport.

Gayle and her co-pilot choose to address this risk by agreeing that each pilot has a veto on continuing the flight, and that they will divert if either becomes uncomfortable with flight conditions. Because the destination airport is still IMC at the time of departure, Gayle establishes a specific point in the trip—an en route VORTAC located between the destination airport and the two alternates—as the logical place for her "final" continue/divert decision. Rather than give her family a specific ETA that might make Gayle feel pressured to meet the schedule, she manages her family's expectations by advising them that she will call when she arrives.

Figure 2-13. A real-world example of how the 3P model guides decisions on a cross-country trip using the TEAM checklist.

alternative airport for every 25–30 nautical mile segment of your route.

- Preflight your passengers by preparing them for the possibility of delay and diversion, and involve them in your evaluation process.

- Another important tool—overlooked by many pilots—is a good post-flight analysis. When you have safely secured the airplane, take the time to review and analyze the flight as objectively as you can. Mistakes and judgment errors are inevitable; the most important thing is for you to recognize, analyze, and learn from them before your next flight.

The DECIDE Model

Using the acronym "DECIDE," the six-step process DECIDE Model is another continuous loop process that provides the pilot with a logical way of making decisions. *[Figure 2-14]* DECIDE means to Detect, Estimate, Choose a course of action, Identify solutions, Do the necessary actions, and Evaluate the effects of the actions.

First, consider a recent accident involving a Piper Apache (PA-23). The aircraft was substantially damaged during impact with terrain at a local airport in Alabama. The certificated airline transport pilot (ATP) received minor injuries and the certificated private pilot was not injured. The private pilot was receiving a checkride from the ATP (who was also a designated examiner) for a commercial pilot certificate with a multi-engine rating. After performing airwork at altitude, they returned to the airport and the private pilot performed a single-engine approach to a full stop landing. He then taxied back for takeoff, performed a short field takeoff, and then joined the traffic pattern to return for another landing. During the approach for the second landing, the ATP simulated a right engine failure by reducing power on the right engine to zero thrust. This caused the aircraft to yaw right.

The procedure to identify the failed engine is a two-step process. First, adjust the power to the maximum controllable level on both engines. Because the left engine is the only engine delivering thrust, the yaw increases to the right, which necessitates application of additional left rudder application.

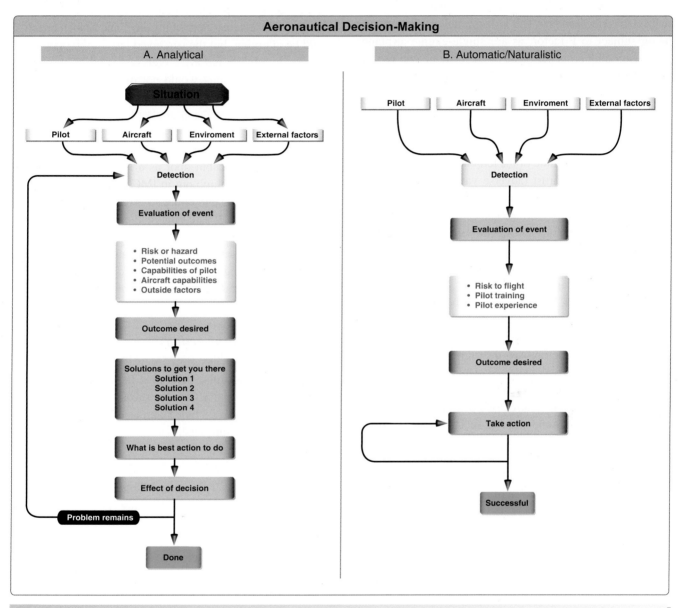

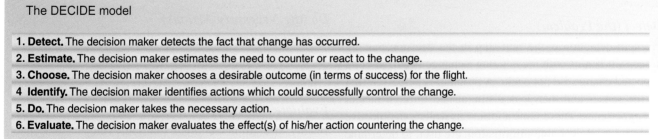

Figure 2-14. *The DECIDE model has been recognized worldwide. Its application is illustrated in column A while automatic/naturalistic decision-making is shown in column B.*

The failed engine is the side that requires no rudder pressure, in this case the right engine. Second, having identified the failed right engine, the procedure is to feather the right engine and adjust power to maintain descent angle to a landing.

However, in this case the pilot feathered the left engine because he assumed the engine failure was a left engine failure. During twin-engine training, the left engine out is emphasized more than the right engine because the left engine on most light twins is the critical engine. This is due to multiengine airplanes being subject to P-factor, as are single-engine airplanes. The descending propeller blade of each engine will produce greater thrust than the ascending blade when the airplane is operated under power and at positive angles of attack. The descending propeller blade of the right engine is also a greater distance from the center of gravity, and therefore has a longer moment arm than the descending propeller blade of the left engine. As a result, failure of the left engine will result in the most asymmetrical thrust (adverse yaw) because the right engine will be providing the remaining thrust. Many twins are designed with a counter-rotating right engine. With this design, the degree of asymmetrical thrust is the same with either engine inoperative. Neither engine is more critical than the other.

Since the pilot never executed the first step of identifying which engine failed, he feathered the left engine and set the right engine at zero thrust. This essentially restricted the aircraft to a controlled glide. Upon realizing that he was not going to make the runway, the pilot increased power to both engines causing an enormous yaw to the left (the left propeller was feathered) whereupon the aircraft started to turn left. In desperation, the instructor closed both throttles and the aircraft hit the ground and was substantially damaged.

This case is interesting because it highlights two particular issues. First, taking action without forethought can be just as dangerous as taking no action at all. In this case, the pilot's actions were incorrect; yet, there was sufficient time to take the necessary steps to analyze the simulated emergency. The second and more subtle issue is that decisions made under pressure are sometimes executed based upon limited experience and the actions taken may be incorrect, incomplete, or insufficient to handle the situation.

Detect (the Problem)

Problem detection is the first step in the decision-making process. It begins with recognizing a change occurred or an expected change did not occur. A problem is perceived first by the senses and then it is distinguished through insight and experience. These same abilities, as well as an objective analysis of all available information, are used to determine the nature and severity of the problem. One critical error made during the decision-making process is incorrectly

detecting the problem. In the previous example, the change that occurred was a yaw.

Estimate (the Need To React)

In the engine-out example, the aircraft yawed right, the pilot was on final approach, and the problem warranted a prompt solution. In many cases, overreaction and fixation excludes a safe outcome. For example, what if the cabin door of a Mooney suddenly opened in flight while the aircraft climbed through 1,500 feet on a clear sunny day? The sudden opening would be alarming, but the perceived hazard the open door presents is quickly and effectively assessed as minor. In fact, the door's opening would not impact safe flight and can almost be disregarded. Most likely, a pilot would return to the airport to secure the door after landing.

The pilot flying on a clear day faced with this minor problem may rank the open cabin door as a low risk. What about the pilot on an IFR climb out in IMC conditions with light intermittent turbulence in rain who is receiving an amended clearance from ATC? The open cabin door now becomes a higher risk factor. The problem has not changed, but the perception of risk a pilot assigns it changes because of the multitude of ongoing tasks and the environment. Experience, discipline, awareness, and knowledge influences how a pilot ranks a problem.

Choose (a Course of Action)

After the problem has been identified and its impact estimated, the pilot must determine the desirable outcome and choose a course of action. In the case of the multiengine pilot given the simulated failed engine, the desired objective is to safely land the airplane.

Identify (Solutions)

The pilot formulates a plan that will take him or her to the objective. Sometimes, there may be only one course of action available. In the case of the engine failure already at 500 feet or below, the pilot solves the problem by identifying one or more solutions that lead to a successful outcome. It is important for the pilot not to become fixated on the process to the exclusion of making a decision.

Do (the Necessary Actions)

Once pathways to resolution are identified, the pilot selects the most suitable one for the situation. The multiengine pilot given the simulated failed engine must now safely land the aircraft.

Evaluate (the Effect of the Action)

Finally, after implementing a solution, evaluate the decision to see if it was correct. If the action taken does not provide the desired results, the process may have to be repeated.

Decision-Making in a Dynamic Environment

A solid approach to decision-making is through the use of analytical models, such as the 5 Ps, 3P, and DECIDE. Good decisions result when pilots gather all available information, review it, analyze the options, rate the options, select a course of action, and evaluate that course of action for correctness.

In some situations, there is not always time to make decisions based on analytical decision-making skills. A good example is a quarterback whose actions are based upon a highly fluid and changing situation. He intends to execute a plan, but new circumstances dictate decision-making on the fly. This type of decision-making is called automatic decision-making or naturalized decision-making. *[Figure 2-14B]*

Automatic Decision-Making

In an emergency situation, a pilot might not survive if he or she rigorously applies analytical models to every decision made as there is not enough time to go through all the options. Under these circumstances he or she should attempt to find the best possible solution to every problem.

For the past several decades, research into how people actually make decisions has revealed that when pressed for time, experts faced with a task loaded with uncertainty first assess whether the situation strikes them as familiar. Rather than comparing the pros and cons of different approaches, they quickly imagine how one or a few possible courses of action in such situations will play out. Experts take the first workable option they can find. While it may not be the best of all possible choices, it often yields remarkably good results.

The terms "naturalistic" and "automatic decision-making" have been coined to describe this type of decision-making. The ability to make automatic decisions holds true for a range of experts from firefighters to chess players. It appears the expert's ability hinges on the recognition of patterns and consistencies that clarify options in complex situations. Experts appear to make provisional sense of a situation, without actually reaching a decision, by launching experience-based actions that in turn trigger creative revisions.

This is a reflexive type of decision-making anchored in training and experience and is most often used in times of emergencies when there is no time to practice analytical decision-making. Naturalistic or automatic decision-making improves with training and experience, and a pilot will find himself or herself using a combination of decision-making tools that correlate with individual experience and training.

Operational Pitfalls

Although more experienced pilots are likely to make more automatic decisions, there are tendencies or operational pitfalls that come with the development of pilot experience. These are classic behavioral traps into which pilots have been known to fall. More experienced pilots, as a rule, try to complete a flight as planned, please passengers, and meet schedules. The desire to meet these goals can have an adverse effect on safety and contribute to an unrealistic assessment of piloting skills. All experienced pilots have fallen prey to, or have been tempted by, one or more of these tendencies in their flying careers. These dangerous tendencies or behavior patterns, which must be identified and eliminated, include the operational pitfalls shown in *Figure 2-15*.

Stress Management

Everyone is stressed to some degree almost all of the time. A certain amount of stress is good since it keeps a person alert and prevents complacency. Effects of stress are cumulative and, if the pilot does not cope with them in an appropriate way, they can eventually add up to an intolerable burden. Performance generally increases with the onset of stress, peaks, and then begins to fall off rapidly as stress levels exceed a person's ability to cope. The ability to make effective decisions during flight can be impaired by stress. There are two categories of stress—acute and chronic. These are both explained in Chapter 17, "Aeromedical Factors."

Factors referred to as stressors can increase a pilot's risk of error in the flight deck. *[Figure 2-16]* Remember the cabin door that suddenly opened in flight on the Mooney climbing through 1,500 feet on a clear sunny day? It may startle the pilot, but the stress would wane when it became apparent the situation was not a serious hazard. Yet, if the cabin door opened in IMC conditions, the stress level makes significant impact on the pilot's ability to cope with simple tasks. The key to stress management is to stop, think, and analyze before jumping to a conclusion. There is usually time to think before drawing unnecessary conclusions.

There are several techniques to help manage the accumulation of life stresses and prevent stress overload. For example, to help reduce stress levels, set aside time for relaxation each day or maintain a program of physical fitness. To prevent stress overload, learn to manage time more effectively to avoid pressures imposed by getting behind schedule and not meeting deadlines.

Use of Resources

To make informed decisions during flight operations, a pilot must also become aware of the resources found inside and outside the flight deck. Since useful tools and sources of information may not always be readily apparent, learning to recognize these resources is an essential part of ADM training. Resources must not only be identified, but a pilot must also develop the skills to evaluate whether there is

Operational Pitfalls
Peer pressure
Poor decision-making may be based upon an emotional response to peers, rather than evaluating a situation objectively.
Mindset
A pilot displays mind set through an inability to recognize and cope with changes in a given situation.
Get-there-itis
This disposition impairs pilot judgment through a fixation on the original goal or destination, combined with a disregard for any alternative course of action.
Duck-under syndrome
A pilot may be tempted to make it into an airport by descending below minimums during an approach. There may be a belief that there is a built-in margin of error in every approach procedure, or a pilot may want to admit that the landing cannot be completed and a missed approach must be initiated.
Scud running
This occurs when a pilot tries to maintain visual contact with the terrain at low altitudes while instrument conditions exist.
Continuing visual flight rules (VFR) into instrument conditions
Spatial disorientation or collision with ground/obstacles may occur when a pilot continues VFR into instrument conditions. This can be even more dangerous if the pilot is not instrument rated or current.
Getting behind the aircraft
This pitfall can be caused by allowing events or the situation to control pilot actions. A constant state of surprise at what happens next may be exhibited when the pilot is getting behind the aircraft.
Loss of positional or situational awareness
In extreme cases, when a pilot gets behind the aircraft, a loss of positional or situational awareness may result. The pilot may not know the aircraft's geographical location or may be unable to recognize deteriorating circumstances.
Operating without adequate fuel reserves
Ignoring minimum fuel reserve requirements is generally the result of overconfidence, lack of flight planning, or disregarding applicable regulations.
Descent below the minimum en route altitude
The duck-under syndrome, as mentioned above, can also occur during the en route portion of an IFR flight.
Flying outside the envelope
The assumed high performance capability of a particular aircraft may cause a mistaken belief that it can meet the demands imposed by a pilot's overestimated flying skills.
Neglect of flight planning, preflight inspections, and checklists
A pilot may rely on short- and long-term memory, regular flying skills, and familiar routes instead of established procedures and published checklists. This can be particularly true of experienced pilots.

Figure 2-15. *Typical operational pitfalls requiring pilot awareness.*

Stressors
Environmental
Conditions associated with the environment, such as temperature and humidity extremes, noise, vibration, and lack of oxygen.
Physiological stress
Physical conditions, such as fatigue, lack of physical fitness, sleep loss, missed meals (leading to low blood sugar levels), and illness.
Psychological stress
Social or emotional factors, such as a death in the family, a divorce, a sick child, or a demotion at work. This type of stress may also be related to mental workload, such as analyzing a problem, navigating an aircraft, or making decisions.

Figure 2-16. *System stressors. Environmental, physiological, and psychological stress are factors that affect decision-making skills. These stressors have a profound impact especially during periods of high workload.*

time to use a particular resource and the impact its use will have upon the safety of flight. For example, the assistance of ATC may be very useful if a pilot becomes lost, but in an emergency situation, there may be no time available to contact ATC.

Internal Resources

One of the most underutilized resources may be the person in the right seat, even if the passenger has no flying experience. When appropriate, the PIC can ask passengers to assist with certain tasks, such as watching for traffic or reading checklist items. The following are some other ways a passenger can assist:

- Provide information in an irregular situation, especially if familiar with flying. A strange smell or sound may alert a passenger to a potential problem.

- Confirm after the pilot that the landing gear is down.

- Learn to look at the altimeter for a given altitude in a descent.

- Listen to logic or lack of logic.

Also, the process of a verbal briefing (which can happen whether or not passengers are aboard) can help the PIC in the decision-making process. For example, assume a pilot provides a lone passenger a briefing of the forecast landing weather before departure. When the Automatic Terminal Information Service (ATIS) is picked up, the weather has significantly changed. The discussion of this forecast change can lead the pilot to reexamine his or her activities and decision-making. [Figure 2-17] Other valuable internal resources include ingenuity, aviation knowledge, and flying skill. Pilots can increase flight deck resources by improving these characteristics.

When flying alone, another internal resource is verbal communication. It has been established that verbal communication reinforces an activity; touching an object while communicating further enhances the probability an activity has been accomplished. For this reason, many solo pilots read the checklist out loud; when they reach critical items, they touch the switch or control. For example, to ascertain the landing gear is down, the pilot can read the checklist. But, if he or she touches the gear handle during the process, a safe extension of the landing gear is confirmed.

It is necessary for a pilot to have a thorough understanding of all the equipment and systems in the aircraft being flown. Lack of knowledge, such as knowing if the oil pressure gauge is direct reading or uses a sensor, is the difference between making a wise decision or poor one that leads to a tragic error.

Figure 2-17. *When possible, have a passenger reconfirm that critical tasks are completed.*

Checklists are essential flight deck internal resources. They are used to verify the aircraft instruments and systems are checked, set, and operating properly, as well as ensuring the proper procedures are performed if there is a system malfunction or in-flight emergency. Students reluctant to use checklists can be reminded that pilots at all levels of experience refer to checklists, and that the more advanced the aircraft is, the more crucial checklists become. In addition, the pilot's operating handbook (POH) is required to be carried on board the aircraft and is essential for accurate flight planning and resolving in-flight equipment malfunctions. However, the most valuable resource a pilot has is the ability to manage workload whether alone or with others.

External Resources

ATC and flight service specialists are the best external resources during flight. In order to promote the safe, orderly flow of air traffic around airports and, along flight routes, the ATC provides pilots with traffic advisories, radar vectors, and assistance in emergency situations. Although it is the PIC's responsibility to make the flight as safe as possible, a pilot with a problem can request assistance from ATC. [Figure 2-18] For example, if a pilot needs to level off, be

Figure 2-18. *Controllers work to make flights as safe as possible.*

given a vector, or decrease speed, ATC assists and becomes integrated as part of the crew. The services provided by ATC can not only decrease pilot workload, but also help pilots make informed in-flight decisions.

The Flight Service Stations (FSSs) are air traffic facilities that provide pilot briefing, en route communications, VFR search and rescue services, assist lost aircraft and aircraft in emergency situations, relay ATC clearances, originate Notices to Airmen (NOTAM), broadcast aviation weather and National Airspace System (NAS) information, receive and process IFR flight plans, and monitor navigational aids (NAVAIDs). In addition, at selected locations, FSSs provide En Route Flight Advisory Service (Flight Watch), issue airport advisories, and advise Customs and Immigration of transborder flights. Selected FSSs in Alaska also provide TWEB recordings and take weather observations.

Situational Awareness

Situational awareness is the accurate perception and understanding of all the factors and conditions within the five fundamental risk elements (flight, pilot, aircraft, environment, and type of operation that comprise any given aviation situation) that affect safety before, during, and after the flight. Monitoring radio communications for traffic, weather discussion, and ATC communication can enhance situational awareness by helping the pilot develop a mental picture of what is happening.

Maintaining situational awareness requires an understanding of the relative significance of all flight related factors and their future impact on the flight. When a pilot understands what is going on and has an overview of the total operation, he or she is not fixated on one perceived significant factor. Not only is it important for a pilot to know the aircraft's geographical location, it is also important he or she understand what is happening. For instance, while flying above Richmond, Virginia, toward Dulles Airport or Leesburg, the pilot should know why he or she is being vectored and be able to anticipate spatial location. A pilot who is simply making turns without understanding why has added an additional burden to his or her management in the event of an emergency. To maintain situational awareness, all of the skills involved in ADM are used.

Obstacles to Maintaining Situational Awareness

Fatigue, stress, and work overload can cause a pilot to fixate on a single perceived important item and reduce an overall situational awareness of the flight. A contributing factor in many accidents is a distraction that diverts the pilot's attention from monitoring the instruments or scanning outside the aircraft. Many flight deck distractions begin as a minor problem, such as a gauge that is not reading correctly, but result in accidents as the pilot diverts attention to the perceived problem and neglects proper control of the aircraft.

Workload Management

Effective workload management ensures essential operations are accomplished by planning, prioritizing, and sequencing tasks to avoid work overload. *[Figure 2-19]* As experience is gained, a pilot learns to recognize future workload requirements and can prepare for high workload periods during times of low workload. Reviewing the appropriate chart and setting radio frequencies well in advance of when they are needed helps reduce workload as the flight nears the airport. In addition, a pilot should listen to ATIS, Automated Surface Observing System (ASOS), or Automated Weather Observing System (AWOS), if available, and then monitor the tower frequency or Common Traffic Advisory Frequency (CTAF) to get a good idea of what traffic conditions to expect. Checklists should be performed well in advance so there is time to focus on traffic and ATC instructions. These procedures are especially important prior to entering a high-density traffic area, such as Class B airspace.

Recognizing a work overload situation is also an important component of managing workload. The first effect of high workload is that the pilot may be working harder but accomplishing less. As workload increases, attention cannot be devoted to several tasks at one time, and the pilot may begin to focus on one item. When a pilot becomes task saturated, there is no awareness of input from various sources, so decisions may be made on incomplete information and the possibility of error increases. *[Figure 2-20]*

When a work overload situation exists, a pilot needs to stop, think, slow down, and prioritize. It is important to understand how to decrease workload. For example, in the case of the cabin door that opened in VFR flight, the impact on workload should be insignificant. If the cabin door opens under IFR different conditions, its impact on workload changes. Therefore, placing a situation in the proper perspective,

Figure 2-19. *Balancing workloads can be a difficult task.*

remaining calm, and thinking rationally are key elements in reducing stress and increasing the capacity to fly safely. This ability depends upon experience, discipline, and training.

Managing Risks

The ability to manage risks begins with preparation. Here are some things a pilot can do to manage risks:

- Assess the flight's risk based upon experience. Use some form of risk assessment. For example, if the weather is marginal and the pilot has little IMC training, it is probably a good idea to cancel the flight.

- Brief passengers using the SAFETY list:

    S   Seat belts fastened for taxi, takeoff, landing

    Shoulder harness fastened for takeoff, landing

    Seat position adjusted and locked in place

    A   Air vents (location and operation)

    All environmental controls (discussed)

    Action in case of any passenger discomfort

    F   Fire extinguisher (location and operation)

    E   Exit doors (how to secure; how to open)

    Emergency evacuation plan

    Emergency/survival kit (location and contents)

    T   Traffic (scanning, spotting, notifying pilot)

    Talking, ("sterile flight deck" expectations)

    Y   Your questions? (Speak up!)

- In addition to the SAFETY list, discuss with passengers whether or not smoking is permitted, flight route altitudes, time en route, destination, weather during flight, expected weather at the destination, controls and what they do, and the general capabilities and limitations of the aircraft.

- Use a sterile flight deck (one that is completely silent with no pilot communication with passengers or by passengers) from the time of departure to the first intermediate altitude and clearance from the local airspace.

- Use a sterile flight deck during arrival from the first radar vector for approach or descent for the approach.

- Keep the passengers informed during times when the workload is low.

- Consider using the passenger in the right seat for simple tasks, such as holding the chart. This relieves the pilot of a task.

Automation

In the GA community, an automated aircraft is generally comprised of an integrated advanced avionics system consisting of a primary flight display (PFD), a multifunction flight display (MFD) including an instrument-certified global positioning system (GPS) with traffic and terrain graphics, and a fully integrated autopilot. This type of aircraft is commonly known as a technically advanced aircraft (TAA). In a TAA aircraft, there are typically two display (computer) screens: PFD (left display screen) and MFD.

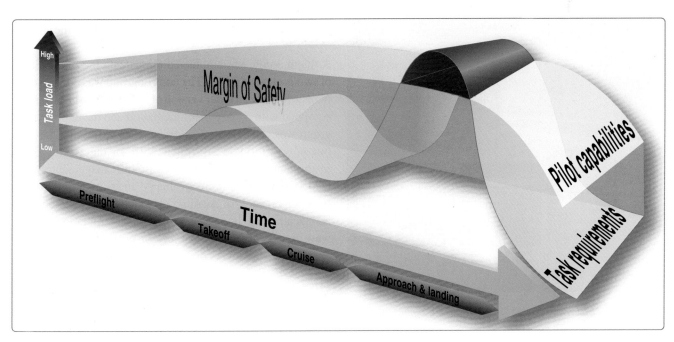

Figure 2-20. *The pilot has a certain capacity of doing work and handling tasks. However, there is a point where the tasking exceeds the pilot's capability. When this happens, tasks are either not performed properly or some are not performed at all.*

Automation is the single most important advance in aviation technologies. Electronic flight displays (EFDs) have made vast improvements in how information is displayed and what information is available to the pilot. Pilots can access electronic databases that contain all of the information traditionally contained in multiple handbooks, reducing clutter in the flight deck. *[Figure 2-21]*

MFDs are capable of displaying moving maps that mirror sectional charts. These detailed displays depict all airspace, including Temporary Flight Restrictions (TFRs). MFDs are so descriptive that many pilots fall into the trap of relying solely on the moving maps for navigation. Pilots also draw upon the database to familiarize themselves with departure and destination airport information.

More pilots now rely on electronic databases for flight planning and use automated flight planning tools rather than planning the flight by the traditional methods of laying out charts, drawing the course, identifying navigation points (assuming a VFR flight), and using the POH to figure out the weight and balance and performance charts. Whichever method a pilot chooses to plan a flight, it is

Figure 2-21. *Electronic flight instrumentation comes in many systems and provides a myriad of information to the pilot.*

important to remember to check and confirm calculations. Always remember that it is up to the pilot to maintain basic airmanship skills and use those skills often to maintain proficiency in all tasks.

Although automation has made flying safer, automated systems can make some errors more evident and sometimes hide other errors or make them less evident. There are concerns about the effect of automation on pilots. In a study published in 1995, the British Airline Pilots Association officially voiced its concern that "Airline pilots increasingly lack 'basic flying skills' as a result of reliance on automation."

This reliance on automation translates into a lack of basic flying skills that may affect the pilot's ability to cope with an in-flight emergency, such as sudden mechanical failure. The worry that pilots are becoming too reliant on automated systems and are not being encouraged or trained to fly manually has grown with the increase in the number of MFD flight decks.

As automated flight decks began entering everyday line operations, instructors and check airmen grew concerned about some of the unanticipated side effects. Despite the promise of reducing human mistakes, the flight managers reported the automation actually created much larger errors at times. In the terminal environment, the workload in an automated flight deck actually seemed higher than in the older analog flight decks. At other times, the automation seemed to lull the flight crews into complacency. Over time, concern surfaced that the manual flying skills of the automated flight crews deteriorated due to over-reliance on computers. The flight crew managers said they worried that pilots would have less "stick-and-rudder" proficiency when those skills were needed to manually resume direct control of the aircraft.

A major study was conducted to evaluate the performance of two groups of pilots. The control group was composed of pilots who flew an older version of a common twin-jet airliner equipped with analog instrumentation and the experimental group was composed of pilots who flew the same aircraft, but newer models equipped with an electronic flight instrument system (EFIS) and a flight management system (FMS). The pilots were evaluated in maintaining aircraft parameters, such as heading, altitude, airspeed, glideslope, and localizer deviations, as well as pilot control inputs. These were recorded during a variety of normal, abnormal, and emergency maneuvers during 4 hours of simulator sessions.

Results of the Study

When pilots who had flown EFIS for several years were required to fly various maneuvers manually, the aircraft parameters and flight control inputs clearly showed some erosion of flying skills. During normal maneuvers, such as turns to headings without a flight director, the EFIS group exhibited somewhat greater deviations than the analog group. Most of the time, the deviations were within the practical test standards (PTS), but the pilots definitely did not keep on the localizer and glideslope as smoothly as the analog group.

The differences in hand-flying skills between the two groups became more significant during abnormal maneuvers, such as accelerated descent profiles known as "slam-dunks." When given close crossing restrictions, the analog crews were more adept at the mental math and usually maneuvered the aircraft in a smoother manner to make the restriction. On the other hand, the EFIS crews tended to go "heads down" and tried to solve the crossing restriction on the FMS. *[Figure 2-22]*

Another situation used in the simulator experiment reflected real world changes in approach that are common and can be assigned on short notice. Once again, the analog crews transitioned more easily to the parallel runway's localizer, whereas the EFIS crews had a much more difficult time with the pilot going head down for a significant amount of time trying to program the new approach into the FMS.

While a pilot's lack of familiarity with the EFIS is often an issue, the approach would have been made easier by disengaging the automated system and manually flying the approach. At the time of this study, the general guidelines in the industry were to let the automated system do as much of the flying as possible. That view has since changed and it is recommended that pilots use their best judgment when choosing which level of automation will most efficiently do the task considering the workload and situational awareness.

Emergency maneuvers clearly broadened the difference in manual flying skills between the two groups. In general, the analog pilots tended to fly raw data, so when they were given an emergency, such as an engine failure, and were instructed to fly the maneuver without a flight director, they performed it expertly. By contrast, SOP for EFIS operations at the time was to use the flight director. When EFIS crews had their flight directors disabled, their eye scan again began a more erratic searching pattern and their manual flying subsequently suffered.

Those who reviewed the data saw that the EFIS pilots who better managed the automation also had better flying skills. While the data did not reveal whether those skills preceded or followed automation, it did indicate that automation management needed to be improved. Recommended "best practices" and procedures have remedied some of the earlier problems with automation.

Pilots must maintain their flight skills and ability to maneuver aircraft manually within the standards set forth in the PTS. It is recommended that pilots of automated aircraft occasionally disengage the automation and manually fly the aircraft to maintain stick-and-rudder proficiency. It is imperative that the pilots understand that the EFD adds to the overall quality of the flight experience, but it can also lead to catastrophe if not utilized properly. At no time is the moving map meant to substitute for a VFR sectional or low altitude en route chart.

Equipment Use

Autopilot Systems

In a single-pilot environment, an autopilot system can greatly reduce workload. *[Figure 2-23]* As a result, the pilot is free to focus his or her attention on other flight deck duties. This can improve situational awareness and reduce the possibility of a CFIT accident. While the addition of an autopilot may certainly be considered a risk control measure, the real challenge comes in determining the impact of an inoperative unit. If the autopilot is known to be inoperative prior to departure, this may factor into the evaluation of other risks.

For example, the pilot may be planning for a VHF omnidirectional range (VOR) approach down to minimums on a dark night into an unfamiliar airport. In such a case, the pilot may have been relying heavily on a functioning autopilot capable of flying a coupled approach. This would free the pilot to monitor aircraft performance. A malfunctioning autopilot could be the single factor that takes this from a medium to a serious risk. At this point, an alternative needs to be considered. On the other hand, if the autopilot were to fail at a critical (high workload) portion of this same flight, the pilot must be prepared to take action. Instead of simply being an inconvenience, this could quickly turn into an emergency if not properly handled. The best way to ensure a pilot is prepared for such an event is to carefully study the issue prior to departure and determine well in advance how an autopilot failure is to be handled.

Familiarity

As previously discussed, pilot familiarity with all equipment is critical in optimizing both safety and efficiency. If a pilot is unfamiliar with any aircraft systems, this will add to workload and may contribute to a loss of situational awareness. This level of proficiency is critical and should be looked upon as a requirement, not unlike carrying an adequate supply of fuel. As a result, pilots should not look upon unfamiliarity with the aircraft and its systems as a risk control measure, but instead as a hazard with high risk potential. Discipline is key to success.

Figure 2-22. *Two similar flight decks equipped with the same information two different ways, analog and digital. What are they indicating? Chances are that the analog pilot will review the top display before the bottom display. Conversely, the digitally trained pilot will review the instrument panel on the bottom first.*

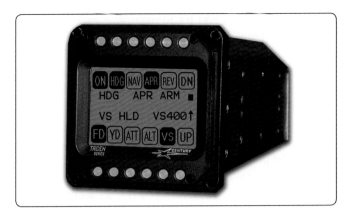

Figure 2-23. *An example of an autopilot system.*

Respect for Onboard Systems

Automation can assist the pilot in many ways, but a thorough understanding of the system(s) in use is essential to gaining the benefits it can offer. Understanding leads to respect, which is achieved through discipline and the mastery of the onboard systems. It is important to fly the aircraft using minimal information from the primary flight display (PFD). This includes turns, climbs, descents, and being able to fly approaches.

Reinforcement of Onboard Suites

The use of an EFD may not seem intuitive, but competency becomes better with understanding and practice. Computer-based software and incremental training help the pilot become comfortable with the onboard suites. Then the pilot needs to practice what was learned in order to gain experience. Reinforcement not only yields dividends in the use of automation, it also reduces workload significantly.

Getting Beyond Rote Workmanship

The key to working effectively with automation is getting beyond the sequential process of executing an action. If a pilot has to analyze what key to push next, or always uses the same sequence of keystrokes when others are available, he or she may be trapped in a rote process. This mechanical process indicates a shallow understanding of the system. Again, the desire is to become competent and know what to do without having to think about, "what keystroke is next." Operating the system with competency and comprehension benefits a pilot when situations become more diverse and tasks increase.

Understand the Platform

Contrary to popular belief, flight in aircraft equipped with different electronic management suites requires the same attention as aircraft equipped with analog instrumentation and a conventional suite of avionics. The pilot should review and understand the different ways in which EFD are used in a particular aircraft. *[Figure 2-24]*

The following are two simple rules for use of an EFD:

- Be able to fly the aircraft to the standards in the PTS. Although this may seem insignificant, knowing how to fly the aircraft to a standard makes a pilot's airmanship smoother and allows him or her more time to attend to the system instead of managing multiple tasks.

- Read and understand the installed electronic flight systems manuals to include the use of the autopilot and the other onboard electronic management tools.

Managing Aircraft Automation

Before any pilot can master aircraft automation, he or she must first know how to fly the aircraft. Maneuvers training remains an important component of flight training because almost 40 percent of all GA accidents take place in the

Figure 2-24. *Examples of different platforms. Top to bottom are the Beechcraft Baron G58, Cirrus SR22, and Cirrus Entega.*

landing phase, one realm of flight that still does not involve programming a computer to execute. Another 15 percent of all GA accidents occurs during takeoff and initial climb.

An advanced avionics safety issue identified by the FAA concerns pilots who apparently develop an unwarranted over-reliance in their avionics and the aircraft, believing that the equipment will compensate for pilot shortcomings. Related to the over-reliance is the role of ADM, which is probably the most significant factor in the GA accident record of high performance aircraft used for cross-country flight. The FAA advanced avionics aircraft safety study found that poor decision-making seems to afflict new advanced avionics pilots at a rate higher than that of GA as a whole. The review of advanced avionics accidents cited in this study shows the majority are not caused by something directly related to the aircraft, but by the pilot's lack of experience and a chain of poor decisions. One consistent theme in many of the fatal accidents is continued VFR flight into IMC.

Thus, pilot skills for normal and emergency operations hinge not only on mechanical manipulation of the stick and rudder, but also include the mental mastery of the EFD. Three key flight management skills are needed to fly the advanced avionics safely: information, automation, and risk.

Information Management

For the newly transitioning pilot, the PFD, MFD, and GPS/VHF navigator screens seem to offer too much information presented in colorful menus and submenus. In fact, the pilot may be drowning in information but unable to find a specific piece of information. It might be helpful to remember these systems are similar to computers that store some folders on a desktop and some within a hierarchy.

The first critical information management skill for flying with advanced avionics is to understand the system at a conceptual level. Remembering how the system is organized helps the pilot manage the available information. It is important to understanding that learning knob-and-dial procedures is not enough. Learning more about how advanced avionics systems work leads to better memory for procedures and allows pilots to solve problems they have not seen before.

There are also limits to understanding. It is generally impossible to understand all of the behaviors of a complex avionics system. Knowing to expect surprises and to continually learn new things is more effective than attempting to memorize mechanical manipulation of the knobs. Simulation software and books on the specific system used are of great value.

The second critical information management skill is stop, look, and read. Pilots new to advanced avionics often become fixated on the knobs and try to memorize each and every sequence of button pushes, pulls, and turns. A far better strategy for accessing and managing the information available in advanced avionics computers is to stop, look, and read. Reading before pushing, pulling, or twisting can often save a pilot some trouble.

Once behind the display screens on an advanced avionics aircraft, the pilot's goal is to meter, manage, and prioritize the information flow to accomplish specific tasks. Certificated flight instructors (CFIs), as well as pilots transitioning to advanced avionics, will find it helpful to corral the information flow. This is possible through such tactics as configuring the aspects of the PFD and MFD screens according to personal preferences. For example, most systems offer map orientation options that include "north up," "track up," "DTK" (desired track up), and "heading up." Another tactic is to decide, when possible, how much (or how little) information to display. Pilots can also tailor the information displayed to suit the needs of a specific flight.

Information flow can also be managed for a specific operation. The pilot has the ability to prioritize information for a timely display of exactly the information needed for any given flight operation. Examples of managing information display for a specific operation include:

- Program map scale settings for en route versus terminal area operation.

- Utilize the terrain awareness page on the MFD for a night or IMC flight in or near the mountains.

- Use the nearest airports inset on the PFD at night or over inhospitable terrain.

- Program the weather datalink set to show echoes and METAR status flags.

Enhanced Situational Awareness

An advanced avionics aircraft offers increased safety with enhanced situational awareness. Although aircraft flight manuals (AFM) explicitly prohibit using the moving map, topography, terrain awareness, traffic, and weather datalink displays as the primary data source, these tools nonetheless give the pilot unprecedented information for enhanced situational awareness. Without a well-planned information management strategy, these tools also make it easy for an unwary pilot to slide into the complacent role of passenger in command.

Consider the pilot whose navigational information management strategy consists solely of following the magenta line on the moving map. He or she can easily fly into geographic or regulatory disaster, if the straight-line GPS course goes through high terrain or prohibited airspace, or if the moving map display fails.

A good strategy for maintaining situational awareness information management should include practices that help ensure that awareness is enhanced, not diminished, by the use of automation. Two basic procedures are to always double-check the system and verbal callouts. At a minimum, ensure the presentation makes sense. Was the correct destination fed into the navigation system? Callouts—even for single-pilot operations—are an excellent way to maintain situational awareness, as well as manage information.

Other ways to maintain situational awareness include:

- Perform verification check of all programming. Before departure, check all information programmed while on the ground.

- Check the flight routing. Before departure, ensure all routing matches the planned flight route. Enter the planned route and legs, to include headings and leg length, on a paper log. Use this log to evaluate what has been programmed. If the two do not match, do not assume the computer data is correct, double check the computer entry.

- Verify waypoints.

- Make use of all onboard navigation equipment. For example, use VOR to back up GPS and vice versa.

- Match the use of the automated system with pilot proficiency. Stay within personal limitations.

- Plan a realistic flight route to maintain situational awareness. For example, although the onboard equipment allows a direct flight from Denver, Colorado, to Destin, Florida, the likelihood of rerouting around Eglin Air Force Base's airspace is high.

- Be ready to verify computer data entries. For example, incorrect keystrokes could lead to loss of situational awareness because the pilot may not recognize errors made during a high workload period.

Automation Management

Advanced avionics offer multiple levels of automation, from strictly manual flight to highly automated flight. No one level of automation is appropriate for all flight situations, but in order to avoid potentially dangerous distractions when flying with advanced avionics, the pilot must know how to manage the course deviation indicator (CDI), the navigation source, and the autopilot. It is important for a pilot to know the peculiarities of the particular automated system being used. This ensures the pilot knows what to expect, how to monitor for proper operation, and promptly take appropriate action if the system does not perform as expected.

For example, at the most basic level, managing the autopilot means knowing at all times which modes are engaged and which modes are armed to engage. The pilot needs to verify that armed functions (e.g., navigation tracking or altitude capture) engage at the appropriate time. Automation management is another good place to practice the callout technique, especially after arming the system to make a change in course or altitude.

In advanced avionics aircraft, proper automation management also requires a thorough understanding of how the autopilot interacts with the other systems. For example, with some autopilots, changing the navigation source on the e-HSI from GPS to LOC or VOR while the autopilot is engaged in NAV (course tracking mode) causes the autopilot's NAV mode to disengage. The autopilot's lateral control will default to ROL (wing level) until the pilot takes action to reengage the NAV mode to track the desired navigation source.

Risk Management

Risk management is the last of the three flight management skills needed for mastery of the glass flight deck aircraft. The enhanced situational awareness and automation capabilities offered by a glass flight deck airplane vastly expand its safety and utility, especially for personal transportation use. At the same time, there is some risk that lighter workloads could lead to complacency.

Humans are characteristically poor monitors of automated systems. When asked to passively monitor an automated system for faults, abnormalities, or other infrequent events, humans perform poorly. The more reliable the system, the poorer the human performance. For example, the pilot only monitors a backup alert system, rather than the situation that the alert system is designed to safeguard. It is a paradox of automation that technically advanced avionics can both increase and decrease pilot awareness.

It is important to remember that EFDs do not replace basic flight knowledge and skills. They are a tool for improving flight safety. Risk increases when the pilot believes the gadgets compensate for lack of skill and knowledge. It is especially important to recognize there are limits to what the electronic systems in any light GA aircraft can do. Being PIC requires sound ADM, which sometimes means saying "no" to a flight.

Risk is also increased when the pilot fails to monitor the systems. By failing to monitor the systems and failing to check the results of the processes, the pilot becomes detached from the aircraft operation and slides into the complacent role of passenger in command. Complacency led to tragedy in a 1999 aircraft accident.

In Colombia, a multi-engine aircraft crewed with two pilots struck the face of the Andes Mountains. Examination of their FMS revealed they entered a waypoint into the FMS incorrectly by one degree resulting in a flight path taking them to a point 60 NM off their intended course. The pilots were equipped with the proper charts, their route was posted on the charts, and they had a paper navigation log indicating the direction of each leg. They had all the tools to manage and monitor their flight, but instead allowed the automation to fly and manage itself. The system did exactly what it was programmed to do; it flew on a programmed course into a mountain resulting in multiple deaths. The pilots simply failed to manage the system and inherently created their own hazard. Although this hazard was self-induced, what is notable is the risk the pilots created through their own inattention. By failing to evaluate each turn made at the direction of automation, the pilots maximized risk instead of minimizing it. In this case, a totally avoidable accident become a tragedy through simple pilot error and complacency.

For the GA pilot transitioning to automated systems, it is helpful to note that all human activity involving technical devices entails some element of risk. Knowledge, experience, and mission requirements tilt the odds in favor of safe and successful flights. The advanced avionics aircraft offers many new capabilities and simplifies the basic flying tasks, but only if the pilot is properly trained and all the equipment is working as advertised.

Chapter Summary

This chapter focused on helping the pilot improve his or her ADM skills with the goal of mitigating the risk factors associated with flight in both classic and automated aircraft. In the end, the discussion is not so much about aircraft, but about the people who fly them.

Chapter 3

Aircraft Construction

Introduction

An aircraft is a device that is used, or intended to be used, for flight according to the current Title 14 of the Code of Federal Regulations (14 CFR) part 1, Definitions and Abbreviations. Categories of aircraft for certification of airmen include airplane, rotorcraft, glider, lighter-than-air, powered-lift, powered parachute, and weight-shift control aircraft. Title 14 CFR part 1 also defines airplane as an engine-driven, fixed-wing aircraft that is supported in flight by the dynamic reaction of air against its wings. Another term, not yet codified in 14 CFR part 1, is advanced avionics aircraft, which refers to an aircraft that contains a global positioning system (GPS) navigation system with a moving map display, in conjunction with another system, such as an autopilot. This chapter provides a brief introduction to the structure of aircraft and uses an airplane for most illustrations. Light Sport Aircraft (LSA), such as weight-shift control aircraft, balloon, glider, powered parachute, and gyroplane, have their own handbooks to include detailed information regarding aerodynamics and control.

Wing

Empennage

Powerplant

Fuselage

Landing gear

Aircraft Design, Certification, and Airworthiness

The FAA certifies three types of aviation products: aircraft, aircraft engines, and propellers. Each of these products has been designed to a set of airworthiness standards. These standards are parts of Title 14 of the Code of Federal Regulations (14 CFR), published by the FAA. The airworthiness standards were developed to help ensure that aviation products are designed with no unsafe features. Different airworthiness standards apply to the different categories of aviation products as follows:

- Normal, Utility, Acrobatic, and Commuter Category Airplanes- 14 CFR part 23

- Transport Category Airplanes—14 CFR part 25

- Normal Category—14 CFR part 27

- Transport Category Rotorcraft—14 CFR part 29

- Manned Free Balloons—14 CFR part 31

- Aircraft Engines—14 CFR part 33

- Propellers—14 CFR part 35

Some aircraft are considered "special classes" of aircraft and do not have their own airworthiness standards, such as gliders and powered lift. The airworthiness standards used for these aircraft are a combination of requirements in 14 CFR parts 23, 25, 27, and 29 that the FAA and the designer have agreed are appropriate for the proposed aircraft.

The FAA issues a Type Certificate (TC) for the product when they are satisfied it complies with the applicable airworthiness standards. When the TC is issued, a Type Certificate Data Sheet (TCDS) is generated that specifies the important design and operational characteristics of the aircraft, aircraft engine, or propeller. The TCDS defines the product and are available to the public from the FAA website at www.faa.gov.

A Note About Light Sport Aircraft

Light sport aircraft are not designed according to FAA airworthiness standards. Instead, they are designed to a consensus of standards agreed upon in the aviation industry. The FAA has agreed the consensus of standards is acceptable as the design criteria for these aircraft. Light sport aircraft do not necessarily have individually type certificated engines and propellers. Instead, a TC is issued to the aircraft as a whole. It includes the airframe, engine, and propeller.

Aircraft, aircraft engines, and propellers can be manufactured one at a time from the design drawings, or through an FAA approved manufacturing process, depending on the size and capabilities of the manufacturer. During the manufacturing process, each part is inspected to ensure that it has been built exactly according to the approved design. This inspection is called a conformity inspection.

When the aircraft is complete, with the airframe, engine, and propeller, it is inspected and the FAA issues an airworthiness certificate for the aircraft. Having an airworthiness certificate means the complete aircraft meets the design and manufacturing standards, and is in a condition for safe flight. This airworthiness certificate must be carried in the aircraft during all flight operations. The airworthiness certificate remains valid as long as the required maintenance and inspections are kept up to date for the aircraft.

Airworthiness certificates are classified as either "Standard" or "Special." Standard airworthiness certificates are white, and are issued for normal, utility, acrobatic, commuter, or transport category aircraft. They are also issued for manned free balloons and aircraft designated as "Special Class."

Special airworthiness certificates are pink, and are issued for primary, restricted, and limited category aircraft, and light sport aircraft. They are also issued as provisional airworthiness certificates, special flight permits (ferry permits), and for experimental aircraft.

More information on airworthiness certificates can be found in Chapter 9, in 14 CFR parts 175-225, and also on the FAA website at www.faa.gov.

Lift and Basic Aerodynamics

In order to understand the operation of the major components and subcomponents of an aircraft, it is important to understand basic aerodynamic concepts. This chapter briefly introduces aerodynamics; a more detailed explanation can be found in Chapter 5, Aerodynamics of Flight.

Four forces act upon an aircraft in relation to straight-and-level, unaccelerated flight. These forces are thrust, lift, weight, and drag. *[Figure 3-1]*

Thrust is the forward force produced by the powerplant/propeller. It opposes or overcomes the force of drag. As a general rule, it is said to act parallel to the longitudinal axis. This is not always the case as explained later.

Drag is a rearward, retarding force and is caused by disruption of airflow by the wing, fuselage, and other protruding objects. Drag opposes thrust and acts rearward parallel to the relative wind.

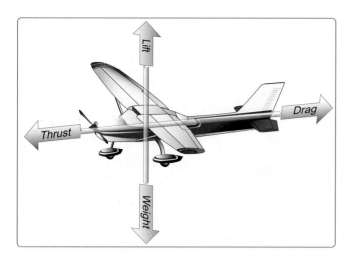

Figure 3-1. *The four forces.*

Weight is the combined load of the aircraft itself, the crew, the fuel, and the cargo or baggage. Weight pulls the aircraft downward because of the force of gravity. It opposes lift and acts vertically downward through the aircraft's center of gravity (CG).

Lift opposes the downward force of weight, is produced by the dynamic effect of the air acting on the wing, and acts perpendicular to the flight path through the wing's center of lift (CL).

An aircraft moves in three dimensions and is controlled by moving it about one or more of its axes. The longitudinal, or roll, axis extends through the aircraft from nose to tail, with the line passing through the CG. The lateral or pitch axis extends across the aircraft on a line through the wing tips, again passing through the CG. The vertical, or yaw, axis passes through the aircraft vertically, intersecting the CG. All control movements cause the aircraft to move around one or more of these axes and allows for the control of the aircraft in flight. *[Figure 3-2]*

One of the most significant components of aircraft design is CG. It is the specific point where the mass or weight of an aircraft may be said to center; that is, a point around which, if the aircraft could be suspended or balanced, the aircraft would remain relatively level. The position of the CG of an aircraft determines the stability of the aircraft in flight. As the CG moves rearward (towards the tail), the aircraft becomes more and more dynamically unstable. In aircraft with fuel tanks situated in front of the CG, it is important that the CG is set with the fuel tank empty. Otherwise, as the fuel is used, the aircraft becomes unstable. *[Figure 3-3]* The CG is computed during initial design and construction and is further affected by the installation of onboard equipment, aircraft loading, and other factors.

Major Components

Although airplanes are designed for a variety of purposes, most of them have the same major components. *[Figure 3-4]* The overall characteristics are largely determined by the original design objectives. Most airplane structures include a fuselage, wings, an empennage, landing gear, and a powerplant.

Fuselage

The fuselage is the central body of an airplane and is designed to accommodate the crew, passengers, and cargo. It also provides the structural connection for the wings and tail assembly. Older types of aircraft design utilized an open truss structure constructed of wood, steel, or aluminum tubing. *[Figure 3-5]* The most popular types of fuselage structures used in today's aircraft are the monocoque (French for "single shell") and semimonocoque. These structure types are discussed in more detail under aircraft construction later in the chapter.

Wings

The wings are airfoils attached to each side of the fuselage and are the main lifting surfaces that support the airplane in

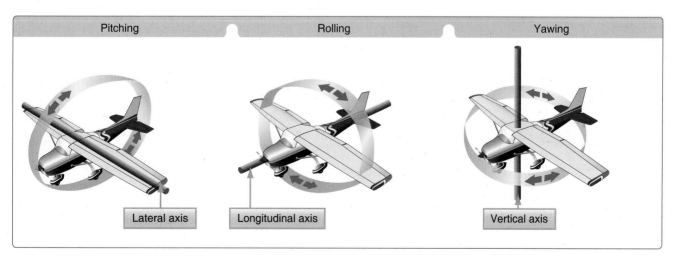

Figure 3-2. *Illustrates the pitch, roll, and yaw motion of the aircraft along the lateral, longitudinal, and vertical axes, respectively.*

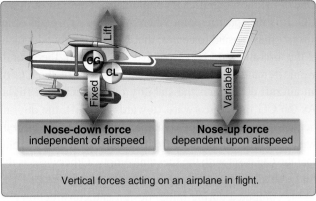

Nose-down force
independent of airspeed

Nose-up force
dependent upon airspeed

Vertical forces acting on an airplane in flight.

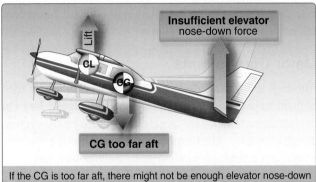

Insufficient elevator
nose-down force

CG too far aft

If the CG is too far aft, there might not be enough elevator nose-down force at the low stall airspeed to get the nose down for recovery.

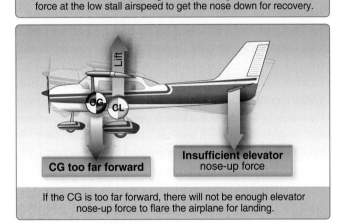

CG too far forward

Insufficient elevator
nose-up force

If the CG is too far forward, there will not be enough elevator nose-up force to flare the airplane for landing.

Figure 3-3. *Center of gravity (CG).*

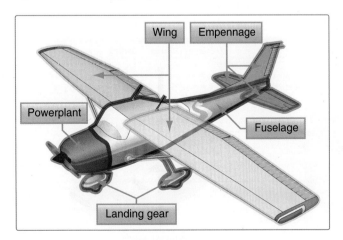

Figure 3-4. *Airplane components.*

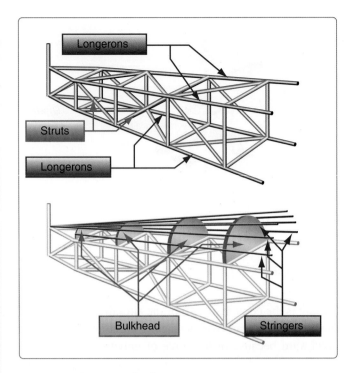

Figure 3-5. *Truss-type fuselage structure.*

flight. There are numerous wing designs, sizes, and shapes used by the various manufacturers. Each fulfills a certain need with respect to the expected performance for the particular airplane. How the wing produces lift is explained in Chapter 5, Aerodynamics of Flight.

Wings may be attached at the top, middle, or lower portion of the fuselage. These designs are referred to as high-, mid-, and low-wing, respectively. The number of wings can also vary. Airplanes with a single set of wings are referred to as monoplanes, while those with two sets are called biplanes. *[Figure 3-6]*

Many high-wing airplanes have external braces, or wing struts that transmit the flight and landing loads through the struts to the main fuselage structure. Since the wing struts are usually attached approximately halfway out on the wing, this type of wing structure is called semi-cantilever. A few high-wing and most low-wing airplanes have a full cantilever wing designed to carry the loads without external struts.

The principal structural parts of the wing are spars, ribs, and stringers. *[Figure 3-7]* These are reinforced by trusses, I-beams, tubing, or other devices, including the skin. The wing ribs determine the shape and thickness of the wing (airfoil). In most modern airplanes, the fuel tanks are either an integral part of the wing's structure or consist of flexible containers mounted inside of the wing.

Figure 3-6. *Monoplane (left) and biplane (right).*

Attached to the rear, or trailing edges, of the wings are two types of control surfaces referred to as ailerons and flaps. Ailerons extend from about the midpoint of each wing outward toward the tip, and move in opposite directions to create aerodynamic forces that cause the airplane to roll. Flaps extend outward from the fuselage to near the midpoint of each wing. The flaps are normally flush with the wing's surface during cruising flight. When extended, the flaps move simultaneously downward to increase the lifting force of the wing for takeoffs and landings. *[Figure 3-8]*

Alternate Types of Wings

Alternate types of wings are often found on aircraft. The shape and design of a wing is dependent upon the type of operation for which an aircraft is intended and is tailored to specific types of flying. These design variations are discussed in Chapter 5, Aerodynamics of Flight, which provides information on the effect controls have on lifting surfaces from traditional wings to wings that use both flexing (due to billowing) and shifting (through the change of the aircraft's CG). For example, the wing of the weight-shift control aircraft is highly swept in an effort to reduce drag and allow for the shifting of weight to provide controlled flight. *[Figure 3-9]* Handbooks specific to most categories of aircraft are available for the interested pilot and can be found on the Federal Aviation Administration (FAA) website at www.faa.gov.

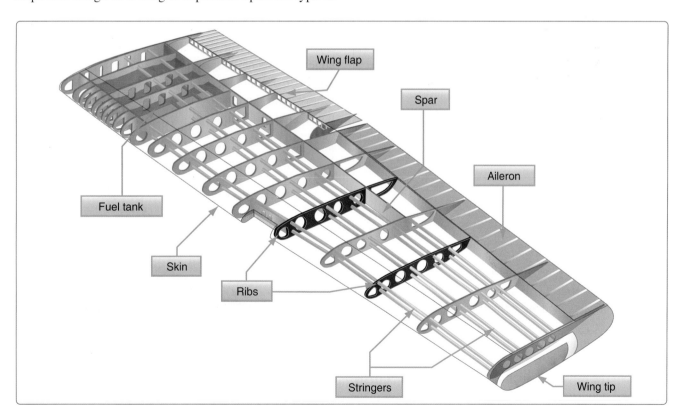

Figure 3-7. *Wing components.*

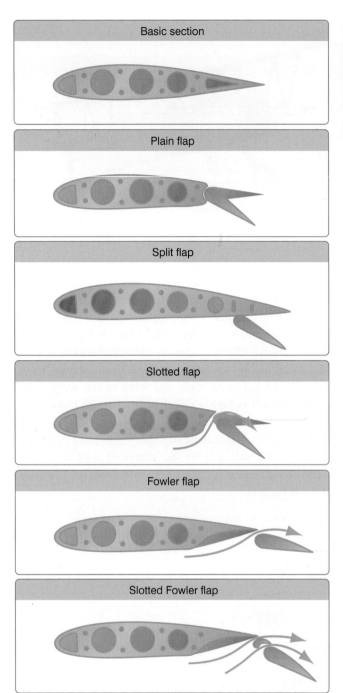

Figure 3-8. *Types of flaps.*

Empennage

The empennage includes the entire tail group and consists of fixed surfaces, such as the vertical stabilizer and the horizontal stabilizer. The movable surfaces include the rudder, the elevator, and one or more trim tabs. *[Figure 3-10]*

The rudder is attached to the back of the vertical stabilizer. During flight, it is used to move the airplane's nose left and right. The elevator, which is attached to the back of the horizontal stabilizer, is used to move the nose of the airplane up and down during flight. Trim tabs are small, movable

Figure 3-9. *Weight-shift control aircraft use the shifting of weight for control.*

portions of the trailing edge of the control surface. These movable trim tabs, which are controlled from the flight deck, reduce control pressures. Trim tabs may be installed on the ailerons, the rudder, and/or the elevator.

A second type of empennage design does not require an elevator. Instead, it incorporates a one-piece horizontal stabilizer that pivots from a central hinge point. This type of design is called a stabilator and is moved using the control wheel, just as the elevator is moved. For example, when a pilot pulls back on the control wheel, the stabilator pivots so the trailing edge moves up. This increases the aerodynamic tail load and causes the nose of the airplane to move up. Stabilators have an antiservo tab extending across their trailing edge. *[Figure 3-11]*

The antiservo tab moves in the same direction as the trailing edge of the stabilator and helps make the stabilator less sensitive. The antiservo tab also functions as a trim tab to relieve control pressures and helps maintain the stabilator in the desired position.

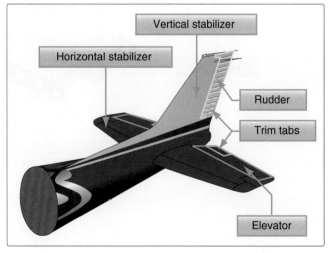

Figure 3-10. *Empennage components.*

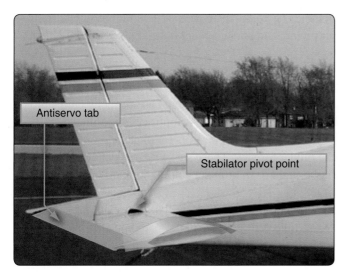

Figure 3-11. *Stabilator components.*

Landing Gear

The landing gear is the principal support of the airplane when parked, taxiing, taking off, or landing. The most common type of landing gear consists of wheels, but airplanes can also be equipped with floats for water operations or skis for landing on snow. *[Figure 3-12]*

Wheeled landing gear consists of three wheels—two main wheels and a third wheel positioned either at the front or rear of the airplane. Landing gear with a rear mounted wheel is called conventional landing gear.

Airplanes with conventional landing gear are sometimes referred to as tailwheel airplanes. When the third wheel is located on the nose, it is called a nosewheel, and the design is referred to as a tricycle gear. A steerable nosewheel or tailwheel permits the airplane to be controlled throughout all operations while on the ground. Most aircraft are steered by moving the rudder pedals, whether nosewheel or tailwheel. Additionally, some aircraft are steered by differential braking.

The Powerplant

The powerplant usually includes both the engine and the propeller. The primary function of the engine is to provide the power to turn the propeller. It also generates electrical power, provides a vacuum source for some flight instruments, and in most single-engine airplanes, provides a source of heat for the pilot and passengers. *[Figure 3-13]* The engine is covered by a cowling, or a nacelle, which are both types of covered housing. The purpose of the cowling or nacelle is to streamline the flow of air around the engine and to help cool the engine by ducting air around the cylinders.

The propeller, mounted on the front of the engine, translates the rotating force of the engine into thrust, a forward acting force that helps move the airplane through the air. A propeller

Figure 3-12. *Types of landing gear: floats (top), skis (middle), and wheels (bottom).*

is a rotating airfoil that produces thrust through aerodynamic action. A high-pressure area is formed at the back of the propeller's airfoil, and low pressure is produced at the face of the propeller, similar to the way lift is generated by an airfoil used as a lifting surface or wing. This pressure differential develops thrust from the propeller, which in turn pulls the airplane forward. Engines may be turned around to be pushers with the propeller at the rear.

There are two significant factors involved in the design of a propeller that impact its effectiveness. The angle of a

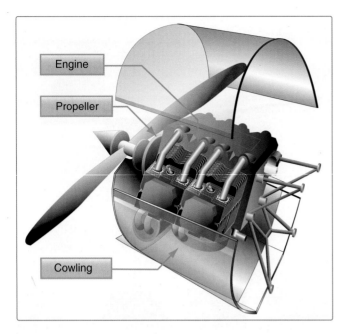

Figure 3-13. *Engine compartment.*

propeller blade, as measured against the hub of the propeller, keeps the angle of attack (AOA) (See definition in Glossary) relatively constant along the span of the propeller blade, reducing or eliminating the possibility of a stall. The amount of lift being produced by the propeller is directly related to the AOA, which is the angle at which the relative wind meets the blade. The AOA continuously changes during the flight depending upon the direction of the aircraft.

The pitch is defined as the distance a propeller would travel in one revolution if it were turning in a solid. These two factors combine to allow a measurement of the propeller's efficiency. Propellers are usually matched to a specific aircraft/powerplant combination to achieve the best efficiency at a particular power setting, and they pull or push depending on how the engine is mounted.

Subcomponents

The subcomponents of an airplane include the airframe, electrical system, flight controls, and brakes.

The airframe is the basic structure of an aircraft and is designed to withstand all aerodynamic forces, as well as the stresses imposed by the weight of the fuel, crew, and payload.

The primary function of an aircraft electrical system is to generate, regulate, and distribute electrical power throughout the aircraft. There are several different power sources on aircraft to power the aircraft electrical systems. These power sources include: engine-driven alternating current (AC) generators, auxiliary power units (APUs), and external power. The aircraft's electrical power system is used to operate the flight instruments, essential systems, such as anti-icing, and passenger services, such as cabin lighting.

The flight controls are the devices and systems that govern the attitude of an aircraft and, as a result, the flight path followed by the aircraft. In the case of many conventional airplanes, the primary flight controls utilize hinged, trailing-edge surfaces called elevators for pitch, ailerons for roll, and the rudder for yaw. These surfaces are operated by the pilot in the flight deck or by an automatic pilot.

In the case of most modern airplanes, airplane brakes consist of multiple pads (called caliper pads) that are hydraulically squeezed toward each other with a rotating disk (called a rotor) between them. The pads place pressure on the rotor which is turning with the wheels. As a result of the increased friction on the rotor, the wheels inherently slow down and stop turning. The disks and brake pads are made either from steel, like those in a car, or from a carbon material that weighs less and can absorb more energy. Because airplane brakes are used principally during landings and must absorb enormous amounts of energy, their life is measured in landings rather than miles.

Types of Aircraft Construction

The construction of aircraft fuselages evolved from the early wood truss structural arrangements to monocoque shell structures to the current semimonocoque shell structures.

Truss Structure

The main drawback of truss structure is its lack of a streamlined shape. In this construction method, lengths of tubing, called longerons, are welded in place to form a well-braced framework. Vertical and horizontal struts are welded to the longerons and give the structure a square or rectangular shape when viewed from the end. Additional struts are needed to resist stress that can come from any direction. Stringers and bulkheads, or formers, are added to shape the fuselage and support the covering.

As technology progressed, aircraft designers began to enclose the truss members to streamline the airplane and improve performance. This was originally accomplished with cloth fabric, which eventually gave way to lightweight metals such as aluminum. In some cases, the outside skin can support all or a major portion of the flight loads. Most modern aircraft use a form of this stressed skin structure known as monocoque or semimonocoque construction. *[Figure 3-14]*

Monocoque

Monocoque construction uses stressed skin to support almost all loads much like an aluminum beverage can. Although very strong, monocoque construction is not highly tolerant

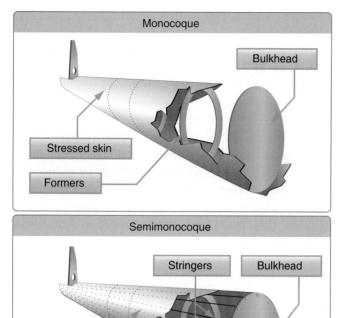

Figure 3-14. *Semimonocoque and monocoque fuselage design.*

to deformation of the surface. For example, an aluminum beverage can supports considerable forces at the ends of the can, but if the side of the can is deformed slightly while supporting a load, it collapses easily.

Because most twisting and bending stresses are carried by the external skin rather than by an open framework, the need for internal bracing was eliminated or reduced, saving weight and maximizing space. One of the notable and innovative methods for using monocoque construction was employed by Jack Northrop. In 1918, he devised a new way to construct a monocoque fuselage used for the Lockheed S-1 Racer. The technique utilized two molded plywood half-shells that were glued together around wooden hoops or stringers. To construct the half shells, rather than gluing many strips of plywood over a form, three large sets of spruce strips were soaked with glue and laid in a semi-circular concrete mold that looked like a bathtub. Then, under a tightly clamped lid, a rubber balloon was inflated in the cavity to press the plywood against the mold. Twenty-four hours later, the smooth half-shell was ready to be joined to another to create the fuselage. The two halves were each less than a quarter inch thick. Although employed in the early aviation period, monocoque construction would not reemerge for several decades due to the complexities involved. Every day examples of monocoque construction can be found in

automobile manufacturing where the unibody is considered standard in manufacturing.

Semimonocoque

Semimonocoque construction, partial or one-half, uses a substructure to which the airplane's skin is attached. The substructure, which consists of bulkheads and/or formers of various sizes and stringers, reinforces the stressed skin by taking some of the bending stress from the fuselage. The main section of the fuselage also includes wing attachment points and a firewall. On single-engine airplanes, the engine is usually attached to the front of the fuselage. There is a fireproof partition between the rear of the engine and the flight deck or cabin to protect the pilot and passengers from accidental engine fires. This partition is called a firewall and is usually made of heat-resistant material such as stainless steel. However, a new emerging process of construction is the integration of composites or aircraft made entirely of composites.

Composite Construction
History

The use of composites in aircraft construction can be dated to World War II aircraft when soft fiberglass insulation was used in B-29 fuselages. By the late 1950s, European high performance sailplane manufacturers were using fiberglass as primary structures. In 1965, the FAA type certified the first all-fiberglass aircraft in the normal category, a Swiss sailplane called a Diamant HBV. Four years later, the FAA certified a four-seat, single-engine Windecker Eagle in the normal category. By 2005, over 35 percent of new aircraft were constructed of composite materials.

Composite is a broad term and can mean materials such as fiberglass, carbon fiber cloth, Kevlar™ cloth, and mixtures of all of the above. Composite construction offers two advantages: extremely smooth skins and the ability to easily form complex curved or streamlined structures. *[Figure 3-15]*

Composite Materials in Aircraft
Composite materials are fiber-reinforced matrix systems. The matrix is the "glue" used to hold the fibers together and, when cured, gives the part its shape, but the fibers carry most of the load. There are many different types of fibers and matrix systems.

In aircraft, the most common matrix is epoxy resin, which is a type of thermosetting plastic. Compared to other choices such as polyester resin, epoxy is stronger and has good high-temperature properties. There are many different types of epoxies available with a wide range of structural properties, cure times and temperatures, and costs.

Figure 3-15. *Composite aircraft.*

The most common reinforcing fibers used in aircraft construction are fiberglass and carbon fiber. Fiberglass has good tensile and compressive strength, good impact resistance, is easy to work with, and is relatively inexpensive and readily available. Its main disadvantage is that it is somewhat heavy, and it is difficult to make a fiberglass load-carrying structure lighter than a well designed equivalent aluminum structure.

Carbon fiber is generally stronger in tensile and compressive strength than fiberglass and has much higher bending stiffness. It is also considerably lighter than fiberglass. However, it is relatively poor in impact resistance; the fibers are brittle and tend to shatter under sharp impact. This can be greatly improved with a "toughened" epoxy resin system, as used in the Boeing 787 horizontal and vertical stabilizers. Carbon fiber is more expensive than fiberglass, but the price has dropped due to innovations driven by the B-2 program in the 1980s and Boeing 777 work in the 1990s. Very well-designed carbon fiber structures can be significantly lighter than an equivalent aluminum structure, sometimes by 30 percent or so.

Advantages of Composites

Composite construction offers several advantages over metal, wood, or fabric, with its lighter weight being the most frequently cited. Lighter weight is not always automatic. It must be remembered that building an aircraft structure out of composites does not guarantee it will be lighter; it depends on the structure, as well as the type of composite being used.

A more important advantage is that a very smooth, compound curved, aerodynamic structure made from composites reduces drag. This is the main reason sailplane designers switched from metal and wood to composites in the 1960s. In aircraft, the use of composites reduces drag for the Cirrus

and Columbia line of production aircraft, leading to their high performance despite their fixed landing gear. Composites also help mask the radar signature of "stealth" aircraft designs, such as the B-2 and the F-22. Today, composites can be found in aircraft as varied as gliders to most new helicopters.

Lack of corrosion is a third advantage of composites. Boeing is designing the 787, with its all-composite fuselage, to have both a higher pressure differential and higher humidity in the cabin than previous airliners. Engineers are no longer as concerned about corrosion from moisture condensation on the hidden areas of the fuselage skins, such as behind insulation blankets. This should lead to lower long-term maintenance costs for the airlines.

Another advantage of composites is their good performance in a flexing environment, such as in helicopter rotor blades. Composites do not suffer from metal fatigue and crack growth as do metals. While it takes careful engineering, composite rotor blades can have considerably higher design lives than metal blades, and most new large helicopter designs have all composite blades, and in many cases, composite rotor hubs.

Disadvantages of Composites

Composite construction comes with its own set of disadvantages, the most important of which is the lack of visual proof of damage. Composites respond differently from other structural materials to impact, and there is often no obvious sign of damage. For example, if a car backs into an aluminum fuselage, it might dent the fuselage. If the fuselage is not dented, there is no damage. If the fuselage is dented, the damage is visible and repairs are made.

In a composite structure, a low energy impact, such as a bump or a tool drop, may not leave any visible sign of the impact on the surface. Underneath the impact site there may be extensive delaminations, spreading in a cone-shaped area from the impact location. The damage on the backside of the structure can be significant and extensive, but it may be hidden from view. Anytime one has reason to think there may have been an impact, even a minor one, it is best to get an inspector familiar with composites to examine the structure to determine underlying damage. The appearance of "whitish" areas in a fiberglass structure is a good tip-off that delaminations of fiber fracture has occurred.

A medium energy impact (perhaps the car backing into the structure) results in local crushing of the surface, which should be visible to the eye. The damaged area is larger than the visible crushed area and will need to be repaired. A high energy impact, such as a bird strike or hail while in flight, results in a puncture and a severely damaged structure. In medium and high energy impacts, the damage is visible to the eye, but low energy impact is difficult to detect. *[Figure 3-16]*

If an impact results in delaminations, crushing of the surface, or a puncture, then a repair is mandatory. While waiting for the repair, the damaged area should be covered and protected from rain. Many composite parts are composed of thin skins over a honeycomb core, creating a "sandwich" structure. While excellent for structural stiffness reasons, such a structure is an easy target for water ingress (entering), leading to further problems later. A piece of "speed tape" over the puncture is a good way to protect it from water, but it is not a structural repair. The use of a paste filler to cover up the damage, while acceptable for cosmetic purposes, is not a structural repair, either.

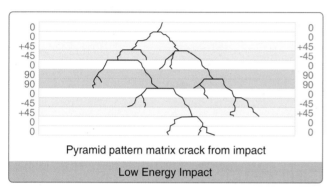

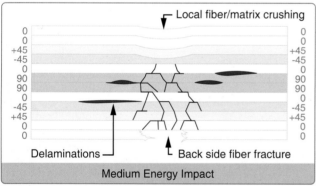

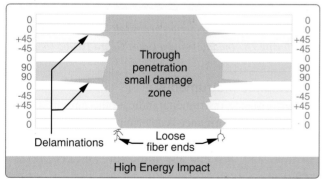

Figure 3-16. *Impact energy affects the visibility, as well as the severity, of damage in composite structures. High and medium energy impacts, while severe, are easy to detect. Low energy impacts can easily cause hidden damage.*

The potential for heat damage to the resin is another disadvantage of using composites. While "too hot" depends on the particular resin system chosen, many epoxies begin to weaken over 150 °F. White paint on composites is often used to minimize this issue. For example, the bottom of a wing that is painted black facing a black asphalt ramp on a hot, sunny day can get as hot as 220 °F. The same structure, painted white, rarely exceeds 140 °F. As a result, composite aircraft often have specific recommendations on allowable paint colors. If the aircraft is repainted, these recommendations must be followed. Heat damage can also occur due to a fire. Even a quickly extinguished small brake fire can damage bottom wing skins, composite landing gear legs, or wheel pants.

Also, chemical paint strippers are very harmful to composites and must not be used on them. If paint needs to be removed from composites, only mechanical methods are allowed, such as gentle grit blasting or sanding. Many expensive composite parts have been ruined by the use of paint stripper and such damage is generally not repairable.

Fluid Spills on Composites

Some owners are concerned about fuel, oil, or hydraulic fluid spills on composite surfaces. These are generally not a problem with modern composites using epoxy resin. Usually, if the spill does not attack the paint, it will not hurt the underlying composite. Some aircraft use fiberglass fuel tanks, for example, in which the fuel rides directly against the composite surface with no sealant being used. If the fiberglass structure is made with some of the more inexpensive types of polyester resin, there can be a problem when using auto gas with ethanol blended into the mixture. The more expensive types of polyester resin, as well as epoxy resin, can be used with auto gas, as well as 100 octane aviation gas (avgas) and jet fuel.

Lightning Strike Protection

Lightning strike protection is an important consideration in aircraft design. When an aircraft is hit by lightning, a very large amount of energy is delivered to the structure. Whether flying a light general aviation (GA) aircraft or a large airliner, the basic principle of lightning strike protection is the same. For any size aircraft, the energy from the strike must be spread over a large surface area to lower the amps per square inch to a harmless level.

If lightning strikes an aluminum airplane, the electrical energy naturally conducts easily through the aluminum structure. The challenge is to keep the energy out of avionics, fuel systems, etc., until it can be safely conducted overboard. The outer skin of the aircraft is the path of least resistance.

In a composite aircraft, fiberglass is an excellent electrical insulator, while carbon fiber conducts electricity, but not as easily as aluminum. Therefore, additional electrical conductivity needs to be added to the outside layer of composite skin. This is done typically with fine metal meshes bonded to the skin surfaces. Aluminum and copper mesh are the two most common types, with aluminum used on fiberglass and copper on carbon fiber. Any structural repairs on lightning-strike protected areas must also include the mesh as well as the underlying structure.

For composite aircraft with internal radio antennas, there must be "windows" in the lightning strike mesh in the area of the antenna. Internal radio antennas may be found in fiberglass composites because fiberglass is transparent to radio frequencies, where carbon fiber is not.

The Future of Composites

In the decades since World War II, composites have earned an important role in aircraft structure design. Their design flexibility and corrosion resistance, as well as the high strength-to-weight ratios possible, will undoubtedly continue to lead to more innovative aircraft designs in the future. From the Cirrus SR-20 to the Boeing 787, it is obvious that composites have found a home in aircraft construction and are here to stay. *[Figure 3-17]*

Instrumentation: Moving into the Future

Until recently, most GA aircraft were equipped with individual instruments utilized collectively to safely operate and maneuver the aircraft. With the release of the electronic flight display (EFD) system, conventional instruments have been replaced by multiple liquid crystal display (LCD) screens. The first screen is installed in front of the pilot position and is referred to as the primary flight display (PFD). The second screen, positioned approximately in the center of the instrument panel, is referred to as the multi-function display (MFD). These two screens de-clutter instrument panels while increasing safety. This has been accomplished through the utilization of solid state instruments that have a failure rate far less than those of conventional analog instrumentation. *[Figure 3-18]*

With today's improvements in avionics and the introduction of EFDs, pilots at any level of experience need an astute knowledge of the onboard flight control systems, as well as an understanding of how automation melds with aeronautical decision-making (ADM). These subjects are covered in detail in Chapter 2, Aeronautical Decision-Making.

Whether an aircraft has analog or digital (glass) instruments, the instrumentation falls into three different categories: performance, control, and navigation.

Figure 3-17. *Composite materials in aircraft, such as Columbia 350 (top), Boeing 787 (middle), and a Coast Guard HH-65 (bottom).*

Performance Instruments

The performance instruments indicate the aircraft's actual performance. Performance is determined by reference to the altimeter, airspeed or vertical speed indicator (VSI), heading indicator, and turn-and-slip indicator. The performance instruments directly reflect the performance the aircraft is achieving. The speed of the aircraft can be referenced on the airspeed indicator. The altitude can be referenced on the altimeter. The aircraft's climb performance can be determined by referencing the VSI. Other performance instruments available are the heading indicator, angle of attack indicator, and the slip-skid indicator. *[Figure 3-19]*

Figure 3-18. *Analog display (top) and digital display (bottom) from a Cessna 172.*

Control Instruments

The control instruments display immediate attitude and power changes and are calibrated to permit adjustments in precise increments. *[Figure 3-20]* The instrument for attitude display is the attitude indicator. The control instruments do not indicate aircraft speed or altitude. In order to determine these variables and others, a pilot must reference the performance instruments.

Navigation Instruments

The navigation instruments indicate the position of the aircraft in relation to a selected navigation facility or fix. This group of instruments includes various types of course indicators, range indicators, glideslope indicators, and bearing pointers. Newer aircraft with more technologically advanced instrumentation provide blended information, giving the pilot more accurate positional information.

Navigation instruments are comprised of indicators that display GPS, very high frequency (VHF) omni-directional radio range (VOR), nondirectional beacon (NDB), and instrument landing system (ILS) information. The instruments indicate the position of the aircraft relative to a selected navigation facility or fix. They also provide pilotage information so the aircraft can be maneuvered to keep it on a predetermined path. The pilotage information can be in either two or three dimensions relative to the ground-based or space-based navigation information. *[Figures 3-21 and 3-22]*

Global Positioning System (GPS)

GPS is a satellite-based navigation system composed of a network of satellites placed into orbit by the United States Department of Defense (DOD). GPS was originally intended for military applications, but in the 1980s the government made the system available for civilian use. GPS works in all weather conditions, anywhere in the world, 24 hours a day. A GPS receiver must be locked onto the signal of at least three satellites to calculate a two-dimensional position (latitude and longitude) and track movement. With four or more satellites in view, the receiver can determine the user's three-dimensional position (latitude, longitude, and altitude). Other satellites must also be in view to offset signal loss and signal ambiguity. The use of the GPS is discussed in more detail in Chapter 16, Navigation. Additionally, GPS is discussed in the Aeronautical Information Manual (AIM).

Chapter Summary

This chapter provides an overview of aircraft structures. A more in-depth understanding of aircraft structures and controls can be gained through the use of flight simulation software or interactive programs available online through aviation organizations, such as the Aircraft Owners and Pilots Association (AOPA). Pilots are also encouraged to subscribe to or review the various aviation periodicals that contain valuable flying information. As discussed in Chapter 1, the National Aeronautics and Space Administration (NASA) and the FAA also offer free information for pilots.

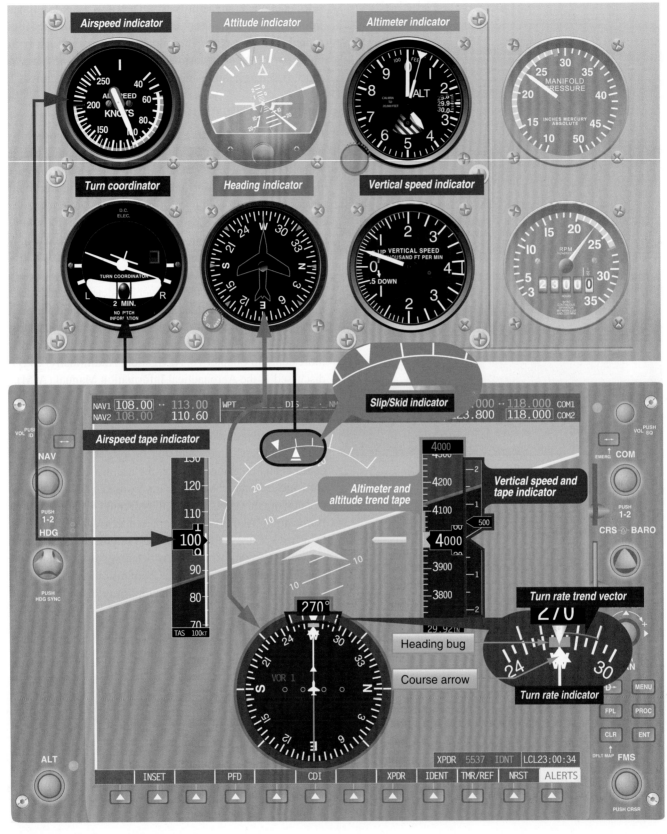

Figure 3-19. *Performance instruments.*

Figure 3-20. *Control instruments.*

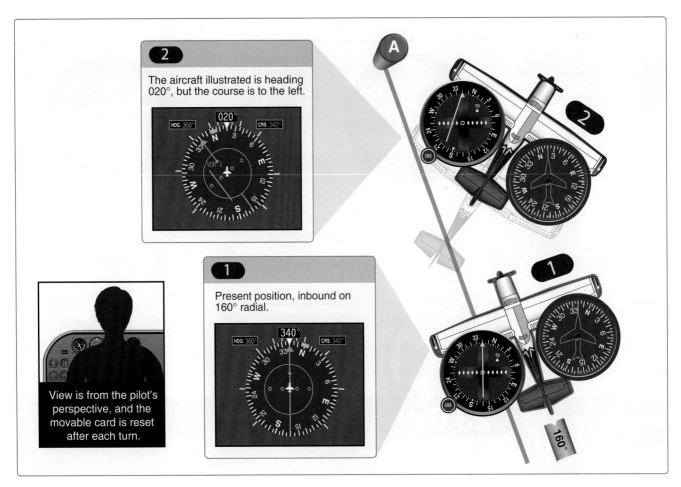

Figure 3-21. *A comparison of navigation information as depicted on both analog and digital displays.*

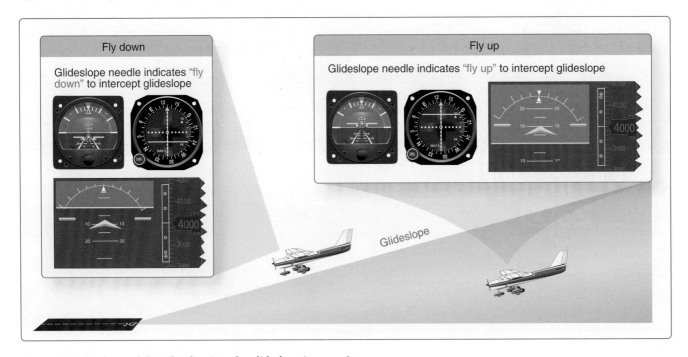

Figure 3-22. *Analog and digital indications for glideslope interception.*

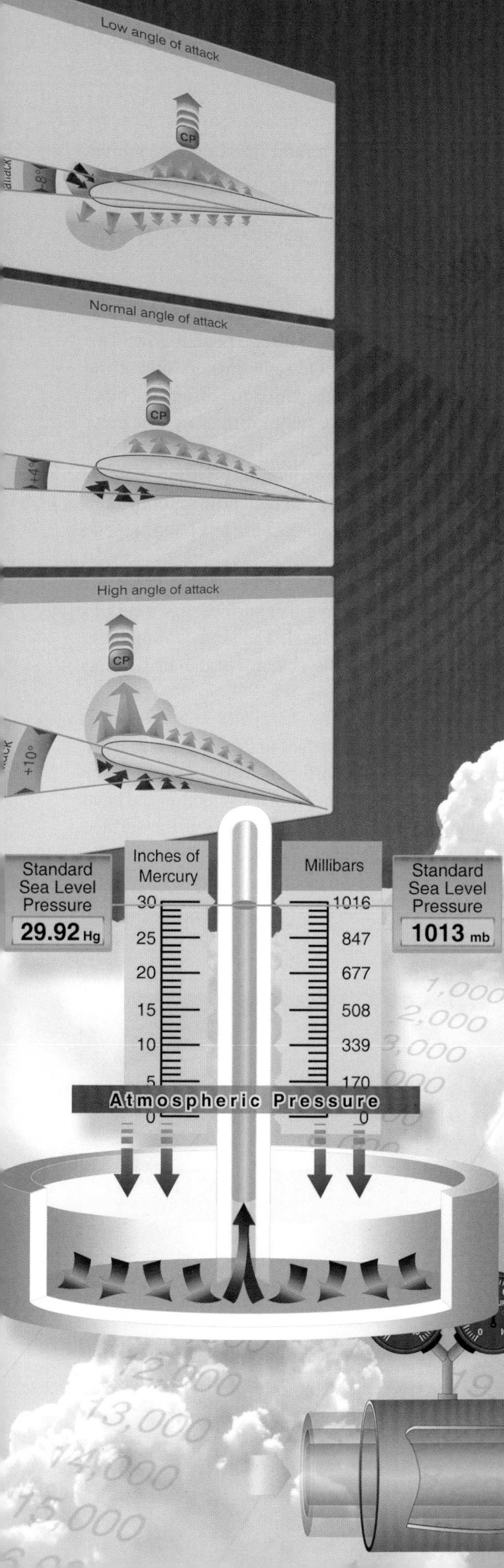

Low angle of attack

CP

Normal angle of attack

CP

High angle of attack

CP

Chapter 4
Principles of Flight

Introduction

This chapter examines the fundamental physical laws governing the forces acting on an aircraft in flight, and what effect these natural laws and forces have on the performance characteristics of aircraft. To control an aircraft, be it an airplane, helicopter, glider, or balloon, the pilot must understand the principles involved and learn to use or counteract these natural forces.

Structure of the Atmosphere

The atmosphere is an envelope of air that surrounds the Earth and rests upon its surface. It is as much a part of the Earth as the seas or the land, but air differs from land and water as it is a mixture of gases. It has mass, weight, and indefinite shape.

The atmosphere is composed of 78 percent nitrogen, 21 percent oxygen, and 1 percent other gases, such as argon or helium. Some of these elements are heavier than others. The heavier elements, such as oxygen, settle to the surface of the Earth, while the lighter elements are lifted up to the region of higher altitude. Most of the atmosphere's oxygen is contained below 35,000 feet altitude.

Standard Sea Level Pressure **29.92 Hg**	Inches of Mercury	Millibars	Standard Sea Level Pressure **1013 mb**
	30	1016	
	25	847	
	20	677	
	15	508	
	10	339	
	5	170	
	0	0	

Atmospheric Pressure

Air is a Fluid

When most people hear the word "fluid," they usually think of liquid. However, gasses, like air, are also fluids. Fluids take on the shape of their containers. Fluids generally do not resist deformation when even the smallest stress is applied, or they resist it only slightly. We call this slight resistance viscosity. Fluids also have the ability to flow. Just as a liquid flows and fills a container, air will expand to fill the available volume of its container. Both liquids and gasses display these unique fluid properties, even though they differ greatly in density. Understanding the fluid properties of air is essential to understanding the principles of flight.

Viscosity

Viscosity is the property of a fluid that causes it to resist flowing. The way individual molecules of the fluid tend to adhere, or stick, to each other determines how much a fluid resists flow. High-viscosity fluids are "thick" and resist flow; low-viscosity fluids are "thin" and flow easily. Air has a low viscosity and flows easily.

Using two liquids as an example, similar amounts of oil and water poured down two identical ramps will flow at different rates due to their different viscosity. The water seems to flow freely while the oil flows much more slowly.

As another example, different types of similar liquids will display different behaviors because of different viscosities. Grease is very viscous. Given time, grease will flow, even though the flow rate will be slow. Motor oil is less viscous than grease and flows much more easily, but it is more viscous and flows more slowly than gasoline.

All fluids are viscous and have a resistance to flow, whether or not we observe this resistance. We cannot easily observe the viscosity of air. However, since air is a fluid and has viscosity properties, it resists flow around any object to some extent.

Friction

Another factor at work when a fluid flows over or around an object is called friction. Friction is the resistance that one surface or object encounters when moving over another. Friction exists between any two materials that contact each other.

The effects of friction can be demonstrated using a similar example as before. If identical fluids are poured down two identical ramps, they flow in the same manner and at the same speed. If the surface of one ramp is rough, and the other smooth, the flow down the two ramps differs significantly. The rough surface ramp impedes the flow of the fluid due to resistance from the surface (friction). It is important to remember that all surfaces, no matter how smooth they appear, are not smooth on a microscopic level and impede the flow of a fluid.

The surface of a wing, like any other surface, has a certain roughness at the microscopic level. The surface roughness causes resistance and slows the velocity of the air flowing over the wing. *[Figure 4-1]*

Molecules of air pass over the surface of the wing and actually adhere (stick, or cling) to the surface because of friction. Air molecules near the surface of the wing resist motion and have a relative velocity near zero. The roughness of the surface impedes their motion. The layer of molecules that adhere to the wing surface is referred to as the boundary layer.

Leading edge of wing under 1,500x magnification

Figure 4-1. *Microscopic surface of a wing.*

Once the boundary layer of the air adheres to the wing by friction, further resistance to the airflow is caused by the viscosity, the tendency of the air to stick to itself. When these two forces act together to resist airflow over a wing, it is called drag.

Pressure

Pressure is the force applied in a perpendicular direction to the surface of an object. Often, pressure is measured in pounds of force exerted per square inch of an object, or PSI. An object completely immersed in a fluid will feel pressure uniformly around the entire surface of the object. If the pressure on one surface of the object becomes less than the pressure exerted on the other surfaces, the object will move in the direction of the lower pressure.

Atmospheric Pressure

Although there are various kinds of pressure, pilots are mainly concerned with atmospheric pressure. It is one of the basic factors in weather changes, helps to lift an aircraft, and actuates some of the important flight instruments. These instruments are the altimeter, airspeed indicator, vertical speed indicator, and manifold pressure gauge.

Air is very light, but it has mass and is affected by the attraction of gravity. Therefore, like any other substance, it has weight, and because of its weight, it has force. Since air is a fluid substance, this force is exerted equally in all directions. Its effect on bodies within the air is called pressure. Under standard conditions at sea level, the average pressure exerted by the weight of the atmosphere is approximately 14.70 pounds per square inch (psi) of surface, or 1,013.2 millibars (mb). The thickness of the atmosphere is limited; therefore, the higher the altitude, the less air there is above. For this reason, the weight of the atmosphere at 18,000 feet is one-half what it is at sea level.

The pressure of the atmosphere varies with time and location. Due to the changing atmospheric pressure, a standard reference was developed. The standard atmosphere at sea level is a surface temperature of 59 °F or 15 °C and a surface pressure of 29.92 inches of mercury ("Hg) or 1,013.2 mb. *[Figure 4-2]*

A standard temperature lapse rate is when the temperature decreases at the rate of approximately 3.5 °F or 2 °C per thousand feet up to 36,000 feet, which is approximately –65 °F or –55 °C. Above this point, the temperature is considered constant up to 80,000 feet. A standard pressure lapse rate is when pressure decreases at a rate of approximately 1 "Hg per 1,000 feet of altitude gain to 10,000 feet. *[Figure 4-3]* The International Civil Aviation Organization (ICAO) has established this as a worldwide standard, and it is often

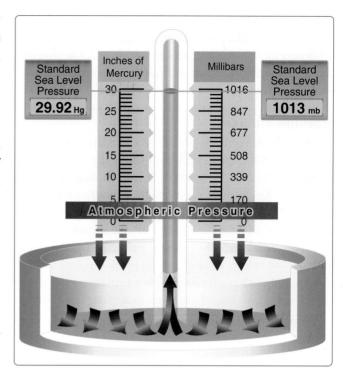

Figure 4-2. *Standard sea level pressure.*

Standard Atmosphere			
Altitude (ft)	Pressure (Hg)	Temperature	
		(°C)	(°F)
0	29.92	15.0	59.0
1,000	28.86	13.0	55.4
2,000	27.82	11.0	51.9
3,000	26.82	9.1	48.3
4,000	25.84	7.1	44.7
5,000	24.89	5.1	41.2
6,000	23.98	3.1	37.6
7,000	23.09	1.1	34.0
8,000	22.22	-0.9	30.5
9,000	21.38	-2.8	26.9
10,000	20.57	-4.8	23.3
11,000	19.79	-6.8	19.8
12,000	19.02	-8.8	16.2
13,000	18.29	-10.8	12.6
14,000	17.57	-12.7	9.1
15,000	16.88	-14.7	5.5
16,000	16.21	-16.7	1.9
17,000	15.56	-18.7	-1.6
18,000	14.94	-20.7	-5.2
19,000	14.33	-22.6	-8.8
20,000	13.74	-24.6	-12.3

Figure 4-3. *Properties of standard atmosphere.*

referred to as International Standard Atmosphere (ISA) or ICAO Standard Atmosphere. Any temperature or pressure that differs from the standard lapse rates is considered nonstandard temperature and pressure.

Since aircraft performance is compared and evaluated with respect to the standard atmosphere, all aircraft instruments are calibrated for the standard atmosphere. In order to properly account for the nonstandard atmosphere, certain related terms must be defined.

Pressure Altitude

Pressure altitude is the height above a standard datum plane (SDP), which is a theoretical level where the weight of the atmosphere is 29.92 "Hg (1,013.2 mb) as measured by a barometer. An altimeter is essentially a sensitive barometer calibrated to indicate altitude in the standard atmosphere. If the altimeter is set for 29.92 "Hg SDP, the altitude indicated is the pressure altitude. As atmospheric pressure changes, the SDP may be below, at, or above sea level. Pressure altitude is important as a basis for determining airplane performance, as well as for assigning flight levels to airplanes operating at or above 18,000 feet.

The pressure altitude can be determined by one of the following methods:

1. Setting the barometric scale of the altimeter to 29.92 and reading the indicated altitude

2. Applying a correction factor to the indicated altitude according to the reported altimeter setting

Density Altitude

SDP is a theoretical pressure altitude, but aircraft operate in a nonstandard atmosphere and the term density altitude is used for correlating aerodynamic performance in the nonstandard atmosphere. Density altitude is the vertical distance above sea level in the standard atmosphere at which a given density is to be found. The density of air has significant effects on the aircraft's performance because as air becomes less dense, it reduces:

• Power because the engine takes in less air

• Thrust because a propeller is less efficient in thin air

• Lift because the thin air exerts less force on the airfoils

Density altitude is pressure altitude corrected for nonstandard temperature. As the density of the air increases (lower density altitude), aircraft performance increases; conversely as air density decreases (higher density altitude), aircraft performance decreases. A decrease in air density means a high density altitude; an increase in air density means a lower density altitude. Density altitude is used in calculating aircraft performance because under standard atmospheric conditions, air at each level in the atmosphere not only has a specific density, its pressure altitude and density altitude identify the same level.

The computation of density altitude involves consideration of pressure (pressure altitude) and temperature. Since aircraft performance data at any level is based upon air density under standard day conditions, such performance data apply to air density levels that may not be identical with altimeter indications. Under conditions higher or lower than standard, these levels cannot be determined directly from the altimeter.

Density altitude is determined by first finding pressure altitude, and then correcting this altitude for nonstandard temperature variations. Since density varies directly with pressure and inversely with temperature, a given pressure altitude may exist for a wide range of temperatures by allowing the density to vary. However, a known density occurs for any one temperature and pressure altitude. The density of the air has a pronounced effect on aircraft and engine performance. Regardless of the actual altitude of the aircraft, it will perform as though it were operating at an altitude equal to the existing density altitude.

Air density is affected by changes in altitude, temperature, and humidity. High density altitude refers to thin air, while low density altitude refers to dense air. The conditions that result in a high density altitude are high elevations, low atmospheric pressures, high temperatures, high humidity, or some combination of these factors. Lower elevations, high atmospheric pressure, low temperatures, and low humidity are more indicative of low density altitude.

Effect of Pressure on Density

Since air is a gas, it can be compressed or expanded. When air is compressed, a greater amount of air can occupy a given volume. Conversely, when pressure on a given volume of air is decreased, the air expands and occupies a greater space. At a lower pressure, the original column of air contains a smaller mass of air. The density is decreased because density is directly proportional to pressure. If the pressure is doubled, the density is doubled; if the pressure is lowered, the density is lowered. This statement is true only at a constant temperature.

Effect of Temperature on Density

Increasing the temperature of a substance decreases its density. Conversely, decreasing the temperature increases the density. Thus, the density of air varies inversely with temperature. This statement is true only at a constant pressure.

In the atmosphere, both temperature and pressure decrease with altitude and have conflicting effects upon density. However, a fairly rapid drop in pressure as altitude increases usually has a dominating effect. Hence, pilots can expect the density to decrease with altitude.

Effect of Humidity (Moisture) on Density

The preceding paragraphs refer to air that is perfectly dry. In reality, it is never completely dry. The small amount of water vapor suspended in the atmosphere may be almost negligible under certain conditions, but in other conditions humidity may become an important factor in the performance of an aircraft. Water vapor is lighter than air; consequently, moist air is lighter than dry air. Therefore, as the water content of the air increases, the air becomes less dense, increasing density altitude and decreasing performance. It is lightest or least dense when, in a given set of conditions, it contains the maximum amount of water vapor.

Humidity, also called relative humidity, refers to the amount of water vapor contained in the atmosphere and is expressed as a percentage of the maximum amount of water vapor the air can hold. This amount varies with temperature. Warm air holds more water vapor, while cold air holds less. Perfectly dry air that contains no water vapor has a relative humidity of zero percent, while saturated air, which cannot hold any more water vapor, has a relative humidity of 100 percent. Humidity alone is usually not considered an important factor in calculating density altitude and aircraft performance, but it is a contributing factor.

As temperature increases, the air can hold greater amounts of water vapor. When comparing two separate air masses, the first warm and moist (both qualities tending to lighten the air) and the second cold and dry (both qualities making it heavier), the first must be less dense than the second. Pressure, temperature, and humidity have a great influence on aircraft performance because of their effect upon density. There are no rules of thumb that can be easily applied, but the affect of humidity can be determined using several online formulas. In the first example, the pressure is needed at the altitude for which density altitude is being sought. Using *Figure 4-2,* select the barometric pressure closest to the associated altitude. As an example, the pressure at 8,000 feet is 22.22 "Hg. Using the National Oceanic and Atmospheric Administration (NOAA) website (www.srh.noaa.gov/epz/?n=wxcalc_densityaltitude) for density altitude, enter the 22.22 for 8,000 feet in the station pressure window. Enter a temperature of 80° and a dew point of 75°. The result is a density altitude of 11,564 feet. With no humidity, the density altitude would be almost 500 feet lower.

Another website (www.wahiduddin.net/calc/density_altitude.htm) provides a more straight forward method of determining the effects of humidity on density altitude without using additional interpretive charts. In any case, the effects of humidity on density altitude include a decrease in overall performance in high humidity conditions.

Theories in the Production of Lift

In order to achieve flight in a machine that is heavier than air, there are several obstacles we must overcome. One of those obstacles, discussed previously, is the resistance to movement called drag. The most challenging obstacle to overcome in aviation, however, is the force of gravity. A wing moving through air generates the force called lift, also previously discussed. Lift from the wing that is greater than the force of gravity, directed opposite to the direction of gravity, enables an aircraft to fly. Generating this force called lift is based on some important principles, Newton's basic laws of motion, and Bernoulli's principle of differential pressure.

Newton's Basic Laws of Motion

The formulation of lift has historically been an adaptation over the past few centuries of basic physical laws. These laws, although seemingly applicable to all aspects of lift, do not explain how lift is formulated. In fact, one must consider the many airfoils that are symmetrical, yet produce significant lift.

The fundamental physical laws governing the forces acting upon an aircraft in flight were adopted from postulated theories developed before any human successfully flew an aircraft. The use of these physical laws grew out of the Scientific Revolution, which began in Europe in the 1600s. Driven by the belief the universe operated in a predictable manner open to human understanding, many philosophers, mathematicians, natural scientists, and inventors spent their lives unlocking the secrets of the universe. One of the most well-known was Sir Isaac Newton, who not only formulated the law of universal gravitation, but also described the three basic laws of motion.

Newton's First Law: "Every object persists in its state of rest or uniform motion in a straight line unless it is compelled to change that state by forces impressed on it."

This means that nothing starts or stops moving until some outside force causes it to do so. An aircraft at rest on the ramp remains at rest unless a force strong enough to overcome its inertia is applied. Once it is moving, its inertia keeps it moving, subject to the various other forces acting on it. These forces may add to its motion, slow it down, or change its direction.

Newton's Second Law: "Force is equal to the change in momentum per change in time. For a constant mass, force equals mass times acceleration."

When a body is acted upon by a constant force, its resulting acceleration is inversely proportional to the mass of the body and is directly proportional to the applied force. This takes into account the factors involved in overcoming Newton's First Law. It covers both changes in direction and speed, including starting up from rest (positive acceleration) and coming to a stop (negative acceleration or deceleration).

Newton's Third Law: "For every action, there is an equal and opposite reaction."

In an airplane, the propeller moves and pushes back the air; consequently, the air pushes the propeller (and thus the airplane) in the opposite direction—forward. In a jet airplane, the engine pushes a blast of hot gases backward; the force of equal and opposite reaction pushes against the engine and forces the airplane forward.

Bernoulli's Principle of Differential Pressure

A half-century after Newton formulated his laws, Daniel Bernoulli, a Swiss mathematician, explained how the pressure of a moving fluid (liquid or gas) varies with its speed of motion. Bernoulli's Principle states that as the velocity of a moving fluid (liquid or gas) increases, the pressure within the fluid decreases. This principle explains what happens to air passing over the curved top of the airplane wing.

A practical application of Bernoulli's Principle is the venturi tube. The venturi tube has an air inlet that narrows to a throat (constricted point) and an outlet section that increases in diameter toward the rear. The diameter of the outlet is the same as that of the inlet. The mass of air entering the tube must exactly equal the mass exiting the tube. At the constriction, the speed must increase to allow the same amount of air to pass in the same amount of time as in all other parts of the tube. When the air speeds up, the pressure also decreases. Past the constriction, the airflow slows and the pressure increases. *[Figure 4-4]*

Since air is recognized as a body, and it is understood that air will follow the above laws, one can begin to see how and why an airplane wing develops lift. As the wing moves through the air, the flow of air across the curved top surface increases in velocity creating a low-pressure area.

Although Newton, Bernoulli, and hundreds of other early scientists who studied the physical laws of the universe did not have the sophisticated laboratories available today, they provided great insight to the contemporary viewpoint of how lift is created.

Airfoil Design

An airfoil is a structure designed to obtain reaction upon its surface from the air through which it moves or that moves past such a structure. Air acts in various ways when submitted to different pressures and velocities; but this discussion is confined to the parts of an aircraft that a pilot is most concerned with in flight—namely, the airfoils designed to produce lift. By looking at a typical airfoil profile, such as the cross section of a wing, one can see several obvious characteristics of design. *[Figure 4-5]* Notice that there is a difference in the curvatures (called cambers) of the upper and lower surfaces of the airfoil. The camber of the upper surface is more pronounced than that of the lower surface, which is usually somewhat flat.

NOTE: The two extremities of the airfoil profile also differ in appearance. The rounded end, which faces forward in flight, is called the leading edge; the other end, the trailing edge, is quite narrow and tapered.

A reference line often used in discussing the airfoil is the chord line, a straight line drawn through the profile connecting the extremities of the leading and trailing edges. The distance from this chord line to the upper and lower surfaces of the wing denotes the magnitude of the upper and lower camber at any point. Another reference line, drawn

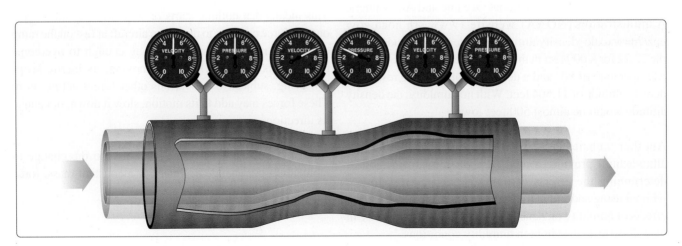

Figure 4-4. *Air pressure decreases in a venturi tube.*

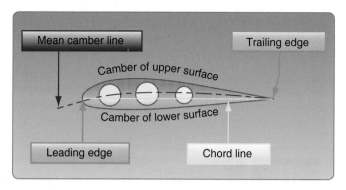

Figure 4-5. *Typical airfoil section.*

from the leading edge to the trailing edge, is the mean camber line. This mean line is equidistant at all points from the upper and lower surfaces.

An airfoil is constructed in such a way that its shape takes advantage of the air's response to certain physical laws. This develops two actions from the air mass: a positive pressure lifting action from the air mass below the wing, and a negative pressure lifting action from lowered pressure above the wing.

As the air stream strikes the relatively flat lower surface of a wing or rotor blade when inclined at a small angle to its direction of motion, the air is forced to rebound downward, causing an upward reaction in positive lift. At the same time, the air stream striking the upper curved section of the leading edge is deflected upward. An airfoil is shaped to cause an action on the air, and forces air downward, which provides an equal reaction from the air, forcing the airfoil upward. If a wing is constructed in such form that it causes a lift force greater than the weight of the aircraft, the aircraft will fly.

If all the lift required were obtained merely from the deflection of air by the lower surface of the wing, an aircraft would only need a flat wing like a kite. However, the balance of the lift needed to support the aircraft comes from the flow of air above the wing. Herein lies the key to flight.

It is neither accurate nor useful to assign specific values to the percentage of lift generated by the upper surface of an airfoil versus that generated by the lower surface. These are not constant values. They vary, not only with flight conditions, but also with different wing designs.

Different airfoils have different flight characteristics. Many thousands of airfoils have been tested in wind tunnels and in actual flight, but no one airfoil has been found that satisfies every flight requirement. The weight, speed, and purpose of each aircraft dictate the shape of its airfoil. The most efficient airfoil for producing the greatest lift is one that has a concave or "scooped out" lower surface. As a fixed design, this type of airfoil sacrifices too much speed while producing

lift and is not suitable for high-speed flight. Advancements in engineering have made it possible for today's high-speed jets to take advantage of the concave airfoil's high lift characteristics. Leading edge (Kreuger) flaps and trailing edge (Fowler) flaps, when extended from the basic wing structure, literally change the airfoil shape into the classic concave form, thereby generating much greater lift during slow flight conditions.

On the other hand, an airfoil that is perfectly streamlined and offers little wind resistance sometimes does not have enough lifting power to take the airplane off the ground. Thus, modern airplanes have airfoils that strike a medium between extremes in design. The shape varies according to the needs of the airplane for which it is designed. *Figure 4-6* shows some of the more common airfoil designs.

Low Pressure Above

In a wind tunnel or in flight, an airfoil is simply a streamlined object inserted into a moving stream of air. If the airfoil profile were in the shape of a teardrop, the speed and the pressure changes of the air passing over the top and bottom would be the same on both sides. But if the teardrop shaped airfoil were cut in half lengthwise, a form resembling the basic airfoil (wing) section would result. If the airfoil were then inclined so the airflow strikes it at an angle, the air moving over the upper surface would be forced to move faster than the air moving along the bottom of the airfoil. This increased velocity reduces the pressure above the airfoil.

Applying Bernoulli's Principle of Pressure, the increase in the speed of the air across the top of an airfoil produces a

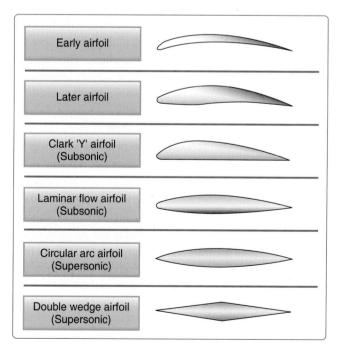

Figure 4-6. *Airfoil designs.*

drop in pressure. This lowered pressure is a component of total lift. The pressure difference between the upper and lower surface of a wing alone does not account for the total lift force produced.

The downward backward flow from the top surface of an airfoil creates a downwash. This downwash meets the flow from the bottom of the airfoil at the trailing edge. Applying Newton's third law, the reaction of this downward backward flow results in an upward forward force on the airfoil.

High Pressure Below

A certain amount of lift is generated by pressure conditions underneath the airfoil. Because of the manner in which air flows underneath the airfoil, a positive pressure results, particularly at higher angles of attack. However, there is another aspect to this airflow that must be considered. At a point close to the leading edge, the airflow is virtually stopped (stagnation point) and then gradually increases speed. At some point near the trailing edge, it again reaches a velocity equal to that on the upper surface. In conformance with Bernoulli's principle, where the airflow was slowed beneath the airfoil, a positive upward pressure was created (i.e., as the fluid speed decreases, the pressure must increase). Since the pressure differential between the upper and lower surface of the airfoil increases, total lift increases. Both Bernoulli's Principle and Newton's Laws are in operation whenever lift is being generated by an airfoil.

Pressure Distribution

From experiments conducted on wind tunnel models and on full size airplanes, it has been determined that as air flows along the surface of a wing at different angles of attack (AOA), there are regions along the surface where the pressure is negative, or less than atmospheric, and regions where the pressure is positive, or greater than atmospheric. This negative pressure on the upper surface creates a relatively larger force on the wing than is caused by the positive pressure resulting from the air striking the lower wing surface. *Figure 4-7* shows the pressure distribution along an airfoil at three different angles of attack. The average of the pressure variation for any given AOA is referred to as the center of pressure (CP). Aerodynamic force acts through this CP. At high angles of attack, the CP moves forward, while at low angles of attack the CP moves aft. In the design of wing structures, this CP travel is very important, since it affects the position of the air loads imposed on the wing structure in both low and high AOA conditions. An airplane's aerodynamic balance and controllability are governed by changes in the CP.

Airfoil Behavior

Although specific examples can be cited in which each of the principles predict and contribute to the formation of lift,

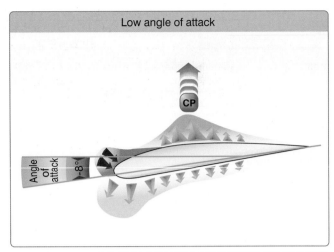

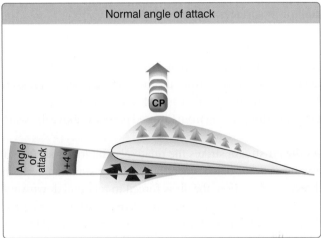

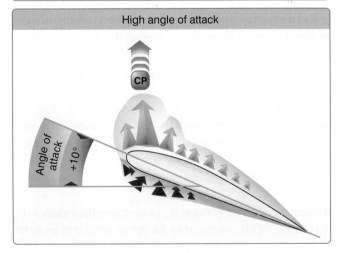

Figure 4-7. *Pressure distribution on an airfoil and CP changes with AOA.*

lift is a complex subject. The production of lift is much more complex than a simple differential pressure between upper and lower airfoil surfaces. In fact, many lifting airfoils do not have an upper surface longer than the bottom, as in the case of symmetrical airfoils. These are seen in high-speed aircraft having symmetrical wings, or on symmetrical rotor blades for many helicopters whose upper and lower surfaces are identical. In both examples, the only difference is the relationship of the airfoil with the oncoming airstream (angle). A paper airplane, which is simply a flat plate, has a bottom and top exactly the same shape and length. Yet, these airfoils do produce lift, and "flow turning" is partly (or fully) responsible for creating lift.

As an airfoil moves through air, the airfoil is inclined against the airflow, producing a different flow caused by the airfoil's relationship to the oncoming air. Think of a hand being placed outside the car window at a high speed. If the hand is inclined in one direction or another, the hand will move upward or downward. This is caused by deflection, which in turn causes the air to turn about the object within the air stream. As a result of this change, the velocity about the object changes in both magnitude and direction, in turn resulting in a measurable velocity force and direction.

A Third Dimension

To this point, the discussion has centered on the flow across the upper and lower surfaces of an airfoil. While most of the lift is produced by these two dimensions, a third dimension, the tip of the airfoil also has an aerodynamic effect. The high-pressure area on the bottom of an airfoil pushes around the tip to the low-pressure area on the top. *[Figure 4-8]* This action creates a rotating flow called a tip vortex. The vortex flows behind the airfoil creating a downwash that extends back to the trailing edge of the airfoil. This downwash results in an overall reduction in lift for the affected portion of the airfoil. Manufacturers have developed different methods to counteract this action. Winglets can be added to the tip of an airfoil to reduce this flow. The winglets act as a dam preventing the vortex from forming. Winglets can be on the top or bottom of the airfoil. Another method of countering the flow is to taper the airfoil tip, reducing the pressure differential and smoothing the airflow around the tip.

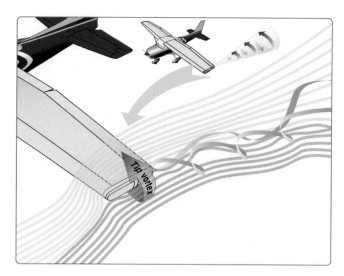

Figure 4-8. *Tip vortex.*

Chapter Summary

Modern general aviation aircraft have what may be considered high performance characteristics. Therefore, it is increasingly necessary that pilots appreciate and understand the principles upon which the art of flying is based. For additional information on the principles discussed in this chapter, visit the National Aeronautics and Space Administration (NASA) Beginner's Guide to Aerodynamics at www.grc.nasa.gov/www/k-12/airplane/bga.html.

Chapter 5

Aerodynamics of Flight

Forces Acting on the Aircraft

Thrust, drag, lift, and weight are forces that act upon all aircraft in flight. Understanding how these forces work and knowing how to control them with the use of power and flight controls are essential to flight. This chapter discusses the aerodynamics of flight—how design, weight, load factors, and gravity affect an aircraft during flight maneuvers.

The four forces acting on an aircraft in straight-and-level, unaccelerated flight are thrust, drag, lift, and weight. They are defined as follows:

- Thrust—the forward force produced by the powerplant/propeller or rotor. It opposes or overcomes the force of drag. As a general rule, it acts parallel to the longitudinal axis. However, this is not always the case, as explained later.

- Drag—a rearward, retarding force caused by disruption of airflow by the wing, rotor, fuselage, and other protruding objects. As a general rule, drag opposes thrust and acts rearward parallel to the relative wind.

- Lift—is a force that is produced by the dynamic effect of the air acting on the airfoil, and acts perpendicular to the flight path through the center of lift (CL) and perpendicular to the lateral axis. In level flight, lift opposes the downward force of weight.

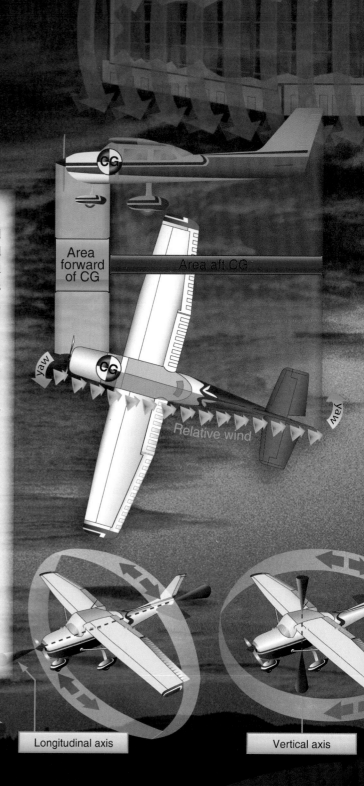

Area forward of CG

Area aft CG

yaw

Relative wind

yaw

Lateral axis

Longitudinal axis

Vertical axis

Applied Force

Applied Force

Applied Force

CG

- Weight—the combined load of the aircraft itself, the crew, the fuel, and the cargo or baggage. Weight is a force that pulls the aircraft downward because of the force of gravity. It opposes lift and acts vertically downward through the aircraft's center of gravity (CG).

In steady flight, the sum of these opposing forces is always zero. There can be no unbalanced forces in steady, straight flight based upon Newton's Third Law, which states that for every action or force there is an equal, but opposite, reaction or force. This is true whether flying level or when climbing or descending.

It does not mean the four forces are equal. It means the opposing forces are equal to, and thereby cancel, the effects of each other. In *Figure 5-1*, the force vectors of thrust, drag, lift, and weight appear to be equal in value. The usual explanation states (without stipulating that thrust and drag do not equal weight and lift) that thrust equals drag and lift equals weight. Although true, this statement can be misleading. It should be understood that in straight, level, unaccelerated flight, it is true that the opposing lift/weight forces are equal. They are also greater than the opposing forces of thrust/drag that are equal only to each other. Therefore, in steady flight:

- The sum of all upward components of forces (not just lift) equals the sum of all downward components of forces (not just weight)

- The sum of all forward components of forces (not just thrust) equals the sum of all backward components of forces (not just drag)

This refinement of the old "thrust equals drag; lift equals weight" formula explains that a portion of thrust is directed upward in climbs and slow flight and acts as if it were lift while a portion of weight is directed backward opposite to the direction of flight and acts as if it were drag. In slow flight, thrust has an upward component. But because the aircraft is in level flight, weight does not contribute to drag. *[Figure 5-2]*

In glides, a portion of the weight vector is directed along the forward flight path and, therefore, acts as thrust. In other words, any time the flight path of the aircraft is not horizontal, lift, weight, thrust, and drag vectors must each be broken down into two components.

Another important concept to understand is angle of attack (AOA). Since the early days of flight, AOA is fundamental to understanding many aspects of airplane performance, stability, and control. The AOA is defined as the acute angle between the chord line of the airfoil and the direction of the relative wind.

Discussions of the preceding concepts are frequently omitted in aeronautical texts/handbooks/manuals. The reason is not that they are inconsequential, but because the main ideas with respect to the aerodynamic forces acting upon an aircraft in flight can be presented in their most essential elements without being involved in the technicalities of the aerodynamicist. In point of fact, considering only level flight, and normal climbs and glides in a steady state, it is still true that lift provided by the wing or rotor is the primary upward force, and weight is the primary downward force.

By using the aerodynamic forces of thrust, drag, lift, and weight, pilots can fly a controlled, safe flight. A more detailed discussion of these forces follows.

Thrust

For an aircraft to start moving, thrust must be exerted and be greater than drag. The aircraft continues to move and gain speed until thrust and drag are equal. In order to maintain a

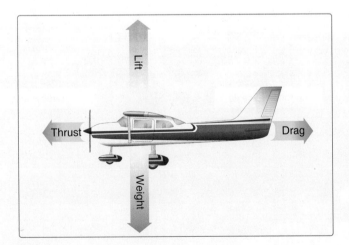

Figure 5-1. *Relationship of forces acting on an aircraft.*

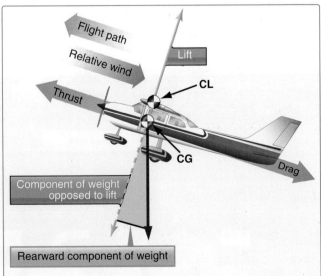

Figure 5-2. *Force vectors during a stabilized climb.*

constant airspeed, thrust and drag must remain equal, just as lift and weight must be equal to maintain a constant altitude. If in level flight, the engine power is reduced, the thrust is lessened, and the aircraft slows down. As long as the thrust is less than the drag, the aircraft continues to decelerate. To a point, as the aircraft slows down, the drag force will also decrease. The aircraft will continue to slow down until thrust again equals drag at which point the airspeed will stabilize.

Likewise, if the engine power is increased, thrust becomes greater than drag and the airspeed increases. As long as the thrust continues to be greater than the drag, the aircraft continues to accelerate. When drag equals thrust, the aircraft flies at a constant airspeed.

Straight-and-level flight may be sustained at a wide range of speeds. The pilot coordinates AOA and thrust in all speed regimes if the aircraft is to be held in level flight. An important fact related to the principal of lift (for a given airfoil shape) is that lift varies with the AOA and airspeed. Therefore, a large AOA at low airspeeds produces an equal amount of lift at high airspeeds with a low AOA. The speed regimes of flight can be grouped in three categories: low-speed flight, cruising flight, and high-speed flight.

When the airspeed is low, the AOA must be relatively high if the balance between lift and weight is to be maintained. *[Figure 5-3]* If thrust decreases and airspeed decreases, lift will become less than weight and the aircraft will start to descend. To maintain level flight, the pilot can increase the AOA an amount that generates a lift force again equal to the weight of the aircraft. While the aircraft will be flying more slowly, it will still maintain level flight. The AOA is adjusted to maintain lift equal weight. The airspeed will naturally adjust until drag equals thrust and then maintain that airspeed (assumes the pilot is not trying to hold an exact speed).

Straight-and-level flight in the slow-speed regime provides some interesting conditions relative to the equilibrium of forces. With the aircraft in a nose-high attitude, there is a vertical component of thrust that helps support it. For one thing, wing loading tends to be less than would be expected.

In level flight, when thrust is increased, the aircraft speeds up and the lift increases. The aircraft will start to climb unless the AOA is decreased just enough to maintain the relationship between lift and weight. The timing of this decrease in AOA needs to be coordinated with the increase in thrust and airspeed. Otherwise, if the AOA is decreased too fast, the aircraft will descend, and if the AOA is decreased too slowly, the aircraft will climb.

As the airspeed varies due to thrust, the AOA must also vary to maintain level flight. At very high speeds and level flight, it is even possible to have a slightly negative AOA. As thrust is reduced and airspeed decreases, the AOA must increase in order to maintain altitude. If speed decreases enough, the required AOA will increase to the critical AOA. Any further increase in the AOA will result in the wing stalling. Therefore, extra vigilance is required at reduced thrust settings and low speeds so as not to exceed the critical angle of attack. If the airplane is equipped with an AOA indicator, it should be referenced to help monitor the proximity to the critical AOA.

Some aircraft have the ability to change the direction of the thrust rather than changing the AOA. This is accomplished either by pivoting the engines or by vectoring the exhaust gases. *[Figure 5-4]*

Lift

The pilot can control the lift. Any time the control yoke or stick is moved fore or aft, the AOA is changed. As the AOA increases, lift increases (all other factors being equal). When the aircraft reaches the maximum AOA, lift begins to diminish rapidly. This is the stalling AOA, known as C_{L-MAX} critical AOA. Examine *Figure 5-5,* noting how the C_L increases until the critical AOA is reached, then decreases rapidly with any further increase in the AOA.

Before proceeding further with the topic of lift and how it can be controlled, velocity must be discussed. The shape of the wing or rotor cannot be effective unless it continually keeps "attacking" new air. If an aircraft is to keep flying, the lift-producing airfoil must keep moving. In a helicopter or gyroplane, this is accomplished by the rotation of the rotor blades. For other types of aircraft, such as airplanes, weight-

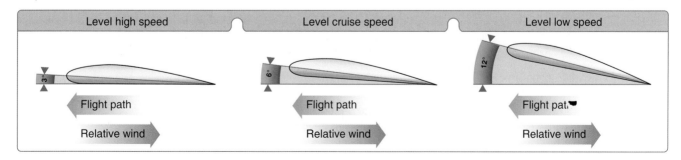

Figure 5-3. *Angle of attack at various speeds.*

Figure 5-4. *Some aircraft have the ability to change the direction of thrust.*

shift control, or gliders, air must be moving across the lifting surface. This is accomplished by the forward speed of the aircraft. Lift is proportional to the square of the aircraft's velocity. For example, an airplane traveling at 200 knots has four times the lift as the same airplane traveling at 100 knots, if the AOA and other factors remain constant.

$$L = \frac{C_L \cdot \rho \cdot V^2 \cdot S}{2}$$

The above lift equation exemplifies this mathematically and supports that doubling of the airspeed will result in four times the lift. As a result, one can see that velocity is an important component to the production of lift, which itself can be affected through varying AOA. When examining the equation, lift (L) is determined through the relationship of the air density (ρ), the airfoil velocity (V), the surface area of the wing (S) and the coefficient of lift (C_L) for a given airfoil.

Taking the equation further, one can see an aircraft could not continue to travel in level flight at a constant altitude and maintain the same AOA if the velocity is increased. The lift would increase and the aircraft would climb as a result of the increased lift force or speed up. Therefore, to keep the aircraft straight and level (not accelerating upward) and in a state of equilibrium, as velocity is increased, lift must be kept constant. This is normally accomplished by reducing the AOA by lowering the nose. Conversely, as the aircraft is slowed, the decreasing velocity requires increasing the AOA to maintain lift sufficient to maintain flight. There is, of course, a limit to how far the AOA can be increased, if a stall is to be avoided.

All other factors being constant, for every AOA there is a corresponding airspeed required to maintain altitude in steady, unaccelerated flight (true only if maintaining level flight). Since an airfoil always stalls at the same AOA, if increasing weight, lift must also be increased. The only

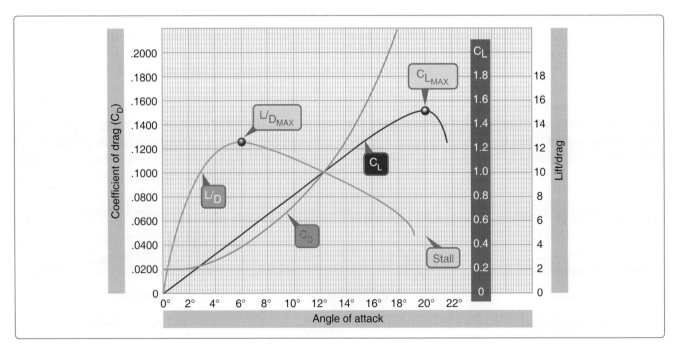

Figure 5-5. *Coefficients of lift and drag at various angles of attack.*

method of increasing lift is by increasing velocity if the AOA is held constant just short of the "critical," or stalling, AOA (assuming no flaps or other high lift devices).

Lift and drag also vary directly with the density of the air. Density is affected by several factors: pressure, temperature, and humidity. At an altitude of 18,000 feet, the density of the air has one-half the density of air at sea level. In order to maintain its lift at a higher altitude, an aircraft must fly at a greater true airspeed for any given AOA.

Warm air is less dense than cool air, and moist air is less dense than dry air. Thus, on a hot humid day, an aircraft must be flown at a greater true airspeed for any given AOA than on a cool, dry day.

If the density factor is decreased and the total lift must equal the total weight to remain in flight, it follows that one of the other factors must be increased. The factor usually increased is the airspeed or the AOA because these are controlled directly by the pilot.

Lift varies directly with the wing area, provided there is no change in the wing's planform. If the wings have the same proportion and airfoil sections, a wing with a planform area of 200 square feet lifts twice as much at the same AOA as a wing with an area of 100 square feet.

Two major aerodynamic factors from the pilot's viewpoint are lift and airspeed because they can be controlled readily and accurately. Of course, the pilot can also control density by adjusting the altitude and can control wing area if the aircraft happens to have flaps of the type that enlarge wing area. However, for most situations, the pilot controls lift and airspeed to maneuver an aircraft. For instance, in straight-and-level flight, cruising along at a constant altitude, altitude is maintained by adjusting lift to match the aircraft's velocity or cruise airspeed, while maintaining a state of equilibrium in which lift equals weight. In an approach to landing, when the pilot wishes to land as slowly as practical, it is necessary to increase AOA near maximum to maintain lift equal to the weight of the aircraft.

Lift/Drag Ratio

The lift-to-drag ratio (L/D) is the amount of lift generated by a wing or airfoil compared to its drag. A ratio of L/D indicates airfoil efficiency. Aircraft with higher L/D ratios are more efficient than those with lower L/D ratios. In unaccelerated flight with the lift and drag data steady, the proportions of the coefficient of lift (C_L) and coefficient of drag (C_D) can be calculated for specific AOA. [Figure 5-5]

The coefficient of lift is dimensionless and relates the lift generated by a lifting body, the dynamic pressure of the fluid

flow around the body, and a reference area associated with the body. The coefficient of drag is also dimensionless and is used to quantify the drag of an object in a fluid environment, such as air, and is always associated with a particular surface area.

The L/D ratio is determined by dividing the C_L by the C_D, which is the same as dividing the lift equation by the drag equation as all of the variables, aside from the coefficients, cancel out. The lift and drag equations are as follows (L = Lift in pounds; D = Drag; C_L = coefficient of lift; ρ = density (expressed in slugs per cubic feet); V = velocity (in feet per second); q = dynamic pressure per square foot ($q = \frac{1}{2} \rho v^2$); S = the area of the lifting body (in square feet); and C_D = Ratio of drag pressure to dynamic pressure):

$$D = \frac{C_D \cdot \rho \cdot V^2 \cdot S}{2}$$

Typically at low AOA, the coefficient of drag is low and small changes in AOA create only slight changes in the coefficient of drag. At high AOA, small changes in the AOA cause significant changes in drag. The shape of an airfoil, as well as changes in the AOA, affects the production of lift.

Notice in *Figure 5-5* that the coefficient of lift curve (red) reaches its maximum for this particular wing section at 20° AOA and then rapidly decreases. 20° AOA is therefore the critical angle of attack. The coefficient of drag curve (orange) increases very rapidly from 14° AOA and completely overcomes the lift curve at 21° AOA. The lift/drag ratio (green) reaches its maximum at 6° AOA, meaning that at this angle, the most lift is obtained for the least amount of drag.

Note that the maximum lift/drag ratio (L/D_{MAX}) occurs at one specific C_L and AOA. If the aircraft is operated in steady flight at L/D_{MAX}, the total drag is at a minimum. Any AOA lower or higher than that for L/D_{MAX} reduces the L/D and consequently increases the total drag for a given aircraft's

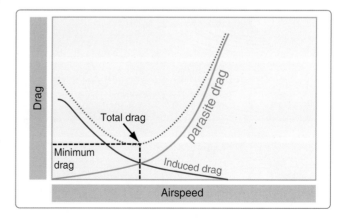

Figure 5-6. *Drag versus speed.*

lift. *Figure 5-6* depicts the L/D$_{MAX}$ by the lowest portion of the blue line labeled "total drag." The configuration of an aircraft has a great effect on the L/D.

Drag

Drag is the force that resists movement of an aircraft through the air. There are two basic types: parasite drag and induced drag. The first is called parasite because it in no way functions to aid flight, while the second, induced drag, is a result of an airfoil developing lift.

Parasite Drag

Parasite drag is comprised of all the forces that work to slow an aircraft's movement. As the term parasite implies, it is the drag that is not associated with the production of lift. This includes the displacement of the air by the aircraft, turbulence generated in the airstream, or a hindrance of air moving over the surface of the aircraft and airfoil. There are three types of parasite drag: form drag, interference drag, and skin friction.

Form Drag

Form drag is the portion of parasite drag generated by the aircraft due to its shape and airflow around it. Examples include the engine cowlings, antennas, and the aerodynamic shape of other components. When the air has to separate to move around a moving aircraft and its components, it eventually rejoins after passing the body. How quickly and smoothly it rejoins is representative of the resistance that it creates, which requires additional force to overcome. *[Figure 5-7]*

Notice how the flat plate in *Figure 5-7* causes the air to swirl around the edges until it eventually rejoins downstream. Form drag is the easiest to reduce when designing an aircraft. The solution is to streamline as many of the parts as possible.

Interference Drag

Interference drag comes from the intersection of airstreams that creates eddy currents, turbulence, or restricts smooth airflow. For example, the intersection of the wing and the fuselage at the wing root has significant interference drag. Air flowing around the fuselage collides with air flowing over the wing, merging into a current of air different from the two original currents. The most interference drag is observed when two surfaces meet at perpendicular angles. Fairings are used to reduce this tendency. If a jet fighter carries two identical wing tanks, the overall drag is greater than the sum of the individual tanks because both of these create and generate interference drag. Fairings and distance between lifting surfaces and external components (such as radar antennas hung from wings) reduce interference drag. *[Figure 5-8]*

Skin Friction Drag

Skin friction drag is the aerodynamic resistance due to the contact of moving air with the surface of an aircraft. Every surface, no matter how apparently smooth, has a rough, ragged surface when viewed under a microscope. The air molecules, which come in direct contact with the surface of the wing, are virtually motionless. Each layer of molecules above the surface moves slightly faster until the molecules are moving at the velocity of the air moving around the aircraft. This speed is called the free-stream velocity. The area between the wing and the free-stream velocity level is about as wide as a playing card and is called the boundary layer. At the top of the boundary layer, the molecules increase velocity and move at the same speed as the molecules outside the boundary layer. The actual speed at which the molecules move depends upon the shape of the wing, the viscosity (stickiness) of the air through which the wing or airfoil is moving, and its compressibility (how much it can be compacted).

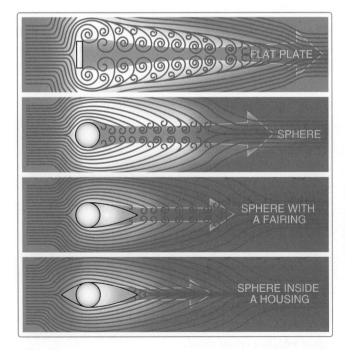

Figure 5-7. *Form drag.*

Figure 5-8. *A wing root can cause interference drag.*

The airflow outside of the boundary layer reacts to the shape of the edge of the boundary layer just as it would to the physical surface of an object. The boundary layer gives any object an "effective" shape that is usually slightly different from the physical shape. The boundary layer may also separate from the body, thus creating an effective shape much different from the physical shape of the object. This change in the physical shape of the boundary layer causes a dramatic decrease in lift and an increase in drag. When this happens, the airfoil has stalled.

In order to reduce the effect of skin friction drag, aircraft designers utilize flush mount rivets and remove any irregularities that may protrude above the wing surface. In addition, a smooth and glossy finish aids in transition of air across the surface of the wing. Since dirt on an aircraft disrupts the free flow of air and increases drag, keep the surfaces of an aircraft clean and waxed.

Induced Drag

The second basic type of drag is induced drag. It is an established physical fact that no system that does work in the mechanical sense can be 100 percent efficient. This means that whatever the nature of the system, the required work is obtained at the expense of certain additional work that is dissipated or lost in the system. The more efficient the system, the smaller this loss.

In level flight, the aerodynamic properties of a wing or rotor produce a required lift, but this can be obtained only at the expense of a certain penalty. The name given to this penalty is induced drag. Induced drag is inherent whenever an airfoil is producing lift and, in fact, this type of drag is inseparable from the production of lift. Consequently, it is always present if lift is produced.

An airfoil (wing or rotor blade) produces the lift force by making use of the energy of the free airstream. Whenever an airfoil is producing lift, the pressure on the lower surface of it is greater than that on the upper surface (Bernoulli's Principle). As a result, the air tends to flow from the high pressure area below the tip upward to the low pressure area on the upper surface. In the vicinity of the tips, there is a tendency for these pressures to equalize, resulting in a lateral flow outward from the underside to the upper surface. This lateral flow imparts a rotational velocity to the air at the tips, creating vortices that trail behind the airfoil.

When the aircraft is viewed from the tail, these vortices circulate counterclockwise about the right tip and clockwise about the left tip. [Figure 5-9] As the air (and vortices) roll off the back of your wing, they angle down, which is known as downwash. Figure 5-10 shows the difference in downwash at

Figure 5-9. *Wingtip vortex from a crop duster.*

altitude versus near the ground. Bearing in mind the direction of rotation of these vortices, it can be seen that they induce an upward flow of air beyond the tip and a downwash flow behind the wing's trailing edge. This induced downwash has nothing in common with the downwash that is necessary to produce lift. It is, in fact, the source of induced drag.

Downwash points the relative wind downward, so the more downwash you have, the more your relative wind points downward. That's important for one very good reason: lift is always perpendicular to the relative wind. In *Figure 5-11,* you can see that when you have less downwash, your lift vector is more vertical, opposing gravity. And when you have more downwash, your lift vector points back more, causing induced drag. On top of that, it takes energy for your wings to create downwash and vortices, and that energy creates drag.

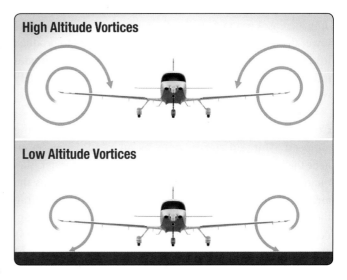

Figure 5-10. *The difference in wingtip vortex size at altitude versus near the ground.*

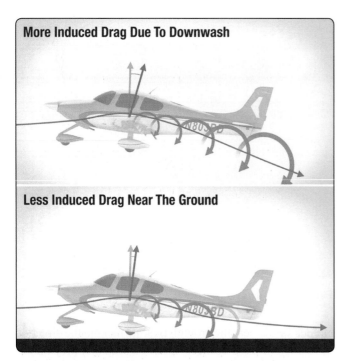

More Induced Drag Due To Downwash

Less Induced Drag Near The Ground

Figure 5-11. *The difference in downwash at altitude versus near the ground.*

The greater the size and strength of the vortices and consequent downwash component on the net airflow over the airfoil, the greater the induced drag effect becomes. This downwash over the top of the airfoil at the tip has the same effect as bending the lift vector rearward; therefore, the lift is slightly aft of perpendicular to the relative wind, creating a rearward lift component. This is induced drag.

In order to create a greater negative pressure on the top of an airfoil, the airfoil can be inclined to a higher AOA. If the AOA of a symmetrical airfoil were zero, there would be no pressure differential, and consequently, no downwash component and no induced drag. In any case, as AOA increases, induced drag increases proportionally. To state this another way—the lower the airspeed, the greater the AOA required to produce lift equal to the aircraft's weight and, therefore, the greater induced drag. The amount of induced drag varies inversely with the square of the airspeed.

Conversely, parasite drag increases as the square of the airspeed. Thus, in steady state, as airspeed decreases to near the stalling speed, the total drag becomes greater, due mainly to the sharp rise in induced drag. Similarly, as the aircraft reaches its never-exceed speed (V_{NE}), the total drag increases rapidly due to the sharp increase of parasite drag. As seen in *Figure 5-6,* at some given airspeed, total drag is at its minimum amount. In figuring the maximum range of aircraft, the thrust required to overcome drag is at a minimum if drag is at a minimum. The minimum power and maximum endurance occur at a different point.

Weight

Gravity is the pulling force that tends to draw all bodies to the center of the earth. The CG may be considered as a point at which all the weight of the aircraft is concentrated. If the aircraft were supported at its exact CG, it would balance in any attitude. It will be noted that CG is of major importance in an aircraft, for its position has a great bearing upon stability. The allowable location of the CG is determined by the general design of each particular aircraft. The designers determine how far the center of pressure (CP) will travel. It is important to understand that an aircraft's weight is concentrated at the CG and the aerodynamic forces of lift occur at the CP. When the CG is forward of the CP, there is a natural tendency for the aircraft to want to pitch nose down. If the CP is forward of the CG, a nose up pitching moment is created. Therefore, designers fix the aft limit of the CG forward of the CP for the corresponding flight speed in order to retain flight equilibrium.

Weight has a definite relationship to lift. This relationship is simple, but important in understanding the aerodynamics of flying. Lift is the upward force on the wing acting perpendicular to the relative wind and perpendicular to the aircraft's lateral axis. Lift is required to counteract the aircraft's weight. In stabilized level flight, when the lift force is equal to the weight force, the aircraft is in a state of equilibrium and neither accelerates upward or downward. If lift becomes less than weight, the vertical speed will decrease. When lift is greater than weight, the vertical speed will increase.

Wingtip Vortices

Formation of Vortices

The action of the airfoil that gives an aircraft lift also causes induced drag. When an airfoil is flown at a positive AOA, a pressure differential exists between the upper and lower surfaces of the airfoil. The pressure above the wing is less than atmospheric pressure and the pressure below the wing is equal to or greater than atmospheric pressure. Since air always moves from high pressure toward low pressure, and the path of least resistance is toward the airfoil's tips, there is a spanwise movement of air from the bottom of the airfoil outward from the fuselage around the tips. This flow of air results in "spillage" over the tips, thereby setting up a whirlpool of air called a vortex. *[Figure 5-12]*

At the same time, the air on the upper surface has a tendency to flow in toward the fuselage and off the trailing edge. This air current forms a similar vortex at the inboard portion of the trailing edge of the airfoil, but because the fuselage limits the inward flow, the vortex is insignificant. Consequently, the deviation in flow direction is greatest at the outer tips where the unrestricted lateral flow is the strongest.

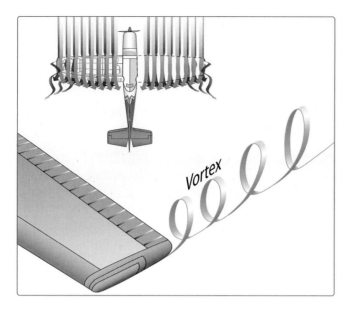

Figure 5-12. *Wingtip vortices.*

As the air curls upward around the tip, it combines with the downwash to form a fast-spinning trailing vortex. These vortices increase drag because of energy spent in producing the turbulence. Whenever an airfoil is producing lift, induced drag occurs and wingtip vortices are created.

Just as lift increases with an increase in AOA, induced drag also increases. This occurs because as the AOA is increased, there is a greater pressure difference between the top and bottom of the airfoil, and a greater lateral flow of air; consequently, this causes more violent vortices to be set up, resulting in more turbulence and more induced drag.

In *Figure 5-12,* it is easy to see the formation of wingtip vortices. The intensity or strength of the vortices is directly proportional to the weight of the aircraft and inversely proportional to the wingspan and speed of the aircraft. The heavier and slower the aircraft, the greater the AOA and the stronger the wingtip vortices. Thus, an aircraft will create wingtip vortices with maximum strength occurring during the takeoff, climb, and landing phases of flight. These

vortices lead to a particularly dangerous hazard to flight, wake turbulence.

Avoiding Wake Turbulence

Wingtip vortices are greatest when the generating aircraft is "heavy, clean, and slow." This condition is most commonly encountered during approaches or departures because an aircraft's AOA is at the highest to produce the lift necessary to land or take off. To minimize the chances of flying through an aircraft's wake turbulence:

- Avoid flying through another aircraft's flight path.

- Rotate prior to the point at which the preceding aircraft rotated when taking off behind another aircraft.

- Avoid following another aircraft on a similar flight path at an altitude within 1,000 feet. *[Figure 5-13]*

- Approach the runway above a preceding aircraft's path when landing behind another aircraft and touch down after the point at which the other aircraft wheels contacted the runway. *[Figure 5-14]*

A hovering helicopter generates a down wash from its main rotor(s) similar to the vortices of an airplane. Pilots of small aircraft should avoid a hovering helicopter by at least three rotor disc diameters to avoid the effects of this down wash. In forward flight, this energy is transformed into a pair of strong, high-speed trailing vortices similar to wing-tip vortices of larger fixed-wing aircraft. Helicopter vortices should be avoided because helicopter forward flight airspeeds are often very slow and can generate exceptionally strong wake turbulence.

Wind is an important factor in avoiding wake turbulence because wingtip vortices drift with the wind at the speed of the wind. For example, a wind speed of 10 knots causes the vortices to drift at about 1,000 feet in a minute in the wind direction. When following another aircraft, a pilot should consider wind speed and direction when selecting an intended takeoff or landing point. If a pilot is unsure of the other aircraft's takeoff or landing point, approximately 3 minutes provides a margin of

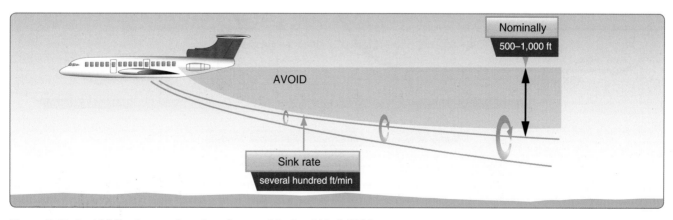

Figure 5-13. *Avoid following another aircraft at an altitude within 1,000 feet.*

Figure 5-14. *Avoid turbulence from another aircraft.*

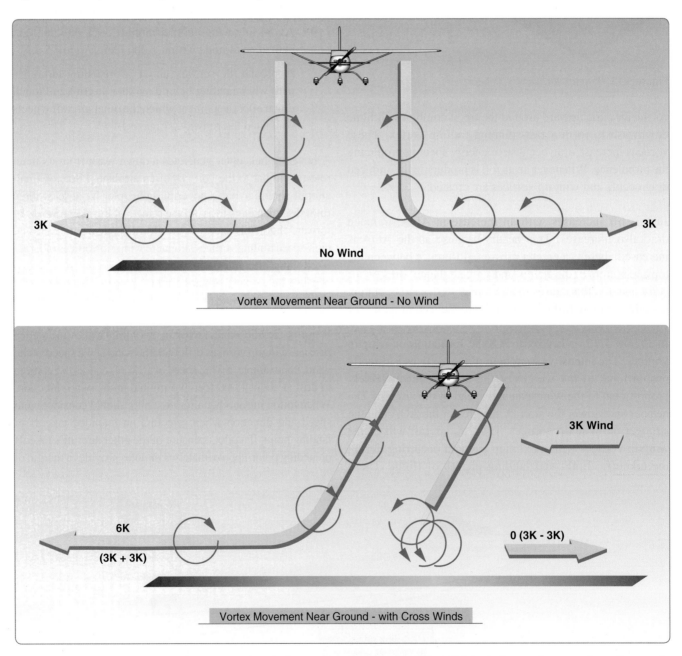

Figure 5-15. *When the vortices of larger aircraft sink close to the ground (within 100 to 200 feet), they tend to move laterally over the ground at a speed of 2 or 3 knots (top). A crosswind will decrease the lateral movement of the upwind vortex and increase the movement of the downwind vortex. Thus a light wind with a cross runway component of 1 to 5 knots could result in the upwind vortex remaining in the touchdown zone for a period of time and hasten the drift of the downwind vortex toward another runway (bottom).*

safety that allows wake turbulence dissipation. *[Figure 5-15]* For more information on wake turbulence, see Advisory Circular (AC) 90-23, Aircraft Wake Turbulence.

Ground Effect

Ever since the beginning of manned flight, pilots realized that just before touchdown it would suddenly feel like the aircraft did not want to go lower, and it would just want to go on and on. This is due to the air that is trapped between the wing and the landing surface, as if there were an air cushion. This phenomenon is called ground effect.

When an aircraft in flight comes within several feet of the surface, ground or water, a change occurs in the three-dimensional flow pattern around the aircraft because the vertical component of the airflow around the wing is restricted by the surface. This alters the wing's upwash, downwash, and wingtip vortices. *[Figure 5-16]* Ground effect, then, is due to the interference of the ground (or water) surface with the airflow patterns about the aircraft in flight. While the aerodynamic characteristics of the tail surfaces and the fuselage are altered by ground effect, the principal effects due to proximity of the ground are the changes in the aerodynamic characteristics of the wing. As the wing encounters ground effect and is maintained at a constant AOA, there is consequent reduction in the upwash, downwash, and wingtip vortices.

Induced drag is a result of the airfoil's work of sustaining the aircraft, and a wing or rotor lifts the aircraft simply by accelerating a mass of air downward. It is true that reduced pressure on top of an airfoil is essential to lift, but that is only one of the things contributing to the overall effect of pushing an air mass downward. The more downwash there is, the harder the wing pushes the mass of air down. At high angles of attack, the amount of induced drag is high; since this corresponds to lower airspeeds in actual flight, it can be said that induced drag predominates at low speed. However, the reduction of the wingtip vortices due to ground effect alters

Figure 5-16. *Ground effect changes airflow.*

the spanwise lift distribution and reduces the induced AOA and induced drag. Therefore, the wing will require a lower AOA in ground effect to produce the same C_L. If a constant AOA is maintained, an increase in C_L results. *[Figure 5-17]*

Ground effect also alters the thrust required versus velocity. Since induced drag predominates at low speeds, the reduction of induced drag due to ground effect will cause a significant reduction of thrust required (parasite plus induced drag) at low speeds. Due to the change in upwash, downwash, and wingtip vortices, there may be a change in position (installation) error of the airspeed system associated with ground effect. In the majority of cases, ground effect causes an increase in the local pressure at the static source and produces a lower indication of airspeed and altitude. Thus, an aircraft may be airborne at an indicated airspeed less than that normally required.

In order for ground effect to be of significant magnitude, the wing must be quite close to the ground. One of the direct results of ground effect is the variation of induced drag with wing height above the ground at a constant C_L. When the wing is at a height equal to its span, the reduction in induced drag is only 1.4 percent. However, when the wing is at a height equal to one-fourth its span, the reduction in induced drag is 23.5 percent and, when the wing is at a height equal to one-tenth its span, the reduction in induced drag is 47.6 percent. Thus, a large reduction in induced drag takes place only when the wing is very close to the ground. Because of this variation, ground effect is most usually recognized during the liftoff for takeoff or just prior to touchdown when landing.

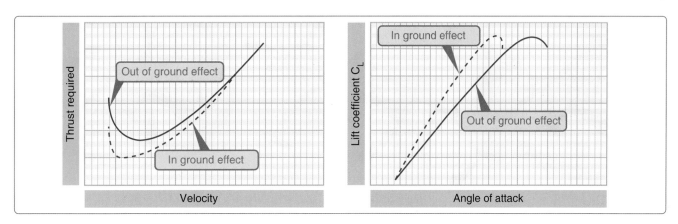

Figure 5-17. *Ground effect changes drag and lift.*

During the takeoff phase of flight, ground effect produces some important relationships. An aircraft leaving ground effect after takeoff encounters just the reverse of an aircraft entering ground effect during landing. The aircraft leaving ground effect will:

- Require an increase in AOA to maintain the same C_L

- Experience an increase in induced drag and thrust required

- Experience a decrease in stability and a nose-up change in moment

- Experience a reduction in static source pressure and increase in indicated airspeed

Ground effect must be considered during takeoffs and landings. For example, if a pilot fails to understand the relationship between the aircraft and ground effect during takeoff, a hazardous situation is possible because the recommended takeoff speed may not be achieved. Due to the reduced drag in ground effect, the aircraft may seem capable of takeoff well below the recommended speed. As the aircraft rises out of ground effect with a deficiency of speed, the greater induced drag may result in marginal initial climb performance. In extreme conditions, such as high gross weight, high density altitude, and high temperature, a deficiency of airspeed during takeoff may permit the aircraft to become airborne but be incapable of sustaining flight out of ground effect. In this case, the aircraft may become airborne initially with a deficiency of speed and then settle back to the runway.

A pilot should not attempt to force an aircraft to become airborne with a deficiency of speed. The manufacturer's recommended takeoff speed is necessary to provide adequate initial climb performance. It is also important that a definite climb be established before a pilot retracts the landing gear or flaps. Never retract the landing gear or flaps prior to

establishing a positive rate of climb and only after achieving a safe altitude.

If, during the landing phase of flight, the aircraft is brought into ground effect with a constant AOA, the aircraft experiences an increase in C_L and a reduction in the thrust required, and a "floating" effect may occur. Because of the reduced drag and lack of power-off deceleration in ground effect, any excess speed at the point of flare may incur a considerable "float" distance. As the aircraft nears the point of touchdown, ground effect is most realized at altitudes less than the wingspan. During the final phases of the approach as the aircraft nears the ground, a reduction of power is necessary to offset the increase in lift caused from ground effect otherwise the aircraft will have a tendency to climb above the desired glidepath (GP).

Axes of an Aircraft

The axes of an aircraft are three imaginary lines that pass through an aircraft's CG. The axes can be considered as imaginary axles around which the aircraft turns. The three axes pass through the CG at 90° angles to each other. The axis passes through the CG and parallel to a line from nose to tail is the longitudinal axis, the axis that passes parallel to a line from wingtip to wingtip is the lateral axis, and the axis that passes through the CG at right angles to the other two axes is the vertical axis. Whenever an aircraft changes its flight attitude or position in flight, it rotates about one or more of the three axes. *[Figure 5-18]*

The aircraft's motion about its longitudinal axis resembles the roll of a ship from side to side. In fact, the names used to describe the motion about an aircraft's three axes were originally nautical terms. They have been adapted to aeronautical terminology due to the similarity of motion of aircraft and seagoing ships. The motion about the aircraft's longitudinal axis is "roll," the motion about its lateral axis is

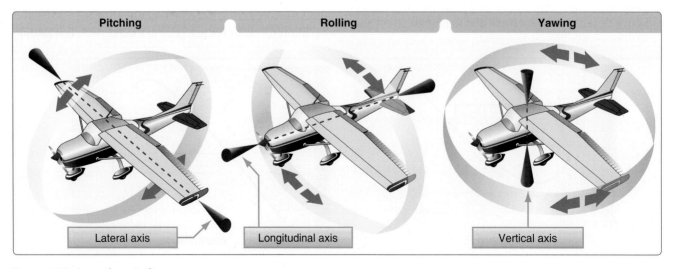

Figure 5-18. *Axes of an airplane.*

"pitch," and the motion about its vertical axis is "yaw." Yaw is the left and right movement of the aircraft's nose.

The three motions of the conventional airplane (roll, pitch, and yaw) are controlled by three control surfaces. Roll is controlled by the ailerons; pitch is controlled by the elevators; yaw is controlled by the rudder. The use of these controls is explained in Chapter 6, Flight Controls. Other types of aircraft may utilize different methods of controlling the movements about the various axes.

For example, weight-shift control aircraft control two axes (roll and pitch) using an "A" frame suspended from the flexible wing attached to a three-wheeled carriage. These aircraft are controlled by moving a horizontal bar (called a control bar) in roughly the same way hang glider pilots fly. [Figure 5-19] They are termed weight-shift control aircraft because the pilot controls the aircraft by shifting the CG. For more information on weight-shift control aircraft, see the Federal Aviation Administration (FAA) Weight-Shift Control Flying Handbook, FAA-H-8083-5. In the case of powered parachutes, aircraft control is accomplished by altering the airfoil via steering lines.

A powered parachute wing is a parachute that has a cambered upper surface and a flatter under surface. The two surfaces are separated by ribs that act as cells, which open to the airflow at the leading edge and have internal ports to allow lateral airflow. The principle at work holds that the cell pressure is greater than the outside pressure, thereby forming a wing that maintains its airfoil shape in flight. The pilot and passenger sit in tandem in front of the engine, which is located at the rear of a vehicle. The airframe is attached to the parachute via two attachment points and lines. Control is accomplished by both power and the changing of the airfoil via the control lines. [Figure 5-20]

Figure 5-19. *A weight-shift control aircraft.*

Figure 5-20. *A powered parachute.*

Moment and Moment Arm

A study of physics shows that a body that is free to rotate will always turn about its CG. In aerodynamic terms, the mathematical measure of an aircraft's tendency to rotate about its CG is called a "moment." A moment is said to be equal to the product of the force applied and the distance at which the force is applied. (A moment arm is the distance from a datum [reference point or line] to the applied force.) For aircraft weight and balance computations, "moments" are expressed in terms of the distance of the arm times the aircraft's weight, or simply, inch-pounds.

Aircraft designers locate the fore and aft position of the aircraft's CG as nearly as possible to the 20 percent point of the mean aerodynamic chord (MAC). If the thrust line is designed to pass horizontally through the CG, it will not cause the aircraft to pitch when power is changed, and there will be no difference in moment due to thrust for a power-on or power-off condition of flight. Although designers have some control over the location of the drag forces, they are not always able to make the resultant drag forces pass through the CG of the aircraft. However, the one item over which they have the greatest control is the size and location of the tail. The objective is to make the moments (due to thrust, drag, and lift) as small as possible and, by proper location of the tail, to provide the means of balancing an aircraft longitudinally for any condition of flight.

The pilot has no direct control over the location of forces acting on the aircraft in flight, except for controlling the center of lift by changing the AOA. The pilot can control the magnitude of the forces. Such a change, however, immediately involves changes in other forces. Therefore, the pilot cannot independently change the location of one force without changing the effect of others. For example, a change in airspeed involves a change in lift, as well as a change in drag and a change in the up or down force on the tail. As forces such as turbulence and gusts act to displace the aircraft, the pilot reacts by providing opposing control forces to counteract this displacement.

Some aircraft are subject to changes in the location of the CG with variations of load. Trimming devices, such as elevator trim tabs and adjustable horizontal stabilizers, are used to counteract the moments set up by fuel burnoff and loading or off-loading of passengers or cargo.

Aircraft Design Characteristics

Each aircraft handles somewhat differently because each resists or responds to control pressures in its own way. For example, a training aircraft is quick to respond to control applications, while a transport aircraft feels heavy on the controls and responds to control pressures more slowly. These features can be designed into an aircraft to facilitate the particular purpose of the aircraft by considering certain stability and maneuvering requirements. The following discussion summarizes the more important aspects of an aircraft's stability, maneuverability, and controllability qualities; how they are analyzed; and their relationship to various flight conditions.

Stability

Stability is the inherent quality of an aircraft to correct for conditions that may disturb its equilibrium and to return to or to continue on the original flight path. It is primarily an aircraft design characteristic. The flight paths and attitudes an aircraft flies are limited by the aerodynamic characteristics of the aircraft, its propulsion system, and its structural strength. These limitations indicate the maximum performance and maneuverability of the aircraft. If the aircraft is to provide maximum utility, it must be safely controllable to the full extent of these limits without exceeding the pilot's strength or requiring exceptional flying ability. If an aircraft is to fly straight and steady along any arbitrary flight path, the forces acting on it must be in static equilibrium. The reaction of any body when its equilibrium is disturbed is referred to as stability. The two types of stability are static and dynamic.

Static Stability

Static stability refers to the initial tendency, or direction of movement, back to equilibrium. In aviation, it refers to the aircraft's initial response when disturbed from a given pitch, yaw, or bank.

- Positive static stability—the initial tendency of the aircraft to return to the original state of equilibrium after being disturbed. *[Figure 5-21]*

- Neutral static stability—the initial tendency of the aircraft to remain in a new condition after its equilibrium has been disturbed. *[Figure 5-21]*

- Negative static stability—the initial tendency of the aircraft to continue away from the original state of equilibrium after being disturbed. *[Figure 5-21]*

Dynamic Stability

Static stability has been defined as the initial tendency to return to equilibrium that the aircraft displays after being disturbed from its trimmed condition. Occasionally, the initial tendency is different or opposite from the overall tendency, so a distinction must be made between the two. Dynamic stability refers to the aircraft response over time

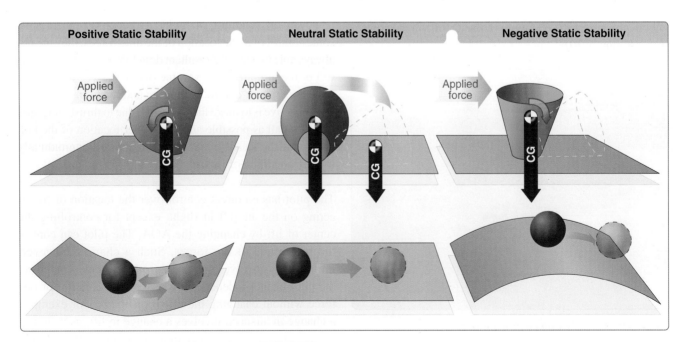

Figure 5-21. *Types of static stability.*

when disturbed from a given pitch, yaw, or bank. This type of stability also has three subtypes: [*Figure 5-22*]

- Positive dynamic stability—over time, the motion of the displaced object decreases in amplitude and, because it is positive, the object displaced returns toward the equilibrium state.

- Neutral dynamic stability—once displaced, the displaced object neither decreases nor increases in amplitude. A worn automobile shock absorber exhibits this tendency.

- Negative dynamic stability—over time, the motion of the displaced object increases and becomes more divergent.

Stability in an aircraft affects two areas significantly:

- Maneuverability—the quality of an aircraft that permits it to be maneuvered easily and to withstand the stresses imposed by maneuvers. It is governed by the aircraft's weight, inertia, size and location of flight controls, structural strength, and powerplant. It too is an aircraft design characteristic.

- Controllability—the capability of an aircraft to respond to the pilot's control, especially with regard to flight path and attitude. It is the quality of the aircraft's response to the pilot's control application when maneuvering the aircraft, regardless of its stability characteristics.

Longitudinal Stability (Pitching)

In designing an aircraft, a great deal of effort is spent in developing the desired degree of stability around all three axes. But longitudinal stability about the lateral axis is considered to be the most affected by certain variables in various flight conditions.

Longitudinal stability is the quality that makes an aircraft stable about its lateral axis. It involves the pitching motion as the aircraft's nose moves up and down in flight. A longitudinally unstable aircraft has a tendency to dive or climb progressively into a very steep dive or climb, or even a stall. Thus, an aircraft with longitudinal instability becomes difficult and sometimes dangerous to fly.

Static longitudinal stability, or instability in an aircraft, is dependent upon three factors:

1. Location of the wing with respect to the CG

2. Location of the horizontal tail surfaces with respect to the CG

3. Area or size of the tail surfaces

In analyzing stability, it should be recalled that a body free to rotate always turns about its CG.

To obtain static longitudinal stability, the relation of the wing and tail moments must be such that, if the moments are initially balanced and the aircraft is suddenly nose up, the wing moments and tail moments change so that the sum of their forces provides an unbalanced but restoring moment which, in turn, brings the nose down again. Similarly, if the aircraft is nose down, the resulting change in moments brings the nose back up.

The Center of Lift (CL) in most asymmetrical airfoils has a tendency to change its fore and aft positions with a change in the AOA. The CL tends to move forward with an increase in AOA and to move aft with a decrease in AOA. This means

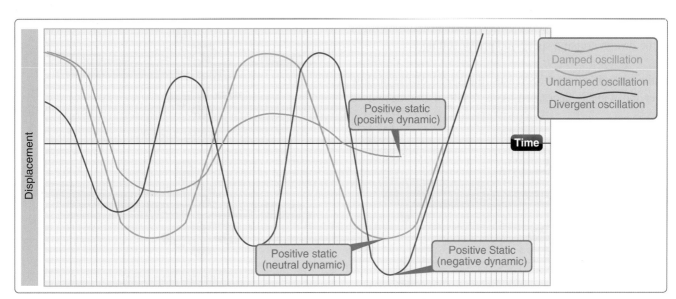

Figure 5-22. *Damped versus undamped stability.*

that when the AOA of an airfoil is increased, the CL, by moving forward, tends to lift the leading edge of the wing still more. This tendency gives the wing an inherent quality of instability. (NOTE: CL is also known as the center of pressure (CP).)

Figure 5-23 shows an aircraft in straight-and-level flight. The line CG-CL-T represents the aircraft's longitudinal axis from the CG to a point T on the horizontal stabilizer.

Most aircraft are designed so that the wing's CL is to the rear of the CG. This makes the aircraft "nose heavy" and requires that there be a slight downward force on the horizontal stabilizer in order to balance the aircraft and keep the nose from continually pitching downward. Compensation for this nose heaviness is provided by setting the horizontal stabilizer at a slight negative AOA. The downward force thus produced holds the tail down, counterbalancing the "heavy" nose. It is as if the line CG-CL-T were a lever with an upward force at CL and two downward forces balancing each other, one a strong force at the CG point and the other, a much lesser force, at point T (downward air pressure on the stabilizer). To better visualize this physics principle: If an iron bar were suspended at point CL, with a heavy weight hanging on it at the CG, it would take downward pressure at point T to keep the "lever" in balance.

Even though the horizontal stabilizer may be level when the aircraft is in level flight, there is a downwash of air from the wings. This downwash strikes the top of the stabilizer and produces a downward pressure, which at a certain speed is just enough to balance the "lever." The faster the aircraft is flying, the greater this downwash and the greater the downward force on the horizontal stabilizer (except T-tails). *[Figure 5-24]* In aircraft with fixed-position horizontal stabilizers, the aircraft manufacturer sets the stabilizer at an angle that provides the best stability (or balance) during flight at the design cruising speed and power setting.

If the aircraft's speed decreases, the speed of the airflow over the wing is decreased. As a result of this decreased

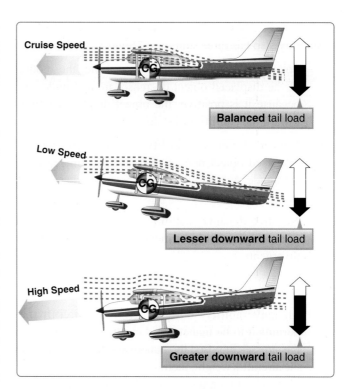

Figure 5-24. *Effect of speed on downwash.*

flow of air over the wing, the downwash is reduced, causing a lesser downward force on the horizontal stabilizer. In turn, the characteristic nose heaviness is accentuated, causing the aircraft's nose to pitch down more. *[Figure 5-25]* This places the aircraft in a nose-low attitude, lessening the wing's AOA and drag and allowing the airspeed to increase. As the aircraft continues in the nose-low attitude and its speed increases, the downward force on the horizontal stabilizer is once again increased. Consequently, the tail is again pushed downward and the nose rises into a climbing attitude.

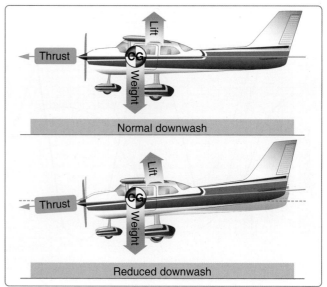

Figure 5-25. *Reduced power allows pitch down.*

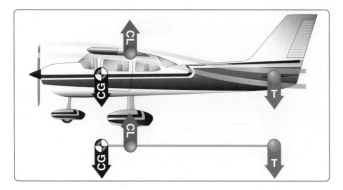

Figure 5-23. *Longitudinal stability.*

As this climb continues, the airspeed again decreases, causing the downward force on the tail to decrease until the nose lowers once more. Because the aircraft is dynamically stable, the nose does not lower as far this time as it did before. The aircraft acquires enough speed in this more gradual dive to start it into another climb, but the climb is not as steep as the preceding one.

After several of these diminishing oscillations, in which the nose alternately rises and lowers, the aircraft finally settles down to a speed at which the downward force on the tail exactly counteracts the tendency of the aircraft to dive. When this condition is attained, the aircraft is once again in balanced flight and continues in stabilized flight as long as this attitude and airspeed are not changed.

A similar effect is noted upon closing the throttle. The downwash of the wings is reduced and the force at T in *Figure 5-23* is not enough to hold the horizontal stabilizer down. It seems as if the force at T on the lever were allowing the force of gravity to pull the nose down. This is a desirable characteristic because the aircraft is inherently trying to regain airspeed and reestablish the proper balance.

Power or thrust can also have a destabilizing effect in that an increase of power may tend to make the nose rise. The aircraft designer can offset this by establishing a "high thrust line" wherein the line of thrust passes above the CG. *[Figures 5-26 and 5-27]* In this case, as power or thrust is

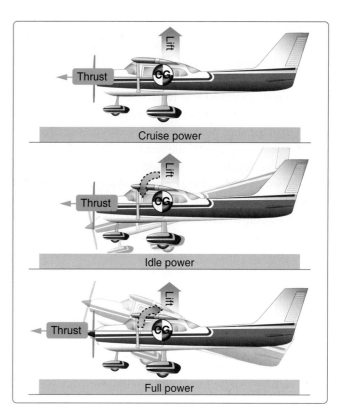

Figure 5-27. *Power changes affect longitudinal stability.*

increased a moment is produced to counteract the down load on the tail. On the other hand, a very "low thrust line" would tend to add to the nose-up effect of the horizontal tail surface. Conclusion: with CG forward of the CL and with an aerodynamic tail-down force, the aircraft usually tries to return to a safe flying attitude.

The following is a simple demonstration of longitudinal stability. Trim the aircraft for "hands off" control in level flight. Then, momentarily give the controls a slight push to nose the aircraft down. If, within a brief period, the nose rises towards the original position, the aircraft is statically stable. Ordinarily, the nose passes the original position (that of level flight) and a series of slow pitching oscillations follows. If the oscillations gradually cease, the aircraft has positive stability; if they continue unevenly, the aircraft has neutral stability; if they increase, the aircraft is unstable.

Lateral Stability (Rolling)

Stability about the aircraft's longitudinal axis, which extends from the nose of the aircraft to its tail, is called lateral stability. Positive lateral stability helps to stabilize the lateral or "rolling effect" when one wing gets lower than the wing on the opposite side of the aircraft. There are four main design factors that make an aircraft laterally stable: dihedral, sweepback, keel effect, and weight distribution.

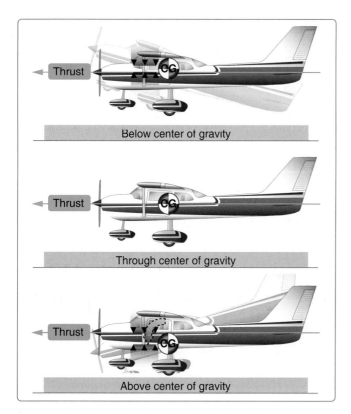

Figure 5-26. *Thrust line affects longitudinal stability.*

Dihedral

Some aircraft are designed so that the outer tips of the wings are higher than the wing roots. The upward angle thus formed by the wings is called dihedral. *[Figure 5-28]* When a gust causes a roll, a sideslip will result. This sideslip causes the relative wind affecting the entire airplane to be from the direction of the slip. When the relative wind comes from the side, the wing slipping into the wind is subject to an increase in AOA and develops an increase in lift. The wing away from the wind is subject to a decrease in angle of attack, and develops a decrease in lift. The changes in lift effect a rolling moment tending to raise the windward wing, hence dihedral contributes to a stable roll due to sideslip. *[Figure 5-29]*

Sweepback and Wing Location

Many aspects of an aircraft's configuration can affect its effective dihedral, but two major components are wing sweepback and the wing location with respect to the fuselage (such as a low wing or high wing). As a rough estimation, 10° of sweepback on a wing provides about 1° of effective dihedral, while a high wing configuration can provide about 5° of effective dihedral over a low wing configuration.

A sweptback wing is one in which the leading edge slopes backward. *[Figure 5-30]* When a disturbance causes an aircraft with sweepback to slip or drop a wing, the low wing presents its leading edge at an angle that is more perpendicular to the relative airflow. As a result, the low wing acquires more lift, rises, and the aircraft is restored to its original flight attitude.

Keel Effect and Weight Distribution

A high wing aircraft always has the tendency to turn the longitudinal axis of the aircraft into the relative wind, which is often referred to as the keel effect. These aircraft are laterally stable simply because the wings are attached in a high position on the fuselage, making the fuselage behave like a keel exerting a steadying influence on the aircraft laterally about the longitudinal axis. When a high-winged aircraft is

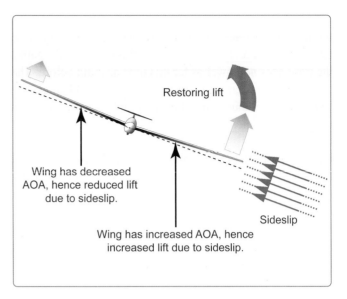

Figure 5-29. *Sideslip causing different AOA on each blade.*

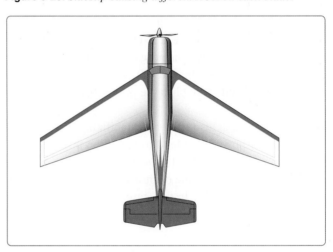

Figure 5-30. *Sweepback wings.*

disturbed and one wing dips, the fuselage weight acts like a pendulum returning the aircraft to the horizontal level.

Laterally stable aircraft are constructed so that the greater portion of the keel area is above the CG. *[Figure 5-31]* Thus, when the aircraft slips to one side, the combination of the

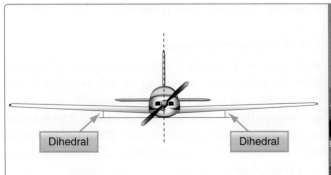

Figure 5-28. *Dihedral is the upward angle of the wings from a horizontal (front/rear view) axis of the plane as shown in the graphic depiction and the rear view of a Ryanair Boeing 737.*

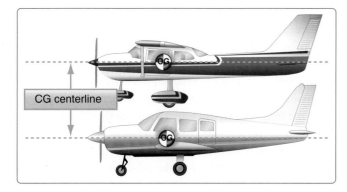

Figure 5-31. *Keel area for lateral stability.*

aircraft's weight and the pressure of the airflow against the upper portion of the keel area (both acting about the CG) tends to roll the aircraft back to wings-level flight.

Directional Stability (Yawing)

Stability about the aircraft's vertical axis (the sideways moment) is called yawing or directional stability. Yawing or directional stability is the most easily achieved stability in aircraft design. The area of the vertical fin and the sides of the fuselage aft of the CG are the prime contributors that make the aircraft act like the well known weather vane or arrow, pointing its nose into the relative wind.

In examining a weather vane, it can be seen that if exactly the same amount of surface were exposed to the wind in front of the pivot point as behind it, the forces fore and aft would be in balance and little or no directional movement would result. Consequently, it is necessary to have a greater surface aft of the pivot point than forward of it.

Similarly, the aircraft designer must ensure positive directional stability by making the side surface greater aft than ahead of the CG. *[Figure 5-32]* To provide additional positive stability to that provided by the fuselage, a vertical fin is added. The fin acts similar to the feather on an arrow in maintaining straight flight. Like the weather vane and the arrow, the farther aft this fin is placed and the larger its size, the greater the aircraft's directional stability.

If an aircraft is flying in a straight line, and a sideward gust of air gives the aircraft a slight rotation about its vertical axis (i.e., the right), the motion is retarded and stopped by the fin because while the aircraft is rotating to the right, the air is striking the left side of the fin at an angle. This causes pressure on the left side of the fin, which resists the turning motion and slows down the aircraft's yaw. In doing so, it acts somewhat like the weather vane by turning the aircraft into the relative wind. The initial change in direction of the aircraft's flight path is generally slightly behind its change of heading. Therefore, after a slight yawing of the aircraft

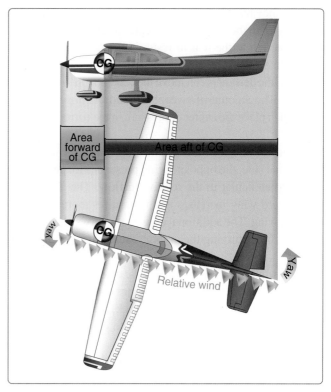

Figure 5-32. *Fuselage and fin for directional stability.*

to the right, there is a brief moment when the aircraft is still moving along its original path, but its longitudinal axis is pointed slightly to the right.

The aircraft is then momentarily skidding sideways and, during that moment (since it is assumed that although the yawing motion has stopped, the excess pressure on the left side of the fin still persists), there is necessarily a tendency for the aircraft to be turned partially back to the left. That is, there is a momentary restoring tendency caused by the fin.

This restoring tendency is relatively slow in developing and ceases when the aircraft stops skidding. When it ceases, the aircraft is flying in a direction slightly different from the original direction. In other words, it will not return of its own accord to the original heading; the pilot must reestablish the initial heading.

A minor improvement of directional stability may be obtained through sweepback. Sweepback is incorporated in the design of the wing primarily to delay the onset of compressibility during high-speed flight. In lighter and slower aircraft, sweepback aids in locating the center of pressure in the correct relationship with the CG. A longitudinally stable aircraft is built with the center of pressure aft of the CG.

Because of structural reasons, aircraft designers sometimes cannot attach the wings to the fuselage at the exact desired point. If they had to mount the wings too far forward, and at right angles to the fuselage, the center of pressure would not be far enough to the rear to result in the desired amount of longitudinal stability. By building sweepback into the wings, however, the designers can move the center of pressure toward the rear. The amount of sweepback and the position of the wings then place the center of pressure in the correct location.

When turbulence or rudder application causes the aircraft to yaw to one side, the opposite wing presents a longer leading edge perpendicular to the relative airflow. The airspeed of the forward wing increases and it acquires more drag than the back wing. The additional drag on the forward wing pulls the wing back, turning the aircraft back to its original path.

The contribution of the wing to static directional stability is usually small. The swept wing provides a stable contribution depending on the amount of sweepback, but the contribution is relatively small when compared with other components.

Free Directional Oscillations (Dutch Roll)

Dutch roll is a coupled lateral/directional oscillation that is usually dynamically stable but is unsafe in an aircraft because of the oscillatory nature. The damping of the oscillatory mode may be weak or strong depending on the properties of the particular aircraft.

If the aircraft has a right wing pushed down, the positive sideslip angle corrects the wing laterally before the nose is realigned with the relative wind. As the wing corrects the position, a lateral directional oscillation can occur resulting in the nose of the aircraft making a figure eight on the horizon as a result of two oscillations (roll and yaw), which, although of about the same magnitude, are out of phase with each other.

In most modern aircraft, except high-speed swept wing designs, these free directional oscillations usually die out automatically in very few cycles unless the air continues to be gusty or turbulent. Those aircraft with continuing Dutch roll tendencies are usually equipped with gyro-stabilized yaw dampers. Manufacturers try to reach a midpoint between too much and too little directional stability. Because it is more desirable for the aircraft to have "spiral instability" than Dutch roll tendencies, most aircraft are designed with that characteristic.

Spiral Instability

Spiral instability exists when the static directional stability of the aircraft is very strong as compared to the effect of its dihedral in maintaining lateral equilibrium. When the lateral equilibrium of the aircraft is disturbed by a gust of air and a sideslip is introduced, the strong directional stability tends to yaw the nose into the resultant relative wind while the comparatively weak dihedral lags in restoring the lateral balance. Due to this yaw, the wing on the outside of the turning moment travels forward faster than the inside wing and, as a consequence, its lift becomes greater. This produces an overbanking tendency which, if not corrected by the pilot, results in the bank angle becoming steeper and steeper. At the same time, the strong directional stability that yaws the aircraft into the relative wind is actually forcing the nose to a lower pitch attitude. A slow downward spiral begins which, if not counteracted by the pilot, gradually increases into a steep spiral dive. Usually the rate of divergence in the spiral motion is so gradual the pilot can control the tendency without any difficulty.

Many aircraft are affected to some degree by this characteristic, although they may be inherently stable in all other normal parameters. This tendency explains why an aircraft cannot be flown "hands off" indefinitely.

Much research has gone into the development of control devices (wing leveler) to correct or eliminate this instability. The pilot must be careful in application of recovery controls during advanced stages of this spiral condition or excessive loads may be imposed on the structure. Improper recovery from spiral instability leading to inflight structural failures has probably contributed to more fatalities in general aviation aircraft than any other factor. Since the airspeed in the spiral condition builds up rapidly, the application of back elevator force to reduce this speed and to pull the nose up only "tightens the turn," increasing the load factor. The results of the prolonged uncontrolled spiral are inflight structural failure, crashing into the ground, or both. Common recorded causes for pilots who get into this situation are loss of horizon reference, inability to control the aircraft by reference to instruments, or a combination of both.

Effect of Wing Planform

Understanding the effects of different wing planforms is important when learning about wing performance and airplane flight characteristics. A planform is the shape of the wing as viewed from directly above and deals with airflow in three dimensions. Aspect ratio, taper ratio, and sweepback are factors in planform design that are very important to the overall aerodynamic characteristic of a wing. [Figure 5-33]

Aspect ratio is the ratio of wing span to wing chord. Taper ratio can be either in planform or thickness, or both. In its simplest terms, it is a decrease from wing root to wingtip in wing chord or wing thickness. Sweepback is the rearward slant of a wing, horizontal tail, or other airfoil surface.

There are two general means by which the designer can change the planform of a wing and both will affect the

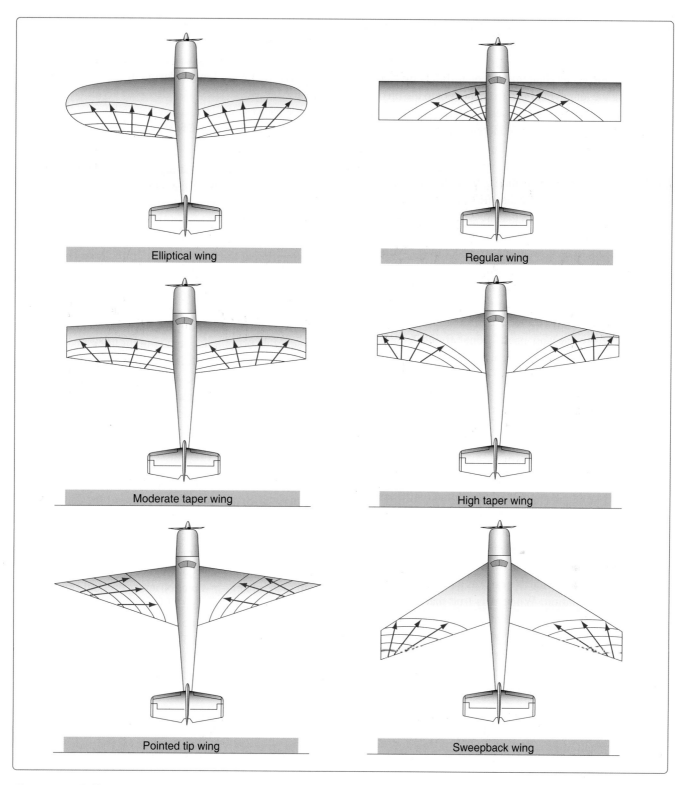

Figure 5-33. *Different types of wing planforms.*

aerodynamic characteristics of the wing. The first is to effect a change in the aspect ratio. Aspect ratio is the primary factor in determining the three dimensional characteristics of the ordinary wing and its lift/drag ratio. An increase in aspect ratio with constant velocity will decrease the drag, especially at high angles of attack, improving the performance of the wing when in a climbing attitude.

A decrease in aspect ratio will give a corresponding increase in drag. It should be noted, however, that with an increase in aspect ratio there is an increase in the length of span, with a corresponding increase in the weight of the wing structure, which means the wing must be heavier to carry the same load. For this reason, part of the gain (due to a decrease in drag) is lost because of the increased weight, and a compromise in design is necessary to obtain the best results from these two conflicting conditions.

The second means of changing the planform is by tapering (decreasing the length of chord from the root to the tip of the wing). In general, tapering causes a decrease in drag (most effective at high speeds) and an increase in lift. There is also a structural benefit due to a saving in weight of the wing.

Most training and general aviation type airplanes are operated at high coefficients of lift, and therefore require comparatively high aspect ratios. Airplanes that are developed to operate at very high speeds demand greater aerodynamic cleanness and greater strength, which require low aspect ratios. Very low aspect ratios result in high wing loadings and high stall speeds. When sweepback is combined with low aspect ratio, it results in flying qualities very different from a more conventional high aspect ratio airplane configuration. Such airplanes require very precise and professional flying techniques, especially at slow speeds, while airplanes with a high aspect ratio are usually more forgiving of improper pilot techniques.

The elliptical wing is the ideal subsonic planform since it provides for a minimum of induced drag for a given aspect ratio, though as we shall see, its stall characteristics in some respects are inferior to the rectangular wing. It is also comparatively difficult to construct. The tapered airfoil is desirable from the standpoint of weight and stiffness, but again is not as efficient aerodynamically as the elliptical wing. In order to preserve the aerodynamic efficiency of the elliptical wing, rectangular and tapered wings are sometimes tailored through use of wing twist and variation in airfoil sections until they provide as nearly as possible the elliptical wing's lift distribution. While it is true that the elliptical wing provides the best coefficients of lift before reaching an incipient stall, it gives little advance warning of a complete stall, and lateral control may be difficult because of poor aileron effectiveness.

In comparison, the rectangular wing has a tendency to stall first at the wing root and provides adequate stall warning, adequate aileron effectiveness, and is usually quite stable. It is, therefore, favored in the design of low cost, low speed airplanes.

Aerodynamic Forces in Flight Maneuvers
Forces in Turns

If an aircraft were viewed in straight-and-level flight from the front [Figure 5-34], and if the forces acting on the aircraft could be seen, lift and weight would be apparent: two forces. If the aircraft were in a bank it would be apparent that lift did not act directly opposite to the weight, rather it now acts in the direction of the bank. A basic truth about turns is that when the aircraft banks, lift acts inward toward the center of the turn, perpendicular to the lateral axis as well as upward.

Newton's First Law of Motion, the Law of Inertia, states that an object at rest or moving in a straight line remains at rest or continues to move in a straight line until acted on by some other force. An aircraft, like any moving object, requires a sideward force to make it turn. In a normal turn, this force is supplied by banking the aircraft so that lift is exerted inward, as well as upward. The force of lift during a turn is separated into two components at right angles to each other. One component, which acts vertically and opposite to the weight (gravity), is called the "vertical component of lift." The other, which acts horizontally toward the center of the turn, is called the "horizontal component of lift" or centripetal force. The horizontal component of lift is the force that pulls the aircraft from a straight flight path to make it turn. Centrifugal force is the "equal and opposite reaction" of the aircraft to the change in direction and acts equal and opposite to the horizontal component of lift. This explains why, in a correctly executed turn, the force that turns the aircraft is not supplied by the rudder. The rudder is used to correct any deviation between the straight track of the nose and tail of the aircraft into the relative wind. A good turn is one in which the

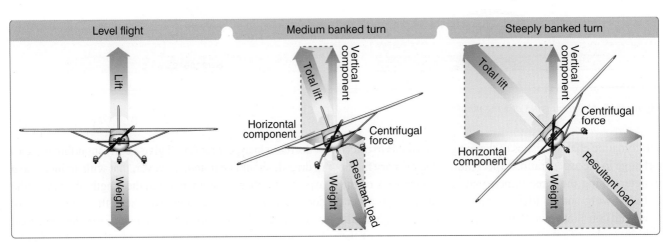

Figure 5-34. *Forces during normal, coordinated turn at constant altitude.*

nose and tail of the aircraft track along the same path. If no rudder is used in a turn, the nose of the aircraft yaws to the outside of the turn. The rudder is used rolling into the turn to bring the nose back in line with the relative wind. Once in the turn, the rudder should not be needed.

An aircraft is not steered like a boat or an automobile. In order for an aircraft to turn, it must be banked. If it is not banked, there is no force available to cause it to deviate from a straight flight path. Conversely, when an aircraft is banked, it turns provided it is not slipping to the inside of the turn. Good directional control is based on the fact that the aircraft attempts to turn whenever it is banked. Pilots should keep this fact in mind when attempting to hold the aircraft in straight-and-level flight.

Merely banking the aircraft into a turn produces no change in the total amount of lift developed. Since the lift during the bank is divided into vertical and horizontal components, the amount of lift opposing gravity and supporting the aircraft's weight is reduced. Consequently, the aircraft loses altitude unless additional lift is created. This is done by increasing the AOA until the vertical component of lift is again equal to the weight. Since the vertical component of lift decreases as the bank angle increases, the AOA must be progressively increased to produce sufficient vertical lift to support the aircraft's weight. An important fact for pilots to remember when making constant altitude turns is that the vertical component of lift must be equal to the weight to maintain altitude.

At a given airspeed, the rate at which an aircraft turns depends upon the magnitude of the horizontal component of lift. It is found that the horizontal component of lift is proportional to the angle of bank—that is, it increases or decreases respectively as the angle of bank increases or decreases. As the angle of bank is increased, the horizontal component of lift increases, thereby increasing the rate of turn (ROT). Consequently, at any given airspeed, the ROT can be controlled by adjusting the angle of bank.

To provide a vertical component of lift sufficient to hold altitude in a level turn, an increase in the AOA is required. Since the drag of the airfoil is directly proportional to its AOA, induced drag increases as the lift is increased. This, in turn, causes a loss of airspeed in proportion to the angle of bank. A small angle of bank results in a small reduction in airspeed while a large angle of bank results in a large reduction in airspeed. Additional thrust (power) must be applied to prevent a reduction in airspeed in level turns. The required amount of additional thrust is proportional to the angle of bank.

To compensate for added lift, which would result if the airspeed were increased during a turn, the AOA must be decreased, or the angle of bank increased, if a constant altitude is to be maintained. If the angle of bank is held constant and the AOA decreased, the ROT decreases. In order to maintain a constant ROT as the airspeed is increased, the AOA must remain constant and the angle of bank increased.

An increase in airspeed results in an increase of the turn radius, and centrifugal force is directly proportional to the radius of the turn. In a correctly executed turn, the horizontal component of lift must be exactly equal and opposite to the centrifugal force. As the airspeed is increased in a constant-rate level turn, the radius of the turn increases. This increase in the radius of turn causes an increase in the centrifugal force, which must be balanced by an increase in the horizontal component of lift, which can only be increased by increasing the angle of bank.

In a slipping turn, the aircraft is not turning at the rate appropriate to the bank being used, since the aircraft is yawed toward the outside of the turning flight path. The aircraft is banked too much for the ROT, so the horizontal lift component is greater than the centrifugal force. *[Figure 5-35]* Equilibrium between the horizontal lift component and centrifugal force is reestablished by either decreasing the bank, increasing the ROT, or a combination of the two changes.

A skidding turn results from an excess of centrifugal force over the horizontal lift component, pulling the aircraft toward the outside of the turn. The ROT is too great for the angle of bank. Correction of a skidding turn thus involves a reduction in the ROT, an increase in bank, or a combination of the two changes.

To maintain a given ROT, the angle of bank must be varied with the airspeed. This becomes particularly important in high-speed aircraft. For instance, at 400 miles per hour (mph), an aircraft must be banked approximately 44° to execute a standard-rate turn (3° per second). At this angle of bank, only about 79 percent of the lift of the aircraft comprises the vertical component of the lift. This causes a loss of altitude unless the AOA is increased sufficiently to compensate for the loss of vertical lift.

Forces in Climbs

For all practical purposes, the wing's lift in a steady state normal climb is the same as it is in a steady level flight at the same airspeed. Although the aircraft's flight path changed when the climb was established, the AOA of the wing with respect to the inclined flight path reverts to practically the same values, as does the lift. There is an initial momentary change as shown in *Figure 5-36*. During the transition from straight-and-level flight to a climb, a change in lift occurs when back elevator pressure is first applied. Raising the aircraft's nose increases the AOA and momentarily increases

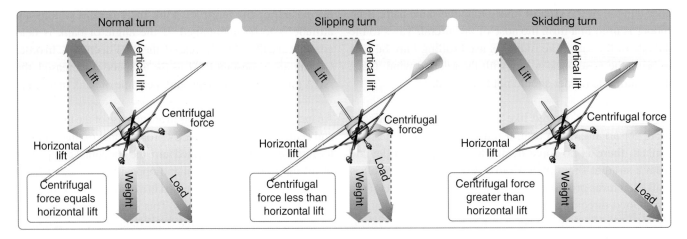

Figure 5-35. *Normal, slipping, and skidding turns at a constant altitude.*

the lift. Lift at this moment is now greater than weight and starts the aircraft climbing. After the flight path is stabilized on the upward incline, the AOA and lift again revert to about the level flight values.

If the climb is entered with no change in power setting, the airspeed gradually diminishes because the thrust required to maintain a given airspeed in level flight is insufficient to maintain the same airspeed in a climb. When the flight path is inclined upward, a component of the aircraft's weight acts in the same direction as, and parallel to, the total drag of the aircraft, thereby increasing the total effective drag. Consequently, the total effective drag is greater than the power, and the airspeed decreases. The reduction in airspeed gradually results in a corresponding decrease in drag until the total drag (including the component of weight acting in the same direction) equals the thrust. *[Figure 5-37]* Due to momentum, the change in airspeed is gradual, varying considerably with differences in aircraft size, weight, total drag, and other factors. Consequently, the total effective drag is greater than the thrust, and the airspeed decreases.

Generally, the forces of thrust and drag, and lift and weight, again become balanced when the airspeed stabilizes but at a value lower than in straight-and-level flight at the same power setting. Since the aircraft's weight is acting not only downward but rearward with drag while in a climb, additional power is required to maintain the same airspeed as in level flight. The amount of power depends on the angle of climb. When the climb is established steep enough that there is insufficient power available, a slower speed results.

The thrust required for a stabilized climb equals drag plus a percentage of weight dependent on the angle of climb. For example, a 10° climb would require thrust to equal drag plus 17 percent of weight. To climb straight up would require thrust to equal all of weight and drag. Therefore, the angle of climb for climb performance is dependent on the amount of excess thrust available to overcome a portion of weight. Note that aircraft are able to sustain a climb due to excess thrust. When the excess thrust is gone, the aircraft is no longer able to climb. At this point, the aircraft has reached its "absolute ceiling."

Forces in Descents

As in climbs, the forces that act on the aircraft go through definite changes when a descent is entered from straight-and-level flight. For the following example, the aircraft

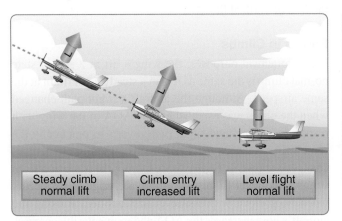

Figure 5-36. *Changes in lift during climb entry.*

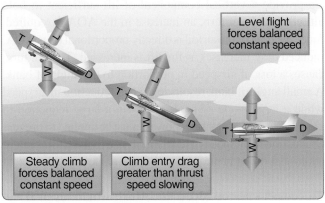

Figure 5-37. *Changes in speed during climb entry.*

is descending at the same power as used in straight-and-level flight.

As forward pressure is applied to the control yoke to initiate the descent, the AOA is decreased momentarily. Initially, the momentum of the aircraft causes the aircraft to briefly continue along the same flight path. For this instant, the AOA decreases causing the total lift to decrease. With weight now being greater than lift, the aircraft begins to descend. At the same time, the flight path goes from level to a descending flight path. Do not confuse a reduction in lift with the inability to generate sufficient lift to maintain level flight. The flight path is being manipulated with available thrust in reserve and with the elevator.

To descend at the same airspeed as used in straight-and-level flight, the power must be reduced as the descent is entered. Entering the descent, the component of weight acting forward along the flight path increases as the angle of descent increases and, conversely, when leveling off, the component of weight acting along the flight path decreases as the angle of descent decreases.

Stalls

An aircraft stall results from a rapid decrease in lift caused by the separation of airflow from the wing's surface brought on by exceeding the critical AOA. A stall can occur at any pitch attitude or airspeed. Stalls are one of the most misunderstood areas of aerodynamics because pilots often believe an airfoil stops producing lift when it stalls. In a stall, the wing does not totally stop producing lift. Rather, it cannot generate adequate lift to sustain level flight.

Since the C_L increases with an increase in AOA, at some point the C_L peaks and then begins to drop off. This peak is called the C_{L-MAX}. The amount of lift the wing produces drops dramatically after exceeding the C_{L-MAX} or critical AOA, but as stated above, it does not completely stop producing lift.

In most straight-wing aircraft, the wing is designed to stall the wing root first. The wing root reaches its critical AOA first making the stall progress outward toward the wingtip. By having the wing root stall first, aileron effectiveness is maintained at the wingtips, maintaining controllability of the aircraft. Various design methods are used to achieve the stalling of the wing root first. In one design, the wing is "twisted" to a higher AOA at the wing root. Installing stall strips on the first 20–25 percent of the wing's leading edge is another method to introduce a stall prematurely.

The wing never completely stops producing lift in a stalled condition. If it did, the aircraft would fall to the Earth. Most training aircraft are designed for the nose of the aircraft to

drop during a stall, reducing the AOA and "unstalling" the wing. The nose-down tendency is due to the CL being aft of the CG. The CG range is very important when it comes to stall recovery characteristics. If an aircraft is allowed to be operated outside of the CG range, the pilot may have difficulty recovering from a stall. The most critical CG violation would occur when operating with a CG that exceeds the rear limit. In this situation, a pilot may not be able to generate sufficient force with the elevator to counteract the excess weight aft of the CG. Without the ability to decrease the AOA, the aircraft continues in a stalled condition until it contacts the ground.

The stalling speed of a particular aircraft is not a fixed value for all flight situations, but a given aircraft always stalls at the same AOA regardless of airspeed, weight, load factor, or density altitude. Each aircraft has a particular AOA where the airflow separates from the upper surface of the wing and the stall occurs. This critical AOA varies from approximately 16° to 20° depending on the aircraft's design. But each aircraft has only one specific AOA where the stall occurs.

There are three flight situations in which the critical AOA is most frequently exceeded: low speed, high speed, and turning.

One way the aircraft can be stalled in straight-and-level flight by flying too slowly. As the airspeed decreases, the AOA must be increased to retain the lift required for maintaining altitude. The lower the airspeed becomes, the more the AOA must be increased. Eventually, an AOA is reached that results in the wing not producing enough lift to support the aircraft, which then starts settling. If the airspeed is reduced further, the aircraft stalls because the AOA has exceeded the critical angle and the airflow over the wing is disrupted.

Low speed is not necessary to produce a stall. The wing can be brought into an excessive AOA at any speed. For example, an aircraft is in a dive with an airspeed of 100 knots when the pilot pulls back sharply on the elevator control. [Figure 5-38] Gravity and centrifugal force prevent an immediate alteration of the flight path, but the aircraft's AOA changes abruptly from quite low to very high. Since the flight path of the aircraft in relation to the oncoming air determines the direction of the relative wind, the AOA is suddenly increased, and the aircraft would reach the stalling angle at a speed much greater than the normal stall speed.

The stalling speed of an aircraft is also higher in a level turn than in straight-and-level flight. [Figure 5-39] Centrifugal force is added to the aircraft's weight and the wing must produce sufficient additional lift to counterbalance the load imposed by the combination of centrifugal force and weight. In a turn, the necessary additional lift is acquired by applying back pressure to the elevator control. This increases the wing's

Figure 5-38. *Forces exerted when pulling out of a dive.*

AOA and results in increased lift. The AOA must increase as the bank angle increases to counteract the increasing load caused by centrifugal force. If at any time during a turn the AOA becomes excessive, the aircraft stalls.

At this point, the action of the aircraft during a stall should be examined. To balance the aircraft aerodynamically, the CL is normally located aft of the CG. Although this makes the aircraft inherently nose-heavy, downwash on the horizontal stabilizer counteracts this condition. At the point of stall, when the upward force of the wing's lift diminishes below that required for sustained flight and the downward tail force decreases to a point of ineffectiveness, or causes it to have an upward force, an unbalanced condition exists. This causes the aircraft to pitch down abruptly, rotating about its CG. During this nose-down attitude, the AOA decreases and the airspeed again increases. The smooth flow of air over the wing begins again, lift returns, and the aircraft begins to fly again. Considerable altitude may be lost before this cycle is complete.

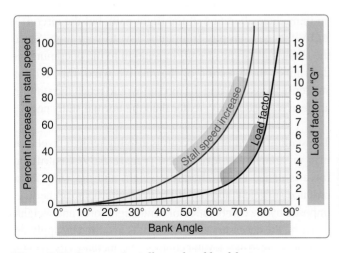

Figure 5-39. *Increase in stall speed and load factor.*

Airfoil shape and degradation of that shape must also be considered in a discussion of stalls. For example, if ice, snow, and frost are allowed to accumulate on the surface of an aircraft, the smooth airflow over the wing is disrupted. This causes the boundary layer to separate at an AOA lower than that of the critical angle. Lift is greatly reduced, altering expected aircraft performance. If ice is allowed to accumulate on the aircraft during flight, the weight of the aircraft is increased while the ability to generate lift is decreased. *[Figure 5-40]* As little as 0.8 millimeter of ice on the upper wing surface increases drag and reduces aircraft lift by 25 percent.

Pilots can encounter icing in any season, anywhere in the country, at altitudes of up to 18,000 feet and sometimes higher. Small aircraft, including commuter planes, are most vulnerable because they fly at lower altitudes where ice is more prevalent. They also lack mechanisms common on jet aircraft that prevent ice buildup by heating the front edges of wings.

Icing can occur in clouds any time the temperature drops below freezing and super-cooled droplets build up on an aircraft and freeze. (Super-cooled droplets are still liquid even though the temperature is below 32 °Fahrenheit (F), or 0 °Celsius (C).

Angle of Attack Indicators

The FAA along with the General Aviation Joint Steering Committee (GAJSC) is promoting AOA indicators as one of the many safety initiatives aimed at reducing the general aviation accident rate. AOA indicators will specifically target Loss of Control (LOC) accidents. Loss of control is the number one root cause of fatalities in both general aviation and commercial aviation. More than 25 percent of general aviation fatal accidents occur during the maneuvering phase of flight. Of those accidents, half involve stall/spin scenarios. Technology such as AOA indicators can have a tremendous impact on reversing this trend and are increasingly affordable for general aviation airplanes. *[Figure 5-41]*

The purpose of an AOA indicator is to give the pilot better situation awareness pertaining to the aerodynamic health

Figure 5-40. *Inflight ice formation.*

of the airfoil. This can also be referred to as stall margin awareness. More simply explained, it is the margin that exists between the current AOA that the airfoil is operating at, and the AOA at which the airfoil will stall (critical AOA).

Angle of attack is taught to student pilots as theory in ground training. When beginning flight training, students typically rely solely on airspeed and the published 1G stall speed to avoid stalls. This creates problems since this speed is only valid when the following conditions are met:

- Unaccelerated flight (a 1G load factor)
- Coordinated flight (inclinometer centered)
- At one weight (typically maximum gross weight)

Speed by itself is not a reliable parameter to avoid a stall. An airplane can stall at any speed. Angle of attack is a better parameter to use to avoid a stall. For a given configuration, the airplane always stalls at the same AOA, referred to as the critical AOA. This critical AOA does not change with:

- Weight
- Bank angle
- Temperature
- Density altitude
- Center of gravity

An AOA indicator can have several benefits when installed in general aviation aircraft, not the least of which is increased situational awareness. Without an AOA indicator, the AOA is "invisible" to pilots. These devices measure several parameters simultaneously and determine the current AOA providing a visual image to the pilot of the current AOA along with representations of the proximity to the critical AOA. [*Figure 5-42*] These devices can give a visual representation of the energy management state of the airplane. The energy

state of an airplane is the balance between airspeed, altitude, drag, and thrust and represents how efficiently the airfoil is operating. The more efficiently the airfoil operates; the larger stall margin that is present. With this increased situational awareness pertaining to the energy condition of the airplane, pilots will have information that they need to aid in preventing a LOC scenario resulting from a stall/spin. Additionally, the less energy that is utilized to maintain flight means greater overall efficiency of the airplane, which is typically realized in fuel savings. This equates to a lower operating cost to the pilot.

Just as training is required for any system on an aircraft, AOA indicators have training considerations also. A more comprehensive understanding of AOA in general should be the goal of this training along with the specific operating characteristics and limitations of the installed AOA indicator. Ground and flight instructors should make every attempt to receive training from an instructor knowledgeable about AOA indicators prior to giving instruction pertaining to or in airplanes equipped with AOA indicators. Pilot schools should incorporate training on AOA indicators in their syllabi, whether their training aircraft are equipped with them or not.

Installation of AOA indicators not required by type certification in general aviation airplanes has recently been streamlined by the FAA. The FAA established policy in February 2014 pertaining to non-required AOA systems and how they may be installed as a minor alteration, depending upon their installation requirements and operational utilization, and the procedures to take for certification of these installations. For updated information, reference the FAA website at www.faa.gov.

While AOA indicators provide a simple visual representation of the current AOA and its proximity to the critical AOA, they are not without their limitations. These limitations should be understood by operators of general aviation airplanes

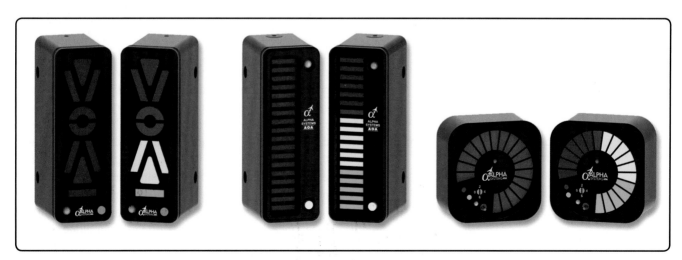

Figure 5-41. *A variety of AOA indicators.*

Coefficient of Lift Curve

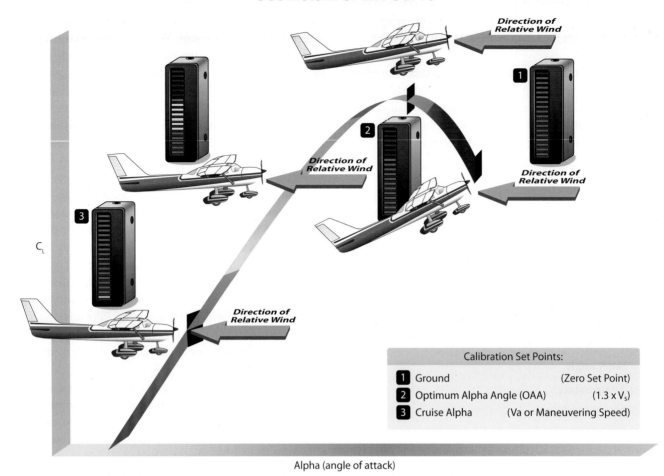

Calibration Set Points:

1	Ground	(Zero Set Point)
2	Optimum Alpha Angle (OAA)	(1.3 x V_s)
3	Cruise Alpha	(Va or Maneuvering Speed)

Alpha (angle of attack)

Figure 5-42. *An AOA indicator has several benefits when installed in general aviation aircraft.*

equipped with these devices. Like advanced automation, such as autopilots and moving maps, the misunderstanding or misuse of the equipment can have disastrous results. Some items which may limit the effectiveness of an AOA indicator are listed below:

- Calibration techniques
- Probes or vanes not being heated
- The type of indicator itself
- Flap setting
- Wing contamination

Pilots of general aviation airplanes equipped with AOA indicators should contact the manufacturer for specific limitations applicable to that installation.

Basic Propeller Principles

The aircraft propeller consists of two or more blades and a central hub to which the blades are attached. Each blade of an aircraft propeller is essentially a rotating wing. As a result of their construction, the propeller blades are like airfoils

and produce forces that create the thrust to pull, or push, the aircraft through the air. The engine furnishes the power needed to rotate the propeller blades through the air at high speeds, and the propeller transforms the rotary power of the engine into forward thrust.

A cross-section of a typical propeller blade is shown in *Figure 5-43*. This section or blade element is an airfoil comparable to a cross-section of an aircraft wing. One surface of the blade is cambered or curved, similar to the upper surface of an aircraft wing, while the other surface is flat like the bottom surface of a wing. The chord line is an imaginary line drawn through the blade from its leading edge to its trailing edge. As in a wing, the leading edge is the thick edge of the blade that meets the air as the propeller rotates. Blade angle, usually measured in degrees, is the angle between the chord of the blade and the plane of rotation and is measured at a specific point along the length of the blade. *[Figure 5-44]* Because most propellers have a flat blade "face," the chord line is often drawn along the face of the propeller blade. Pitch is not blade angle, but because pitch is largely determined by blade angle, the two terms are

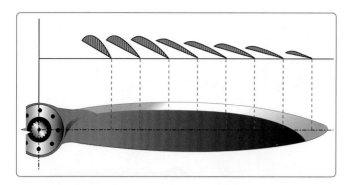

Figure 5-43. *Airfoil sections of propeller blade.*

often used interchangeably. An increase or decrease in one is usually associated with an increase or decrease in the other. The pitch of a propeller may be designated in inches. A propeller designated as a "74–48" would be 74 inches in length and have an effective pitch of 48 inches. The pitch is the distance in inches, which the propeller would screw through the air in one revolution if there were no slippage.

When specifying a fixed-pitch propeller for a new type of aircraft, the manufacturer usually selects one with a pitch that operates efficiently at the expected cruising speed of the aircraft. Every fixed-pitch propeller must be a compromise because it can be efficient at only a given combination of airspeed and revolutions per minute (rpm). Pilots cannot change this combination in flight.

When the aircraft is at rest on the ground with the engine operating, or moving slowly at the beginning of takeoff, the propeller efficiency is very low because the propeller is restrained from advancing with sufficient speed to permit its fixed-pitch blades to reach their full efficiency. In this situation, each propeller blade is turning through the air at an AOA that produces relatively little thrust for the amount of power required to turn it.

To understand the action of a propeller, consider first its motion, which is both rotational and forward. As shown by

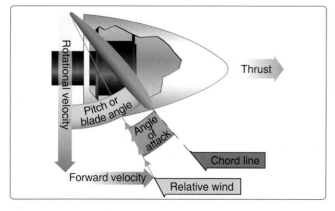

Figure 5-44. *Propeller blade angle.*

the vectors of propeller forces in *Figure 5-44*, each section of a propeller blade moves downward and forward. The angle at which this air (relative wind) strikes the propeller blade is its AOA. The air deflection produced by this angle causes the dynamic pressure at the engine side of the propeller blade to be greater than atmospheric pressure, thus creating thrust.

The shape of the blade also creates thrust because it is cambered like the airfoil shape of a wing. As the air flows past the propeller, the pressure on one side is less than that on the other. As in a wing, a reaction force is produced in the direction of the lesser pressure. The airflow over the wing has less pressure, and the force (lift) is upward. In the case of the propeller, which is mounted in a vertical instead of a horizontal plane, the area of decreased pressure is in front of the propeller, and the force (thrust) is in a forward direction. Aerodynamically, thrust is the result of the propeller shape and the AOA of the blade.

Thrust can be considered also in terms of the mass of air handled by the propeller. In these terms, thrust equals mass of air handled multiplied by slipstream velocity minus velocity of the aircraft. The power expended in producing thrust depends on the rate of air mass movement. On average, thrust constitutes approximately 80 percent of the torque (total horsepower absorbed by the propeller). The other 20 percent is lost in friction and slippage. For any speed of rotation, the horsepower absorbed by the propeller balances the horsepower delivered by the engine. For any single revolution of the propeller, the amount of air handled depends on the blade angle, which determines how big a "bite" of air the propeller takes. Thus, the blade angle is an excellent means of adjusting the load on the propeller to control the engine rpm.

The blade angle is also an excellent method of adjusting the AOA of the propeller. On constant-speed propellers, the blade angle must be adjusted to provide the most efficient AOA at all engine and aircraft speeds. Lift versus drag curves, which are drawn for propellers as well as wings, indicate that the most efficient AOA is small, varying from +2° to +4°. The actual blade angle necessary to maintain this small AOA varies with the forward speed of the aircraft.

Fixed-pitch and ground-adjustable propellers are designed for best efficiency at one rotation and forward speed. They are designed for a given aircraft and engine combination. A propeller may be used that provides the maximum efficiency for takeoff, climb, cruise, or high-speed flight. Any change in these conditions results in lowering the efficiency of both the propeller and the engine. Since the efficiency of any machine is the ratio of the useful power output to the actual power input, propeller efficiency is the ratio of thrust horsepower to brake horsepower. Propeller efficiency varies from 50 to

87 percent, depending on how much the propeller "slips." Propeller slip is the difference between the geometric pitch of the propeller and its effective pitch. *[Figure 5-45]* Geometric pitch is the theoretical distance a propeller should advance in one revolution; effective pitch is the distance it actually advances. Thus, geometric or theoretical pitch is based on no slippage, but actual or effective pitch includes propeller slippage in the air.

The reason a propeller is "twisted" is that the outer parts of the propeller blades, like all things that turn about a central point, travel faster than the portions near the hub. *[Figure 5-46]* If the blades had the same geometric pitch throughout their lengths, portions near the hub could have negative AOAs while the propeller tips would be stalled at cruise speed. Twisting or variations in the geometric pitch of the blades permits the propeller to operate with a relatively constant AOA along its length when in cruising flight. Propeller blades are twisted to change the blade angle in proportion to the differences in speed of rotation along the length of the propeller, keeping thrust more nearly equalized along this length.

Usually 1° to 4° provides the most efficient lift/drag ratio, but in flight the propeller AOA of a fixed-pitch propeller varies—normally from 0° to 15°. This variation is caused by changes in the relative airstream, which in turn results from changes in aircraft speed. Thus, propeller AOA is the product of two motions: propeller rotation about its axis and its forward motion.

A constant-speed propeller automatically keeps the blade angle adjusted for maximum efficiency for most conditions encountered in flight. During takeoff, when maximum power and thrust are required, the constant-speed propeller is at a low propeller blade angle or pitch. The low blade angle keeps the AOA small and efficient with respect to the relative wind. At the same time, it allows the propeller to handle a smaller mass of air per revolution. This light load allows the engine to turn at high rpm and to convert the maximum amount of fuel into heat energy in a given time. The high rpm also creates maximum thrust because, although the mass of air handled per revolution is small, the rpm and slipstream velocity are

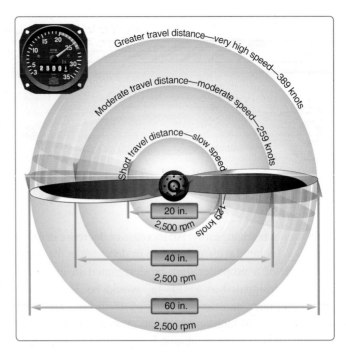

Figure 5-46. *Propeller tips travel faster than the hub.*

high, and with the low aircraft speed, there is maximum thrust. After liftoff, as the speed of the aircraft increases, the constant-speed propeller automatically changes to a higher angle (or pitch). Again, the higher blade angle keeps the AOA small and efficient with respect to the relative wind. The higher blade angle increases the mass of air handled per revolution. This decreases the engine rpm, reducing fuel consumption and engine wear, and keeps thrust at a maximum.

After the takeoff climb is established in an aircraft having a controllable-pitch propeller, the pilot reduces the power output of the engine to climb power by first decreasing the manifold pressure and then increasing the blade angle to lower the rpm.

At cruising altitude, when the aircraft is in level flight and less power is required than is used in takeoff or climb, the pilot again reduces engine power by reducing the manifold pressure and then increasing the blade angle to decrease the rpm. Again, this provides a torque requirement to match the reduced engine power. Although the mass of air handled per revolution is greater, it is more than offset by a decrease in slipstream velocity and an increase in airspeed. The AOA is still small because the blade angle has been increased with an increase in airspeed.

Torque and P-Factor

To the pilot, "torque" (the left turning tendency of the airplane) is made up of four elements that cause or produce a twisting or rotating motion around at least one of the airplane's three axes. These four elements are:

1. Torque reaction from engine and propeller

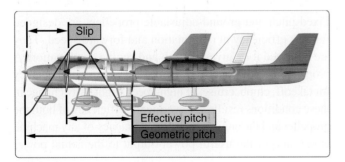

Figure 5-45. *Propeller slippage.*

2. Corkscrewing effect of the slipstream

3. Gyroscopic action of the propeller

4. Asymmetric loading of the propeller (P-factor)

Torque Reaction

Torque reaction involves Newton's Third Law of Physics—for every action, there is an equal and opposite reaction. As applied to the aircraft, this means that as the internal engine parts and propeller are revolving in one direction, an equal force is trying to rotate the aircraft in the opposite direction. [Figure 5-47]

When the aircraft is airborne, this force is acting around the longitudinal axis, tending to make the aircraft roll. To compensate for roll tendency, some of the older aircraft are rigged in a manner to create more lift on the wing that is being forced downward. The more modern aircraft are designed with the engine offset to counteract this effect of torque.

NOTE: Most United States built aircraft engines rotate the propeller clockwise, as viewed from the pilot's seat. The discussion here is with reference to those engines.

Generally, the compensating factors are permanently set so that they compensate for this force at cruising speed, since most of the aircraft's operating time is at that speed. However, aileron trim tabs permit further adjustment for other speeds. When the aircraft's wheels are on the ground during the takeoff roll, an additional turning moment around the vertical axis is induced by torque reaction. As the left side of the aircraft is being forced down by torque reaction, more weight is being placed on the left main landing gear. This results in more ground friction, or drag, on the left tire than on the right, causing a further turning moment to the left. The magnitude of this moment is dependent on many variables. Some of these variables are:

1. Size and horsepower of engine

2. Size of propeller and the rpm

3. Size of the aircraft

4. Condition of the ground surface

This yawing moment on the takeoff roll is corrected by the pilot's proper use of the rudder or rudder trim.

Corkscrew Effect

The high-speed rotation of an aircraft propeller gives a corkscrew or spiraling rotation to the slipstream. At high propeller speeds and low forward speed (as in the takeoffs and approaches to power-on stalls), this spiraling rotation is very compact and exerts a strong sideward force on the aircraft's vertical tail surface. [Figure 5-48]

When this spiraling slipstream strikes the vertical fin, it causes a yawing moment about the aircraft's vertical axis. The more compact the spiral, the more prominent this force is. As the forward speed increases, however, the spiral elongates and becomes less effective. The corkscrew flow of the slipstream also causes a rolling moment around the longitudinal axis.

Note that this rolling moment caused by the corkscrew flow of the slipstream is to the right, while the yawing moment caused by torque reaction is to the left—in effect one may be counteracting the other. However, these forces vary greatly and it is the pilot's responsibility to apply proper corrective action by use of the flight controls at all times. These forces must be counteracted regardless of which is the most prominent at the time.

Gyroscopic Action

Before the gyroscopic effects of the propeller can be understood, it is necessary to understand the basic principle of a gyroscope. All practical applications of the gyroscope are based upon two fundamental properties of gyroscopic action: rigidity in space and precession. The one of interest for this discussion is precession.

Precession is the resultant action, or deflection, of a spinning rotor when a deflecting force is applied to its rim. As can be seen in Figure 5-49, when a force is applied, the resulting force takes effect 90° ahead of and in the direction of rotation. The rotating propeller of an airplane makes a very good

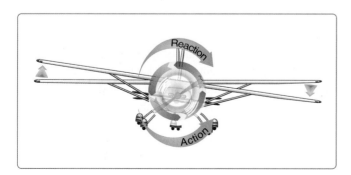

Figure 5-47. Torque reaction.

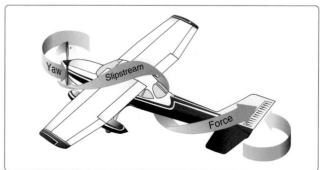

Figure 5-48. Corkscrewing slipstream.

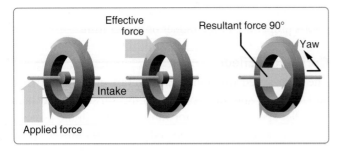

Figure 5-49. *Gyroscopic precession.*

gyroscope and thus has similar properties. Any time a force is applied to deflect the propeller out of its plane of rotation, the resulting force is 90° ahead of and in the direction of rotation and in the direction of application, causing a pitching moment, a yawing moment, or a combination of the two depending upon the point at which the force was applied.

This element of torque effect has always been associated with and considered more prominent in tailwheel-type aircraft and most often occurs when the tail is being raised during the takeoff roll. *[Figure 5-50]* This change in pitch attitude has the same effect as applying a force to the top of the propeller's plane of rotation. The resultant force acting 90° ahead causes a yawing moment to the left around the vertical axis. The magnitude of this moment depends on several variables, one of which is the abruptness with which the tail is raised (amount of force applied). However, precession, or gyroscopic action, occurs when a force is applied to any point on the rim of the propeller's plane of rotation; the resultant force will still be 90° from the point of application in the direction of rotation. Depending on where the force is applied, the airplane is caused to yaw left or right, to pitch up or down, or a combination of pitching and yawing.

It can be said that, as a result of gyroscopic action, any yawing around the vertical axis results in a pitching moment, and any pitching around the lateral axis results in a yawing moment. To correct for the effect of gyroscopic action, it is necessary for the pilot to properly use elevator and rudder to prevent undesired pitching and yawing.

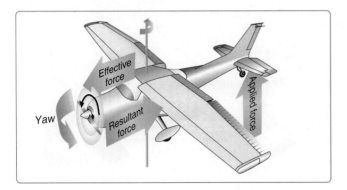

Figure 5-50. *Raising tail produces gyroscopic precession.*

Asymmetric Loading (P-Factor)

When an aircraft is flying with a high AOA, the "bite" of the downward moving blade is greater than the "bite" of the upward moving blade. This moves the center of thrust to the right of the prop disc area, causing a yawing moment toward the left around the vertical axis. Proving this explanation is complex because it would be necessary to work wind vector problems on each blade while considering both the AOA of the aircraft and the AOA of each blade.

This asymmetric loading is caused by the resultant velocity, which is generated by the combination of the velocity of the propeller blade in its plane of rotation and the velocity of the air passing horizontally through the propeller disc. With the aircraft being flown at positive AOAs, the right (viewed from the rear) or downswinging blade, is passing through an area of resultant velocity, which is greater than that affecting the left or upswinging blade. Since the propeller blade is an airfoil, increased velocity means increased lift. The downswinging blade has more lift and tends to pull (yaw) the aircraft's nose to the left.

When the aircraft is flying at a high AOA, the downward moving blade has a higher resultant velocity, creating more lift than the upward moving blade. *[Figure 5-51]* This might be easier to visualize if the propeller shaft was mounted perpendicular to the ground (like a helicopter). If there were no air movement at all, except that generated by the propeller itself, identical sections of each blade would have the same airspeed. With air moving horizontally across this vertically mounted propeller, the blade proceeding forward into the flow of air has a higher airspeed than the blade retreating with the airflow. Thus, the blade proceeding into the horizontal airflow is creating more lift, or thrust, moving the center of thrust toward that blade. Visualize rotating the vertically mounted propeller shaft to shallower angles relative to the moving air (as on an aircraft). This unbalanced thrust then becomes proportionately smaller and continues getting smaller until it reaches the value of zero when the propeller shaft is exactly horizontal in relation to the moving air.

The effects of each of these four elements of torque vary in value with changes in flight situations. In one phase of flight, one of these elements may be more prominent than another. In another phase of flight, another element may be more prominent. The relationship of these values to each other varies with different aircraft depending on the airframe, engine, and propeller combinations, as well as other design features. To maintain positive control of the aircraft in all flight conditions, the pilot must apply the flight controls as necessary to compensate for these varying values.

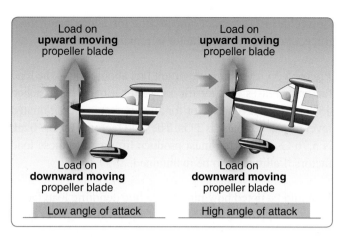

Figure 5-51. *Asymmetrical loading of propeller (P-factor).*

Load Factors

In aerodynamics, the maximum load factor (at given bank angle) is a proportion between lift and weight and has a trigonometric relationship. The load factor is measured in Gs (acceleration of gravity), a unit of force equal to the force exerted by gravity on a body at rest and indicates the force to which a body is subjected when it is accelerated. Any force applied to an aircraft to deflect its flight from a straight line produces a stress on its structure. The amount of this force is the load factor. While a course in aerodynamics is not a prerequisite for obtaining a pilot's license, the competent pilot should have a solid understanding of the forces that act on the aircraft, the advantageous use of these forces, and the operating limitations of the aircraft being flown.

For example, a load factor of 3 means the total load on an aircraft's structure is three times its weight. Since load factors are expressed in terms of Gs, a load factor of 3 may be spoken of as 3 Gs, or a load factor of 4 as 4 Gs.

If an aircraft is pulled up from a dive, subjecting the pilot to 3 Gs, he or she would be pressed down into the seat with a force equal to three times his or her weight. Since modern aircraft operate at significantly higher speeds than older aircraft, increasing the potential for large load factors, this effect has become a primary consideration in the design of the structure of all aircraft.

With the structural design of aircraft planned to withstand only a certain amount of overload, a knowledge of load factors has become essential for all pilots. Load factors are important for two reasons:

1. It is possible for a pilot to impose a dangerous overload on the aircraft structures.

2. An increased load factor increases the stalling speed and makes stalls possible at seemingly safe flight speeds.

Load Factors in Aircraft Design

The answer to the question "How strong should an aircraft be?" is determined largely by the use to which the aircraft is subjected. This is a difficult problem because the maximum possible loads are much too high for use in efficient design. It is true that any pilot can make a very hard landing or an extremely sharp pull up from a dive, which would result in abnormal loads. However, such extremely abnormal loads must be dismissed somewhat if aircraft are built that take off quickly, land slowly, and carry worthwhile payloads.

The problem of load factors in aircraft design becomes how to determine the highest load factors that can be expected in normal operation under various operational situations. These load factors are called "limit load factors." For reasons of safety, it is required that the aircraft be designed to withstand these load factors without any structural damage. Although the Code of Federal Regulations (CFR) requires the aircraft structure be capable of supporting one and one-half times these limit load factors without failure, it is accepted that parts of the aircraft may bend or twist under these loads and that some structural damage may occur.

This 1.5 load limit factor is called the "factor of safety" and provides, to some extent, for loads higher than those expected under normal and reasonable operation. This strength reserve is not something that pilots should willfully abuse; rather, it is there for protection when encountering unexpected conditions.

The above considerations apply to all loading conditions, whether they be due to gusts, maneuvers, or landings. The gust load factor requirements now in effect are substantially the same as those that have been in existence for years. Hundreds of thousands of operational hours have proven them adequate for safety. Since the pilot has little control over gust load factors (except to reduce the aircraft's speed when rough air is encountered), the gust loading requirements are substantially the same for most general aviation type aircraft regardless of their operational use. Generally, the gust load factors control the design of aircraft which are intended for strictly nonacrobatic usage.

An entirely different situation exists in aircraft design with maneuvering load factors. It is necessary to discuss this matter separately with respect to: (1) aircraft designed in accordance with the category system (i.e., normal, utility, acrobatic); and (2) older designs built according to requirements that did not provide for operational categories.

Aircraft designed under the category system are readily identified by a placard in the flight deck, which states the operational category (or categories) in which the aircraft is certificated. The maximum safe load factors (limit load factors) specified for aircraft in the various categories are:

CATEGORY	LIMIT LOAD FACTOR
Normal[1]	3.8 to −1.52
Utility (mild acrobatics, including spins)	4.4 to −1.76
Acrobatic	6.0 to −3.00

[1] For aircraft with gross weight of more than 4,000 pounds, the limit load factor is reduced. To the limit loads given above, a safety factor of 50 percent is added.

There is an upward graduation in load factor with the increasing severity of maneuvers. The category system provides for maximum utility of an aircraft. If normal operation alone is intended, the required load factor (and consequently the weight of the aircraft) is less than if the aircraft is to be employed in training or acrobatic maneuvers as they result in higher maneuvering loads.

Aircraft that do not have the category placard are designs that were constructed under earlier engineering requirements in which no operational restrictions were specifically given to the pilots. For aircraft of this type (up to weights of about 4,000 pounds), the required strength is comparable to present-day utility category aircraft, and the same types of operation are permissible. For aircraft of this type over 4,000 pounds, the load factors decrease with weight. These aircraft should be regarded as being comparable to the normal category aircraft designed under the category system, and they should be operated accordingly.

Load Factors in Steep Turns

At a constant altitude, during a coordinated turn in any aircraft, the load factor is the result of two forces: centrifugal force and weight. *[Figure 5-52]* For any given bank angle, the ROT varies with the airspeed—the higher the speed, the slower the ROT. This compensates for added centrifugal force, allowing the load factor to remain the same.

Figure 5-53 reveals an important fact about turns—the load factor increases at a terrific rate after a bank has reached 45° or 50°. The load factor for any aircraft in a coordinated level turn at 60° bank is 2 Gs. The load factor in an 80° bank is 5.76 Gs. The wing must produce lift equal to these load factors if altitude is to be maintained.

It should be noted how rapidly the line denoting load factor rises as it approaches the 90° bank line, which it never quite reaches because a 90° banked, constant altitude turn is not mathematically possible. An aircraft may be banked to 90° in a coordinated turn if not trying to hold altitude. An aircraft that can be held in a 90° banked slipping turn is capable of straight knife-edged flight. At slightly more than 80°, the load factor exceeds the limit of 6 Gs, the limit load factor of an acrobatic aircraft.

For a coordinated, constant altitude turn, the approximate maximum bank for the average general aviation aircraft is 60°. This bank and its resultant necessary power setting reach the limit of this type of aircraft. An additional 10° bank increases the load factor by approximately 1 G, bringing it close to the yield point established for these aircraft. *[Figure 5-54]*

Load Factors and Stalling Speeds

Any aircraft, within the limits of its structure, may be stalled at any airspeed. When a sufficiently high AOA is imposed, the smooth flow of air over an airfoil breaks up and separates, producing an abrupt change of flight characteristics and a sudden loss of lift, which results in a stall.

A study of this effect has revealed that an aircraft's stalling speed increases in proportion to the square root of the

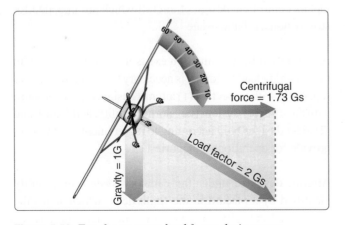

Figure 5-52. *Two forces cause load factor during turns.*

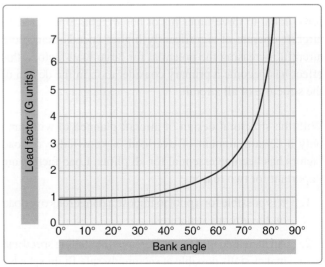

Figure 5-53. *Angle of bank changes load factor in level flight.*

load factor. This means that an aircraft with a normal unaccelerated stalling speed of 50 knots can be stalled at 100 knots by inducing a load factor of 4 Gs. If it were possible for this aircraft to withstand a load factor of nine, it could be stalled at a speed of 150 knots. A pilot should be aware of the following:

- The danger of inadvertently stalling the aircraft by increasing the load factor, as in a steep turn or spiral;

- When intentionally stalling an aircraft above its design maneuvering speed, a tremendous load factor is imposed.

Figures 5-53 and *5-54* show that banking an aircraft greater than 72° in a steep turn produces a load factor of 3, and the stalling speed is increased significantly. If this turn is made in an aircraft with a normal unaccelerated stalling speed of 45 knots, the airspeed must be kept greater than 75 knots to prevent inducing a stall. A similar effect is experienced in a quick pull up or any maneuver producing load factors above 1 G. This sudden, unexpected loss of control, particularly in a steep turn or abrupt application of the back elevator control near the ground, has caused many accidents.

Since the load factor is squared as the stalling speed doubles, tremendous loads may be imposed on structures by stalling an aircraft at relatively high airspeeds.

The following information primarily applies to fixed-wing airplanes. The maximum speed at which an airplane may be stalled safely is now determined for all new designs.

This speed is called the "design maneuvering speed" (V_A), which is the speed below which you can move a single flight control, one time, to its full deflection, for one axis of airplane rotation only (pitch, roll or yaw), in smooth air, without risk of damage to the airplane. V_A must be entered in the FAA-approved Airplane Flight Manual/Pilot's Operating Handbook (AFM/POH) of all recently designed airplanes. For older general aviation airplanes, this speed is approximately 1.7 times the normal stalling speed. Thus, an older airplane that normally stalls at 60 knots must never be stalled at above 102 knots (60 knots × 1.7 = 102 knots). An airplane with a normal stalling speed of 60 knots stalled at 102 knots undergoes a load factor equal to the square of the increase in speed, or 2.89 Gs (1.7 × 1.7 = 2.89 Gs). (The above figures are approximations to be considered as a guide, and are not the exact answers to any set of problems. The design maneuvering speed should be determined from the particular airplane's operating limitations provided by the manufacturer.) Operating at or below design maneuvering speed does not provide structural protection against multiple full control inputs in one axis or full control inputs in more than one axis at the same time.

Since the leverage in the control system varies with different aircraft (some types employ "balanced" control surfaces while others do not), the pressure exerted by the pilot on the controls cannot be accepted as an index of the load factors produced in different aircraft. In most cases, load factors can be judged by the experienced pilot from the feel of seat pressure. Load factors can also be measured by an instrument called an "accelerometer," but this instrument is not common in general

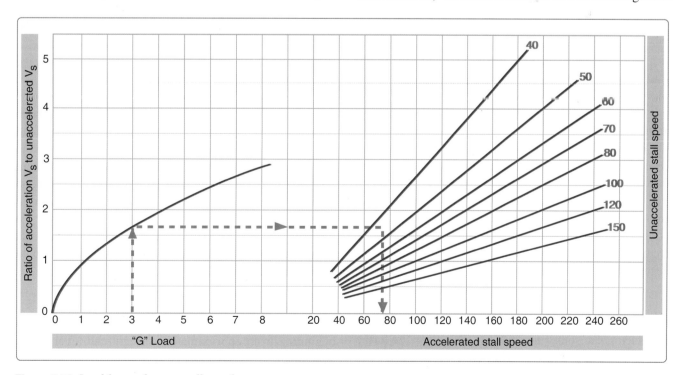

Figure 5-54. *Load factor changes stall speed.*

aviation training aircraft. The development of the ability to judge load factors from the feel of their effect on the body is important. A knowledge of these principles is essential to the development of the ability to estimate load factors.

A thorough knowledge of load factors induced by varying degrees of bank and the V_A aids in the prevention of two of the most serious types of accidents:

1. Stalls from steep turns or excessive maneuvering near the ground

2. Structural failures during acrobatics or other violent maneuvers resulting from loss of control

Load Factors and Flight Maneuvers

Critical load factors apply to all flight maneuvers except unaccelerated straight flight where a load factor of 1 G is always present. Certain maneuvers considered in this section are known to involve relatively high load factors. Full application of pitch, roll, or yaw controls should be confined to speeds below the maneuvering speed. Avoid rapid and large alternating control inputs, especially in combination with large changes in pitch, roll, or yaw (e.g., large sideslip angles) as they may result in structural failures at any speed, including below V_A.

Turns

Increased load factors are a characteristic of all banked turns. As noted in the section on load factors in steep turns, load factors become significant to both flight performance and load on wing structure as the bank increases beyond approximately 45°.

The yield factor of the average light plane is reached at a bank of approximately 70° to 75°, and the stalling speed is increased by approximately one-half at a bank of approximately 63°.

Stalls

The normal stall entered from straight-and-level flight, or an unaccelerated straight climb, does not produce added load factors beyond the 1 G of straight-and-level flight. As the stall occurs, however, this load factor may be reduced toward zero, the factor at which nothing seems to have weight. The pilot experiences a sensation of "floating free in space." If recovery is effected by snapping the elevator control forward, negative load factors (or those that impose a down load on the wings and raise the pilot from the seat) may be produced.

During the pull up following stall recovery, significant load factors are sometimes induced. These may be further increased inadvertently during excessive diving (and consequently high airspeed) and abrupt pull ups to level flight. One usually leads to the other, thus increasing the load

factor. Abrupt pull ups at high diving speeds may impose critical loads on aircraft structures and may produce recurrent or secondary stalls by increasing the AOA to that of stalling.

As a generalization, a recovery from a stall made by diving only to cruising or design maneuvering airspeed, with a gradual pull up as soon as the airspeed is safely above stalling, can be effected with a load factor not to exceed 2 or 2.5 Gs. A higher load factor should never be necessary unless recovery has been effected with the aircraft's nose near or beyond the vertical attitude or at extremely low altitudes to avoid diving into the ground.

Spins

A stabilized spin is not different from a stall in any element other than rotation and the same load factor considerations apply to spin recovery as apply to stall recovery. Since spin recoveries are usually effected with the nose much lower than is common in stall recoveries, higher airspeeds and consequently higher load factors are to be expected. The load factor in a proper spin recovery usually is found to be about 2.5 Gs.

The load factor during a spin varies with the spin characteristics of each aircraft, but is usually found to be slightly above the 1 G of level flight. There are two reasons for this:

1. Airspeed in a spin is very low, usually within 2 knots of the unaccelerated stalling speeds.

2. An aircraft pivots, rather than turns, while it is in a spin.

High Speed Stalls

The average light plane is not built to withstand the repeated application of load factors common to high speed stalls. The load factor necessary for these maneuvers produces a stress on the wings and tail structure, which does not leave a reasonable margin of safety in most light aircraft.

The only way this stall can be induced at an airspeed above normal stalling involves the imposition of an added load factor, which may be accomplished by a severe pull on the elevator control. A speed of 1.7 times stalling speed (about 102 knots in a light aircraft with a stalling speed of 60 knots) produces a load factor of 3 Gs. Only a very narrow margin for error can be allowed for acrobatics in light aircraft. To illustrate how rapidly the load factor increases with airspeed, a high-speed stall at 112 knots in the same aircraft would produce a load factor of 4 Gs.

Chandelles and Lazy Eights

A chandelle is a maximum performance climbing turn beginning from approximately straight-and-level flight, and ending at the completion of a precise 180° turn in a wings-level, nose-high attitude at the minimum controllable airspeed. In this flight maneuver, the aircraft is in a steep climbing turn and almost stalls to gain altitude while changing direction. A lazy eight derives its name from the manner in which the extended longitudinal axis of the aircraft is made to trace a flight pattern in the form of a figure "8" lying on its side. It would be difficult to make a definite statement concerning load factors in these maneuvers as both involve smooth, shallow dives and pull-ups. The load factors incurred depend directly on the speed of the dives and the abruptness of the pull-ups during these maneuvers.

Generally, the better the maneuver is performed, the less extreme the load factor induced. A chandelle or lazy eight in which the pull-up produces a load factor greater than 2 Gs will not result in as great a gain in altitude; in low-powered aircraft, it may result in a net loss of altitude.

The smoothest pull-up possible, with a moderate load factor, delivers the greatest gain in altitude in a chandelle and results in a better overall performance in both chandelles and lazy eights. The recommended entry speed for these maneuvers is generally near the manufacturer's design maneuvering speed, which allows maximum development of load factors without exceeding the load limits.

Rough Air

All standard certificated aircraft are designed to withstand loads imposed by gusts of considerable intensity. Gust load factors increase with increasing airspeed, and the strength used for design purposes usually corresponds to the highest level flight speed. In extremely rough air, as in thunderstorms or frontal conditions, it is wise to reduce the speed to the design maneuvering speed. Regardless of the speed held, there may be gusts that can produce loads that exceed the load limits.

Each specific aircraft is designed with a specific G loading that can be imposed on the aircraft without causing structural damage. There are two types of load factors factored into aircraft design: limit load and ultimate load. The limit load is a force applied to an aircraft that causes a bending of the aircraft structure that does not return to the original shape. The ultimate load is the load factor applied to the aircraft beyond the limit load and at which point the aircraft material experiences structural failure (breakage). Load factors lower than the limit load can be sustained without compromising the integrity of the aircraft structure.

Speeds up to, but not exceeding, the maneuvering speed allow an aircraft to stall prior to experiencing an increase in load factor that would exceed the limit load of the aircraft.

Most AFM/POH now include turbulent air penetration information, which help today's pilots safely fly aircraft capable of a wide range of speeds and altitudes. It is important for the pilot to remember that the maximum "never-exceed" placard dive speeds are determined for smooth air only. High speed dives or acrobatics involving speed above the known maneuvering speed should never be practiced in rough or turbulent air.

Vg Diagram

The flight operating strength of an aircraft is presented on a graph whose vertical scale is based on load factor. [Figure 5-55] The diagram is called a Vg diagram—velocity versus G loads or load factor. Each aircraft has its own Vg diagram that is valid at a certain weight and altitude.

The lines of maximum lift capability (curved lines) are the first items of importance on the Vg diagram. The aircraft in Figure 5-53 is capable of developing no more than +1 G at 64 mph, the wing level stall speed of the aircraft. Since the maximum load factor varies with the square of the airspeed, the maximum positive lift capability of this aircraft is 2 G at 92 mph, 3 G at 112 mph, 4.4 G at 137 mph, and so forth. Any load factor above this line is unavailable aerodynamically (i.e., the aircraft cannot fly above the line of maximum lift capability because it stalls). The same situation exists for negative lift flight with the exception that the speed necessary to produce a given negative load factor is higher than that to produce the same positive load factor.

If the aircraft is flown at a positive load factor greater than the positive limit load factor of 4.4, structural damage is possible. When the aircraft is operated in this region, objectionable permanent deformation of the primary structure may take place and a high rate of fatigue damage is incurred. Operation above the limit load factor must be avoided in normal operation.

There are two other points of importance on the Vg diagram. One point is the intersection of the positive limit load factor and the line of maximum positive lift capability. The airspeed at this point is the minimum airspeed at which the limit load can be developed aerodynamically. Any airspeed greater than this provides a positive lift capability sufficient to damage the aircraft. Conversely, any airspeed less than this does not provide positive lift capability sufficient to cause damage from excessive flight loads. The usual term given to this speed is "maneuvering speed," since consideration of subsonic aerodynamics would predict minimum usable turn radius or maneuverability to occur at this condition. The maneuver speed is a valuable reference point, since an aircraft operating below this point cannot produce a damaging positive flight load. Any combination of maneuver and gust cannot create damage due to excess airload when the aircraft is below the maneuver speed.

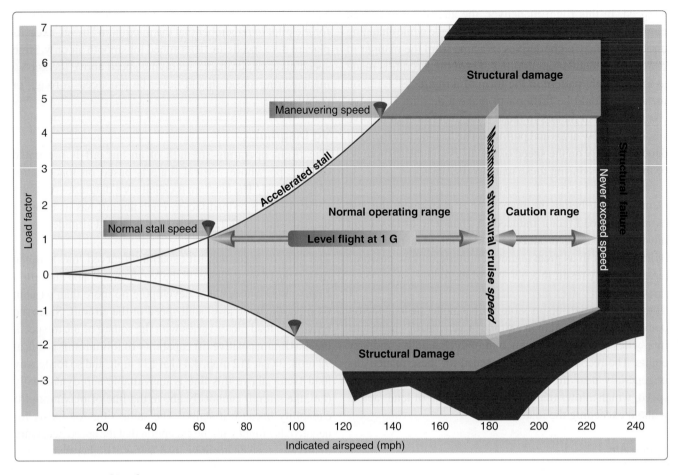

Figure 5-55. *Typical Vg diagram.*

The other point of importance on the Vg diagram is the intersection of the negative limit load factor and line of maximum negative lift capability. Any airspeed greater than this provides a negative lift capability sufficient to damage the aircraft; any airspeed less than this does not provide negative lift capability sufficient to damage the aircraft from excessive flight loads.

The limit airspeed (or redline speed) is a design reference point for the aircraft—this aircraft is limited to 225 mph. If flight is attempted beyond the limit airspeed, structural damage or structural failure may result from a variety of phenomena.

The aircraft in flight is limited to a regime of airspeeds and Gs that do not exceed the limit (or redline) speed, do not exceed the limit load factor, and cannot exceed the maximum lift capability. The aircraft must be operated within this "envelope" to prevent structural damage and ensure the anticipated service lift of the aircraft is obtained. The pilot must appreciate the Vg diagram as describing the allowable combination of airspeeds and load factors for safe operation. Any maneuver, gust, or gust plus maneuver outside the structural envelope can cause structural damage and effectively shorten the service life of the aircraft.

Rate of Turn

The rate of turn (ROT) is the number of degrees (expressed in degrees per second) of heading change that an aircraft makes. The ROT can be determined by taking the constant of 1,091, multiplying it by the tangent of any bank angle and dividing that product by a given airspeed in knots as illustrated in *Figure 5-55*. If the airspeed is increased and the ROT desired is to be constant, the angle of bank must be increased, otherwise, the ROT decreases. Likewise, if the airspeed is held constant, an aircraft's ROT increases if the bank angle is increased. The formula in *Figures 5-56* through *5-58* depicts the relationship between bank angle and airspeed as they affect the ROT.

NOTE: All airspeed discussed in this section is true airspeed (TAS).

Airspeed significantly effects an aircraft's ROT. If airspeed is increased, the ROT is reduced if using the same angle of bank used at the lower speed. Therefore, if airspeed is increased as illustrated in *Figure 5-57*, it can be inferred that the angle of bank must be increased in order to achieve the same ROT achieved in *Figure 5-58*.

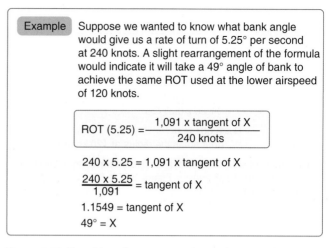

$$ROT = \frac{1{,}091 \times \text{tangent of the bank angle}}{\text{airspeed (in knots)}}$$

Example The rate of turn for an aircraft in a coordinated turn of 30° and traveling at 120 knots would have a ROT as follows.

$$ROT = \frac{1{,}091 \times \text{tangent of } 30°}{120 \text{ knots}}$$

$$ROT = \frac{1{,}091 \times 0.5773 \text{ (tangent of } 30°)}{120 \text{ knots}}$$

$$ROT = 5.25 \text{ degrees per second}$$

Figure 5-56. *Rate of turn for a given airspeed (knots, TAS) and bank angle.*

What does this mean on a practicable side? If a given airspeed and bank angle produces a specific ROT, additional conclusions can be made. Knowing the ROT is a given number of degrees of change per second, the number of seconds it takes to travel 360° (a circle) can be determined by simple division. For example, if moving at 120 knots with a 30° bank angle, the ROT is 5.25° per second and it takes 68.6 seconds (360° divided by 5.25 = 68.6 seconds) to make a complete circle. Likewise, if flying at 240 knots TAS and using a 30° angle of bank, the ROT is only about 2.63° per second and it takes about 137 seconds to complete a 360° circle. Looking at the formula, any increase in airspeed is directly proportional to the time the aircraft takes to travel an arc.

So why is this important to understand? Once the ROT is understood, a pilot can determine the distance required to make that particular turn, which is explained in radius of turn.

Radius of Turn

The radius of turn is directly linked to the ROT, which explained earlier is a function of both bank angle and airspeed. If the bank angle is held constant and the airspeed is increased, the radius of the turn changes (increases). A higher airspeed causes the aircraft to travel through a longer

Example Suppose we were to increase the speed to 240 knots, what is the ROT? Using the same formula from above we see that:

$$ROT = \frac{1{,}091 \times \text{tangent of } 30°}{240 \text{ knots}}$$

$$ROT = 2.62 \text{ degrees per second}$$

An increase in speed causes a decrease in the ROT when using the same bank angle.

Figure 5-57. *Rate of turn when increasing speed.*

Example Suppose we wanted to know what bank angle would give us a rate of turn of 5.25° per second at 240 knots. A slight rearrangement of the formula would indicate it will take a 49° angle of bank to achieve the same ROT used at the lower airspeed of 120 knots.

$$ROT (5.25) = \frac{1{,}091 \times \text{tangent of } X}{240 \text{ knots}}$$

$$240 \times 5.25 = 1{,}091 \times \text{tangent of } X$$

$$\frac{240 \times 5.25}{1{,}091} = \text{tangent of } X$$

$$1.1549 = \text{tangent of } X$$

$$49° = X$$

Figure 5-58. *To achieve the same rate of turn of an aircraft traveling at 120 knots, an increase of bank angle is required.*

arc due to a greater speed. An aircraft traveling at 120 knots is able to turn a 360° circle in a tighter radius than an aircraft traveling at 240 knots. In order to compensate for the increase in airspeed, the bank angle would need to be increased.

The radius of turn (R) can be computed using a simple formula. The radius of turn is equal to the velocity squared (V^2) divided by 11.26 times the tangent of the bank angle.

$$R = \frac{V^2}{11.26 \times \text{tangent of bank angle}}$$

Using the examples provided in *Figures 5-56* through *5-58*, the turn radius for each of the two speeds can be computed.

Note that if the speed is doubled, the radius is quadrupled. *[Figures 5-59 and 5-60]*

Another way to determine the radius of turn is speed using feet per second (fps), π (3.1415), and the ROT. In one of the previous examples, it was determined that an aircraft with a ROT of 5.25 degrees per second required 68.6 seconds to make a complete circle. An aircraft's speed (in knots) can

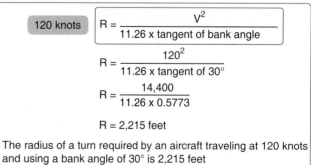

120 knots

$$R = \frac{V^2}{11.26 \times \text{tangent of bank angle}}$$

$$R = \frac{120^2}{11.26 \times \text{tangent of } 30°}$$

$$R = \frac{14{,}400}{11.26 \times 0.5773}$$

$$R = 2{,}215 \text{ feet}$$

The radius of a turn required by an aircraft traveling at 120 knots and using a bank angle of 30° is 2,215 feet

Figure 5-59. *Radius at 120 knots with bank angle of 30°.*

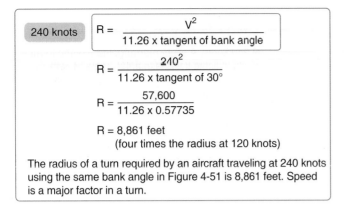

$$R = \frac{V^2}{11.26 \times \text{tangent of bank angle}}$$

240 knots

$$R = \frac{240^2}{11.26 \times \text{tangent of } 30°}$$

$$R = \frac{57,600}{11.26 \times 0.57735}$$

$$R = 8,861 \text{ feet}$$
(four times the radius at 120 knots)

The radius of a turn required by an aircraft traveling at 240 knots using the same bank angle in Figure 4-51 is 8,861 feet. Speed is a major factor in a turn.

Figure 5-60. *Radius at 240 knots.*

be converted to fps by multiplying it by a constant of 1.69. Therefore, an aircraft traveling at 120 knots (TAS) travels at 202.8 fps. Knowing the speed in fps (202.8) multiplied by the time an aircraft takes to complete a circle (68.6 seconds) can determine the size of the circle; 202.8 times 68.6 equals 13,912 feet. Dividing by π yields a diameter of 4,428 feet, which when divided by 2 equals a radius of 2,214 feet [*Figure 5-61*], a foot within that determined through use of the formula in *Figure 5-59*.

In *Figure 5-62*, the pilot enters a canyon and decides to turn 180° to exit. The pilot uses a 30° bank angle in his turn.

Weight and Balance

The aircraft's weight and balance data is important information for a pilot that must be frequently reevaluated. Although the aircraft was weighed during the certification process, this information is not valid indefinitely. Equipment changes or modifications affect the weight and balance data. Too often pilots reduce the aircraft weight and balance into a rule of thumb, such as: "If I have three passengers, I can load only 100 gallons of fuel; four passengers, 70 gallons."

Weight and balance computations should be part of every preflight briefing. Never assume three passengers are always

$$r = \frac{\text{speed (fps)} \times \dfrac{360}{ROT}}{\dfrac{Pi (\pi)}{2}}$$

$$r = \frac{\dfrac{202.8 \times 68.6}{\pi}}{2}$$

$$r = \frac{\dfrac{13,912}{\pi}}{2}$$

$$r = \frac{4,428}{2} = 2,214 \text{ feet}$$

Figure 5-61. *Another formula that can be used for radius.*

of equal weight. Instead, do a full computation of all items to be loaded on the aircraft, including baggage, as well as the pilot and passenger. It is recommended that all bags be weighed to make a precise computation of how the aircraft CG is positioned.

The importance of the CG was stressed in the discussion of stability, controllability, and performance. Unequal load distribution causes accidents. A competent pilot understands and respects the effects of CG on an aircraft.

Weight and balance are critical components in the utilization of an aircraft to its fullest potential. The pilot must know how much fuel can be loaded onto the aircraft without violating CG limits, as well as weight limits to conduct long or short flights with or without a full complement of allowable passengers. For example, an aircraft has four seats and can carry 60 gallons of fuel. How many passengers can the aircraft safely carry? Can all those seats be occupied at all times with the varying fuel loads? Four people who each weigh 150 pounds leads to a different weight and balance computation than four people who each weigh 200 pounds. The second scenario loads an additional 200 pounds onto the aircraft and is equal to about 30 gallons of fuel.

The additional weight may or may not place the CG outside of the CG envelope, but the maximum gross weight could be exceeded. The excess weight can overstress the aircraft and degrade the performance.

Aircraft are certificated for weight and balance for two principal reasons:

1. The effect of the weight on the aircraft's primary structure and its performance characteristics

2. The effect of the location of this weight on flight characteristics, particularly in stall and spin recovery and stability

Aircraft, such as balloons and weight-shift control, do not require weight and balance computations because the load is suspended below the lifting mechanism. The CG range in these types of aircraft is such that it is difficult to exceed loading limits. For example, the rear seat position and fuel of a weight-shift control aircraft are as close as possible to the hang point with the aircraft in a suspended attitude. Thus, load variations have little effect on the CG. This also holds true for the balloon basket or gondola. While it is difficult to exceed CG limits in these aircraft, pilots should never overload an aircraft because overloading causes structural damage and failures. Weight and balance computations are not required, but pilots should calculate weight and remain within the manufacturer's established limit.

120 knots

4,430 feet

2,215 feet

5,000 feet

$$R = \frac{V^2}{11.26 \times \text{tangent of the bank angle } 30°}$$

$$R = \frac{120^2}{11.26 \times 0.5773} = \frac{14,400}{6.50096} = 2,215 \text{ feet}$$

140 knots

6,028 feet

3,014 feet

5,000 feet

$$R = \frac{V^2}{11.26 \times \text{tangent of the bank angle } 30°}$$

$$R = \frac{140^2}{11.26 \times 0.5773} = \frac{19,600}{6.50096} = 3,014 \text{ feet}$$

Figure 5-62. *Two aircraft have flown into a canyon by error. The canyon is 5,000 feet across and has sheer cliffs on both sides. The pilot in the top image is flying at 120 knots. After realizing the error, the pilot banks hard and uses a 30° bank angle to reverse course. This aircraft requires about 4,000 feet to turn 180°, and makes it out of the canyon safely. The pilot in the bottom image is flying at 140 knots and also uses a 30° angle of bank in an attempt to reverse course. The aircraft, although flying just 20 knots faster than the aircraft in the top image, requires over 6,000 feet to reverse course to safety. Unfortunately, the canyon is only 5,000 feet across and the aircraft will hit the canyon wall. The point is that airspeed is the most influential factor in determining how much distance is required to turn. Many pilots have made the error of increasing the steepness of their bank angle when a simple reduction of speed would have been more appropriate.*

Effect of Weight on Flight Performance

The takeoff/climb and landing performance of an aircraft are determined on the basis of its maximum allowable takeoff and landing weights. A heavier gross weight results in a longer takeoff run and shallower climb, and a faster touchdown speed and longer landing roll. Even a minor overload may make it impossible for the aircraft to clear an obstacle that normally would not be a problem during takeoff under more favorable conditions.

The detrimental effects of overloading on performance are not limited to the immediate hazards involved with takeoffs and landings. Overloading has an adverse effect on all climb and cruise performance, which leads to overheating during climbs, added wear on engine parts, increased fuel consumption, slower cruising speeds, and reduced range.

The manufacturers of modern aircraft furnish weight and balance data with each aircraft produced. Generally, this information may be found in the FAA-approved AFM/POH and easy-to-read charts for determining weight and balance data are now provided. Increased performance and load-carrying capability of these aircraft require strict adherence to the operating limitations prescribed by the manufacturer. Deviations from the recommendations can result in structural damage or complete failure of the aircraft's structure. Even if an aircraft is loaded well within the maximum weight limitations, it is imperative that weight distribution be within the limits of CG location. The preceding brief study of aerodynamics and load factors points out the reasons for this precaution. The following discussion is background information into some of the reasons why weight and balance conditions are important to the safe flight of an aircraft.

In some aircraft, it is not possible to fill all seats, baggage compartments, and fuel tanks, and still remain within approved weight or balance limits. For example, in several popular four-place aircraft, the fuel tanks may not be filled to capacity when four occupants and their baggage are carried. In a certain two-place aircraft, no baggage may be carried in the compartment aft of the seats when spins are to be practiced. It is important for a pilot to be aware of the weight and balance limitations of the aircraft being flown and the reasons for these limitations.

Effect of Weight on Aircraft Structure

The effect of additional weight on the wing structure of an aircraft is not readily apparent. Airworthiness requirements prescribe that the structure of an aircraft certificated in the normal category (in which acrobatics are prohibited) must be strong enough to withstand a load factor of 3.8 Gs to take care of dynamic loads caused by maneuvering and gusts. This means that the primary structure of the aircraft can withstand a load of 3.8 times the approved gross weight of the aircraft without structural failure occurring. If this is accepted as indicative of the load factors that may be imposed during operations for which the aircraft is intended, a 100-pound overload imposes a potential structural overload of 380 pounds. The same consideration is even more impressive in the case of utility and acrobatic category aircraft, which have load factor requirements of 4.4 and 6.0, respectively.

Structural failures that result from overloading may be dramatic and catastrophic, but more often they affect structural components progressively in a manner that is difficult to detect and expensive to repair. Habitual overloading tends to cause cumulative stress and damage that may not be detected during preflight inspections and result in structural failure later during completely normal operations. The additional stress placed on structural parts by overloading is believed to accelerate the occurrence of metallic fatigue failures.

A knowledge of load factors imposed by flight maneuvers and gusts emphasizes the consequences of an increase in the gross weight of an aircraft. The structure of an aircraft about to undergo a load factor of 3 Gs, as in recovery from a steep dive, must be prepared to withstand an added load of 300 pounds for each 100-pound increase in weight. It should be noted that this would be imposed by the addition of about 16 gallons of unneeded fuel in a particular aircraft. FAA-certificated civil aircraft have been analyzed structurally and tested for flight at the maximum gross weight authorized and within the speeds posted for the type of flights to be performed. Flights at weights in excess of this amount are quite possible and often are well within the performance capabilities of an aircraft. This fact should not mislead the pilot, as the pilot may not realize that loads for which the aircraft was not designed are being imposed on all or some part of the structure.

In loading an aircraft with either passengers or cargo, the structure must be considered. Seats, baggage compartments, and cabin floors are designed for a certain load or concentration of load and no more. For example, a light plane baggage compartment may be placarded for 20 pounds because of the limited strength of its supporting structure even though the aircraft may not be overloaded or out of CG limits with more weight at that location.

Effect of Weight on Stability and Controllability

Overloading also affects stability. An aircraft that is stable and controllable when loaded normally may have very different flight characteristics when overloaded. Although the distribution of weight has the most direct effect on this, an increase in the aircraft's gross weight may be expected to have an adverse effect on stability, regardless of location of the CG. The stability of many certificated aircraft is completely unsatisfactory if the gross weight is exceeded.

Effect of Load Distribution

The effect of the position of the CG on the load imposed on an aircraft's wing in flight is significant to climb and cruising performance. An aircraft with forward loading is "heavier" and consequently, slower than the same aircraft with the CG further aft.

Figure 5-63 illustrates why this is true. With forward loading, "nose-up" trim is required in most aircraft to maintain level cruising flight. Nose-up trim involves setting the tail surfaces to produce a greater down load on the aft portion of the fuselage, which adds to the wing loading and the total lift required from the wing if altitude is to be maintained. This requires a higher AOA of the wing, which results in more drag and, in turn, produces a higher stalling speed.

With aft loading and "nose-down" trim, the tail surfaces exert less down load, relieving the wing of that much wing loading and lift required to maintain altitude. The required AOA of the wing is less, so the drag is less, allowing for a faster cruise speed. Theoretically, a neutral load on the tail surfaces in cruising flight would produce the most efficient overall performance and fastest cruising speed, but it would also result in instability. Modern aircraft are designed to require a down load on the tail for stability and controllability. A zero indication on the trim tab control is not necessarily the same as "neutral trim" because of the force exerted by downwash from the wings and the fuselage on the tail surfaces.

The effects of the distribution of the aircraft's useful load have a significant influence on its flight characteristics, even when the load is within the CG limits and the maximum

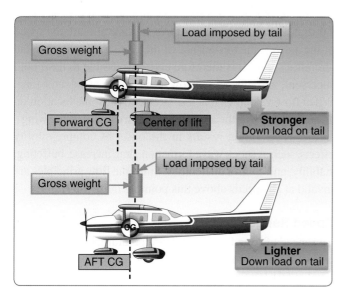

Figure 5-63. *Effect of load distribution on balance.*

permissible gross weight. Important among these effects are changes in controllability, stability, and the actual load imposed on the wing.

Generally, an aircraft becomes less controllable, especially at slow flight speeds, as the CG is moved further aft. An aircraft that cleanly recovers from a prolonged spin with the CG at one position may fail completely to respond to normal recovery attempts when the CG is moved aft by one or two inches.

It is common practice for aircraft designers to establish an aft CG limit that is within one inch of the maximum, which allows normal recovery from a one-turn spin. When certificating an aircraft in the utility category to permit intentional spins, the aft CG limit is usually established at a point several inches forward of that permissible for certification in the normal category.

Another factor affecting controllability, which has become more important in current designs of large aircraft, is the effect of long moment arms to the positions of heavy equipment and cargo. The same aircraft may be loaded to maximum gross weight within its CG limits by concentrating fuel, passengers, and cargo near the design CG, or by dispersing fuel and cargo loads in wingtip tanks and cargo bins forward and aft of the cabin.

With the same total weight and CG, maneuvering the aircraft or maintaining level flight in turbulent air requires the application of greater control forces when the load is dispersed. The longer moment arms to the positions of the heavy fuel and cargo loads must be overcome by the action of the control surfaces. An aircraft with full outboard wing tanks or tip tanks tends to be sluggish in roll when control situations are marginal, while one with full nose and aft cargo bins tends to be less responsive to the elevator controls.

The rearward CG limit of an aircraft is determined largely by considerations of stability. The original airworthiness requirements for a type certificate specify that an aircraft in flight at a certain speed dampens out vertical displacement of the nose within a certain number of oscillations. An aircraft loaded too far rearward may not do this. Instead, when the nose is momentarily pulled up, it may alternately climb and dive becoming steeper with each oscillation. This instability is not only uncomfortable to occupants, but it could even become dangerous by making the aircraft unmanageable under certain conditions.

The recovery from a stall in any aircraft becomes progressively more difficult as its CG moves aft. This is particularly important in spin recovery, as there is a point in rearward loading of any aircraft at which a "flat" spin develops. A flat spin is one in which centrifugal force, acting through a CG located well to the rear, pulls the tail of the aircraft out away from the axis of the spin, making it impossible to get the nose down and recover.

An aircraft loaded to the rear limit of its permissible CG range handles differently in turns and stall maneuvers and has different landing characteristics than when it is loaded near the forward limit.

The forward CG limit is determined by a number of considerations. As a safety measure, it is required that the trimming device, whether tab or adjustable stabilizer, be capable of holding the aircraft in a normal glide with the power off. A conventional aircraft must be capable of a full stall, power-off landing in order to ensure minimum landing speed in emergencies. A tailwheel-type aircraft loaded excessively nose-heavy is difficult to taxi, particularly in high winds. It can be nosed over easily by use of the brakes, and it is difficult to land without bouncing since it tends to pitch down on the wheels as it is slowed down and flared for landing. Steering difficulties on the ground may occur in nosewheel-type aircraft, particularly during the landing roll and takeoff. The effects of load distribution are summarized as follows:

- The CG position influences the lift and AOA of the wing, the amount and direction of force on the tail, and the degree of deflection of the stabilizer needed to supply the proper tail force for equilibrium. The latter is very important because of its relationship to elevator control force.

- The aircraft stalls at a higher speed with a forward CG location. This is because the stalling AOA is reached at a higher speed due to increased wing loading.

- Higher elevator control forces normally exist with a forward CG location due to the increased stabilizer deflection required to balance the aircraft.

- The aircraft cruises faster with an aft CG location because of reduced drag. The drag is reduced because a smaller AOA and less downward deflection of the stabilizer are required to support the aircraft and overcome the nose-down pitching tendency.

- The aircraft becomes less stable as the CG is moved rearward. This is because when the CG is moved rearward, it causes a decrease in the AOA. Therefore, the wing contribution to the aircraft's stability is now decreased, while the tail contribution is still stabilizing. When the point is reached that the wing and tail contributions balance, then neutral stability exists. Any CG movement further aft results in an unstable aircraft.

- A forward CG location increases the need for greater back elevator pressure. The elevator may no longer be able to oppose any increase in nose-down pitching. Adequate elevator control is needed to control the aircraft throughout the airspeed range down to the stall.

A detailed discussion and additional information relating to weight and balance can be found in Chapter 10, Weight and Balance.

High Speed Flight

Subsonic Versus Supersonic Flow

In subsonic aerodynamics, the theory of lift is based upon the forces generated on a body and a moving gas (air) in which it is immersed. At speeds of approximately 260 knots or less, air can be considered incompressible in that, at a fixed altitude, its density remains nearly constant while its pressure varies. Under this assumption, air acts the same as water and is classified as a fluid. Subsonic aerodynamic theory also assumes the effects of viscosity (the property of a fluid that tends to prevent motion of one part of the fluid with respect to another) are negligible and classifies air as an ideal fluid conforming to the principles of ideal-fluid aerodynamics such as continuity, Bernoulli's principle, and circulation.

In reality, air is compressible and viscous. While the effects of these properties are negligible at low speeds, compressibility effects in particular become increasingly important as speed increases. Compressibility (and to a lesser extent viscosity) is of paramount importance at speeds approaching the speed of sound. In these speed ranges, compressibility causes a change in the density of the air around an aircraft.

During flight, a wing produces lift by accelerating the airflow over the upper surface. This accelerated air can, and does, reach sonic speeds even though the aircraft itself may be flying subsonic. At some extreme AOAs, in some aircraft, the speed of the air over the top surface of the wing may be double the aircraft's speed. It is therefore entirely possible to have both supersonic and subsonic airflow on an aircraft at the same time. When flow velocities reach sonic speeds at some location on an aircraft (such as the area of maximum camber on the wing), further acceleration results in the onset of compressibility effects, such as shock wave formation, drag increase, buffeting, stability, and control difficulties. Subsonic flow principles are invalid at all speeds above this point. *[Figure 5-64]*

Speed Ranges

The speed of sound varies with temperature. Under standard temperature conditions of 15 °C, the speed of sound at sea level is 661 knots. At 40,000 feet, where the temperature is –55 °C, the speed of sound decreases to 574 knots. In high-

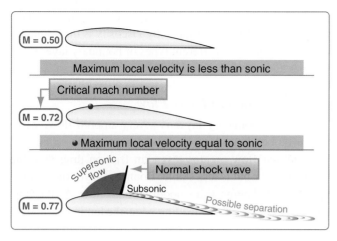

Figure 5-64. *Wing airflow.*

speed flight and/or high-altitude flight, the measurement of speed is expressed in terms of a "Mach number"—the ratio of the true airspeed of the aircraft to the speed of sound in the same atmospheric conditions. An aircraft traveling at the speed of sound is traveling at Mach 1.0. Aircraft speed regimes are defined approximately as follows:

Subsonic—Mach numbers below 0.75

Transonic—Mach numbers from 0.75 to 1.20

Supersonic—Mach numbers from 1.20 to 5.00

Hypersonic—Mach numbers above 5.00

While flights in the transonic and supersonic ranges are common occurrences for military aircraft, civilian jet aircraft normally operate in a cruise speed range of Mach 0.7 to Mach 0.90.

The speed of an aircraft in which airflow over any part of the aircraft or structure under consideration first reaches (but does not exceed) Mach 1.0 is termed "critical Mach number" or "Mach Crit." Thus, critical Mach number is the boundary between subsonic and transonic flight and is largely dependent on the wing and airfoil design. Critical Mach number is an important point in transonic flight. When shock waves form on the aircraft, airflow separation followed by buffet and aircraft control difficulties can occur. Shock waves, buffet, and airflow separation take place above critical Mach number. A jet aircraft typically is most efficient when cruising at or near its critical Mach number. At speeds 5–10 percent above the critical Mach number, compressibility effects begin. Drag begins to rise sharply. Associated with the "drag rise" are buffet, trim, and stability changes and a decrease in control surface effectiveness. This is the point of "drag divergence." *[Figure 5-65]*

V_{MO}/M_{MO} is defined as the maximum operating limit speed. V_{MO} is expressed in knots calibrated airspeed (KCAS), while

M_{MO} is expressed in Mach number. The V_{MO} limit is usually associated with operations at lower altitudes and deals with structural loads and flutter. The M_{MO} limit is associated with operations at higher altitudes and is usually more concerned with compressibility effects and flutter. At lower altitudes, structural loads and flutter are of concern; at higher altitudes, compressibility effects and flutter are of concern.

Adherence to these speeds prevents structural problems due to dynamic pressure or flutter, degradation in aircraft control response due to compressibility effects (e.g., Mach Tuck, aileron reversal, or buzz), and separated airflow due to shock waves resulting in loss of lift or vibration and buffet. Any of these phenomena could prevent the pilot from being able to adequately control the aircraft.

For example, an early civilian jet aircraft had a V_{MO} limit of 306 KCAS up to approximately FL 310 (on a standard day). At this altitude (FL 310), an M_{MO} of 0.82 was approximately equal to 306 KCAS. Above this altitude, an M_{MO} of 0.82 always equaled a KCAS less than 306 KCAS and, thus, became the operating limit as you could not reach the V_{MO} limit without first reaching the M_{MO} limit. For example, at FL 380, an M_{MO} of 0.82 is equal to 261 KCAS.

Mach Number Versus Airspeed

It is important to understand how airspeed varies with Mach number. As an example, consider how the stall speed of a jet transport aircraft varies with an increase in altitude. The increase in altitude results in a corresponding drop in air density and outside temperature. Suppose this jet transport is in the clean configuration (gear and flaps up) and weighs 550,000 pounds. The aircraft might stall at approximately 152 KCAS at sea level. This is equal to (on a standard day) a true velocity of 152 KTAS and a Mach number of 0.23. At FL 380, the aircraft will still stall at approximately 152 KCAS, but the true velocity is about 287 KTAS with a Mach number of 0.50.

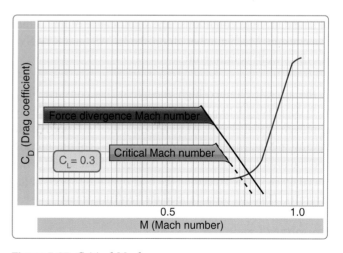

Figure 5-65. *Critical Mach.*

Although the stalling speed has remained the same for our purposes, both the Mach number and TAS have increased. With increasing altitude, the air density has decreased; this requires a faster true airspeed in order to have the same pressure sensed by the pitot tube for the same KCAS, or KIAS (for our purposes, KCAS and KIAS are relatively close to each other). The dynamic pressure the wing experiences at FL 380 at 287 KTAS is the same as at sea level at 152 KTAS. However, it is flying at higher Mach number.

Another factor to consider is the speed of sound. A decrease in temperature in a gas results in a decrease in the speed of sound. Thus, as the aircraft climbs in altitude with outside temperature dropping, the speed of sound is dropping. At sea level, the speed of sound is approximately 661 KCAS, while at FL 380 it is 574 KCAS. Thus, for our jet transport aircraft, the stall speed (in KTAS) has gone from 152 at sea level to 287 at FL 380. Simultaneously, the speed of sound (in KCAS) has decreased from 661 to 574 and the Mach number has increased from 0.23 (152 KTAS divided by 661 KTAS) to 0.50 (287 KTAS divided by 574 KTAS). All the while, the KCAS for stall has remained constant at 152. This describes what happens when the aircraft is at a constant KCAS with increasing altitude, but what happens when the pilot keeps Mach constant during the climb? In normal jet flight operations, the climb is at 250 KIAS (or higher (e.g. heavy)) to 10,000 feet and then at a specified en route climb airspeed (about 330 if a DC10) until reaching an altitude in the "mid-twenties" where the pilot then climbs at a constant Mach number to cruise altitude.

Assuming for illustration purposes that the pilot climbs at a M_{MO} of 0.82 from sea level up to FL 380. KCAS goes from 543 to 261. The KIAS at each altitude would follow the same behavior and just differ by a few knots. Recall from the earlier discussion that the speed of sound is decreasing with the drop in temperature as the aircraft climbs. The Mach number is simply the ratio of the true airspeed to the speed of sound at flight conditions. The significance of this is that at a constant Mach number climb, the KCAS (and KTAS or KIAS as well) is falling off.

If the aircraft climbed high enough at this constant M_{MO} with decreasing KIAS, KCAS, and KTAS, it would begin to approach its stall speed. At some point, the stall speed of the aircraft in Mach number could equal the M_{MO} of the aircraft, and the pilot could neither slow down (without stalling) nor speed up (without exceeding the max operating speed of the aircraft). This has been dubbed the "coffin corner."

Boundary Layer

The viscous nature of airflow reduces the local velocities on a surface and is responsible for skin friction. As discussed earlier in the chapter, the layer of air over the wing's surface that is slowed down or stopped by viscosity is the boundary layer. There are two different types of boundary layer flow: laminar and turbulent.

Laminar Boundary Layer Flow

The laminar boundary layer is a very smooth flow, while the turbulent boundary layer contains swirls or eddies. The laminar flow creates less skin friction drag than the turbulent flow but is less stable. Boundary layer flow over a wing surface begins as a smooth laminar flow. As the flow continues back from the leading edge, the laminar boundary layer increases in thickness.

Turbulent Boundary Layer Flow

At some distance back from the leading edge, the smooth laminar flow breaks down and transitions to a turbulent flow. From a drag standpoint, it is advisable to have the transition from laminar to turbulent flow as far aft on the wing as possible or have a large amount of the wing surface within the laminar portion of the boundary layer. The low energy laminar flow, however, tends to break down more suddenly than the turbulent layer.

Boundary Layer Separation

Another phenomenon associated with viscous flow is separation. Separation occurs when the airflow breaks away from an airfoil. The natural progression is from laminar boundary layer to turbulent boundary layer and then to airflow separation. Airflow separation produces high drag and ultimately destroys lift. The boundary layer separation point moves forward on the wing as the AOA is increased. [Figure 5-66]

Vortex generators are used to delay or prevent shock wave induced boundary layer separation encountered in transonic flight. They are small low aspect ratio airfoils placed at a 12° to 15° AOA to the airstream. Usually spaced a few inches apart along the wing ahead of the ailerons or other control surfaces, vortex generators create a vortex that mixes the boundary airflow with the high energy airflow just above the surface. This produces higher surface velocities and increases the energy of the boundary layer. Thus, a stronger shock wave is necessary to produce airflow separation.

Shock Waves

When an airplane flies at subsonic speeds, the air ahead is "warned" of the airplane's coming by a pressure change transmitted ahead of the airplane at the speed of sound. Because of this warning, the air begins to move aside before the airplane arrives and is prepared to let it pass easily. When the airplane's speed reaches the speed of sound, the pressure

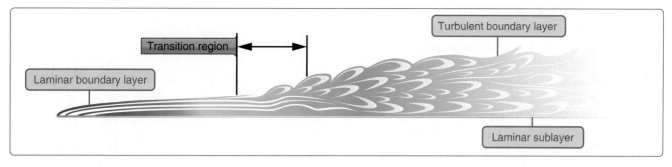

Figure 5-66. *Boundary layer.*

change can no longer warn the air ahead because the airplane is keeping up with its own pressure waves. Rather, the air particles pile up in front of the airplane causing a sharp decrease in the flow velocity directly in front of the airplane with a corresponding increase in air pressure and density.

As the airplane's speed increases beyond the speed of sound, the pressure and density of the compressed air ahead of it increase, the area of compression extending some distance ahead of the airplane. At some point in the airstream, the air particles are completely undisturbed, having had no advanced warning of the airplane's approach, and in the next instant the same air particles are forced to undergo sudden and drastic changes in temperature, pressure, density, and velocity. The boundary between the undisturbed air and the region of compressed air is called a shock or "compression" wave. This same type of wave is formed whenever a supersonic airstream is slowed to subsonic without a change in direction, such as when the airstream is accelerated to sonic speed over the cambered portion of a wing, and then decelerated to subsonic speed as the area of maximum camber is passed. A shock wave forms as a boundary between the supersonic and subsonic ranges.

Whenever a shock wave forms perpendicular to the airflow, it is termed a "normal" shock wave, and the flow immediately behind the wave is subsonic. A supersonic airstream passing through a normal shock wave experiences these changes:

- The airstream is slowed to subsonic.

- The airflow immediately behind the shock wave does not change direction.

- The static pressure and density of the airstream behind the wave is greatly increased.

- The energy of the airstream (indicated by total pressure—dynamic plus static) is greatly reduced.

Shock wave formation causes an increase in drag. One of the principal effects of a shock wave is the formation of a dense high pressure region immediately behind the wave. The instability of the high pressure region, and the fact that

part of the velocity energy of the airstream is converted to heat as it flows through the wave, is a contributing factor in the drag increase, but the drag resulting from airflow separation is much greater. If the shock wave is strong, the boundary layer may not have sufficient kinetic energy to withstand airflow separation. The drag incurred in the transonic region due to shock wave formation and airflow separation is known as "wave drag." When speed exceeds the critical Mach number by about 10 percent, wave drag increases sharply. A considerable increase in thrust (power) is required to increase flight speed beyond this point into the supersonic range where, depending on the airfoil shape and the AOA, the boundary layer may reattach.

Normal shock waves form on the wing's upper surface and form an additional area of supersonic flow and a normal shock wave on the lower surface. As flight speed approaches the speed of sound, the areas of supersonic flow enlarge and the shock waves move nearer the trailing edge. *[Figure 5-67]*

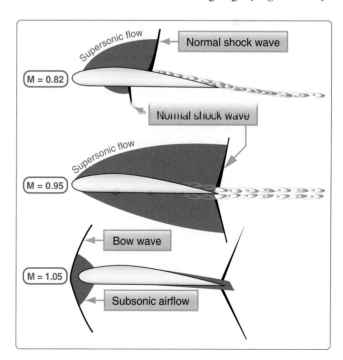

Figure 5-67. *Shock waves.*

Associated with "drag rise" are buffet (known as Mach buffet), trim, and stability changes and a decrease in control force effectiveness. The loss of lift due to airflow separation results in a loss of downwash and a change in the position of the center pressure on the wing. Airflow separation produces a turbulent wake behind the wing, which causes the tail surfaces to buffet (vibrate). The nose-up and nose-down pitch control provided by the horizontal tail is dependent on the downwash behind the wing. Thus, an increase in downwash decreases the horizontal tail's pitch control effectiveness since it effectively increases the AOA that the tail surface is seeing. Movement of the wing CP affects the wing pitching moment. If the CP moves aft, a diving moment referred to as "Mach tuck" or "tuck under" is produced, and if it moves forward, a nose-up moment is produced. This is the primary reason for the development of the T-tail configuration on many turbine-powered aircraft, which places the horizontal stabilizer as far as practical from the turbulence of the wings.

Sweepback

Most of the difficulties of transonic flight are associated with shock wave induced flow separation. Therefore, any means of delaying or alleviating the shock induced separation improves aerodynamic performance. One method is wing sweepback. Sweepback theory is based upon the concept that it is only the component of the airflow perpendicular to the leading edge of the wing that affects pressure distribution and formation of shock waves. *[Figure 5-68]*

On a straight wing aircraft, the airflow strikes the wing leading edge at 90°, and its full impact produces pressure and lift. A wing with sweepback is struck by the same airflow at an angle smaller than 90°. This airflow on the swept wing has the effect of persuading the wing into believing that it is flying slower than it really is; thus the formation of shock waves is delayed. Advantages of wing sweep include an increase in critical Mach number, force divergence Mach number, and the Mach number at which drag rise peaks. In other words, sweep delays the onset of compressibility effects.

The Mach number that produces a sharp change in coefficient of drag is termed the "force divergence" Mach number and, for most airfoils, usually exceeds the critical Mach number by 5 to 10 percent. At this speed, the airflow separation induced by shock wave formation can create significant variations in the drag, lift, or pitching moment coefficients. In addition to the delay of the onset of compressibility effects, sweepback reduces the magnitude in the changes of drag, lift, or moment coefficients. In other words, the use of sweepback "softens" the force divergence.

A disadvantage of swept wings is that they tend to stall at the wingtips rather than at the wing roots. *[Figure 5-69]* This is

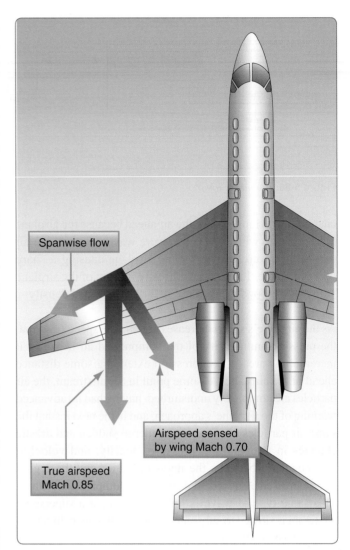

Figure 5-68. *Sweepback effect.*

because the boundary layer tends to flow spanwise toward the tips and to separate near the leading edges. Because the tips of a swept wing are on the aft part of the wing (behind the CL), a wingtip stall causes the CL to move forward on the wing, forcing the nose to rise further. The tendency for tip stall is greatest when wing sweep and taper are combined.

The stall situation can be aggravated by a T-tail configuration, which affords little or no pre-stall warning in the form of tail control surface buffet. *[Figure 5-70]* The T-tail, being above the wing wake remains effective even after the wing has begun to stall, allowing the pilot to inadvertently drive the wing into a deeper stall at a much greater AOA. If the horizontal tail surfaces then become buried in the wing's wake, the elevator may lose all effectiveness, making it impossible to reduce pitch attitude and break the stall. In the pre-stall and immediate post-stall regimes, the lift/drag qualities of a swept wing aircraft (specifically the enormous increase in drag at low speeds) can cause an increasingly descending flight path with no change in pitch attitude, further increasing the

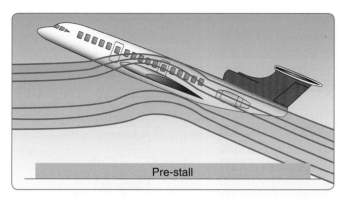

Figure 5-69. *Wingtip pre-stall.*

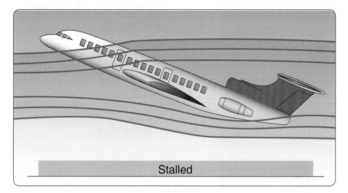

Figure 5-70. *T-tail stall.*

AOA. In this situation, without reliable AOA information, a nose-down pitch attitude with an increasing airspeed is no guarantee that recovery has been affected, and up-elevator movement at this stage may merely keep the aircraft stalled.

It is a characteristic of T-tail aircraft to pitch up viciously when stalled in extreme nose-high attitudes, making recovery difficult or violent. The stick pusher inhibits this type of stall. At approximately one knot above stall speed, pre-programmed stick forces automatically move the stick forward, preventing the stall from developing. A G-limiter may also be incorporated into the system to prevent the pitch down generated by the stick pusher from imposing excessive loads on the aircraft. A "stick shaker," on the other hand, provides stall warning when the airspeed is five to seven percent above stall speed.

Mach Buffet Boundaries

Mach buffet is a function of the speed of the airflow over the wing—not necessarily the speed of the aircraft. Any time that too great a lift demand is made on the wing, whether from too fast an airspeed or from too high an AOA near the M_{MO}, the "high-speed" buffet occurs. There are also occasions when the buffet can be experienced at much lower speeds known as the "low-speed Mach buffet."

An aircraft flown at a speed too slow for its weight and altitude necessitating a high AOA is the most likely situation to cause a low-speed Mach buffet. This very high AOA has the effect of increasing airflow velocity over the upper surface of the wing until the same effects of the shock waves and buffet occur as in the high-speed buffet situation. The AOA of the wing has the greatest effect on inducing the Mach buffet at either the high-speed or low-speed boundaries for the aircraft. The conditions that increase the AOA, the speed of the airflow over the wing, and chances of Mach buffet are:

- High altitudes—the higher an aircraft flies, the thinner the air and the greater the AOA required to produce the lift needed to maintain level flight.

- Heavy weights—the heavier the aircraft, the greater the lift required of the wing, and all other factors being equal, the greater the AOA.

- G loading—an increase in the G loading on the aircraft has the same effect as increasing the weight of the aircraft. Whether the increase in G forces is caused by turns, rough control usage, or turbulence, the effect of increasing the wing's AOA is the same.

High Speed Flight Controls

On high-speed aircraft, flight controls are divided into primary flight controls and secondary or auxiliary flight controls. The primary flight controls maneuver the aircraft about the pitch, roll, and yaw axes. They include the ailerons, elevator, and rudder. Secondary or auxiliary flight controls include tabs, leading edge flaps, trailing edge flaps, spoilers, and slats.

Spoilers are used on the upper surface of the wing to spoil or reduce lift. High speed aircraft, due to their clean low drag design, use spoilers as speed brakes to slow them down. Spoilers are extended immediately after touchdown to dump lift and thus transfer the weight of the aircraft from the wings onto the wheels for better braking performance. *[Figure 5-71]*

Jet transport aircraft have small ailerons. The space for ailerons is limited because as much of the wing trailing edge as possible is needed for flaps. Also, a conventional size aileron would cause wing twist at high speed. For that reason, spoilers are used in unison with ailerons to provide additional roll control.

Some jet transports have two sets of ailerons, a pair of outboard low-speed ailerons and a pair of high-speed inboard ailerons. When the flaps are fully retracted after takeoff, the outboard ailerons are automatically locked out in the faired position.

When used for roll control, the spoiler on the side of the up-going aileron extends and reduces the lift on that side, causing the wing to drop. If the spoilers are extended as speed brakes, they can still be used for roll control. If they are the

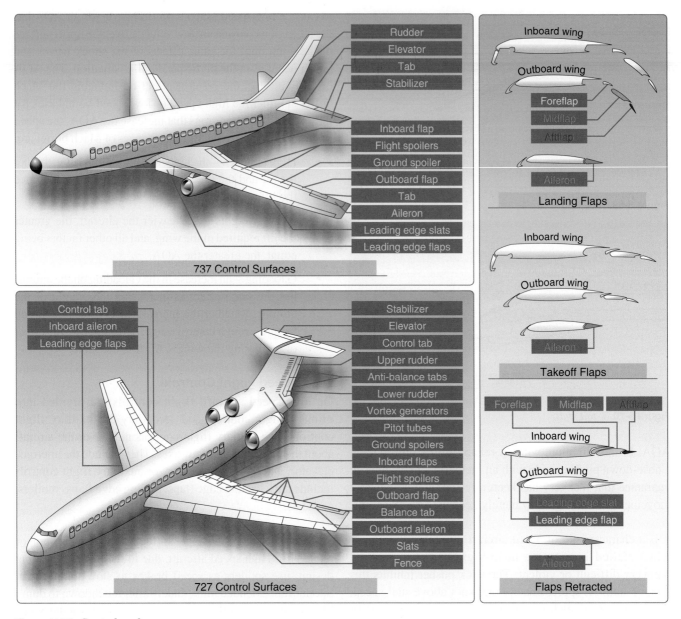

Figure 5-71. *Control surfaces.*

differential type, they extend further on one side and retract on the other side. If they are the non-differential type, they extend further on one side but do not retract on the other side. When fully extended as speed brakes, the non-differential spoilers remain extended and do not supplement the ailerons.

To obtain a smooth stall and a higher AOA without airflow separation, the wing's leading edge should have a well-rounded almost blunt shape that the airflow can adhere to at the higher AOA. With this shape, the airflow separation starts at the trailing edge and progresses forward gradually as AOA is increased.

The pointed leading edge necessary for high-speed flight results in an abrupt stall and restricts the use of trailing edge flaps because the airflow cannot follow the sharp curve around the wing leading edge. The airflow tends to tear loose rather suddenly from the upper surface at a moderate AOA. To utilize trailing edge flaps, and thus increase the C_{L-MAX}, the wing must go to a higher AOA without airflow separation. Therefore, leading edge slots, slats, and flaps are used to improve the low-speed characteristics during takeoff, climb, and landing. Although these devices are not as powerful as trailing edge flaps, they are effective when used full span in combination with high-lift trailing edge flaps. With the aid of these sophisticated high-lift devices, airflow separation is delayed and the C_{L-MAX} is increased considerably. In fact, a 50 knot reduction in stall speed is not uncommon.

The operational requirements of a large jet transport aircraft necessitate large pitch trim changes. Some requirements are:

- A large CG range

- A large speed range

- The ability to perform large trim changes due to wing leading edge and trailing edge high-lift devices without limiting the amount of elevator remaining

- Maintaining trim drag to a minimum

These requirements are met by the use of a variable incidence horizontal stabilizer. Large trim changes on a fixed-tail aircraft require large elevator deflections. At these large deflections, little further elevator movement remains in the same direction. A variable incidence horizontal stabilizer is designed to take out the trim changes. The stabilizer is larger than the elevator, and consequently does not need to be moved through as large an angle. This leaves the elevator streamlining the tail plane with a full range of movement up and down. The variable incidence horizontal stabilizer can be set to handle the bulk of the pitch control demand, with the elevator handling the rest. On aircraft equipped with a variable incidence horizontal stabilizer, the elevator is smaller and less effective in isolation than it is on a fixed-tail aircraft. In comparison to other flight controls, the variable incidence horizontal stabilizer is enormously powerful in its effect.

Because of the size and high speeds of jet transport aircraft, the forces required to move the control surfaces can be beyond the strength of the pilot. Consequently, the control surfaces are actuated by hydraulic or electrical power units. Moving the controls in the flight deck signals the control angle required, and the power unit positions the actual control surface. In the event of complete power unit failure, movement of the control surface can be affected by manually controlling the control tabs. Moving the control tab upsets the aerodynamic balance, which causes the control surface to move.

Chapter Summary

In order to sustain an aircraft in flight, a pilot must understand how thrust, drag, lift, and weight act on the aircraft. By understanding the aerodynamics of flight, how design, weight, load factors, and gravity affect an aircraft during flight maneuvers from stalls to high speed flight, the pilot learns how to control the balance between these forces. For information on stall speeds, load factors, and other important aircraft data, always consult the AFM/POH for specific information pertaining to the aircraft being flown.

Chapter 6
Flight Controls

Introduction

This chapter focuses on the flight control systems a pilot uses to control the forces of flight and the aircraft's direction and attitude. It should be noted that flight control systems and characteristics can vary greatly depending on the type of aircraft flown. The most basic flight control system designs are mechanical and date back to early aircraft. They operate with a collection of mechanical parts, such as rods, cables, pulleys, and sometimes chains to transmit the forces of the flight deck controls to the control surfaces. Mechanical flight control systems are still used today in small general and sport category aircraft where the aerodynamic forces are not excessive. [Figure 6-1]

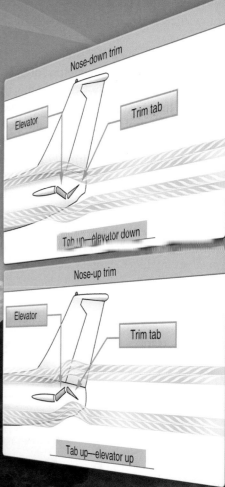

Nose-down trim

Elevator

Trim tab

Tab up—elevator down

Nose-up trim

Elevator

Trim tab

Tab up—elevator up

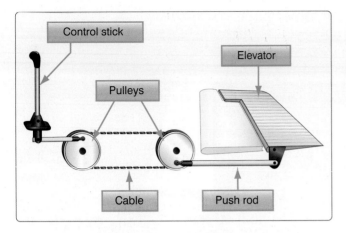

Figure 6-1. *Mechanical flight control system.*

As aviation matured and aircraft designers learned more about aerodynamics, the industry produced larger and faster aircraft. Therefore, the aerodynamic forces acting upon the control surfaces increased exponentially. To make the control force required by pilots manageable, aircraft engineers designed more complex systems. At first, hydromechanical designs, consisting of a mechanical circuit and a hydraulic circuit, were used to reduce the complexity, weight, and limitations of mechanical flight controls systems. *[Figure 6-2]*

As aircraft became more sophisticated, the control surfaces were actuated by electric motors, digital computers, or fiber optic cables. Called "fly-by-wire," this flight control system replaces the physical connection between pilot controls and the flight control surfaces with an electrical interface. In addition, in some large and fast aircraft, controls are boosted by hydraulically or electrically actuated systems. In both the fly-by-wire and boosted controls, the feel of the control reaction is fed back to the pilot by simulated means.

Current research at the National Aeronautics and Space Administration (NASA) Dryden Flight Research Center involves Intelligent Flight Control Systems (IFCS). The goal of this project is to develop an adaptive neural network-based flight control system. Applied directly to flight control system feedback errors, IFCS provides adjustments to improve aircraft performance in normal flight, as well as with system failures. With IFCS, a pilot is able to maintain control and safely land an aircraft that has suffered a failure to a control surface or damage to the airframe. It also improves mission capability, increases the reliability and safety of flight, and eases the pilot workload.

Today's aircraft employ a variety of flight control systems. For example, some aircraft in the sport pilot category rely on weight-shift control to fly while balloons use a standard burn technique. Helicopters utilize a cyclic to tilt the rotor in the desired direction along with a collective to manipulate rotor pitch and anti-torque pedals to control yaw. *[Figure 6-3]*

For additional information on flight control systems, refer to the appropriate handbook for information related to the flight control systems and characteristics of specific types of aircraft.

Flight Control Systems

Flight Controls

Aircraft flight control systems consist of primary and secondary systems. The ailerons, elevator (or stabilator), and rudder constitute the primary control system and are required to control an aircraft safely during flight. Wing flaps, leading edge devices, spoilers, and trim systems constitute the secondary control system and improve the performance characteristics of the airplane or relieve the pilot of excessive control forces.

Primary Flight Controls

Aircraft control systems are carefully designed to provide adequate responsiveness to control inputs while allowing a

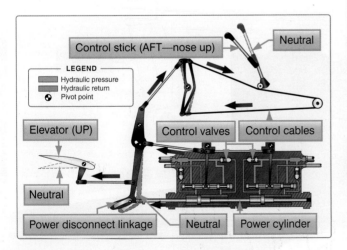

Figure 6-2. *Hydromechanical flight control system.*

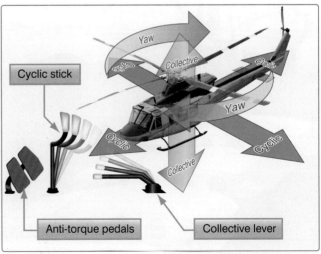

Figure 6-3. *Helicopter flight control system.*

natural feel. At low airspeeds, the controls usually feel soft and sluggish, and the aircraft responds slowly to control applications. At higher airspeeds, the controls become increasingly firm and aircraft response is more rapid.

Movement of any of the three primary flight control surfaces (ailerons, elevator or stabilator, or rudder), changes the airflow and pressure distribution over and around the airfoil. These changes affect the lift and drag produced by the airfoil/control surface combination, and allow a pilot to control the aircraft about its three axes of rotation.

Design features limit the amount of deflection of flight control surfaces. For example, control-stop mechanisms may be incorporated into the flight control linkages, or movement of the control column and/or rudder pedals may be limited. The purpose of these design limits is to prevent the pilot from inadvertently overcontrolling and overstressing the aircraft during normal maneuvers.

A properly designed aircraft is stable and easily controlled during normal maneuvering. Control surface inputs cause movement about the three axes of rotation. The types of stability an aircraft exhibits also relate to the three axes of rotation. *[Figure 6-4]*

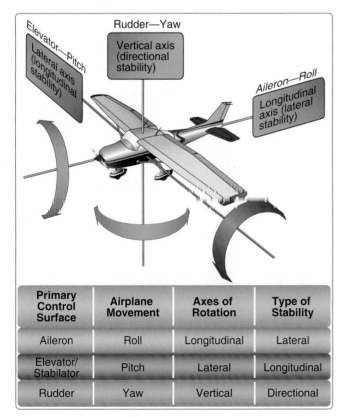

Primary Control Surface	Airplane Movement	Axes of Rotation	Type of Stability
Aileron	Roll	Longitudinal	Lateral
Elevator/ Stabilator	Pitch	Lateral	Longitudinal
Rudder	Yaw	Vertical	Directional

Figure 6-4. *Airplane controls, movement, axes of rotation, and type of stability.*

Ailerons

Ailerons control roll about the longitudinal axis. The ailerons are attached to the outboard trailing edge of each wing and move in the opposite direction from each other. Ailerons are connected by cables, bellcranks, pulleys, and/or push-pull tubes to a control wheel or control stick.

Moving the control wheel, or control stick, to the right causes the right aileron to deflect upward and the left aileron to deflect downward. The upward deflection of the right aileron decreases the camber resulting in decreased lift on the right wing. The corresponding downward deflection of the left aileron increases the camber resulting in increased lift on the left wing. Thus, the increased lift on the left wing and the decreased lift on the right wing causes the aircraft to roll to the right.

Adverse Yaw

Since the downward deflected aileron produces more lift as evidenced by the wing raising, it also produces more drag. This added drag causes the wing to slow down slightly. This results in the aircraft yawing toward the wing which had experienced an increase in lift (and drag). From the pilot's perspective, the yaw is opposite the direction of the bank. The adverse yaw is a result of differential drag and the slight difference in the velocity of the left and right wings. *[Figure 6-5]*

Adverse yaw becomes more pronounced at low airspeeds. At these slower airspeeds, aerodynamic pressure on control surfaces are low, and larger control inputs are required to

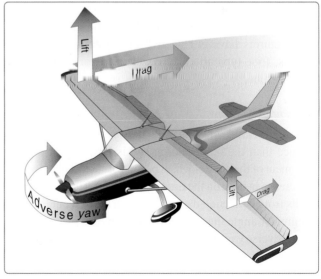

Figure 6-5. *Adverse yaw is caused by higher drag on the outside wing that is producing more lift.*

effectively maneuver the aircraft. As a result, the increase in aileron deflection causes an increase in adverse yaw. The yaw is especially evident in aircraft with long wing spans.

Application of the rudder is used to counteract adverse yaw. The amount of rudder control required is greatest at low airspeeds, high angles of attack, and with large aileron deflections. Like all control surfaces at lower airspeeds, the vertical stabilizer/rudder becomes less effective and magnifies the control problems associated with adverse yaw.

All turns are coordinated by use of ailerons, rudder, and elevator. Applying aileron pressure is necessary to place the aircraft in the desired angle of bank, while simultaneous application of rudder pressure is necessary to counteract the resultant adverse yaw. Additionally, because more lift is required during a turn than during straight-and-level flight, the angle of attack (AOA) must be increased by applying elevator back pressure. The steeper the turn, the more elevator back pressure that is needed.

As the desired angle of bank is established, aileron and rudder pressures should be relaxed. This stops the angle of bank from increasing, because the aileron and rudder control surfaces are in a neutral and streamlined position. Elevator back pressure should be held constant to maintain altitude. The roll-out from a turn is similar to the roll-in, except the flight controls are applied in the opposite direction. The aileron and rudder are applied in the direction of the roll-out or toward the high wing. As the angle of bank decreases, the elevator back pressure should be relaxed as necessary to maintain altitude.

In an attempt to reduce the effects of adverse yaw, manufacturers have engineered four systems: differential ailerons, frise-type ailerons, coupled ailerons and rudder, and flaperons.

Differential Ailerons

With differential ailerons, one aileron is raised a greater distance than the other aileron and is lowered for a given movement of the control wheel or control stick. This produces an increase in drag on the descending wing. The greater drag results from deflecting the up aileron on the descending wing to a greater angle than the down aileron on the rising wing. While adverse yaw is reduced, it is not eliminated completely. [Figure 6-6]

Frise-Type Ailerons

With a frise-type aileron, when pressure is applied to the control wheel, or control stick, the aileron that is being raised pivots on an offset hinge. This projects the leading edge of the aileron into the airflow and creates drag. It helps equalize

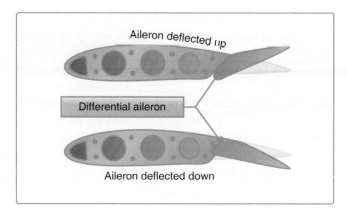

Figure 6-6. *Differential ailerons.*

the drag created by the lowered aileron on the opposite wing and reduces adverse yaw. *[Figure 6-7]*

The frise-type aileron also forms a slot so air flows smoothly over the lowered aileron, making it more effective at high angles of attack. Frise-type ailerons may also be designed to function differentially. Like the differential aileron, the frise-type aileron does not eliminate adverse yaw entirely. Coordinated rudder application is still needed when ailerons are applied.

Coupled Ailerons and Rudder

Coupled ailerons and rudder are linked controls. This is accomplished with rudder-aileron interconnect springs, which help correct for aileron drag by automatically deflecting the rudder at the same time the ailerons are deflected. For

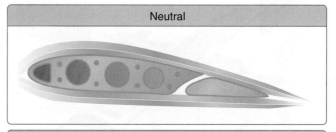

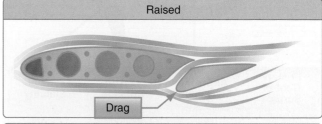

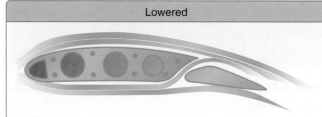

Figure 6-7. *Frise-type ailerons.*

example, when the control wheel, or control stick, is moved to produce a left roll, the interconnect cable and spring pulls forward on the left rudder pedal just enough to prevent the nose of the aircraft from yawing to the right. The force applied to the rudder by the springs can be overridden if it becomes necessary to slip the aircraft. *[Figure 6-8]*

Flaperons

Flaperons combine both aspects of flaps and ailerons. In addition to controlling the bank angle of an aircraft like conventional ailerons, flaperons can be lowered together to function much the same as a dedicated set of flaps. The pilot retains separate controls for ailerons and flaps. A mixer is used to combine the separate pilot inputs into this single set of control surfaces called flaperons. Many designs that incorporate flaperons mount the control surfaces away from the wing to provide undisturbed airflow at high angles of attack and/or low airspeeds. *[Figure 6-9]*

Elevator

The elevator controls pitch about the lateral axis. Like the ailerons on small aircraft, the elevator is connected to the control column in the flight deck by a series of mechanical

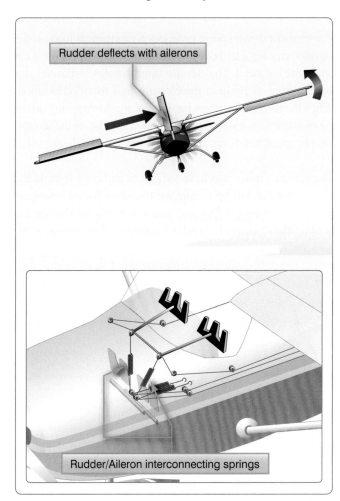

Figure 6-8. *Coupled ailerons and rudder.*

Figure 6-9. *Flaperons on a Skystar Kitfox MK 7.*

linkages. Aft movement of the control column deflects the trailing edge of the elevator surface up. This is usually referred to as the up-elevator position. *[Figure 6-10]*

The up-elevator position decreases the camber of the elevator and creates a downward aerodynamic force, which is greater than the normal tail-down force that exists in straight-and-level flight. The overall effect causes the tail of the aircraft to move down and the nose to pitch up. The pitching moment occurs about the center of gravity (CG). The strength of the pitching moment is determined by the distance between the CG and the horizontal tail surface, as well as by the aerodynamic effectiveness of the horizontal tail surface. Moving the control column forward has the opposite effect. In this case, elevator camber increases, creating more lift (less tail-down force) on the horizontal stabilizer/elevator. This moves the tail upward and pitches the nose down. Again, the pitching moment occurs about the CG.

As mentioned earlier, stability, power, thrustline, and the position of the horizontal tail surfaces on the empennage are factors in elevator effectiveness controlling pitch. For

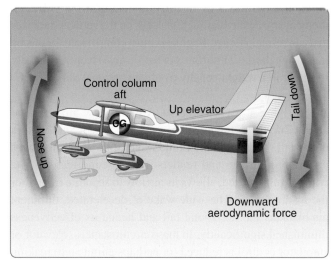

Figure 6-10. *The elevator is the primary control for changing the pitch attitude of an aircraft.*

example, the horizontal tail surfaces may be attached near the lower part of the vertical stabilizer, at the midpoint, or at the high point, as in the T-tail design.

T-Tail

In a T-tail configuration, the elevator is above most of the effects of downwash from the propeller, as well as airflow around the fuselage and/or wings during normal flight conditions. Operation of the elevators in this undisturbed air allows control movements that are consistent throughout most flight regimes. T-tail designs have become popular on many light and large aircraft, especially those with aft fuselage-mounted engines because the T-tail configuration removes the tail from the exhaust blast of the engines. Seaplanes and amphibians often have T-tails in order to keep the horizontal surfaces as far from the water as possible. An additional benefit is reduced noise and vibration inside the aircraft.

In comparison with conventional-tail aircraft, the elevator on a T-tail aircraft must be moved a greater distance to raise the nose a given amount when traveling at slow speeds. This is because the conventional-tail aircraft has the downwash from the propeller pushing down on the tail to assist in raising the nose.

Aircraft controls are rigged so that an increase in control force is required to increase control travel. The forces required to raise the nose of a T-tail aircraft are greater than the forces required to raise the nose of a conventional-tail aircraft. Longitudinal stability of a trimmed aircraft is the same for both types of configuration, but the pilot must be aware that the required control forces are greater at slow speeds during takeoffs, landings, or stalls than for similar size aircraft equipped with conventional tails.

T-tail aircraft also require additional design considerations to counter the problem of flutter. Since the weight of the horizontal surfaces is at the top of the vertical stabilizer, the moment arm created causes high loads on the vertical stabilizer that can result in flutter. Engineers must compensate for this by increasing the design stiffness of the vertical stabilizer, usually resulting in a weight penalty over conventional tail designs.

When flying at a very high AOA with a low airspeed and an aft CG, the T-tail aircraft may be more susceptible to a deep stall. In this condition, the wake of the wing impinges on the tail surface and renders it almost ineffective. The wing, if fully stalled, allows its airflow to separate right after the leading edge. The wide wake of decelerated, turbulent air blankets the horizontal tail and hence its effectiveness diminished significantly. In these circumstances, elevator or stabilator control is reduced (or perhaps eliminated) making it difficult to recover from the stall. It should be noted that an aft CG is often a contributing factor in these incidents, since

similar recovery problems are also found with conventional tail aircraft with an aft CG. *[Figure 6-11]* Deep stalls can occur on any aircraft but are more likely to occur on aircraft with "T" tails as a high AOA may be more likely to place the wings separated airflow into the path of the horizontal surface of the tail. Additionally, the distance between the wings and the tail, the position of the engines (such as being mounted on the tail) may increase the susceptibility of deep stall events. Therefore a deep stall may be more prevalent on transport versus general aviation aircraft.

Since flight at a high AOA with a low airspeed and an aft CG position can be dangerous, many aircraft have systems to compensate for this situation. The systems range from control stops to elevator down springs. On transport category jets, stick pushers are commonly used. An elevator down spring assists in lowering the nose of the aircraft to prevent a stall caused by the aft CG position. The stall occurs because the properly trimmed airplane is flying with the elevator in a trailing edge down position, forcing the tail up and the nose down. In this unstable condition, if the aircraft encounters turbulence and slows down further, the trim tab no longer positions the elevator in the nose-down position. The elevator then streamlines, and the nose of the aircraft pitches upward, possibly resulting in a stall.

The elevator down spring produces a mechanical load on the elevator, causing it to move toward the nose-down position if not otherwise balanced. The elevator trim tab balances the elevator down spring to position the elevator in a trimmed position. When the trim tab becomes ineffective, the down spring drives the elevator to a nose-down position. The nose of the aircraft lowers, speed builds up, and a stall is prevented. *[Figure 6-12]*

The elevator must also have sufficient authority to hold the nose of the aircraft up during the roundout for a landing. In this case, a forward CG may cause a problem. During the landing flare, power is usually reduced, which decreases the

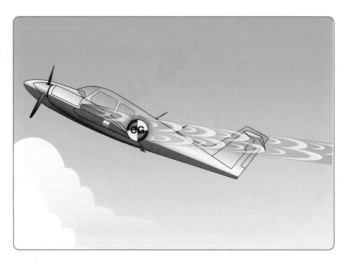

Figure 6-11. *Aircraft with a T-tail design at a high AOA and an aft CG.*

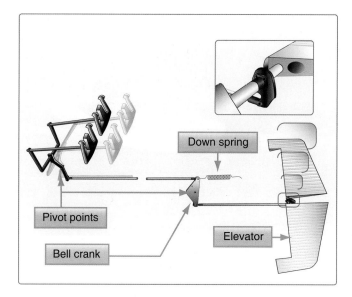

Figure 6-12. *When the aerodynamic efficiency of the horizontal tail surface is inadequate due to an aft CG condition, an elevator down spring may be used to supply a mechanical load to lower the nose.*

airflow over the empennage. This, coupled with the reduced landing speed, makes the elevator less effective.

As this discussion demonstrates, pilots must understand and follow proper loading procedures, particularly with regard to the CG position. More information on aircraft loading, as well as weight and balance, is included in Chapter 10, Weight and Balance.

Stabilator

As mentioned in Chapter 3, Aircraft Structure, a stabilator is essentially a one-piece horizontal stabilizer that pivots from a central hinge point. When the control column is pulled back, it raises the stabilator's trailing edge, pulling the nose of the aircraft. Pushing the control column forward lowers the trailing edge of the stabilator and pitches the nose of the aircraft down.

Because stabilators pivot around a central hinge point, they are extremely sensitive to control inputs and aerodynamic loads. Antiservo tabs are incorporated on the trailing edge to decrease sensitivity. They deflect in the same direction as the stabilator. This results in an increase in the force required to move the stabilator, thus making it less prone to pilot-induced overcontrolling. In addition, a balance weight is usually incorporated in front of the main spar. The balance weight may project into the empennage or may be incorporated on the forward portion of the stabilator tips. *[Figure 6-13]*

Canard

The canard design utilizes the concept of two lifting surfaces. The canard functions as a horizontal stabilizer located in front

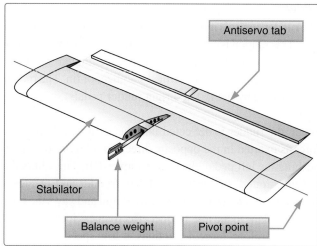

Figure 6-13. *The stabilator is a one-piece horizontal tail surface that pivots up and down about a central hinge point.*

of the main wings. In effect, the canard is an airfoil similar to the horizontal surface on a conventional aft-tail design. The difference is that the canard actually creates lift and holds the nose up, as opposed to the aft-tail design which exerts downward force on the tail to prevent the nose from rotating downward. *[Figure 6-14]*

The canard design dates back to the pioneer days of aviation. Most notably, it was used on the Wright Flyer. Recently, the canard configuration has regained popularity and is appearing on newer aircraft. Canard designs include two types–one with a horizontal surface of about the same size as a normal aft-tail design, and the other with a surface of the same approximate size and airfoil of the aft-mounted wing known as a tandem wing configuration. Theoretically, the canard is considered more efficient because using the horizontal surface to help lift the weight of the aircraft should result in less drag for a given amount of lift.

Figure 6-14. *The Piaggio P180 includes a variable-sweep canard design that provides longitudinal stability about the lateral axis.*

Rudder

The rudder controls movement of the aircraft about its vertical axis. This motion is called yaw. Like the other primary control surfaces, the rudder is a movable surface hinged to a fixed surface in this case, to the vertical stabilizer or fin. The rudder is controlled by the left and right rudder pedals.

When the rudder is deflected into the airflow, a horizontal force is exerted in the opposite direction. *[Figure 6-15]* By pushing the left pedal, the rudder moves left. This alters the airflow around the vertical stabilizer/rudder and creates a sideward lift that moves the tail to the right and yaws the nose of the airplane to the left. Rudder effectiveness increases with speed; therefore, large deflections at low speeds and small deflections at high speeds may be required to provide the desired reaction. In propeller-driven aircraft, any slipstream flowing over the rudder increases its effectiveness.

V-Tail

The V-tail design utilizes two slanted tail surfaces to perform the same functions as the surfaces of a conventional elevator and rudder configuration. The fixed surfaces act as both horizontal and vertical stabilizers. *[Figure 6-16]*

The movable surfaces, which are usually called ruddervators, are connected through a special linkage that allows the control wheel to move both surfaces simultaneously. On the other hand, displacement of the rudder pedals moves the surfaces differentially, thereby providing directional control.

When both rudder and elevator controls are moved by the pilot, a control mixing mechanism moves each surface the

Figure 6-16. *Beechcraft Bonanza V35.*

appropriate amount. The control system for the V-tail is more complex than the control system for a conventional tail. In addition, the V-tail design is more susceptible to Dutch roll tendencies than a conventional tail, and total reduction in drag is minimal.

Secondary Flight Controls

Secondary flight control systems may consist of wing flaps, leading edge devices, spoilers, and trim systems.

Flaps

Flaps are the most common high-lift devices used on aircraft. These surfaces, which are attached to the trailing edge of the wing, increase both lift and induced drag for any given AOA. Flaps allow a compromise between high cruising speed and low landing speed because they may be extended when needed and retracted into the wing's structure when not needed. There are four common types of flaps: plain, split, slotted, and Fowler flaps. *[Figure 6-17]*

The plain flap is the simplest of the four types. It increases the airfoil camber, resulting in a significant increase in the coefficient of lift (C_L) at a given AOA. At the same time, it greatly increases drag and moves the center of pressure (CP) aft on the airfoil, resulting in a nose-down pitching moment.

The split flap is deflected from the lower surface of the airfoil and produces a slightly greater increase in lift than the plain flap. More drag is created because of the turbulent air pattern produced behind the airfoil. When fully extended, both plain and split flaps produce high drag with little additional lift.

The most popular flap on aircraft today is the slotted flap. Variations of this design are used for small aircraft, as well as for large ones. Slotted flaps increase the lift coefficient significantly more than plain or split flaps. On small aircraft, the hinge is located below the lower surface of the flap, and

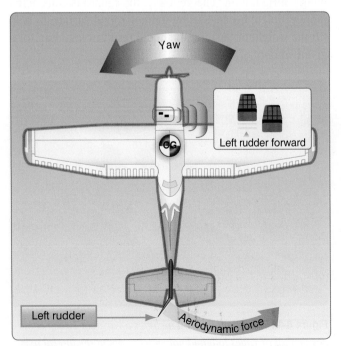

Figure 6-15. *The effect of left rudder pressure.*

when the flap is lowered, a duct forms between the flap well in the wing and the leading edge of the flap. When the slotted flap is lowered, high energy air from the lower surface is ducted to the flap's upper surface. The high energy air from the slot accelerates the upper surface boundary layer and delays airflow separation, providing a higher C_L. Thus, the slotted flap produces much greater increases in maximum coefficient of lift (C_{L-MAX}) than the plain or split flap. While there are many types of slotted flaps, large aircraft often have double- and even triple-slotted flaps. These allow the maximum increase in drag without the airflow over the flaps separating and destroying the lift they produce.

Fowler flaps are a type of slotted flap. This flap design not only changes the camber of the wing, it also increases the wing area. Instead of rotating down on a hinge, it slides backwards on tracks. In the first portion of its extension, it increases the drag very little, but increases the lift a great deal as it increases both the area and camber. Pilots should be aware that flap extension may cause a nose-up or down pitching moment, depending on the type of aircraft, which the pilot will need to compensate for, usually with a trim adjustment. As the extension continues, the flap deflects downward. During the last portion of its travel, the flap increases the drag with little additional increase in lift.

Leading Edge Devices

High-lift devices also can be applied to the leading edge of the airfoil. The most common types are fixed slots, movable slats, leading edge flaps, and cuffs. *[Figure 6-18]*

Fixed slots direct airflow to the upper wing surface and delay airflow separation at higher angles of attack. The slot does not

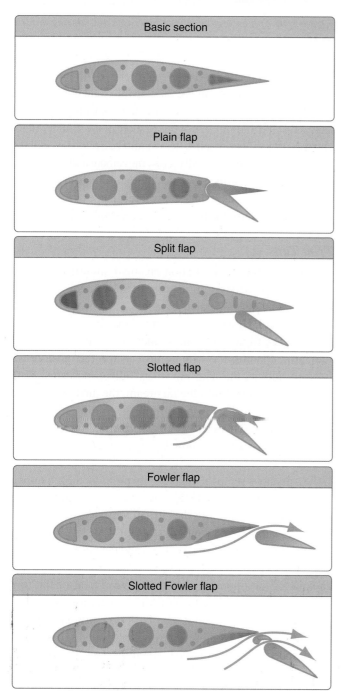

Figure 6-17. *Five common types of flaps.*

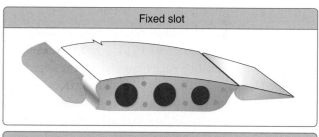
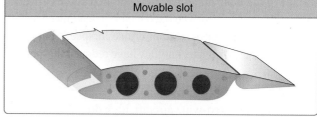
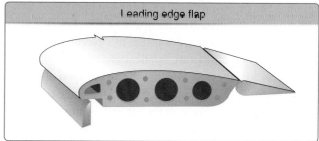
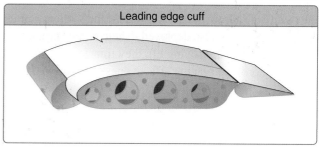

Figure 6-18. *Leading edge high lift devices.*

6-9

increase the wing camber, but allows a higher maximum C_L because the stall is delayed until the wing reaches a greater AOA.

Movable slats consist of leading edge segments that move on tracks. At low angles of attack, each slat is held flush against the wing's leading edge by the high pressure that forms at the wing's leading edge. As the AOA increases, the high-pressure area moves aft below the lower surface of the wing, allowing the slats to move forward. Some slats, however, are pilot operated and can be deployed at any AOA. Opening a slat allows the air below the wing to flow over the wing's upper surface, delaying airflow separation.

Leading edge flaps, like trailing edge flaps, are used to increase both C_{L-MAX} and the camber of the wings. This type of leading edge device is frequently used in conjunction with trailing edge flaps and can reduce the nose-down pitching movement produced by the latter. As is true with trailing edge flaps, a small increment of leading edge flaps increases lift to a much greater extent than drag. As flaps are extended, drag increases at a greater rate than lift.

Leading edge cuffs, like leading edge flaps and trailing edge flaps are used to increase both C_{L-MAX} and the camber of the wings. Unlike leading edge flaps and trailing edge flaps, leading edge cuffs are fixed aerodynamic devices. In most cases, leading edge cuffs extend the leading edge down and forward. This causes the airflow to attach better to the upper surface of the wing at higher angles of attack, thus lowering an aircraft's stall speed. The fixed nature of leading edge cuffs extracts a penalty in maximum cruise airspeed, but recent advances in design and technology have reduced this penalty.

Spoilers

Found on some fixed-wing aircraft, high drag devices called spoilers are deployed from the wings to spoil the smooth airflow, reducing lift and increasing drag. On gliders, spoilers are most often used to control rate of descent for accurate landings. On other aircraft, spoilers are often used for roll control, an advantage of which is the elimination of adverse yaw. To turn right, for example, the spoiler on the right wing is raised, destroying some of the lift and creating more drag on the right. The right wing drops, and the aircraft banks and yaws to the right. Deploying spoilers on both wings at the same time allows the aircraft to descend without gaining speed. Spoilers are also deployed to help reduce ground roll after landing. By destroying lift, they transfer weight to the wheels, improving braking effectiveness. *[Figure 6-19]*

Trim Systems

Although an aircraft can be operated throughout a wide range of attitudes, airspeeds, and power settings, it can be designed to fly hands-off within only a very limited combination of these

Figure 6-19. *Spoilers reduce lift and increase drag during descent and landing.*

variables. Trim systems are used to relieve the pilot of the need to maintain constant pressure on the flight controls, and usually consist of flight deck controls and small hinged devices attached to the trailing edge of one or more of the primary flight control surfaces. Designed to help minimize a pilot's workload, trim systems aerodynamically assist movement and position of the flight control surface to which they are attached. Common types of trim systems include trim tabs, balance tabs, antiservo tabs, ground adjustable tabs, and an adjustable stabilizer.

Trim Tabs

The most common installation on small aircraft is a single trim tab attached to the trailing edge of the elevator. Most trim tabs are manually operated by a small, vertically mounted control wheel. However, a trim crank may be found in some aircraft. The flight deck control includes a trim tab position indicator. Placing the trim control in the full nose-down position moves the trim tab to its full up position. With the trim tab up and into the airstream, the airflow over the horizontal tail surface tends to force the trailing edge of the elevator down. This causes the tail of the aircraft to move up and the nose to move down. *[Figure 6-20]*

If the trim tab is set to the full nose-up position, the tab moves to its full down position. In this case, the air flowing under the horizontal tail surface hits the tab and forces the trailing edge of the elevator up, reducing the elevator's AOA. This causes the tail of the aircraft to move down and the nose to move up.

In spite of the opposing directional movement of the trim tab and the elevator, control of trim is natural to a pilot. If the pilot needs to exert constant back pressure on a control column, the need for nose-up trim is indicated. The normal trim procedure is to continue trimming until the aircraft is balanced and the nose-heavy condition is no longer apparent. Pilots normally establish the desired power, pitch attitude, and configuration first, and then trim the aircraft to relieve

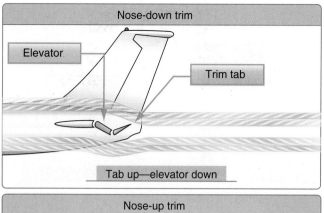

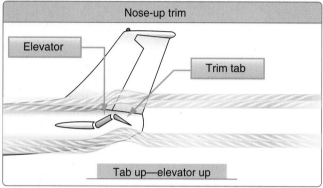

Figure 6-20. *The movement of the elevator is opposite to the direction of movement of the elevator trim tab.*

control pressures that may exist for that flight condition. As power, pitch attitude, or configuration changes, retrimming is necessary to relieve the control pressures for the new flight condition.

Balance Tabs

The control forces may be excessively high in some aircraft, and, in order to decrease them, the manufacturer may use balance tabs. They look like trim tabs and are hinged in approximately the same places as trim tabs. The essential difference between the two in that the balancing tab is coupled to the control surface rod so that when the primary control surface is moved in any direction, the tab automatically moves in the opposite direction. The airflow striking the tab counterbalances some of the air pressure against the primary control surface and enables the pilot to move the control more easily and hold the control surface in position.

If the linkage between the balance tab and the fixed surface is adjustable from the flight deck, the tab acts as a combination trim and balance tab that can be adjusted to a desired deflection.

Servo Tabs

Servo tabs are very similar in operation and appearance to the trim tabs previously discussed. A servo tab is a small portion of a flight control surface that deploys in such a way that it helps to move the entire flight control surface in the direction that the pilot wishes it to go. A servo tab is a dynamic device that deploys to decrease the pilots work load and de-stabilize the aircraft. Servo tabs are sometimes referred to as flight tabs and are used primarily on large aircraft. They aid the pilot in moving the control surface and in holding it in the desired position. Only the servo tab moves in response to movement of the pilot's flight control, and the force of the airflow on the servo tab then moves the primary control surface.

Antiservo Tabs

Antiservo tabs work in the same manner as balance tabs except, instead of moving in the opposite direction, they move in the same direction as the trailing edge of the stabilator. In addition to decreasing the sensitivity of the stabilator, an antiservo tab also functions as a trim device to relieve control pressure and maintain the stabilator in the desired position. The fixed end of the linkage is on the opposite side of the surface from the horn on the tab; when the trailing edge of the stabilator moves up, the linkage forces the trailing edge of the tab up. When the stabilator moves down, the tab also moves down. Conversely, trim tabs on elevators move opposite of the control surface. *[Figure 6-21]*

Ground Adjustable Tabs

Many small aircraft have a nonmovable metal trim tab on the rudder. This tab is bent in one direction or the other while on the ground to apply a trim force to the rudder. The correct displacement is determined by trial and error. Usually, small

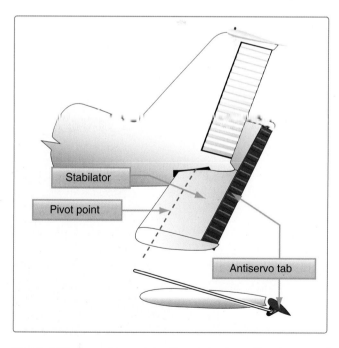

Figure 6-21. *An antiservo tab attempts to streamline the control surface and is used to make the stabilator less sensitive by opposing the force exerted by the pilot.*

adjustments are necessary until the aircraft no longer skids left or right during normal cruising flight. *[Figure 6-22]*

Adjustable Stabilizer

Rather than using a movable tab on the trailing edge of the elevator, some aircraft have an adjustable stabilizer. With this arrangement, linkages pivot the horizontal stabilizer about its rear spar. This is accomplished by the use of a jackscrew mounted on the leading edge of the stabilator. *[Figure 6-23]* On small aircraft, the jackscrew is cable operated with a trim wheel or crank. On larger aircraft, it is motor driven. The trimming effect and flight deck indications for an adjustable stabilizer are similar to those of a trim tab.

Autopilot

Autopilot is an automatic flight control system that keeps an aircraft in level flight or on a set course. It can be directed by the pilot, or it may be coupled to a radio navigation signal. Autopilot reduces the physical and mental demands on a pilot and increases safety. The common features available on an autopilot are altitude and heading hold.

The simplest systems use gyroscopic attitude indicators and magnetic compasses to control servos connected to the flight control system. *[Figure 6-24]* The number and location of these servos depends on the complexity of the system. For example, a single-axis autopilot controls the aircraft about the longitudinal axis and a servo actuates the ailerons. A three-axis autopilot controls the aircraft about the longitudinal, lateral, and vertical axes. Three different servos actuate ailerons, elevator, and rudder. More advanced systems often include a vertical speed and/or indicated airspeed hold mode. Advanced autopilot systems are coupled to navigational aids through a flight director.

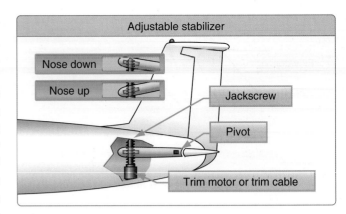

Figure 6-23. *Some aircraft, including most jet transports, use an adjustable stabilizer to provide the required pitch trim forces.*

The autopilot system also incorporates a disconnect safety feature to disengage the system automatically or manually. These autopilots work with inertial navigation systems, global positioning systems (GPS), and flight computers to control the aircraft. In fly-by-wire systems, the autopilot is an integrated component.

Additionally, autopilots can be manually overridden. Because autopilot systems differ widely in their operation, refer to the autopilot operating instructions in the Airplane Flight Manual (AFM) or the Pilot's Operating Handbook (POH).

Chapter Summary

Because flight control systems and aerodynamic characteristics vary greatly between aircraft, it is essential that a pilot become familiar with the primary and secondary flight control systems of the aircraft being flown. The primary source of this information is the AFM or the POH. Various manufacturer and owner group websites can also be a valuable source of additional information.

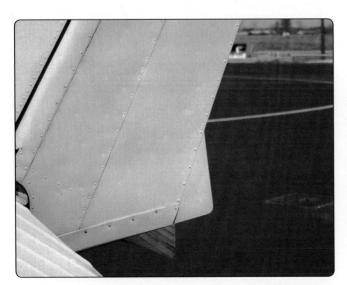

Figure 6-22. *A ground adjustable tab is used on the rudder of many small airplanes to correct for a tendency to fly with the fuselage slightly misaligned with the relative wind.*

Figure 6-24. *Basic autopilot system integrated into the flight control system.*

Chapter 7

Aircraft Systems

Introduction

This chapter covers the primary systems found on most aircraft. These include the engine, propeller, induction, ignition, as well as the fuel, lubrication, cooling, electrical, landing gear, and environmental control systems.

Powerplant

An aircraft engine, or powerplant, produces thrust to propel an aircraft. Reciprocating engines and turboprop engines work in combination with a propeller to produce thrust. Turbojet and turbofan engines produce thrust by increasing the velocity of air flowing through the engine. All of these powerplants also drive the various systems that support the operation of an aircraft.

Reciprocating Engines

Most small aircraft are designed with reciprocating engines. The name is derived from the back-and-forth, or reciprocating, movement of the pistons that produces the mechanical energy necessary to accomplish work.

Driven by a revitalization of the general aviation (GA) industry and advances in both material and engine design, reciprocating engine technology has improved dramatically over the past two decades. The integration of computerized engine management systems has improved fuel efficiency, decreased emissions, and reduced pilot workload.

Reciprocating engines operate on the basic principle of converting chemical energy (fuel) into mechanical energy. This conversion occurs within the cylinders of the engine through the process of combustion. The two primary reciprocating engine designs are the spark ignition and the compression ignition. The spark ignition reciprocating engine has served as the powerplant of choice for many years. In an effort to reduce operating costs, simplify design, and improve reliability, several engine manufacturers are turning to compression ignition as a viable alternative. Often referred to as jet fuel piston engines, compression ignition engines have the added advantage of utilizing readily available and lower cost diesel or jet fuel.

The main mechanical components of the spark ignition and the compression ignition engine are essentially the same. Both use cylindrical combustion chambers and pistons that travel the length of the cylinders to convert linear motion into the rotary motion of the crankshaft. The main difference between spark ignition and compression ignition is the process of igniting the fuel. Spark ignition engines use a spark plug to ignite a pre-mixed fuel-air mixture. (Fuel-air mixture is the ratio of the "weight" of fuel to the "weight" of air in the mixture to be burned.) A compression ignition engine first compresses the air in the cylinder, raising its temperature to a degree necessary for automatic ignition when fuel is injected into the cylinder.

These two engine designs can be further classified as:

1. Cylinder arrangement with respect to the crankshaft—radial, in-line, v-type, or opposed

2. Operating cycle—two or four

3. Method of cooling—liquid or air

Radial engines were widely used during World War II and many are still in service today. With these engines, a row or rows of cylinders are arranged in a circular pattern around the crankcase. The main advantage of a radial engine is the favorable power-to-weight ratio. [Figure 7-1]

Figure 7-1. *Radial engine.*

In-line engines have a comparatively small frontal area, but their power-to-weight ratios are relatively low. In addition, the rearmost cylinders of an air-cooled, in-line engine receive very little cooling air, so these engines are normally limited to four or six cylinders. V-type engines provide more horsepower than in-line engines and still retain a small frontal area.

Continued improvements in engine design led to the development of the horizontally-opposed engine, which remains the most popular reciprocating engines used on smaller aircraft. These engines always have an even number of cylinders, since a cylinder on one side of the crankcase "opposes" a cylinder on the other side. [Figure 7-2] The majority of these engines are air cooled and usually are mounted in a horizontal position when installed on fixed-wing airplanes. Opposed-type engines have high power-to-weight ratios because they have a comparatively small, lightweight crankcase. In addition, the compact cylinder arrangement reduces the engine's frontal area and allows a streamlined installation that minimizes aerodynamic drag.

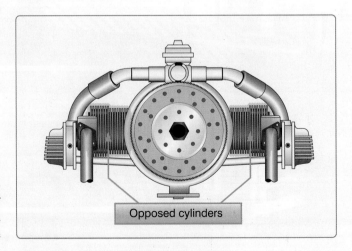

Opposed cylinders

Figure 7-2. *Horizontally opposed engine.*

Depending on the engine manufacturer, all of these arrangements can be designed to utilize spark or compression ignition and operate on either a two- or four-stroke cycle.

In a two-stroke engine, the conversion of chemical energy into mechanical energy occurs over a two-stroke operating cycle. The intake, compression, power, and exhaust processes occur in only two strokes of the piston rather than the more common four strokes. Because a two-stroke engine has a power stroke upon each revolution of the crankshaft, it typically has higher power-to-weight ratio than a comparable four-stroke engine. Due to the inherent inefficiency and disproportionate emissions of the earliest designs, use of the two-stroke engine has been limited in aviation.

Recent advances in material and engine design have reduced many of the negative characteristics associated with two-stroke engines. Modern two-stroke engines often use conventional oil sumps, oil pumps, and full pressure fed lubrication systems. The use of direct fuel injection and pressurized air, characteristic of advanced compression ignition engines, make two-stroke compression ignition engines a viable alternative to the more common four-stroke spark ignition designs. *[Figure 7-3]*

Spark ignition four-stroke engines remain the most common design used in GA today. *[Figure 7-4]* The main parts of a spark ignition reciprocating engine include the cylinders, crankcase, and accessory housing. The intake/exhaust valves, spark plugs, and pistons are located in the cylinders. The crankshaft and connecting rods are located in the crankcase. The magnetos are normally located on the engine accessory housing.

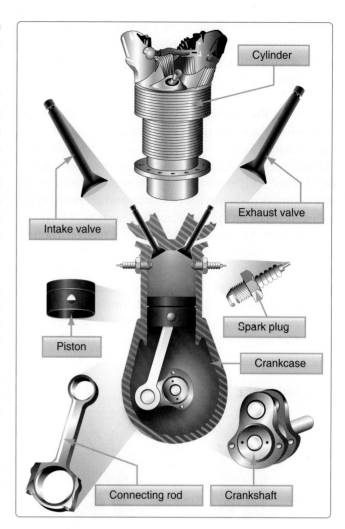

Figure 7-4. *Main components of a spark ignition reciprocating engine.*

In a four-stroke engine, the conversion of chemical energy into mechanical energy occurs over a four-stroke operating cycle. The intake, compression, power, and exhaust processes occur in four separate strokes of the piston in the following order.

1. The intake stroke begins as the piston starts its downward travel. When this happens, the intake valve opens and the fuel-air mixture is drawn into the cylinder.

2. The compression stroke begins when the intake valve closes, and the piston starts moving back to the top of the cylinder. This phase of the cycle is used to obtain a much greater power output from the fuel-air mixture once it is ignited.

3. The power stroke begins when the fuel-air mixture is ignited. This causes a tremendous pressure increase in the cylinder and forces the piston downward away from the cylinder head, creating the power that turns the crankshaft.

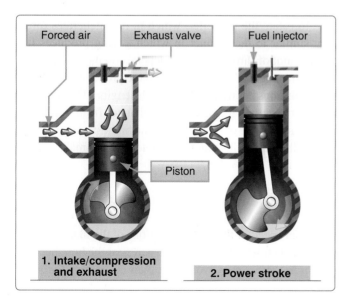

Figure 7-3. *Two-stroke compression ignition.*

4. The exhaust stroke is used to purge the cylinder of burned gases. It begins when the exhaust valve opens, and the piston starts to move toward the cylinder head once again.

Even when the engine is operated at a fairly low speed, the four-stroke cycle takes place several hundred times each minute. *[Figure 7-5]* In a four-cylinder engine, each cylinder operates on a different stroke. Continuous rotation of a crankshaft is maintained by the precise timing of the power strokes in each cylinder. Continuous operation of the engine depends on the simultaneous function of auxiliary systems, including the induction, ignition, fuel, oil, cooling, and exhaust systems.

The latest advance in aircraft reciprocating engines was pioneered in the mid-1960s by Frank Thielert, who looked to the automotive industry for answers on how to integrate diesel technology into an aircraft engine. The advantage

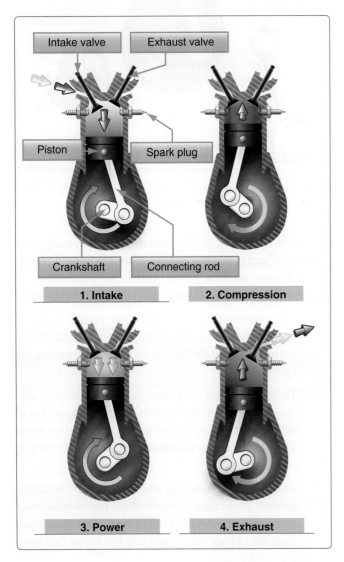

Figure 7-5. *The arrows in this illustration indicate the direction of motion of the crankshaft and piston during the four-stroke cycle.*

of a diesel-fueled reciprocating engine lies in the physical similarity of diesel and kerosene. Aircraft equipped with a diesel piston engine runs on standard aviation fuel kerosene, which provides more independence, higher reliability, lower consumption, and operational cost saving.

In 1999, Thielert formed Thielert Aircraft Engines (TAE) to design, develop, certify, and manufacture a brand-new Jet-A-burning diesel cycle engine (also known as jet-fueled piston engine) for the GA industry. By March 2001, the first prototype engine became the first certified diesel engine since World War II. TAE continues to design and develop diesel cycle engines and other engine manufacturers, such as Société de Motorisations Aéronautiques (SMA), now offer jet-fueled piston engines as well. TAE engines can be found on the Diamond DA40 single and the DA42 Twin Star; the first diesel engine to be part of the type certificate of a new original equipment manufacturer (OEM) aircraft.

These engines have also gained a toehold in the retrofit market with a supplemental type certificate (STC) to re-engine the Cessna 172 models and the Piper PA-28 family. The jet-fueled piston engine's technology has continued to progress and a full authority digital engine control (FADEC, discussed more fully later in the chapter) is standard on such equipped aircraft, which minimizes complication of engine control. By 2007, various jet-fueled piston aircraft had logged well over 600,000 hours of service.

Propeller

The propeller is a rotating airfoil, subject to induced drag, stalls, and other aerodynamic principles that apply to any airfoil. It provides the necessary thrust to pull, or in some cases push, the aircraft through the air. The engine power is used to rotate the propeller, which in turn generates thrust very similar to the manner in which a wing produces lift. The amount of thrust produced depends on the shape of the airfoil, the angle of attack (AOA) of the propeller blade, and the revolutions per minute (rpm) of the engine. The propeller itself is twisted so the blade angle changes from hub to tip. The greatest angle of incidence, or the highest pitch, is at the hub while the smallest angle of incidence or smallest pitch is at the tip. *[Figure 7-6]*

The reason for the twist is to produce uniform lift from the hub to the tip. As the blade rotates, there is a difference in the actual speed of the various portions of the blade. The tip of the blade travels faster than the part near the hub, because the tip travels a greater distance than the hub in the same length of time. *[Figure 7-7]* Changing the angle of incidence (pitch) from the hub to the tip to correspond with the speed produces uniform lift throughout the length of the blade. A propeller blade designed with the same angle of incidence

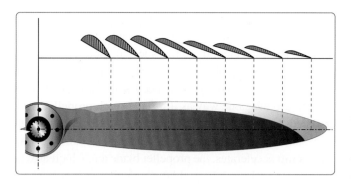

Figure 7-6. *Changes in propeller blade angle from hub to tip.*

throughout its entire length would be inefficient because as airspeed increases in flight, the portion near the hub would have a negative AOA while the blade tip would be stalled. Small aircraft are equipped with either one of two types of propellers: fixed-pitch or adjustable-pitch.

Fixed-Pitch Propeller

A propeller with fixed blade angles is a fixed-pitch propeller. The pitch of this propeller is set by the manufacturer and cannot be changed. Since a fixed-pitch propeller achieves the best efficiency only at a given combination of airspeed and rpm, the pitch setting is ideal for neither cruise nor climb. Thus, the aircraft suffers a bit in each performance category. The fixed-pitch propeller is used when low weight, simplicity, and low cost are needed.

There are two types of fixed-pitch propellers: climb and cruise. Whether the airplane has a climb or cruise propeller installed depends upon its intended use. The climb propeller has a lower pitch, therefore less drag. Less drag results in higher rpm and more horsepower capability, which increases performance during takeoffs and climbs but decreases performance during cruising flight.

The cruise propeller has a higher pitch, therefore more drag. More drag results in lower rpm and less horsepower capability, which decreases performance during takeoffs and climbs but increases efficiency during cruising flight.

The propeller is usually mounted on a shaft, which may be an extension of the engine crankshaft. In this case, the rpm of the propeller would be the same as the crankshaft rpm. On some engines, the propeller is mounted on a shaft geared to the engine crankshaft. In this type, the rpm of the propeller is different than that of the engine.

In a fixed-pitch propeller, the tachometer is the indicator of engine power. *[Figure 7-8]* A tachometer is calibrated in hundreds of rpm and gives a direct indication of the engine and propeller rpm. The instrument is color coded with a green arc denoting the maximum continuous operating rpm. Some tachometers have additional markings to reflect engine and/or propeller limitations. The manufacturer's recommendations should be used as a reference to clarify any misunderstanding of tachometer markings.

The rpm is regulated by the throttle, which controls the fuel-air flow to the engine. At a given altitude, the higher the tachometer reading, the higher the power output of the engine.

When operating altitude increases, the tachometer may not show correct power output of the engine. For example, 2,300 rpm at 5,000 feet produces less horsepower than 2,300 rpm at sea level because power output depends on air density. Air density decreases as altitude increases and a decrease in an

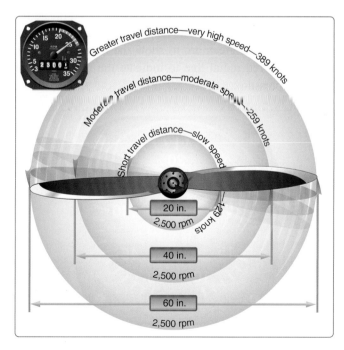

Figure 7-7. *Relationship of travel distance and speed of various portions of propeller blade.*

Figure 7-8. *Engine rpm is indicated on the tachometer.*

density (higher density altitude) decreases the power output of the engine. As altitude changes, the position of the throttle must be changed to maintain the same rpm. As altitude is increased, the throttle must be opened further to indicate the same rpm as at a lower altitude.

Adjustable-Pitch Propeller

The adjustable-pitch propeller was the forerunner of the constant-speed propeller. It is a propeller with blades whose pitch can be adjusted on the ground with the engine not running, but which cannot be adjusted in flight. It is also referred to as a ground adjustable propeller. By the 1930s, pioneer aviation inventors were laying the ground work for automatic pitch-change mechanisms, which is why the term sometimes refers to modern constant-speed propellers that are adjustable in flight.

The first adjustable-pitch propeller systems provided only two pitch settings: low and high. Today, most adjustable-pitch propeller systems are capable of a range of pitch settings.

A constant-speed propeller is a controllable-pitch propeller whose pitch is automatically varied in flight by a governor maintaining constant rpm despite varying air loads. It is the most common type of adjustable-pitch propeller. The main advantage of a constant-speed propeller is that it converts a high percentage of brake horsepower (BHP) into thrust horsepower (THP) over a wide range of rpm and airspeed combinations. A constant-speed propeller is more efficient than other propellers because it allows selection of the most efficient engine rpm for the given conditions.

An aircraft with a constant-speed propeller has two controls: the throttle and the propeller control. The throttle controls power output, and the propeller control regulates engine rpm. This regulates propeller rpm, which is registered on the tachometer.

Once a specific rpm is selected, a governor automatically adjusts the propeller blade angle as necessary to maintain the selected rpm. For example, after setting the desired rpm during cruising flight, an increase in airspeed or decrease in propeller load causes the propeller blade angle to increase as necessary to maintain the selected rpm. A reduction in airspeed or increase in propeller load causes the propeller blade angle to decrease.

The propeller's constant-speed range, defined by the high and low pitch stops, is the range of possible blade angles for a constant-speed propeller. As long as the propeller blade angle is within the constant-speed range and not against either pitch stop, a constant engine rpm is maintained. If the propeller blades contact a pitch stop, the engine rpm

will increase or decrease as appropriate, with changes in airspeed and propeller load. For example, once a specific rpm has been selected, if aircraft speed decreases enough to rotate the propeller blades until they contact the low pitch stop, any further decrease in airspeed will cause engine rpm to decrease the same way as if a fixed-pitch propeller were installed. The same holds true when an aircraft equipped with a constant-speed propeller accelerates to a faster airspeed. As the aircraft accelerates, the propeller blade angle increases to maintain the selected rpm until the high pitch stop is reached. Once this occurs, the blade angle cannot increase any further and engine rpm increases.

On aircraft equipped with a constant-speed propeller, power output is controlled by the throttle and indicated by a manifold pressure gauge. The gauge measures the absolute pressure of the fuel-air mixture inside the intake manifold and is more correctly a measure of manifold absolute pressure (MAP). At a constant rpm and altitude, the amount of power produced is directly related to the fuel-air mixture being delivered to the combustion chamber. As the throttle setting is increased, more fuel and air flows to the engine and MAP increases. When the engine is not running, the manifold pressure gauge indicates ambient air pressure (i.e., 29.92 inches mercury (29.92 "Hg)). When the engine is started, the manifold pressure indication decreases to a value less than ambient pressure (i.e., idle at 12 "Hg). Engine failure or power loss is indicated on the manifold gauge as an increase in manifold pressure to a value corresponding to the ambient air pressure at the altitude where the failure occurred. [Figure 7-9]

The manifold pressure gauge is color coded to indicate the engine's operating range. The face of the manifold pressure gauge contains a green arc to show the normal operating range and a red radial line to indicate the upper limit of manifold pressure.

Figure 7-9. *Engine power output is indicated on the manifold pressure gauge.*

For any given rpm, there is a manifold pressure that should not be exceeded. If manifold pressure is excessive for a given rpm, the pressure within the cylinders could be exceeded, placing undue stress on the cylinders. If repeated too frequently, this stress can weaken the cylinder components and eventually cause engine failure.

A pilot can avoid conditions that overstress the cylinders by being constantly aware of the rpm, especially when increasing the manifold pressure. Consult the manufacturer's recommendations for power settings of a particular engine to maintain the proper relationship between manifold pressure and rpm.

When both manifold pressure and rpm need to be changed, avoid engine overstress by making power adjustments in the proper order:

- When power settings are being decreased, reduce manifold pressure before reducing rpm. If rpm is reduced before manifold pressure, manifold pressure automatically increases, possibly exceeding the manufacturer's tolerances.

- When power settings are being increased, reverse the order—increase rpm first, then manifold pressure.

- To prevent damage to radial engines, minimize operating time at maximum rpm and manifold pressure, and avoid operation at maximum rpm and low manifold pressure.

The engine and/or airframe manufacturer's recommendations should be followed to prevent severe wear, fatigue, and damage to high-performance reciprocating engines.

Propeller Overspeed in Piston Engine Aircraft

On March 17, 2010, the Federal Aviation Administration (FAA) issued Special Airworthiness Information Bulletin (SAIB) CE-10-21. The subject was Propellers/Propulsers; Propeller Overspeed in Piston Engine Aircraft to alert operators, pilots, and aircraft manufacturers of concerns for an optimum response to a propeller overspeed in piston engine aircraft with variable pitch propellers. Although a SAIB is not regulatory in nature, the FAA recommends that the information be read and taken into consideration for the safety of flight.

The document explains that a single-engine aircraft experienced a propeller overspeed during cruise flight at 7,000 feet. The pilot reported that the application of throttle resulted in a propeller overspeed with no appreciable thrust. The pilot attempted to glide to a nearby airport and established the "best glide" speed of 110 knots, as published in the Pilot's Operating Handbook (POH), but was unable to reach the airport and was forced to conduct an off-field landing.

It was further explained that a determination was made that the propeller experienced a failure causing the blade pitch change mechanism to move to the low pitch stop position. This caused the propeller to operate as a fixed-pitch propeller such that it changes rpm with changes in power and airspeed. The low pitch setting allows for maximum power during takeoff but can result in a propeller overspeed at a higher airspeed.

A performance evaluation of the flight condition was performed for the particular aircraft model involved in this incident. This evaluation indicated that an airspeed lower than the best glide speed would have resulted in increased thrust enabling the pilot to maintain level flight. There are numerous variables in aircraft, engines, and propellers that affect aircraft performance. For some aircraft models, the published best glide speed may not be low enough to generate adequate thrust for a given propeller installation in this situation (propeller blades at low pitch stop position).

The operators of aircraft with variable pitch propellers should be aware that in certain instances of propeller overspeed, the airspeed necessary to maintain level flight may be different than the speed associated with engine-out best glide speed. The appropriate emergency procedures should be followed to mitigate the emergency situation in the event of a propeller overspeed; however, pilots should be aware that some reduction in airspeed may result in the ability for continued safe flight and landing. The determination of an airspeed that is more suitable than engine-out best glide speed should only be conducted at a safe altitude when the pilot has time to determine an alternative course of action other than landing immediately.

Induction Systems

The induction system brings in air from the outside, mixes it with fuel, and delivers the fuel-air mixture to the cylinder where combustion occurs. Outside air enters the induction system through an intake port on the front of the engine cowling. This port normally contains an air filter that inhibits the entry of dust and other foreign objects. Since the filter may occasionally become clogged, an alternate source of air must be available. Usually, the alternate air comes from inside the engine cowling, where it bypasses a clogged air filter. Some alternate air sources function automatically, while others operate manually.

Two types of induction systems are commonly used in small aircraft engines:

1. The carburetor system mixes the fuel and air in the carburetor before this mixture enters the intake manifold.

2. The fuel injection system mixes the fuel and air immediately before entry into each cylinder or injects fuel directly into each cylinder.

Carburetor Systems

Aircraft carburetors are separated into two categories: float-type carburetors and pressure-type carburetors. Float-type carburetors, complete with idling, accelerating, mixture control, idle cutoff, and power enrichment systems, are the most common of the two carburetor types. Pressure-type carburetors are usually not found on small aircraft. The basic difference between a float-type and a pressure-type carburetor is the delivery of fuel. The pressure-type carburetor delivers fuel under pressure by a fuel pump.

In the operation of the float-type carburetor system, the outside air first flows through an air filter, usually located at an air intake in the front part of the engine cowling. This filtered air flows into the carburetor and through a venturi, a narrow throat in the carburetor. When the air flows through the venturi, a low-pressure area is created that forces the fuel to flow through a main fuel jet located at the throat. The fuel then flows into the airstream where it is mixed with the flowing air. *[Figure 7-10]*

The fuel-air mixture is then drawn through the intake manifold and into the combustion chambers where it is ignited. The float-type carburetor acquires its name from a float that rests on fuel within the float chamber. A needle attached to the float opens and closes an opening at the bottom of the carburetor bowl. This meters the amount of fuel entering into the carburetor, depending upon the position of the float, which is controlled by the level of fuel in the float chamber. When the level of the fuel forces the float to rise, the needle valve closes the fuel opening and shuts off the fuel flow to the carburetor. The needle valve opens again when the engine requires additional fuel. The flow of the fuel-air mixture to the combustion chambers is regulated by the throttle valve, which is controlled by the throttle in the flight deck.

The float-type carburetor has several distinct disadvantages. First, they do not function well during abrupt maneuvers. Secondly, the discharge of fuel at low pressure leads to incomplete vaporization and difficulty in discharging fuel into some types of supercharged systems. The chief disadvantage of the float-type carburetor, however, is its icing tendency. Since the float-type carburetor must discharge

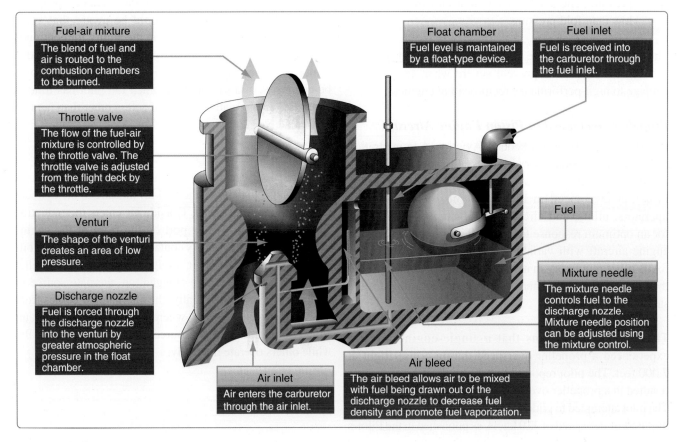

Fuel-air mixture
The blend of fuel and air is routed to the combustion chambers to be burned.

Throttle valve
The flow of the fuel-air mixture is controlled by the throttle valve. The throttle valve is adjusted from the flight deck by the throttle.

Venturi
The shape of the venturi creates an area of low pressure.

Discharge nozzle
Fuel is forced through the discharge nozzle into the venturi by greater atmospheric pressure in the float chamber.

Air inlet
Air enters the carburetor through the air inlet.

Air bleed
The air bleed allows air to be mixed with fuel being drawn out of the discharge nozzle to decrease fuel density and promote fuel vaporization.

Float chamber
Fuel level is maintained by a float-type device.

Fuel inlet
Fuel is received into the carburetor through the fuel inlet.

Fuel

Mixture needle
The mixture needle controls fuel to the discharge nozzle. Mixture needle position can be adjusted using the mixture control.

Figure 7-10. *Float-type carburetor.*

fuel at a point of low pressure, the discharge nozzle must be located at the venturi throat, and the throttle valve must be on the engine side of the discharge nozzle. This means that the drop in temperature due to fuel vaporization takes place within the venturi. As a result, ice readily forms in the venturi and on the throttle valve.

A pressure-type carburetor discharges fuel into the airstream at a pressure well above atmospheric pressure. This results in better vaporization and permits the discharge of fuel into the airstream on the engine side of the throttle valve. With the discharge nozzle in this position fuel vaporization takes place after the air has passed through the throttle valve and at a point where the drop in temperature is offset by heat from the engine. Thus, the danger of fuel vaporization icing is practically eliminated. The effects of rapid maneuvers and rough air on the pressure-type carburetors are negligible, since their fuel chambers remain filled under all operating conditions.

Mixture Control

Carburetors are normally calibrated at sea-level air pressure where the correct fuel-air mixture ratio is established with the mixture control set in the FULL RICH position. However, as altitude increases, the density of air entering the carburetor decreases, while the density of the fuel remains the same. This creates a progressively richer mixture that can result in engine roughness and an appreciable loss of power. The roughness normally is due to spark plug fouling from excessive carbon buildup on the plugs. Carbon buildup occurs because the rich mixture lowers the temperature inside the cylinder, inhibiting complete combustion of the fuel. This condition may occur during the runup prior to takeoff at high-elevation airports and during climbs or cruise flight at high altitudes. To maintain the correct fuel-air mixture, the mixture must be leaned using the mixture control. Leaning the mixture decreases fuel flow, which compensates for the decreased air density at high altitude.

During a descent from high altitude, the fuel-air mixture must be enriched, or it may become too lean. An overly lean mixture causes detonation, which may result in rough engine operation, overheating, and/or a loss of power. The best way to maintain the proper fuel-air mixture is to monitor the engine temperature and enrich the mixture as needed. Proper mixture control and better fuel economy for fuel-injected engines can be achieved by using an exhaust gas temperature (EGT) gauge. Since the process of adjusting the mixture can vary from one aircraft to another, it is important to refer to the airplane flight manual (AFM) or the POH to determine the specific procedures for a given aircraft.

Carburetor Icing

As mentioned earlier, one disadvantage of the float-type carburetor is its icing tendency. Carburetor ice occurs due to the effect of fuel vaporization and the decrease in air pressure in the venturi, which causes a sharp temperature drop in the carburetor. If water vapor in the air condenses when the carburetor temperature is at or below freezing, ice may form on internal surfaces of the carburetor, including the throttle valve. [Figure 7-11]

The reduced air pressure, as well as the vaporization of fuel, contributes to the temperature decrease in the carburetor. Ice generally forms in the vicinity of the throttle valve and in the venturi throat. This restricts the flow of the fuel-air mixture and reduces power. If enough ice builds up, the engine may cease to operate. Carburetor ice is most likely to occur when temperatures are below 70 degrees Fahrenheit (°F) or 21 degrees Celsius (°C) and the relative humidity is above 80 percent. Due to the sudden cooling that takes place in the carburetor, icing can occur even in outside air temperatures as high as 100 °F (38 °C) and humidity as low as 50 percent. This temperature drop can be as much as 60 to 70 absolute (versus relative) Fahrenheit degrees (70 x 100/180 = 38.89

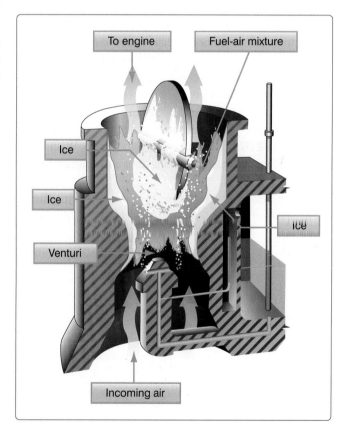

Figure 7-11. *The formation of carburetor ice may reduce or block fuel-air flow to the engine.*

Celsius degrees) (Remember there are 180 Fahrenheit degrees from freezing to boiling versus 100 degrees for the Celsius scale.) Therefore, an outside air temperature of 100 F (38 C), a temperature drop of an absolute 70 F degrees (38.89 Celsius degrees) results in an air temperature in the carburetor of 30 F (-1 C). [Figure 7-12]

The first indication of carburetor icing in an aircraft with a fixed-pitch propeller is a decrease in engine rpm, which may be followed by engine roughness. In an aircraft with a constant-speed propeller, carburetor icing is usually indicated by a decrease in manifold pressure, but no reduction in rpm. Propeller pitch is automatically adjusted to compensate for loss of power. Thus, a constant rpm is maintained. Although carburetor ice can occur during any phase of flight, it is particularly dangerous when using reduced power during a descent. Under certain conditions, carburetor ice could build unnoticed until power is added. To combat the effects of carburetor ice, engines with float-type carburetors employ a carburetor heat system.

Carburetor Heat

Carburetor heat is an anti-icing system that preheats the air before it reaches the carburetor and is intended to keep the fuel-air mixture above freezing to prevent the formation of carburetor ice. Carburetor heat can be used to melt ice that has already formed in the carburetor if the accumulation is not too great, but using carburetor heat as a preventative measure is the better option. Additionally, carburetor heat may be used as an alternate air source if the intake filter clogs, such as in sudden or unexpected airframe icing conditions. The carburetor heat should be checked during the engine runup. When using carburetor heat, follow the manufacturer's recommendations.

When conditions are conducive to carburetor icing during flight, periodic checks should be made to detect its presence. If detected, full carburetor heat should be applied immediately, and it should be left in the ON position until the pilot is certain that all the ice has been removed. If ice is present, applying partial heat or leaving heat on for an insufficient time might aggravate the situation. In extreme cases of carburetor icing, even after the ice has been removed, full carburetor heat should be used to prevent further ice formation. If installed, a carburetor temperature gauge is useful in determining when to use carburetor heat.

Whenever the throttle is closed during flight, the engine cools rapidly and vaporization of the fuel is less complete than if the engine is warm. Also, in this condition, the engine is more susceptible to carburetor icing. If carburetor icing conditions are suspected and closed-throttle operation anticipated, adjust the carburetor heat to the full ON position before closing the throttle and leave it on during the closed-throttle operation. The heat aids in vaporizing the fuel and helps prevent the formation of carburetor ice. Periodically, open the throttle smoothly for a few seconds to keep the engine warm; otherwise, the carburetor heater may not provide enough heat to prevent icing.

The use of carburetor heat causes a decrease in engine power, sometimes up to 15 percent, because the heated air is less dense than the outside air that had been entering the engine. This enriches the mixture. When ice is present in an aircraft with a fixed-pitch propeller and carburetor heat is being used, there is a decrease in rpm, followed by a gradual increase in rpm as the ice melts. The engine also should run more smoothly after the ice has been removed. If ice is not present, the rpm decreases and then remains constant. When carburetor heat is used on an aircraft with a constant-speed propeller and ice is present, a decrease in the manifold pressure is noticed, followed by a gradual increase. If carburetor icing is not present, the gradual increase in manifold pressure is not apparent until the carburetor heat is turned off.

It is imperative for a pilot to recognize carburetor ice when it forms during flight to prevent a loss in power, altitude, and/or airspeed. These symptoms may sometimes be accompanied by vibration or engine roughness. Once a power loss is noticed, immediate action should be taken to eliminate ice already formed in the carburetor and to prevent further ice formation. This is accomplished by applying full carburetor heat, which will further reduce power and may cause engine roughness as melted ice goes through the engine. These symptoms may last from 30 seconds to several minutes, depending on the severity of the icing. During this period, the pilot must resist the temptation to decrease the carburetor heat usage. Carburetor heat must remain in the full-hot position until normal power returns.

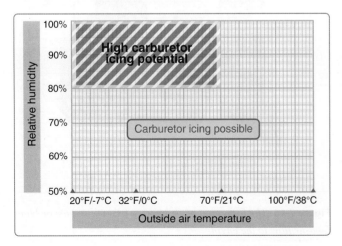

Figure 7-12. *Although carburetor ice is most likely to form when the temperature and humidity are in ranges indicated by this chart, carburetor icing is possible under conditions not depicted.*

Since the use of carburetor heat tends to reduce the output of the engine and to increase the operating temperature, carburetor heat should not be used when full power is required (as during takeoff) or during normal engine operation, except to check for the presence of, or to remove, carburetor ice.

Carburetor Air Temperature Gauge
Some aircraft are equipped with a carburetor air temperature gauge, which is useful in detecting potential icing conditions. Usually, the face of the gauge is calibrated in degrees Celsius with a yellow arc indicating the carburetor air temperatures where icing may occur. This yellow arc typically ranges between –15 °C and +5 °C (5 °F and 41 °F). If the air temperature and moisture content of the air are such that carburetor icing is improbable, the engine can be operated with the indicator in the yellow range with no adverse effects. If the atmospheric conditions are conducive to carburetor icing, the indicator must be kept outside the yellow arc by application of carburetor heat.

Certain carburetor air temperature gauges have a red radial that indicates the maximum permissible carburetor inlet air temperature recommended by the engine manufacturer. If present, a green arc indicates the normal operating range.

Outside Air Temperature Gauge
Most aircraft are also equipped with an outside air temperature (OAT) gauge calibrated in both degrees Celsius and Fahrenheit. It provides the outside or ambient air temperature for calculating true airspeed and is useful in detecting potential icing conditions.

Fuel Injection Systems
In a fuel injection system, the fuel is injected directly into the cylinders, or just ahead of the intake valve. The air intake for the fuel injection system is similar to that used in a carburetor system, with an alternate air source located within the engine cowling. This source is used if the external air source is obstructed. The alternate air source is usually operated automatically, with a backup manual system that can be used if the automatic feature malfunctions.

A fuel injection system usually incorporates six basic components: an engine-driven fuel pump, a fuel-air control unit, a fuel manifold (fuel distributor), discharge nozzles, an auxiliary fuel pump, and fuel pressure/flow indicators. *[Figure 7-13]*

The auxiliary fuel pump provides fuel under pressure to the fuel-air control unit for engine starting and/or emergency use. After starting, the engine-driven fuel pump provides fuel under pressure from the fuel tank to the fuel-air control unit.

This control unit, which essentially replaces the carburetor, meters fuel based on the mixture control setting and sends it to the fuel manifold valve at a rate controlled by the throttle.

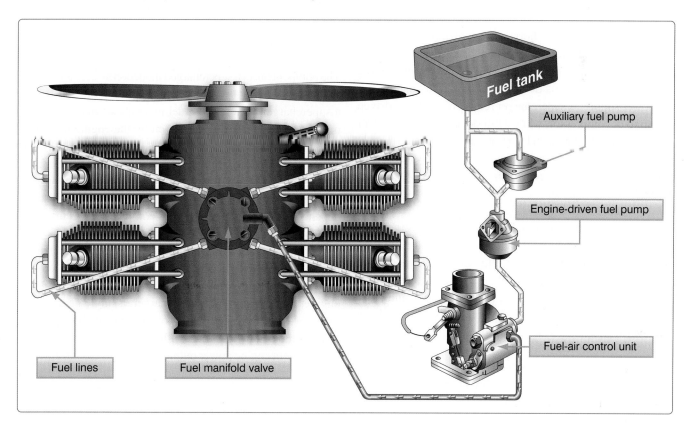

Figure 7-13. *Fuel injection system.*

After reaching the fuel manifold valve, the fuel is distributed to the individual fuel discharge nozzles. The discharge nozzles, which are located in each cylinder head, inject the fuel-air mixture directly into each cylinder intake port.

A fuel injection system is considered to be less susceptible to icing than a carburetor system, but impact icing on the air intake is a possibility in either system. Impact icing occurs when ice forms on the exterior of the aircraft and blocks openings, such as the air intake for the injection system.

The following are advantages of using fuel injection:

- Reduction in evaporative icing
- Better fuel flow
- Faster throttle response
- Precise control of mixture
- Better fuel distribution
- Easier cold weather starts

The following are disadvantages of using fuel injection:

- Difficulty in starting a hot engine
- Vapor locks during ground operations on hot days
- Problems associated with restarting an engine that quits because of fuel starvation

Superchargers and Turbosuperchargers

To increase an engine's horsepower, manufacturers have developed forced induction systems called supercharger and turbosupercharger systems. They both compress the intake air to increase its density. The key difference lies in the power supply. A supercharger relies on an engine-driven air pump or compressor, while a turbocharger gets its power from the exhaust stream that runs through a turbine, which in turn spins the compressor. Aircraft with these systems have a manifold pressure gauge, which displays MAP within the engine's intake manifold.

On a standard day at sea level with the engine shut down, the manifold pressure gauge indicates the ambient absolute air pressure of 29.92 "Hg. Because atmospheric pressure decreases approximately 1 "Hg per 1,000 feet of altitude increase, the manifold pressure gauge indicates approximately 24.92 "Hg at an airport that is 5,000 feet above sea level with standard day conditions.

As a normally aspirated aircraft climbs, it eventually reaches an altitude where the MAP is insufficient for a normal climb. This altitude limit is known as the aircraft's service ceiling, and it is directly affected by the engine's ability to produce power. If the induction air entering the engine is pressurized,

or boosted, by either a supercharger or a turbosupercharger, the aircraft's service ceiling can be increased. With these systems, an aircraft can fly at higher altitudes with the advantage of higher true airspeeds and the increased ability to circumnavigate adverse weather.

Superchargers

A supercharger is an engine-driven air pump or compressor that provides compressed air to the engine to provide additional pressure to the induction air so that the engine can produce additional power. It increases manifold pressure and forces the fuel-air mixture into the cylinders. Higher manifold pressure increases the density of the fuel-air mixture and increases the power an engine can produce. With a normally aspirated engine, it is not possible to have manifold pressure higher than the existing atmospheric pressure. A supercharger is capable of boosting manifold pressure above 30 "Hg.

For example, at 8,000 feet, a typical engine may be able to produce 75 percent of the power it could produce at mean sea level (MSL) because the air is less dense at the higher altitude. The supercharger compresses the air to a higher density allowing a supercharged engine to produce the same manifold pressure at higher altitudes as it could produce at sea level. Thus, an engine at 8,000 feet MSL could still produce 25 "Hg of manifold pressure whereas, without a supercharger, it could only produce 22 "Hg. Superchargers are especially valuable at high altitudes (such as 18,000 feet) where the air density is 50 percent that of sea level. The use of a supercharger in many cases will supply air to the engine at the same density it did at sea level.

The components in a supercharged induction system are similar to those in a normally aspirated system, with the addition of a supercharger between the fuel metering device and intake manifold. A supercharger is driven by the engine through a gear train at one speed, two speeds, or variable speeds. In addition, superchargers can have one or more stages. Each stage also provides an increase in pressure and superchargers may be classified as single stage, two stage, or multistage, depending on the number of times compression occurs.

An early version of a single-stage, single-speed supercharger may be referred to as a sea-level supercharger. An engine equipped with this type of supercharger is called a sea-level engine. With this type of supercharger, a single gear-driven impeller is used to increase the power produced by an engine at all altitudes. The drawback with this type of supercharger is a decrease in engine power output with an increase in altitude.

Single-stage, single-speed superchargers are found on many high-powered radial engines and use an air intake that faces forward so the induction system can take full advantage of the ram air. Intake air passes through ducts to a carburetor, where fuel is metered in proportion to the airflow. The fuel-air charge is then ducted to the supercharger, or blower impeller, which accelerates the fuel-air mixture outward. Once accelerated, the fuel-air mixture passes through a diffuser, where air velocity is traded for pressure energy. After compression, the resulting high pressure fuel-air mixture is directed to the cylinders.

Some of the large radial engines developed during World War II have a single-stage, two-speed supercharger. With this type of supercharger, a single impeller may be operated at two speeds. The low impeller speed is often referred to as the low blower setting, while the high impeller speed is called the high blower setting. On engines equipped with a two-speed supercharger, a lever or switch in the flight deck activates an oil-operated clutch that switches from one speed to the other.

Under normal operations, takeoff is made with the supercharger in the low blower position. In this mode, the engine performs as a ground-boosted engine, and the power output decreases as the aircraft gains altitude. However, once the aircraft reaches a specified altitude, a power reduction is made, and the supercharger control is switched to the high blower position. The throttle is then reset to the desired manifold pressure. An engine equipped with this type of supercharger is called an altitude engine. *[Figure 7-14]*

Turbosuperchargers

The most efficient method of increasing horsepower in an engine is by using a turbosupercharger or turbocharger.

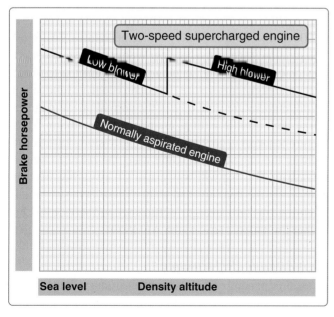

Figure 7-14. *Power output of normally aspirated engine compared to a single-stage, two-speed supercharged engine.*

Installed on an engine, this booster uses the engine's exhaust gases to drive an air compressor to increase the pressure of the air going into the engine through the carburetor or fuel injection system to boost power at higher altitude.

The major disadvantage of the gear-driven supercharger—use of a large amount of the engine's power output for the amount of power increase produced—is avoided with a turbocharger because turbochargers are powered by an engine's exhaust gases. This means a turbocharger recovers energy from hot exhaust gases that would otherwise be lost.

A second advantage of turbochargers over superchargers is the ability to maintain control over an engine's rated sea-level horsepower from sea level up to the engine's critical altitude. Critical altitude is the maximum altitude at which a turbocharged engine can produce its rated horsepower. Above the critical altitude, power output begins to decrease like it does for a normally aspirated engine.

Turbochargers increase the pressure of the engine's induction air, which allows the engine to develop sea level or greater horsepower at higher altitudes. A turbocharger is comprised of two main elements: a compressor and turbine. The compressor section houses an impeller that turns at a high rate of speed. As induction air is drawn across the impeller blades, the impeller accelerates the air, allowing a large volume of air to be drawn into the compressor housing. The impeller's action subsequently produces high-pressure, high-density air that is delivered to the engine. To turn the impeller, the engine's exhaust gases are used to drive a turbine wheel that is mounted on the opposite end of the impeller's drive shaft. By directing different amounts of exhaust gases to flow over the turbine, more energy can be extracted, causing the impeller to deliver more compressed air to the engine. The waste gate, essentially an adjustable butterfly valve installed in the exhaust system, is used to vary the mass of exhaust gas flowing into the turbine. When closed, most of the exhaust gases from the engine are forced to flow through the turbine. When open, the exhaust gases are allowed to bypass the turbine by flowing directly out through the engine's exhaust pipe. *[Figure 7-15]*

Since the temperature of a gas rises when it is compressed, turbocharging causes the temperature of the induction air to increase. To reduce this temperature and lower the risk of detonation, many turbocharged engines use an intercooler. This small heat exchanger uses outside air to cool the hot compressed air before it enters the fuel metering device.

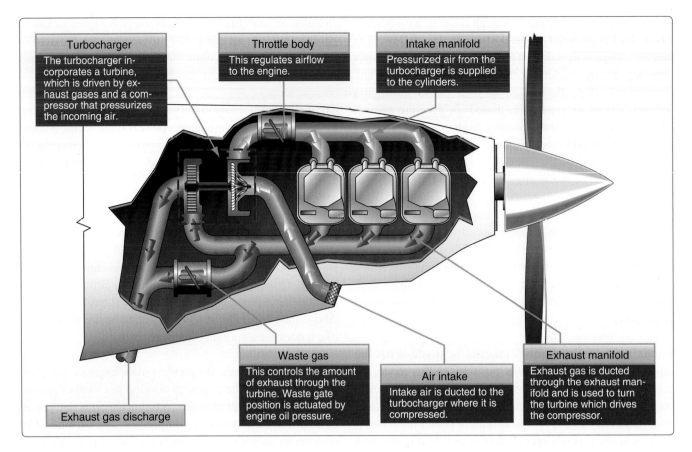

Figure 7-15. *Turbocharger components.*

System Operation

On most modern turbocharged engines, the position of the waste gate is governed by a pressure-sensing control mechanism coupled to an actuator. Engine oil directed into or away from this actuator moves the waste gate position. On these systems, the actuator is automatically positioned to produce the desired MAP simply by changing the position of the throttle control.

Other turbocharging system designs use a separate manual control to position the waste gate. With manual control, the manifold pressure gauge must be closely monitored to determine when the desired MAP has been achieved. Manual systems are often found on aircraft that have been modified with aftermarket turbocharging systems. These systems require special operating considerations. For example, if the waste gate is left closed after descending from a high altitude, it is possible to produce a manifold pressure that exceeds the engine's limitations. This condition, called an overboost, may produce severe detonation because of the leaning effect resulting from increased air density during descent.

Although an automatic waste gate system is less likely to experience an overboost condition, it can still occur. If takeoff power is applied while the engine oil temperature is below its normal operating range, the cold oil may not flow out of the waste gate actuator quickly enough to prevent an overboost. To help prevent overboosting, advance the throttle cautiously to prevent exceeding the maximum manifold pressure limits.

A pilot flying an aircraft with a turbocharger should be aware of system limitations. For example, a turbocharger turbine and impeller can operate at rotational speeds in excess of 80,000 rpm while at extremely high temperatures. To achieve high rotational speed, the bearings within the system must be constantly supplied with engine oil to reduce the frictional forces and high temperature. To obtain adequate lubrication, the oil temperature should be in the normal operating range before high throttle settings are applied. In addition, allow the turbocharger to cool and the turbine to slow down before shutting the engine down. Otherwise, the oil remaining in the bearing housing will boil, causing hard carbon deposits to form on the bearings and shaft. These deposits rapidly deteriorate the turbocharger's efficiency and service life. For further limitations, refer to the AFM/POH.

High Altitude Performance

As an aircraft equipped with a turbocharging system climbs, the waste gate is gradually closed to maintain the maximum allowable manifold pressure. At some point, the waste gate is fully closed and further increases in altitude cause the manifold pressure to decrease. This is the critical altitude, which is established by the aircraft or engine manufacturer. When evaluating the performance of the turbocharging system, be aware that if the manifold pressure begins decreasing before the specified critical altitude, the engine and turbocharging system should be inspected by a qualified aviation maintenance technician (AMT) to verify that the system is operating properly.

Ignition System

In a spark ignition engine, the ignition system provides a spark that ignites the fuel-air mixture in the cylinders and is made up of magnetos, spark plugs, high-tension leads, and an ignition switch. *[Figure 7-16]*

A magneto uses a permanent magnet to generate an electrical current completely independent of the aircraft's electrical system. The magneto generates sufficiently high voltage to jump a spark across the spark plug gap in each cylinder. The system begins to fire when the starter is engaged and the crankshaft begins to turn. It continues to operate whenever the crankshaft is rotating.

Most standard certificated aircraft incorporate a dual ignition system with two individual magnetos, separate sets of wires, and spark plugs to increase reliability of the ignition system. Each magneto operates independently to fire one of the two spark plugs in each cylinder. The firing of two spark plugs improves combustion of the fuel-air mixture and results in a

slightly higher power output. If one of the magnetos fails, the other is unaffected. The engine continues to operate normally, although a slight decrease in engine power can be expected. The same is true if one of the two spark plugs in a cylinder fails.

The operation of the magneto is controlled in the flight deck by the ignition switch. The switch has five positions:

1.   OFF
2.   R (right)
3.   L (left)
4.   BOTH
5.   START

With RIGHT or LEFT selected, only the associated magneto is activated. The system operates on both magnetos when BOTH is selected.

A malfunctioning ignition system can be identified during the pretakeoff check by observing the decrease in rpm that occurs when the ignition switch is first moved from BOTH to RIGHT and then from BOTH to LEFT. A small decrease in engine rpm is normal during this check. The permissible decrease is listed in the AFM or POH. If the engine stops running when switched to one magneto or if the rpm drop exceeds the allowable limit, do not fly the aircraft until the problem is corrected. The cause could be fouled plugs,

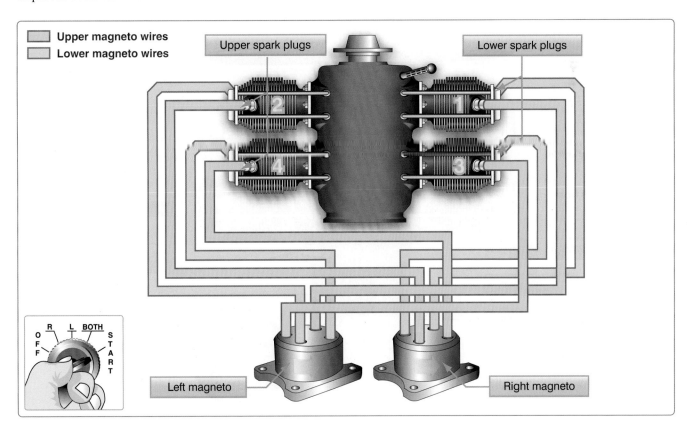

Figure 7-16. *Ignition system components.*

broken or shorted wires between the magneto and the plugs, or improperly timed firing of the plugs. It should be noted that "no drop" in rpm is not normal, and in that instance, the aircraft should not be flown.

Following engine shutdown, turn the ignition switch to the OFF position. Even with the battery and master switches OFF, the engine can fire and turn over if the ignition switch is left ON and the propeller is moved because the magneto requires no outside source of electrical power. Be aware of the potential for serious injury in this situation.

Even with the ignition switch in the OFF position, if the ground wire between the magneto and the ignition switch becomes disconnected or broken, the engine could accidentally start if the propeller is moved with residual fuel in the cylinder. If this occurs, the only way to stop the engine is to move the mixture lever to the idle cutoff position, then have the system checked by a qualified AMT.

Oil Systems

The engine oil system performs several important functions:

- Lubrication of the engine's moving parts
- Cooling of the engine by reducing friction
- Removing heat from the cylinders
- Providing a seal between the cylinder walls and pistons
- Carrying away contaminants

Reciprocating engines use either a wet-sump or a dry-sump oil system. In a wet-sump system, the oil is located in a sump that is an integral part of the engine. In a dry-sump system, the oil is contained in a separate tank and circulated through the engine by pumps. *[Figure 7-17]*

The main component of a wet-sump system is the oil pump, which draws oil from the sump and routes it to the engine. After the oil passes through the engine, it returns to the sump. In some engines, additional lubrication is supplied by the rotating crankshaft, which splashes oil onto portions of the engine.

An oil pump also supplies oil pressure in a dry-sump system, but the source of the oil is located external to the engine in a separate oil tank. After oil is routed through the engine, it is pumped from the various locations in the engine back to the oil tank by scavenge pumps. Dry-sump systems allow for a greater volume of oil to be supplied to the engine, which makes them more suitable for very large reciprocating engines.

The oil pressure gauge provides a direct indication of the oil system operation. It ensures the pressure in pounds per square inch (psi) of the oil supplied to the engine. Green indicates the normal operating range, while red indicates the minimum and maximum pressures. There should be an indication of oil pressure during engine start. Refer to the AFM/POH for manufacturer limitations.

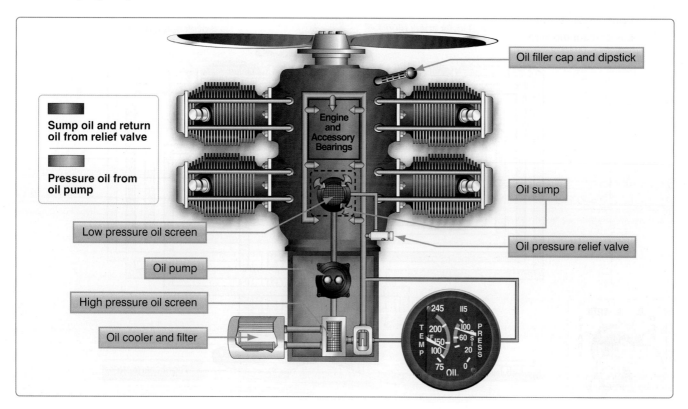

Figure 7-17. *Wet-sump oil system.*

The oil temperature gauge measures the temperature of oil. A green area shows the normal operating range, and the red line indicates the maximum allowable temperature. Unlike oil pressure, changes in oil temperature occur more slowly. This is particularly noticeable after starting a cold engine, when it may take several minutes or longer for the gauge to show any increase in oil temperature.

Check oil temperature periodically during flight especially when operating in high or low ambient air temperature. High oil temperature indications may signal a plugged oil line, a low oil quantity, a blocked oil cooler, or a defective temperature gauge. Low oil temperature indications may signal improper oil viscosity during cold weather operations.

The oil filler cap and dipstick (for measuring the oil quantity) are usually accessible through a panel in the engine cowling. If the quantity does not meet the manufacturer's recommended operating levels, oil should be added. The AFM/POH or placards near the access panel provide information about the correct oil type and weight, as well as the minimum and maximum oil quantity. [Figure 7-18]

Engine Cooling Systems

The burning fuel within the cylinders produces intense heat, most of which is expelled through the exhaust system. Much of the remaining heat, however, must be removed, or at least dissipated, to prevent the engine from overheating. Otherwise, the extremely high engine temperatures can lead to loss of power, excessive oil consumption, detonation, and serious engine damage.

While the oil system is vital to the internal cooling of the engine, an additional method of cooling is necessary for the engine's external surface. Most small aircraft are air cooled, although some are liquid cooled.

Figure 7-18. *Always check the engine oil level during the preflight inspection.*

Air cooling is accomplished by air flowing into the engine compartment through openings in front of the engine cowling. Baffles route this air over fins attached to the engine cylinders, and other parts of the engine, where the air absorbs the engine heat. Expulsion of the hot air takes place through one or more openings in the lower, aft portion of the engine cowling. [Figure 7-19]

The outside air enters the engine compartment through an inlet behind the propeller hub. Baffles direct it to the hottest parts of the engine, primarily the cylinders, which have fins that increase the area exposed to the airflow.

The air cooling system is less effective during ground operations, takeoffs, go-arounds, and other periods of high-power, low-airspeed operation. Conversely, high-speed descents provide excess air and can shock cool the engine, subjecting it to abrupt temperature fluctuations.

Operating the engine at higher than its designed temperature can cause loss of power, excessive oil consumption, and detonation. It will also lead to serious permanent damage, such as scoring the cylinder walls, damaging the pistons and rings, and burning and warping the valves. Monitoring the flight deck engine temperature instruments aids in avoiding high operating temperature.

Under normal operating conditions in aircraft not equipped with cowl flaps, the engine temperature can be controlled

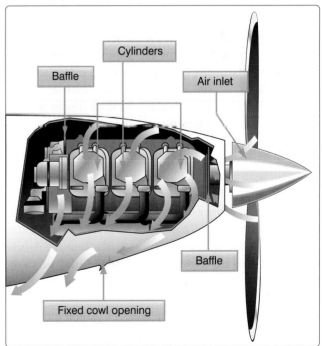

Figure 7-19. *Outside air aids in cooling the engine.*

by changing the airspeed or the power output of the engine. High engine temperatures can be decreased by increasing the airspeed and/or reducing the power.

The oil temperature gauge gives an indirect and delayed indication of rising engine temperature, but can be used for determining engine temperature if this is the only means available.

Most aircraft are equipped with a cylinder-head temperature gauge that indicates a direct and immediate cylinder temperature change. This instrument is calibrated in degrees Celsius or Fahrenheit and is usually color coded with a green arc to indicate the normal operating range. A red line on the instrument indicates maximum allowable cylinder head temperature.

To avoid excessive cylinder head temperatures, increase airspeed, enrich the fuel-air mixture, and/or reduce power. Any of these procedures help to reduce the engine temperature. On aircraft equipped with cowl flaps, use the cowl flap positions to control the temperature. Cowl flaps are hinged covers that fit over the opening through which the hot air is expelled. If the engine temperature is low, the cowl flaps can be closed, thereby restricting the flow of expelled hot air and increasing engine temperature. If the engine temperature is high, the cowl flaps can be opened to permit a greater flow of air through the system, thereby decreasing the engine temperature.

Exhaust Systems

Engine exhaust systems vent the burned combustion gases overboard, provide heat for the cabin, and defrost the windscreen. An exhaust system has exhaust piping attached to the cylinders, as well as a muffler and a muffler shroud. The exhaust gases are pushed out of the cylinder through the exhaust valve and then through the exhaust pipe system to the atmosphere.

For cabin heat, outside air is drawn into the air inlet and is ducted through a shroud around the muffler. The muffler is heated by the exiting exhaust gases and, in turn, heats the air around the muffler. This heated air is then ducted to the cabin for heat and defrost applications. The heat and defrost are controlled in the flight deck and can be adjusted to the desired level.

Exhaust gases contain large amounts of carbon monoxide, which is odorless and colorless. Carbon monoxide is deadly, and its presence is virtually impossible to detect. To ensure that exhaust gases are properly expelled, the exhaust system must be in good condition and free of cracks.

Some exhaust systems have an EGT probe. This probe transmits the EGT to an instrument in the flight deck. The EGT gauge measures the temperature of the gases at the exhaust manifold. This temperature varies with the ratio of fuel to air entering the cylinders and can be used as a basis for regulating the fuel-air mixture. The EGT gauge is highly accurate in indicating the correct fuel-air mixture setting. When using the EGT to aid in leaning the fuel-air mixture, fuel consumption can be reduced. For specific procedures, refer to the manufacturer's recommendations for leaning the fuel-air mixture.

Starting System

Most small aircraft use a direct-cranking electric starter system. This system consists of a source of electricity, wiring, switches, and solenoids to operate the starter and a starter motor. Most aircraft have starters that automatically engage and disengage when operated, but some older aircraft have starters that are mechanically engaged by a lever actuated by the pilot. The starter engages the aircraft flywheel, rotating the engine at a speed that allows the engine to start and maintain operation.

Electrical power for starting is usually supplied by an onboard battery, but can also be supplied by external power through an external power receptacle. When the battery switch is turned on, electricity is supplied to the main power bus bar through the battery solenoid. Both the starter and the starter switch draw current from the main bus bar, but the starter will not operate until the starting solenoid is energized by the starter switch being turned to the "start" position. When the starter switch is released from the "start" position, the solenoid removes power from the starter motor. The starter motor is protected from being driven by the engine through a clutch in the starter drive that allows the engine to run faster than the starter motor. *[Figure 7-20]*

When starting an engine, the rules of safety and courtesy should be strictly observed. One of the most important safety rules is to ensure there is no one near the propeller prior to starting the engine. In addition, the wheels should be chocked and the brakes set to avoid hazards caused by unintentional movement. To avoid damage to the propeller and property, the aircraft should be in an area where the propeller will not stir up gravel or dust.

Combustion

During normal combustion, the fuel-air mixture burns in a very controlled and predictable manner. In a spark ignition engine, the process occurs in a fraction of a second. The mixture actually begins to burn at the point where it is ignited

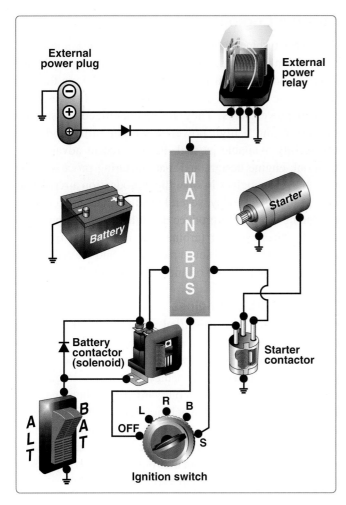

Figure 7-20. *Typical starting circuit.*

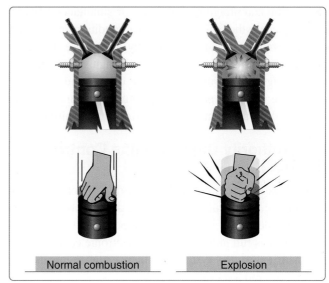

Figure 7-21. *Normal combustion and explosive combustion.*

by the spark plugs. It then burns away from the plugs until it is completely consumed. This type of combustion causes a smooth build-up of temperature and pressure and ensures that the expanding gases deliver the maximum force to the piston at exactly the right time in the power stroke. *[Figure 7-21]* Detonation is an uncontrolled, explosive ignition of the fuel-air mixture within the cylinder's combustion chamber. It causes excessive temperatures and pressures which, if not corrected, can quickly lead to failure of the piston, cylinder, or valves. In less severe cases, detonation causes engine overheating, roughness, or loss of power.

Detonation is characterized by high cylinder head temperatures and is most likely to occur when operating at high power settings. Common operational causes of detonation are:

- Use of a lower fuel grade than that specified by the aircraft manufacturer

- Operation of the engine with extremely high manifold pressures in conjunction with low rpm

- Operation of the engine at high power settings with an excessively lean mixture

- Maintaining extended ground operations or steep climbs in which cylinder cooling is reduced

Detonation may be avoided by following these basic guidelines during the various phases of ground and flight operations:

- Ensure that the proper grade of fuel is used.

- Keep the cowl flaps (if available) in the full-open position while on the ground to provide the maximum airflow through the cowling.

- Use an enriched fuel mixture, as well as a shallow climb angle, to increase cylinder cooling during takeoff and initial climb.

- Avoid extended, high power, steep climbs.

- Develop the habit of monitoring the engine instruments to verify proper operation according to procedures established by the manufacturer.

Preignition occurs when the fuel-air mixture ignites prior to the engine's normal ignition event. Premature burning is usually caused by a residual hot spot in the combustion chamber, often created by a small carbon deposit on a spark plug, a cracked spark plug insulator, or other damage in the cylinder that causes a part to heat sufficiently to ignite the fuel-air charge. Preignition causes the engine to lose power and produces high operating temperature. As with detonation, preignition may also cause severe engine damage because the expanding gases exert excessive pressure on the piston while still on its compression stroke.

Detonation and preignition often occur simultaneously and one may cause the other. Since either condition causes high engine temperature accompanied by a decrease in engine performance, it is often difficult to distinguish between the two. Using the recommended grade of fuel and operating the engine within its proper temperature, pressure, and rpm ranges reduce the chance of detonation or preignition.

Full Authority Digital Engine Control (FADEC)

FADEC is a system consisting of a digital computer and ancillary components that control an aircraft's engine and propeller. First used in turbine-powered aircraft, and referred to as full authority digital electronic control, these sophisticated control systems are increasingly being used in piston powered aircraft.

In a spark-ignition reciprocating engine, the FADEC uses speed, temperature, and pressure sensors to monitor the status of each cylinder. A digital computer calculates the ideal pulse for each injector and adjusts ignition timing as necessary to achieve optimal performance. In a compression-ignition engine, the FADEC operates similarly and performs all of the same functions, excluding those specifically related to the spark ignition process.

FADEC systems eliminate the need for magnetos, carburetor heat, mixture controls, and engine priming. A single throttle lever is characteristic of an aircraft equipped with a FADEC system. The pilot simply positions the throttle lever to a desired detent, such as start, idle, cruise power, or max power, and the FADEC system adjusts the engine and propeller automatically for the mode selected. There is no need for the pilot to monitor or control the fuel-air mixture.

During aircraft starting, the FADEC primes the cylinders, adjusts the mixture, and positions the throttle based on engine temperature and ambient pressure. During cruise flight, the FADEC constantly monitors the engine and adjusts fuel flow and ignition timing individually in each cylinder. This precise control of the combustion process often results in decreased fuel consumption and increased horsepower.

FADEC systems are considered an essential part of the engine and propeller control and may be powered by the aircraft's main electrical system. In many aircraft, FADEC uses power from a separate generator connected to the engine. In either case, there must be a backup electrical source available because failure of a FADEC system could result in a complete loss of engine thrust. To prevent loss of thrust, two separate and identical digital channels are incorporated for redundancy. Each channel is capable of providing all engine and propeller functions without limitations.

Turbine Engines

An aircraft turbine engine consists of an air inlet, compressor, combustion chambers, a turbine section, and exhaust. Thrust is produced by increasing the velocity of the air flowing through the engine. Turbine engines are highly desirable aircraft powerplants. They are characterized by smooth operation and a high power-to-weight ratio, and they use readily available jet fuel. Prior to recent advances in material, engine design, and manufacturing processes, the use of turbine engines in small/light production aircraft was cost prohibitive. Today, several aviation manufacturers are producing or plan to produce small/light turbine-powered aircraft. These smaller turbine-powered aircraft typically seat between three and seven passengers and are referred to as very light jets (VLJs) or microjets. *[Figure 7-22]*

Types of Turbine Engines

Turbine engines are classified according to the type of compressors they use. There are three types of compressors—centrifugal flow, axial flow, and centrifugal-axial flow. Compression of inlet air is achieved in a centrifugal flow engine by accelerating air outward perpendicular to the longitudinal axis of the machine. The axial-flow engine compresses air by a series of rotating and stationary airfoils moving the air parallel to the longitudinal axis. The centrifugal-axial flow design uses both kinds of compressors to achieve the desired compression.

The path the air takes through the engine and how power is produced determines the type of engine. There are four types of aircraft turbine engines—turbojet, turboprop, turbofan, and turboshaft.

Turbojet

The turbojet engine consists of four sections—compressor, combustion chamber, turbine section, and exhaust. The compressor section passes inlet air at a high rate of speed to

Figure 7-22. *Eclipse 500 VLJ.*

the combustion chamber. The combustion chamber contains the fuel inlet and igniter for combustion. The expanding air drives a turbine, which is connected by a shaft to the compressor, sustaining engine operation. The accelerated exhaust gases from the engine provide thrust. This is a basic application of compressing air, igniting the fuel-air mixture, producing power to self-sustain the engine operation, and exhaust for propulsion. [Figure 7-23]

Turbojet engines are limited in range and endurance. They are also slow to respond to throttle applications at slow compressor speeds.

Turboprop

A turboprop engine is a turbine engine that drives a propeller through a reduction gear. The exhaust gases drive a power turbine connected by a shaft that drives the reduction gear assembly. Reduction gearing is necessary in turboprop engines because optimum propeller performance is achieved at much slower speeds than the engine's operating rpm. Turboprop engines are a compromise between turbojet engines and reciprocating powerplants. Turboprop engines are most efficient at speeds between 250 and 400 mph and altitudes between 18,000 and 30,000 feet. They also perform well at the slow airspeeds required for takeoff and landing and are fuel efficient. The minimum specific fuel consumption

of the turboprop engine is normally available in the altitude range of 25,000 feet to the tropopause. [Figure 7-24]

Turbofan

Turbofans were developed to combine some of the best features of the turbojet and the turboprop. Turbofan engines are designed to create additional thrust by diverting a secondary airflow around the combustion chamber. The turbofan bypass air generates increased thrust, cools the engine, and aids in exhaust noise suppression. This provides turbojet-type cruise speed and lower fuel consumption.

The inlet air that passes through a turbofan engine is usually divided into two separate streams of air. One stream passes through the engine core, while a second stream bypasses the engine core. It is this bypass stream of air that is responsible for the term "bypass engine." A turbofan's bypass ratio refers to the ratio of the mass airflow that passes through the fan divided by the mass airflow that passes through the engine core. [Figure 7-25]

Turboshaft

The fourth common type of jet engine is the turboshaft. [Figure 7-26] It delivers power to a shaft that drives something other than a propeller. The biggest difference between a turbojet and turboshaft engine is that on a

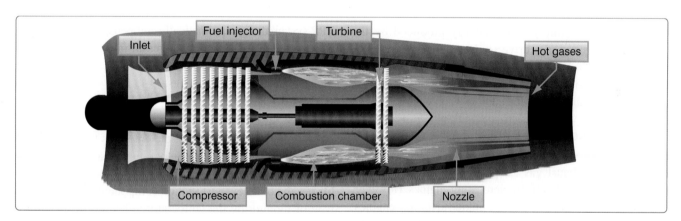

Figure 7-23. *Turbojet engine.*

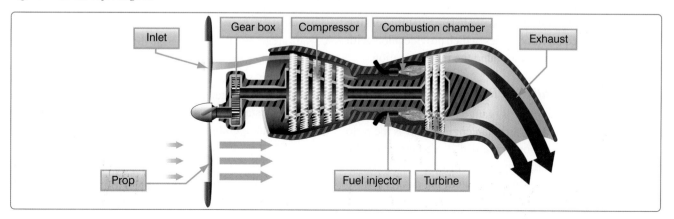

Figure 7-24. *Turboprop engine.*

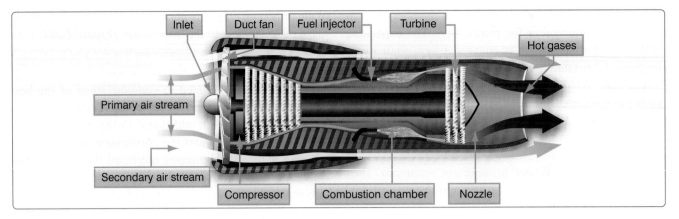

Figure 7-25. *Turbofan engine.*

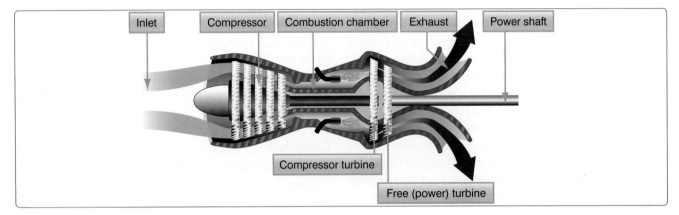

Figure 7-26. *Turboshaft engine.*

turboshaft engine, most of the energy produced by the expanding gases is used to drive a turbine rather than produce thrust. Many helicopters use a turboshaft gas turbine engine. In addition, turboshaft engines are widely used as auxiliary power units on large aircraft.

Turbine Engine Instruments

Engine instruments that indicate oil pressure, oil temperature, engine speed, exhaust gas temperature, and fuel flow are common to both turbine and reciprocating engines. However, there are some instruments that are unique to turbine engines. These instruments provide indications of engine pressure ratio, turbine discharge pressure, and torque. In addition, most gas turbine engines have multiple temperature-sensing instruments, called thermocouples, which provide pilots with temperature readings in and around the turbine section.

Engine Pressure Ratio (EPR)

An engine pressure ratio (EPR) gauge is used to indicate the power output of a turbojet/turbofan engine. EPR is the ratio of turbine discharge to compressor inlet pressure. Pressure measurements are recorded by probes installed in the engine inlet and at the exhaust. Once collected, the data is sent to a differential pressure transducer, which is indicated on a flight deck EPR gauge.

EPR system design automatically compensates for the effects of airspeed and altitude. Changes in ambient temperature require a correction be applied to EPR indications to provide accurate engine power settings.

Exhaust Gas Temperature (EGT)

A limiting factor in a gas turbine engine is the temperature of the turbine section. The temperature of a turbine section must be monitored closely to prevent overheating the turbine blades and other exhaust section components. One common way of monitoring the temperature of a turbine section is with an EGT gauge. EGT is an engine operating limit used to monitor overall engine operating conditions.

Variations of EGT systems bear different names based on the location of the temperature sensors. Common turbine temperature sensing gauges include the turbine inlet temperature (TIT) gauge, turbine outlet temperature (TOT) gauge, interstage turbine temperature (ITT) gauge, and turbine gas temperature (TGT) gauge.

Torquemeter

Turboprop/turboshaft engine power output is measured by the torquemeter. Torque is a twisting force applied to a shaft. The torquemeter measures power applied to the shaft.

Turboprop and turboshaft engines are designed to produce torque for driving a propeller. Torquemeters are calibrated in percentage units, foot-pounds, or psi.

N_1 *Indicator*

N_1 represents the rotational speed of the low pressure compressor and is presented on the indicator as a percentage of design rpm. After start, the speed of the low pressure compressor is governed by the N_1 turbine wheel. The N_1 turbine wheel is connected to the low pressure compressor through a concentric shaft.

N_2 *Indicator*

N_2 represents the rotational speed of the high pressure compressor and is presented on the indicator as a percentage of design rpm. The high pressure compressor is governed by the N_2 turbine wheel. The N_2 turbine wheel is connected to the high pressure compressor through a concentric shaft. *[Figure 7-27]*

Turbine Engine Operational Considerations

The great variety of turbine engines makes it impractical to cover specific operational procedures, but there are certain operational considerations common to all turbine engines. They are engine temperature limits, foreign object damage, hot start, compressor stall, and flameout.

Engine Temperature Limitations

The highest temperature in any turbine engine occurs at the turbine inlet. TIT is therefore usually the limiting factor in turbine engine operation.

Thrust Variations

Turbine engine thrust varies directly with air density. As air density decreases, so does thrust. Additionally, because air density decreases with an increase in temperature, increased

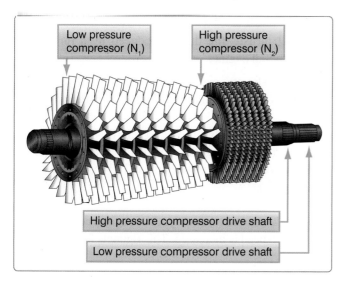

Low pressure compressor (N_1)

High pressure compressor (N_2)

High pressure compressor drive shaft

Low pressure compressor drive shaft

Figure 7-27. *Dual-spool axial-flow compressor.*

temperatures also results in decreased thrust. While both turbine and reciprocating powered engines are affected to some degree by high relative humidity, turbine engines will experience a negligible loss of thrust, while reciprocating engines a significant loss of brake horsepower.

Foreign Object Damage (FOD)

Due to the design and function of a turbine engine's air inlet, the possibility of ingestion of debris always exists. This causes significant damage, particularly to the compressor and turbine sections. When ingestion of debris occurs, it is called foreign object damage (FOD). Typical FOD consists of small nicks and dents caused by ingestion of small objects from the ramp, taxiway, or runway, but FOD damage caused by bird strikes or ice ingestion also occur. Sometimes FOD results in total destruction of an engine.

Prevention of FOD is a high priority. Some engine inlets have a tendency to form a vortex between the ground and the inlet during ground operations. A vortex dissipater may be installed on these engines. Other devices, such as screens and/or deflectors, may also be utilized. Preflight procedures include a visual inspection for any sign of FOD.

Turbine Engine Hot/Hung Start

When the EGT exceeds the safe limit of an aircraft, it experiences a "hot start." This is caused by too much fuel entering the combustion chamber or insufficient turbine rpm. Any time an engine has a hot start, refer to the AFM/POH or an appropriate maintenance manual for inspection requirements.

If the engine fails to accelerate to the proper speed after ignition or does not accelerate to idle rpm, a hung or false start has occurred. A hung start may be caused by an insufficient starting power source or fuel control malfunction.

Compressor Stalls

Compressor blades are small airfoils and are subject to the same aerodynamic principles that apply to any airfoil. A compressor blade has an AOA that is a result of inlet air velocity and the compressor's rotational velocity. These two forces combine to form a vector, which defines the airfoil's actual AOA to the approaching inlet air.

A compressor stall is an imbalance between the two vector quantities, inlet velocity, and compressor rotational speed. Compressor stalls occur when the compressor blades' AOA exceeds the critical AOA. At this point, smooth airflow is interrupted and turbulence is created with pressure fluctuations. Compressor stalls cause air flowing in the compressor to slow down and stagnate, sometimes reversing direction. *[Figure 7-28]*

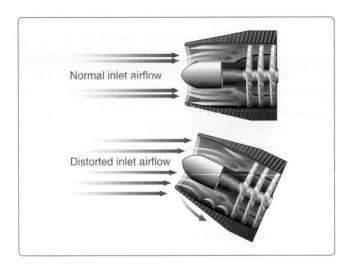

Figure 7-28. *Comparison of normal and distorted airflow into the compressor section.*

Compressor stalls can be transient and intermittent or steady and severe. Indications of a transient/intermittent stall are usually an intermittent "bang" as backfire and flow reversal take place. If the stall develops and becomes steady, strong vibration and a loud roar may develop from the continuous flow reversal. Often, the flight deck gauges do not show a mild or transient stall, but they do indicate a developed stall. Typical instrument indications include fluctuations in rpm and an increase in exhaust gas temperature. Most transient stalls are not harmful to the engine and often correct themselves after one or two pulsations. The possibility of severe engine damage from a steady state stall is immediate. Recovery must be accomplished by quickly reducing power, decreasing the aircraft's AOA, and increasing airspeed.

Although all gas turbine engines are subject to compressor stalls, most models have systems that inhibit them. One system uses a variable inlet guide vane (VIGV) and variable stator vanes that direct the incoming air into the rotor blades at an appropriate angle. To prevent air pressure stalls, operate the aircraft within the parameters established by the manufacturer. If a compressor stall does develop, follow the procedures recommended in the AFM/POH.

Flameout

A flameout occurs in the operation of a gas turbine engine in which the fire in the engine unintentionally goes out. If the rich limit of the fuel-air ratio is exceeded in the combustion chamber, the flame will blow out. This condition is often referred to as a rich flameout. It generally results from very fast engine acceleration where an overly rich mixture causes the fuel temperature to drop below the combustion temperature. It may also be caused by insufficient airflow to support combustion.

A more common flameout occurrence is due to low fuel pressure and low engine speeds, which typically are associated with high-altitude flight. This situation may also occur with the engine throttled back during a descent, which can set up the lean-condition flameout. A weak mixture can easily cause the flame to die out, even with a normal airflow through the engine.

Any interruption of the fuel supply can result in a flameout. This may be due to prolonged unusual attitudes, a malfunctioning fuel control system, turbulence, icing, or running out of fuel.

Symptoms of a flameout normally are the same as those following an engine failure. If the flameout is due to a transitory condition, such as an imbalance between fuel flow and engine speed, an airstart may be attempted once the condition is corrected. In any case, pilots must follow the applicable emergency procedures outlined in the AFM/POH. Generally these procedures contain recommendations concerning altitude and airspeed where the airstart is most likely to be successful.

Performance Comparison

It is possible to compare the performance of a reciprocating powerplant and different types of turbine engines. For the comparison to be accurate, thrust horsepower (usable horsepower) for the reciprocating powerplant must be used rather than brake horsepower, and net thrust must be used for the turbine-powered engines. In addition, aircraft design configuration and size must be approximately the same.

When comparing performance, the following definitions are useful:

- Brake horsepower (BHP)—the horsepower actually delivered to the output shaft. Brake horsepower is the actual usable horsepower.

- Net thrust—the thrust produced by a turbojet or turbofan engine.

- Thrust horsepower (THP)—the horsepower equivalent of the thrust produced by a turbojet or turbofan engine.

Equivalent shaft horsepower (ESHP)—with respect to turboprop engines, the sum of the shaft horsepower (SHP) delivered to the propeller and THP produced by the exhaust gases.

Figure 7-29 shows how four types of engines compare in net thrust as airspeed is increased. This figure is for explanatory

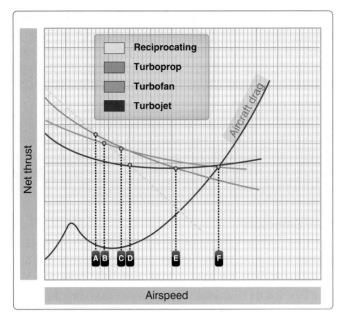

Figure 7-29. *Engine net thrust versus aircraft speed and drag. Points* **A** *through* **F** *are explained in the text below.*

purposes only and is not for specific models of engines. The following are the four types of engines:

- Reciprocating powerplant
- Turbine, propeller combination (turboprop)
- Turbine engine incorporating a fan (turbofan)
- Turbojet (pure jet)

By plotting the performance curve for each engine, a comparison can be made of maximum aircraft speed variation with the type of engine used. Since the graph is only a means of comparison, numerical values for net thrust, aircraft speed, and drag are not included.

Comparison of the four powerplants on the basis of net thrust makes certain performance capabilities evident. In the speed range shown to the left of line A, the reciprocating powerplant outperforms the other three types. The turboprop outperforms the turbofan in the range to the left of line C. The turbofan engine outperforms the turbojet in the range to the left of line F. The turbofan engine outperforms the reciprocating powerplant to the right of line B and the turboprop to the right of line C. The turbojet outperforms the reciprocating powerplant to the right of line D, the turboprop to the right of line E, and the turbofan to the right of line F.

The points where the aircraft drag curve intersects the net thrust curves are the maximum aircraft speeds. The vertical lines from each of the points to the baseline of the graph indicate that the turbojet aircraft can attain a higher maximum speed than aircraft equipped with the other types of engines. Aircraft equipped with the turbofan engine attains a higher

maximum speed than aircraft equipped with a turboprop or reciprocating powerplant.

Airframe Systems

Fuel, electrical, hydraulic, and oxygen systems make up the airframe systems.

Fuel Systems

The fuel system is designed to provide an uninterrupted flow of clean fuel from the fuel tanks to the engine. The fuel must be available to the engine under all conditions of engine power, altitude, attitude, and during all approved flight maneuvers. Two common classifications apply to fuel systems in small aircraft: gravity-feed and fuel-pump systems.

Gravity-Feed System

The gravity-feed system utilizes the force of gravity to transfer the fuel from the tanks to the engine. For example, on high-wing airplanes, the fuel tanks are installed in the wings. This places the fuel tanks above the carburetor, and the fuel is gravity fed through the system and into the carburetor. If the design of the aircraft is such that gravity cannot be used to transfer fuel, fuel pumps are installed. For example, on low-wing airplanes, the fuel tanks in the wings are located below the carburetor. *[Figure 7-30]*

Fuel-Pump System

Aircraft with fuel-pump systems have two fuel pumps. The main pump system is engine driven with an electrically-driven auxiliary pump provided for use in engine starting and in the event the engine pump fails. The auxiliary pump, also known as a boost pump, provides added reliability to the fuel system. The electrically-driven auxiliary pump is controlled by a switch in the flight deck.

Fuel Primer

Both gravity-feed and fuel-pump systems may incorporate a fuel primer into the system. The fuel primer is used to draw fuel from the tanks to vaporize fuel directly into the cylinders prior to starting the engine. During cold weather, when engines are difficult to start, the fuel primer helps because there is not enough heat available to vaporize the fuel in the carburetor. It is important to lock the primer in place when it is not in use. If the knob is free to move, it may vibrate out of position during flight which may cause an excessively rich fuel-air mixture. To avoid overpriming, read the priming instructions for the aircraft.

Fuel Tanks

The fuel tanks, normally located inside the wings of an airplane, have a filler opening on top of the wing through which they can be filled. A filler cap covers this opening.

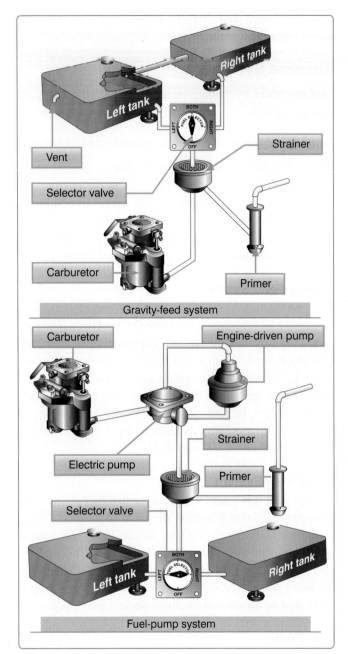

Figure 7-30. *Gravity-feed and fuel-pump systems.*

The tanks are vented to the outside to maintain atmospheric pressure inside the tank. They may be vented through the filler cap or through a tube extending through the surface of the wing. Fuel tanks also include an overflow drain that may stand alone or be collocated with the fuel tank vent. This allows fuel to expand with increases in temperature without damage to the tank itself. If the tanks have been filled on a hot day, it is not unusual to see fuel coming from the overflow drain.

Fuel Gauges

The fuel quantity gauges indicate the amount of fuel measured by a sensing unit in each fuel tank and is displayed in gallons or pounds. Aircraft certification rules require accuracy in fuel gauges only when they read "empty." Any reading other than "empty" should be verified. Do not depend solely on the accuracy of the fuel quantity gauges. Always visually check the fuel level in each tank during the preflight inspection, and then compare it with the corresponding fuel quantity indication.

If a fuel pump is installed in the fuel system, a fuel pressure gauge is also included. This gauge indicates the pressure in the fuel lines. The normal operating pressure can be found in the AFM/POH or on the gauge by color coding.

Fuel Selectors

The fuel selector valve allows selection of fuel from various tanks. A common type of selector valve contains four positions: LEFT, RIGHT, BOTH, and OFF. Selecting the LEFT or RIGHT position allows fuel to feed only from the respective tank, while selecting the BOTH position feeds fuel from both tanks. The LEFT or RIGHT position may be used to balance the amount of fuel remaining in each wing tank. *[Figure 7-31]*

Fuel placards show any limitations on fuel tank usage, such as "level flight only" and/or "both" for landings and takeoffs.

Regardless of the type of fuel selector in use, fuel consumption should be monitored closely to ensure that a tank does not run completely out of fuel. Running a fuel tank dry does not only cause the engine to stop, but running for prolonged periods on one tank causes an unbalanced fuel load between tanks. Running a tank completely dry may allow air to enter the fuel system and cause vapor lock, which makes it difficult to restart the engine. On fuel-injected engines, the fuel becomes so hot it vaporizes in the fuel line, not allowing fuel to reach the cylinders.

Figure 7-31. *Fuel selector valve.*

Fuel Strainers, Sumps, and Drains

After leaving the fuel tank and before it enters the carburetor, the fuel passes through a strainer that removes any moisture and other sediments in the system. Since these contaminants are heavier than aviation fuel, they settle in a sump at the bottom of the strainer assembly. A sump is a low point in a fuel system and/or fuel tank. The fuel system may contain a sump, a fuel strainer, and fuel tank drains, which may be collocated.

The fuel strainer should be drained before each flight. Fuel samples should be drained and checked visually for water and contaminants.

Water in the sump is hazardous because in cold weather the water can freeze and block fuel lines. In warm weather, it can flow into the carburetor and stop the engine. If water is present in the sump, more water in the fuel tanks is probable, and they should be drained until there is no evidence of water. Never take off until all water and contaminants have been removed from the engine fuel system.

Because of the variation in fuel systems, become thoroughly familiar with the systems that apply to the aircraft being flown. Consult the AFM/POH for specific operating procedures.

Fuel Grades

Aviation gasoline (AVGAS) is identified by an octane or performance number (grade), which designates the antiknock value or knock resistance of the fuel mixture in the engine cylinder. The higher the grade of gasoline, the more pressure the fuel can withstand without detonating. Lower grades of fuel are used in lower-compression engines because these fuels ignite at a lower temperature. Higher grades are used in higher-compression engines because they ignite at higher temperatures, but not prematurely. If the proper grade of fuel is not available, use the next higher grade as a substitute. Never use a grade lower than recommended. This can cause the cylinder head temperature and engine oil temperature to exceed their normal operating ranges, which may result in detonation.

Several grades of AVGAS are available. Care must be exercised to ensure that the correct aviation grade is being used for the specific type of engine. The proper fuel grade is stated in the AFM/POH, on placards in the flight deck, and next to the filler caps. Automobile gas should NEVER be used in aircraft engines unless the aircraft has been modified with a Supplemental Type Certificate (STC) issued by the Federal Aviation Administration (FAA).

The current method identifies AVGAS for aircraft with reciprocating engines by the octane and performance number, along with the abbreviation AVGAS. These aircraft use AVGAS 80, 100, and 100LL. Although AVGAS 100LL performs the same as grade 100, the "LL" indicates it has a low lead content. Fuel for aircraft with turbine engines is classified as JET A, JET A-1, and JET B. Jet fuel is basically kerosene and has a distinctive kerosene smell. Since use of the correct fuel is critical, dyes are added to help identify the type and grade of fuel. *[Figure 7-32]*

In addition to the color of the fuel itself, the color-coding system extends to decals and various airport fuel handling equipment. For example, all AVGAS is identified by name, using white letters on a red background. In contrast, turbine fuels are identified by white letters on a black background.

Special Airworthiness Information Bulleting (SAIB) NE-11-15 advises that grade 100VLL AVGAS is acceptable for use on aircraft and engines. 100VLL meets all performance requirements of grades 80, 91, 100, and 100LL; meets the approved operating limitations for aircraft and engines certificated to operate with these other grades of AVGAS; and is basically identical to 100LL AVGAS. The lead content of 100VLL is reduced by about 19 percent. 100VLL is blue like 100LL and virtually indistinguishable.

Fuel Contamination

Accidents attributed to powerplant failure from fuel contamination have often been traced to:

- Inadequate preflight inspection by the pilot
- Servicing aircraft with improperly filtered fuel from small tanks or drums
- Storing aircraft with partially filled fuel tanks
- Lack of proper maintenance

Fuel should be drained from the fuel strainer quick drain and from each fuel tank sump into a transparent container and then checked for dirt and water. When the fuel strainer is being drained, water in the tank may not appear until all the fuel has been drained from the lines leading to the tank. This indicates that water remains in the tank and is not forcing the fuel out of the fuel lines leading to the fuel strainer. Therefore, drain enough fuel from the fuel strainer to be certain that fuel is being drained from the tank. The amount depends on

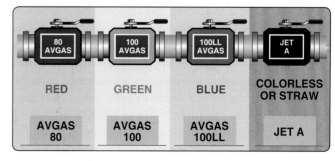

Figure 7-32. *Aviation fuel color-coding system.*

the length of fuel line from the tank to the drain. If water or other contaminants are found in the first sample, drain further samples until no trace appears.

Water may also remain in the fuel tanks after the drainage from the fuel strainer has ceased to show any trace of water. This residual water can be removed only by draining the fuel tank sump drains.

Water is the principal fuel contaminant. Suspended water droplets in the fuel can be identified by a cloudy appearance of the fuel, or by the clear separation of water from the colored fuel, which occurs after the water has settled to the bottom of the tank. As a safety measure, the fuel sumps should be drained before every flight during the preflight inspection.

Fuel tanks should be filled after each flight or after the last flight of the day to prevent moisture condensation within the tank. To prevent fuel contamination, avoid refueling from cans and drums.

In remote areas or in emergency situations, there may be no alternative to refueling from sources with inadequate anti-contamination systems. While a chamois skin and funnel may be the only possible means of filtering fuel, using them is hazardous. Remember, the use of a chamois does not always ensure decontaminated fuel. Worn-out chamois do not filter water; neither will a new, clean chamois that is already water-wet or damp. Most imitation chamois skins do not filter water.

Fuel System Icing

Ice formation in the aircraft fuel system results from the presence of water in the fuel system. This water may be undissolved or dissolved. One condition of undissolved water is entrained water that consists of minute water particles suspended in the fuel. This may occur as a result of mechanical agitation of free water or conversion of dissolved water through temperature reduction. Entrained water settles out in time under static conditions and may or may not be drained during normal servicing, depending on the rate at which it is converted to free water. In general, it is not likely that all entrained water can ever be separated from fuel under field conditions. The settling rate depends on a series of factors including temperature, quiescence, and droplet size.

The droplet size varies depending upon the mechanics of formation. Usually, the particles are so small as to be invisible to the naked eye, but in extreme cases, can cause slight haziness in the fuel. Water in solution cannot be removed except by dehydration or by converting it through temperature reduction to entrained, then to free water.

Another condition of undissolved water is free water that may be introduced as a result of refueling or the settling of entrained water that collects at the bottom of a fuel tank. Free water is usually present in easily detected quantities at the bottom of the tank, separated by a continuous interface from the fuel above. Free water can be drained from a fuel tank through the sump drains, which are provided for that purpose. Free water, frozen on the bottom of reservoirs, such as the fuel tanks and fuel filter, may render water drains useless and can later melt releasing the water into the system thereby causing engine malfunction or stoppage. If such a condition is detected, the aircraft may be placed in a warm hangar to reestablish proper draining of these reservoirs, and all sumps and drains should be activated and checked prior to flight.

Entrained water (i.e., water in solution with petroleum fuels) constitutes a relatively small part of the total potential water in a particular system, the quantity dissolved being dependent on fuel temperature and the existing pressure and the water volubility characteristics of the fuel. Entrained water freezes in mid fuel and tends to stay in suspension longer since the specific gravity of ice is approximately the same as that of AVGAS.

Water in suspension may freeze and form ice crystals of sufficient size such that fuel screens, strainers, and filters may be blocked. Some of this water may be cooled further as the fuel enters carburetor air passages and causes carburetor metering component icing, when conditions are not otherwise conducive to this form of icing.

Prevention Procedures

The use of anti-icing additives for some aircraft has been approved as a means of preventing problems with water and ice in AVGAS. Some laboratory and flight testing indicates that the use of hexylene glycol, certain methanol derivatives, and ethylene glycol monoethyl ether (EGME) in small concentrations inhibit fuel system icing. These tests indicate that the use of EGME at a maximum 0.15 percent by volume concentration substantially inhibits fuel system icing under most operating conditions. The concentration of additives in the fuel is critical. Marked deterioration in additive effectiveness may result from too little or too much additive. Pilots should recognize that anti-icing additives are in no way a substitute or replacement for carburetor heat. Aircraft operating instructions involving the use of carburetor heat should be adhered to at all times when operating under atmospheric conditions conducive to icing.

Refueling Procedures

Static electricity is formed by the friction of air passing over the surfaces of an aircraft in flight and by the flow of fuel through the hose and nozzle during refueling. Nylon, Dacron, or wool clothing is especially prone to accumulate and discharge static electricity from the person to the funnel or nozzle. To guard against the possibility of static electricity igniting fuel fumes, a ground wire should be attached to the aircraft before the fuel cap is removed from the tank. Because both the aircraft and refueler have different static charges, bonding both components to each other is critical. By bonding both components to each other, the static differential charge is equalized. The refueling nozzle should be bonded to the aircraft before refueling begins and should remain bonded throughout the refueling process. When a fuel truck is used, it should be grounded prior to the fuel nozzle contacting the aircraft.

If fueling from drums or cans is necessary, proper bonding and grounding connections are important. Drums should be placed near grounding posts, and the following sequence of connections observed:

1. Drum to ground

2. Ground to aircraft

3. Drum to aircraft or nozzle to aircraft before removing the fuel cap

When disconnecting, reverse the order.

The passage of fuel through a chamois increases the charge of static electricity and the danger of sparks. The aircraft must be properly grounded and the nozzle, chamois filter, and funnel bonded to the aircraft. If a can is used, it should be connected to either the grounding post or the funnel. Under no circumstances should a plastic bucket or similar nonconductive container be used in this operation.

Heating System

There are many different types of aircraft heating systems that are available depending on the type of aircraft. Regardless of which type or the safety features that accompany them, it is always important to reference the specific aircraft operator's manual and become knowledgeable about the heating system. Each has different repair and inspection criteria that should be precisely followed.

Fuel Fired Heaters

A fuel fired heater is a small mounted or portable space-heating device. The fuel is brought to the heater by using piping from a fuel tank, or taps into the aircraft's fuel system. A fan blows air into a combustion chamber, and a spark plug or ignition device lights the fuel-air mixture. A built-in safety switch prevents fuel from flowing unless the fan is working. Outside the combustion chamber, a second, larger diameter tube conducts air around the combustion tube's outer surface, and a second fan blows the warmed air into tubing to direct it towards the interior of the aircraft. Most gasoline heaters can produce between 5,000 and 50,000 British Thermal Units (BTU) per hour.

Fuel fired heaters require electricity to operate and are compatible with a 12-volt and 24-volt aircraft electrical system. The heater requires routine maintenance, such as regular inspection of the combustion tube and replacement of the igniter at periodic intervals. Because gasoline heaters are required to be vented, special care must be made to ensure the vents do not leak into the interior of the aircraft. Combustion byproducts include soot, sulfur dioxide, carbon dioxide, and some carbon monoxide. An improperly adjusted, fueled, or poorly maintained fuel heater can be dangerous.

Exhaust Heating Systems

Exhaust heating systems are the simplest type of aircraft heating system and are used on most light aircraft. Exhaust heating systems are used to route exhaust gases away from the engine and fuselage while reducing engine noise. The exhaust systems also serve as a heat source for the cabin and carburetor.

The risks of operating an aircraft with a defective exhaust heating system include carbon monoxide poisoning, a decrease in engine performance, and an increased potential for fire. Because of these risks, technicians should be aware of the rate of exhaust heating system deterioration and should thoroughly inspect all areas of the exhaust heating system to look for deficiencies inside and out.

Combustion Heater Systems

Combustion heaters or surface combustion heaters are often used to heat the cabin of larger, more expensive aircraft. This type of heater burns the aircraft's fuel in a combustion chamber or tube to develop required heat, and the air flowing around the tube is heated and ducted to the cabin. A combustion heater is an airtight burner chamber with a stainless-steel jacket. Fuel from the aircraft fuel system is ignited and burns to provide heat. Ventilation air is forced over the airtight burn chamber picking up heat, which is then dispersed into the cabin area.

When the heater control switch is turned on, airflow, ignition, and fuel are supplied to the heater. Airflow and ignition are constant within the burner chamber while the heater control switch is on. When heat is required, the temperature control is advanced, activating the thermostat. The thermostat (which senses ventilation air temperature) turns on the fuel solenoid allowing fuel to spray into the burner chamber. Fuel mixes with air inside the chamber and is ignited by the spark plug, producing heat.

The by-product, carbon monoxide, leaves the aircraft through the heater exhaust pipe. Air flowing over the outside of the burner chamber and inside the jacket of the heater absorbs the heat and carries it through ducts into the cabin. As the thermostat reaches its preset temperature, it turns off the fuel solenoid and stops the flow of fuel into the burner chamber. When ventilation air cools to the point that the thermostat again turns the fuel solenoid on, the burner starts again.

This method of heat is very safe as an overheat switch is provided on all combustion heaters, which is wired into the heater's electrical system to shut off the fuel in the case of malfunction. In the unlikely event that the heater fuel solenoid, located at the heater, remains open or the control switches fail, the remote fuel solenoid and/or fuel pump is shut off by the mechanical overheat switch, stopping all fuel flow to the system.

As opposed to the fuel fired cabin heaters that are used on most single-engine aircraft, it is unlikely for carbon monoxide poisoning to occur in combustion heaters. Combustion heaters have low pressure in the combustion tube that is vented through its exhaust into the air stream. The ventilation air on the outside of the combustion chamber is of higher pressure than on the inside, and ram air increases the pressure on the outside of the combustion tube. In the event a leak would develop in the combustion chamber, the higher-pressure air outside the chamber would travel into the chamber and out the exhaust.

Bleed Air Heating Systems

Bleed air heating systems are used on turbine-engine aircraft. Extremely hot compressor bleed air is ducted into a chamber where it is mixed with ambient or re-circulated air to cool the air to a useable temperature. The air mixture is then ducted into the cabin. This type of system contains several safety features to include temperature sensors that prevent excessive heat from entering the cabin, check valves to prevent a loss of compressor bleed air when starting the engine and when full power is required, and engine sensors to eliminate the bleed system if the engine becomes inoperative.

Electrical System

Most aircraft are equipped with either a 14- or a 28-volt direct current (DC) electrical system. A basic aircraft electrical system consists of the following components:

- Alternator/generator

- Battery
- Master/battery switch
- Alternator/generator switch
- Bus bar, fuses, and circuit breakers
- Voltage regulator
- Ammeter/loadmeter
- Associated electrical wiring

Engine-driven alternators or generators supply electric current to the electrical system. They also maintain a sufficient electrical charge in the battery. Electrical energy stored in a battery provides a source of electrical power for starting the engine and a limited supply of electrical power for use in the event the alternator or generator fails.

Most DC generators do not produce a sufficient amount of electrical current at low engine rpm to operate the entire electrical system. During operations at low engine rpm, the electrical needs must be drawn from the battery, which can quickly be depleted.

Alternators have several advantages over generators. Alternators produce sufficient current to operate the entire electrical system, even at slower engine speeds, by producing alternating current (AC), which is converted to DC. The electrical output of an alternator is more constant throughout a wide range of engine speeds.

Some aircraft have receptacles to which an external ground power unit (GPU) may be connected to provide electrical energy for starting. These are very useful, especially during cold weather starting. Follow the manufacturer's recommendations for engine starting using a GPU.

The electrical system is turned on or off with a master switch. Turning the master switch to the ON position provides electrical energy to all the electrical equipment circuits except the ignition system. Equipment that commonly uses the electrical system for its source of energy includes:

- Position lights
- Anticollision lights
- Landing lights
- Taxi lights
- Interior cabin lights
- Instrument lights
- Radio equipment
- Turn indicator
- Fuel gauges

- Electric fuel pump
- Stall warning system
- Pitot heat
- Starting motor

Many aircraft are equipped with a battery switch that controls the electrical power to the aircraft in a manner similar to the master switch. In addition, an alternator switch is installed that permits the pilot to exclude the alternator from the electrical system in the event of alternator failure. *[Figure 7-33]*

With the alternator half of the switch in the OFF position, the entire electrical load is placed on the battery. All nonessential electrical equipment should be turned off to conserve battery power.

A bus bar is used as a terminal in the aircraft electrical system to connect the main electrical system to the equipment using electricity as a source of power. This simplifies the wiring system and provides a common point from which voltage can be distributed throughout the system. *[Figure 7-34]*

Fuses or circuit breakers are used in the electrical system to protect the circuits and equipment from electrical overload. Spare fuses of the proper amperage limit should be carried in the aircraft to replace defective or blown fuses. Circuit breakers have the same function as a fuse but can be manually reset, rather than replaced, if an overload condition occurs in the electrical system. Placards at the fuse or circuit breaker panel identify the circuit by name and show the amperage limit.

Figure 7-33. *On this master switch, the left half is for the alternator and the right half is for the battery.*

An ammeter is used to monitor the performance of the aircraft electrical system. The ammeter shows if the alternator/generator is producing an adequate supply of electrical power. It also indicates whether or not the battery is receiving an electrical charge.

Ammeters are designed with the zero point in the center of the face and a negative or positive indication on either side. *[Figure 7-35]* When the pointer of the ammeter is on the plus side, it shows the charging rate of the battery. A minus indication means more current is being drawn from the battery than is being replaced. A full-scale minus deflection indicates a malfunction of the alternator/generator. A full-scale positive deflection indicates a malfunction of the regulator. In either case, consult the AFM/POH for appropriate action to be taken.

Not all aircraft are equipped with an ammeter. Some have a warning light that, when lighted, indicates a discharge in the system as a generator/alternator malfunction. Refer to the AFM/POH for appropriate action to be taken.

Another electrical monitoring indicator is a loadmeter. This type of gauge has a scale beginning with zero and shows the load being placed on the alternator/generator. *[Figure 7-35]* The loadmeter reflects the total percentage of the load placed on the generating capacity of the electrical system by the electrical accessories and battery. When all electrical components are turned off, it reflects only the amount of charging current demanded by the battery.

A voltage regulator controls the rate of charge to the battery by stabilizing the generator or alternator electrical output. The generator/alternator voltage output should be higher than the battery voltage. For example, a 12-volt battery would be fed by a generator/alternator system of approximately 14 volts. The difference in voltage keeps the battery charged.

Hydraulic Systems

There are multiple applications for hydraulic use in aircraft, depending on the complexity of the aircraft. For example, a hydraulic system is often used on small airplanes to operate wheel brakes, retractable landing gear, and some constant-speed propellers. On large airplanes, a hydraulic system is used for flight control surfaces, wing flaps, spoilers, and other systems.

A basic hydraulic system consists of a reservoir, pump (either hand, electric, or engine-driven), a filter to keep the fluid clean, a selector valve to control the direction of flow, a relief valve to relieve excess pressure, and an actuator. *[Figure 7-36]*

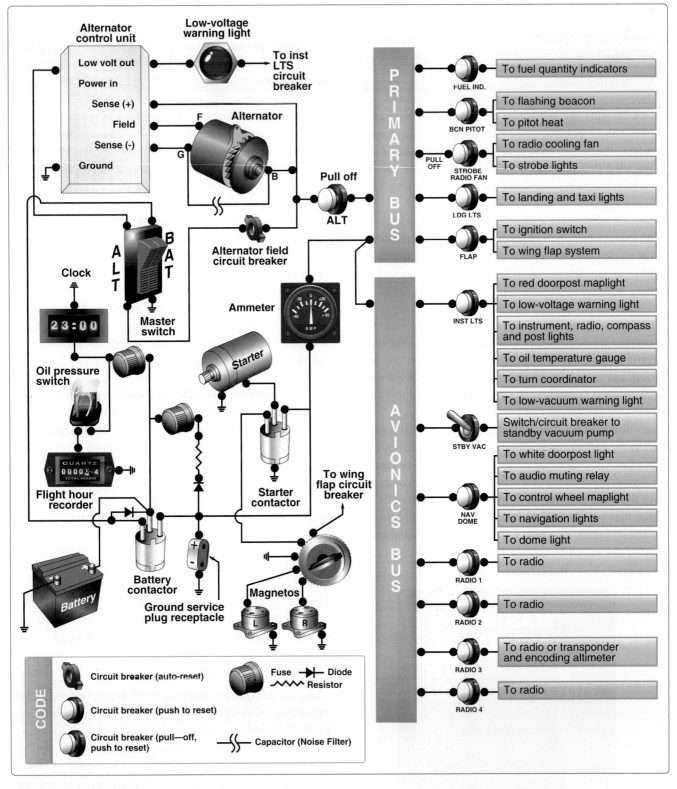

Figure 7-34. *Electrical system schematic.*

The hydraulic fluid is pumped through the system to an actuator or servo. A servo is a cylinder with a piston inside that turns fluid power into work and creates the power needed to move an aircraft system or flight control. Servos can be either single-acting or double-acting, based on the needs of the system. This means that the fluid can be applied to one or both sides of the servo, depending on the servo type. A single-acting servo provides power in one direction. The selector valve allows the fluid direction to be controlled. This is necessary for operations such as the extension and retraction of landing gear during which the fluid must work in two different directions. The relief valve provides an outlet

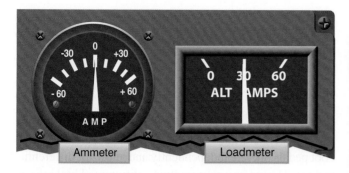

Figure 7-35. *Ammeter and loadmeter.*

for the system in the event of excessive fluid pressure in the system. Each system incorporates different components to meet the individual needs of different aircraft.

A mineral-based hydraulic fluid is the most widely used type for small aircraft. This type of hydraulic fluid, a kerosene-like petroleum product, has good lubricating properties, as well as additives to inhibit foaming and prevent the formation of corrosion. It is chemically stable, has very little viscosity change with temperature, and is dyed for identification. Since several types of hydraulic fluids are commonly used, an aircraft must be serviced with the type specified by the manufacturer. Refer to the AFM/POH or the Maintenance Manual.

Landing Gear

The landing gear forms the principal support of an aircraft on the surface. The most common type of landing gear consists of wheels, but aircraft can also be equipped with floats for water operations or skis for landing on snow. *[Figure 7-37]* The landing gear on small aircraft consists of three wheels: two main wheels (one located on each side of the fuselage) and a third wheel positioned either at the front or rear of the

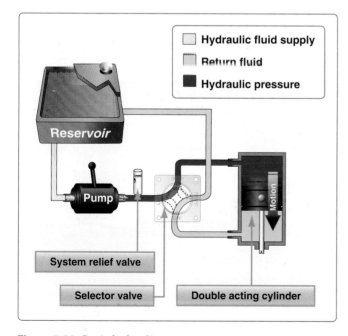

Figure 7-36. *Basic hydraulic system.*

Figure 7-37. *The landing gear supports the airplane during the takeoff run, landing, taxiing, and when parked.*

airplane. Landing gear employing a rear-mounted wheel is called conventional landing gear. Airplanes with conventional landing gear are often referred to as tailwheel airplanes. When the third wheel is located on the nose, it is called a nosewheel, and the design is referred to as a tricycle gear. A steerable nosewheel or tailwheel permits the airplane to be controlled throughout all operations while on the ground.

Tricycle Landing Gear

There are three advantages to using tricycle landing gear:

1. It allows more forceful application of the brakes during landings at high speeds without causing the aircraft to nose over.

2. It permits better forward visibility for the pilot during takeoff, landing, and taxiing.

3. It tends to prevent ground looping (swerving) by providing more directional stability during ground operation since the aircraft's center of gravity (CG) is forward of the main wheels. The forward CG keeps the airplane moving forward in a straight line rather than ground looping.

Nosewheels are either steerable or castering. Steerable nosewheels are linked to the rudders by cables or rods, while castering nosewheels are free to swivel. In both cases, the aircraft is steered using the rudder pedals. Airplanes with a castering nosewheel may require the pilot to combine the use of the rudder pedals with independent use of the brakes.

Tailwheel Landing Gear

Tailwheel landing gear airplanes have two main wheels attached to the airframe ahead of its CG that support most of the weight of the structure. A tailwheel at the very back of the fuselage provides a third point of support. This arrangement allows adequate ground clearance for a larger propeller and is more desirable for operations on unimproved fields. *[Figure 7-38]*

With the CG located behind the main landing gear, directional control using this type of landing gear is more difficult while on the ground. This is the main disadvantage of the tailwheel landing gear. For example, if the pilot allows the aircraft to swerve while rolling on the ground at a low speed, he or she may not have sufficient rudder control and the CG will attempt to get ahead of the main gear, which may cause the airplane to ground loop.

Diminished forward visibility when the tailwheel is on or near the ground is a second disadvantage of tailwheel landing gear airplanes. Because of these disadvantages, specific training is required to operate tailwheel airplanes.

Fixed and Retractable Landing Gear
Landing gear can also be classified as either fixed or retractable. Fixed landing gear always remains extended and has the advantage of simplicity combined with low

Figure 7-38. *Tailwheel landing gear.*

maintenance. Retractable landing gear is designed to streamline the airplane by allowing the landing gear to be stowed inside the structure during cruising flight. *[Figure 7-39]*

Brakes
Airplane brakes are located on the main wheels and are applied by either a hand control or by foot pedals (toe or heel). Foot pedals operate independently and allow for differential braking. During ground operations, differential braking can supplement nosewheel/tailwheel steering.

Pressurized Aircraft
Aircraft are flown at high altitudes for two reasons. First, an aircraft flown at high altitude consumes less fuel for a given airspeed than it does for the same speed at a lower altitude because the aircraft is more efficient at a high altitude. Second, bad weather and turbulence may be avoided by flying in relatively smooth air above the storms. Many modern aircraft are being designed to operate at high altitudes, taking advantage of that environment. In order to fly at higher altitudes, the aircraft must be pressurized or suitable supplemental oxygen must be provided for each occupant. It is important for pilots who fly these aircraft to be familiar with the basic operating principles.

In a typical pressurization system, the cabin, flight compartment, and baggage compartments are incorporated into a sealed unit capable of containing air under a pressure higher than outside atmospheric pressure. On aircraft powered by turbine engines, bleed air from the engine compressor section is used to pressurize the cabin. Superchargers may be used on older model turbine-powered aircraft to pump air into the sealed fuselage. Piston-powered aircraft may use air supplied from each engine turbocharger through a sonic venturi (flow limiter). Air is released from the fuselage by

Figure 7-39. *Fixed (left) and retractable (right) gear airplanes.*

a device called an outflow valve. By regulating the air exit, the outflow valve allows for a constant inflow of air to the pressurized area. *[Figure 7-40]*

A cabin pressurization system typically maintains a cabin pressure altitude of approximately 8,000 feet at the maximum designed cruising altitude of an aircraft. This prevents rapid changes of cabin altitude that may be uncomfortable or cause injury to passengers and crew. In addition, the pressurization system permits a reasonably fast exchange of air from the inside to the outside of the cabin. This is necessary to eliminate odors and to remove stale air. *[Figure 7-41]*

Pressurization of the aircraft cabin is necessary in order to protect occupants against hypoxia. Within a pressurized cabin, occupants can be transported comfortably and safely for long periods of time, particularly if the cabin altitude is maintained at 8,000 feet or below, where the use of oxygen equipment is not required. The flight crew in this type of aircraft must be aware of the danger of accidental loss of cabin pressure and be prepared to deal with such an emergency whenever it occurs.

Atmosphere pressure	
Altitude (ft)	Pressure (psi)
Sea level	14.7
2,000	13.7
4,000	12.7
6,000	11.8
8,000	10.9
10,000	10.1
12,000	9.4
14,000	8.6
16,000	8.0
18,000	7.3
20,000	6.8
22,000	6.2
24,000	5.7
26,000	5.2
28,000	4.8
30,000	4.4

The altitude at which the standard air pressure is equal to 10.9 psi can be found at 8,000 feet.

At an altitude of 28,000 feet, standard atmospheric pressure is 4.8 psi. By adding this pressure to the cabin pressure differential of 6.1 psi difference (psid), a total air pressure of 10.9 psi is obtained.

Figure 7-41. *Standard atmospheric pressure chart.*

The following terms will aid in understanding the operating principles of pressurization and air conditioning systems:

- Aircraft altitude—the actual height above sea level at which the aircraft is flying

- Ambient temperature—the temperature in the area immediately surrounding the aircraft

- Ambient pressure—the pressure in the area immediately surrounding the aircraft

- Cabin altitude—cabin pressure in terms of equivalent altitude above sea level

- Differential pressure—the difference in pressure between the pressure acting on one side of a wall and the pressure acting on the other side of the wall. In aircraft air-conditioning and pressurizing systems, it is the difference between cabin pressure and atmospheric pressure.

The cabin pressure control system provides cabin pressure regulation, pressure relief, vacuum relief, and the means for selecting the desired cabin altitude in the isobaric and differential range. In addition, dumping of the cabin pressure is a function of the pressure control system. A cabin pressure regulator, an outflow valve, and a safety valve are used to accomplish these functions.

The cabin pressure regulator controls cabin pressure to a selected value in the isobaric range and limits cabin pressure to a preset differential value in the differential range. When an aircraft reaches the altitude at which the difference between the pressure inside and outside the cabin is equal to the highest differential pressure for which the fuselage structure is designed, a further increase in aircraft altitude will result

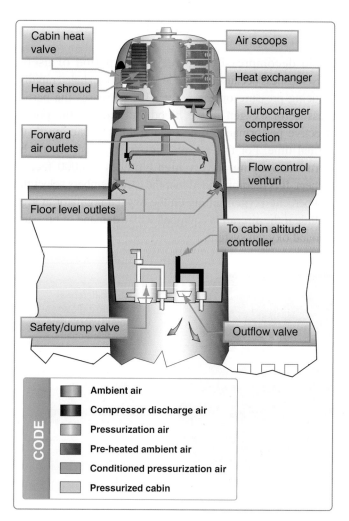

Figure 7-40. *High performance airplane pressurization system.*

in a corresponding increase in cabin altitude. Differential control is used to prevent the maximum differential pressure, for which the fuselage was designed, from being exceeded. This differential pressure is determined by the structural strength of the cabin and often by the relationship of the cabin size to the probable areas of rupture, such as window areas and doors.

The cabin air pressure safety valve is a combination pressure relief, vacuum relief, and dump valve. The pressure relief valve prevents cabin pressure from exceeding a predetermined differential pressure above ambient pressure. The vacuum relief prevents ambient pressure from exceeding cabin pressure by allowing external air to enter the cabin when ambient pressure exceeds cabin pressure. The flight deck control switch actuates the dump valve. When this switch is positioned to ram, a solenoid valve opens, causing the valve to dump cabin air into the atmosphere.

The degree of pressurization and the operating altitude of the aircraft are limited by several critical design factors. Primarily, the fuselage is designed to withstand a particular maximum cabin differential pressure.

Several instruments are used in conjunction with the pressurization controller. The cabin differential pressure gauge indicates the difference between inside and outside pressure. This gauge should be monitored to assure that the cabin does not exceed the maximum allowable differential pressure. A cabin altimeter is also provided as a check on the performance of the system. In some cases, these two instruments are combined into one. A third instrument indicates the cabin rate of climb or descent. A cabin rate-of-climb instrument and a cabin altimeter are illustrated in *Figure 7-42*.

Decompression is defined as the inability of the aircraft's pressurization system to maintain its designed pressure differential. This can be caused by a malfunction in the pressurization system or structural damage to the aircraft.

Physiologically, decompressions fall into the following two categories:

- Explosive decompression—a change in cabin pressure faster than the lungs can decompress, possibly resulting in lung damage. Normally, the time required to release air from the lungs without restrictions, such as masks, is 0.2 seconds. Most authorities consider any decompression that occurs in less than 0.5 seconds to be explosive and potentially dangerous.

- Rapid decompression—a change in cabin pressure in which the lungs decompress faster than the cabin.

During an explosive decompression, there may be noise, and one may feel dazed for a moment. The cabin air fills with fog, dust, or flying debris. Fog occurs due to the rapid drop in temperature and the change of relative humidity. Normally, the ears clear automatically. Air rushes from the mouth and nose due to the escape of air from the lungs and may be noticed by some individuals.

Rapid decompression decreases the period of useful consciousness because oxygen in the lungs is exhaled rapidly, reducing pressure on the body. This decreases the partial pressure of oxygen in the blood and reduces the pilot's effective performance time by one-third to one-fourth its normal time. For this reason, an oxygen mask should be worn when flying at very high altitudes (35,000 feet or higher). It is recommended that the crewmembers select the 100 percent oxygen setting on the oxygen regulator at high altitude if the aircraft is equipped with a demand or pressure demand oxygen system.

Figure 7-42. *Cabin pressurization instruments.*

The primary danger of decompression is hypoxia. Quick, proper utilization of oxygen equipment is necessary to avoid unconsciousness. Another potential danger that pilots, crew, and passengers face during high altitude decompressions is evolved gas decompression sickness. This occurs when the pressure on the body drops sufficiently, nitrogen comes out of solution, and forms bubbles inside the person that can have adverse effects on some body tissues.

Decompression caused by structural damage to the aircraft presents another type of danger to pilots, crew, and passengers—being tossed or blown out of the aircraft if they are located near openings. Individuals near openings should wear safety harnesses or seatbelts at all times when the aircraft is pressurized and they are seated. Structural damage also has the potential to expose them to wind blasts and extremely cold temperatures.

Rapid descent from altitude is necessary in order to minimize these problems. Automatic visual and aural warning systems are included in the equipment of all pressurized aircraft.

Oxygen Systems

Crew and passengers use oxygen systems, in conjunction with pressurization systems, to prevent hypoxia. Regulations require, at a minimum, flight crews have and use supplemental oxygen after 30 minutes exposure to cabin pressure altitudes between 12,500 and 14,000 feet. Use of supplemental oxygen is required immediately upon exposure to cabin pressure altitudes above 14,000 feet. Every aircraft occupant, above 15,000 feet cabin pressure altitude, must have supplemental oxygen. However, based on a person's physical characteristics and condition, a person may feel the effects of oxygen deprivation at much lower altitudes. Some people flying above 10,000 feet during the day may experience disorientation due to the lack of adequate oxygen. At night, especially when fatigued, these effects may occur as low as 5,000 feet. Therefore, for optimum protection, pilots are encouraged to use supplemental oxygen above 10,000 feet cabin altitude during the day and above 5,000 feet at night.

Most high altitude aircraft come equipped with some type of fixed oxygen installation. If the aircraft does not have a fixed installation, portable oxygen equipment must be readily accessible during flight. The portable equipment usually consists of a container, regulator, mask outlet, and pressure gauge. Aircraft oxygen is usually stored in high pressure system containers of 1,800–2,200 psi. When the ambient temperature surrounding an oxygen cylinder decreases, pressure within that cylinder decreases because pressure varies directly with temperature if the volume of a gas remains constant. A drop in the indicated pressure of a supplemental oxygen cylinder may be due to the container being stored in an unheated area of the aircraft rather than an actual depletion of the oxygen supply. High pressure oxygen containers should be marked with the psi tolerance (i.e., 1,800 psi) before filling the container to that pressure. The containers should be supplied with oxygen that meets or exceeds SAE AS8010 (as revised), Aviator's Breathing Oxygen Purity Standard. To assure safety, periodic inspection and servicing of the oxygen system should be performed.

An oxygen system consists of a mask or cannula and a regulator that supplies a flow of oxygen dependent upon cabin altitude. Most regulators approved for use up to 40,000 feet are designed to provide zero percent cylinder oxygen and 100 percent cabin air at cabin altitudes of 8,000 feet or less, with the ratio changing to 100 percent oxygen and zero percent cabin air at approximately 34,000 feet cabin altitude. [Figure 7-43] Most regulators approved up to 45,000 feet are designed to provide 40 percent cylinder oxygen and 60 percent cabin air at lower altitudes, with the ratio changing to 100 percent at the higher altitude.

Pilots should be aware of the danger of fire when using oxygen. Materials that are nearly fireproof in ordinary air may be susceptible to combustion in oxygen. Oils and greases may ignite if exposed to oxygen and cannot be used for sealing the valves and fittings of oxygen equipment. Smoking during any kind of oxygen equipment use is prohibited. Before each flight, the pilot should thoroughly inspect and test all oxygen equipment. The inspection should include a thorough examination of the aircraft oxygen equipment, including available supply, an operational check of the system, and assurance that the supplemental oxygen is readily accessible. The inspection should be accomplished with clean hands and should include a visual inspection of the mask and tubing for tears, cracks, or deterioration; the regulator for valve and lever condition and positions; oxygen quantity; and the location and functioning of oxygen pressure gauges, flow indicators, and connections. The mask should be donned and the system should be tested. After any oxygen use, verify that all components and valves are shut off.

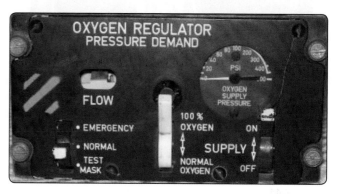

Figure 7-43. *Oxygen system regulator.*

Oxygen Masks

There are numerous types and designs of oxygen masks in use. The most important factor in oxygen mask use is to ensure that the masks and oxygen system are compatible. Crew masks are fitted to the user's face with a minimum of leakage and usually contain a microphone. Most masks are the oronasal type that covers only the mouth and nose.

A passenger mask may be a simple, cup-shaped rubber molding sufficiently flexible to obviate individual fitting. It may have a simple elastic head strap or the passenger may hold it to his or her face.

All oxygen masks should be kept clean to reduce the danger of infection and prolong the life of the mask. To clean the mask, wash it with a mild soap and water solution and rinse it with clear water. If a microphone is installed, use a clean swab, instead of running water, to wipe off the soapy solution. The mask should also be disinfected. A gauze pad that has been soaked in a water solution of Merthiolate can be used to swab out the mask. This solution used should contain one-fifth teaspoon of Merthiolate per quart of water. Wipe the mask with a clean cloth and air dry.

Cannula

A cannula is an ergonomic piece of plastic tubing that runs under the nose to administer oxygen to the user. *[Figure 7-44]* Cannulas are typically more comfortable than masks, but may not provide an adequate flow of oxygen as reliably as masks when operating at higher altitudes. Airplanes certified to older regulations had cannulas installed with an on-board

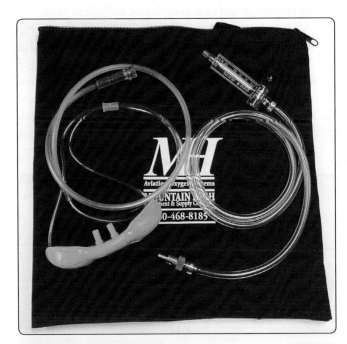

Figure 7-44. *Cannula with green flow detector.*

oxygen system. However, current regulations require aircraft with oxygen systems installed and certified for operations above 18,000 feet to be equipped with oxygen masks instead of cannulas. Many cannulas have a flow meter in the oxygen supply line. If equipped, a periodic check of the green flow detector should be a part of the pilot's regular scan.

Diluter-Demand Oxygen Systems

Diluter-demand oxygen systems supply oxygen only when the user inhales through the mask. An automix lever allows the regulators to automatically mix cabin air and oxygen or supply 100 percent oxygen, depending on the altitude. The demand mask provides a tight seal over the face to prevent dilution with outside air and can be used safely up to 40,000 feet. A pilot who has a beard or mustache should be sure it is trimmed in a manner that will not interfere with the sealing of the oxygen mask. The fit of the mask around the beard or mustache should be checked on the ground for proper sealing.

Pressure-Demand Oxygen Systems

Pressure-demand oxygen systems are similar to diluter demand oxygen equipment, except that oxygen is supplied to the mask under pressure at cabin altitudes above 34,000 feet. Pressure-demand regulators create airtight and oxygen-tight seals, but they also provide a positive pressure application of oxygen to the mask face piece that allows the user's lungs to be pressurized with oxygen. This feature makes pressure demand regulators safe at altitudes above 40,000 feet. Some systems may have a pressure demand mask with the regulator attached directly to the mask, rather than mounted on the instrument panel or other area within the flight deck. The mask-mounted regulator eliminates the problem of a long hose that must be purged of air before 100 percent oxygen begins flowing into the mask.

Continuous-Flow Oxygen System

Continuous-flow oxygen systems are usually provided for passengers. The passenger mask typically has a reservoir bag that collects oxygen from the continuous-flow oxygen system during the time when the mask user is exhaling. The oxygen collected in the reservoir bag allows a higher aspiratory flow rate during the inhalation cycle, which reduces the amount of air dilution. Ambient air is added to the supplied oxygen during inhalation after the reservoir bag oxygen supply is depleted. The exhaled air is released to the cabin. *[Figure 7-45]*

Electrical Pulse-Demand Oxygen System

Portable electrical pulse-demand oxygen systems deliver oxygen by detecting an individual's inhalation effort and provide oxygen flow during the initial portion of inhalation. Pulse demand systems do not waste oxygen during the

Figure 7-45. *Continuous flow mask and rebreather bag.*

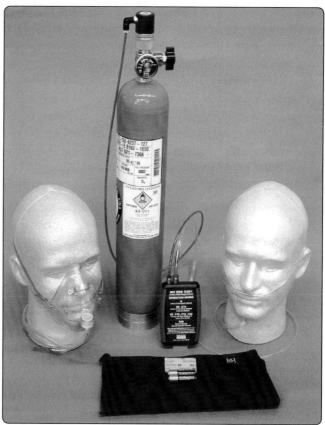

Figure 7-46. *EDS-011 portable pulse-demand oxygen system.*

breathing cycle because oxygen is only delivered during inhalation. Compared to continuous-flow systems, the pulse-demand method of oxygen delivery can reduce the amount of oxygen needed by 50–85 percent. Most pulse-demand oxygen systems also incorporate an internal barometer that automatically compensates for changes in altitude by increasing the amount of oxygen delivered for each pulse as altitude is increased. *[Figure 7-46]*

Pulse Oximeters

A pulse oximeter is a device that measures the amount of oxygen in an individual's blood, in addition to heart rate. This non-invasive device measures the color changes that red blood cells undergo when they become saturated with oxygen. By transmitting a special light beam through a fingertip to evaluate the color of the red cells, a pulse oximeter can calculate the degree of oxygen saturation within one percent of directly measured blood oxygen. Because of their portability and speed, pulse oximeters are very useful for pilots operating in nonpressurized aircraft above 12,500 feet where supplemental oxygen is required. A pulse oximeter permits crewmembers and passengers of an aircraft to evaluate their actual need for supplemental oxygen. *[Figure 7-47]*

Servicing of Oxygen Systems

Before servicing any aircraft with oxygen, consult the specific aircraft service manual to determine the type of equipment required and procedures to be used. Certain precautions should be observed whenever aircraft oxygen

systems are to be serviced. Oxygen system servicing should be accomplished only when the aircraft is located outside of the hangars. Personal cleanliness and good housekeeping are imperative when working with oxygen. Oxygen under pressure creates spontaneous results when brought in contact with petroleum products. Service people should be certain to wash dirt, oil, and grease (including lip salves and hair oil) from their hands before working around oxygen equipment. It is also essential that clothing and tools are free of oil, grease,

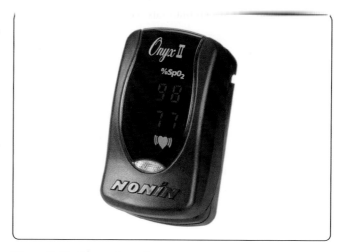

Figure 7-47. *Onyx pulse oximeter.*

and dirt. Aircraft with permanently installed oxygen tanks usually require two persons to accomplish servicing of the system. One should be stationed at the service equipment control valves, and the other stationed where he or she can observe the aircraft system pressure gauges. Oxygen system servicing is not recommended during aircraft fueling operations or while other work is performed that could provide a source of ignition. Oxygen system servicing while passengers are on board the aircraft is not recommended.

Anti-Ice and Deice Systems

Anti-icing equipment is designed to prevent the formation of ice, while deicing equipment is designed to remove ice once it has formed. These systems protect the leading edge of wing and tail surfaces, pitot and static port openings, fuel tank vents, stall warning devices, windshields, and propeller blades. Ice detection lighting may also be installed on some aircraft to determine the extent of structural icing during night flights.

Most light aircraft have only a heated pitot tube and are not certified for flight in icing. These light aircraft have limited cross-country capability in the cooler climates during late fall, winter, and early spring. Noncertificated aircraft must exit icing conditions immediately. Refer to the AFM/POH for details.

Airfoil Anti-Ice and Deice

Inflatable deicing boots consist of a rubber sheet bonded to the leading edge of the airfoil. When ice builds up on the leading edge, an engine-driven pneumatic pump inflates the rubber boots. Many turboprop aircraft divert engine bleed air to the wing to inflate the rubber boots. Upon inflation, the ice is cracked and should fall off the leading edge of the wing. Deicing boots are controlled from the flight deck by a switch and can be operated in a single cycle or allowed to cycle at automatic, timed intervals. *[Figure 7-48]*

In the past, it was believed that if the boots were cycled too soon after encountering ice, the ice layer would expand instead of breaking off, resulting in a condition referred to as ice "bridging." Consequently, subsequent deice boot cycles would be ineffective at removing the ice buildup. Although some residual ice may remain after a boot cycle, "bridging" does not occur with any modern boots. Pilots can cycle the boots as soon as an ice accumulation is observed. Consult the AFM/POH for information on the operation of deice boots on an aircraft.

Many deicing boot systems use the instrument system suction gauge and a pneumatic pressure gauge to indicate proper boot operation. These gauges have range markings that indicate the operating limits for boot operation. Some systems may

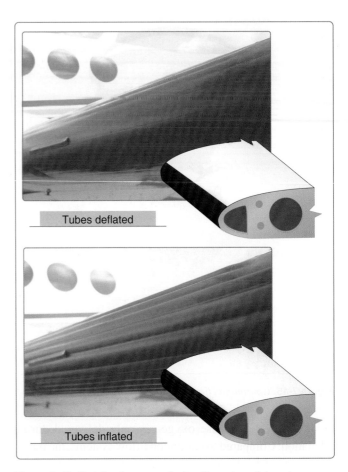

Figure 6-48. *Deicing boots on the leading edge of the wing.*

also incorporate an annunciator light to indicate proper boot operation.

Proper maintenance and care of deicing boots are important for continued operation of this system. They need to be carefully inspected during preflight.

Another type of leading edge protection is the thermal anti-ice system. Heat provides one of the most effective methods for preventing ice accumulation on an airfoil. High performance turbine aircraft often direct hot air from the compressor section of the engine to the leading edge surfaces. The hot air heats the leading edge surfaces sufficiently to prevent the formation of ice. A newer type of thermal anti-ice system referred to as ThermaWing uses electrically heated graphite foil laminate applied to the leading edge of the wing and horizontal stabilizer. ThermaWing systems typically have two zones of heat application. One zone on the leading edge receives continuous heat; the second zone further aft receives heat in cycles to dislodge the ice allowing aerodynamic forces to remove it. Thermal anti-ice systems should be activated prior to entering icing conditions.

An alternate type of leading edge protection that is not as common as thermal anti-ice and deicing boots is known as a weeping wing. The weeping-wing design uses small holes located in the leading edge of the wing to prevent the formation and build-up of ice. An antifreeze solution is pumped to the leading edge and weeps out through the holes. Additionally, the weeping wing is capable of deicing an aircraft. When ice has accumulated on the leading edges, application of the antifreeze solution chemically breaks down the bond between the ice and airframe, allowing aerodynamic forces to remove the ice. *[Figure 7-49]*

Windscreen Anti-Ice

There are two main types of windscreen anti-ice systems. The first system directs a flow of alcohol to the windscreen. If used early enough, the alcohol prevents ice from building up on the windscreen. The rate of alcohol flow can be controlled by a dial in the flight deck according to procedures recommended by the aircraft manufacturer.

Another effective method of anti-icing equipment is the electric heating method. Small wires or other conductive material is imbedded in the windscreen. The heater can be turned on by a switch in the flight deck, causing an electrical current to be passed across the shield through the wires to provide sufficient heat to prevent the formation of ice on the windscreen. The heated windscreen should only be used during flight. Do not leave it on during ground operations, as it can overheat and cause damage to the windscreen. Warning: the electrical current can cause compass deviation errors by as much as 40°.

Propeller Anti-Ice

Propellers are protected from icing by the use of alcohol or electrically heated elements. Some propellers are equipped with a discharge nozzle that is pointed toward the root of the blade. Alcohol is discharged from the nozzles, and centrifugal force drives the alcohol down the leading edge of the blade. The boots are also grooved to help direct the flow of alcohol. This prevents ice from forming on the leading edge of the propeller. Propellers can also be fitted with propeller anti-ice boots. The propeller boot is divided into two sections—the

inboard and the outboard sections. The boots are imbedded with electrical wires that carry current for heating the propeller. The prop anti-ice system can be monitored for proper operation by monitoring the prop anti-ice ammeter. During the preflight inspection, check the propeller boots for proper operation. If a boot fails to heat one blade, an unequal blade loading can result and may cause severe propeller vibration. *[Figure 7-50]*

Other Anti-Ice and Deice Systems

Pitot and static ports, fuel vents, stall-warning sensors, and other optional equipment may be heated by electrical elements. Operational checks of the electrically heated systems are to be checked in accordance with the AFM /POH.

Operation of aircraft anti-icing and deicing systems should be checked prior to encountering icing conditions. Encounters with structural ice require immediate action. Anti-icing and deicing equipment are not intended to sustain long-term flight in icing conditions.

Chapter Summary

All aircraft have a requirement for essential systems such as the engine, propeller, induction, ignition systems as well as the fuel, lubrication, cooling, electrical, landing gear, and

Figure 7-49. *TKS weeping wing anti-ice/deicing system.*

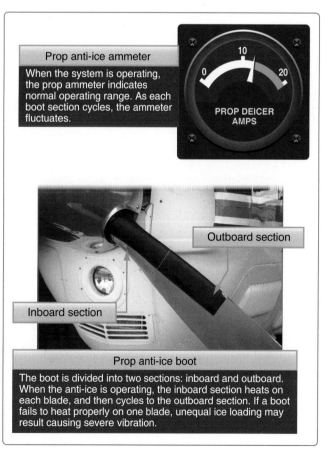

Figure 7-50. *Prop ammeter and anti-ice boots.*

environmental control systems to support flight. Understanding the aircraft systems of the aircraft being flown is critical to its safe operation and proper maintenance. Consult the AFM/POH for specific information pertaining to the aircraft being flown. Various manufacturer and owners group websites can also be a valuable source of additional information.

Chapter 8
Flight Instruments

Introduction

In order to safely fly any aircraft, a pilot must understand how to interpret and operate the flight instruments. The pilot also needs to be able to recognize associated errors and malfunctions of these instruments. This chapter addresses the pitot-static system and associated instruments, the vacuum system and related instruments, gyroscopic instruments, and the magnetic compass. When a pilot understands how each instrument works and recognizes when an instrument is malfunctioning, he or she can safely utilize the instruments to their fullest potential.

Pitot-Static Flight Instruments

The pitot-static system is a combined system that utilizes the static air pressure and the dynamic pressure due to the motion of the aircraft through the air. These combined pressures are utilized for the operation of the airspeed indicator (ASI), altimeter, and vertical speed indicator (VSI). [Figure 8-1]

Climbing left bank

Level left bank

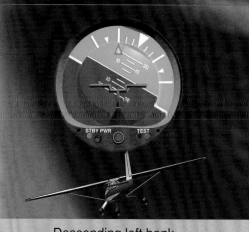

Descending left bank

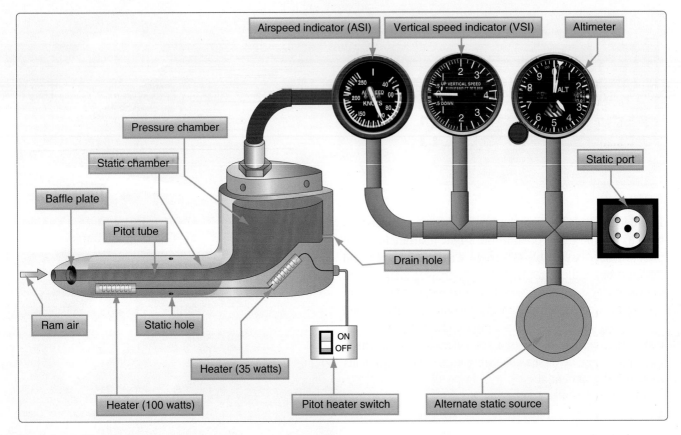

Figure 8-1. *Pitot-static system and instruments.*

Impact Pressure Chamber and Lines

The pitot tube is utilized to measure the total combined pressures that are present when an aircraft moves through the air. Static pressure, also known as ambient pressure, is always present whether an aircraft is moving or at rest. It is simply the barometric pressure in the local area. Dynamic pressure is present only when an aircraft is in motion; therefore, it can be thought of as a pressure due to motion. Wind also generates dynamic pressure. It does not matter if the aircraft is moving through still air at 70 knots or if the aircraft is facing a wind with a speed of 70 knots, the same dynamic pressure is generated.

When the wind blows from an angle less than 90° off the nose of the aircraft, dynamic pressure can be depicted on the ASI. The wind moving across the airfoil at 20 knots is the same as the aircraft moving through calm air at 20 knots. The pitot tube captures the dynamic pressure, as well as the static pressure that is always present.

The pitot tube has a small opening at the front that allows the total pressure to enter the pressure chamber. The total pressure is made up of dynamic pressure plus static pressure. In addition to the larger hole in the front of the pitot tube, there is a small hole in the back of the chamber that allows moisture to drain from the system should the aircraft enter precipitation. Both openings in the pitot tube must be checked prior to flight to ensure that neither is blocked. Many aircraft have pitot tube covers installed when they sit for extended periods of time. This helps to keep bugs and other objects from becoming lodged in the opening of the pitot tube.

The one instrument that utilizes the pitot tube is the ASI. The total pressure is transmitted to the ASI from the pitot tube's pressure chamber via a small tube. The static pressure is also delivered to the opposite side of the ASI, which serves to cancel out the two static pressures, thereby leaving the dynamic pressure to be indicated on the instrument. When the dynamic pressure changes, the ASI shows either increase or decrease, corresponding to the direction of change. The two remaining instruments (altimeter and VSI) utilize only the static pressure that is derived from the static port.

Static Pressure Chamber and Lines

The static chamber is vented through small holes to the free undisturbed air on the side(s) of the aircraft. As the atmospheric pressure changes, the pressure is able to move freely in and out of the instruments through the small lines that connect the instruments to the static system. An alternate static source is provided in some aircraft to provide static pressure should the primary static source become blocked. The alternate static source is normally found inside the flight deck. Due to the venturi effect of the air flowing around the fuselage, the air pressure inside the flight deck is lower than the exterior pressure.

When the alternate static source pressure is used, the following instrument indications are observed:

1. The altimeter indicates a slightly higher altitude than actual.

2. The ASI indicates an airspeed greater than the actual airspeed.

3. The VSI shows a momentary climb and then stabilizes if the altitude is held constant.

Each pilot is responsible for consulting the Aircraft Flight Manual (AFM) or the Pilot's Operating Handbook (POH) to determine the amount of error that is introduced into the system when utilizing the alternate static source. In an aircraft not equipped with an alternate static source, an alternate method of introducing static pressure into the system should a blockage occur is to break the glass face of the VSI. This most likely renders the VSI inoperative. The reason for choosing the VSI as the instrument to break is that it is the least important static source instrument for flight.

Altimeter

The altimeter is an instrument that measures the height of an aircraft above a given pressure level. Pressure levels are discussed later in detail. Since the altimeter is the only instrument that is capable of indicating altitude, this is one of the most vital instruments installed in the aircraft. To use the altimeter effectively, the pilot must understand the operation of the instrument, as well as the errors associated with the altimeter and how each affect the indication.

A stack of sealed aneroid wafers comprise the main component of the altimeter. An aneroid wafer is a sealed wafer that is evacuated to an internal pressure of 29.92 inches of mercury ("Hg). These wafers are free to expand and contract with changes to the static pressure. A higher static pressure presses down on the wafers and causes them to collapse. A lower static pressure (less than 29.92 "Hg) allows the wafers to expand. A mechanical linkage connects the wafer movement to the needles on the indicator face, which translates compression of the wafers into a decrease in altitude and translates an expansion of the wafers into an increase in altitude. *[Figure 8-2]*

Notice how the static pressure is introduced into the rear of the sealed altimeter case. The altimeter's outer chamber is sealed, which allows the static pressure to surround the aneroid wafers. If the static pressure is higher than the pressure in the aneroid wafers (29.92 "Hg), then the wafers are compressed

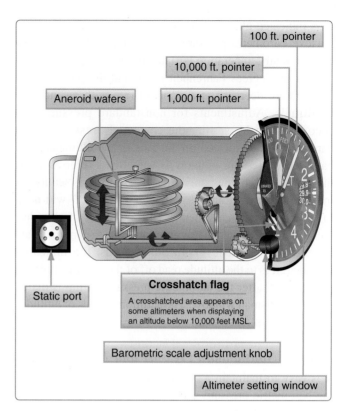

Figure 8-2. *Altimeter.*

until the pressure inside the wafers is equal to the surrounding static pressure. Conversely, if the static pressure is less than the pressure inside of the wafers, the wafers are able to expand which increases the volume. The expansion and contraction of the wafers moves the mechanical linkage which drives the needles on the face of the altimeter.

Principle of Operation

The pressure altimeter is an aneroid barometer that measures the pressure of the atmosphere at the level where the altimeter is located and presents an altitude indication in feet. The altimeter uses static pressure as its source of operation. Air is denser at sea level than aloft—as altitude increases, atmospheric pressure decreases. This difference in pressure at various levels causes the altimeter to indicate changes in altitude.

The presentation of altitude varies considerably between different types of altimeters. Some have one pointer while others have two or more. Only the multipointer type is discussed in this handbook. The dial of a typical altimeter is graduated with numerals arranged clockwise from zero to nine. Movement of the aneroid element is transmitted through gears to the three hands that indicate altitude. In *Figure 8-2*, the long, thin needle with the inverted triangle at the end indicates tens of thousands of feet; the short, wide needle indicates thousands of feet; and the long needle on top indicates hundreds of feet.

This indicated altitude is correct, however, only when the sea level barometric pressure is standard (29.92 "Hg), the sea level free air temperature is standard (+15 degrees Celsius (°C) or 59 degrees Fahrenheit (°F)), and the pressure and temperature decrease at a standard rate with an increase in altitude. Adjustments for nonstandard pressures are accomplished by setting the corrected pressure into a barometric scale located on the face of the altimeter. The barometric pressure window is sometimes referred to as the Kollsman window; only after the altimeter is set does it indicate the correct altitude. The word "correct" will need to be better explained when referring to types of altitudes, but is commonly used in this case to denote the approximate altitude above sea level. In other words, the indicated altitude refers to the altitude read off of the altitude which is uncorrected, after the barometric pressure setting is dialed into the Kollsman window. The additional types of altitudes are further explained later.

Effect of Nonstandard Pressure and Temperature

It is easy to maintain a consistent height above ground if the barometric pressure and temperature remain constant, but this is rarely the case. The pressure and temperature can change between takeoff and landing even on a local flight. If these changes are not taken into consideration, flight becomes dangerous.

If altimeters could not be adjusted for nonstandard pressure, a hazardous situation could occur. For example, if an aircraft is flown from a high pressure area to a low pressure area without adjusting the altimeter, a constant altitude will be displayed, but the actual height of the aircraft above the ground would be lower then the indicated altitude. There is an old aviation axiom: "GOING FROM A HIGH TO A LOW, LOOK OUT BELOW." Conversely, if an aircraft is flown from a low pressure area to a high pressure area without an adjustment of the altimeter, the actual altitude of the aircraft is higher than the indicated altitude. Once in flight, it is important to frequently obtain current altimeter settings en route to ensure terrain and obstruction clearance.

Many altimeters do not have an accurate means of being adjusted for barometric pressures in excess of 31.00 "Hg. When the altimeter cannot be set to the higher pressure setting, the aircraft actual altitude is higher than the altimeter indicates. When low barometric pressure conditions occur (below 28.00), flight operations by aircraft unable to set the actual altimeter setting are not recommended.

Adjustments to compensate for nonstandard pressure do not compensate for nonstandard temperature. Since cold air is denser than warm air, when operating in temperatures that are colder than standard, the altitude is lower than the altimeter indication. *[Figure 8-3]* It is the magnitude of this "difference" that determines the magnitude of the error. It is the difference due to colder temperatures that concerns the pilot. When flying into a cooler air mass while maintaining a constant indicated altitude, true altitude is lower. If terrain or obstacle clearance is a factor in selecting a cruising altitude, particularly in mountainous terrain, remember to anticipate that a colder-than-standard temperature places the aircraft lower than the altimeter indicates. Therefore, a higher indicated altitude may be required to provide adequate terrain clearance. A variation of the memory aid used for pressure

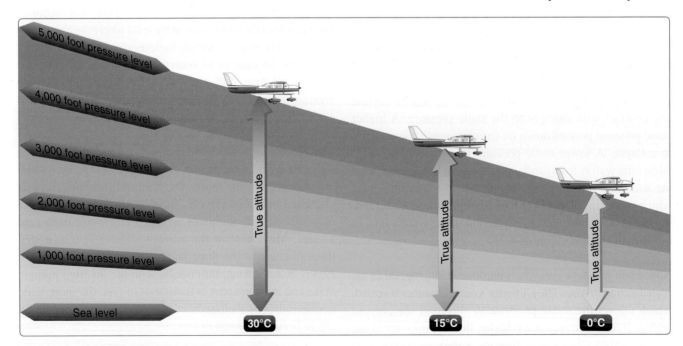

Figure 8-3. *Effects of nonstandard temperature on an altimeter.*

can be employed: "FROM HOT TO COLD, LOOK OUT BELOW." When the air is warmer than standard, the aircraft is higher than the altimeter indicates. Altitude corrections for temperature can be computed on the navigation computer.

Extremely cold temperatures also affect altimeter indications. *Figure 8-4,* which was derived from ICAO formulas, indicates how much error can exist when the temperature is extremely cold.

Setting the Altimeter

Most altimeters are equipped with a barometric pressure setting window (or Kollsman window) providing a means to adjust the altimeter. A knob is located at the bottom of the instrument for this adjustment.

To adjust the altimeter for variation in atmospheric pressure, the pressure scale in the altimeter setting window, calibrated in inches of mercury ("Hg) and/or millibars (mb), is adjusted to match the given altimeter setting. Altimeter setting is defined as station pressure reduced to sea level, but an altimeter setting is accurate only in the vicinity of the reporting station. Therefore, the altimeter must be adjusted as the flight progresses from one station to the next. Air traffic control (ATC) will advise when updated altimeter settings are available. If a pilot is not utilizing ATC assistance, local altimeter settings can be obtained by monitoring local automated weather observing system/automated surface observation system (AWOS/ASOS) or automatic terminal information service (ATIS) broadcasts.

Many pilots confidently expect the current altimeter setting will compensate for irregularities in atmospheric pressure at all altitudes, but this is not always true. The altimeter setting broadcast by ground stations is the station pressure corrected to mean sea level. It does not account for the irregularities

at higher levels, particularly the effect of nonstandard temperature. If each pilot in a given area is using the same altimeter setting, each altimeter should be equally affected by temperature and pressure variation errors, making it possible to maintain the desired vertical separation between aircraft. This does not guarantee vertical separation though. It is still imperative to maintain a regimented visual scan for intruding air traffic.

When flying over high, mountainous terrain, certain atmospheric conditions cause the altimeter to indicate an altitude of 1,000 feet or more higher than the actual altitude. For this reason, a generous margin of altitude should be allowed—not only for possible altimeter error, but also for possible downdrafts that might be associated with high winds.

To illustrate the use of the altimeter setting system, follow a flight from Dallas Love Field, Texas, to Abilene Municipal Airport, Texas, via Mineral Wells. Before taking off from Love Field, the pilot receives a current altimeter setting of 29.85 "Hg from the control tower or ATIS and sets this value in the altimeter setting window. The altimeter indication should then be compared with the known airport elevation of 487 feet. Since most altimeters are not perfectly calibrated, an error may exist.

When over Mineral Wells, assume the pilot receives a current altimeter setting of 29.94 "Hg and sets this in the altimeter window. Before entering the traffic pattern at Abilene Municipal Airport, a new altimeter setting of 29.69 "Hg is received from the Abilene Control Tower and set in the altimeter setting window. If the pilot desires to fly the traffic pattern at approximately 800 feet above the terrain, and the field elevation of Abilene is 1,791 feet, an indicated altitude of 2,600 feet should be maintained (1,791 feet + 800 feet = 2,591 feet, rounded to 2,600 feet).

The importance of properly setting the altimeter cannot be overemphasized. Assume the pilot did not adjust the altimeter at Abilene to the current setting and continued using the Mineral Wells setting of 29.94 "Hg. When entering the Abilene traffic pattern at an indicated altitude of 2,600 feet, the aircraft would be approximately 250 feet below the proper traffic pattern altitude. Upon landing, the altimeter would indicate approximately 250 feet higher than the field elevation.

Mineral Wells altimeter setting	29.94
Abilene altimeter setting	29.69
Difference	0.25

(Since 1 inch of pressure is equal to approximately 1,000 feet of altitude, 0.25 × 1,000 feet = 250 feet.)

Reported Temp 0 °C	\multicolumn{14}{c}{Height Above Airport in Feet}
	200	300	400	500	600	700	800	900	1,000	1,500	2,000	3,000	4,000	5,000
+10	10	10	10	10	20	20	20	20	20	30	40	60	80	90
0	20	20	30	30	40	40	50	50	60	90	120	170	230	280
-10	20	30	40	50	60	70	80	90	100	150	200	290	390	490
-20	30	50	60	70	90	100	120	130	140	210	280	420	570	710
-30	40	60	80	100	120	140	150	170	190	280	380	570	760	950
-40	50	80	100	120	150	170	190	220	240	360	480	720	970	1,210
-50	60	90	120	150	180	210	240	270	300	450	590	890	1,190	1,500

Figure 8-4. *Look at the chart using a temperature of –10 °C and an aircraft altitude of 1,000 feet above the airport elevation. The chart shows that the reported current altimeter setting may place the aircraft as much as 100 feet below the altitude indicated by the altimeter.*

When determining whether to add or subtract the amount of altimeter error, remember that when the actual pressure is lower than what is set in the altimeter window, the actual altitude of the aircraft is lower than what is indicated on the altimeter.

The following is another method of computing the altitude deviation. Start by subtracting the current altimeter setting from 29.94 "Hg. Always remember to place the original setting as the top number. Then subtract the current altimeter setting.

Mineral Wells altimeter setting	29.94
Abilene altimeter setting	29.69
29.94 – 29.69 = Difference	0.25

(Since 1 inch of pressure is equal to approximately 1,000 feet of altitude, 0.25 × 1,000 feet = 250 feet.) Always subtract the number from the indicated altitude.

$$2,600 - 250 = 2,350$$

Now, try a lower pressure setting. Adjust from altimeter setting 29.94 to 30.56 "Hg.

Mineral Wells altimeter setting	29.94
Altimeter setting	30.56
29.94 – 30.56 = Difference	–0.62

(Since 1 inch of pressure is equal to approximately 1,000 feet of altitude, 0.62 × 1,000 feet = 620 feet.) Always subtract the number from the indicated altitude.

$$2,600 - (-620) = 3,220$$

The pilot will be 620 feet high.

Notice the difference is a negative number. Starting with the current indicated altitude of 2,600 feet, subtracting a negative number is the same as adding the two numbers. By utilizing this method, a pilot will better understand the importance of using the current altimeter setting (miscalculation of where and in what direction an error lies can affect safety; if altitude is lower than indicated altitude, an aircraft could be in danger of colliding with an obstacle).

Altimeter Operation

There are two means by which the altimeter pointers can be moved. The first is a change in air pressure, while the other is an adjustment to the barometric scale. When the aircraft climbs or descends, changing pressure within the altimeter case expands or contracts the aneroid barometer. This movement is transmitted through mechanical linkage to rotate the pointers.

A decrease in pressure causes the altimeter to indicate an increase in altitude, and an increase in pressure causes the altimeter to indicate a decrease in altitude. Accordingly, if the aircraft is sitting on the ground with a pressure level of 29.98 "Hg and the pressure level changes to 29.68 "Hg, the altimeter would show an increase of approximately 300 feet in altitude. This pressure change is most noticeable when the aircraft is left parked over night. As the pressure falls, the altimeter interprets this as a climb. The altimeter indicates an altitude above the actual field elevation. If the barometric pressure setting is reset to the current altimeter setting of 29.68 "Hg, then the field elevation is again indicated on the altimeter.

This pressure change is not as easily noticed in flight since aircraft fly at specific altitudes. The aircraft steadily decreases true altitude while the altimeter is held constant through pilot action as discussed in the previous section.

Knowing the aircraft's altitude is vitally important to a pilot. The pilot must be sure that the aircraft is flying high enough to clear the highest terrain or obstruction along the intended route. It is especially important to have accurate altitude information when visibility is restricted. To clear obstructions, the pilot must constantly be aware of the altitude of the aircraft and the elevation of the surrounding terrain. To reduce the possibility of a midair collision, it is essential to maintain altitude in accordance with air traffic rules.

Types of Altitude

Altitude in itself is a relevant term only when it is specifically stated to which type of altitude a pilot is referring. Normally when the term "altitude" is used, it is referring to altitude above sea level since this is the altitude which is used to depict obstacles and airspace, as well as to separate air traffic.

Altitude is vertical distance above some point or level used as a reference. There are as many kinds of altitude as there are reference levels from which altitude is measured, and each may be used for specific reasons. Pilots are mainly concerned with five types of altitudes:

1. Indicated altitude—read directly from the altimeter (uncorrected) when it is set to the current altimeter setting.

2. True altitude—the vertical distance of the aircraft above sea level—the actual altitude. It is often expressed as feet above mean sea level (MSL). Airport, terrain, and obstacle elevations on aeronautical charts are true altitudes.

3. Absolute altitude—the vertical distance of an aircraft above the terrain, or above ground level (AGL).

4. Pressure altitude—the altitude indicated when the altimeter setting window (barometric scale) is adjusted to 29.92 "Hg. This is the altitude above the standard datum plane, which is a theoretical plane where air pressure (corrected to 15 °C) equals 29.92 "Hg. Pressure altitude is used to compute density altitude, true altitude, true airspeed (TAS), and other performance data.

5. Density altitude—pressure altitude corrected for variations from standard temperature. When conditions are standard, pressure altitude and density altitude are the same. If the temperature is above standard, the density altitude is higher than pressure altitude. If the temperature is below standard, the density altitude is lower than pressure altitude. This is an important altitude because it is directly related to the aircraft's performance.

A pilot must understand how the performance of the aircraft is directly related to the density of the air. The density of the air affects how much power a naturally aspirated engine produces, as well as how efficient the airfoils are. If there are fewer air molecules (lower pressure) to accelerate through the propeller, the acceleration to rotation speed is longer and thus produces a longer takeoff roll, which translates to a decrease in performance.

As an example, consider an airport with a field elevation of 5,048 feet MSL where the standard temperature is 5 °C. Under these conditions, pressure altitude and density altitude are the same—5,048 feet. If the temperature changes to 30 °C, the density altitude increases to 7,855 feet. This means an aircraft would perform on takeoff as though the field elevation were 7,855 feet at standard temperature. Conversely, a temperature of –25 °C would result in a density altitude of 1,232 feet. An aircraft would perform much better under these conditions.

Instrument Check

Prior to each flight, a pilot should examine the altimeter for proper indications in order to verify its validity. To determine the condition of an altimeter, set the barometric scale to the current reported altimeter setting transmitted by the local airport traffic control tower, flight service station (FSS), or any other reliable source, such as ATIS, AWOS, or ASOS. The altimeter pointers should indicate the surveyed field elevation of the airport. If the indication is off more than 75 feet from the surveyed field elevation, the instrument should be referred to a certificated instrument repair station for recalibration.

Vertical Speed Indicator (VSI)

The VSI, which is sometimes called a vertical velocity indicator (VVI), indicates whether the aircraft is climbing, descending, or in level flight. The rate of climb or descent is indicated in feet per minute (fpm). If properly calibrated, the VSI indicates zero in level flight. *[Figure 8-5]*

Principle of Operation

Although the VSI operates solely from static pressure, it is a differential pressure instrument. It contains a diaphragm with connecting linkage and gearing to the indicator pointer inside an airtight case. The inside of the diaphragm is connected directly to the static line of the pitot-static system. The area outside the diaphragm, which is inside the instrument case, is also connected to the static line but through a restricted orifice (calibrated leak).

Both the diaphragm and the case receive air from the static line at existing atmospheric pressure. The diaphragm receives unrestricted air, while the case receives the static pressure via the metered leak. When the aircraft is on the ground or in level flight, the pressures inside the diaphragm and the instrument case are equal, and the pointer is at the zero indication. When the aircraft climbs or descends, the pressure inside the diaphragm changes immediately, but due to the metering action of the restricted passage, the case pressure remains higher or lower for a short time, causing the diaphragm to contract or expand. This causes a pressure differential that is indicated on the instrument needle as a climb or descent. When the pressure differential stabilizes at a definite ratio, the needle indicates the rate of altitude change.

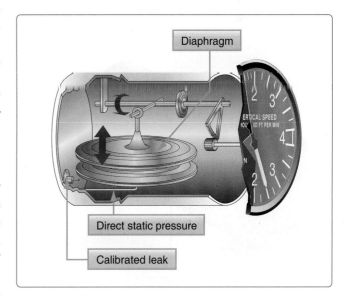

Figure 8-5. *Vertical speed indicator (VSI).*

The VSI displays two different types of information:

- Trend information shows an immediate indication of an increase or decrease in the aircraft's rate of climb or descent.

- Rate information shows a stabilized rate of change in altitude.

The trend information is the direction of movement of the VSI needle. For example, if an aircraft is maintaining level flight and the pilot pulls back on the control yoke causing the nose of the aircraft to pitch up, the VSI needle moves upward to indicate a climb. If the pitch attitude is held constant, the needle stabilizes after a short period (6–9 seconds) and indicates the rate of climb in hundreds of fpm. The time period from the initial change in the rate of climb, until the VSI displays an accurate indication of the new rate, is called the lag. Rough control technique and turbulence can extend the lag period and cause erratic and unstable rate indications. Some aircraft are equipped with an instantaneous vertical speed indicator (IVSI), which incorporates accelerometers to compensate for the lag in the typical VSI. *[Figure 8-6]*

Instrument Check

As part of a preflight check, proper operation of the VSI must be established. Make sure the VSI indicates a near zero reading prior to leaving the ramp area and again just before takeoff. If the VSI indicates anything other than zero, that indication can be referenced as the zero mark. Normally, if the needle is not exactly zero, it is only slightly above or below the zero line. After takeoff, the VSI should trend upward to indicate a positive rate of climb and then, once a stabilized climb is established, a rate of climb can be referenced.

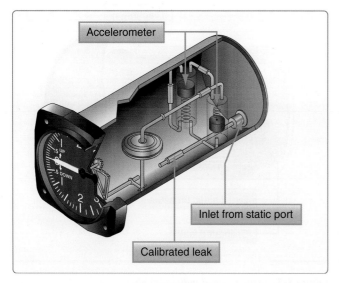

Figure 8-6. *An IVSI incorporates accelerometers to help the instrument immediately indicate changes in vertical speed.*

Airspeed Indicator (ASI)

The ASI is a sensitive, differential pressure gauge that measures and promptly indicates the difference between pitot (impact/dynamic pressure) and static pressure. These two pressures are equal when the aircraft is parked on the ground in calm air. When the aircraft moves through the air, the pressure on the pitot line becomes greater than the pressure in the static lines. This difference in pressure is registered by the airspeed pointer on the face of the instrument, which is calibrated in miles per hour, knots (nautical miles per hour), or both. *[Figure 8-7]*

The ASI is the one instrument that utilizes both the pitot, as well as the static system. The ASI introduces the static pressure into the airspeed case while the pitot pressure (dynamic) is introduced into the diaphragm. The dynamic pressure expands or contracts one side of the diaphragm, which is attached to an indicating system. The system drives the mechanical linkage and the airspeed needle.

Just as in altitudes, there are multiple types of airspeeds. Pilots need to be very familiar with each type.

- Indicated airspeed (IAS)—the direct instrument reading obtained from the ASI, uncorrected for variations in atmospheric density, installation error, or instrument error. Manufacturers use this airspeed as the basis for determining aircraft performance. Takeoff, landing, and stall speeds listed in the AFM/POH are IAS and do not normally vary with altitude or temperature.

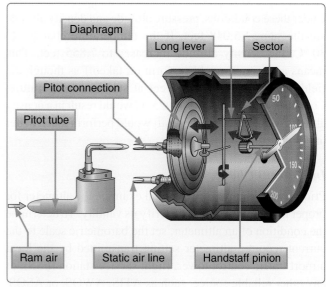

Figure 8-7. *Airspeed indicator (ASI).*

- Calibrated airspeed (CAS)—IAS corrected for installation error and instrument error. Although manufacturers attempt to keep airspeed errors to a minimum, it is not possible to eliminate all errors throughout the airspeed operating range. At certain airspeeds and with certain flap settings, the installation and instrument errors may total several knots. This error is generally greatest at low airspeeds. In the cruising and higher airspeed ranges, IAS and CAS are approximately the same. Refer to the airspeed calibration chart to correct for possible airspeed errors.

- True airspeed (TAS)—CAS corrected for altitude and nonstandard temperature. Because air density decreases with an increase in altitude, an aircraft has to be flown faster at higher altitudes to cause the same pressure difference between pitot impact pressure and static pressure. Therefore, for a given CAS, TAS increases as altitude increases; or for a given TAS, CAS decreases as altitude increases. A pilot can find TAS by two methods. The most accurate method is to use a flight computer. With this method, the CAS is corrected for temperature and pressure variation by using the airspeed correction scale on the computer. Extremely accurate electronic flight computers are also available. Just enter the CAS, pressure altitude, and temperature, and the computer calculates the TAS. A second method, which is a rule of thumb, provides the approximate TAS. Simply add 2 percent to the CAS for each 1,000 feet of altitude. The TAS is the speed that is used for flight planning and is used when filing a flight plan.

- Groundspeed (GS)—the actual speed of the airplane over the ground. It is TAS adjusted for wind. GS decreases with a headwind and increases with a tailwind.

Airspeed Indicator Markings

Aircraft weighing 12,500 pounds or less, manufactured after 1945, and certified by the FAA are required to have ASIs marked in accordance with a standard color-coded marking system. This system of color-coded markings enables a pilot to determine at a glance certain airspeed limitations that are important to the safe operation of the aircraft. For example, if during the execution of a maneuver, it is noted that the airspeed needle is in the yellow arc and rapidly approaching the red line, the immediate reaction should be to reduce airspeed.

As shown in *Figure 8-8,* ASIs on single-engine small aircraft include the following standard color-coded markings:

- White arc—commonly referred to as the flap operating range since its lower limit represents the full flap stall speed and its upper limit provides the maximum flap

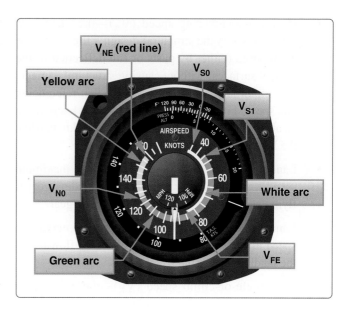

Figure 8-8. *Single engine airspeed indicator (ASI).*

speed. Approaches and landings are usually flown at speeds within the white arc.

- Lower limit of white arc (V_{S0})—the stalling speed or the minimum steady flight speed in the landing configuration. In small aircraft, this is the power-off stall speed at the maximum landing weight in the landing configuration (gear and flaps down).

- Upper limit of the white arc (V_{FE})—the maximum speed with the flaps extended.

- Green arc—the normal operating range of the aircraft. Most flying occurs within this range.

- Lower limit of green arc (V_{S1})—the stalling speed or the minimum steady flight speed obtained in a specified configuration. For most aircraft, this is the power-off stall speed at the maximum takeoff weight in the clean configuration (gear up, if retractable, and flaps up).

- Upper limit of green arc (V_{NO})—the maximum structural cruising speed. Do not exceed this speed except in smooth air.

- Yellow arc—caution range. Fly within this range only in smooth air and then only with caution.

- Red line (V_{NE})—never exceed speed. Operating above this speed is prohibited since it may result in damage or structural failure.

Other Airspeed Limitations

Some important airspeed limitations are not marked on the face of the ASI, but are found on placards and in the AFM/POH. These airspeeds include:

- Design maneuvering speed (V_A)—the maximum speed at which the structural design's limit load can be imposed (either by gusts or full deflection of the control surfaces) without causing structural damage. It is important to consider weight when referencing this speed. For example, V_A may be 100 knots when an airplane is heavily loaded, but only 90 knots when the load is light.

- Landing gear operating speed (V_{LO})—the maximum speed for extending or retracting the landing gear if flying an aircraft with retractable landing gear.

- Landing gear extended speed (V_{LE})—the maximum speed at which an aircraft can be safely flown with the landing gear extended.

- Best angle-of-climb speed (V_X)—the airspeed at which an aircraft gains the greatest amount of altitude in a given distance. It is used during a short-field takeoff to clear an obstacle.

- Best rate-of-climb speed (V_Y)—the airspeed that provides the most altitude gain in a given period of time.

- Single-engine best rate-of-climb (V_{YSE})—the best rate-of-climb or minimum rate-of-sink in a light twin-engine aircraft with one engine inoperative. It is marked on the ASI with a blue line. V_{YSE} is commonly referred to as "Blue Line."

- Minimum control speed (V_{MC})—the minimum flight speed at which a light, twin-engine aircraft can be satisfactorily controlled when an engine suddenly becomes inoperative and the remaining engine is at takeoff power.

Instrument Check

Prior to takeoff, the ASI should read zero. However, if there is a strong wind blowing directly into the pitot tube, the ASI may read higher than zero. When beginning the takeoff, make sure the airspeed is increasing at an appropriate rate.

Blockage of the Pitot-Static System

Errors almost always indicate blockage of the pitot tube, the static port(s), or both. Blockage may be caused by moisture (including ice), dirt, or even insects. During preflight, make sure the pitot tube cover is removed. Then, check the pitot and static port openings. A blocked pitot tube affects the accuracy of the ASI, but a blockage of the static port not only affects the ASI, but also causes errors in the altimeter and VSI.

Blocked Pitot System

The pitot system can become blocked completely or only partially if the pitot tube drain hole remains open. If the pitot tube becomes blocked and its associated drain hole remains clear, ram air is no longer able to enter the pitot system. Air

already in the system vents through the drain hole, and the remaining pressure drops to ambient (outside) air pressure. Under these circumstances, the ASI reading decreases to zero because the ASI senses no difference between ram and static air pressure. The ASI no longer operates since dynamic pressure cannot enter the pitot tube opening. Static pressure is able to equalize on both sides since the pitot drain hole is still open. The apparent loss of airspeed is not usually instantaneous but happens very quickly. *[Figure 8-9]*

If both the pitot tube opening and the drain hole should become clogged simultaneously, then the pressure in the pitot tube is trapped. No change is noted on the airspeed indication should the airspeed increase or decrease. If the static port is unblocked and the aircraft should change altitude, then a change is noted on the ASI. The change is not related to a change in airspeed but a change in static pressure. The total pressure in the pitot tube does not change due to the blockage; however, the static pressure will change.

Because airspeed indications rely upon both static and dynamic pressure together, the blockage of either of these systems affects the ASI reading. Remember that the ASI has a diaphragm in which dynamic air pressure is entered. Behind this diaphragm is a reference pressure called static pressure that comes from the static ports. The diaphragm pressurizes against this static pressure and as a result changes the airspeed indication via levers and indicators. *[Figure 8-10]*

For example, take an aircraft and slow it down to zero knots at a given altitude. If the static port (providing static pressure) and the pitot tube (providing dynamic pressure) are both unobstructed, the following claims can be made:

1. The ASI would be zero.

2. Dynamic pressure and static pressure are equal.

3. Because both dynamic and static air pressure are equal at zero speed with increased speed, dynamic pressure

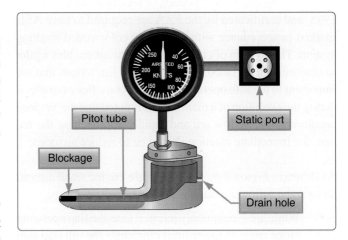

Figure 8-9. *A blocked pitot tube, but clear drain hole.*

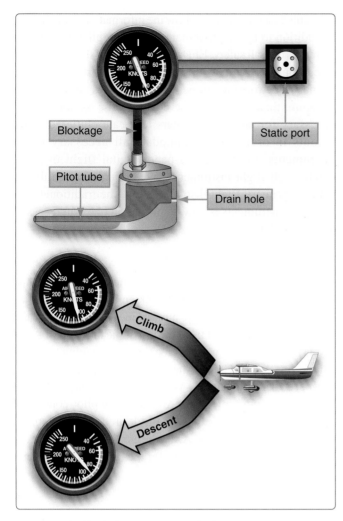

Figure 8-10. *Blocked pitot system with clear static system.*

must include two components: static pressure and dynamic pressure.

It can be inferred that airspeed indication must be based upon a relationship between these two pressures, and indeed it is. An ASI uses the static pressure as a reference pressure and as a result, the ASI's case is kept at this pressure behind the diaphragm. On the other hand, the dynamic pressure through the pitot tube is connected to a highly sensitive diaphragm within the ASI case. Because an aircraft in zero motion (regardless of altitude) results in a zero airspeed, the pitot tube always provides static pressure in addition to dynamic pressure.

Therefore, the airspeed indication is the result of two pressures: the pitot tube static and dynamic pressure within the diaphragm as measured against the static pressure in the ASI's case.

If the aircraft were to descend while the pitot tube is obstructed, the pressure in the pitot system, including the diaphragm, would remain constant. But as the descent is made, the static pressure would increase against the

diaphragm causing it to compress, thereby resulting in an indication of decreased airspeed. Conversely, if the aircraft were to climb, the static pressure would decrease allowing the diaphragm to expand, thereby showing an indication of greater airspeed. *[Figure 8-10]*

The pitot tube may become blocked during flight due to visible moisture. Some aircraft may be equipped with pitot heat for flight in visible moisture. Consult the AFM/POH for specific procedures regarding the use of pitot heat.

Blocked Static System

If the static system becomes blocked but the pitot tube remains clear, the ASI continues to operate; however, it is inaccurate. The airspeed indicates lower than the actual airspeed when the aircraft is operated above the altitude where the static ports became blocked because the trapped static pressure is higher than normal for that altitude. When operating at a lower altitude, a faster than actual airspeed is displayed due to the relatively low static pressure trapped in the system.

Revisiting the ratios that were used to explain a blocked pitot tube, the same principle applies for a blocked static port. If the aircraft descends, the static pressure increases on the pitot side showing an increase on the ASI. This assumes that the aircraft does not actually increase its speed. The increase in static pressure on the pitot side is equivalent to an increase in dynamic pressure since the pressure cannot change on the static side.

If an aircraft begins to climb after a static port becomes blocked, the airspeed begins to show a decrease as the aircraft continues to climb. This is due to the decrease in static pressure on the pitot side, while the pressure on the static side is held constant.

A blockage of the static system also affects the altimeter and VSI. Trapped static pressure causes the altimeter to freeze at the altitude where the blockage occurred. In the case of the VSI, a blocked static system produces a continuous zero indication. *[Figure 8-11]*

Some aircraft are equipped with an alternate static source in the flight deck. In the case of a blocked static source, opening the alternate static source introduces static pressure from the flight deck into the system. Flight deck static pressure is lower than outside static pressure. Check the aircraft AOM/POH for airspeed corrections when utilizing alternate static pressure.

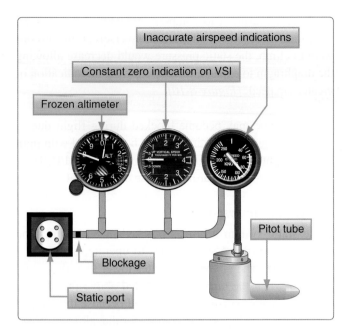

Figure 8-11. *Blocked static system.*

Electronic Flight Display (EFD)

Advances in technology have brought about changes in the instrumentation found in all types of aircraft; for example, Electronic Flight Displays (EFDs) commonly referred to as "glass cockpits." EFDs include flight displays such as primary flight displays (PFD) and multi-function displays (MFD). This has changed not only what information is available to a pilot, but also how the information is displayed. In addition to the improvement in system reliability, which increases overall safety, EFDs have decreased the overall cost of equipping aircraft with state-of-the-art instrumentation. Primary electronic instrumentation packages are less prone to failure than their analogue counterparts. No longer is it necessary for aircraft designers to create cluttered panel layouts in order to accommodate all necessary flight instruments. Instead, multi-panel digital flight displays combine all flight instruments onto a single screen that is called a primary flight display (PFD). The traditional "six pack" of instruments is now displayed on one liquid crystal display (LCD) screen.

Airspeed Tape

Configured similarly to traditional panel layouts, the ASI is located on the left side of the screen and is displayed as a vertical speed tape. As the aircraft increases in speed, the larger numbers descend from the top of the tape. The TAS is displayed at the bottom of the tape through the input to the air data computer (ADC) from the outside air temperature probe. Airspeed markings for V_X, V_Y, and rotation speed (V_R) are displayed for pilot reference. An additional pilot-controlled airspeed bug is available to set at any desired reference speed. As on traditional analogue ASIs, the electronic airspeed tape displays the color-coded ranges for the flap operating range,

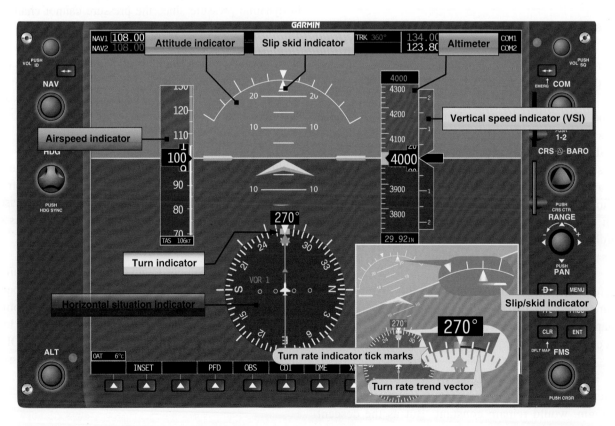

Figure 8-12. *Primary flight display (PFD). Note that the actual location of indications vary depending on manufacturers.*

normal range, and caution range. *[Figure 8-12]* The number value changes color to red when the airspeed exceeds V_{NE} to warn the pilot of exceeding the maximum speed limitation.

Attitude Indicator

One improvement over analogue instrumentation is the larger attitude indicator on EFD. The artificial horizon spans the entire width of the PFD. *[Figure 8-12]* This expanded instrumentation offers better reference through all phases of flight and all flight maneuvers. The attitude indicator receives its information from the Attitude Heading and Reference System (AHRS).

Altimeter

The altimeter is located on the right side of the PFD. *[Figure 8-12]* As the altitude increases, the larger numbers descend from the top of the display tape, with the current altitude being displayed in the black box in the center of the display tape. The altitude is displayed in increments of 20 feet.

Vertical Speed Indicator (VSI)

The VSI is displayed to the right of the altimeter tape and can take the form of an arced indicator or a vertical speed tape. *[Figure 8-12]* Both are equipped with a vertical speed bug.

Heading Indicator

The heading indicator is located below the artificial horizon and is normally modeled after a Horizontal Situation Indicator (HSI). *[Figure 8-12]* As in the case of the attitude indicator, the heading indicator receives its information from the magnetometer, which feeds information to the AHRS unit and then out to the PFD.

Turn Indicator

The turn indicator takes a slightly different form than the traditional instrumentation. A sliding bar moves left and right below the triangle to indicate deflection from coordinated flight. *[Figure 8-12]* Reference for coordinated flight comes from accelerometers contained in the AHRS unit.

Tachometer

The sixth instrument normally associated with the "six pack" package is the tachometer. This is the only instrument that is not located on the PFD. The tachometer is normally located on the multi-function display (MFD). In the event of a display screen failure, it is displayed on the remaining screen with the PFD flight instrumentation. *[Figure 8-13]*

Slip/Skid Indicator

The slip/skid indicator is the horizontal line below the roll pointer. *[Figure 8-12]* Like a ball in a turn-and-slip indicator, a bar width off center is equal to one ball width displacement.

Turn Rate Indicator

The turn rate indicator, illustrated in *Figure 8-12*, is typically found directly above the rotating compass card. Tick marks to the left and right of the lubber line denote the turn (standard-rate versus half standard-rate). Typically denoted by a trend line, if the trend vector is extended to the second tick mark the aircraft is in a standard-rate turn.

Individual panel displays can be configured for a variety of aircraft by installing different software packages. *[Figure 8-14]* Manufacturers are also able to upgrade existing instrument displays in a similar manner, eliminating the need to replace individual gauges in order to upgrade.

Figure 8-13. *Multi-function display (MFD).*

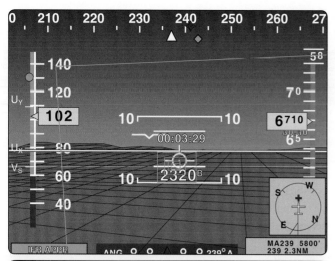

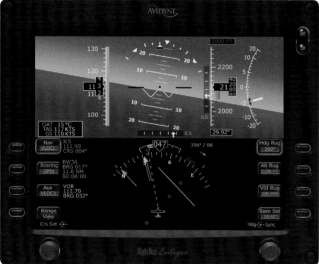

Figure 8-14. *Chelton's FlightLogic (top) and Avidyne's Entegra (bottom) are examples of panel displays that are configurable.*

Air Data Computer (ADC)

EFDs utilize the same type of instrument inputs as traditional analogue gauges; however, the processing system is different. The pitot static inputs are received by an ADC. The ADC computes the difference between the total pressure and the static pressure and generates the information necessary to display the airspeed on the PFD. Outside air temperatures are also monitored and introduced into various components within the system, as well as being displayed on the PFD screen. *[Figure 8-15]*

The ADC is a separate solid state device that, in addition to providing data to the PFD, is capable of providing data to the autopilot control system. In the event of system malfunction, the ADC can quickly be removed and replaced in order to decrease downtime and decrease maintenance turn-around times.

Altitude information is derived from the static pressure port just as an analogue system does; however, the static pressure

Figure 8-15. *Teledyne's 90004 TAS/Plus Air Data Computer (ADC) computes air data information from the pitot-static pneumatic system, aircraft temperature probe, and barometric correction device to help create a clear picture of flight characteristics.*

does not enter a diaphragm. The ADC computes the received barometric pressure and sends a digital signal to the PFD to display the proper altitude readout. EFDs also show trend vectors, which show the pilot how the altitude and airspeed are progressing.

Trend Vectors

Trend vectors are magenta lines that move up and down both the ASI and the altimeter. *[Figures 8-16 and 8-17]* The ADC computes the rate of change and displays the 6-second projection of where the aircraft will be. Pilots can utilize the trend vectors to better control the aircraft's attitude. By including the trend vectors in the instrument scan, pilots are able to precisely control airspeed and altitude. Additional information can be obtained by referencing the Instrument Flying Handbook or specific avionics manufacturer's training material.

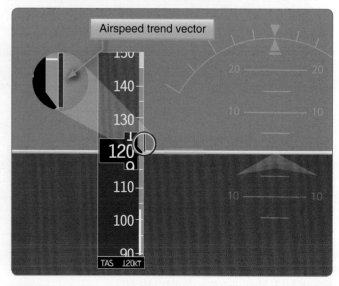

Figure 8-16. *Airspeed trend vector.*

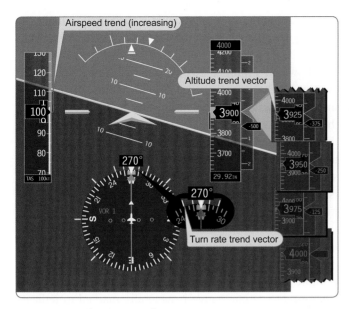

Figure 8-17. *Altimeter trend vector.*

Gyroscopic Flight Instruments

Several flight instruments utilize the properties of a gyroscope for their operation. The most common instruments containing gyroscopes are the turn coordinator, heading indicator, and the attitude indicator. To understand how these instruments operate requires knowledge of the instrument power systems, gyroscopic principles, and the operating principles of each instrument.

Gyroscopic Principles

Any spinning object exhibits gyroscopic properties. A wheel or rotor designed and mounted to utilize these properties is called a gyroscope. Two important design characteristics of an instrument gyro are great weight for its size, or high density, and rotation at high speed with low friction bearings.

There are two general types of mountings; the type used depends upon which property of the gyro is utilized. A freely or universally mounted gyroscope is free to rotate in any direction about its center of gravity. Such a wheel is said to have three planes of freedom. The wheel or rotor is free to rotate in any plane in relation to the base and is balanced so that, with the gyro wheel at rest, it remains in the position in which it is placed. Restricted or semi-rigidly mounted gyroscopes are those mounted so that one of the planes of freedom is held fixed in relation to the base.

There are two fundamental properties of gyroscopic action: rigidity in space and precession.

Rigidity in Space

Rigidity in space refers to the principle that a gyroscope remains in a fixed position in the plane in which it is spinning. An example of rigidity in space is that of a bicycle wheel. As the bicycle wheels increase speed, they become more stable in their plane of rotation. This is why a bicycle is unstable and maneuverable at low speeds and stable and less maneuverable at higher speeds.

By mounting this wheel, or gyroscope, on a set of gimbal rings, the gyro is able to rotate freely in any direction. Thus, if the gimbal rings are tilted, twisted, or otherwise moved, the gyro remains in the plane in which it was originally spinning. *[Figure 8-18]*

Precession

Precession is the tilting or turning of a gyro in response to a deflective force. The reaction to this force does not occur at the point at which it was applied; rather, it occurs at a point that is 90° later in the direction of rotation. This principle allows the gyro to determine a rate of turn by sensing the amount of pressure created by a change in direction. The rate at which the gyro precesses is inversely proportional to the speed of the rotor and proportional to the deflective force.

Using the example of the bicycle, precession acts on the wheels in order to allow the bicycle to turn. While riding at normal speed, it is not necessary to turn the handle bars in the direction of the desired turn. A rider simply leans in the direction that he or she wishes to go. Since the wheels are rotating in a clockwise direction when viewed from the right side of the bicycle, if a rider leans to the left, a force is applied to the top of the wheel to the left. The force actually acts 90° in the direction of rotation, which has the effect of applying a force to the front of the tire, causing the bicycle

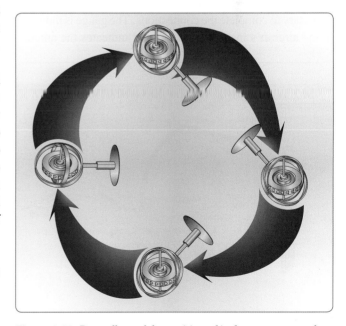

Figure 8-18. *Regardless of the position of its base, a gyro tends to remain rigid in space, with its axis of rotation pointed in a constant direction.*

to move to the left. There is a need to turn the handlebars at low speeds because of the instability of the slowly turning gyros and also to increase the rate of turn.

Precession can also create some minor errors in some instruments. *[Figure 8-19]* Precession can cause a freely spinning gyro to become displaced from its intended plane of rotation through bearing friction, etc. Certain instruments may require corrective realignment during flight, such as the heading indicator.

Sources of Power

In some aircraft, all the gyros are vacuum, pressure, or electrically operated. In other aircraft, vacuum or pressure systems provide the power for the heading and attitude indicators, while the electrical system provides the power for the turn coordinator. Most aircraft have at least two sources of power to ensure at least one source of bank information is available if one power source fails. The vacuum or pressure system spins the gyro by drawing a stream of air against the rotor vanes to spin the rotor at high speed, much like the operation of a waterwheel or turbine. The amount of vacuum or pressure required for instrument operation varies, but is usually between 4.5 "Hg and 5.5 "Hg.

One source of vacuum for the gyros is a vane-type engine-driven pump that is mounted on the accessory case of the engine. Pump capacity varies in different aircraft, depending on the number of gyros.

A typical vacuum system consists of an engine-driven vacuum pump, relief valve, air filter, gauge, and tubing necessary to complete the connections. The gauge is mounted in the aircraft's instrument panel and indicates the amount

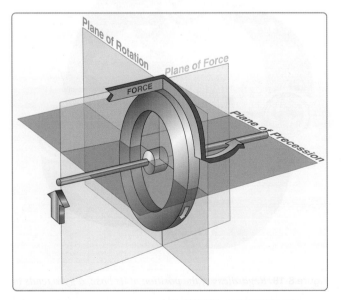

Figure 8-19. *Precession of a gyroscope resulting from an applied deflective force.*

of pressure in the system (vacuum is measured in inches of mercury less than ambient pressure).

As shown in *Figure 8-20,* air is drawn into the vacuum system by the engine-driven vacuum pump. It first goes through a filter, which prevents foreign matter from entering the vacuum or pressure system. The air then moves through the attitude and heading indicators where it causes the gyros to spin. A relief valve prevents the vacuum pressure, or suction, from exceeding prescribed limits. After that, the air is expelled overboard or used in other systems, such as for inflating pneumatic deicing boots.

It is important to monitor vacuum pressure during flight, because the attitude and heading indicators may not provide reliable information when suction pressure is low. The vacuum, or suction, gauge is generally marked to indicate the normal range. Some aircraft are equipped with a warning light that illuminates when the vacuum pressure drops below the acceptable level.

When the vacuum pressure drops below the normal operating range, the gyroscopic instruments may become unstable and inaccurate. Cross-checking the instruments routinely is a good habit to develop.

Turn Indicators

Aircraft use two types of turn indicators: turn-and-slip indicators and turn coordinators. Because of the way the gyro is mounted, the turn-and-slip indicator shows only the rate of turn in degrees per second. The turn coordinator is mounted at an angle, or canted, so it can initially show roll rate. When the roll stabilizes, it indicates rate of turn. Both instruments indicate turn direction and quality (coordination), and also serve as a backup source of bank information in the event an attitude indicator fails. Coordination is achieved by referring to the inclinometer, which consists of a liquid-filled curved tube with a ball inside. *[Figure 8-21]*

Turn-and-Slip Indicator

The gyro in the turn-and-slip indicator rotates in the vertical plane corresponding to the aircraft's longitudinal axis. A single gimbal limits the planes in which the gyro can tilt, and a spring works to maintain a center position. Because of precession, a yawing force causes the gyro to tilt left or right, as viewed from the pilot seat. The turn-and-slip indicator uses a pointer, called the turn needle, to show the direction and rate of turn. The turn-and-slip indicator is incapable of "tumbling" off its rotational axis because of the restraining springs. When extreme forces are applied to a gyro, the gyro is displaced from its normal plane of rotation, rendering its indications invalid. Certain instruments have specific pitch and bank limits that induce a tumble of the gyro.

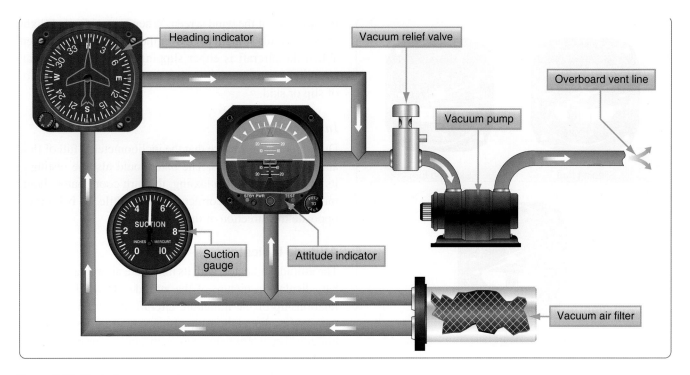

Figure 8-20. *Typical vacuum system.*

Turn Coordinator

The gimbal in the turn coordinator is canted; therefore, its gyro can sense both rate of roll and rate of turn. Since turn coordinators are more prevalent in training aircraft, this discussion concentrates on that instrument. When rolling into or out of a turn, the miniature aircraft banks in the direction the aircraft is rolled. A rapid roll rate causes the miniature aircraft to bank more steeply than a slow roll rate.

The turn coordinator can be used to establish and maintain a standard-rate turn by aligning the wing of the miniature aircraft with the turn index. *Figure 8-22* shows a picture of a turn coordinator. There are two marks on each side (left and right) of the face of the instrument. The first mark is used to reference a wings level zero rate of turn. The second mark on the left and right side of the instrument serve to indicate a standard rate of turn. A standard-rate turn is defined as a turn rate of 3° per second. The turn coordinator indicates only the rate and direction of turn; it does not display a specific angle of bank.

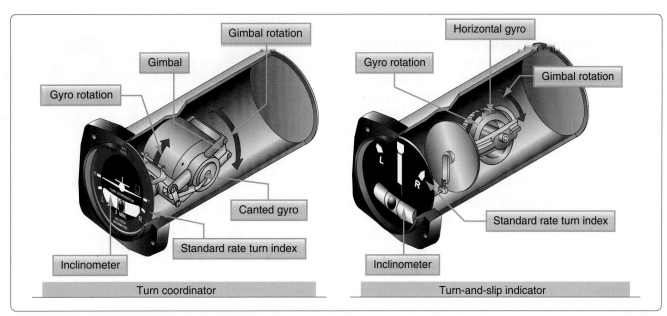

Figure 8-21. *Turn indicators rely on controlled precession for their operation.*

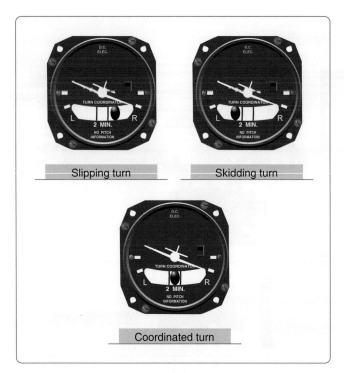

Slipping turn

Skidding turn

Coordinated turn

Figure 8-22. *If inadequate right rudder is applied in a right turn, a slip results. Too much right rudder causes the aircraft to skid through the turn. Centering the ball results in a coordinated turn.*

Inclinometer

The inclinometer is used to depict aircraft yaw, which is the side-to-side movement of the aircraft's nose. During coordinated, straight-and-level flight, the force of gravity causes the ball to rest in the lowest part of the tube, centered between the reference lines. Coordinated flight is maintained by keeping the ball centered. If the ball is not centered, it can be centered by using the rudder.

To center the ball, apply rudder pressure on the side to which the ball is deflected. Use the simple rule, "step on the ball," to remember which rudder pedal to press. If aileron and rudder are coordinated during a turn, the ball remains centered in the tube. If aerodynamic forces are unbalanced, the ball moves away from the center of the tube. As shown in *Figure 8-22,* in a slip, the rate of turn is too slow for the angle of bank, and the ball moves to the inside of the turn. In a skid, the rate of turn is too great for the angle of bank, and the ball moves to the outside of the turn. To correct for these conditions, and improve the quality of the turn, remember to "step on the ball." Varying the angle of bank can also help restore coordinated flight from a slip or skid. To correct for a slip, decrease bank and/or increase the rate of turn. To correct for a skid, increase the bank and/or decrease the rate of turn.

Yaw String

One additional tool that can be added to the aircraft is a yaw string. A yaw string is simply a string or piece of yarn attached

to the center of the wind screen. When in coordinated flight, the string trails straight back over the top of the wind screen. When the aircraft is either slipping or skidding, the yaw string moves to the right or left depending on the direction of slip or skid.

Instrument Check

During preflight, ensure that the inclinometer is full of fluid and has no air bubbles. The ball should also be resting at its lowest point. When taxiing, the turn coordinator should indicate a turn in the correct direction while the ball moves opposite the direction of the turn.

Attitude Indicator

The attitude indicator, with its miniature aircraft and horizon bar, displays a picture of the attitude of the aircraft. The relationship of the miniature aircraft to the horizon bar is the same as the relationship of the real aircraft to the actual horizon. The instrument gives an instantaneous indication of even the smallest changes in attitude.

The gyro in the attitude indicator is mounted in a horizontal plane and depends upon rigidity in space for its operation. The horizon bar represents the true horizon. This bar is fixed to the gyro and remains in a horizontal plane as the aircraft is pitched or banked about its lateral or longitudinal axis, indicating the attitude of the aircraft relative to the true horizon. *[Figure 8-23]*

The gyro spins in the horizontal plane and resists deflection of the rotational path. Since the gyro relies on rigidity in space, the aircraft actually rotates around the spinning gyro.

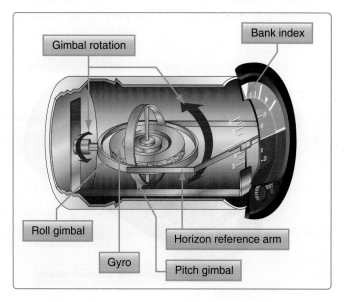

Figure 8-23. *Attitude indicator.*

An adjustment knob is provided with which the pilot may move the miniature aircraft up or down to align the miniature aircraft with the horizon bar to suit the pilot's line of vision. Normally, the miniature aircraft is adjusted so that the wings overlap the horizon bar when the aircraft is in straight-and-level cruising flight.

The pitch and bank limits of an attitude indicator depend upon the make and model of the instrument. Some attitude indicators have limits in the banking plane from 100° to 110°, and the pitch limits can be from 60° to 70°. For those attitude indicators that display only pitch information of +/- 25° vertically, the instrument could "peg" (stop) and remain at this pitch indication until the pitch no longer exceeds limitation or "tumble" and provide erroneous pitch and bank indications when the aircraft exceeds these limits. This may be extremely hazardous when the aircraft is operating in instrument meteorological conditions or confuse a pilot during an unusual attitude recovery. A number of modern attitude indicators do not have this problem.

Every pilot should be able to interpret the banking scale illustrated in Figure 8-24. Most banking scale indicators on the top of the instrument move in the same direction from that in which the aircraft is actually banked. Some other models move in the opposite direction from that in which the aircraft is actually banked. This may confuse the pilot if the indicator is used to determine the direction of bank. This scale should be used only to control the degree of desired bank. The relationship of the miniature aircraft to the horizon bar should be used for an indication of the direction of bank. The attitude indicator is reliable and the most realistic flight instrument on the instrument panel. Its indications are very close approximations of the actual attitude of the aircraft.

Heading Indicator

The heading indicator is fundamentally a mechanical instrument designed to facilitate the use of the magnetic compass. Errors in the magnetic compass are numerous,

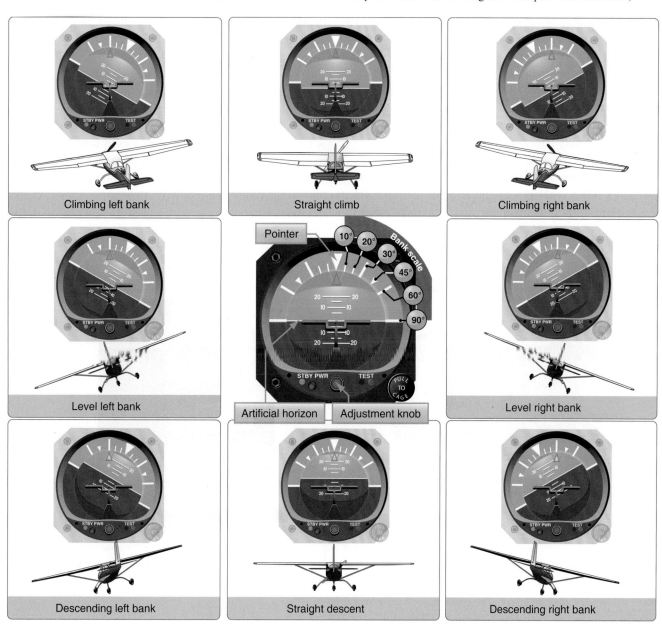

Figure 8-24. *Attitude representation by the attitude indicator corresponds to the relation of the aircraft to the real horizon.*

making straight flight and precision turns to headings difficult to accomplish, particularly in turbulent air. A heading indicator, however, is not affected by the forces that make the magnetic compass difficult to interpret. *[Figure 8-25]*

The operation of the heading indicator depends upon the principle of rigidity in space. The rotor turns in a vertical plane and fixed to the rotor is a compass card. Since the rotor remains rigid in space, the points on the card hold the same position in space relative to the vertical plane of the gyro. The aircraft actually rotates around the rotating gyro, not the other way around. As the instrument case and the aircraft revolve around the vertical axis of the gyro, the card provides clear and accurate heading information.

Because of precession caused by friction, the heading indicator creeps or drifts from its set position. Among other factors, the amount of drift depends largely upon the condition of the instrument. If the bearings are worn, dirty, or improperly lubricated, the drift may be excessive. Another error in the heading indicator is caused by the fact that the gyro is oriented in space, and the Earth rotates in space at a rate of 15° in 1 hour. Thus, discounting precession caused by friction, the heading indicator may indicate as much as 15° error per every hour of operation.

Some heading indicators referred to as horizontal situation indicators (HSI) receive a magnetic north reference from a magnetic slaving transmitter and generally need no adjustment. The magnetic slaving transmitter is called a magnetometer.

Attitude and Heading Reference System (AHRS)

Electronic flight displays have replaced free-spinning gyros with solid-state laser systems that are capable of flight at any attitude without tumbling. This capability is the result of the development of the Attitude and Heading Reference System (AHRS).

The AHRS sends attitude information to the PFD in order to generate the pitch and bank information of the attitude indicator. The heading information is derived from a magnetometer that senses the earth's lines of magnetic flux. This information is then processed and sent out to the PFD to generate the heading display. *[Figure 8-26]*

The Flux Gate Compass System

As mentioned earlier, the lines of flux in the Earth's magnetic field have two basic characteristics: a magnet aligns with them, and an electrical current is induced, or generated, in any wire crossed by them.

The flux gate compass that drives slaved gyros uses the characteristic of current induction. The flux valve is a small, segmented ring, like the one in *Figure 8-27,* made of soft iron that readily accepts lines of magnetic flux. An electrical coil is wound around each of the three legs to accept the current induced in this ring by the Earth's magnetic field. A coil wound around the iron spacer in the center of the frame has 400 Hz alternating current (AC) flowing through it. During the times when this current reaches its peak, twice during each cycle, there is so much magnetism produced by this coil that the frame cannot accept the lines of flux from the Earth's field.

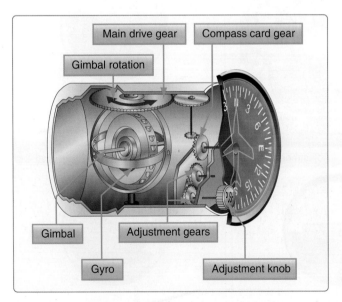

Figure 8-25. *A heading indicator displays headings based on a 360° azimuth, with the final zero omitted. For example, "6" represents 060°, while "21" indicates 210°. The adjustment knob is used to align the heading indicator with the magnetic compass.*

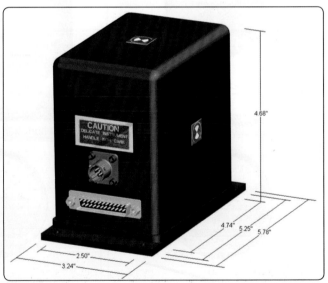

Figure 8-26. *Attitude and heading reference system (AHRS).*

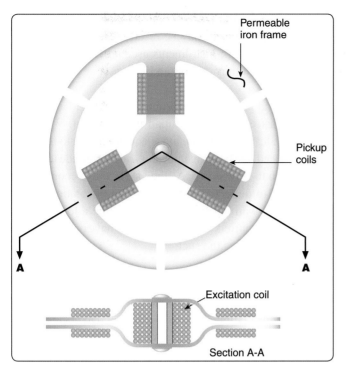

Figure 8-27. *The soft iron frame of the flux valve accepts the flux from the Earth's magnetic field each time the current in the center coil reverses. This flux causes current to flow in the three pickup coils.*

As the current reverses between the peaks, it demagnetizes the frame so it can accept the flux from the Earth's field. As this flux cuts across the windings in the three coils, it causes current to flow in them. These three coils are connected in such a way that the current flowing in them changes as the heading of the aircraft changes. *[Figure 8-28]*

The three coils are connected to three similar but smaller coils in a synchro inside the instrument case. The synchro rotates the dial of a radio magnetic indicator (RMI) or a HSI.

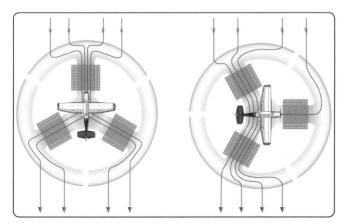

Figure 8-28. *The current in each of the three pickup coils changes with the heading of the aircraft.*

Remote Indicating Compass

Remote indicating compasses were developed to compensate for the errors and limitations of the older type of heading indicators. The two panel-mounted components of a typical system are the pictorial navigation indicator and the slaving control and compensator unit. *[Figure 8-29]* The pictorial navigation indicator is commonly referred to as an HSI.

The slaving control and compensator unit has a push button that provides a means of selecting either the "slaved gyro" or "free gyro" mode. This unit also has a slaving meter and two manual heading-drive buttons. The slaving meter indicates the difference between the displayed heading and the magnetic heading. A right deflection indicates a clockwise error of the compass card; a left deflection indicates a counterclockwise error. Whenever the aircraft is in a turn and the card rotates, the slaving meter shows a full deflection to one side or the other. When the system is in "free gyro" mode, the compass card may be adjusted by depressing the appropriate heading-drive button.

A separate unit, the magnetic slaving transmitter, is mounted remotely, usually in a wingtip to eliminate the possibility of magnetic interference. It contains the flux valve, which is the direction-sensing device of the system. A concentration of lines of magnetic force, after being amplified, becomes

Figure 8-29. *Pictorial navigation indicator (HSI, top), slaving meter (lower right), and slaving control compensator unit (lower left).*

a signal relayed to the heading indicator unit, which is also remotely mounted. This signal operates a torque motor in the heading indicator unit that processes the gyro unit until it is aligned with the transmitter signal. The magnetic slaving transmitter is connected electrically to the HSI.

There are a number of designs of the remote indicating compass; therefore, only the basic features of the system are covered here. Instrument pilots must become familiar with the characteristics of the equipment in their aircraft.

As instrument panels become more crowded and the pilot's available scan time is reduced by a heavier flight deck workload, instrument manufacturers have worked toward combining instruments. One good example of this is the RMI in *Figure 8-30*. The compass card is driven by signals from the flux valve, and the two pointers are driven by an automatic direction finder (ADF) and a very high frequency (VHF) omni-directional radio range (VOR).

Heading indicators that do not have this automatic northseeking capability are called "free" gyros and require periodic adjustment. It is important to check the indications frequently (approximately every 15 minutes) and reset the heading indicator to align it with the magnetic compass when required. Adjust the heading indicator to the magnetic compass heading when the aircraft is straight and level at a constant speed to avoid compass errors.

The bank and pitch limits of the heading indicator vary with the particular design and make of instrument. On some heading indicators found in light aircraft, the limits are approximately 55° of pitch and 55° of bank. When either of these attitude limits is exceeded, the instrument "tumbles"

Figure 8-30. *Driven by signals from a flux valve, the compass card in this RMI indicates the heading of the aircraft opposite the upper center index mark. The green pointer is driven by the ADF.*

or "spills" and no longer gives the correct indication until reset. After spilling, it may be reset with the caging knob. Many of the modern instruments used are designed in such a manner so that they do not tumble.

An additional precession error may occur due to a gyro not spinning fast enough to maintain its alignment. When the vacuum system stops producing adequate suction to maintain the gyro speed, the heading indicator and the attitude indicator gyros begin to slow down. As they slow, they become more susceptible to deflection from the plane of rotation. Some aircraft have warning lights to indicate that a low vacuum situation has occurred. Other aircraft may have only a vacuum gauge that indicates the suction.

Instrument Check

As the gyro spools up, make sure there are no abnormal sounds. While taxiing, the instrument should indicate turns in the correct direction, and precession should be normal. At idle power settings, the gyroscopic instruments using the vacuum system might not be up to operating speeds and precession might occur more rapidly than during flight.

Angle of Attack Indicators

The purpose of an AOA indicator is to give the pilot better situational awareness pertaining to the aerodynamic health of the airfoil. This can also be referred to as stall margin awareness. More simply explained, it is the margin that exists between the current AOA that the airfoil is operating at, and the AOA at which the airfoil will stall (critical AOA).

Speed by itself is not a reliable parameter to avoid a stall. An airplane can stall at any speed. Angle of attack is a better parameter to use to avoid a stall. For a given configuration, the airplane always stalls at the same AOA, referred to as the critical AOA. This critical AOA does not change with:

- Weight
- Bank Angle
- Temperature
- Density Altitude
- Center of Gravity

An AOA indicator can have several benefits when installed in General Aviation aircraft, not the least of which is increased situational awareness. Without an AOA indicator, the AOA is "invisible" to pilots. These devices measure several parameters simultaneously and determine the current AOA providing a visual image to the pilot of the current AOA along with representations of the proximity to the critical AOA. *[Figure 8-31]* These devices can give a visual representation of the energy management state of the airplane. The energy

Figure 8-31. *Angle of attack indicators.*

state of an airplane is the balance between airspeed, altitude, drag, and thrust and represents how efficiently the airfoil is operating.

Compass Systems

The Earth is a huge magnet, spinning in space, surrounded by a magnetic field made up of invisible lines of flux. These lines leave the surface at the magnetic North Pole and reenter at the magnetic South Pole.

Lines of magnetic flux have two important characteristics: any magnet that is free to rotate will align with them, and an electrical current is induced into any conductor that cuts across them. Most direction indicators installed in aircraft make use of one of these two characteristics.

Magnetic Compass

One of the oldest and simplest instruments for indicating direction is the magnetic compass. It is also one of the basic instruments required by Title 14 of the Code of Federal Regulations (14 CFR) part 91 for both VFR and IFR flight.

A magnet is a piece of material, usually a metal containing iron, that attracts and holds lines of magnetic flux. Regardless of size, every magnet has two poles: north and south. When one magnet is placed in the field of another, the unlike poles attract each other, and like poles repel.

An aircraft magnetic compass, such as the one in *Figure 8-32,* has two small magnets attached to a metal float sealed inside a bowl of clear compass fluid similar to kerosene. A graduated scale, called a card, is wrapped around the float and viewed through a glass window with a lubber line across it. The card is marked with letters representing the cardinal directions, north, east, south, and west, and a number for each 30° between these letters. The final "0" is omitted from these directions. For example, 3 = 30°, 6 = 60°, and 33 = 330°.

There are long and short graduation marks between the letters and numbers, each long mark representing 10° and each short mark representing 5°.

The float and card assembly has a hardened steel pivot in its center that rides inside a special, spring-loaded, hard glass jewel cup. The buoyancy of the float takes most of the weight off of the pivot, and the fluid damps the oscillation of the float and card. This jewel-and-pivot type mounting allows the float freedom to rotate and tilt up to approximately 18° angle of bank. At steeper bank angles, the compass indications are erratic and unpredictable.

The compass housing is entirely full of compass fluid. To prevent damage or leakage when the fluid expands and contracts with temperature changes, the rear of the compass case is sealed with a flexible diaphragm, or with a metal bellows in some compasses.

The magnets align with the Earth's magnetic field and the pilot reads the direction on the scale opposite the lubber line. Note that in *Figure 8-32,* the pilot views the compass

Figure 8-32. *A magnetic compass. The vertical line is called the lubber line.*

8-23

card from its backside. When the pilot is flying north, as the compass indicates, east is to the pilot's right. On the card, "33," which represents 330° (west of north), is to the right of north. The reason for this apparent backward graduation is that the card remains stationary, and the compass housing and the pilot rotate around it. Because of this setup, the magnetic compass can be confusing to read.

Magnetic Compass Induced Errors

The magnetic compass is the simplest instrument in the panel, but it is subject to a number of errors that must be considered.

Variation

The Earth rotates about its geographic axis; maps and charts are drawn using meridians of longitude that pass through the geographic poles. Directions measured from the geographic poles are called true directions. The magnetic North Pole to which the magnetic compass points is not collocated with the geographic North Pole, but is some 1,300 miles away; directions measured from the magnetic poles are called magnetic directions. In aerial navigation, the difference between true and magnetic directions is called variation. This same angular difference in surveying and land navigation is called declination.

Figure 8-33 shows the isogonic lines that identify the number of degrees of variation in their area. The line that passes near Chicago is called the agonic line. Anywhere along this line

the two poles are aligned, and there is no variation. East of this line, the magnetic North Pole is to the west of the geographic North Pole and a correction must be applied to a compass indication to get a true direction.

Flying in the Washington, D.C., area, for example, the variation is 10° west. If a pilot wants to fly a true course of south (180°), the variation must be added to this, resulting in a magnetic course of 190° to fly. Flying in the Los Angeles, California area, the variation is 14° east. To fly a true course of 180° there, the pilot would have to subtract the variation and fly a magnetic course of 166°. The variation error does not change with the heading of the aircraft; it is the same anywhere along the isogonic line.

Deviation

The magnets in a compass align with any magnetic field. Some causes for magnetic fields in aircraft include flowing electrical current, magnetized parts, and conflict with the Earth's magnetic field. These aircraft magnetic fields create a compass error called deviation.

Deviation, unlike variation, depends on the aircraft heading. Also unlike variation, the aircraft's geographic location does not affect deviation. While no one can reduce or change variation error, an aviation maintenance technician (AMT) can provide the means to minimize deviation error by performing the maintenance task known as "swinging the compass."

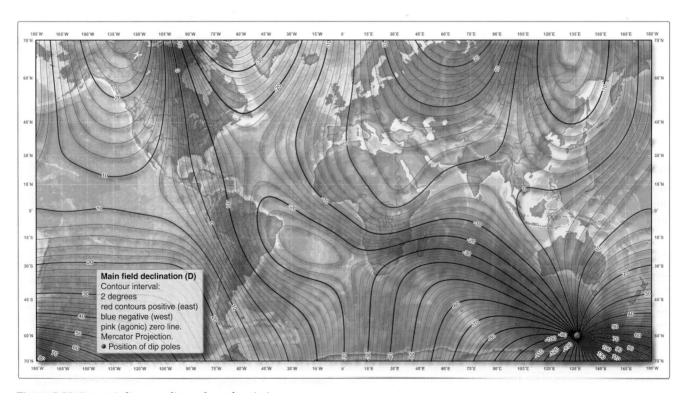

Figure 8-33. *Isogonic lines are lines of equal variation.*

To swing the compass, an AMT positions the aircraft on a series of known headings, usually at a compass rose. *[Figure 8-34]* A compass rose consists of a series of lines marked every 30° on an airport ramp, oriented to magnetic north. There is minimal magnetic interference at the compass rose. The pilot or the AMT, if authorized, can taxi the aircraft to the compass rose and maneuver the aircraft to the headings prescribed by the AMT.

As the aircraft is "swung" or aligned to each compass rose heading, the AMT adjusts the compensator assembly located on the top or bottom of the compass. The compensator assembly has two shafts whose ends have screwdriver slots accessible from the front of the compass. Each shaft rotates one or two small compensating magnets. The end of one shaft is marked E-W, and its magnets affect the compass when the aircraft is pointed east or west. The other shaft is marked N-S and its magnets affect the compass when the aircraft is pointed north or south.

The adjustments position the compensating magnets to minimize the difference between the compass indication and the actual aircraft magnetic heading. The AMT records any remaining error on a compass correction card like the one in *Figure 8-35* and places it in a holder near the compass. Only AMTs can adjust the compass or complete the compass correction card. Pilots determine and fly compass headings using the deviation errors noted on the card. Pilots must also note the use of any equipment causing operational magnetic interference such as radios, deicing equipment, pitot heat, radar, or magnetic cargo.

The corrections for variation and deviation must be applied in the correct sequence as shown below, starting from the true course desired.

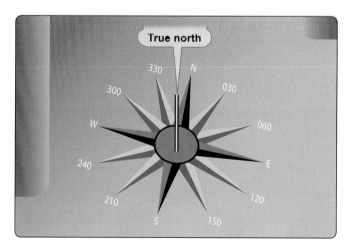

Figure 8-34. *Utilization of a compass rose aids compensation for deviation errors.*

FOR STEER	000	030	060	090	120	150
RDO. ON	001	032	062	095	123	155
RDO. OFF	002	031	064	094	125	157

FOR STEER	180	210	240	270	300	330
RDO. ON	176	210	243	271	296	325
RDO. OFF	174	210	240	273	298	327

Figure 8-35. *A compass correction card shows the deviation correction for any heading.*

Step 1: Determine the Magnetic Course
True Course (180°) ± Variation (+10°) = Magnetic Course (190°)

The magnetic course (190°) is steered if there is no deviation error to be applied. The compass card must now be considered for the compass course of 190°.

Step 2: Determine the Compass Course
Magnetic Course (190°, from step 1) ± Deviation (−2°, from correction card) = Compass Course (188°)

NOTE: Intermediate magnetic courses between those listed on the compass card need to be interpreted. Therefore, to steer a true course of 180°, the pilot would follow a compass course of 188°.

To find the true course that is being flown when the compass course is known:
Compass Course ± Deviation = Magnetic Course ± Variation= True Course

Dip Errors

The Earth's magnetic field runs parallel to its surface only at the Magnetic Equator, which is the point halfway between the Magnetic North and South Poles. As you move away from the Magnetic Equator towards the magnetic poles, the angle created by the vertical pull of the Earth's magnetic field in relation to the Earth's surface increases gradually. This angle is known as the dip angle. The dip angle increases in a downward direction as you move towards the Magnetic North Pole and increases in an upward direction as you move towards the Magnetic South Pole.

If the compass needle were mounted so that it could pivot freely in three dimensions, it would align itself with the magnetic field, pointing up or down at the dip angle in the direction of local Magnetic North. Because the dip angle is of no navigational interest, the compass is made so that it can rotate only in the horizontal plane. This is done by lowering the center of gravity below the pivot point and making the assembly heavy enough that the vertical component of the magnetic force is too weak to tilt it significantly out of the horizontal plane. The compass can then work effectively at all latitudes without specific compensation for dip. However, close to the magnetic poles, the horizontal component of the Earth's field is too small to align the compass which makes the compass unusable for navigation. Because of this constraint, the compass only indicates correctly if the card is horizontal. Once tilted out of the horizontal plane, it will be affected by the vertical component of the Earth's field which leads to the following discussions on northerly and southerly turning errors.

Northerly Turning Errors

The center of gravity of the float assembly is located lower than the pivotal point. As the aircraft turns, the force that results from the magnetic dip causes the float assembly to swing in the same direction that the float turns. The result is a false northerly turn indication. Because of this lead of the compass card, or float assembly, a northerly turn should be stopped prior to arrival at the desired heading. This compass error is amplified with the proximity to either magnetic pole. One rule of thumb to correct for this leading error is to stop the turn 15 degrees plus half of the latitude (i.e., if the aircraft is being operated in a position near 40 degrees latitude, the turn should be stopped 15+20=35 degrees prior to the desired heading). [Figure 8-36A]

Southerly Turning Errors

When turning in a southerly direction, the forces are such that the compass float assembly lags rather than leads. The result is a false southerly turn indication. The compass card, or float assembly, should be allowed to pass the desired heading prior to stopping the turn. As with the northerly error, this error is amplified with the proximity to either magnetic pole. To correct this lagging error, the aircraft should be allowed to pass the desired heading prior to stopping the turn. The same rule of 15 degrees plus half of the latitude applies here (i.e., if the aircraft is being operated in a position near 30 degrees latitude, the turn should be stopped 15+15=30 degrees after passing the desired heading). [Figure 8-36B]

Acceleration Error

The magnetic dip and the forces of inertia cause magnetic compass errors when accelerating and decelerating on easterly and westerly headings. Because of the pendulous-type mounting, the aft end of the compass card is tilted upward when accelerating and downward when decelerating during changes of airspeed. When accelerating on either an easterly or westerly heading, the error appears as a turn indication toward north. When decelerating on either of these headings, the compass indicates a turn toward south. A mnemonic, or memory jogger, for the effect of acceleration error is the word "ANDS" (Acceleration-North/Deceleration-South) may help you to remember the acceleration error. [Figure 8-37] Acceleration causes an

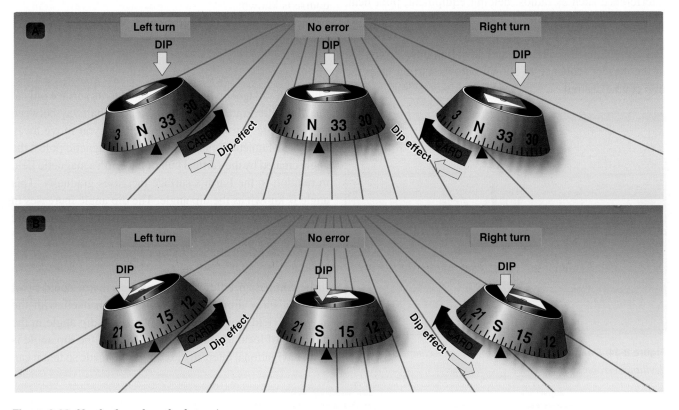

Figure 8-36. *Northerly and southerly turning errors.*

Figure 8-37. *The effects of acceleration error.*

indication toward north; deceleration causes an indication toward south.

Oscillation Error

Oscillation is a combination of all of the errors previously mentioned and results in fluctuation of the compass card in relation to the actual heading direction of the aircraft. When setting the gyroscopic heading indicator to agree with the magnetic compass, use the average indication between the swings.

The Vertical Card Magnetic Compass

The vertical card magnetic compass eliminates some of the errors and confusion encountered with the magnetic compass. The dial of this compass is graduated with letters representing the cardinal directions, numbers every 30°, and tick marks every 5°. The dial is rotated by a set of gears from the shaft-mounted magnet, and the nose of the symbolic aircraft on the instrument glass represents the lubber line for reading the heading of the aircraft from the dial. *[Figure 8-38]*

Lags or Leads

When starting a turn from a northerly heading, the compass lags behind the turn. When starting a turn from a southerly heading, the compass leads the turn.

Figure 8-38. *Vertical card magnetic compass.*

Eddy Current Damping

In the case of a vertical card magnetic compass, flux from the oscillating permanent magnet produces eddy currents in a damping disk or cup. The magnetic flux produced by the eddy currents opposes the flux from the permanent magnet and decreases the oscillations.

Outside Air Temperature (OAT) Gauge

The outside air temperature (OAT) gauge is a simple and effective device mounted so that the sensing element is exposed to the outside air. The sensing element consists of a bimetallic-type thermometer in which two dissimilar materials are welded together in a single strip and twisted into a helix. One end is anchored into protective tube and the other end is affixed to the pointer, which reads against the calibration on a circular face. OAT gauges are calibrated in degrees °C, °F, or both. An accurate air temperature provides the pilot with useful information about temperature lapse rate with altitude change. *[Figure 8-39]*

Figure 8-39. *Outside air temperature (OAT) gauge.*

Chapter Summary

Flight instruments enable an aircraft to be operated with maximum performance and enhanced safety, especially when flying long distances. Manufacturers provide the necessary flight instruments, but to use them effectively, pilots need to understand how they operate. As a pilot, it is important to become very familiar with the operational aspects of the pitot-static system and associated instruments, the vacuum system and associated instruments, the gyroscopic instruments, and the magnetic compass.

Chapter 9
Flight Manuals and Other Documents

Introduction

Each aircraft comes with documentation and a set of manuals with which a pilot must be familiar in order to fly that aircraft. This chapter covers airplane flight manuals (AFM), the pilot's operating handbook (POH), and aircraft documents pertaining to ownership, airworthiness, maintenance, and operations with inoperative equipment. Knowledge of these required documents and manuals is essential for a pilot to conduct a safe flight.

Airplane Flight Manuals (AFM)

Flight manuals and operating handbooks are concise reference books that provide specific information about a particular aircraft or subject. They contain basic facts, information, and/or instructions for the pilot about the operation of an aircraft, flying techniques, etc., and are intended to be kept on hand for ready reference.

The aircraft owner/information manual is a document developed by the aircraft manufacturer and contains general information about the make and model of the aircraft. The manual is not approved by the Federal Aviation Administration (FAA) and is not specific to an individual aircraft. The manual provides general information about the operation of an aircraft, is not kept current, and cannot be substituted for the AFM/POH.

An AFM is a document developed by the aircraft manufacturer and approved by the FAA. This book contains the information and instructions required to operate an aircraft safely. A pilot must comply with this information which is specific to a particular make and model of aircraft, usually by serial number. An AFM contains the operating procedures and limitations of that aircraft. Title 14 of the Code of Federal Regulations (14 CFR) part 91 requires that pilots comply with the operating limitations specified in the approved flight manuals, markings, and placards.

Originally, flight manuals followed whatever format and content the manufacturer felt was appropriate, but this changed with the acceptance of Specification No. 1 prepared by the General Aviation Manufacturers Association (GAMA). Specification No. 1 established a standardized format for all general aviation airplane and helicopter flight manuals.

The POH is a document developed by the aircraft manufacturer and contains FAA-approved AFM information. If "POH" is used in the main title, a statement must be included on the title page indicating that sections of the document are FAA approved as the AFM.

The POH for most light aircraft built after 1975 is also designated as the FAA-approved flight manual. The typical AFM/POH contains the following nine sections: General; Limitations; Emergency Procedures; Normal Procedures; Performance; Weight and Balance/Equipment List; Systems Description; Handling, Service, and Maintenance; and Supplements. Manufacturers also have the option of including additional sections, such as one on Safety and Operational Tips or an alphabetical index at the end of the POH.

Preliminary Pages

While the AFM/POH may appear similar for the same make and model of aircraft, each manual is unique and contains specific information about a particular aircraft, such as the equipment installed and weight and balance information. Manufacturers are required to include the serial number and registration on the title page to identify the aircraft to which the manual belongs. If a manual does not indicate a specific aircraft registration and serial number, it is limited to general study purposes only.

Most manufacturers include a table of contents that identifies the order of the entire manual by section number and title. Usually, each section also contains a table of contents for that section. Page numbers reflect the section and page within that section (1 1, 1 2, 2 1, 3 1, etc.). If the manual is published in loose-leaf form, each section is usually marked with a divider tab indicating the section number, title, or both. The Emergency Procedures section may have a red tab for quick identification and reference.

General (Section 1)

The General section provides the basic descriptive information on the airframe and powerplant(s). Some manuals include a three-dimensional drawing of the aircraft that provides dimensions of various components. Included are such items as wingspan, maximum height, overall length, wheelbase length, main landing gear track width, diameter of the rotor system, maximum propeller diameter, propeller ground clearance, minimum turning radius, and wing area. This section serves as a quick reference and helps a pilot become familiar with the aircraft.

The last segment of the General section contains definitions, abbreviations, explanations of symbology, and some of the terminology used in the POH. At the discretion of the manufacturer, metric and other conversion tables may also be included.

Limitations (Section 2)

The Limitations section contains only those limitations required by regulation or that are necessary for the safe operation of the aircraft, powerplant, systems, and equipment. It includes operating limitations, instrument markings, color-coding, and basic placards. Some of the limitation areas are airspeed, powerplant, weight and loading distribution, and flight.

Airspeed

Airspeed limitations are shown on the airspeed indicator (ASI) by color coding and on placards or graphs in the aircraft. [Figure 9-1] A red line on the ASI shows the airspeed limit beyond which structural damage could occur. This is called the never-exceed speed (V_{NE}). A yellow arc indicates the speed range between maximum structural cruising speed (V_{N0}) and V_{NE}. Operation of an aircraft in the yellow airspeed arc is for smooth air only and then only with caution. A green arc depicts the normal operating speed range, with the upper end at V_{N0} and the lower end at stalling speed at maximum weight with the landing gear and flaps retracted (V_{S1}). For airplanes, the flap operating range is depicted by the white arc, with the upper end at the maximum flap extended speed (V_{FE}), and the lower end at the stalling speed with the landing gear and flaps in the landing configuration (V_{S0}).

Figure 9-1. *Single-engine airspeed indicator.*

In addition to the markings listed above, small multi-engine airplanes have a red radial line to indicate single-engine minimum controllable airspeed (V_{MC}). A blue radial line is used to indicate single-engine best rate of climb speed at maximum weight at sea level (V_{YSE}). *[Figure 9-2]*

Powerplant

The Powerplant Limitations portion describes operating limitations on an aircraft's reciprocating or turbine engine(s). These include limitations for takeoff power, maximum continuous power, and maximum normal operating power, which is the maximum power the engine can produce without any restrictions and is depicted by a green arc. Other items that can be included in this area are the minimum and maximum oil and fuel pressures, oil and fuel grades, and propeller operating limits. *[Figure 9-3]*

All reciprocating-engine powered aircraft must have a revolutions per minute (rpm) indicator for each engine. Aircraft equipped with a constant-speed propeller or rotor system use a manifold pressure gauge to monitor power output and a tachometer to monitor propeller or rotor speed. Both instruments depict the maximum operating limit with

Figure 9-2. *Multi-engine airspeed indicator.*

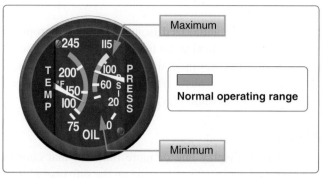

Figure 9-3. *Minimum, maximum, and normal operating range markings on oil gauge.*

a red radial line and the normal operating range with a green arc. *[Figure 9-4]* Some instruments may have a yellow arc to indicate a caution area.

Weight and Loading Distribution

Weight and Loading Distribution contains the maximum certificated weights, as well as the center of gravity (CG) range. The location of the reference datum used in balance computations is included in this section. Weight and balance computations are not provided in this area, but rather in the weight and balance section of the AFM/POH.

Figure 9-4. *Manifold pressure gauge (top) and tachometer (bottom).*

Flight Limits

Flight Limits list authorized maneuvers with appropriate entry speeds, flight load factor limits, and types of operation limits. It also indicates those maneuvers that are prohibited, such as spins or acrobatic flight, as well as operational limitations such as flight into known icing conditions.

Placards

Most aircraft display one or more placards that contain information having a direct bearing on the safe operation of the aircraft. These placards are located in conspicuous places and are reproduced in the Limitations section or as directed by an Airworthiness Directive (AD). *[Figure 9-5]* Airworthiness Directives are explained in detail later in this chapter.

Emergency Procedures (Section 3)

Checklists describing the recommended procedures and airspeeds for coping with various types of emergencies or critical situations are located in the Emergency Procedures section. Some of the emergencies covered include: engine failure, fire, and system failure. The procedures for inflight engine restarting and ditching may also be included. Manufacturers may first show an emergency checklist in an abbreviated form with the order of items reflecting the sequence of action. Amplified checklists that provide additional information on the procedures follow the abbreviated checklist. To be prepared for emergency situations, memorize the immediate action items and, after completion, refer to the appropriate checklist.

Figure 9-5. *Placards are used to depict aircraft limitations.*

Manufacturers may include an optional subsection entitled Abnormal Procedures. This subsection describes recommended procedures for handling malfunctions that are not considered emergencies.

Normal Procedures (Section 4)

This section begins with a list of the airspeeds for normal operations. The next area consists of several checklists that may include preflight inspection, before starting procedures, starting engine, before taxiing, taxiing, before takeoff, climb, cruise, descent, before landing, balked landing, after landing, and post flight procedures. An Amplified Procedures area follows the checklists to provide more detailed information about the various previously mentioned procedures.

To avoid missing important steps, always use the appropriate checklists when available. Consistent adherence to approved checklists is a sign of a disciplined and competent pilot.

Performance (Section 5)

The Performance section contains all the information required by the aircraft certification regulations and any additional performance information the manufacturer deems important to pilot ability to safely operate the aircraft. Performance charts, tables, and graphs vary in style, but all contain the same basic information. Examples of the performance information found in most flight manuals include a graph or table for converting calibrated airspeed to true airspeed; stall speeds in various configurations; and data for determining takeoff and climb performance, cruise performance, and landing performance. *Figure 9-6* is an example of a typical performance graph. For more information on use of the charts, graphs, and tables, refer to Chapter 10, Aircraft Performance.

Weight and Balance/Equipment List (Section 6)

The Weight and Balance/Equipment List section contains all the information required by the FAA to calculate the weight and balance of an aircraft. Manufacturers include sample weight and balance problems. Weight and balance is discussed in greater detail in Chapter 10, Weight and Balance.

Systems Description (Section 7)

This section describes the aircraft systems in a manner appropriate to the pilot most likely to operate the aircraft. For example, a manufacturer might assume an experienced pilot will be reading the information for an advanced aircraft. For more information on aircraft systems, refer to Chapter 7, Aircraft Systems.

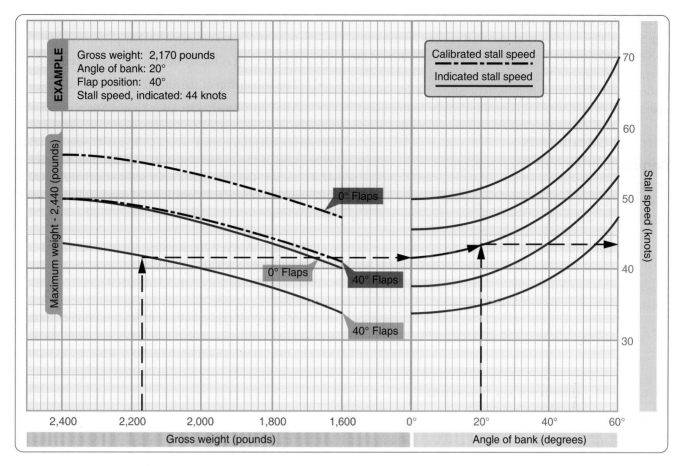

Figure 9-6. *Stall speed chart.*

Handling, Service, and Maintenance (Section 8)

The Handling, Service, and Maintenance section describes the maintenance and inspections recommended by the manufacturer (and the regulations). Additional maintenance or inspections may be required by the issuance of ADs applicable to the airframe, engine, propeller, or components. This section also describes preventive maintenance that may be accomplished by certificated pilots, as well as the manufacturer's recommended ground handling procedures. It includes considerations for hangaring, tie-down, and general storage procedures for the aircraft.

Supplements (Section 9)

The Supplements section contains information necessary to safely and efficiently operate the aircraft when equipped with optional systems and equipment (not provided with the standard aircraft). Some of this information may be supplied by the aircraft manufacturer or by the manufacturer of the optional equipment. The appropriate information is inserted into the flight manual at the time the equipment is installed. Autopilots, navigation systems, and air-conditioning systems are examples of equipment described in this section. *[Figure 9-7]*

Figure 9-7. *Supplements provide information on optional equipment.*

Safety Tips (Section 10)

The Safety Tips section is an optional section containing a review of information that enhances the safe operation of the aircraft. For example, physiological factors, general weather information, fuel conservation procedures, high altitude operations, or cold weather operations might be discussed.

Aircraft Documents

Certificate of Aircraft Registration

Before an aircraft can be flown legally, it must be registered with the FAA Aircraft Registry. The Certificate of Aircraft Registration, which is issued to the owner as evidence of the registration, must be carried in the aircraft at all times. [Figure 9-8]

The Certificate of Aircraft Registration cannot be used for operations when:

- The aircraft is registered under the laws of a foreign country

- The aircraft's registration is canceled upon written request of the certificate holder

- The aircraft is totally destroyed or scrapped

- The ownership of the aircraft is transferred

- The certificate holder loses United States citizenship

For additional information, see 14 CFR part 47, section 47.41. When one of the events listed in 14 CFR part 47, section 47.41 occurs, the previous owner must notify the FAA by filling in the back of the Certificate of Aircraft Registration, and mailing it to:

FAA Aircraft Registration Branch, AFS-750
P.O. Box 25504
Oklahoma City, OK 73125-0504

A dealer's aircraft registration certificate is another form of registration certificate, but is valid only for required flight tests by the manufacturer or in flights that are necessary for

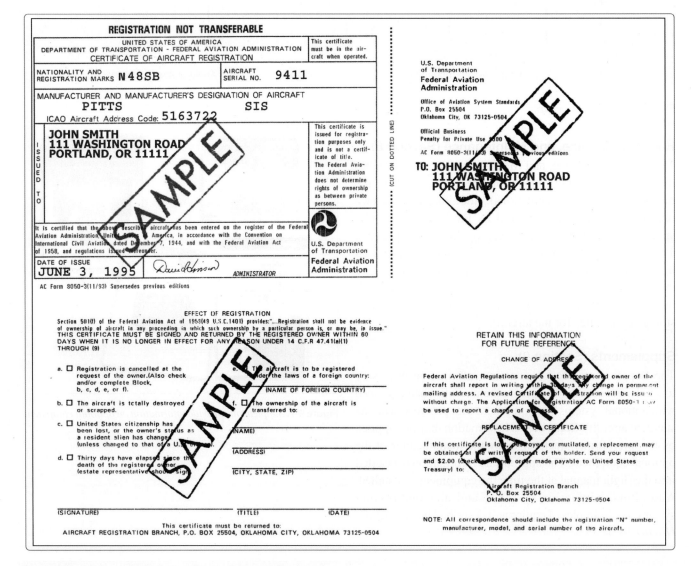

Figure 9-8. *AC Form 8050-3, Certificate of Aircraft Registration.*

the sale of the aircraft by the manufacturer or a dealer. The dealer must remove the certificate when the aircraft is sold.

Upon complying with 14 CFR part 47, section 47.31, the pink copy of the application for an Aircraft Registration Application, Aeronautical Center (AC) Form 8050-1, provides authorization to operate an unregistered aircraft for a period not to exceed 90 days. Since the aircraft is unregistered, it cannot be operated outside of the United States until a permanent Certificate of Aircraft Registration is received and placed in the aircraft.

The FAA does not issue any certificate of ownership or endorse any information with respect to ownership on a Certificate of Aircraft Registration.

NOTE: For additional information concerning the Aircraft Registration Application or the Aircraft Bill of Sale, contact the nearest FAA Flight Standards District Office (FSDO).

Airworthiness Certificate

An Airworthiness Certificate is issued by a representative of the FAA after the aircraft has been inspected, is found to meet the requirements of 14 CFR part 21, and is in condition for safe operation. The Airworthiness Certificate must be displayed in the aircraft so it is legible to the passengers and crew whenever it is operated. The Airworthiness Certificate must remain with the aircraft unless it is sold to a foreign purchaser.

A Standard Airworthiness Certificate is issued for aircraft type certificated in the normal, utility, acrobatic, commuter, transport categories, and manned free balloons. *Figure 9-9* illustrates a Standard Airworthiness Certificate, and an explanation of each item in the certificate follows.

1. *Nationality and Registration Marks.* The "N" indicates the aircraft is registered in the United States. Registration marks consist of a series of up to five numbers or numbers and letters. In this case, N2631A is the registration number assigned to this aircraft.

2. *Manufacturer and Model.* Indicates the manufacturer, make, and model of the aircraft.

3. *Aircraft Serial Number.* Indicates the manufacturer's serial number assigned to the aircraft, as noted on the aircraft data plate.

4. *Category.* Indicates the category in which the aircraft must be operated. In this case, it must be operated in accordance with the limitations specified for the "NORMAL" category.

5. *Authority and Basis for Issuance.* Indicates the aircraft conforms to its type certificate and is considered in condition for safe operation at the time of inspection and issuance of the certificate. Any exemptions from the applicable airworthiness standards are briefly noted here and the exemption number given. The word "NONE" is entered if no exemption exists.

UNITED STATES OF AMERICA
DEPARTMENT OF TRANSPORTATION-FEDERAL AVIATION ADMINISTRATION
STANDARD AIRWORTHINESS CERTIFICATE

1 NATIONALITY AND REGISTRATION MARKS	2 MANUFACTURER AND MODEL	3 AIRCRAFT SERIAL NUMBER	4 CATEGORY
N12345	Douglas DC-6A	43219	Transport

5 AUTHORITY AND BASIS FOR ISSUANCE

This airworthiness certificate is issued pursuant to the Federal Aviation Act of 1958 and certifies that, as of the date of issuance, the aircraft to which issued has been inspected and found to conform to the type certificate therefor, to be in condition for safe operation, and has been shown to meet the requirements of the applicable comprehensive and detailed airworthiness code as provided by Annex 8 to the Convention on International Civil Aviation, except as noted herein. Exceptions:

None

6 TERMS AND CONDITIONS

Unless sooner surrendered, suspended, revoked, or a termination date is otherwise established by the Administrator, this airworthiness certificate is effective as long as the maintenance, preventative maintenance, and alterations are performed in accordance with Parts 21, 43, and 91 of the Federal Aviation Regulations, as appropriate, and the aircraft is registered in the United States.

DATE OF ISSUANCE	FAA REPRESENTATIVE	DESIGNATION NUMBER
01/20/00	E.R. White *E.R. White*	NE-XX

Any iteration, reproduction, or misuse of this certificate may be punishable by a fine not exceeding $1,000 or imprisonment not exceeding 3 years or both. THIS CERTIFICATE MUST BE DISPLAYED IN THE AIRCRAFT IN ACCORDANCE WITH APPLICABLE FEDERAL AVIATION REGULATIONS.

FAA Form 8100-2 (04-11) Supersedes Previous Edition

Figure 9-9. *FAA Form 8100-2, Standard Airworthiness Certificate.*

6. *Terms and Conditions*. Indicates the Airworthiness Certificate is in effect indefinitely if the aircraft is maintained in accordance with 14 CFR parts 21, 43, and 91, and the aircraft is registered in the United States.

Also included are the date the certificate was issued and the signature and office identification of the FAA representative.

A Standard Airworthiness Certificate remains in effect if the aircraft receives the required maintenance and is properly registered in the United States. Flight safety relies in part on the condition of the aircraft, which is determined by inspections performed by mechanics, approved repair stations, or manufacturers that meet specific requirements of 14 CFR part 43.

A Special Airworthiness Certificate is issued for all aircraft certificated in other than the Standard classifications, such as Experimental, Restricted, Limited, Provisional, and Light-Sport Aircraft (LSA). LSA receive a pink special airworthiness certificate; however, there are exceptions. For example, the Piper Cub is in the LSA category, but it was certificated as a normal aircraft during its manufacture. When purchasing an aircraft classified as other than Standard, it is recommended that the local FSDO be contacted for an explanation of the pertinent airworthiness requirements and the limitations of such a certificate.

Aircraft Maintenance

Maintenance is defined as the preservation, inspection, overhaul, and repair of an aircraft, including the replacement of parts. Regular and proper maintenance ensures that an aircraft meets an acceptable standard of airworthiness throughout its operational life.

Although maintenance requirements vary for different types of aircraft, experience shows that aircraft need some type of preventive maintenance every 25 hours of flying time or less and minor maintenance at least every 100 hours. This is influenced by the kind of operation, climatic conditions, storage facilities, age, and construction of the aircraft. Manufacturers supply maintenance manuals, parts catalogs, and other service information that should be used in maintaining the aircraft.

Aircraft Inspections

Under 14 CFR part 91, the primary responsibility for maintaining an aircraft in an airworthy condition falls on the owner or operator of the aircraft. Certain inspections must be performed on the aircraft, and the owner must maintain the airworthiness of the aircraft during the time between required inspections by having any defects corrected.

Under 14 CFR, part 91, subpart E, all civil aircraft are required to be inspected at specific intervals to determine the overall condition. The interval depends upon the type of operations in which the aircraft is engaged. All aircraft need to be inspected at least once every 12 calendar months, while inspection is required for others after every 100 hours of operation. Some aircraft are inspected in accordance with an inspection system set up to provide for total inspection of the aircraft on the basis of calendar time, time in service, number of system operations, or any combination of these.

All inspections should follow the current manufacturer's maintenance manual, including the Instructions for Continued Airworthiness concerning inspection intervals, parts replacement, and life-limited items as applicable to the aircraft.

Annual Inspection

Any reciprocating engine or single-engine turbojet/turbopropeller-powered small aircraft (weighing 12,500 pounds or less) flown for business or pleasure and not flown for compensation or hire is required to be inspected at least annually. The inspection shall be performed by a certificated airframe and powerplant (A&P) mechanic who holds an inspection authorization (IA) by the manufacturer of the aircraft or by a certificated and appropriately rated repair station. The aircraft may not be operated unless the annual inspection has been performed within the preceding 12 calendar months. A period of 12 calendar months extends from any day of a month to the last day of the same month the following year. An aircraft overdue for an annual inspection may be operated under a Special Flight Permit issued by the FAA for the purpose of flying the aircraft to a location where the annual inspection can be performed. However, all applicable ADs that are due must be complied with before the flight.

100-Hour Inspection

All aircraft under 12,500 pounds (except turbojet/turbopropeller-powered multi-engine airplanes and turbine powered rotorcraft), used to carry passengers for hire, must receive a 100-hour inspection within the preceding 100 hours of time in service and must be approved for return to service. Additionally, an aircraft used for flight instruction for hire, when provided by the person giving the flight instruction, must also have received a 100-hour inspection. This inspection must be performed by an FAA-certificated A&P mechanic, an appropriately rated FAA-certificated repair station, or by the aircraft manufacturer. An annual inspection, or an inspection for the issuance of an Airworthiness Certificate, may be substituted for a required 100-hour inspection. The 100-hour limitation may be exceeded by no more than 10 hours for the purpose of traveling to a location at which the required inspection can be performed. Any excess time used for this purpose must be included in computing the next 100 hours of time in service.

Other Inspection Programs

The annual and 100-hour inspection requirements do not apply to large (over 12,500 pounds) airplanes, turbojets, or turbopropeller-powered multi-engine airplanes or to aircraft for which the owner complies with a progressive inspection program. Details of these requirements may be determined by referencing 14 CFR, part 43, section 43.11 and 14 CFR part 91, subpart E, or by inquiring at a local FSDO.

Altimeter System Inspection

Under 14 CFR, part 91, section 91.411, requires that the altimeter, encoding altimeter, and related system must be tested and inspected within the 24 months prior to operating in controlled airspace under instrument flight rules (IFR). This applies to all aircraft being operated in controlled airspace.

Transponder Inspection

Title 14 CFR, part 91, section 91.413, requires that before a transponder can be used under 14 CFR, part 91, section 91.215(a), it shall be tested and inspected within the 24 months prior to operation of the aircraft regardless of airspace restrictions.

Emergency Locator Transmitter

An emergency locator transmitter (ELT) is required by 14 CFR, part 91, section 91.207, and must be inspected within 12 calendar months after the last inspection for the following:

- Proper installation

- Battery corrosion

- Operation of the controls and crash sensor

- The presence of a sufficient signal radiated from its antenna

The ELT must be attached to the airplane in such a manner that the probability of damage to the transmitter in the event of crash impact is minimized. Fixed and deployable automatic type transmitters must be attached to the airplane as far aft as practicable. Batteries used in the ELTs must be replaced (or recharged, if the batteries are rechargeable):

- When the transmitter has been in use for more than 1 cumulative hour

- When 50 percent of the battery useful life or, for rechargeable batteries, 50 percent of useful life of the charge has expired

An expiration date for replacing (or recharging) the battery must be legibly marked on the outside of the transmitter and entered in the aircraft maintenance record. This does not apply to batteries that are essentially unaffected during storage intervals, such as water-activated batteries.

Preflight Inspections

The preflight inspection is a thorough and systematic means by which a pilot determines if an aircraft is airworthy and in condition for safe operation. POHs and owner/information manuals contain a section devoted to a systematic method of performing a preflight inspection.

Minimum Equipment Lists (MEL) and Operations With Inoperative Equipment

Under 14 CFR, all aircraft instruments and installed equipment are required to be operative prior to each departure. When the FAA adopted the minimum equipment list (MEL) concept for 14 CFR part 91 operations, it allowed aircraft to be operated with inoperative equipment determined to be nonessential for safe flight. At the same time, it allowed part 91 operators, without an MEL, to defer repairs on nonessential equipment within the guidelines of part 91.

The FAA has two acceptable methods of deferring maintenance on small rotorcraft, non-turbine powered airplanes, gliders, or lighter-than-air aircraft operated under part 91. They are the deferral provision of 14 CFR, part 91, section 91.213(d) and an FAA-approved MEL.

The deferral provision of 14 CFR, part 91, section 91.213(d) is widely used by most pilot/operators. Its popularity is due to simplicity and minimal paperwork. When inoperative equipment is found during a preflight inspection or prior to departure, the decision should be to cancel the flight, obtain maintenance prior to flight, or to defer the item or equipment.

Maintenance deferrals are not used for inflight discrepancies. The manufacturer's AFM/POH procedures are to be used in those situations. The discussion that follows assumes that the pilot wishes to defer maintenance that would ordinarily be required prior to flight.

Using the deferral provision of 14 CFR, part 91, section 91.213(d), the pilot determines whether the inoperative equipment is required by type design, 14 CFR, or ADs. If the inoperative item is not required, and the aircraft can be safely operated without it, the deferral may be made. The inoperative item shall be deactivated or removed and an INOPERATIVE placard placed near the appropriate switch, control, or indicator. If deactivation or removal involves maintenance (removal always will), it must be accomplished by certificated maintenance personnel and recorded in accordance with 14 CFR part 43.

For example, if the position lights (installed equipment) were discovered to be inoperative prior to a daytime flight, the pilot would follow the requirements of 14 CFR, part 91, section 91.213(d).

The deactivation may be a process as simple as the pilot positioning a circuit breaker to the OFF position or as complex as rendering instruments or equipment totally inoperable. Complex maintenance tasks require a certificated and appropriately rated maintenance person to perform the deactivation. In all cases, the item or equipment must be placarded INOPERATIVE.

All small rotorcraft, non-turbine powered airplanes, gliders, or lighter-than-air aircraft operated under 14 CFR part 91 are eligible to use the maintenance deferral provisions of 14 CFR, part 91, section 91.213(d). However, once an operator requests an MEL, and a Letter of Authorization (LOA) is issued by the FAA, then the use of the MEL becomes mandatory for that aircraft. All maintenance deferrals must be accomplished in accordance with the terms and conditions of the MEL and the operator-generated procedures document.

The use of an MEL for an aircraft operated under 14 CFR part 91 also allows for the deferral of inoperative items or equipment. The primary guidance becomes the FAA-approved MEL issued to that specific operator and N-numbered aircraft.

The FAA has developed master minimum equipment lists (MMELs) for aircraft in current use. Upon written request by an operator, the local FSDO may issue the appropriate make and model MMEL, along with an LOA, and the preamble. The operator then develops operations and maintenance (O&M) procedures from the MMEL. This MMEL with O&M procedures now becomes the operator's MEL. The MEL, LOA, preamble, and procedures document developed by the operator must be on board the aircraft during each operation. The FAA considers an approved MEL to be a supplemental type certificate (STC) issued to an aircraft by serial number and registration number. It, therefore, becomes the authority to operate that aircraft in a condition other than originally type certificated.

With an approved MEL, if the position lights were discovered inoperative prior to a daytime flight, the pilot would make an entry in the maintenance record or discrepancy record provided for that purpose. The item would then either be repaired or deferred in accordance with the MEL. Upon

confirming that daytime flight with inoperative position lights is acceptable in accordance with the provisions of the MEL, the pilot would leave the position lights switch OFF, open the circuit breaker (or whatever action is called for in the procedures document), and placard the position light switch as INOPERATIVE.

There are exceptions to the use of the MEL for deferral. For example, should a component fail that is not listed in the MEL as deferrable (the tachometer, flaps, or stall warning device, for example), then repairs are required to be performed prior to departure. If maintenance or parts are not readily available at that location, a special flight permit can be obtained from the nearest FSDO. This permit allows the aircraft to be flown to another location for maintenance. This allows an aircraft that may not currently meet applicable airworthiness requirements, but is capable of safe flight, to be operated under the restrictive special terms and conditions attached to the special flight permit.

Deferral of maintenance is not to be taken lightly, and due consideration should be given to the effect an inoperative component may have on the operation of an aircraft, particularly if other items are inoperative. Further information regarding MELs and operations with inoperative equipment can be found in AC 91-67, Minimum Equipment Requirements for General Aviation Operations Under FAR Part 91.

Preventive Maintenance

Preventive maintenance is regarded as simple or minor preservation operations and the replacement of small standard parts, not involving complex assembly operations. Allowed items of preventative maintenance are listed and limited to the items of 14 CFR part 43, appendix A(c).

Maintenance Entries

All pilots who perform preventive maintenance must make an entry in the maintenance record of the aircraft. The entry must include the following information:

1. A description of the work, such as "changed oil (Shell Aero-50) at 2,345 hours"

2. The date of completion of the work performed

3. The pilot's name, signature, certificate number, and type of certificate held

Examples of Preventive Maintenance

The following examples of preventive maintenance are taken from 14 CFR, part 43, Maintenance, Preventive Maintenance, Rebuilding, and Alternation, which should be consulted for a more in-depth look at the preventive maintenance a pilot can perform on an aircraft. Remember, preventive maintenance is limited to work that does not involve complex assembly operations including the following:

- Removal, installation, and repair of landing gear tires and shock cords; servicing landing gear shock struts by adding oil, air, or both; servicing gear wheel bearings; replacing defective safety wiring or cotter keys; lubrication not requiring disassembly other than removal of nonstructural items, such as cover plates, cowlings, and fairings; making simple fabric patches not requiring rib stitching or the removal of structural parts or control surfaces. In the case of balloons, the making of small fabric repairs to envelopes (as defined in, and in accordance with, the balloon manufacturer's instructions) not requiring load tape repair or replacement.

- Replenishing hydraulic fluid in the hydraulic reservoir; refinishing decorative coating of fuselage, balloon baskets, wings, tail group surfaces (excluding balanced control surfaces), fairings, cowlings, landing gear, cabin, or flight deck interior when removal or disassembly of any primary structure or operating system is not required; applying preservative or protective material to components where no disassembly of any primary structure or operating system is involved and where such coating is not prohibited or is not contrary to good practices; repairing upholstery and decorative furnishings of the cabin, flight deck, or balloon basket interior when the repair does not require disassembly of any primary structure or operating system or interfere with an operating system or affect the primary structure of the aircraft; making small, simple repairs to fairings, nonstructural cover plates, cowlings, and small patches and reinforcements not changing the contour to interfere with proper air flow; replacing side windows where that work does not interfere with the structure or any operating system, such as controls, electrical equipment, etc.

- Replacing safety belts, seats or seat parts with replacement parts approved for the aircraft, not involving disassembly of any primary structure or operating system, bulbs, reflectors, and lenses of position and landing lights.

- Replacing wheels and skis where no weight-and-balance computation is involved; replacing any cowling not requiring removal of the propeller or disconnection of flight controls; replacing or cleaning spark plugs and setting of spark plug gap clearance; replacing any hose connection, except hydraulic connections; however, prefabricated fuel lines may be replaced.

- Cleaning or replacing fuel and oil strainers or filter elements; servicing batteries, cleaning balloon burner pilot and main nozzles in accordance with the balloon manufacturer's instructions.

- The interchange of balloon baskets and burners on envelopes when the basket or burner is designated as interchangeable in the balloon type certificate data and the baskets and burners are specifically designed for quick removal and installation; adjustment of nonstructural standard fasteners incidental to operations.

- The installations of anti-misfueling devices to reduce the diameter of fuel tank filler openings only if the specific device has been made a part of the aircraft type certificate data by the aircraft manufacturer, the aircraft manufacturer has provided FAA-approved instructions for installation of the specific device, and installation does not involve the disassembly of the existing tank filler opening; troubleshooting and repairing broken circuits in landing light wiring circuits.

- Removing and replacing self-contained, front instrument panel-mounted navigation and communication devices that employ tray-mounted connectors which connect the unit when the unit is installed into the instrument panel (excluding automatic flight control systems, transponders, and microwave frequency distance measuring equipment (DME)). The approved unit must be designed to be readily and repeatedly removed and replaced, and pertinent instructions must be provided. Prior to the unit's intended use, an operational check must be performed in accordance with the applicable sections of 14 CFR part 91 on checking, removing, and replacing magnetic chip detectors.

- Inspection and maintenance tasks prescribed and specifically identified as preventive maintenance in a primary category aircraft type certificate or STC holder's approved special inspection and preventive maintenance program when accomplished on a primary category aircraft.

- Updating self-contained, front instrument panel-mounted air traffic control (ATC) navigational software databases (excluding those of automatic flight control systems, transponders, and microwave frequency DME), only if no disassembly of the unit is required and pertinent instructions are provided; prior to the unit's intended use, an operational check must be performed in accordance with applicable sections of 14 CFR part 91.

Certificated pilots, excluding student pilots, sport pilots, and recreational pilots, may perform preventive maintenance on any aircraft that is owned or operated by them provided that the aircraft is not used in air carrier service and does not qualify under 14 CFR parts 121, 129, or 135. A pilot holding a sport pilot certificate may perform preventive maintenance on an aircraft owned or operated by that pilot if that aircraft is issued a special airworthiness certificate in the LSA category. (Sport pilots operating LSA should refer to 14 CFR part 65 for maintenance privileges.) 14 CFR part 43, appendix A, contains a list of the operations that are considered to be preventive maintenance.

Repairs and Alterations

Repairs and alterations are classified as either major or minor. 14 CFR part 43, appendix A, describes the alterations and repairs considered major. Major repairs or alterations shall be approved for return to service on FAA Form 337, Major Repair and Alteration, by an appropriately rated certificated repair station, an FAA-certificated A&P mechanic holding an IA, or a representative of the Administrator. Minor repairs and minor alterations may be approved for return to service with a proper entry in the maintenance records by an FAA-certificated A&P mechanic or an appropriately certificated repair station.

For modifications of experimental aircraft, refer to the operating limitations issued to that aircraft. Modifications in accordance with FAA Order 8130.2, Airworthiness Certification of Aircraft and Related Products, may require the notification of the issuing authority.

Special Flight Permits

A special flight permit is a Special Airworthiness Certificate authorizing operation of an aircraft that does not currently meet applicable airworthiness requirements but is safe for a specific flight. Before the permit is issued, an FAA inspector may personally inspect the aircraft or require it to be inspected by an FAA-certificated A&P mechanic or an appropriately certificated repair station to determine its safety for the intended flight. The inspection shall be recorded in the aircraft records.

The special flight permit is issued to allow the aircraft to be flown to a base where repairs, alterations, or maintenance can be performed; for delivering or exporting the aircraft; or for evacuating an aircraft from an area of impending danger. A special flight permit may be issued to allow the operation of an overweight aircraft for flight beyond its normal range over water or land areas where adequate landing facilities or fuel is not available.

If a special flight permit is needed, assistance and the necessary forms may be obtained from the local FSDO or Designated Airworthiness Representative (DAR). *[Figure 9-10]*

Airworthiness Directives (ADs)

A primary safety function of the FAA is to require correction of unsafe conditions found in an aircraft, aircraft engine, propeller, or appliance when such conditions exist and are likely to exist or develop in other products of the same design. The unsafe condition may exist because of a design defect, maintenance, or other causes. Airworthiness Directives (ADs), under 14 CFR, part 39, define the authority and responsibility of the Administrator for requiring the necessary corrective action. ADs are used to notify aircraft owners and other interested persons of unsafe conditions and to specify the conditions under which the product may continue to be operated. ADs are divided into two categories:

1. Those of an emergency nature requiring immediate compliance prior to further flight

2. Those of a less urgent nature requiring compliance within a specified period of time

ADs are regulatory and shall be complied with unless a specific exemption is granted. It is the responsibility of the aircraft owner or operator to ensure compliance with all pertinent ADs, including those ADs that require recurrent or continuing action. For example, an AD may require a repetitive inspection each 50 hours of operation, meaning the particular inspection shall be accomplished and recorded every 50 hours of time in service. Owners/operators are reminded that there is no provision to overfly the maximum hour requirement of an AD unless it is specifically written into the AD. To help determine if an AD applies to an amateur-built aircraft, contact the local FSDO.

14 CFR, part 91, section 91.417 requires a record to be maintained that shows the current status of applicable ADs, including the method of compliance; the AD number and revision date, if recurring; next due date and time; the signature; type of certificate; and certificate number of the repair station or mechanic who performed the work. For ready reference, many aircraft owners have a chronological listing of the pertinent ADs in the back of their aircraft, engine, and propeller maintenance records.

All ADs and the AD Biweekly are free on the Internet at http://rgl.faa.gov and are available through e-mail. Individuals can enroll for the e-mail service at the website above. Paper copies of the Summary of Airworthiness Directives and the AD Biweekly may be purchased from the Superintendent of Documents. The Summary contains all the valid ADs previously published and is divided into

SPECIAL AIRWORTHINESS CERTIFICATE

UNITED STATES OF AMERICA
DEPARTMENT OF TRANSPORTATION - FEDERAL AVIATION ADMINISTRATION

A	CATEGORY/DESIGNATION	Special Flight Permit	
	PURPOSE	Production Flight Testing or Customer Demonstration	
B	MANU-FACTURER	NAME The Boeing Company	
		ADDRESS P.O. Box 767, Renton WA 13567	
C	FLIGHT	FROM N/A	
		TO N/A	
D	N- N/A		SERIAL NO. N/A
	BUILDER N/A		MODEL N/A
E	DATE OF ISSUANCE 01/31/2001		EXPIRY 01/31/2001
	OPERATING LIMITATIONS DATED 01/31/2001 ARE PART OF THIS CERTIFICATE		
	SIGNATURE OF FAA REPRESENTATIVE Sam T. Smith *Sam T. Smith*		DESIGNATION OR OFFICE NO. NM-XX

Any alteration, reproduction or misuse of this certificate may be punishable by a fine not exceeding $1,000 or imprisonment not exceeding 3 years, or both. THIS CERTIFICATE MUST BE DISPLAYED IN THE AIRCRAFT IN ACCORDANCE WITH APPLICABLE TITLE 14, CODE OF FEDERAL REGULATIONS (CFR).

FAA Form 8130-7 (07/04)   *SEE REVERSE SIDE*

Figure 9-10. *FAA Form 8130-7, Special Airworthiness Certificate.*

two areas. The small aircraft and helicopter books contain all ADs applicable to small aircraft (12,500 pounds or less maximum certificated takeoff weight) and ADs applicable to all helicopters. The large aircraft books contain all ADs applicable to large aircraft.

For current information on how to order paper copies of AD books and the AD Biweekly, visit the FAA online regulatory and guidance library at: http://rgl.faa.gov.

Aircraft Owner/Operator Responsibilities

The registered owner/operator of an aircraft is responsible for:

- Having a current Airworthiness Certificate and a Certificate of Aircraft Registration in the aircraft.

- Maintaining the aircraft in an airworthy condition, including compliance with all applicable ADs and assuring that maintenance is properly recorded.

- Keeping abreast of current regulations concerning the operation and maintenance of the aircraft.

- Notifying the FAA Aircraft Registry immediately of any change of permanent mailing address, of the sale or export of the aircraft, or of the loss of the eligibility to register an aircraft. (Refer to 14 CFR, part 47, section 47.41.)

- Having a current Federal Communications Commission (FCC) radio station license if equipped with radios, including emergency locator transmitter (ELT), if operated outside of the United States.

Chapter Summary

Knowledge of an aircraft's AFM/POH and documents, such as ADs, provide pilots with ready access to pertinent information needed to safely fly a particular aircraft. By understanding the operations, limitations, and performance characteristics of the aircraft, the pilot can make educated flight decisions. By learning what preventive maintenance is allowed on the aircraft, a pilot can maintain his or her aircraft in an airworthy condition. The goal of every pilot is a safe flight. Flight manuals and aircraft documentation are essential tools used to reach that goal.

Chapter 10
Weight and Balance

Introduction

Compliance with the weight and balance limits of any aircraft is critical to flight safety. Operating above the maximum weight limitation compromises the structural integrity of an aircraft and adversely affects its performance. Operation with the center of gravity (CG) outside the approved limits results in control difficulty.

Weight Control

As discussed in Chapter 5, Aerodynamics of Flight, weight is the force with which gravity attracts a body toward the center of the Earth. It is a product of the mass of a body and the acceleration acting on the body. Weight is a major factor in aircraft construction and operation and demands respect from all pilots.

The force of gravity continuously attempts to pull an aircraft down toward Earth. The force of lift is the only force that counteracts weight and sustains an aircraft in flight. The amount of lift produced by an airfoil is limited by the airfoil design, angle of attack (AOA), airspeed, and air density. To assure that the lift generated is sufficient to counteract weight, loading an aircraft beyond the manufacturer's recommended weight must be avoided. If the weight is greater than the lift generated, the aircraft may be incapable of flight.

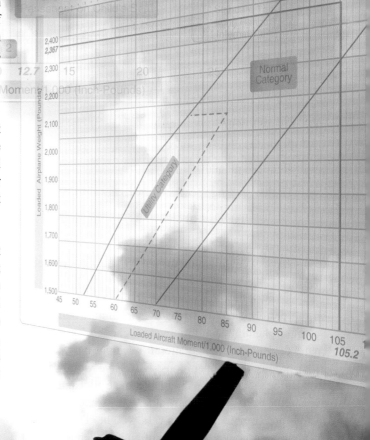

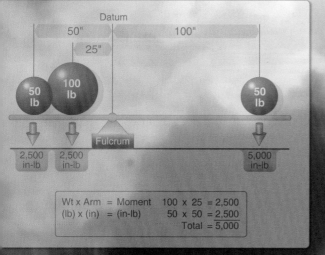

Wt x Arm	= Moment	100 x 25	= 2,500
(lb) x (in)	= (in-lb)	50 x 50	= 2,500
		Total	= 5,000

Effects of Weight

Any item aboard an aircraft that increases the total weight is undesirable for performance. Manufacturers attempt to make an aircraft as light as possible without sacrificing strength or safety.

The pilot should always be aware of the consequences of overloading. An overloaded aircraft may not be able to leave the ground, or if it does become airborne, it may exhibit unexpected and unusually poor flight characteristics. If not properly loaded, the initial indication of poor performance usually takes place during takeoff.

Excessive weight reduces the flight performance in almost every respect. For example, the most important performance deficiencies of an overloaded aircraft are:

- Higher takeoff speed
- Longer takeoff run
- Reduced rate and angle of climb
- Lower maximum altitude
- Shorter range
- Reduced cruising speed
- Reduced maneuverability
- Higher stalling speed
- Higher approach and landing speed
- Longer landing roll
- Excessive weight on the nose wheel or tail wheel

The pilot must be knowledgeable about the effect of weight on the performance of the particular aircraft being flown. Preflight planning should include a check of performance charts to determine if the aircraft's weight may contribute to hazardous flight operations. Excessive weight in itself reduces the safety margins available to the pilot and becomes even more hazardous when other performance-reducing factors are combined with excess weight. The pilot must also consider the consequences of an overweight aircraft if an emergency condition arises. If an engine fails on takeoff or airframe ice forms at low altitude, it is usually too late to reduce an aircraft's weight to keep it in the air.

Weight Changes

The operating weight of an aircraft can be changed by simply altering the fuel load. Gasoline has considerable weight—6 pounds per gallon. Thirty gallons of fuel may weigh more than one passenger. If a pilot lowers airplane weight by reducing fuel, the resulting decrease in the range of the airplane must be taken into consideration during flight planning. During flight, fuel burn is normally the only weight change that takes place. As fuel is used, an aircraft becomes lighter and performance is improved.

Changes of fixed equipment have a major effect upon the weight of an aircraft. The installation of extra radios or instruments, as well as repairs or modifications, may also affect the weight of an aircraft.

Balance, Stability, and Center of Gravity

Balance refers to the location of the CG of an aircraft, and is important to stability and safety in flight. The CG is a point at which the aircraft would balance if it were suspended at that point.

The primary concern in balancing an aircraft is the fore and aft location of the CG along the longitudinal axis. The CG is not necessarily a fixed point; its location depends on the distribution of weight in the aircraft. As variable load items are shifted or expended, there is a resultant shift in CG location. The distance between the forward and back limits for the position of the center for gravity or CG range is certified for an aircraft by the manufacturer. The pilot should realize that if the CG is displaced too far forward on the longitudinal axis, a nose-heavy condition will result. Conversely, if the CG is displaced too far aft on the longitudinal axis, a tail heavy condition results. It is possible that the pilot could not control the aircraft if the CG location produced an unstable condition. *[Figure 10-1]*

Location of the CG with reference to the lateral axis is also important. For each item of weight existing to the left of

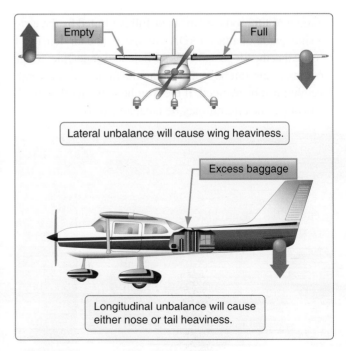

Figure 10-1. *Lateral and longitudinal unbalance.*

the fuselage centerline, there is an equal weight existing at a corresponding location on the right. This may be upset by unbalanced lateral loading. The position of the lateral CG is not computed in all aircraft, but the pilot must be aware that adverse effects arise as a result of a laterally unbalanced condition. In an airplane, lateral unbalance occurs if the fuel load is mismanaged by supplying the engine(s) unevenly from tanks on one side of the airplane. The pilot can compensate for the resulting wing-heavy condition by adjusting the trim or by holding a constant control pressure. This action places the aircraft controls in an out-of-streamline condition, increases drag, and results in decreased operating efficiency. Since lateral balance is addressed when needed in the aircraft flight manual (AFM) and longitudinal balance is more critical, further reference to balance in this handbook means longitudinal location of the CG.

Flying an aircraft that is out of balance can produce increased pilot fatigue with obvious effects on the safety and efficiency of flight. The pilot's natural correction for longitudinal unbalance is a change of trim to remove the excessive control pressure. Excessive trim, however, has the effect of reducing not only aerodynamic efficiency but also primary control travel distance in the direction the trim is applied.

Effects of Adverse Balance

Adverse balance conditions affect flight characteristics in much the same manner as those mentioned for an excess weight condition. It is vital to comply with weight and balance limits established for all aircraft. Operating above the maximum weight limitation compromises the structural integrity of the aircraft and can adversely affect performance. Stability and control are also affected by improper balance.

Stability

Loading in a nose-heavy condition causes problems in controlling and raising the nose, especially during takeoff and landing. Loading in a tail heavy condition has a serious effect upon longitudinal stability, and reduces the capability to recover from stalls and spins. Tail heavy loading also produces very light control forces, another undesirable characteristic. This makes it easy for the pilot to inadvertently overstress an aircraft.

Stability and Center of Gravity

Limits for the location of the CG are established by the manufacturer. These are the fore and aft limits beyond which the CG should not be located for flight. These limits are published for each aircraft in the Type Certificate Data Sheet (TCDS), or aircraft specification and the AFM or pilot's operating handbook (POH). If the CG is not within the allowable limits after loading, it will be necessary to relocate some items before flight is attempted.

The forward CG limit is often established at a location that is determined by the landing characteristics of an aircraft. During landing, one of the most critical phases of flight, exceeding the forward CG limit may result in excessive loads on the nosewheel, a tendency to nose over on tailwheel type airplanes, decreased performance, higher stalling speeds, and higher control forces.

Control

In extreme cases, a CG location that is beyond the forward limit may result in nose heaviness, making it difficult or impossible to flare for landing. Manufacturers purposely place the forward CG limit as far rearward as possible to aid pilots in avoiding damage when landing. In addition to decreased static and dynamic longitudinal stability, other undesirable effects caused by a CG location aft of the allowable range may include extreme control difficulty, violent stall characteristics, and very light control forces which make it easy to overstress an aircraft inadvertently.

A restricted forward CG limit is also specified to assure that sufficient elevator/control deflection is available at minimum airspeed. When structural limitations do not limit the forward CG position, it is located at the position where full-up elevator/control deflection is required to obtain a high AOA for landing.

The aft CG limit is the most rearward position at which the CG can be located for the most critical maneuver or operation. As the CG moves aft, a less stable condition occurs, which decreases the ability of the aircraft to right itself after maneuvering or turbulence.

For some aircraft, both fore and aft CG limits may be specified to vary as gross weight changes. They may also be changed for certain operations, such as acrobatic flight, retraction of the landing gear, or the installation of special loads and devices that change the flight characteristics.

The actual location of the CG can be altered by many variable factors and is usually controlled by the pilot. Placement of baggage and cargo items determines the CG location. The assignment of seats to passengers can also be used as a means of obtaining a favorable balance. If an aircraft is tail heavy, it is only logical to place heavy passengers in forward seats. Fuel burn can also affect the CG based on the location of the fuel tanks. For example, most small aircraft carry fuel in the wings very near the CG and burning off fuel has little effect on the loaded CG.

Management of Weight and Balance Control

Title 14 of the Code of Federal Regulations (14 CFR) part 23, section 23.23 requires establishment of the ranges of weights and CGs within which an aircraft may be operated safely. The manufacturer provides this information, which is included in the approved AFM, TCDS, or aircraft specifications.

While there are no specified requirements for a pilot operating under 14 CFR part 91 to conduct weight and balance calculations prior to each flight, 14 CFR part 91, section 91.9 requires the pilot in command (PIC) to comply with the operating limits in the approved AFM. These limits include the weight and balance of the aircraft. To enable pilots to make weight and balance computations, charts and graphs are provided in the approved AFM.

Weight and balance control should be a matter of concern to all pilots. The pilot controls loading and fuel management (the two variable factors that can change both total weight and CG location) of a particular aircraft. The aircraft owner or operator should make certain that up-to-date information is available for pilot use, and should ensure that appropriate entries are made in the records when repairs or modifications have been accomplished. The removal or addition of equipment results in changes to the CG.

Weight changes must be accounted for and the proper notations made in weight and balance records. The equipment list must be updated, if appropriate. Without such information, the pilot has no foundation upon which to base the necessary calculations and decisions.

Standard parts with negligible weight or the addition of minor items of equipment such as nuts, bolts, washers, rivets, and similar standard parts of negligible weight on fixed-wing aircraft do not require a weight and balance check. The following criteria for negligible weight change is outlined in Advisory Circular (AC) 43.13-1 (as revised), Methods Techniques and Practices—Aircraft Inspection and Repair:

- One pound or less for an aircraft whose weight empty is less than 5,000 pounds

- Two pounds or less for aircraft with an empty weight of more than 5,000 pounds to 50,000 pounds

- Five pounds or less for aircraft with an empty weight of more than 50,000 pounds

Negligible CG change is any change of less than 0.05 percent Mean Aerodynamic Chord (MAC) for fixed-wing aircraft or 0.2 percent for rotary wing aircraft. MAC is the average distance from the leading edge to the trailing edge of the wing. Exceeding these limits would require a weight and balance check.

Before any flight, the pilot should determine the weight and balance condition of the aircraft. Simple and orderly procedures based on sound principles have been devised by the manufacturer for the determination of loading conditions. The pilot uses these procedures and exercises good judgment when determining weight and balance. In many modern aircraft, it is not possible to fill all seats, baggage compartments, and fuel tanks, and still remain within the approved weight and balance limits. If the maximum passenger load is carried, the pilot must often reduce the fuel load or reduce the amount of baggage.

14 CFR part 125 requires aircraft with 20 or more seats or maximum payload capacity of 6,000 pounds or more to be weighed every 36 calendar months. Multi-engine aircraft operated under 14 CFR part 135 are also required to be weighed every 36 months. Aircraft operated under 14 CFR part 135 are exempt from the 36 month requirement if operated under a weight and balance system approved in the operations specifications of the certificate holder. For additional information on approved weight and balance control programs for operations under parts 121 and 135, reference the current edition of AC 120-27, Aircraft Weight and Balance Control. AC 43.13-1, Acceptable Methods, Techniques and Practices—Aircraft Inspection and Repair also requires that the aircraft mechanic ensure that the weight and balance data in the aircraft records is current and accurate after a 100-hour or annual inspection.

Terms and Definitions

The pilot should be familiar with the appropriate terms regarding weight and balance. The following list of terms and their definitions is standardized, and knowledge of these terms aids the pilot to better understand weight and balance calculations of any aircraft. Terms defined by the General Aviation Manufacturers Association (GAMA) as industry standard are marked in the titles with GAMA.

- Arm (moment arm)—the horizontal distance in inches from the reference datum line to the CG of an item. The algebraic sign is plus (+) if measured aft of the datum and minus (–) if measured forward of the datum.

- Basic empty weight (GAMA)—the standard empty weight plus the weight of optional and special equipment that have been installed.

- Center of gravity (CG)—the point about which an aircraft would balance if it were possible to suspend it at that point. It is the mass center of the aircraft or the theoretical point at which the entire weight of the aircraft is assumed to be concentrated. It may be expressed in inches from the reference datum or in percent of MAC. The CG is a three-dimensional point with longitudinal, lateral, and vertical positioning in the aircraft.

- CG limits—the specified forward and aft points within which the CG must be located during flight. These limits are indicated on pertinent aircraft specifications.

- CG range—the distance between the forward and aft CG limits indicated on pertinent aircraft specifications.

- Datum (reference datum)—an imaginary vertical plane or line from which all measurements of arm are taken. The datum is established by the manufacturer. Once the datum has been selected, all moment arms and the location of CG range are measured from this point.

- Delta—a Greek letter expressed by the symbol $\triangle$ to indicate a change of values. As an example, $\triangle CG$ indicates a change (or movement) of the CG.

- Floor load limit—the maximum weight the floor can sustain per square inch/foot as provided by the manufacturer.

- Fuel load—the expendable part of the load of the aircraft. It includes only usable fuel, not fuel required to fill the lines or that which remains trapped in the tank sumps.

- Licensed empty weight—the empty weight that consists of the airframe, engine(s), unusable fuel, and undrainable oil plus standard and optional equipment as specified in the equipment list. Some manufacturers used this term prior to GAMA standardization.

- Maximum landing weight—the greatest weight that an aircraft is normally allowed to have at landing.

- Maximum ramp weight—the total weight of a loaded aircraft including all fuel. It is greater than the takeoff weight due to the fuel that will be burned during the taxi and run-up operations. Ramp weight may also be referred to as taxi weight.

- Maximum takeoff weight—the maximum allowable weight for takeoff.

- Maximum weight—the maximum authorized weight of the aircraft and all of its equipment as specified in the TCDS for the aircraft.

- Maximum zero fuel weight (GAMA)—the maximum weight, exclusive of usable fuel.

- Mean aerodynamic chord (MAC)—the average distance from the leading edge to the trailing edge of the wing.

- Moment—the product of the weight of an item multiplied by its arm. Moments are expressed in pound-inches (in-lb). Total moment is the weight of the airplane multiplied by the distance between the datum and the CG.

- Moment index (or index)—a moment divided by a constant such as 100, 1,000, or 10,000. The purpose of using a moment index is to simplify weight and balance computations of aircraft where heavy items and long arms result in large, unmanageable numbers.

- Payload (GAMA)—the weight of occupants, cargo, and baggage.

- Standard empty weight (GAMA)—aircraft weight that consists of the airframe, engines, and all items of operating equipment that have fixed locations and are permanently installed in the aircraft, including fixed ballast, hydraulic fluid, unusable fuel, and full engine oil.

- Standard weights—established weights for numerous items involved in weight and balance computations. These weights should not be used if actual weights are available. Some of the standard weights are:

Gasoline.. 6 lb/US gal

Jet A, Jet A-1 .... 6.8 lb/US gal

Jet B..6.5 lb/US gal

Oil...7.5 lb/US gal

Water... 8.35 lb/US gal

- Station—a location in the aircraft that is identified by a number designating its distance in inches from the datum. The datum is, therefore, identified as station zero. An item located at station +50 would have an arm of 50 inches.

- Useful load—the weight of the pilot, copilot, passengers, baggage, usable fuel, and drainable oil. It is the basic empty weight subtracted from the maximum allowable gross weight. This term applies to general aviation (GA) aircraft only.

Principles of Weight and Balance Computations

It is imperative that all pilots understand the basic principles of weight and balance determination. The following methods of computation can be applied to any object or vehicle for which weight and balance information is essential.

By determining the weight of the empty aircraft and adding the weight of everything loaded on the aircraft, a total weight can be determined—a simple concept. A greater problem, particularly if the basic principles of weight and balance are not understood, is distributing this weight in such a manner that the entire mass of the loaded aircraft is balanced around a point (CG) that must be located within specified limits.

The point at which an aircraft balances can be determined by locating the CG, which is, as stated in the definitions of terms, the imaginary point at which all the weight is concentrated. To provide the necessary balance between longitudinal stability and elevator control, the CG is usually located slightly forward of the center of lift. This loading condition causes a nose-down tendency in flight, which is desirable during flight at a high AOA and slow speeds.

As mentioned earlier, a safe zone within which the balance point (CG) must fall is called the CG range. The extremities of the range are called the forward CG limits and aft CG limits. These limits are usually specified in inches, along the longitudinal axis of the airplane, measured from a reference point called a datum reference. The datum is an arbitrary point, established by aircraft designers that may vary in location between different aircraft. [Figure 10-2]

The distance from the datum to any component part or any object loaded on the aircraft is called the arm. When the object or component is located aft of the datum, it is measured in positive inches; if located forward of the datum, it is measured as negative inches or minus inches. The location of the object or part is often referred to as the station. If the weight of any object or component is multiplied by the distance from the datum (arm), the product is the moment. The moment is the measurement of the gravitational force that causes a tendency of the weight to rotate about a point or axis and is expressed in inch-pounds (in-lb).

To illustrate, assume a weight of 50 pounds is placed on the board at a station or point 100 inches from the datum. The downward force of the weight can be determined by multiplying 50 pounds by 100 inches, which produces a moment of 5,000 in-lb. [Figure 10-3]

To establish a balance, a total of 5,000 in-lb must be applied to the other end of the board. Any combination of weight and distance which, when multiplied, produces a 5,000 in-lb moment will balance the board. For example (illustrated

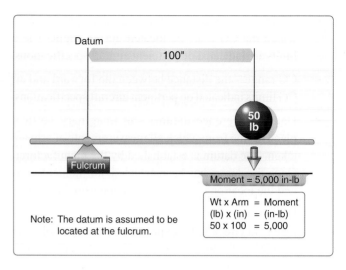

Figure 10-3. *Determining moment.*

in *Figure 10-4*), if a 100-pound weight is placed at a point (station) 25 inches from the datum, and another 50-pound weight is placed at a point (station) 50 inches from the datum, the sum of the product of the two weights and their distances total a moment of 5,000 in-lb, which will balance the board.

Weight and Balance Restrictions

An aircraft's weight and balance restrictions should be closely followed. The loading conditions and empty weight of a particular aircraft may differ from that found in the AFM/POH because modifications or equipment changes may have been made. Sample loading problems in the AFM/POH are intended for guidance only; therefore, each aircraft must be treated separately. Although an aircraft is certified for a specified maximum gross takeoff weight, it may not safely take off at this weight under all conditions. Conditions that affect takeoff and climb performance, such as high elevations, high temperatures, and high humidity (high density altitudes), may require a reduction in weight before flight is attempted. Other factors to consider when computing

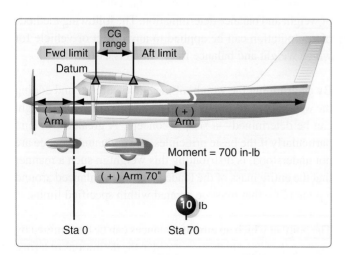

Figure 10-2. *Weight and balance.*

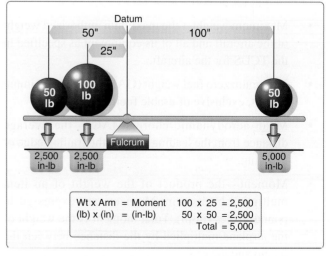

Figure 10-4. *Establishing a balance.*

weight and balance distribution prior to takeoff are runway length, runway surface, runway slope, surface wind, and the presence of obstacles. These factors may require a reduction in or redistribution of weight prior to flight.

Some aircraft are designed so that it is difficult to load them in a manner that places the CG out of limits. These are usually small aircraft with the seats, fuel, and baggage areas located near the CG limit. Pilots must be aware that while within CG limits these aircraft can be overloaded in weight. Other aircraft can be loaded in such a manner that they will be out of CG limits even though the useful load has not been exceeded. Because of the effects of an out-of-balance or overweight condition, a pilot should always be sure that an aircraft is properly loaded.

Determining Loaded Weight and CG

There are various methods for determining the loaded weight and CG of an aircraft. There is the computational method as well as methods that utilize graphs and tables provided by the aircraft manufacturer.

Computational Method

The following is an example of the computational method involving the application of basic math functions.

Aircraft Allowances:

Maximum gross weight.......................3,400 pounds

CG range...78–86 inches

Given:

Weight of front seat occupants..............340 pounds

Weight of rear seat occupants..............350 pounds

Fuel...75 gallons

Weight of baggage in area 1....................80 pounds

1. List the weight of the aircraft, occupants, fuel, and baggage. Remember that aviation gas (AVGAS) weighs 6 pounds per gallon and is used in this example.

2. Enter the moment for each item listed. Remember "weight x arm = moment."

3. Find the total weight and total moment.

4. To determine the CG, divide the total moment by the total weight.

NOTE: The weight and balance records for a particular aircraft provide the empty weight and moment, as well as the information on the arm distance. [Figure 10-5]

The total loaded weight of 3,320 pounds does not exceed the maximum gross weight of 3,400 pounds, and the CG of

Item	Weight	Arm	Moment
Aircraft Empty Weight	2,100	78.3	164,430
Front Seat Occupants	340	85.0	28,900
Rear Seat Occupants	350	121.0	42,350
Fuel	450	75.0	33,750
Baggage Area 1	80	150.0	12,000
Total	3,320		281,430
		281,430 ÷ 3,320 = 84.8	

Figure 10-5. *Example of weight and balance computations.*

84.8 is within the 78–86 inch range; therefore, the aircraft is loaded within limits.

Graph Method

Another method for determining the loaded weight and CG is the use of graphs provided by the manufacturers. To simplify calculations, the moment may sometimes be divided by 100, 1,000, or 10,000. [Figures 10-6, 10-7, and 10-8]

Front seat occupants.....................................340 pounds

Rear seat occupants ....300 pounds

Fuel...40 gallons

Baggage area 1 ....20 pounds

The same steps should be followed in the graph method as were used in the computational method except the graphs provided will calculate the moments and allow the pilot to determine if the aircraft is loaded within limits. To determine the moment using the loading graph, find the weight and draw a line straight across until it intercepts the item for which the moment is to be calculated. Then draw a line straight down to determine the moment. (The red line on the loading graph in *Figure 10-7* represents the moment for the pilot and front passenger. All other moments were determined the same

Sample Loading Problem	Weight (lb)	Moment (in-lb/1,000)
1. Basic empty weight (Use data pertaining to aircraft as it is presently equipped) includes unusable fuel and full oil	1,467	57.3
2. Usable fuel (At 6 lb/gal) ■ Standard tanks (40 gal maximum) ■ Long range tanks (50 gal maximum) ■ Integral tanks (62 gal maximum) ■ Integral reduced fuel (42 gal)	240	11.5
3. Pilot and front passenger (Station 34 to 46)	340	12.7
4. Rear passengers	300	21.8
5. Baggage area 1 or passenger on child's seat (Station 82 to 108, 120 lb maximum)	20	1.9
6. Baggage area 2 (Station 108 to 142, 50 lb maximum)		
7. Weight and moment	2,367	105.2

Figure 10-6. *Weight and balance data.*

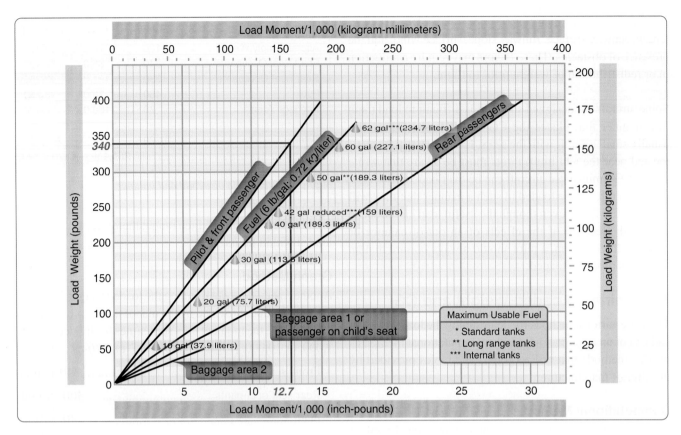

Figure 10-7. *Loading graph.*

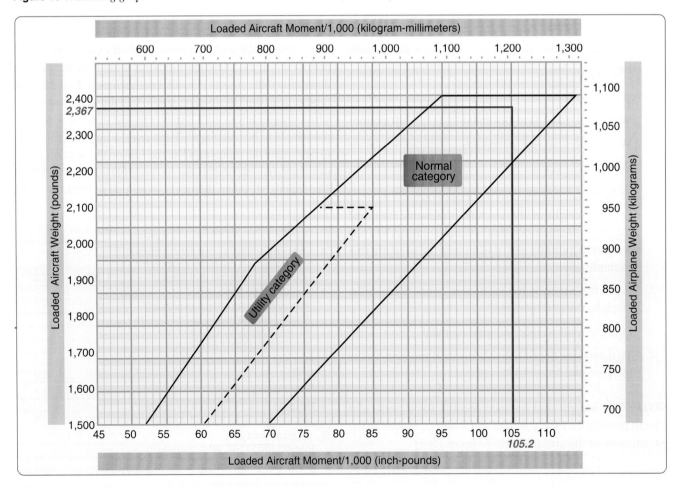

Figure 10-8. *CG moment envelope.*

way.) Once this has been done for each item, total the weight and moments and draw a line for both weight and moment on the CG envelope graph. If the lines intersect within the envelope, the aircraft is loaded within limits. In this sample loading problem, the aircraft is loaded within limits.

Table Method

The table method applies the same principles as the computational and graph methods. The information and limitations are contained in tables provided by the manufacturer. *Figure 10-9* is an example of a table and a

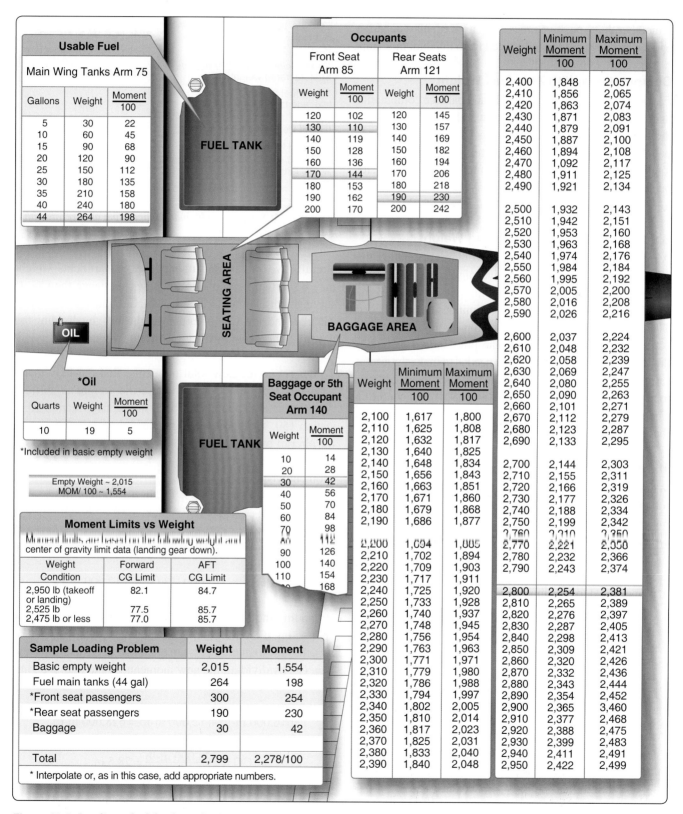

Usable Fuel

Main Wing Tanks Arm 75

Gallons	Weight	Moment 100
5	30	22
10	60	45
15	90	68
20	120	90
25	150	112
30	180	135
35	210	158
40	240	180
44	264	198

Occupants

Front Seat Arm 85		Rear Seats Arm 121	
Weight	Moment 100	Weight	Moment 100
120	102	120	145
130	110	130	157
140	119	140	169
150	128	150	182
160	136	160	194
170	144	170	206
180	153	180	218
190	162	190	230
200	170	200	242

Weight	Minimum Moment 100	Maximum Moment 100
2,400	1,848	2,057
2,410	1,856	2,065
2,420	1,863	2,074
2,430	1,871	2,083
2,440	1,879	2,091
2,450	1,887	2,100
2,460	1,894	2,108
2,470	1,092	2,117
2,480	1,911	2,125
2,490	1,921	2,134
2,500	1,932	2,143
2,510	1,942	2,151
2,520	1,953	2,160
2,530	1,963	2,168
2,540	1,974	2,176
2,550	1,984	2,184
2,560	1,995	2,192
2,570	2,005	2,200
2,580	2,016	2,208
2,590	2,026	2,216
2,600	2,037	2,224
2,610	2,048	2,232
2,620	2,058	2,239
2,630	2,069	2,247
2,640	2,080	2,255
2,650	2,090	2,263
2,660	2,101	2,271
2,670	2,112	2,279
2,680	2,123	2,287
2,690	2,133	2,295
2,700	2,144	2,303
2,710	2,155	2,311
2,720	2,166	2,319
2,730	2,177	2,326
2,740	2,188	2,334
2,750	2,199	2,342
2,760	2,210	2,350
2,770	2,221	2,358
2,780	2,232	2,366
2,790	2,243	2,374
2,800	2,254	2,381
2,810	2,265	2,389
2,820	2,276	2,397
2,830	2,287	2,405
2,840	2,298	2,413
2,850	2,309	2,421
2,860	2,320	2,426
2,870	2,332	2,436
2,880	2,343	2,444
2,890	2,354	2,452
2,900	2,365	3,460
2,910	2,377	2,468
2,920	2,388	2,475
2,930	2,399	2,483
2,940	2,411	2,491
2,950	2,422	2,499

*Oil

Quarts	Weight	Moment 100
10	19	5

*Included in basic empty weight

Empty Weight ~ 2,015
MOM/ 100 ~ 1,554

Baggage or 5th Seat Occupant Arm 140

Weight	Moment 100
10	14
20	28
30	42
40	56
50	70
60	84
70	98
80	112
90	126
100	140
110	154
120	168

Weight	Minimum Moment 100	Maximum Moment 100
2,100	1,617	1,800
2,110	1,625	1,808
2,120	1,632	1,817
2,130	1,640	1,825
2,140	1,648	1,834
2,150	1,656	1,843
2,160	1,663	1,851
2,170	1,671	1,860
2,180	1,679	1,868
2,190	1,686	1,877
2,200	1,694	1,885
2,210	1,702	1,894
2,220	1,709	1,903
2,230	1,717	1,911
2,240	1,725	1,920
2,250	1,733	1,928
2,260	1,740	1,937
2,270	1,748	1,945
2,280	1,756	1,954
2,290	1,763	1,963
2,300	1,771	1,971
2,310	1,779	1,980
2,320	1,786	1,988
2,330	1,794	1,997
2,340	1,802	2,005
2,350	1,810	2,014
2,360	1,817	2,023
2,370	1,825	2,031
2,380	1,833	2,040
2,390	1,840	2,048

Moment Limits vs Weight

Moment limits are based on the following weight and center of gravity limit data (landing gear down).

Weight Condition	Forward CG Limit	AFT CG Limit
2,950 lb (takeoff or landing)	82.1	84.7
2,525 lb	77.5	85.7
2,475 lb or less	77.0	85.7

Sample Loading Problem

	Weight	Moment
Basic empty weight	2,015	1,554
Fuel main tanks (44 gal)	264	198
*Front seat passengers	300	254
*Rear seat passengers	190	230
Baggage	30	42
Total	2,799	2,278/100

* Interpolate or, as in this case, add appropriate numbers.

Figure 10-9. *Loading schedule placard.*

weight and balance calculation based on that table. In this problem, the total weight of 2,799 pounds and moment of 2,278/100 are within the limits of the table.

Computations With a Negative Arm

Figure 10-10 is a sample of weight and balance computation using an aircraft with a negative arm. It is important to remember that a positive times a negative equals a negative, and a negative would be subtracted from the total moments.

Computations With Zero Fuel Weight

Figure 10-11 is a sample of weight and balance computation using an aircraft with a zero fuel weight. In this example, the total weight of the aircraft less fuel is 4,240 pounds, which is under the zero fuel weight of 4,400 pounds. If the total weight of the aircraft without fuel had exceeded 4,400 pounds, passengers or cargo would have needed to be reduced to bring the weight at or below the max zero fuel weight.

Shifting, Adding, and Removing Weight

A pilot must be able to solve any problems accurately that involve the shift, addition, or removal of weight. For example, the pilot may load the aircraft within the allowable takeoff weight limit, then find that the CG limit has been exceeded. The most satisfactory solution to this problem is to shift baggage, passengers, or both. The pilot should be able to determine the minimum load shift needed to make the aircraft safe for flight. Pilots should be able to determine if shifting a load to a new location will correct an out-of-limit condition. There are some standardized calculations that can help make these determinations.

Weight Shifting

When weight is shifted from one location to another, the total weight of the aircraft is unchanged. The total moments, however, do change in relation and proportion to the direction and distance the weight is moved. When weight is moved forward, the total moments decrease; when weight is moved aft, total moments increase. The moment change is proportional to the amount of weight moved. Since many

Item	Weight	Arm	Moment
Licensed empty weight	1,011.9	68.6	69,393.0
Oil (6 quarts)	11.0	−31.0	−341.0
Fuel (18 gallons)	108.0	84.0	9,072.0
Fuel, auxiliary (18 gallons)	108.0	84.0	9,072.0
Pilot	170.0	81.0	13,770.0
Passenger	170.0	81.0	13,770.0
Baggage	70.0	105.0	7,350.0
Total	1,648.9		122,086.0
CG		74.0	

Figure 10-10. *Sample weight and balance using a negative.*

Item	Weight	Arm	Moment
Basic empty weight	3,230	CG 90.5	292,315.0
Front seat occupants	335	89.0	29,815.0
3rd & 4th seat occupants forward facing	350	126.0	44,100.0
5th & 6th seat occupants	200	157.0	31,400.0
Nose baggage	100	10.0	1,000.0
Aft baggage	25	183.0	4,575.0
Zero fuel weight max 4,400 pounds Subtotal	4,240	CG 95.1	403,205.0
Fuel	822	113.0	92,886.0
Ramp weight max 5,224 pounds Subtotal ramp weight	5,062	CG 98.0	496,091.0
* Less fuel for start, taxi, and takeoff	−24	113.0	−2,712.0
Subtotal takeoff weight	5,038	CG 97.9	493,379.0
Less fuel to destination	−450	113.0	−50,850.0
Max landing weight 4,940 pounds Actual landing weight	4,588	CG 96.5	442,529.0
Fuel for start, taxi, and takeoff is normally 24 pounds.			

Figure 10-11. *Sample weight and balance using an aircraft with a published zero fuel weight.*

aircraft have forward and aft baggage compartments, weight may be shifted from one to the other to change the CG. If starting with a known aircraft weight, CG, and total moments, calculate the new CG (after the weight shift) by dividing the new total moments by the total aircraft weight.

To determine the new total moments, find out how many moments are gained or lost when the weight is shifted. Assume that 100 pounds has been shifted from station 30 to station 150. This movement increases the total moments of the aircraft by 12,000 in-lb.

Moment when
at station 150 $= 100 \text{ lb} \times 150 \text{ in} = 15,000 \text{ in-lb}$

Moment when
at station 30 $= 100 \text{ lb} \times 30 \text{ in} = 3,000 \text{ in-lb}$

Moment change $= [15,000 - 3,000] = 12,000 \text{ in-lb}$

By adding the moment change to the original moment (or subtracting if the weight has been moved forward instead of aft), the new total moments are obtained. Then determine the new CG by dividing the new moments by the total weight:

$$\text{Total moments} = 616,000 \text{ in-lb} + 12,000 \text{ in-lb} = 628,000 \text{ in-lb}$$

$$CG = \frac{628,000 \text{ in-lb}}{8,000 \text{ lb}} = 78.5 \text{ in}$$

The shift has caused the CG to shift to station 78.5.

$$\frac{\text{Weight shifted}}{\text{Total weight}} = \frac{\Delta CG \text{ (change of CG)}}{\text{Distance weight is shifted}}$$

$$\frac{100}{8,000} = \frac{\Delta CG}{120}$$

$$\Delta CG = 1.5 \text{ in}$$

The change of CG is added to (or subtracted from when appropriate) the original CG to determine the new CG:
77 + 1.5 = 78.5 inches aft of datum

The shifting weight proportion formula can also be used to determine how much weight must be shifted to achieve a particular shift of the CG. The following problem illustrates a solution of this type.

Given:
Aircraft total weight . . . . 7,800 lb
CG station. . . . . 81.5 in
Aft CG limit . . . . 80.5 in

Determine how much cargo must be shifted from the aft cargo compartment at station 150 to the forward cargo compartment at station 30 to move the CG to exactly the aft limit.

Solution:

$$\frac{\text{Weight to be shifted}}{\text{Total weight}} = \frac{\Delta CG}{\text{Distance weight is shifted}}$$

$$\frac{\text{Weight to be shifted}}{7,800 \text{ lb}} = \frac{1.0 \text{ in}}{120 \text{ in}}$$

$$\text{Weight to be shifted} = 65 \text{ lb}$$

A simpler solution may be obtained by using a computer or calculator and a proportional formula. This can be done because the CG will shift a distance that is proportional to the distance the weight is shifted.

Weight Addition or Removal

In many instances, the weight and balance of the aircraft will be changed by the addition or removal of weight. When this happens, a new CG must be calculated and checked against the limitations to see if the location is acceptable. This type of weight and balance problem is commonly encountered when the aircraft burns fuel in flight, thereby reducing the weight located at the fuel tanks. Most small aircraft are designed with the fuel tanks positioned close to the CG; therefore, the consumption of fuel does not affect the CG to any great extent.

The addition or removal of cargo presents a CG change problem that must be calculated before flight. The problem may always be solved by calculations involving total moments. A typical problem may involve the calculation of a new CG for an aircraft which, when loaded and ready for flight, receives some additional cargo or passengers just before departure time.

Given:
Aircraft total weight . . . . 6,860 lb
CG station. . . . . 80.0 in

Determine the location of the CG if 140 pounds of baggage is added to station 150.

Solution:

$$\frac{\text{Added weight}}{\text{New total weight}} = \frac{\Delta CG}{\text{Distance between weight and old CG}}$$

$$\frac{140 \text{ lb}}{6,860 \text{ lb} + 140 \text{ lb}} = \frac{\Delta CG}{150 \text{ in} - 80 \text{ in}}$$

$$\frac{140 \text{ lb}}{7,000 \text{ lb}} = \frac{\Delta CG}{70 \text{ in}}$$

$$CG = 1.4 \text{ in aft}$$

Add ΔCG to old CG
New CG = 80 in + 1.4 in = 81.4 in

Given:
Aircraft total weight . . . . 6,100 lb
CG station. . . . . 80.0 in

Determine the location of the CG if 100 pounds is removed from station 150.

Solution:

$$\frac{\text{Weight removed}}{\text{New total weight}} = \frac{\Delta CG}{\text{Distance between weight and old CG}}$$

$$\frac{100 \text{ lb}}{6,100 \text{ lb} - 100 \text{ lb}} = \frac{\Delta CG}{150 \text{ in} - 80 \text{ in}}$$

$$\frac{100 \text{ lb}}{6,000 \text{ lb}} = \frac{\Delta CG}{70 \text{ in}}$$

$$CG = 1.2 \text{ in forward}$$

Subtract ΔCG from old CG
New CG = 80 in − 1.2 in = 78.8 in

In the previous examples, the △CG is either added or subtracted from the old CG. Deciding which to accomplish is best handled by mentally calculating which way the CG will shift for the particular weight change. If the CG is shifting aft, the △CG is added to the old CG; if the CG is shifting forward, the △CG is subtracted from the old CG.

Chapter Summary

Operating an aircraft within the weight and balance limits is critical to flight safety. Pilots must ensure that the CG is and remains within approved limits throughout all phases of a flight. For additional information on weight, balance, CG, and aircraft stability refer to the FAA handbook appropriate to the specific aircraft category.

Chapter 11
Aircraft Performance

Introduction

This chapter discusses the factors that affect aircraft performance, which include the aircraft weight, atmospheric conditions, runway environment, and the fundamental physical laws governing the forces acting on an aircraft.

Importance of Performance Data

The performance or operational information section of the Aircraft Flight Manual/Pilot's Operating Handbook (AFM/POH) contains the operating data for the aircraft; that is, the data pertaining to takeoff, climb, range, endurance, descent, and landing. The use of this data in flying operations is mandatory for safe and efficient operation. Considerable knowledge and familiarity of the aircraft can be gained by studying this material.

Method for Determining Pressure Altitude

Altimeter setting	Altitude correction
28.0	1,824
28.1	1,727
28.2	1,630
28.3	1,533
28.4	1,436
28.5	1,340
28.6	1,244
28.7	1,148
28.8	1,053
28.9	957
29.0	863
29.1	768
29.2	673
29.3	579
29.4	485
29.5	392
29.6	298
29.7	205
29.8	112
29.9	20
29.92	0
30.0	−73
30.1	−165
30.2	−257
30.3	−348
30.4	−440
30.5	−531

Field elevation is sea level

To field elevation

To get pressure altitude

From field elevation

LANDING DISTANCE

Condition:
Flaps lowered to 40°
Power off
Hard surface runway
Zero wind

Gross weight lb	Approach speed IAS, MPH	At sea level & 59 °F		At 2,500 ft & 50 °F		At 5,000 ft & 41 °F		At 7,500 ft & 32 °F	
		Ground roll	Total to clear 50 ft OBS	Ground roll	Total to clear 50 ft OBS	Ground roll	Total to clear 50 ft OBS	Ground roll	Total to clear 50 ft OBS
1,600	60	445	1,075	470	1,135	495	1,195	520	1,255

Note
1. Decrease the distances shown by 10% for each 4 knots of headwind.
2. Increase the distance by 10% for each 60 °F temperature increase above standard.
3. For operation on a dry, grass runway, increase distances (both "ground roll" and "total to clear 50 ft obstacle") by 20% of the "total to clear 50 ft obstacle" figure.

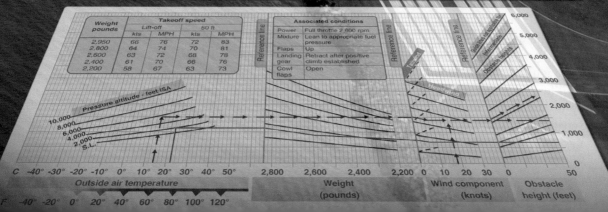

It must be emphasized that the manufacturers' information and data furnished in the AFM/POH is not standardized. Some provide the data in tabular form, while others use graphs. In addition, the performance data may be presented on the basis of standard atmospheric conditions, pressure altitude, or density altitude. The performance information in the AFM/POH has little or no value unless the user recognizes those variations and makes the necessary adjustments.

To be able to make practical use of the aircraft's capabilities and limitations, it is essential to understand the significance of the operational data. The pilot must be cognizant of the basis for the performance data, as well as the meanings of the various terms used in expressing performance capabilities and limitations.

Since the characteristics of the atmosphere have a major effect on performance, it is necessary to review two dominant factors—pressure and temperature.

Structure of the Atmosphere

The atmosphere is an envelope of air that surrounds the Earth and rests upon its surface. It is as much a part of the Earth as is land and water. However, air differs from land and water in that it is a mixture of gases. It has mass, weight, and indefinite shape.

Air, like any other fluid, is able to flow and change its shape when subjected to even minute pressures because of the lack of strong molecular cohesion. For example, gas will completely fill any container into which it is placed, expanding or contracting to adjust its shape to the limits of the container.

The atmosphere is composed of 78 percent nitrogen, 21 percent oxygen, and 1 percent other gases, such as argon or helium. Most of the oxygen is contained below 35,000 feet altitude.

Atmospheric Pressure

Though there are various kinds of pressure, pilots are mainly concerned with atmospheric pressure. It is one of the basic factors in weather changes, helps to lift the aircraft, and actuates some of the most important flight instruments in the aircraft. These instruments often include the altimeter, the airspeed indicator (ASI), the vertical speed indicator (VSI), and the manifold pressure gauge.

Though air is very light, it has mass and is affected by the attraction of gravity. Therefore, like any other substance, it has weight; because it has weight, it has force. Since it is a fluid substance, this force is exerted equally in all directions, and its effect on bodies within the air is called pressure. Under standard conditions at sea level, the average pressure

exerted by the weight of the atmosphere is approximately 14.7 pounds per square inch (psi). The density of air has significant effects on the aircraft's performance. As air becomes less dense, it reduces:

- Power, because the engine takes in less air
- Thrust, because the propeller is less efficient in thin air
- Lift, because the thin air exerts less force on the airfoils

The pressure of the atmosphere may vary with time but more importantly, it varies with altitude and temperature. Due to the changing atmospheric pressure, a standard reference was developed. The standard atmosphere at sea level has a surface temperature of 59 degrees Fahrenheit (°F) or 15 degrees Celsius (°C) and a surface pressure of 29.92 inches of mercury ("Hg) or 1013.2 millibars (mb). *[Figure 11-1]*

A standard temperature lapse rate is one in which the temperature decreases at the rate of approximately 3.5 °F or 2 °C per thousand feet up to 36,000 feet. Above this point, the temperature is considered constant up to 80,000 feet. A standard pressure lapse rate is one in which pressure decreases at a rate of approximately 1 "Hg per 1,000 feet of altitude gain to 10,000 feet. *[Figure 11-2]* The International Civil Aviation Organization (ICAO) has established this as a worldwide standard, and it is often referred to as International Standard Atmosphere (ISA) or ICAO Standard Atmosphere. Any temperature or pressure that differs from the standard lapse rates is considered nonstandard temperature and pressure. Adjustments for nonstandard temperatures and pressures are provided on the manufacturer's performance charts.

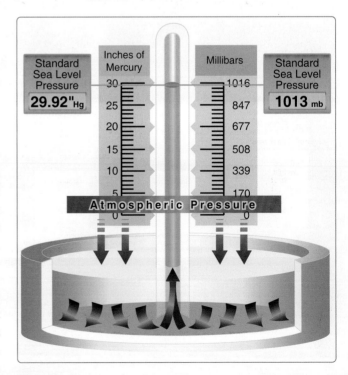

Figure 11-1. *Standard sea level pressure.*

Altitude (ft)	Pressure ("Hg)	Temperature	
		(°C)	(°F)
0	29.92	15.0	59.0
1,000	28.86	13.0	55.4
2,000	27.82	11.0	51.9
3,000	26.82	9.1	48.3
4,000	25.84	7.1	44.7
5,000	24.89	5.1	41.2
6,000	23.98	3.1	37.6
7,000	23.09	1.1	34.0
8,000	22.22	−0.9	30.5
9,000	21.38	−2.8	26.9
10,000	20.57	−4.8	23.3
11,000	19.79	−6.8	19.8
12,000	19.02	−8.8	16.2
13,000	18.29	−10.8	12.6
14,000	17.57	−12.7	9.1
15,000	16.88	−14.7	5.5
16,000	16.21	−16.7	1.9
17,000	15.56	−18.7	−1.6
18,000	14.94	−20.7	−5.2
19,000	14.33	−22.6	−8.8
20,000	13.74	−24.6	−12.3

Figure 11-2. *Properties of standard atmosphere.*

Since all aircraft performance is compared and evaluated using the standard atmosphere, all aircraft instruments are calibrated for the standard atmosphere. Thus, certain corrections must apply to the instrumentation, as well as the aircraft performance, if the actual operating conditions do not fit the standard atmosphere. In order to account properly for the nonstandard atmosphere, certain related terms must be defined.

Pressure Altitude

Pressure altitude is the height above the standard datum plane (SDP). The aircraft altimeter is essentially a sensitive barometer calibrated to indicate altitude in the standard atmosphere. If the altimeter is set for 29.92 "Hg SDP, the altitude indicated is the pressure altitude—the altitude in the standard atmosphere corresponding to the sensed pressure.

The SDP is a theoretical level at which the pressure of the atmosphere is 29.92 "Hg and the weight of air is 14.7 psi. As atmospheric pressure changes, the SDP may be below, at, or above sea level. Pressure altitude is important as a basis for determining aircraft performance, as well as for assigning flight levels to aircraft operating at above 18,000 feet.

The pressure altitude can be determined by any of the three following methods:

1. By setting the barometric scale of the altimeter to 29.92 "Hg and reading the indicated altitude,

2. By applying a correction factor to the indicated altitude according to the reported "altimeter setting," *[Figure 11-3]*

3. By using a flight computer

Density Altitude

The more appropriate term for correlating aerodynamic performance in the nonstandard atmosphere is density altitude—the altitude in the standard atmosphere corresponding to a particular value of air density.

Density altitude is pressure altitude corrected for nonstandard temperature. As the density of the air increases (lower density altitude), aircraft performance increases. Conversely,

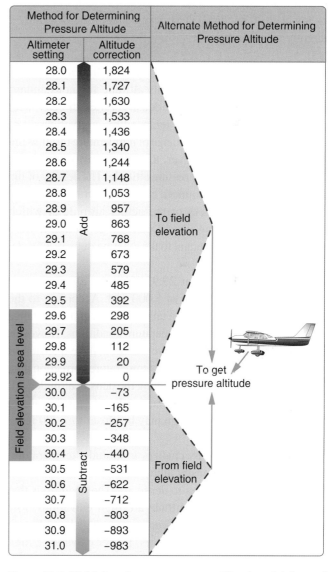

Figure 11-3. *Field elevation versus pressure. The aircraft is located on a field that happens to be at sea level. Set the altimeter to the current altimeter setting (29.7). The difference of 205 feet is added to the elevation or a PA of 205 feet.*

as air density decreases (higher density altitude), aircraft performance decreases. A decrease in air density means a high density altitude; an increase in air density means a lower density altitude. Density altitude is used in calculating aircraft performance. Under standard atmospheric condition, air at each level in the atmosphere has a specific density; under standard conditions, pressure altitude and density altitude identify the same level. Density altitude, then, is the vertical distance above sea level in the standard atmosphere at which a given density is to be found.

Density altitude is computed using pressure altitude and temperature. Since aircraft performance data at any level is based upon air density under standard day conditions, such performance data apply to air density levels that may not be identical to altimeter indications. Under conditions higher or lower than standard, these levels cannot be determined directly from the altimeter.

Density altitude is determined by first finding pressure altitude and then correcting this altitude for nonstandard temperature variations. Since density varies directly with pressure, and inversely with temperature, a given pressure altitude may exist for a wide range of temperature by allowing the density to vary. However, a known density occurs for any one temperature and pressure altitude. The density of the air, of course, has a pronounced effect on aircraft and engine performance. Regardless of the actual altitude at which the aircraft is operating, it will perform as though it were operating at an altitude equal to the existing density altitude.

For example, when set at 29.92 "Hg, the altimeter may indicate a pressure altitude of 5,000 feet. According to the AFM/POH, the ground run on takeoff may require a distance of 790 feet under standard temperature conditions. However, if the temperature is 20 °C above standard, the expansion of air raises the density level. Using temperature correction data from tables or graphs, or by deriving the density altitude with a computer, it may be found that the density level is above 7,000 feet, and the ground run may be closer to 1,000 feet.

Air density is affected by changes in altitude, temperature, and humidity. High density altitude refers to thin air while low density altitude refers to dense air. The conditions that result in a high density altitude are high elevations, low atmospheric pressures, high temperatures, high humidity, or some combination of these factors. Lower elevations, high atmospheric pressure, low temperatures, and low humidity are more indicative of low density altitude.

Using a flight computer, density altitude can be computed by inputting the pressure altitude and outside air temperature at flight level. Density altitude can also be determined by referring to the table and chart in *Figures 11-3 and 11-4* respectively.

Effects of Pressure on Density

Since air is a gas, it can be compressed or expanded. When air is compressed, a greater amount of air can occupy a given volume. Conversely, when pressure on a given volume of air is decreased, the air expands and occupies a greater space. That is, the original column of air at a lower pressure

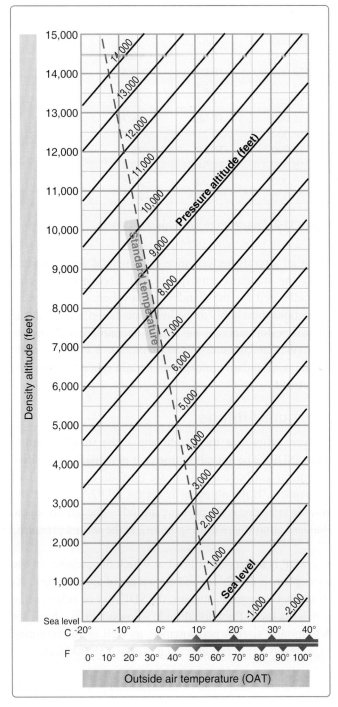

Figure 11-4. *Density altitude chart.*

contains a smaller mass of air. In other words, the density is decreased. In fact, density is directly proportional to pressure. If the pressure is doubled, the density is doubled, and if the pressure is lowered, so is the density. This statement is true only at a constant temperature.

Effects of Temperature on Density

Increasing the temperature of a substance decreases its density. Conversely, decreasing the temperature increases the density. Thus, the density of air varies inversely with temperature. This statement is true only at a constant pressure.

In the atmosphere, both temperature and pressure decrease with altitude and have conflicting effects upon density. However, the fairly rapid drop in pressure as altitude is increased usually has the dominant effect. Hence, pilots can expect the density to decrease with altitude.

Effects of Humidity (Moisture) on Density

The preceding paragraphs are based on the presupposition of perfectly dry air. In reality, it is never completely dry. The small amount of water vapor suspended in the atmosphere may be negligible under certain conditions, but in other conditions humidity may become an important factor in the performance of an aircraft. Water vapor is lighter than air; consequently, moist air is lighter than dry air. Therefore, as the water content of the air increases, the air becomes less dense, increasing density altitude and decreasing performance. It is lightest or least dense when, in a given set of conditions, it contains the maximum amount of water vapor.

Humidity, also called relative humidity, refers to the amount of water vapor contained in the atmosphere and is expressed as a percentage of the maximum amount of water vapor the air can hold. This amount varies with the temperature; warm air can hold more water vapor, while colder air can hold less. Perfectly dry air that contains no water vapor has a relative humidity of zero percent, while saturated air that cannot hold any more water vapor has a relative humidity of 100 percent. Humidity alone is usually not considered an essential factor in calculating density altitude and aircraft performance; however, it does contribute.

The higher the temperature, the greater amount of water vapor that the air can hold. When comparing two separate air masses, the first warm and moist (both qualities making air lighter) and the second cold and dry (both qualities making it heavier), the first must be less dense than the second. Pressure, temperature, and humidity have a great influence on aircraft performance because of their effect upon density. There is no rule-of-thumb or chart used to compute the effects of humidity on density altitude, but it must be taken into consideration. Expect a decrease in overall performance in high humidity conditions.

Performance

Performance is a term used to describe the ability of an aircraft to accomplish certain things that make it useful for certain purposes. For example, the ability of an aircraft to land and take off in a very short distance is an important factor to the pilot who operates in and out of short, unimproved airfields. The ability to carry heavy loads, fly at high altitudes at fast speeds, and/or travel long distances is essential for the performance of airline and executive type aircraft.

The primary factors most affected by performance are the takeoff and landing distance, rate of climb, ceiling, payload, range, speed, maneuverability, stability, and fuel economy. Some of these factors are often directly opposed: for example, high speed versus short landing distance, long range versus great payload, and high rate of climb versus fuel economy. It is the preeminence of one or more of these factors that dictates differences between aircraft and explains the high degree of specialization found in modern aircraft.

The various items of aircraft performance result from the combination of aircraft and powerplant characteristics. The aerodynamic characteristics of the aircraft generally define the power and thrust requirements at various conditions of flight, while powerplant characteristics generally define the power and thrust available at various conditions of flight. The matching of the aerodynamic configuration with the powerplant is accomplished by the manufacturer to provide maximum performance at the specific design condition (e.g., range, endurance, and climb).

Straight-and-Level Flight

All of the principal components of flight performance involve steady-state flight conditions and equilibrium of the aircraft. For the aircraft to remain in steady, level flight, equilibrium must be obtained by a lift equal to the aircraft weight and a powerplant thrust equal to the aircraft drag. Thus, the aircraft drag defines the thrust required to maintain steady, level flight. As presented in Chapter 4, Aerodynamics of Flight, all parts of an aircraft contribute to the drag, either induced (from lifting surfaces) or parasite drag.

While parasite drag predominates at high speed, induced drag predominates at low speed. *[Figure 11-5]* For example, if an aircraft in a steady flight condition at 100 knots is then accelerated to 200 knots, the parasite drag becomes four times as great, but the power required to overcome that drag is eight times the original value. Conversely, when the

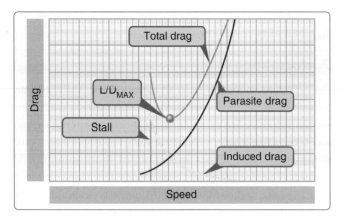

Figure 11-5. *Drag versus speed.*

aircraft is operated in steady, level flight at twice as great a speed, the induced drag is one-fourth the original value, and the power required to overcome that drag is only one-half the original value.

When an aircraft is in steady, level flight, the condition of equilibrium must prevail. The unaccelerated condition of flight is achieved with the aircraft trimmed for lift equal to weight and the powerplant set for a thrust to equal the aircraft drag.

The maximum level flight speed for the aircraft is obtained when the power or thrust required equals the maximum power or thrust available from the powerplant. *[Figure 11-6]* The minimum level flight airspeed is not usually defined by thrust or power requirement since conditions of stall or stability and control problems generally predominate.

Climb Performance

If an aircraft is to move, fly, and perform, work must act upon it. Work involves force moving the aircraft. The aircraft acquires mechanical energy when it moves. Mechanical

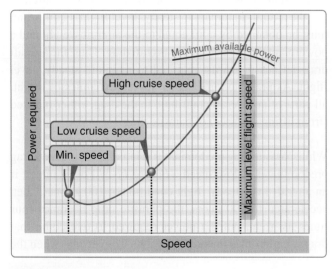

Figure 11-6. *Power versus speed.*

energy comes in two forms: (1) Kinetic Energy (KE), the energy of speed; (2) Potential Energy (PE), the stored energy of position.

Aircraft motion (KE) is described by its velocity (airspeed). Aircraft position (PE) is described by its height (altitude). Both KE and PE are directly proportional to the object's mass. KE is directly proportional to the square of the object's velocity (airspeed). PE is directly proportional to the object's height (altitude). The formulas below summarize these energy relationships:

$$KE = \frac{1}{2} \times m \times v^2$$

m = object mass
v = object velocity

$$PE = m \times g \times h$$

m = object mass
g = gravity field strength
h = object height

We sometimes use the terms "power" and "thrust" interchangeably when discussing climb performance. This erroneously implies the terms are synonymous. It is important to distinguish between these terms. Thrust is a force or pressure exerted on an object. Thrust is measured in pounds (lb) or newtons (N). Power, however, is a measurement of the rate of performing work or transferring energy (KE and PE). Power is typically measured in horsepower (hp) or kilowatts (kw). We can think of power as the motion (KE and PE) a force (thrust) creates when exerted on an object over a period of time.

Positive climb performance occurs when an aircraft gains PE by increasing altitude. Two basic factors, or a combination of the two factors, contribute to positive climb performance in most aircraft:

1. The aircraft climbs (gains PE) using excess power above that required to maintain level flight, or

2. The aircraft climbs by converting airspeed (KE) to altitude (PE).

As an example of factor 1 above, an aircraft with an engine capable of producing 200 horsepower (at a given altitude) is using only 130 horsepower to maintain level flight at that altitude. This leaves 70 horsepower available to climb. The pilot holds airspeed constant and increases power to perform the climb.

As an example of factor 2, an aircraft is flying level at 120 knots. The pilot leaves the engine power setting constant but applies other control inputs to perform a climb. The climb, sometimes called a zoom climb, converts the airspeed (KE) to altitude (PE); the airspeed decreases to something less than 120 knots as the altitude increases.

There are two primary reasons to evaluate climb performance. First, aircraft must climb over obstacles to avoid hitting them. Second, climbing to higher altitudes can provide better weather, fuel economy, and other benefits. Maximum Angle of Climb (AOC), obtained at V_X, may provide climb performance to ensure an aircraft will clear obstacles. Maximum Rate of Climb (ROC), obtained at V_Y, provides climb performance to achieve the greatest altitude gain over time. Maximum ROC may not be sufficient to avoid obstacles in some situations, while maximum AOC may be sufficient to avoid the same obstacles. *[Figure 11-7]*

Angle of Climb (AOC)

AOC is a comparison of altitude gained relative to distance traveled. AOC is the inclination (angle) of the flight path. For maximum AOC performance, a pilot flies the aircraft at V_X

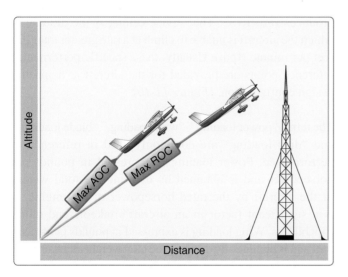

Figure 11-7. *Maximum angle of climb (AOC) versus maximum rate of climb (ROC).*

so as to achieve maximum altitude increase with minimum horizontal travel over the ground. A good use of maximum AOC is when taking off from a short airfield surrounded by high obstacles, such as trees or power lines. The objective is to gain sufficient altitude to clear the obstacle while traveling the least horizontal distance over the surface.

One method to climb (have positive AOC performance) is to have excess thrust available. Essentially, the greater the force that pushes the aircraft upward, the steeper it can climb. Maximum AOC occurs at the airspeed and angle of attack (AOA) combination which allows the maximum excess thrust. The airspeed and AOA combination where excess thrust exists varies amongst aircraft types. As an example, *Figure 11-8* provides a comparison between jet and propeller airplanes as to where maximum excess thrust (for maximum AOC) occurs. In a jet, maximum excess thrust normally occurs at the airspeed where the thrust required is at a minimum (approximately L/D_{MAX}). In a propeller airplane, maximum excess thrust normally occurs at an airspeed below L/D_{MAX} and frequently just above stall speed.

Rate of Climb (ROC)

ROC is a comparison of altitude gained relative to the time needed to reach that altitude. ROC is simply the vertical component of the aircraft's flight path velocity vector. For maximum ROC performance, a pilot flies the aircraft at V_Y so as to achieve a maximum gain in altitude over a given period of time.

Maximum ROC expedites a climb to an assigned altitude. This gains the greatest vertical distance over a period of time. For example, in a maximum AOC profile, a certain aircraft takes 30 seconds to reach 1,000 feet AGL, but covers only 3,000 feet over the ground. By comparison, using its maximum ROC profile, the same aircraft climbs

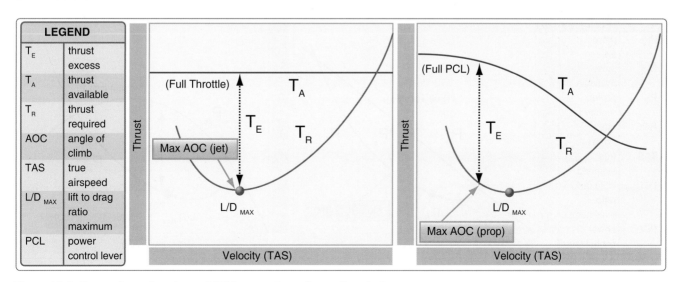

Figure 11-8. *Comparison of maximum AOC between jet and propeller airplanes.*

to 1,500 feet in 30 seconds but covers 6,000 feet across the ground. Note that both ROC and AOC maximum climb profiles use the aircraft's maximum throttle setting. Any differences between max ROC and max AOC lie primarily in the velocity (airspeed) and AOA combination the aircraft manual specifies. *[Figure 11-7]*

ROC performance depends upon excess power. Since climbing is work and power is the rate of performing work, a pilot can increase the climb rate by using any power not used to maintain level flight. Maximum ROC occurs at an airspeed and AOA combination that produces the maximum excess power. Therefore, maximum ROC for a typical jet airplane occurs at an airspeed greater than L/D_{MAX} and at an AOA less than L/D_{MAX} AOA. In contrast, maximum ROC for a typical propeller airplane occurs at an airspeed and AOA combination closer to L/D_{MAX}. *[Figure 11-9]*

Climb Performance Factors

Since weight, altitude and configuration changes affect excess thrust and power, they also affect climb performance. Climb performance is directly dependent upon the ability to produce either excess thrust or excess power. Earlier in the book it was shown that an increase in weight, an increase in altitude, lowering the landing gear, or lowering the flaps all decrease both excess thrust and excess power for all aircraft. Therefore, maximum AOC and maximum ROC performance decreases under any of these conditions.

Weight has a very pronounced effect on aircraft performance. If weight is added to an aircraft, it must fly at a higher AOA to maintain a given altitude and speed. This increases the induced drag of the wings, as well as the parasite drag of the aircraft. Increased drag means that additional thrust is needed to overcome it, which in turn means that less reserve thrust is available for climbing. Aircraft designers go to great lengths

to minimize the weight, since it has such a marked effect on the factors pertaining to performance.

A change in an aircraft's weight produces a twofold effect on climb performance. First, a change in weight changes the drag and the power required. This alters the reserve power available, which in turn, affects both the climb angle and the climb rate. Secondly, an increase in weight reduces the maximum ROC, but the aircraft must be operated at a higher climb speed to achieve the smaller peak climb rate.

An increase in altitude also increases the power required and decreases the power available. Therefore, the climb performance of an aircraft diminishes with altitude. The speeds for maximum ROC, maximum AOC, and maximum and minimum level flight airspeeds vary with altitude. As altitude is increased, these various speeds finally converge at the absolute ceiling of the aircraft. At the absolute ceiling, there is no excess of power and only one speed allows steady, level flight. Consequently, the absolute ceiling of an aircraft produces zero ROC. The service ceiling is the altitude at which the aircraft is unable to climb at a rate greater than 100 feet per minute (fpm). Usually, these specific performance reference points are provided for the aircraft at a specific design configuration. *[Figure 11-10]*

The terms "power loading," "wing loading," "blade loading," and "disk loading" are commonly used in reference to performance. Power loading is expressed in pounds per horsepower and is obtained by dividing the total weight of the aircraft by the rated horsepower of the engine. It is a significant factor in an aircraft's takeoff and climb capabilities. Wing loading is expressed in pounds per square foot and is obtained by dividing the total weight of an airplane in pounds by the wing area (including ailerons) in square feet. It is the airplane's wing loading that determines the landing

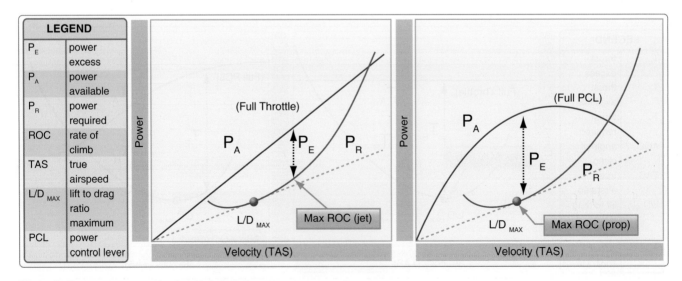

Figure 11-9. *Comparison of maximum ROC between jet and propeller airplanes.*

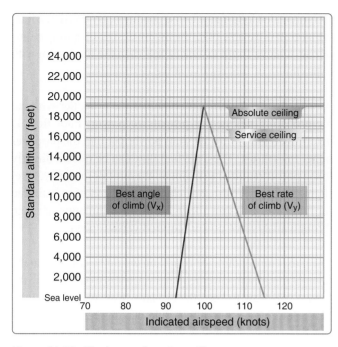

Figure 11-10. *Absolute and service ceiling.*

A common element for each of these operating problems is the specific range; that is, nautical miles (NM) of flying distance versus the amount of fuel consumed. Range must be clearly distinguished from the item of endurance. Range involves consideration of flying distance, while endurance involves consideration of flying time. Thus, it is appropriate to define a separate term, specific endurance.

$$\text{specific endurance} = \frac{\text{flight hours}}{\text{pounds of fuel}}$$

or

$$\text{specific endurance} = \frac{\text{flight hours/hour}}{\text{pounds of fuel/hour}}$$

or

$$\text{specific endurance} = \frac{1}{\text{fuel flow}}$$

speed. Blade loading is expressed in pounds per square foot and is obtained by dividing the total weight of a helicopter by the area of the rotor blades. Blade loading is not to be confused with disk loading, which is the total weight of a helicopter divided by the area of the disk swept by the rotor blades.

Range Performance

The ability of an aircraft to convert fuel energy into flying distance is one of the most important items of aircraft performance. In flying operations, the problem of efficient range operation of an aircraft appears in two general forms:

1. To extract the maximum flying distance from a given fuel load

2. To fly a specified distance with a minimum expenditure of fuel

Fuel flow can be defined in either pounds or gallons. If maximum endurance is desired, the flight condition must provide a minimum fuel flow. In *Figure 11-11* at point A, the airspeed is low and fuel flow is high. This would occur during ground operations or when taking off and climbing. As airspeed is increased, power requirements decrease due to aerodynamic factors, and fuel flow decreases to point B. This is the point of maximum endurance. Beyond this point, increases in airspeed come at a cost. Airspeed increases require additional power and fuel flow increases with additional power.

Cruise flight operations for maximum range should be conducted so that the aircraft obtains maximum specific range throughout the flight. The specific range can be defined by the following relationship

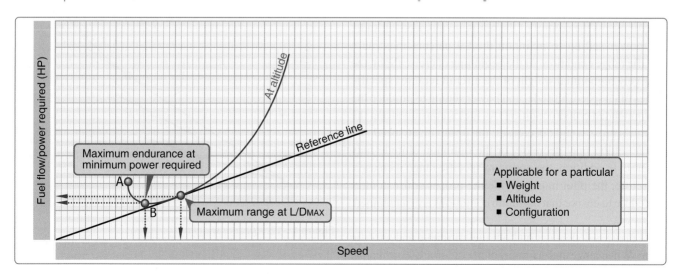

Figure 11-11. *Airspeed for maximum endurance.*

$$specific\ range = \frac{NM}{pounds\ of\ fuel}$$

or

$$specific\ range = \frac{NM/hour}{pounds\ of\ fuel/hour}$$

or

$$specific\ range = \frac{knots}{fuel\ flow}$$

If maximum specific range is desired, the flight condition must provide a maximum of speed per fuel flow. While the peak value of specific range would provide maximum range operation, long-range cruise operation is generally recommended at a slightly higher airspeed. Most long-range cruise operations are conducted at the flight condition that provides 99 percent of the absolute maximum specific range. The advantage of such operation is that one percent of range is traded for three to five percent higher cruise speed. Since the higher cruise speed has a great number of advantages, the small sacrifice of range is a fair bargain. The values of specific range versus speed are affected by three principal variables:

1. Aircraft gross weight

2. Altitude

3. The external aerodynamic configuration of the aircraft.

These are the source of range and endurance operating data included in the performance section of the AFM/POH.

Cruise control of an aircraft implies that the aircraft is operated to maintain the recommended long-range cruise condition throughout the flight. Since fuel is consumed during cruise, the gross weight of the aircraft varies and optimum airspeed, altitude, and power setting can also vary. Cruise control means the control of the optimum airspeed, altitude, and power setting to maintain the 99 percent maximum specific range condition. At the beginning of cruise flight, the relatively high initial weight of the aircraft requires specific values of airspeed, altitude, and power setting to produce the recommended cruise condition. As fuel is consumed and the aircraft's gross weight decreases, the optimum airspeed and power setting may decrease, or the optimum altitude may increase. In addition, the optimum specific range increases. Therefore, the pilot must provide the proper cruise control procedure to ensure that optimum conditions are maintained.

Total range is dependent on both fuel available and specific range. When range and economy of operation are the principal goals, the pilot must ensure that the aircraft is operated at the recommended long-range cruise condition. By this procedure, the aircraft is capable of its maximum design-operating radius or can achieve flight distances less than the maximum with a maximum of fuel reserve at the destination.

A propeller-driven aircraft combines the propeller with the reciprocating engine for propulsive power. Fuel flow is determined mainly by the shaft power put into the propeller rather than thrust. Thus, the fuel flow can be related directly to the power required to maintain the aircraft in steady, level flight, and on performance charts power can be substituted for fuel flow. This fact allows for the determination of range through analysis of power required versus speed.

The maximum endurance condition would be obtained at the point of minimum power required since this would require the lowest fuel flow to keep the airplane in steady, level flight. Maximum range condition would occur where the ratio of speed to power required is greatest. *[Figure 11-11]*

The maximum range condition is obtained at maximum lift/drag ratio (L/D_{MAX}), and it is important to note that for a given aircraft configuration, the L/D_{MAX} occurs at a particular AOA and lift coefficient and is unaffected by weight or altitude. A variation in weight alters the values of airspeed and power required to obtain the L/D_{MAX}. *[Figure 11-12]* Different theories exist on how to achieve max range when there is a headwind or tailwind present. Many say that speeding up in a headwind or slowing down in a tail wind helps to achieve max range. While this theory may be true in a lot of cases, it is not always true as there are different variables to every situation. Each aircraft configuration is different, and there is not a rule of thumb that encompasses all of them as to how to achieve the max range.

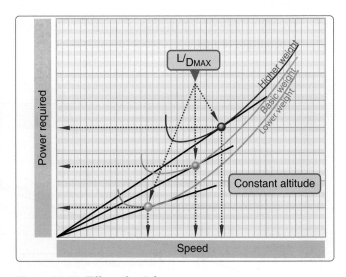

Figure 11-12. *Effect of weight.*

The variations of speed and power required must be monitored by the pilot as part of the cruise control procedure to maintain the L/D$_{MAX}$. When the aircraft's fuel weight is a small part of the gross weight and the aircraft's range is small, the cruise control procedure can be simplified to essentially maintaining a constant speed and power setting throughout the time of cruise flight. However, a long-range aircraft has a fuel weight that is a considerable part of the gross weight, and cruise control procedures must employ scheduled airspeed and power changes to maintain optimum range conditions.

The effect of altitude on the range of a propeller-driven aircraft is illustrated in *Figure 11-13*. A flight conducted at high altitude has a greater true airspeed (TAS), and the power required is proportionately greater than when conducted at sea level. The drag of the aircraft at altitude is the same as the drag at sea level, but the higher TAS causes a proportionately greater power required.

NOTE: The straight line that is tangent to the sea level power curve is also tangent to the altitude power curve.

The effect of altitude on specific range can also be appreciated from the previous relationships. If a change in altitude causes identical changes in speed and power required, the proportion of speed to power required would be unchanged. The fact implies that the specific range of a propeller-driven aircraft would be unaffected by altitude. Actually, this is true to the extent that specific fuel consumption and propeller efficiency are the principal factors that could cause a variation of specific range with altitude. If compressibility effects are negligible, any variation of specific range with altitude is strictly a function of engine/propeller performance.

An aircraft equipped with a reciprocating engine experiences very little, if any, variation of specific range up to its absolute altitude. There is negligible variation of brake specific fuel consumption for values of brake horsepower below the maximum cruise power rating of the engine that is the lean range of engine operation. Thus, an increase in altitude produces a decrease in specific range only when the increased power requirement exceeds the maximum cruise power rating of the engine. One advantage of supercharging is that the cruise power may be maintained at high altitude, and the aircraft may achieve the range at high altitude with the corresponding increase in TAS. The principal differences in the high altitude cruise and low altitude cruise are the TAS and climb fuel requirements.

Region of Reversed Command

The aerodynamic properties of an aircraft generally determine the power requirements at various conditions of flight, while the powerplant capabilities generally determine the power available at various conditions of flight. When an aircraft is in steady, level flight, a condition of equilibrium must prevail. An unaccelerated condition of flight is achieved when lift equals weight, and the powerplant is set for thrust equal to drag. The power required to achieve equilibrium in constant-altitude flight at various airspeeds is depicted on a power required curve. The power required curve illustrates the fact that at low airspeeds near the stall or minimum controllable airspeed, the power setting required for steady, level flight is quite high.

Flight in the region of normal command means that while holding a constant altitude, a higher airspeed requires a higher power setting and a lower airspeed requires a lower power setting. The majority of aircraft flying (climb, cruise, and maneuvers) is conducted in the region of normal command.

Flight in the region of reversed command means flight in which a higher airspeed requires a lower power setting and a lower airspeed requires a higher power setting to hold altitude. It does not imply that a decrease in power produces lower airspeed. The region of reversed command is encountered in the low speed phases of flight. Flight speeds below the speed for maximum endurance (lowest point on the power curve) require higher power settings with a decrease in airspeed. Since the need to increase the required power setting with decreased speed is contrary to the normal command of flight, the regime of flight speeds between the speed for minimum required power setting and the stall speed (or minimum control speed) is termed the region of reversed command. In the region of reversed command, a decrease in airspeed must be accompanied by an increased power setting in order to maintain steady flight.

Figure 11-14 shows the maximum power available as a curved line. Lower power settings, such as cruise power, would also appear in a similar curve. The lowest point on

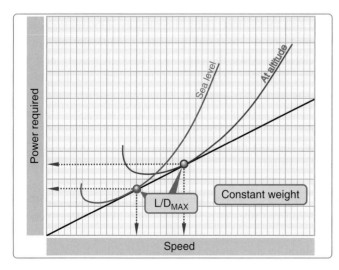

Figure 11-13. *Effect of altitude on range.*

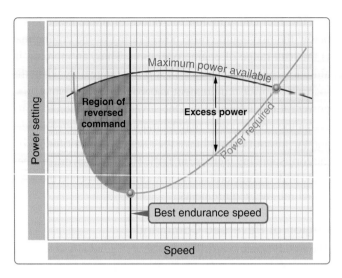

Figure 11-14. *Power required curve.*

the power required curve represents the speed at which the lowest brake horsepower sustains level flight. This is termed the best endurance airspeed.

An airplane performing a low airspeed, high pitch attitude power approach for a short-field landing is an example of operating in the region of reversed command. If an unacceptably high sink rate should develop, it may be possible for the pilot to reduce or stop the descent by applying power. But without further use of power, the airplane would probably stall or be incapable of flaring for the landing. Merely lowering the nose of the airplane to regain flying speed in this situation, without the use of power, would result in a rapid sink rate and corresponding loss of altitude.

If during a soft-field takeoff and climb, for example, the pilot attempts to climb out of ground effect without first attaining normal climb pitch attitude and airspeed, the airplane may inadvertently enter the region of reversed command at a dangerously low altitude. Even with full power, the airplane may be incapable of climbing or even maintaining altitude. The pilot's only recourse in this situation is to lower the pitch attitude in order to increase airspeed, which inevitably results in a loss of altitude.

Airplane pilots must give particular attention to precise control of airspeed when operating in the low flight speeds of the region of reversed command.

Takeoff and Landing Performance

The majority of pilot-caused aircraft accidents occur during the takeoff and landing phase of flight. Because of this fact, the pilot must be familiar with all the variables that influence the takeoff and landing performance of an aircraft and must strive for exacting, professional procedures of operation during these phases of flight.

Takeoff and landing performance is a condition of accelerated and decelerated motion. For instance, during takeoff an aircraft starts at zero speed and accelerates to the takeoff speed to become airborne. During landing, the aircraft touches down at the landing speed and decelerates to zero speed. The important factors of takeoff or landing performance are:

- The takeoff or landing speed is generally a function of the stall speed or minimum flying speed.

- The rate of acceleration/deceleration during the takeoff or landing roll. The speed (acceleration and deceleration) experienced by any object varies directly with the imbalance of force and inversely with the mass of the object. An airplane on the runway moving at 75 knots has four times the energy it has traveling at 37 knots. Thus, an airplane requires four times as much distance to stop as required at half the speed.

- The takeoff or landing roll distance is a function of both acceleration/deceleration and speed.

Runway Surface and Gradient

Runway conditions affect takeoff and landing performance. Typically, performance chart information assumes paved, level, smooth, and dry runway surfaces. Since no two runways are alike, the runway surface differs from one runway to another, as does the runway gradient or slope. *[Figure 11-15]*

Runway surfaces vary widely from one airport to another. The runway surface encountered may be concrete, asphalt, gravel, dirt, or grass. The runway surface for a specific airport is noted in the Chart Supplement U.S. (formerly Airport/Facility Directory). Any surface that is not hard and smooth increases the ground roll during takeoff. This is due to the inability of the tires to roll smoothly along the runway. Tires can sink into soft, grassy, or muddy runways. Potholes or other ruts in the pavement can be the cause of poor tire movement along the runway. Obstructions such as mud, snow, or standing water reduce the airplane's acceleration down the runway. Although muddy and wet surface conditions can reduce friction between the runway and the tires, they can also act as obstructions and reduce the landing distance. *[Figure 11-16]* Braking effectiveness is another consideration when dealing with various runway types. The condition of the surface affects the braking ability of the aircraft.

The amount of power that is applied to the brakes without skidding the tires is referred to as braking effectiveness. Ensure that runways are adequate in length for takeoff acceleration and landing deceleration when less than ideal surface conditions are being reported.

TAKEOFF DISTANCE
MAXIMUM WEIGHT 3800 LBS

SHORT FIELD

CONDITIONS:
Flaps 10°
2850 RPM, Full Throttle and Mixture Set at Placard Fuel Flow Prior to Brake Release
Cowl Flaps Open
Paved, Level, Dry Runway
Zero Wind

MIXTURE SETTING	
PRESS ALT	PPH
S.L.	144
2000	138
4000	132
6000	126
8000	120

NOTES:
1. Short field technique as specified in Section 4.
2. Landing gear extended until takeoff obstacle is cleared.
3. Where distance value has been deleted, climb performance after lift-off is less than 150 fpm. Rate of climb is based on landing gear extended and flaps 10° at takeoff speed.
4. Decrease distances 10% for each 10 knots headwind. For operation with tailwinds up to 10 knots, increase distances by 10% for each 2.5 knots.
5. For operation on a dry, grass runway, increase distances by 15% of the "ground roll" figure.

WEIGHT LBS	TAKEOFF SPEED KIAS		PRESS ALT FT	0°C		10°C		20°C		30°C		40°C	
	LIFT	AT		GRND	TOTAL TO CLEAR	GRND	TOTAL TO CLEAR	GRND	TOTAL TO CLEAR	GRND	TOTAL TO CLEAR	GRND	TOTAL TO CLEAR

Figure 11-15. *Takeoff distance chart.*

The gradient or slope of the runway is the amount of change in runway height over the length of the runway. The gradient is expressed as a percentage, such as a 3 percent gradient. This means that for every 100 feet of runway length, the runway height changes by 3 feet. A positive gradient indicates the runway height increases, and a negative gradient indicates the runway decreases in height. An upsloping runway impedes acceleration and results in a longer ground run during takeoff. However, landing on an upsloping runway typically reduces the landing roll. A downsloping runway aids in acceleration on takeoff resulting in shorter takeoff distances. The opposite is true when landing, as landing on a downsloping runway increases landing distances. Runway slope information is contained in the Chart Supplement U.S. (formerly Airport/ Facility Directory). *[Figure 11-17]*

Water on the Runway and Dynamic Hydroplaning

Water on the runways reduces the friction between the tires and the ground and can reduce braking effectiveness. The ability to brake can be completely lost when the tires are hydroplaning because a layer of water separates the tires from the runway surface. This is also true of braking effectiveness when runways are covered in ice.

When the runway is wet, the pilot may be confronted with dynamic hydroplaning. Dynamic hydroplaning is a condition in which the aircraft tires ride on a thin sheet of water rather than on the runway's surface. Because hydroplaning wheels are not touching the runway, braking and directional control are almost nil. To help minimize dynamic hydroplaning, some runways are grooved to help drain off water; most runways are not.

Figure 11-16. *An aircraft's performance during takeoff depends greatly on the runway surface.*

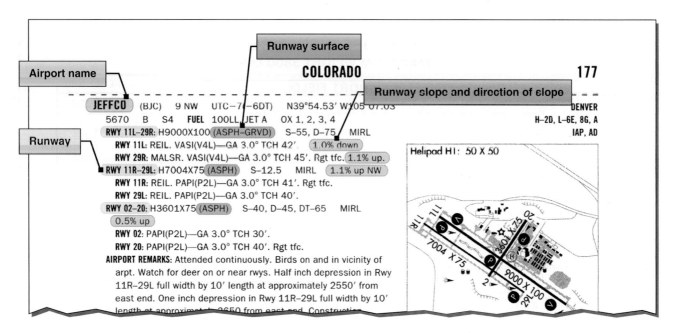

Figure 11-17. *Chart Supplement U.S. (formerly Airport/Facility Directory) information.*

Tire pressure is a factor in dynamic hydroplaning. Using the simple formula in *Figure 11-18,* a pilot can calculate the minimum speed, in knots, at which hydroplaning begins. In plain language, the minimum hydroplaning speed is determined by multiplying the square root of the main gear tire pressure in psi by nine. For example, if the main gear tire pressure is at 36 psi, the aircraft would begin hydroplaning at 54 knots.

Landing at higher than recommended touchdown speeds exposes the aircraft to a greater potential for hydroplaning. And once hydroplaning starts, it can continue well below the minimum initial hydroplaning speed.

On wet runways, directional control can be maximized by landing into the wind. Abrupt control inputs should be avoided. When the runway is wet, anticipate braking problems well before landing and be prepared for hydroplaning. Opt for a suitable runway most aligned with the wind. Mechanical braking may be ineffective, so aerodynamic braking should be used to its fullest advantage.

Takeoff Performance

The minimum takeoff distance is of primary interest in the operation of any aircraft because it defines the runway requirements. The minimum takeoff distance is obtained by taking off at some minimum safe speed that allows sufficient margin above stall and provides satisfactory control and initial ROC. Generally, the lift-off speed is some fixed percentage of the stall speed or minimum control speed for the aircraft in the takeoff configuration. As such, the lift-off is accomplished at some particular value of lift coefficient and AOA. Depending on the aircraft characteristics, the lift-off speed is anywhere from 1.05 to 1.25 times the stall speed or minimum control speed.

To obtain minimum takeoff distance at the specific lift-off speed, the forces that act on the aircraft must provide the maximum acceleration during the takeoff roll. The various forces acting on the aircraft may or may not be under the control of the pilot, and various procedures may be necessary in certain aircraft to maintain takeoff acceleration at the highest value.

The powerplant thrust is the principal force to provide the acceleration and, for minimum takeoff distance, the output thrust should be at a maximum. Lift and drag are produced as soon as the aircraft has speed, and the values of lift and drag depend on the AOA and dynamic pressure.

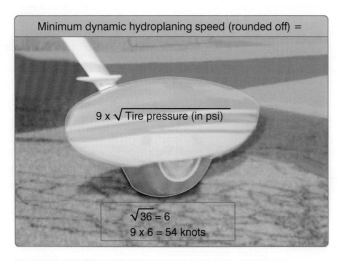

Figure 11-18. *Tire pressure.*

As discussed in Chapter 6, engine pressure ratio (EPR) is the ratio between exhaust pressure (jet blast) and inlet (static) pressure on a turbo jet or turbo fan engine. An EPR gauge tells the pilot how much power the engines are generating. The higher the EPR, the higher the engine thrust. EPR is used to avoid over-boosting an engine and to set takeoff and go around power if needed. This information is important to know before taking off as it helps determine the performance of the aircraft.

In addition to the important factors of proper procedures, many other variables affect the takeoff performance of an aircraft. Any item that alters the takeoff speed or acceleration rate during the takeoff roll affects the takeoff distance.

For example, the effect of gross weight on takeoff distance is significant, and proper consideration of this item must be made in predicting the aircraft's takeoff distance. Increased gross weight can be considered to produce a threefold effect on takeoff performance:

1. Higher lift-off speed

2. Greater mass to accelerate

3. Increased retarding force (drag and ground friction)

If the gross weight increases, a greater speed is necessary to produce the greater lift necessary to get the aircraft airborne at the takeoff lift coefficient. As an example of the effect of a change in gross weight, a 21 percent increase in takeoff weight requires a 10 percent increase in lift-off speed to support the greater weight.

A change in gross weight changes the net accelerating force and changes the mass that is being accelerated. If the aircraft has a relatively high thrust-to-weight ratio, the change in the net accelerating force is slight and the principal effect on acceleration is due to the change in mass.

For example, a 10 percent increase in takeoff gross weight would cause:

•   A 5 percent increase in takeoff velocity

•   At least a 9 percent decrease in rate of acceleration

•   At least a 21 percent increase in takeoff distance

With ISA conditions, increasing the takeoff weight of the average Cessna 182 from 2,400 pounds to 2,700 pounds (11 percent increase) results in an increased takeoff distance from 440 feet to 575 feet (23 percent increase).

For the aircraft with a high thrust-to-weight ratio, the increase in takeoff distance might be approximately 21 to 22 percent, but for the aircraft with a relatively low thrust-to-weight ratio, the increase in takeoff distance would be approximately 25 to 30 percent. Such a powerful effect requires proper consideration of gross weight in predicting takeoff distance.

The effect of wind on takeoff distance is large, and proper consideration must also be provided when predicting takeoff distance. The effect of a headwind is to allow the aircraft to reach the lift-off speed at a lower groundspeed, while the effect of a tailwind is to require the aircraft to achieve a greater groundspeed to attain the lift-off speed.

A headwind that is 10 percent of the takeoff airspeed reduces the takeoff distance approximately 19 percent. However, a tailwind that is 10 percent of the takeoff airspeed increases the takeoff distance approximately 21 percent. In the case where the headwind speed is 50 percent of the takeoff speed, the takeoff distance would be approximately 25 percent of the zero wind takeoff distance (75 percent reduction).

The effect of wind on landing distance is identical to its effect on takeoff distance. *Figure 11-19* illustrates the general effect of wind by the percent change in takeoff or landing distance as a function of the ratio of wind velocity to takeoff or landing speed.

The effect of proper takeoff speed is especially important when runway lengths and takeoff distances are critical. The takeoff speeds specified in the AFM/POH are generally the minimum safe speeds at which the aircraft can become airborne. Any attempt to take off below the recommended speed means that the aircraft could stall, be difficult to control, or have a very low initial ROC. In some cases, an

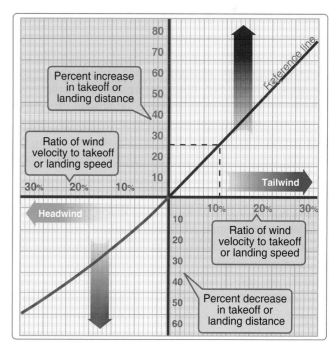

Figure 11-19. *Effect of wind on takeoff and landing.*

excessive AOA may not allow the aircraft to climb out of ground effect. On the other hand, an excessive airspeed at takeoff may improve the initial ROC and "feel" of the aircraft but produces an undesirable increase in takeoff distance. Assuming that the acceleration is essentially unaffected, the takeoff distance varies with the square of the takeoff velocity.

Thus, ten percent excess airspeed would increase the takeoff distance 21 percent. In most critical takeoff conditions, such an increase in takeoff distance would be prohibitive, and the pilot must adhere to the recommended takeoff speeds.

The effect of pressure altitude and ambient temperature is to define the density altitude and its effect on takeoff performance. While subsequent corrections are appropriate for the effect of temperature on certain items of powerplant performance, density altitude defines specific effects on takeoff performance. An increase in density altitude can produce a twofold effect on takeoff performance:

1. Greater takeoff speed

2. Decreased thrust and reduced net accelerating force

If an aircraft of given weight and configuration is operated at greater heights above standard sea level, the aircraft requires the same dynamic pressure to become airborne at the takeoff lift coefficient. Thus, the aircraft at altitude takes off at the same indicated airspeed (IAS) as at sea level, but because of the reduced air density, the TAS is greater.

The effect of density altitude on powerplant thrust depends much on the type of powerplant. An increase in altitude above standard sea level brings an immediate decrease in power output for the unsupercharged reciprocating engine. However, an increase in altitude above standard sea level does not cause a decrease in power output for the supercharged reciprocating engine until the altitude exceeds the critical operating altitude. For those powerplants that experience a decay in thrust with an increase in altitude, the effect on the net accelerating force and acceleration rate can be approximated by assuming a direct variation with density. Actually, this assumed variation would closely approximate the effect on aircraft with high thrust-to-weight ratios.

Proper accounting of pressure altitude and temperature is mandatory for accurate prediction of takeoff roll distance. The most critical conditions of takeoff performance are the result of some combination of high gross weight, altitude, temperature, and unfavorable wind. In all cases, the pilot must make an accurate prediction of takeoff distance from the performance data of the AFM/POH, regardless of the runway available, and strive for a polished, professional takeoff procedure.

In the prediction of takeoff distance from the AFM/POH data, the following primary considerations must be given:

- Pressure altitude and temperature—to define the effect of density altitude on distance

- Gross weight—a large effect on distance

- Wind—a large effect due to the wind or wind component along the runway

- Runway slope and condition—the effect of an incline and retarding effect of factors such as snow or ice

Landing Performance

In many cases, the landing distance of an aircraft defines the runway requirements for flight operations. The minimum landing distance is obtained by landing at some minimum safe speed, that allows sufficient margin above stall and provides satisfactory control and capability for a go-around. Generally, the landing speed is some fixed percentage of the stall speed or minimum control speed for the aircraft in the landing configuration. As such, the landing is accomplished at some particular value of lift coefficient and AOA. The exact values depend on the aircraft characteristics but, once defined, the values are independent of weight, altitude, and wind.

To obtain minimum landing distance at the specified landing speed, the forces that act on the aircraft must provide maximum deceleration during the landing roll. The forces acting on the aircraft during the landing roll may require various procedures to maintain landing deceleration at the peak value.

A distinction should be made between the procedures for minimum landing distance and an ordinary landing roll with considerable excess runway available. Minimum landing distance is obtained by creating a continuous peak deceleration of the aircraft; that is, extensive use of the brakes for maximum deceleration. On the other hand, an ordinary landing roll with considerable excess runway may allow extensive use of aerodynamic drag to minimize wear and tear on the tires and brakes. If aerodynamic drag is sufficient to cause deceleration, it can be used in deference to the brakes in the early stages of the landing roll (i.e., brakes and tires suffer from continuous hard use, but aircraft aerodynamic drag is free and does not wear out with use). The use of aerodynamic drag is applicable only for deceleration to 60 or 70 percent of the touchdown speed. At speeds less than 60 to 70 percent of the touchdown speed, aerodynamic drag is so slight as to be of little use, and braking must be utilized to produce continued deceleration. Since the objective during the landing roll is to decelerate, the powerplant thrust should be the smallest possible positive value (or largest possible negative value in the case of thrust reversers).

In addition to the important factors of proper procedures, many other variables affect the landing performance. Any item that alters the landing speed or deceleration rate during the landing roll affects the landing distance.

The effect of gross weight on landing distance is one of the principal items determining the landing distance. One effect of an increased gross weight is that a greater speed is required to support the aircraft at the landing AOA and lift coefficient. For an example of the effect of a change in gross weight, a 21 percent increase in landing weight requires a ten percent increase in landing speed to support the greater weight.

When minimum landing distances are considered, braking friction forces predominate during the landing roll and, for the majority of aircraft configurations, braking friction is the main source of deceleration.

The minimum landing distance varies in direct proportion to the gross weight. For example, a ten percent increase in gross weight at landing would cause a:

- Five percent increase in landing velocity
- Ten percent increase in landing distance

A contingency of this is the relationship between weight and braking friction force.

The effect of wind on landing distance is large and deserves proper consideration when predicting landing distance. Since the aircraft lands at a particular airspeed independent of the wind, the principal effect of wind on landing distance is the change in the groundspeed at which the aircraft touches down. The effect of wind on deceleration during the landing is identical to the effect on acceleration during the takeoff.

The effect of pressure altitude and ambient temperature is to define density altitude and its effect on landing performance. An increase in density altitude increases the landing speed but does not alter the net retarding force. Thus, the aircraft at altitude lands at the same IAS as at sea level but, because of the reduced density, the TAS is greater. Since the aircraft lands at altitude with the same weight and dynamic pressure, the drag and braking friction throughout the landing roll have the same values as at sea level. As long as the condition is within the capability of the brakes, the net retarding force is unchanged, and the deceleration is the same as with the landing at sea level. Since an increase in altitude does not alter deceleration, the effect of density altitude on landing distance is due to the greater TAS.

The minimum landing distance at 5,000 feet is 16 percent greater than the minimum landing distance at sea level. The approximate increase in landing distance with altitude is approximately three and one-half percent for each 1,000 feet of altitude. Proper accounting of density altitude is necessary to accurately predict landing distance.

The effect of proper landing speed is important when runway lengths and landing distances are critical. The landing speeds specified in the AFM/POH are generally the minimum safe speeds at which the aircraft can be landed. Any attempt to land at below the specified speed may mean that the aircraft may stall, be difficult to control, or develop high rates of descent. On the other hand, an excessive speed at landing may improve the controllability slightly (especially in crosswinds) but causes an undesirable increase in landing distance.

A ten percent excess landing speed causes at least a 21 percent increase in landing distance. The excess speed places a greater working load on the brakes because of the additional kinetic energy to be dissipated. Also, the additional speed causes increased drag and lift in the normal ground attitude, and the increased lift reduces the normal force on the braking surfaces. The deceleration during this range of speed immediately after touchdown may suffer, and it is more probable for a tire to be blown out from braking at this point.

The most critical conditions of landing performance are combinations of high gross weight, high density altitude, and unfavorable wind. These conditions produce the greatest required landing distances and critical levels of energy dissipation on the brakes. In all cases, it is necessary to make an accurate prediction of minimum landing distance to compare with the available runway. A polished, professional landing procedure is necessary because the landing phase of flight accounts for more pilot-caused aircraft accidents than any other single phase of flight.

In the prediction of minimum landing distance from the AFM/POH data, the following considerations must be given:

- Pressure altitude and temperature—to define the effect of density altitude
- Gross weight—which defines the CAS for landing
- Wind—a large effect due to wind or wind component along the runway
- Runway slope and condition—relatively small correction for ordinary values of runway slope, but a significant effect of snow, ice, or soft ground

A tail wind of ten knots increases the landing distance by about 21 percent. An increase of landing speed by ten percent increases the landing distance by 20 percent. Hydroplaning makes braking ineffective until a decrease of speed that can be determined by using *Figure 11-18.*

For instance, a pilot is downwind for runway 18, and the tower asks if runway 27 could be accepted. There is a light rain and the winds are out of the east at ten knots. The pilot accepts because he or she is approaching the extended centerline of runway 27. The turn is tight and the pilot must descend (dive) to get to runway 27. After becoming aligned with the runway and at 50 feet AGL, the pilot is already 1,000 feet down the 3,500 feet runway. The airspeed is still high by about ten percent (should be at 70 knots and is at about 80 knots). The wind of ten knots is blowing from behind.

First, the airspeed being high by about ten percent (80 knots versus 70 knots), as presented in the performance chapter, results in a 20 percent increase in the landing distance. In performance planning, the pilot determined that at 70 knots the distance would be 1,600 feet. However, now it is increased by 20 percent and the required distance is now 1,920 feet.

The newly revised landing distance of 1,920 feet is also affected by the wind. In looking at *Figure 11-19*, the affect of the wind is an additional 20 percent for every ten miles per hour (mph) in wind. This is computed not on the original estimate but on the estimate based upon the increased airspeed. Now the landing distance is increased by another 320 feet for a total requirement of 2,240 feet to land the airplane after reaching 50 feet AGL.

That is the original estimate of 1,600 under planned conditions plus the additional 640 feet for excess speed and the tailwind. Given the pilot overshot the threshhold by 1,000 feet, the total length required is 3,240 on a 3,500 foot runway; 260 feet to spare. But this is in a perfect environment. Most pilots become fearful as the end of the runway is facing them just ahead. A typical pilot reaction is to brake—and brake hard. Because the aircraft does not have antilock braking features like a car, the brakes lock, and the aircraft hydroplanes on the wet surface of the runway until decreasing to a speed of about 54 knots (the square root of the tire pressure ($\sqrt{36}$) × 9). Braking is ineffective when hydroplaning.

The 260 feet that a pilot might feel is left over has long since evaporated as the aircraft hydroplaned the first 300–500 feet when the brakes locked. This is an example of a true story, but one which only changes from year to year because of new participants and aircraft with different N-numbers.

In this example, the pilot actually made many bad decisions. Bad decisions, when combined, have a synergy greater than the individual errors. Therefore, the corrective actions become larger and larger until correction is almost impossible. Aeronautical decision-making is discussed more fully in Chapter 2, Aeronautical Decision-Making (ADM).

Performance Speeds

True airspeed (TAS)—the speed of the aircraft in relation to the air mass in which it is flying.

Indicated airspeed (IAS)—the speed of the aircraft as observed on the ASI. It is the airspeed without correction for indicator, position (or installation), or compressibility errors.

Calibrated airspeed (CAS)—the ASI reading corrected for position (or installation) and instrument errors. (CAS is equal to TAS at sea level in standard atmosphere.) The color coding for various design speeds marked on ASIs may be IAS or CAS.

Equivalent airspeed (EAS)—the ASI reading corrected for position (or installation), for instrument error, and for adiabatic compressible flow for the particular altitude. (EAS is equal to CAS at sea level in standard atmosphere.)

V_{S0}—the calibrated power-off stalling speed or the minimum steady flight speed at which the aircraft is controllable in the landing configuration.

V_{S1}—the calibrated power-off stalling speed or the minimum steady flight speed at which the aircraft is controllable in a specified configuration.

V_Y—the speed at which the aircraft obtains the maximum increase in altitude per unit of time. This best ROC speed normally decreases slightly with altitude.

V_X—the speed at which the aircraft obtains the highest altitude in a given horizontal distance. This best AOC speed normally increases slightly with altitude.

V_{LE}—the maximum speed at which the aircraft can be safely flown with the landing gear extended. This is a problem involving stability and controllability.

V_{LO}—the maximum speed at which the landing gear can be safely extended or retracted. This is a problem involving the air loads imposed on the operating mechanism during extension or retraction of the gear.

V_{FE}—the highest speed permissible with the wing flaps in a prescribed extended position. This is because of the air loads imposed on the structure of the flaps.

V_A—the calibrated design maneuvering airspeed. This is the maximum speed at which the limit load can be imposed (either by gusts or full deflection of the control surfaces) without causing structural damage. Operating at or below maneuvering speed does not provide structural protection against multiple full control inputs in one axis or full control inputs in more than one axis at the same time.

V_{N0}—the maximum speed for normal operation or the maximum structural cruising speed. This is the speed at which exceeding the limit load factor may cause permanent deformation of the aircraft structure.

V_{NE}—the speed that should *never* be exceeded. If flight is attempted above this speed, structural damage or structural failure may result.

Performance Charts

Performance charts allow a pilot to predict the takeoff, climb, cruise, and landing performance of an aircraft. These charts, provided by the manufacturer, are included in the AFM/POH. Information the manufacturer provides on these charts has been gathered from test flights conducted in a new aircraft, under normal operating conditions while using average piloting skills, and with the aircraft and engine in good working order. Engineers record the flight data and create performance charts based on the behavior of the aircraft during the test flights. By using these performance charts, a pilot can determine the runway length needed to take off and land, the amount of fuel to be used during flight, and the time required to arrive at the destination. It is important to remember that the data from the charts will not be accurate if the aircraft is not in good working order or when operating under adverse conditions. Always consider the necessity to compensate for the performance numbers if the aircraft is not in good working order or piloting skills are below average.

Each aircraft performs differently and, therefore, has different performance numbers. Compute the performance of the aircraft prior to every flight, as every flight is different. (See appendix for examples of performance charts for a Cessna Model 172R and Challenger 605.)

Every chart is based on certain conditions and contains notes on how to adapt the information for flight conditions. It is important to read every chart and understand how to use it. Read the instructions provided by the manufacturer. For an explanation on how to use the charts, refer to the example provided by the manufacturer for that specific chart. *[Figure 11-20]*

The information manufacturers furnish is not standardized. Information may be contained in a table format and other information may be contained in a graph format. Sometimes combined graphs incorporate two or more graphs into one chart to compensate for multiple conditions of flight. Combined graphs allow the pilot to predict aircraft performance for variations in density altitude, weight, and winds all on one chart. Because of the vast amount of information that can be extracted from this type of chart, it is important to be very accurate in reading the chart. A small error in the beginning can lead to a large error at the end.

The remainder of this section covers performance information for aircraft in general and discusses what information the charts contain and how to extract information from the charts by direct reading and interpolation methods. Every chart contains a wealth of information that should be used when flight planning. Examples of the table, graph, and combined graph formats for all aspects of flight are discussed.

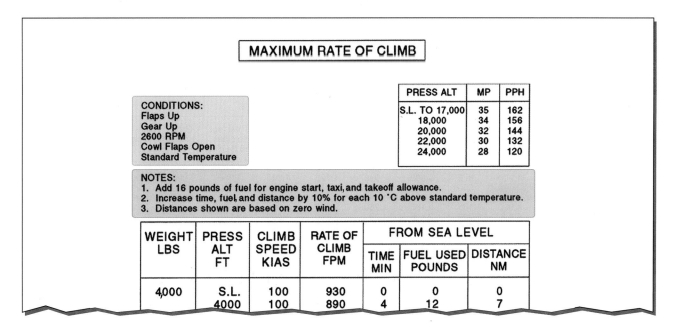

Figure 11-20. *Conditions notes chart.*

Interpolation

Not all of the information on the charts is easily extracted. Some charts require interpolation to find the information for specific flight conditions. Interpolating information means that by taking the known information, a pilot can compute intermediate information. However, pilots sometimes round off values from charts to a more conservative figure.

Using values that reflect slightly more adverse conditions provides a reasonable estimate of performance information and gives a slight margin of safety. The following illustration is an example of interpolating information from a takeoff distance chart. [Figure 11-21]

Density Altitude Charts

Use a density altitude chart to figure the density altitude at the departing airport. Using Figure 11-22, determine the density altitude based on the given information.

Sample Problem 1

Airport Elevation..5,883 feet

OAT..70 °F

Altimeter...30.10 "Hg

First, compute the pressure altitude conversion. Find 30.10 under the altimeter heading. Read across to the second column. It reads "–165." Therefore, it is necessary to subtract 165 from the airport elevation giving a pressure altitude of 5,718 feet. Next, locate the outside air temperature on the scale along the bottom of the graph. From 70°, draw a line up to the 5,718 feet pressure altitude line, which is about two-thirds of the way up between the 5,000 and 6,000 foot lines. Draw a line straight across to the far left side of the graph

and read the approximate density altitude. The approximate density altitude in thousands of feet is 7,700 feet.

Takeoff Charts

Takeoff charts are typically provided in several forms and allow a pilot to compute the takeoff distance of the aircraft with no flaps or with a specific flap configuration. A pilot can also compute distances for a no flap takeoff over a 50 foot obstacle scenario, as well as with flaps over a 50 foot obstacle. The takeoff distance chart provides for various aircraft weights, altitudes, temperatures, winds, and obstacle heights.

Sample Problem 2

Pressure Altitude..2,000 feet

OAT...22 °C

Takeoff Weight..2,600 pounds

Headwind..6 knots

Obstacle Height....................................50 foot obstacle

Refer to Figure 11-23. This chart is an example of a combined takeoff distance graph. It takes into consideration pressure altitude, temperature, weight, wind, and obstacles all on one chart. First, find the correct temperature on the bottom left side of the graph. Follow the line from 22 °C straight up until it intersects the 2,000 foot altitude line. From that point, draw a line straight across to the first dark reference line. Continue to draw the line from the reference point in a diagonal direction following the surrounding lines until it intersects the corresponding weight line. From the intersection of 2,600 pounds, draw a line straight across until it reaches the second reference line. Once again, follow the lines in a diagonal manner until it reaches the six knot headwind mark. Follow

Conditions	Flaps 10° Full throttle prior to brake release Paved level runway Zero wind			TAKEOFF DISTANCE MAXIMUM WEIGHT 2,400 LB										
Weight (lb)	Takeoff speed KIAS		Press ALT (ft)	0 °C		10 °C		20 °C		30 °C		40 °C	
	Lift off	AT 50 ft		Grnd roll (ft)	Total feet to clear 50 ft OBS	Grnd roll (ft)	Total feet to clear 50 ft OBS	Grnd roll (ft)	Total feet to clear 50 ft OBS	Grnd roll (ft)	Total feet to clear 50 ft OBS	Grnd roll (ft)	Total feet to clear 50 ft OBS
2,400	51	56	S.L.	795	1,460	860	1,570	925	1,685	995	1,810	1,065	1,945
			1,000	875	1,605	940	1,725	1,015	1,860	1,090	2,000	1,170	2,155
			2,000	960	1,770	1,035	1,910	1,115	2,060	1,200	2,220	1,290	2,395
			3,000	1,055	1,960	1,140	2,120	1,230	2,295	1,325	2,480	1,425	2,685
			4,000	1,165	2,185	1,260	2,365	1,355	2,570	1,465	2,790	1,575	3,030
			5,000	1,285	2,445	1,390	2,660	1,500	2,895	1,620	3,160	1,745	3,455
			6,000	1,425	2,755	1,540	3,015	1,665	3,300	1,800	3,620	1,940	3,990
			7,000	1,580	3,140	1,710	3,450	1,850	3,805	2,000	4,220	- - -	- - -
			8,000	1,755	3,615	1,905	4,015	2,060	4,480	- - -	- - -	- - -	- - -

To find the takeoff distance for a pressure altitude of 2,500 feet at 20 °C, average the ground roll for 2,000 feet and 3,000 feet.

$$\frac{1{,}115 + 1{,}230}{2} = 1{,}173 \text{ feet}$$

Figure 11-21. *Interpolating charts.*

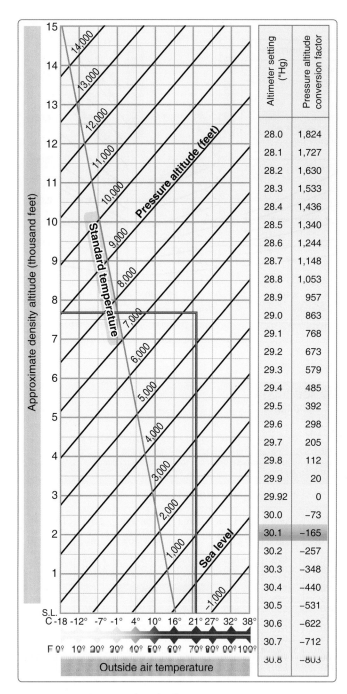

Figure 11-22. *Density altitude chart.*

straight across to the third reference line and from here, draw a line in two directions. First, draw a line straight across to figure the ground roll distance. Next, follow the diagonal lines again until they reach the corresponding obstacle height. In this case, it is a 50 foot obstacle. Therefore, draw the diagonal line to the far edge of the chart. This results in a 700 foot ground roll distance and a total distance of 1,400 feet over a 50 foot obstacle. To find the corresponding takeoff speeds at lift-off and over the 50 foot obstacle, refer to the table on the top of the chart. In this case, the lift-off speed at 2,600 pounds would be 63 knots and over the 50 foot obstacle would be 68 knots.

Sample Problem 3

Pressure Altitude..3,000 feet

OAT..30 °C

Takeoff Weight...2,400 pounds

Headwind..18 knots

Refer to *Figure 11-24*. This chart is an example of a takeoff distance table for short-field takeoffs. For this table, first find the takeoff weight. Once at 2,400 pounds, begin reading from left to right across the table. The takeoff speed is in the second column and, in the third column under pressure altitude, find the pressure altitude of 3,000 feet. Carefully follow that line to the right until it is under the correct temperature column of 30 °C. The ground roll total reads 1,325 feet and the total required to clear a 50 foot obstacle is 2,480 feet. At this point, there is an 18 knot headwind. According to the notes section under point number two, decrease the distances by ten percent for each 9 knots of headwind. With an 18 knot headwind, it is necessary to decrease the distance by 20 percent. Multiply 1,325 feet by 20 percent (1,325 × .20 = 265), subtract the product from the total distance (1,325 − 265 = 1,060). Repeat this process for the total distance over a 50 foot obstacle. The ground roll distance is 1,060 feet and the total distance over a 50 foot obstacle is 1,984 feet.

Climb and Cruise Charts

Climb and cruise chart information is based on actual flight tests conducted in an aircraft of the same type. This information is extremely useful when planning a cross-country flight to predict the performance and fuel consumption of the aircraft. Manufacturers produce several different charts for climb and cruise performance. These charts include everything from fuel, time, and distance to climb to best power setting during cruise to cruise range performance.

The first chart to check for climb performance is a fuel, time, and distance-to-climb chart. This chart gives the fuel amount used during the climb, the time it takes to accomplish the climb, and the ground distance that is covered during the climb. To use this chart, obtain the information for the departing airport and for the cruise altitude. Using *Figure 11-25,* calculate the fuel, time, and distance to climb based on the information provided.

Sample Problem 4

Departing Airport Pressure Altitude.................6,000 feet

Departing Airport OAT...25 °C

Cruise Pressure Altitude.................................10,000 feet

Cruise OAT...10 °C

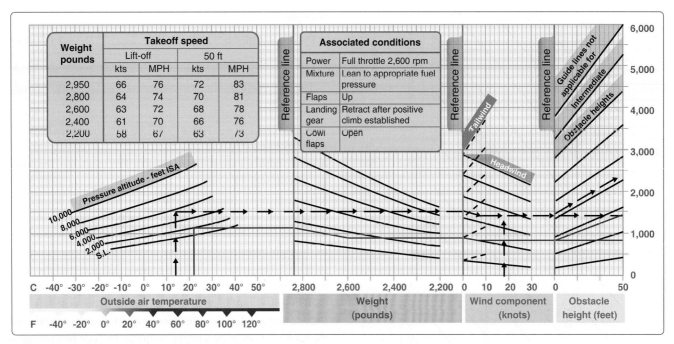

Figure 11-23. *Takeoff distance graph.*

Conditions	TAKEOFF DISTANCE

Flaps 10°
Full throttle prior to brake release
Paved level runway
Zero wind

TAKEOFF DISTANCE
MAXIMUM WEIGHT 2,400 LB

SHORT FIELD

Notes:
1. Prior to takeoff from fields above 3,000 feet elevation, the mixture should be leaned to give maximum rpm in a full throttle, static runup.
2. Decrease distances 10% for each 9 knots headwind. For operation with tailwind up to 10 knots, increase distances by 10% for each 2 knots.
3. For operation on a dry, grass runway, increase distances by 15% of the "ground roll" figure.

Weight (lb)	Takeoff speed KIAS Lift off	Takeoff speed KIAS AT 50 ft	Press ALT (ft)	0 °C Grnd roll (ft)	0 °C Total feet to clear 50 ft OBS	10 °C Grnd roll (ft)	10 °C Total feet to clear 50 ft OBS	20 °C Grnd roll (ft)	20 °C Total feet to clear 50 ft OBS	30 °C Grnd roll (ft)	30 °C Total feet to clear 50 ft OBS	40 °C Grnd roll (ft)	40 °C Total feet to clear 50 ft OBS
2,400	51	56	S.L.	795	1,460	860	1,570	925	1,685	995	1,810	1,065	1,945
			1,000	875	1,605	940	1,725	1,015	1,860	1,090	2,000	1,170	2,155
			2,000	960	1,770	1,035	1,910	1,115	2,060	1,200	2,220	1,290	2,395
			3,000	1,055	1,960	1,140	2,120	1,230	2,295	1,325	2,480	1,425	2,685
			4,000	1,165	2,185	1,260	2,365	1,355	2,570	1,465	2,790	1,575	3,030
			5,000	1,285	2,445	1,390	2,660	1,500	2,895	1,620	3,160	1,745	3,455
			6,000	1,425	2,755	1,540	3,015	1,665	3,300	1,800	3,620	1,940	3,990
			7,000	1,580	3,140	1,710	3,450	1,850	3,805	2,000	4,220	- - -	- - -
			8,000	1,755	3,615	1,905	4,015	2,060	4,480	- - -	- - -	- - -	- - -
2,200	49	54	S.L.	650	1,195	700	1,280	750	1,375	805	1,470	865	1,575
			1,000	710	1,310	765	1,405	825	1,510	885	1,615	950	1,735
			2,000	780	1,440	840	1,545	905	1,660	975	1,785	1,045	1,915
			3,000	855	1,585	925	1,705	995	1,835	1,070	1,975	1,150	2,130
			4,000	945	1,750	1,020	1,890	1,100	2,040	1,180	2,200	1,270	2,375
			5,000	1,040	1,945	1,125	2,105	1,210	2,275	1,305	2,465	1,405	2,665
			6,000	1,150	2,170	1,240	2,355	1,340	2,555	1,445	2,775	1,555	3,020
			7,000	1,270	2,440	1,375	2,655	1,485	2,890	1,605	3,155	1,730	3,450
			8,000	1,410	2,760	1,525	3,015	1,650	3,305	1,785	3,630	1,925	4,005
2,000	46	51	S.L.	525	970	565	1,035	605	1,110	650	1,185	695	1,265
			1,000	570	1,060	615	1,135	665	1,215	710	1,295	765	1,385
			2,000	625	1,160	675	1,240	725	1,330	780	1,425	840	1,525
			3,000	690	1,270	740	1,365	800	1,465	860	1,570	920	1,685
			4,000	755	1,400	815	1,500	880	1,615	945	1,735	1,015	1,865
			5,000	830	1,545	900	1,660	970	1,790	2,145	1,925	1,120	2,070
			6,000	920	1,710	990	1,845	1,070	1,990	2,405	2,145	1,235	2,315
			7,000	1,015	1,900	1,095	2,055	1,180	2,225	2,715	2,405	1,370	2,605
			8,000	1,125	2,125	1,215	2,305	1,310	2,500	1,410	2,715	1,520	2,950

Figure 11-24. *Takeoff distance short field charts.*

11-22

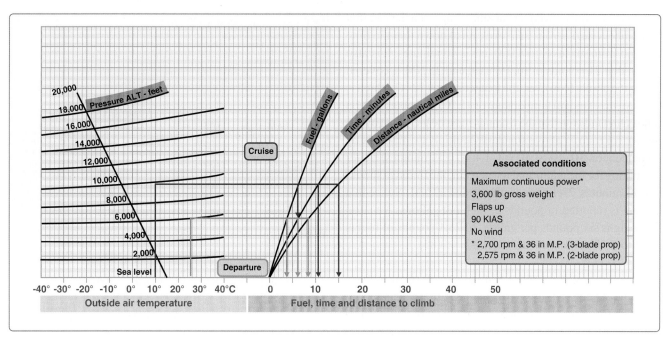

Figure 11-25. *Fuel, time, and distance climb chart.*

First, find the information for the departing airport. Find the OAT for the departing airport along the bottom, left side of the graph. Follow the line from 25 °C straight up until it intersects the line corresponding to the pressure altitude of 6,000 feet. Continue this line straight across until it intersects all three lines for fuel, time, and distance. Draw a line straight down from the intersection of altitude and fuel, altitude and time, and a third line at altitude and distance. It should read three and one-half gallons of fuel, 6 minutes of time, and nine NM. Next, repeat the steps to find the information for the cruise altitude. It should read six gallons of fuel, 10.5 minutes of time, and 15 NM. Take each set of numbers for fuel, time, and distance and subtract them from one another (6.0 – 3.5 = 2.5 gallons of fuel). It takes two and one-half gallons of fuel and 4 minutes of time to climb to 10,000 feet. During that climb, the distance covered is six NM. Remember, according to the notes at the top of the chart, these numbers do not take into account wind, and it is assumed maximum continuous power is being used.

The next example is a fuel, time, and distance-to-climb table. For this table, use the same basic criteria as for the previous chart. However, it is necessary to figure the information in a different manner. Refer to *Figure 11-26* to work the following sample problem.

Sample Problem 5

Departing Airport Pressure Altitude..................Sea level

Departing Airport OAT...22 °C

Cruise Pressure Altitude....................................8,000 feet

Takeoff Weight...3,400 pounds

To begin, find the given weight of 3,400 in the first column of the chart. Move across to the pressure altitude column to find the sea level altitude numbers. At sea level, the numbers read zero. Next, read the line that corresponds with the cruising altitude of 8,000 feet. Normally, a pilot would subtract these

Conditions	Flaps up Gear up 2,500 rpm 30 "Hg 120 PPH fuel flow Cowl flaps open Standard temperature	**NORMAL CLIMB 110 KIAS**		
Notes	1. Add 16 pounds of fuel for engine start, taxi, and takeoff allowance. 2. Increase time, fuel, and distance by 10% for each 7 °C above standard temperature. 3. Distances shown are based on zero wind.			

Weight (pounds)	Press ALT (feet)	Rate of climb fpm	From sea level		
			Time (minutes)	Fuel used (pounds)	Distance (nautical miles)
4,000	S.L.	605	0	0	0
	4,000	570	7	14	13
	8,000	530	14	28	27
	12,000	485	22	44	43
	16,000	430	31	62	63
	20,000	365	41	82	87
3,700	S.L.	700	0	0	0
	4,000	665	6	12	11
	8,000	625	12	24	23
	12,000	580	19	37	37
	16,000	525	26	52	53
	20,000	460	34	68	72
	S.L.	810	0	0	0
	4,000	775	5	10	9
3,400	8,000	735	10	21	20
	12,000	690	16	32	31
	16,000	635	22	44	45
	20,000	565	29	57	61

Figure 11-26. *Fuel time distance climb.*

two sets of numbers from one another, but given the fact that the numbers read zero at sea level, it is known that the time to climb from sea level to 8,000 feet is 10 minutes. It is also known that 21 pounds of fuel is used and 20 NM is covered during the climb. However, the temperature is 22 °C, which is 7° above the standard temperature of 15 °C. The notes section of this chart indicate that the findings must be increased by ten percent for each 7° above standard. Multiply the findings by ten percent or .10 (10 × .10 = 1, 1 + 10 = 11 minutes). After accounting for the additional ten percent, the findings should read 11 minutes, 23.1 pounds of fuel, and 22 NM. Notice that the fuel is reported in pounds of fuel, not gallons. Aviation fuel weighs six pounds per gallon, so 23.1 pounds of fuel is equal to 3.85 gallons of fuel (23.1 ÷ 6 = 3.85).

The next example is a cruise and range performance chart. This type of table is designed to give TAS, fuel consumption, endurance in hours, and range in miles at specific cruise configurations. Use *Figure 11-27* to determine the cruise and range performance under the given conditions.

Sample Problem 6

Pressure Altitude...5,000 feet

RPM...2,400 rpm

Fuel Carrying Capacity..................38 gallons, no reserve

Find 5,000 feet pressure altitude in the first column on the left side of the table. Next, find the correct rpm of 2,400 in the second column. Follow that line straight across and read the TAS of 116 mph and a fuel burn rate of 6.9 gallons per hour. As per the example, the aircraft is equipped with a fuel carrying capacity of 38 gallons. Under this column, read that the endurance in hours is 5.5 hours and the range in miles is 635 miles.

Cruise power setting tables are useful when planning cross-country flights. The table gives the correct cruise power settings, as well as the fuel flow and airspeed performance numbers at that altitude and airspeed.

Sample Problem 7

Pressure Altitude at Cruise................................6,000 feet

OAT...36 °F above standard

Refer to *Figure 11-28* for this sample problem. First, locate the pressure altitude of 6,000 feet on the far left side of the table. Follow that line across to the far right side of the table under the 20 °C (or 36 °F) column. At 6,000 feet, the rpm setting of 2,450 will maintain 65 percent continuous power at 21.0 "Hg with a fuel flow rate of 11.5 gallons per hour and airspeed of 161 knots.

Conditions	Gross weight—2,300 lb. Standard conditions Zero wind Lean mixture
Notes	Maximum cruise is normally limited to 75% power.

ALT	RPM	% BHP	TAS MPH	GAL/ Hour	38 gal (no reserve)		48 gal (no reserve)	
					Endr. hours	Range miles	Endr. hours	Range miles
2,500	2,700	86	134	9.7	3.9	525	4.9	660
	2,600	79	129	8.6	4.4	570	5.6	720
	2,500	72	123	7.8	4.9	600	6.2	760
	2,400	65	117	7.2	5.3	620	6.7	780
	2,300	58	111	6.7	5.7	630	7.2	795
	2,200	52	103	6.3	6.1	625	7.7	790
5,000	2,700	82	134	9.0	4.2	565	5.3	710
	2,600	75	128	8.1	4.7	600	5.9	760
	2,500	68	122	7.4	5.1	625	6.4	790
	2,400	61	116	6.9	5.5	635	6.9	805
	2,300	55	108	6.5	5.9	635	7.4	805
	2,200	49	100	6.0	6.3	630	7.9	795
7,500	2,700	78	133	8.4	4.5	600	5.7	755
	2,600	71	127	7.7	4.9	625	6.2	790
	2,500	64	121	7.1	5.3	645	6.7	810
	2,400	58	113	6.7	5.7	645	7.2	820
	2,300	52	105	6.2	6.1	640	7.7	810
10,000	2,650	70	129	7.6	5.0	640	6.3	810
	2,600	67	125	7.3	5.2	650	6.5	820
	2,500	61	118	6.9	5.5	655	7.0	830
	2,400	55	110	6.4	5.9	650	7.5	825
	2,300	49	100	6.0	6.3	635	8.0	800

Figure 11-27. *Cruise and range performance.*

Another type of cruise chart is a best power mixture range graph. This graph gives the best range based on power setting and altitude. Using *Figure 11-29,* find the range at 65 percent power with and without a reserve based on the provided conditions.

Sample Problem 8

OAT..Standard

Pressure Altitude...5,000 feet

First, move up the left side of the graph to 5,000 feet and standard temperature. Follow the line straight across the graph until it intersects the 65 percent line under both the reserve and no reserve categories. Draw a line straight down from both intersections to the bottom of the graph. At 65 percent power with a reserve, the range is approximately 522 miles. At 65 percent power with no reserve, the range should be 581 miles.

The last cruise chart referenced is a cruise performance graph. This graph is designed to tell the TAS performance of the airplane depending on the altitude, temperature, and power setting. Using *Figure 11-30,* find the TAS performance based on the given information.

Press ALT	ISA −20° (−36 °F)								Standard day (ISA)								ISA +20° (+36 °F)							
	IOAT		Engine speed	Man. press	Fuel flow per engine		TAS		IOAT		Engine speed	Man. press	Fuel flow per engine		TAS		IOAT		Engine speed	Man. press	Fuel flow per engine		TAS	
	°F	°C	RPM	"HG	PSI	GPH	kts	MPH	°F	°C	RPM	"HG	PSI	GPH	kts	MPH	°F	°C	RPM	"HG	PSI	GPH	kts	MPH
S.L.	27	−3	2,450	20.7	6.6	11.5	147	169	63	17	2,450	21.2	6.6	11.5	150	173	99	37	2,450	21.8	6.6	11.5	153	176
2,000	19	−7	2,450	20.4	6.6	11.5	149	171	55	13	2,450	21.0	6.6	11.5	153	176	91	33	2,450	21.5	6.6	11.5	156	180
4,000	12	−11	2,450	20.1	6.6	11.5	152	175	48	9	2,450	20.7	6.6	11.5	156	180	84	29	2,450	21.3	6.6	11.5	159	183
6,000	5	−15	2,450	19.8	6.6	11.5	155	178	41	5	2,450	20.4	6.6	11.5	158	182	79	26	2,450	21.0	6.6	11.5	161	185
8,000	−2	−19	2,450	19.5	6.6	11.5	157	181	36	2	2,450	20.2	6.6	11.5	161	185	72	22	2,450	20.8	6.6	11.5	164	189
10,000	−8	−22	2,450	19.2	6.6	11.5	160	184	28	−2	2,450	19.9	6.6	11.5	163	188	64	18	2,450	20.3	6.5	11.4	166	191
12,000	−15	−26	2,450	18.8	6.4	11.3	162	186	21	−6	2,450	18.8	6.1	10.9	163	188	57	14	2,450	18.8	5.9	10.6	163	188
14,000	−22	−30	2,450	17.4	5.8	10.5	159	183	14	−10	2,450	17.4	5.6	10.1	160	184	50	10	2,450	17.4	5.4	9.8	160	184
16,000	−29	−34	2,450	16.1	5.3	9.7	156	180	7	−14	2,450	16.1	5.1	9.4	156	180	43	6	2,450	16.1	4.9	9.1	155	178

CRUISE POWER SETTING
65% MAXIMUM CONTINUOUS POWER (OR FULL THROTTLE)
2,800 POUNDS

Notes:
1. Full throttle manifold pressure settings are approximate.
2. Shaded area represents operation with full throttle.

Figure 11-28. *Cruise power setting.*

Sample Problem 9

OAT..16 °C

Pressure Altitude..6,000 feet

Power Setting................................65 percent, best power

Wheel Fairings...Not installed

Begin by finding the correct OAT on the bottom left side of the graph. Move up that line until it intersects the pressure altitude of 6,000 feet. Draw a line straight across to the 65 percent, best power line. This is the solid line, that represents best economy. Draw a line straight down from this intersection to the bottom of the graph. The TAS at 65 percent best power is 140 knots. However, it is necessary to subtract 8 knots from the speed since there are no wheel fairings. This note is listed under the title and conditions. The TAS is 132 knots.

Crosswind and Headwind Component Chart

Every aircraft is tested according to Federal Aviation Administration (FAA) regulations prior to certification. The aircraft is tested by a pilot with average piloting skills in 90° crosswinds with a velocity up to 0.2 V_{S0} or two-tenths of the aircraft's stalling speed with power off, gear down, and flaps down. This means that if the stalling speed of the aircraft is 45 knots, it must be capable of landing in a 9-knot, 90° crosswind. The maximum demonstrated crosswind component is published in the AFM/POH. The crosswind and headwind component chart allows for figuring the headwind and crosswind component for any given wind direction and velocity.

Sample Problem 10

Runway..17

Wind...140° at 25 knots

Refer to *Figure 11-31* to solve this problem. First, determine how many degrees difference there is between the runway and the wind direction. It is known that runway 17 means a direction of 170°; from that subtract the wind direction of 140°. This gives a 30° angular difference or wind angle. Next, locate the 30° mark and draw a line from there until it intersects the correct wind velocity of 25 knots. From

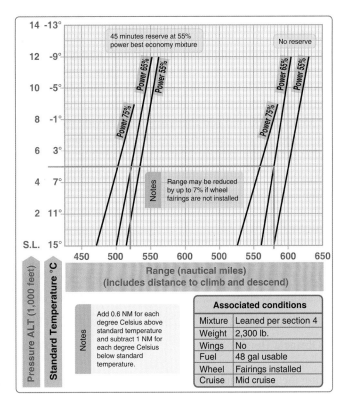

Figure 11-29. *Best power mixture range.*

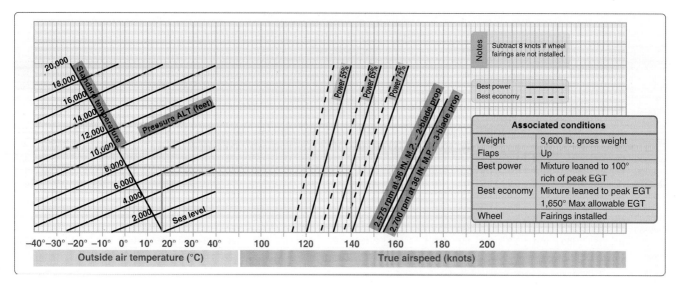

Figure 11-30. *Cruise performance graph.*

there, draw a line straight down and a line straight across. The headwind component is 22 knots and the crosswind component is 13 knots. This information is important when taking off and landing so that, first of all, the appropriate runway can be picked if more than one exists at a particular airport, but also so that the aircraft is not pushed beyond its tested limits.

Landing Charts

Landing performance is affected by variables similar to those affecting takeoff performance. It is necessary to compensate for differences in density altitude, weight of the airplane, and headwinds. Like takeoff performance charts, landing distance information is available as normal landing information, as well as landing distance over a 50 foot obstacle. As

usual, read the associated conditions and notes in order to ascertain the basis of the chart information. Remember, when calculating landing distance that the landing weight is not the same as the takeoff weight. The weight must be recalculated to compensate for the fuel that was used during the flight.

Sample Problem 11

Pressure Altitude...1,250 feet

Temperature...Standard

Refer to *Figure 10-32.* This example makes use of a landing distance table. Notice that the altitude of 1,250 feet is not on this table. It is, therefore, necessary to interpolate to find the correct landing distance. The pressure altitude of 1,250 is halfway between sea level and 2,500 feet. First, find the column for sea level and the column for 2,500 feet. Take the total distance of 1,075 for sea level and the total distance of 1,135 for 2,500 and add them together. Divide the total by two to obtain the distance for 1,250 feet. The distance is 1,105 feet total landing distance to clear a 50 foot obstacle. Repeat this process to obtain the ground roll distance for the pressure altitude. The ground roll should be 457.5 feet.

Sample Problem 12

OAT... 57 °F

Pressure Altitude.. 4,000 feet

Landing Weight...2,400 pounds

Headwind.. 6 knots

Obstacle Height.. 50 feet

Using the given conditions and *Figure 11-33*, determine the landing distance for the aircraft. This graph is an example of

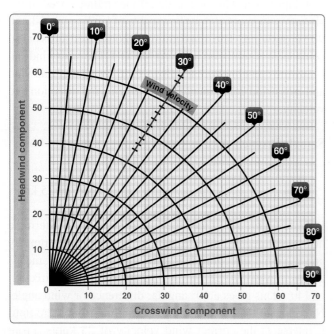

Figure 10-31. *Crosswind component chart.*

Conditions: Flaps lowered to 40° Power off Hard surface runway Zero wind		LANDING DISTANCE							
Gross weight lb	Approach speed IAS, MPH	At sea level & 59 °F		At 2,500 ft & 50 °F		At 5,000 ft & 41 °F		At 7,500 ft & 32 °F	
		Ground roll	Total to clear 50 ft OBS	Ground roll	Total to clear 50 ft OBS	Ground roll	Total to clear 50 ft OBS	Ground roll	Total to clear 50 ft OBS
1,600	60	445	1,075	470	1,135	495	1,195	520	1,255

Note:
1. Decrease the distances shown by 10% for each 4 knots of headwind.
2. Increase the distance by 10% for each 60 °F temperature increase above standard.
3. For operation on a dry, grass runway, increase distances (both "ground roll" and "total to clear 50 ft obstacle") by 20% of the "total to clear 50 ft obstacle" figure.

Figure 11-32. *Landing distance table.*

a combined landing distance graph and allows compensation for temperature, weight, headwinds, tailwinds, and varying obstacle height. Begin by finding the correct OAT on the scale on the left side of the chart. Move up in a straight line to the correct pressure altitude of 4,000 feet. From this intersection, move straight across to the first dark reference line. Follow the lines in the same diagonal fashion until the correct landing weight is reached. At 2,400 pounds, continue in a straight line across to the second dark reference line. Once again, draw a line in a diagonal manner to the correct wind component and then straight across to the third dark reference line. From this point, draw a line in two separate directions: one straight across to figure the ground roll and one in a diagonal manner to the correct obstacle height. This should be 975 feet for the total ground roll and 1,500 feet for the total distance over a 50 foot obstacle.

Stall Speed Performance Charts

Stall speed performance charts are designed to give an understanding of the speed at which the aircraft stalls in a given configuration. This type of chart typically takes into account the angle of bank, the position of the gear and flaps, and the throttle position. Use *Figure 11-34* and the accompanying conditions to find the speed at which the airplane stalls.

Sample Problem 13

Power... OFF

Flaps... Down

Gear.. Down

Angle of Bank... 45°

First, locate the correct flap and gear configuration. The bottom half of the chart should be used since the gear and

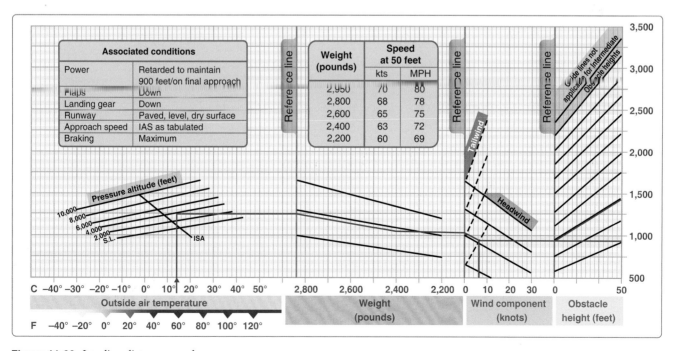

Figure 11-33. *Landing distance graph.*

Gross weight 2,750 lb			Angle of bank			
			Level	30°	45°	60°
			Gear and flaps up			
Power	On	MPH	62	67	74	88
		knots	54	58	64	76
	Off	MPH	75	81	89	106
		knots	65	70	77	92
			Gear and flaps down			
Power	On	MPH	54	58	64	76
		knots	47	50	56	66
	Off	MPH	66	71	78	93
		knots	57	62	68	81

Figure 11-34. *Stall speed table.*

flaps are down. Next, choose the row corresponding to a power-off situation. Now, find the correct angle of bank column, which is 45°. The stall speed is 78 mph, and the stall speed in knots would be 68 knots.

Performance charts provide valuable information to the pilot. By using these charts, a pilot can predict the performance of the aircraft under most flying conditions, providing a better plan for every flight. The Code of Federal Regulations (CFR) requires that a pilot be familiar with all information available prior to any flight. Pilots should use the information to their advantage as it can only contribute to safety in flight.

Transport Category Aircraft Performance

Transport category aircraft are certificated under Title 14 of the CFR (14 CFR) part 25. For additional information concerning transport category airplanes, consult the Airplane Flying Handbook, FAA-H-8083-3 (as revised).

Transport category helicopters are certificated under 14 CFR part 29.

Air Carrier Obstacle Clearance Requirements

For information on air carrier obstacle clearance requirements consult the Instrument Procedures Handbook, FAA-H-8083-16 (as revised).

Chapter Summary

Performance characteristics and capabilities vary greatly among aircraft. As transport aircraft become more capable and more complex, most operators find themselves having to rely increasingly on computerized flight mission planning systems. These systems may be on board or used during the planning phase of the flight. Moreover, aircraft weight, atmospheric conditions, and external environmental factors can significantly affect aircraft performance. It is essential that a pilot become intimately familiar with the mission planning programs, performance characteristics, and capabilities of the aircraft being flown, as well as all of the onboard computerized systems in today's complex aircraft. The primary source of this information is the AFM/POH.

Chapter 12

Weather Theory

introduction

Weather is an important factor that influences aircraft performance and flying safety. It is the state of the atmosphere at a given time and place with respect to variables, such as temperature (heat or cold), moisture (wetness or dryness), wind velocity (calm or storm), visibility (clearness or cloudiness), and barometric pressure (high or low). The term "weather" can also apply to adverse or destructive atmospheric conditions, such as high winds.

This chapter explains basic weather theory and offers pilots background knowledge of weather principles. It is designed to help them gain a good understanding of how weather affects daily flying activities. Understanding the theories behind weather helps a pilot make sound weather decisions based on the reports and forecasts obtained from a Flight Service Station (FSS) weather specialist and other aviation weather services.

Be it a local flight or a long cross-country flight, decisions based on weather can dramatically affect the safety of the flight.

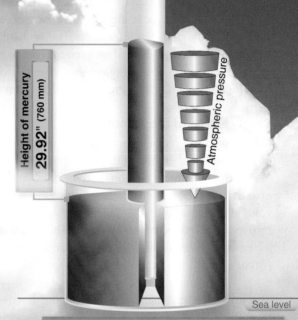

Height of mercury 29.92" (760 mm)

Atmospheric pressure

Sea level

29.92 "Hg = 1,013.2 mb (hPa) = 14.7 lb/in^2

Atmosphere

The atmosphere is a blanket of air made up of a mixture of gases that surrounds the Earth and reaches almost 350 miles from the surface of the Earth. This mixture is in constant motion. If the atmosphere were visible, it might look like an ocean with swirls and eddies, rising and falling air, and waves that travel for great distances.

Life on Earth is supported by the atmosphere, solar energy, and the planet's magnetic fields. The atmosphere absorbs energy from the sun, recycles water and other chemicals, and works with the electrical and magnetic forces to provide a moderate climate. The atmosphere also protects life on Earth from high energy radiation and the frigid vacuum of space.

Composition of the Atmosphere

In any given volume of air, nitrogen accounts for 78 percent of the gases that comprise the atmosphere, while oxygen makes up 21 percent. Argon, carbon dioxide, and traces of other gases make up the remaining one percent. This volume of air also contains some water vapor, varying from zero to about five percent by volume. This small amount of water vapor is responsible for major changes in the weather. *[Figure 12-1]*

The envelope of gases surrounding the Earth changes from the ground up. Four distinct layers or spheres of the

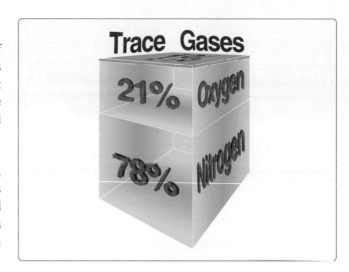

Figure 12-1. *Composition of the atmosphere.*

atmosphere have been identified using thermal characteristics (temperature changes), chemical composition, movement, and density. *[Figure 12-2]*

The first layer, known as the troposphere, extends from 6 to 20 kilometers (km) (4 to 12 miles) over the northern and southern poles and up to 48,000 feet (14.5 km) over the equatorial regions. The vast majority of weather, clouds, storms, and temperature variances occur within this first layer of the atmosphere. Inside the troposphere, the average temperature decreases at a rate of about 2 °Celsius (C) every

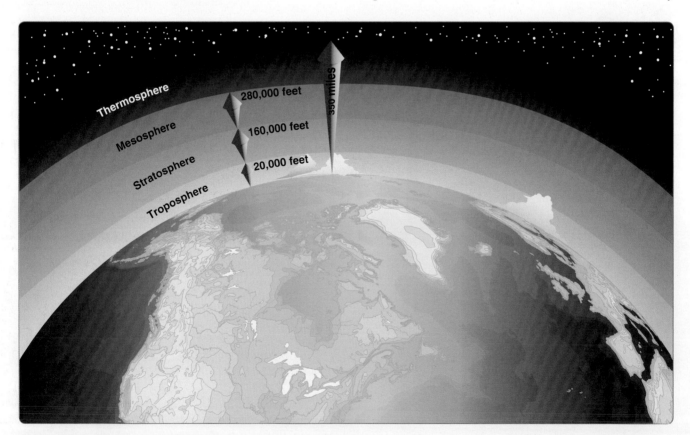

Figure 12-2. *Layers of the atmosphere.*

1,000 feet of altitude gain, and the pressure decreases at a rate of about one inch per 1,000 feet of altitude gain.

At the top of the troposphere is a boundary known as the tropopause, which traps moisture and the associated weather in the troposphere. The altitude of the tropopause varies with latitude and with the season of the year; therefore, it takes on an elliptical shape as opposed to round. Location of the tropopause is important because it is commonly associated with the location of the jet stream and possible clear air turbulence.

Above the tropopause are three more atmospheric levels. The first is the stratosphere, which extends from the tropopause to a height of about 160,000 feet (50 km). Little weather exists in this layer and the air remains stable, although certain types of clouds occasionally extend in it. Above the stratosphere are the mesosphere and thermosphere, which have little influence over weather.

Atmospheric Circulation

As noted earlier, the atmosphere is in constant motion. Certain factors combine to set the atmosphere in motion, but a major factor is the uneven heating of the Earth's surface. This heating upsets the equilibrium of the atmosphere, creating changes in air movement and atmospheric pressure. The movement of air around the surface of the Earth is called atmospheric circulation.

Heating of the Earth's surface is accomplished by several processes, but in the simple convection-only model used for this discussion, the Earth is warmed by energy radiating from the sun. The process causes a circular motion that results when warm air rises and is replaced by cooler air.

Warm air rises because heat causes air molecules to spread apart. As the air expands, it becomes less dense and lighter than the surrounding air. As air cools, the molecules pack together more closely, becoming denser and heavier than warm air. As a result, cool, heavy air tends to sink and replace warmer, rising air.

Because the Earth has a curved surface that rotates on a tilted axis while orbiting the sun, the equatorial regions of the Earth receive a greater amount of heat from the sun than the polar regions. The amount of solar energy that heats the Earth depends on the time of year and the latitude of the specific region. All of these factors affect the length of time and the angle at which sunlight strikes the surface.

Solar heating causes higher temperatures in equatorial areas, which causes the air to be less dense and rise. As the warm air flows toward the poles, it cools, becoming denser and sinks back toward the surface. *[Figure 12-3]*

Figure 12-3. *Circulation pattern in a static environment.*

Atmospheric Pressure

The unequal heating of the Earth's surface not only modifies air density and creates circulation patterns; it also causes changes in air pressure or the force exerted by the weight of air molecules. Although air molecules are invisible, they still have weight and take up space.

Imagine a sealed column of air that has a footprint of one square inch and is 350 miles high. It would take 14.7 pounds of effort to lift that column. This represents the air's weight; if the column is shortened, the pressure exerted at the bottom (and its weight) would be less.

The weight of the shortened column of air at 18,000 feet is approximately 7.4 pounds; almost 50 percent that at sea level. For instance, if a bathroom scale (calibrated for sea level) were raised to 18,000 feet, the column of air weighing 14.7 pounds at sea level would be 18,000 feet shorter and would weigh approximately 7.3 pounds (50 percent) less than at sea level. *[Figure 12-4]*

The actual pressure at a given place and time differs with altitude, temperature, and density of the air. These conditions also affect aircraft performance, especially with regard to takeoff, rate of climb, and landings.

Coriolis Force

In general atmospheric circulation theory, areas of low pressure exist over the equatorial regions and areas of high pressure exist over the polar regions due to a difference in temperature. The resulting low pressure allows the high-pressure air at the poles to flow along the planet's surface toward the equator. While this pattern of air circulation is

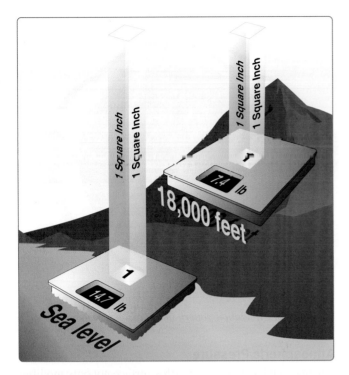

Figure 12-4. *Atmosphere weights.*

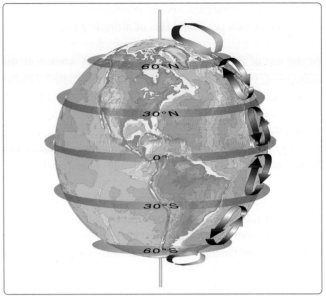

Figure 12-5. *Three-cell circulation pattern due to the rotation of the Earth.*

correct in theory, the circulation of air is modified by several forces, the most important of which is the rotation of the Earth.

The force created by the rotation of the Earth is known as the Coriolis force. This force is not perceptible to humans as they walk around because humans move slowly and travel relatively short distances compared to the size and rotation rate of the Earth. However, the Coriolis force significantly affects motion over large distances, such as an air mass or body of water.

The Coriolis force deflects air to the right in the Northern Hemisphere, causing it to follow a curved path instead of a straight line. The amount of deflection differs depending on the latitude. It is greatest at the poles and diminishes to zero at the equator. The magnitude of Coriolis force also differs with the speed of the moving body—the greater the speed, the greater the deviation. In the Northern Hemisphere, the rotation of the Earth deflects moving air to the right and changes the general circulation pattern of the air.

The Coriolis force causes the general flow to break up into three distinct cells in each hemisphere. *[Figure 12-5]* In the Northern Hemisphere, the warm air at the equator rises upward from the surface, travels northward, and is deflected eastward by the rotation of the Earth. By the time it has traveled one-third of the distance from the equator to the North Pole, it is no longer moving northward, but eastward. This air cools and sinks in a belt-like area at about 30° latitude, creating an area of high pressure as it sinks toward

the surface. Then, it flows southward along the surface back toward the equator. Coriolis force bends the flow to the right, thus creating the northeasterly trade winds that prevail from 30° latitude to the equator. Similar forces create circulation cells that encircle the Earth between 30° and 60° latitude and between 60° and the poles. This circulation pattern results in the prevailing upper level westerly winds in the conterminous United States.

Circulation patterns are further complicated by seasonal changes, differences between the surfaces of continents and oceans, and other factors such as frictional forces caused by the topography of the Earth's surface that modify the movement of the air in the atmosphere. For example, within 2,000 feet of the ground, the friction between the surface and the atmosphere slows the moving air. The wind is diverted from its path because of the frictional force. Thus, the wind direction at the surface varies somewhat from the wind direction just a few thousand feet above the Earth.

Measurement of Atmosphere Pressure

Atmospheric pressure historically was measured in inches of mercury ("Hg) by a mercurial barometer. *[Figure 12-6]* The barometer measures the height of a column of mercury inside a glass tube. A section of the mercury is exposed to the pressure of the atmosphere, which exerts a force on the mercury. An increase in pressure forces the mercury to rise inside the tube. When the pressure drops, mercury drains out of the tube decreasing the height of the column. This type of barometer is typically used in a laboratory or weather observation station, is not easily transported, and difficult to read.

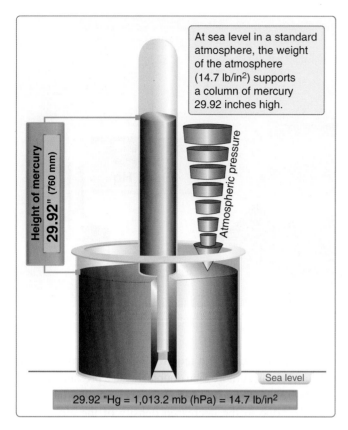

At sea level in a standard atmosphere, the weight of the atmosphere (14.7 lb/in²) supports a column of mercury 29.92 inches high.

Height of mercury 29.92" (760 mm)

Atmospheric pressure

Sea level

29.92 "Hg = 1,013.2 mb (hPa) = 14.7 lb/in²

Figure 12-6. *Although mercurial barometers are no longer used in the U. S., they are still a good historical reference for where the altimeter setting came from (inches of mercury).*

An aneroid barometer is the standard instrument used to measure pressure; it is easier to read and transport. *[Figure 12-7]* The aneroid barometer contains a closed vessel called an aneroid cell that contracts or expands with changes in pressure. The aneroid cell attaches to a pressure indicator with a mechanical linkage to provide pressure readings. The pressure sensing part of an aircraft altimeter is essentially an aneroid barometer. It is important to note that due to

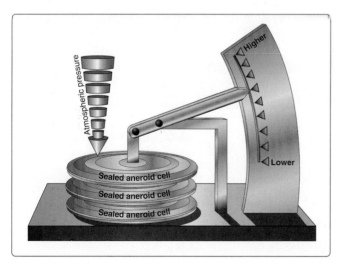

Atmospheric pressure

Higher

Lower

Sealed aneroid cell
Sealed aneroid cell
Sealed aneroid cell

Figure 12-7. *Aneroid barometer.*

the linkage mechanism of an aneroid barometer, it is not as accurate as a mercurial barometer.

To provide a common reference, the International Standard Atmosphere (ISA) has been established. These standard conditions are the basis for certain flight instruments and most aircraft performance data. Standard sea level pressure is defined as 29.92 "Hg and a standard temperature of 59 °F (15 °C). Atmospheric pressure is also reported in millibars (mb), with 1 "Hg equal to approximately 34 mb. Standard sea level pressure is 1,013.2 mb. Typical mb pressure readings range from 950.0 to 1,040.0 mb. Surface charts, high and low pressure centers, and hurricane data are reported using mb.

Since weather stations are located around the globe, all local barometric pressure readings are converted to a sea level pressure to provide a standard for records and reports. To achieve this, each station converts its barometric pressure by adding approximately 1 "Hg for every 1,000 feet of elevation. For example, a station at 5,000 feet above sea level, with a reading of 24.92 "Hg, reports a sea level pressure reading of 29.92 "Hg. *[Figure 12-8]* Using common sea level pressure readings helps ensure aircraft altimeters are set correctly, based on the current pressure readings.

By tracking barometric pressure trends across a large area, weather forecasters can more accurately predict movement of pressure systems and the associated weather. For example, tracking a pattern of rising pressure at a single weather station generally indicates the approach of fair weather. Conversely, decreasing or rapidly falling pressure usually indicates approaching bad weather and, possibly, severe storms.

Altitude and Atmospheric Pressure

As altitude increases, atmospheric pressure decreases. On average, with every 1,000 feet of increase in altitude, the atmospheric pressure decreases 1 "Hg. As pressure decreases, the air becomes less dense or thinner. This is the equivalent of being at a higher altitude and is referred to as density altitude. As pressure decreases, density altitude increases and has a pronounced effect on aircraft performance.

Differences in air density caused by changes in temperature result in a change in pressure. This, in turn, creates motion in the atmosphere, both vertically and horizontally, in the form of currents and wind. The atmosphere is almost constantly in motion as it strives to reach equilibrium. These never-ending air movements set up chain reactions that cause a continuing variety in the weather.

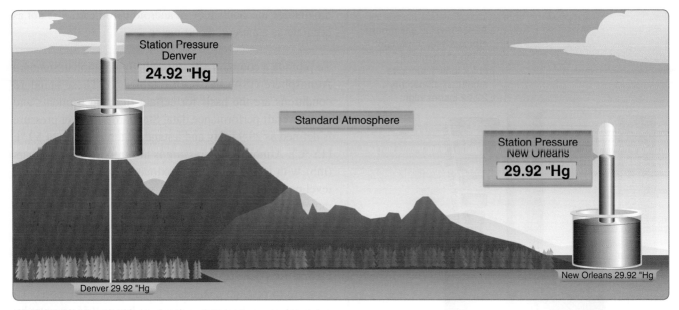

Figure 12-8. *Station pressure is converted to and reported in sea level pressure.*

Altitude and Flight

Altitude affects every aspect of flight from aircraft performance to human performance. At higher altitudes, with a decreased atmospheric pressure, takeoff and landing distances are increased, while climb rates decrease.

When an aircraft takes off, lift is created by the flow of air around the wings. If the air is thin, more speed is required to obtain enough lift for takeoff; therefore, the ground run is longer. An aircraft that requires 745 feet of ground run at sea level requires more than double that at a pressure altitude of 8,000 feet. *[Figure 12-9].* It is also true that at higher altitudes, due to the decreased density of the air, aircraft engines and propellers are less efficient. This leads to reduced rates of climb and a greater ground run for obstacle clearance.

Altitude and the Human Body

As discussed earlier, nitrogen and other trace gases make up 79 percent of the atmosphere, while the remaining 21

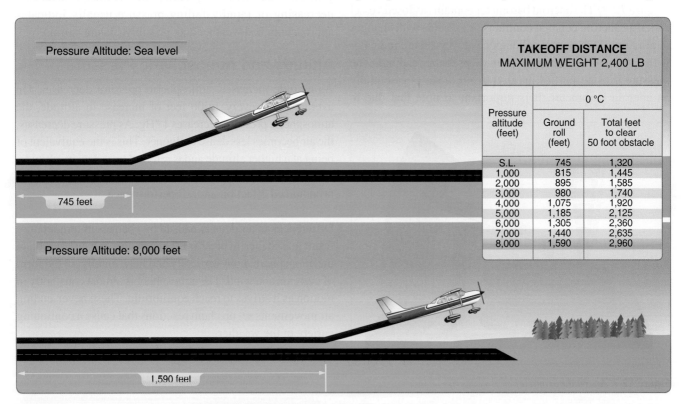

Figure 12-9. *Takeoff distances increase with increased altitude.*

percent is life sustaining atmospheric oxygen. At sea level, atmospheric pressure is great enough to support normal growth, activity, and life. By 18,000 feet, the partial pressure of oxygen is reduced and adversely affects the normal activities and functions of the human body.

The reactions of the average person become impaired at an altitude of about 10,000 feet, but for some people impairment can occur at an altitude as low as 5,000 feet. The physiological reactions to hypoxia or oxygen deprivation are insidious and affect people in different ways. These symptoms range from mild disorientation to total incapacitation, depending on body tolerance and altitude. Supplemental oxygen or cabin pressurization systems help pilots fly at higher altitudes and overcome the effects of oxygen deprivation.

Wind and Currents

Air flows from areas of high pressure into areas of low pressure because air always seeks out lower pressure. The combination of atmospheric pressure differences, Coriolis force, friction, and temperature differences of the air near the earth cause two kinds of atmospheric motion: convective currents (upward and downward motion) and wind (horizontal motion). Currents and winds are important as they affect takeoff, landing, and cruise flight operations. Most importantly, currents and winds or atmospheric circulation cause weather changes.

Wind Patterns

In the Northern Hemisphere, the flow of air from areas of high to low pressure is deflected to the right and produces a clockwise circulation around an area of high pressure. This is known as anticyclonic circulation. The opposite is true of low-pressure areas; the air flows toward a low and is deflected to create a counterclockwise or cyclonic circulation. *[Figure 12-10]*

High-pressure systems are generally areas of dry, descending air. Good weather is typically associated with high-pressure systems for this reason. Conversely, air flows into a low-pressure area to replace rising air. This air usually brings increasing cloudiness and precipitation. Thus, bad weather is commonly associated with areas of low pressure.

A good understanding of high- and low-pressure wind patterns can be of great help when planning a flight because a pilot can take advantage of beneficial tailwinds. *[Figure 12-11]* When planning a flight from west to east, favorable winds would be encountered along the northern side of a high-pressure system or the southern side of a low-pressure system. On the return flight, the most favorable winds would be along the southern side of the same high-pressure system or the northern side of a low-pressure system. An added advantage

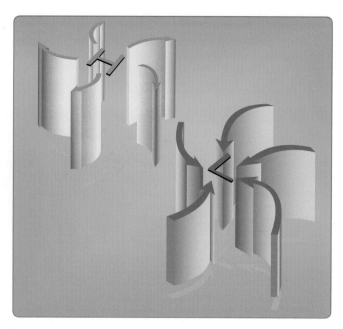

Figure 12-10. *Circulation pattern about areas of high and low pressure.*

is a better understanding of what type of weather to expect in a given area along a route of flight based on the prevailing areas of highs and lows.

While the theory of circulation and wind patterns is accurate for large scale atmospheric circulation, it does not take into account changes to the circulation on a local scale. Local conditions, geological features, and other anomalies can change the wind direction and speed close to the Earth's surface.

Convective Currents

Plowed ground, rocks, sand, and barren land absorb solar energy quickly and can therefore give off a large amount of heat; whereas, water, trees, and other areas of vegetation tend to more slowly absorb heat and give off heat. The resulting uneven heating of the air creates small areas of local circulation called convective currents.

Convective currents cause the bumpy, turbulent air sometimes experienced when flying at lower altitudes during warmer weather. On a low-altitude flight over varying surfaces, updrafts are likely to occur over pavement or barren places, and downdrafts often occur over water or expansive areas of vegetation like a group of trees. Typically, these turbulent conditions can be avoided by flying at higher altitudes, even above cumulus cloud layers. *[Figure 12-12]*

Convective currents are particularly noticeable in areas with a land mass directly adjacent to a large body of water, such as an ocean, large lake, or other appreciable area of water. During the day, land heats faster than water, so the air over the land becomes warmer and less dense. It rises and is replaced by

Figure 12-11. *Favorable winds near a high pressure system.*

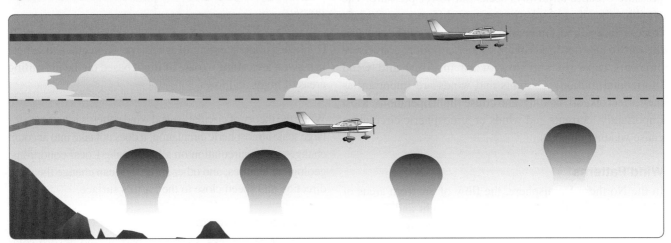

Figure 12-12. *Convective turbulence avoidance.*

cooler, denser air flowing in from over the water. This causes an onshore wind called a sea breeze. Conversely, at night land cools faster than water, as does the corresponding air. In this case, the warmer air over the water rises and is replaced by the cooler, denser air from the land, creating an offshore wind called a land breeze. This reverses the local wind circulation pattern. Convective currents can occur anywhere there is an uneven heating of the Earth's surface. *[Figure 12-13]*

Convective currents close to the ground can affect a pilot's ability to control the aircraft. For example, on final approach, the rising air from terrain devoid of vegetation sometimes produces a ballooning effect that can cause a pilot to overshoot the intended landing spot. On the other hand, an approach over a large body of water or an area of thick vegetation tends to create a sinking effect that can cause an unwary pilot to land short of the intended landing spot. *[Figure 12-14]*

Effect of Obstructions on Wind

Another atmospheric hazard exists that can create problems for pilots. Obstructions on the ground affect the flow of wind and can be an unseen danger. Ground topography and large buildings can break up the flow of the wind and create wind gusts that change rapidly in direction and speed. These obstructions range from man-made structures, like hangars, to large natural obstructions, such as mountains, bluffs, or canyons. It is especially important to be vigilant when flying in or out of airports that have large buildings or natural obstructions located near the runway. *[Figure 12-15]*

The intensity of the turbulence associated with ground obstructions depends on the size of the obstacle and the primary velocity of the wind. This can affect the takeoff and landing performance of any aircraft and can present a very serious hazard. During the landing phase of flight, an aircraft

Figure 12-13. *Sea breeze and land breeze wind circulation patterns.*

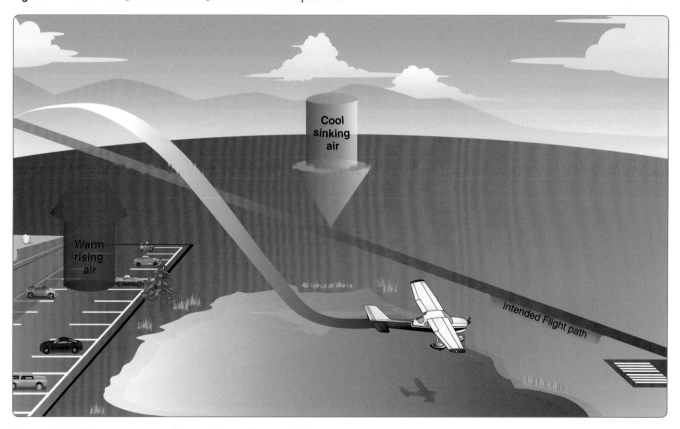

Figure 12-14. *Currents generated by varying surface conditions.*

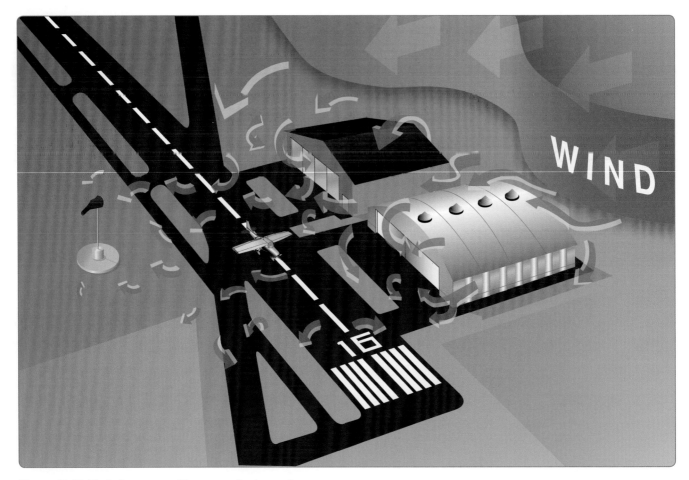

Figure 12-15. *Turbulence caused by manmade obstructions.*

may "drop in" due to the turbulent air and be too low to clear obstacles during the approach.

This same condition is even more noticeable when flying in mountainous regions. *[Figure 12-16]* While the wind flows smoothly up the windward side of the mountain and the upward currents help to carry an aircraft over the peak of the mountain, the wind on the leeward side does not act in

a similar manner. As the air flows down the leeward side of the mountain, the air follows the contour of the terrain and is increasingly turbulent. This tends to push an aircraft into the side of a mountain. The stronger the wind, the greater the downward pressure and turbulence become.

Due to the effect terrain has on the wind in valleys or canyons, downdrafts can be severe. Before conducting a flight in or

Figure 12-16. *Turbulence in mountainous regions.*

near mountainous terrain, it is helpful for a pilot unfamiliar with a mountainous area to get a checkout with a mountain qualified flight instructor.

Low-Level Wind Shear

Wind shear is a sudden, drastic change in wind speed and/or direction over a very small area. Wind shear can subject an aircraft to violent updrafts and downdrafts, as well as abrupt changes to the horizontal movement of the aircraft. While wind shear can occur at any altitude, low-level wind shear is especially hazardous due to the proximity of an aircraft to the ground. Low-level wind shear is commonly associated with passing frontal systems, thunderstorms, temperature inversions, and strong upper level winds (greater than 25 knots).

Wind shear is dangerous to an aircraft. It can rapidly change the performance of the aircraft and disrupt the normal flight attitude. For example, a tailwind quickly changing to a headwind causes an increase in airspeed and performance. Conversely, a headwind changing to a tailwind causes a decrease in airspeed and performance. In either case, a pilot must be prepared to react immediately to these changes to maintain control of the aircraft.

The most severe type of low-level wind shear, a microburst, is associated with convective precipitation into dry air at cloud base. Microburst activity may be indicated by an intense rain shaft at the surface but virga at cloud base and a ring of blowing dust is often the only visible clue. A typical microburst has a horizontal diameter of 1–2 miles and a nominal depth of 1,000 feet. The lifespan of a microburst is about 5–15 minutes during which time it can produce downdrafts of up to 6,000 feet per minute (fpm)

and headwind losses of 30–90 knots, seriously degrading performance. It can also produce strong turbulence and hazardous wind direction changes. Consider *Figure 12-17*: During an inadvertent takeoff into a microburst, the plane may first experience a performance-increasing headwind (1), followed by performance-decreasing downdrafts (2), followed by a rapidly increasing tailwind (3). This can result in terrain impact or flight dangerously close to the ground (4). An encounter during approach involves the same sequence of wind changes and could force the plane to the ground short of the runway.

The FAA has made a substantial investment in microburst accident prevention. The totally redesigned LLWAS-NE, the TDWR, and the ASR-9 WSP are skillful microburst alerting systems installed at major airports. These three systems were extensively evaluated over a 3-year period. Each was seen to issue very few false alerts and to detect microbursts well above the 90 percent detection requirement established by Congress. Many flights involve airports that lack microburst alert equipment, so the FAA has also prepared wind shear training material: Advisory Circular (AC) 00-54, FAA Pilot Wind Shear Guide. Included is information on how to recognize the risk of a microburst encounter, how to avoid an encounter, and the best flight strategy for successful escape should an encounter occur.

It is important to remember that wind shear can affect any flight and any pilot at any altitude. While wind shear may be reported, it often remains undetected and is a silent danger to aviation. Always be alert to the possibility of wind shear, especially when flying in and around thunderstorms and frontal systems.

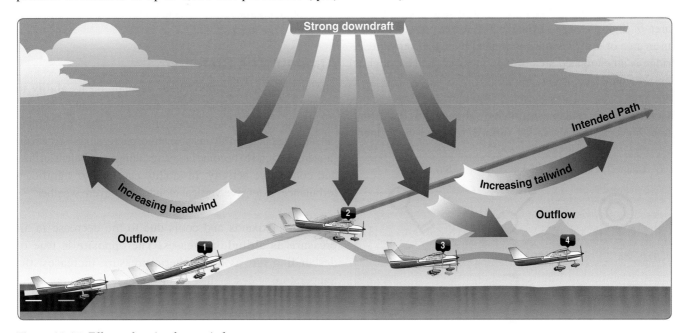

Figure 12-17. *Effects of a microburst wind.*

Wind and Pressure Representation on Surface Weather Maps

Surface weather maps provide information about fronts, areas of high and low pressure, and surface winds and pressures for each station. This type of weather map allows pilots to see the locations of fronts and pressure systems, but more importantly, it depicts the wind and pressure at the surface for each location. For more information on surface analysis and weather depiction charts, see Chapter 13, Aviation Weather Services.

Wind conditions are reported by an arrow attached to the station location circle. *[Figure 12-18]* The station circle represents the head of the arrow, with the arrow pointing in the direction from which the wind is blowing. Winds are described by the direction from which they blow, thus a northwest wind means that the wind is blowing from the northwest toward the southeast. The speed of the wind is depicted by barbs or pennants placed on the wind line. Each barb represents a speed of ten knots, while half a barb is equal to five knots, and a pennant is equal to 50 knots.

The pressure for each station is recorded on the weather chart and is shown in mb. Isobars are lines drawn on the chart to depict lines of equal pressure. These lines result in a pattern that reveals the pressure gradient or change in pressure over distance. *[Figure 12-19]* Isobars are similar to contour lines on a topographic map that indicate terrain altitudes and slope steepness. For example, isobars that are closely spaced indicate a steep pressure gradient and strong winds prevail. Shallow gradients, on the other hand, are represented by isobars that are spaced far apart and are indicative of light winds. Isobars help identify low- and high-pressure systems, as well as the location of ridges and troughs. A high is an area of high pressure surrounded by lower pressure; a low is an area of low pressure surrounded by higher pressure. A ridge is an elongated area of high pressure, and a trough is an elongated area of low pressure.

Isobars furnish valuable information about winds in the first few thousand feet above the surface. Close to the ground,

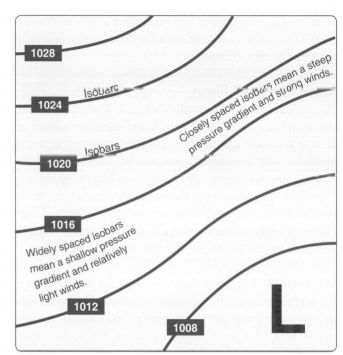

Figure 12-19. *Isobars reveal the pressure gradient of an area of high- or low-pressure areas.*

wind direction is modified by the friction and wind speed decreases due to friction with the surface. At levels 2,000 to 3,000 feet above the surface, however, the speed is greater and the direction becomes more parallel to the isobars.

Generally, the wind 2,000 feet above ground level (AGL) is 20° to 40° to the right of surface winds, and the wind speed is greater. The change of wind direction is greatest over rough terrain and least over flat surfaces, such as open water. In the absence of winds aloft information, this rule of thumb allows for a rough estimate of the wind conditions a few thousand feet above the surface.

Atmospheric Stability

The stability of the atmosphere depends on its ability to resist vertical motion. A stable atmosphere makes vertical movement difficult, and small vertical disturbances dampen out and disappear. In an unstable atmosphere, small vertical air movements tend to become larger, resulting in turbulent airflow and convective activity. Instability can lead to significant turbulence, extensive vertical clouds, and severe weather.

Rising air expands and cools due to the decrease in air pressure as altitude increases. The opposite is true of descending air; as atmospheric pressure increases, the temperature of descending air increases as it is compressed. Adiabatic heating and adiabatic cooling are terms used to describe this temperature change.

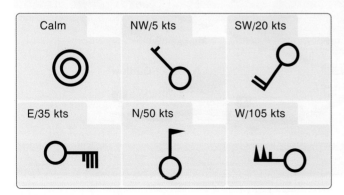

Figure 12-18. *Depiction of winds on a surface weather chart.*

The adiabatic process takes place in all upward and downward moving air. When air rises into an area of lower pressure, it expands to a larger volume. As the molecules of air expand, the temperature of the air lowers. As a result, when a parcel of air rises, pressure decreases, volume increases, and temperature decreases. When air descends, the opposite is true. The rate at which temperature decreases with an increase in altitude is referred to as its lapse rate. As air ascends through the atmosphere, the average rate of temperature change is 2 °C (3.5 °F) per 1,000 feet.

Since water vapor is lighter than air, moisture decreases air density, causing it to rise. Conversely, as moisture decreases, air becomes denser and tends to sink. Since moist air cools at a slower rate, it is generally less stable than dry air since the moist air must rise higher before its temperature cools to that of the surrounding air. The dry adiabatic lapse rate (unsaturated air) is 3 °C (5.4 °F) per 1,000 feet. The moist adiabatic lapse rate varies from 1.1 °C to 2.8 °C (2 °F to 5 °F) per 1,000 feet.

The combination of moisture and temperature determine the stability of the air and the resulting weather. Cool, dry air is very stable and resists vertical movement, which leads to good and generally clear weather. The greatest instability occurs when the air is moist and warm, as it is in the tropical regions in the summer. Typically, thunderstorms appear on a daily basis in these regions due to the instability of the surrounding air.

Inversion

As air rises and expands in the atmosphere, the temperature decreases. There is an atmospheric anomaly that can occur; however, that changes this typical pattern of atmospheric behavior. When the temperature of the air rises with altitude, a temperature inversion exists. Inversion layers are commonly shallow layers of smooth, stable air close to the ground. The temperature of the air increases with altitude to a certain point, which is the top of the inversion. The air at the top of the layer acts as a lid, keeping weather and pollutants trapped below. If the relative humidity of the air is high, it can contribute to the formation of clouds, fog, haze, or smoke resulting in diminished visibility in the inversion layer.

Surface-based temperature inversions occur on clear, cool nights when the air close to the ground is cooled by the lowering temperature of the ground. The air within a few hundred feet of the surface becomes cooler than the air above it. Frontal inversions occur when warm air spreads over a layer of cooler air, or cooler air is forced under a layer of warmer air.

Moisture and Temperature

The atmosphere, by nature, contains moisture in the form of water vapor. The amount of moisture present in the atmosphere is dependent upon the temperature of the air. Every 20 °F increase in temperature doubles the amount of moisture the air can hold. Conversely, a decrease of 20 °F cuts the capacity in half.

Water is present in the atmosphere in three states: liquid, solid, and gaseous. All three forms can readily change to another, and all are present within the temperature ranges of the atmosphere. As water changes from one state to another, an exchange of heat takes place. These changes occur through the processes of evaporation, sublimation, condensation, deposition, melting, or freezing. However, water vapor is added into the atmosphere only by the processes of evaporation and sublimation.

Evaporation is the changing of liquid water to water vapor. As water vapor forms, it absorbs heat from the nearest available source. This heat exchange is known as the latent heat of evaporation. A good example is the evaporation of human perspiration. The net effect is a cooling sensation as heat is extracted from the body. Similarly, sublimation is the changing of ice directly to water vapor, completely bypassing the liquid stage. Though dry ice is not made of water, but rather carbon dioxide, it demonstrates the principle of sublimation when a solid turns directly into vapor.

Relative Humidity

Humidity refers to the amount of water vapor present in the atmosphere at a given time. Relative humidity is the actual amount of moisture in the air compared to the total amount of moisture the air could hold at that temperature. For example, if the current relative humidity is 65 percent, the air is holding 65 percent of the total amount of moisture that it is capable of holding at that temperature and pressure. While much of the western United States rarely sees days of high humidity, relative humidity readings of 75 to 90 percent are not uncommon in the southern United States during warmer months. *[Figure 12-20]*

Temperature/Dew Point Relationship

The relationship between dew point and temperature defines the concept of relative humidity. The dew point, given in degrees, is the temperature at which the air can hold no more moisture. When the temperature of the air is reduced to the dew point, the air is completely saturated and moisture begins to condense out of the air in the form of fog, dew, frost, clouds, rain, or snow.

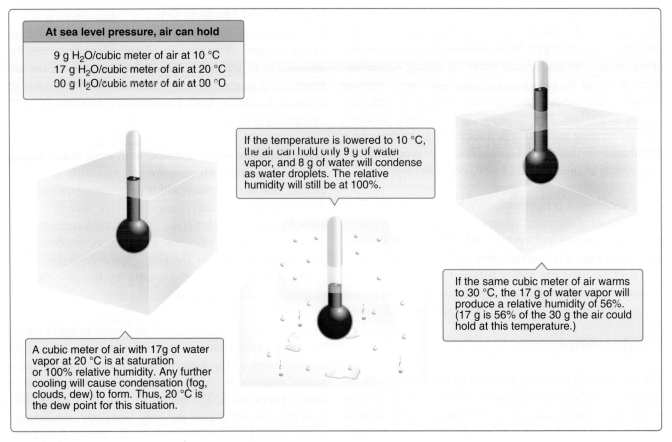

Figure 12-20. *Relationship between relative humidity, temperature, and dewpoint.*

As moist, unstable air rises, clouds often form at the altitude where temperature and dew point reach the same value. When lifted, unsaturated air cools at a rate of 5.4 °F per 1,000 feet and the dew point temperature decreases at a rate of 1 °F per 1,000 feet. This results in a convergence of temperature and dew point at a rate of 4.4 °F. Apply the convergence rate to the reported temperature and dew point to determine the height of the cloud base.

Given:

Temperature (T) = 85 °F

Dew point (DP) = 71 °F

Convergence Rate (CR) = 4.4°

T – DP = Temperature Dew Point Spread (TDS)

TDS ÷ CR = X

X × 1,000 feet = height of cloud base AGL

Example:

85 °F – 71 °F = 14 °F

14 °F ÷ 4.4 °F = 3.18

3.18 × 1,000 = 3,180 feet AGL

The height of the cloud base is 3,180 feet AGL.

Explanation:

With an outside air temperature (OAT) of 85 °F at the surface and dew point at the surface of 71 °F, the spread is 14°. Divide the temperature dew point spread by the convergence rate of 4.4 °F, and multiply by 1,000 to determine the approximate height of the cloud base.

Methods by Which Air Reaches the Saturation Point

If air reaches the saturation point while temperature and dew point are close together, it is highly likely that fog, low clouds, and precipitation will form. There are four methods by which air can reach the saturation point. First, when warm air moves over a cold surface, the air temperature drops and reaches the saturation point. Second, the saturation point may be reached when cold air and warm air mix. Third, when air cools at night through contact with the cooler ground, air reaches its saturation point. The fourth method occurs when air is lifted or is forced upward in the atmosphere.

As air rises, it uses heat energy to expand. As a result, the rising air loses heat rapidly. Unsaturated air loses heat at a rate of 3.0 °C (5.4 °F) for every 1,000 feet of altitude gain. No matter what causes the air to reach its saturation point, saturated air brings clouds, rain, and other critical weather situations.

Dew and Frost

On cool, clear, calm nights, the temperature of the ground and objects on the surface can cause temperatures of the surrounding air to drop below the dew point. When this occurs, the moisture in the air condenses and deposits itself on the ground, buildings, and other objects like cars and aircraft. This moisture is known as dew and sometimes can be seen on grass and other objects in the morning. If the temperature is below freezing, the moisture is deposited in the form of frost. While dew poses no threat to an aircraft, frost poses a definite flight safety hazard. Frost disrupts the flow of air over the wing and can drastically reduce the production of lift. It also increases drag, which when combined with lowered lift production, can adversely affect the ability to take off. An aircraft must be thoroughly cleaned and free of frost prior to beginning a flight.

Fog

Fog is a cloud that is on the surface. It typically occurs when the temperature of air near the ground is cooled to the air's dew point. At this point, water vapor in the air condenses and becomes visible in the form of fog. Fog is classified according to the manner in which it forms and is dependent upon the current temperature and the amount of water vapor in the air.

On clear nights, with relatively little to no wind present, radiation fog may develop. *[Figure 12-21]* Usually, it forms in low-lying areas like mountain valleys. This type of fog occurs when the ground cools rapidly due to terrestrial radiation, and the surrounding air temperature reaches its dew point. As the sun rises and the temperature increases, radiation fog lifts and eventually burns off. Any increase in wind also speeds the dissipation of radiation fog. If radiation fog is less than 20 feet thick, it is known as ground fog.

When a layer of warm, moist air moves over a cold surface, advection fog is likely to occur. Unlike radiation fog, wind is required to form advection fog. Winds of up to 15 knots

Figure 12-21. *Radiation fog.*

allow the fog to form and intensify; above a speed of 15 knots, the fog usually lifts and forms low stratus clouds. Advection fog is common in coastal areas where sea breezes can blow the air over cooler landmasses.

Upslope fog occurs when moist, stable air is forced up sloping land features like a mountain range. This type of fog also requires wind for formation and continued existence. Upslope and advection fog, unlike radiation fog, may not burn off with the morning sun but instead can persist for days. They can also extend to greater heights than radiation fog.

Steam fog, or sea smoke, forms when cold, dry air moves over warm water. As the water evaporates, it rises and resembles smoke. This type of fog is common over bodies of water during the coldest times of the year. Low-level turbulence and icing are commonly associated with steam fog.

Ice fog occurs in cold weather when the temperature is much below freezing and water vapor forms directly into ice crystals. Conditions favorable for its formation are the same as for radiation fog except for cold temperature, usually –25 °F or colder. It occurs mostly in the arctic regions but is not unknown in middle latitudes during the cold season.

Clouds

Clouds are visible indicators and are often indicative of future weather. For clouds to form, there must be adequate water vapor and condensation nuclei, as well as a method by which the air can be cooled. When the air cools and reaches its saturation point, the invisible water vapor changes into a visible state. Through the processes of deposition (also referred to as sublimation) and condensation, moisture condenses or sublimates onto miniscule particles of matter like dust, salt, and smoke known as condensation nuclei. The nuclei are important because they provide a means for the moisture to change from one state to another.

Cloud type is determined by its height, shape, and characteristics. They are classified according to the height of their bases as low, middle, or high clouds, as well as clouds with vertical development. *[Figure 12-22]*

Low clouds are those that form near the Earth's surface and extend up to about 6,500 feet AGL. They are made primarily of water droplets but can include supercooled water droplets that induce hazardous aircraft icing. Typical low clouds are stratus, stratocumulus, and nimbostratus. Fog is also classified as a type of low cloud formation. Clouds in this family create low ceilings, hamper visibility, and can change rapidly. Because of this, they influence flight planning and can make visual flight rules (VFR) flight impossible.

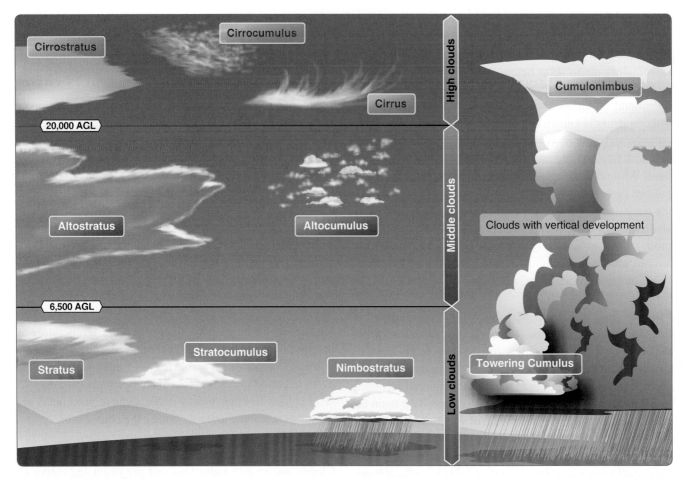

Figure 12-22. *Basic cloud types.*

Middle clouds form around 6,500 feet AGL and extend up to 20,000 feet AGL. They are composed of water, ice crystals, and supercooled water droplets. Typical middle-level clouds include altostratus and altocumulus. These types of clouds may be encountered on cross-country flights at higher altitudes. Altostratus clouds can produce turbulence and may contain moderate icing. Altocumulus clouds, which usually form when altostratus clouds are breaking apart, also may contain light turbulence and icing.

High clouds form above 20,000 feet AGL and usually form only in stable air. They are made up of ice crystals and pose no real threat of turbulence or aircraft icing. Typical high level clouds are cirrus, cirrostratus, and cirrocumulus.

Clouds with extensive vertical development are cumulus clouds that build vertically into towering cumulus or cumulonimbus clouds. The bases of these clouds form in the low to middle cloud base region but can extend into high altitude cloud levels. Towering cumulus clouds indicate areas of instability in the atmosphere, and the air around and inside them is turbulent. These types of clouds often develop into cumulonimbus clouds or thunderstorms. Cumulonimbus clouds contain large amounts of moisture and unstable air

and usually produce hazardous weather phenomena, such as lightning, hail, tornadoes, gusty winds, and wind shear. These extensive vertical clouds can be obscured by other cloud formations and are not always visible from the ground or while in flight. When this happens, these clouds are said to be embedded, hence the term, embedded thunderstorms.

To pilots, the cumulonimbus cloud is perhaps the most dangerous cloud type. It appears individually or in groups and is known as either an air mass or orographic thunderstorm. Heating of the air near the Earth's surface creates an air mass thunderstorm; the upslope motion of air in the mountainous regions causes orographic thunderstorms. Cumulonimbus clouds that form in a continuous line are nonfrontal bands of thunderstorms or squall lines.

Since rising air currents cause cumulonimbus clouds, they are extremely turbulent and pose a significant hazard to flight safety. For example, if an aircraft enters a thunderstorm, the aircraft could experience updrafts and downdrafts that exceed 3,000 fpm. In addition, thunderstorms can produce large hailstones, damaging lightning, tornadoes, and large quantities of water, all of which are potentially hazardous to aircraft.

Cloud classification can be further broken down into specific cloud types according to the outward appearance and cloud composition. Knowing these terms can help a pilot identify visible clouds.

The following is a list of cloud classifications:

- Cumulus—heaped or piled clouds
- Stratus—formed in layers
- Cirrus—ringlets, fibrous clouds, also high level clouds above 20,000 feet
- Castellanus—common base with separate vertical development, castle-like
- Lenticularus—lens-shaped, formed over mountains in strong winds
- Nimbus—rain-bearing clouds
- Fracto—ragged or broken
- Alto—middle level clouds existing at 5,000 to 20,000 feet

Ceiling

For aviation purposes, a ceiling is the lowest layer of clouds reported as being broken or overcast, or the vertical visibility into an obscuration like fog or haze. Clouds are reported as broken when five-eighths to seven-eighths of the sky is covered with clouds. Overcast means the entire sky is covered with clouds. Current ceiling information is reported by the aviation routine weather report (METAR) and automated weather stations of various types.

Visibility

Closely related to cloud cover and reported ceilings is visibility information. Visibility refers to the greatest horizontal distance at which prominent objects can be viewed with the naked eye. Current visibility is also reported in METAR and other aviation weather reports, as well as by automated weather systems. Visibility information, as predicted by meteorologists, is available for a pilot during a preflight weather briefing.

Precipitation

Precipitation refers to any type of water particles that form in the atmosphere and fall to the ground. It has a profound impact on flight safety. Depending on the form of precipitation, it can reduce visibility, create icing situations, and affect landing and takeoff performance of an aircraft.

Precipitation occurs because water or ice particles in clouds grow in size until the atmosphere can no longer support them. It can occur in several forms as it falls toward the Earth, including drizzle, rain, ice pellets, hail, snow, and ice.

Drizzle is classified as very small water droplets, smaller than 0.02 inches in diameter. Drizzle usually accompanies fog or low stratus clouds. Water droplets of larger size are referred to as rain. Rain that falls through the atmosphere but evaporates prior to striking the ground is known as virga. Freezing rain and freezing drizzle occur when the temperature of the surface is below freezing; the rain freezes on contact with the cooler surface.

If rain falls through a temperature inversion, it may freeze as it passes through the underlying cold air and fall to the ground in the form of ice pellets. Ice pellets are an indication of a temperature inversion and that freezing rain exists at a higher altitude. In the case of hail, freezing water droplets are carried up and down by drafts inside cumulonimbus clouds, growing larger in size as they come in contact with more moisture. Once the updrafts can no longer hold the freezing water, it falls to the Earth in the form of hail. Hail can be pea sized, or it can grow as large as five inches in diameter, larger than a softball.

Snow is precipitation in the form of ice crystals that falls at a steady rate or in snow showers that begin, change in intensity, and end rapidly. Snow also varies in size, from very small grains to large flakes. Snow grains are the equivalent of drizzle in size.

Precipitation in any form poses a threat to safety of flight. Often, precipitation is accompanied by low ceilings and reduced visibility. Aircraft that have ice, snow, or frost on their surfaces must be carefully cleaned prior to beginning a flight because of the possible airflow disruption and loss of lift. Rain can contribute to water in the fuel tanks. Precipitation can create hazards on the runway surface itself, making takeoffs and landings difficult, if not impossible, due to snow, ice, or pooling water and very slick surfaces.

Air Masses

Air masses are classified according to the regions where they originate. They are large bodies of air that take on the characteristics of the surrounding area or source region. A source region is typically an area in which the air remains relatively stagnant for a period of days or longer. During this time of stagnation, the air mass takes on the temperature and moisture characteristics of the source region. Areas of stagnation can be found in polar regions, tropical oceans, and dry deserts. Air masses are generally identified as polar or tropical based on temperature characteristics and maritime or continental based on moisture content.

A continental polar air mass forms over a polar region and brings cool, dry air with it. Maritime tropical air masses form over warm tropical waters like the Caribbean Sea and bring warm, moist air. As the air mass moves from its source region and passes over land or water, the air mass is subjected to the varying conditions of the land or water which modify the nature of the air mass. *[Figure 12-23]*

An air mass passing over a warmer surface is warmed from below, and convective currents form, causing the air to rise. This creates an unstable air mass with good surface visibility. Moist, unstable air causes cumulus clouds, showers, and turbulence to form.

Conversely, an air mass passing over a colder surface does not form convective currents but instead creates a stable air mass with poor surface visibility. The poor surface visibility is due to the fact that smoke, dust, and other particles cannot rise out of the air mass and are instead trapped near the surface. A stable air mass can produce low stratus clouds and fog.

Fronts

As an air mass moves across bodies of water and land, it eventually comes in contact with another air mass with different characteristics. The boundary layer between two types of air masses is known as a front. An approaching front of any type always means changes to the weather are imminent.

There are four types of fronts that are named according to the temperature of the advancing air relative to the temperature of the air it is replacing: *[Figure 12-24]*

- Warm
- Cold
- Stationary
- Occluded

Any discussion of frontal systems must be tempered with the knowledge that no two fronts are the same. However, generalized weather conditions are associated with a specific type of front that helps identify the front.

Warm Front

A warm front occurs when a warm mass of air advances and replaces a body of colder air. Warm fronts move slowly, typically 10 to 25 miles per hour (mph). The slope of the advancing front slides over the top of the cooler air and gradually pushes it out of the area. Warm fronts contain warm air that often has very high humidity. As the warm air is lifted, the temperature drops and condensation occurs.

Generally, prior to the passage of a warm front, cirriform or stratiform clouds, along with fog, can be expected to form along the frontal boundary. In the summer months, cumulonimbus clouds (thunderstorms) are likely to develop.

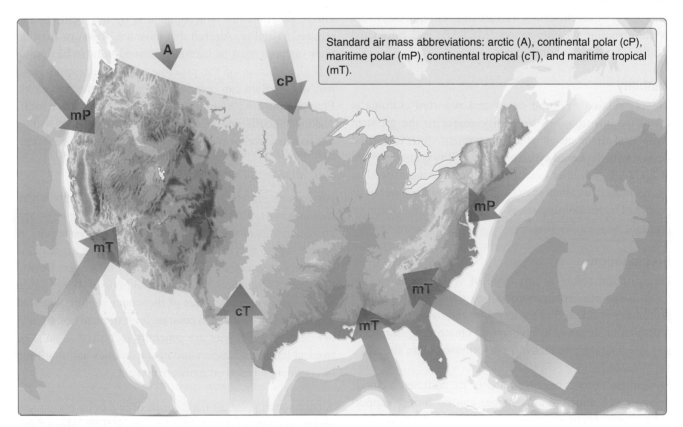

Standard air mass abbreviations: arctic (A), continental polar (cP), maritime polar (mP), continental tropical (cT), and maritime tropical (mT).

Figure 12-23. *North American air mass source regions.*

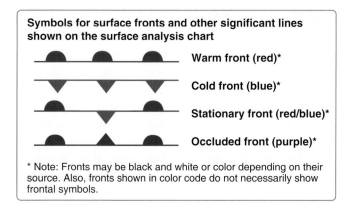

Figure 12-24. *Common chart symbology to depict weather front location.*

Light to moderate precipitation is probable, usually in the form of rain, sleet, snow, or drizzle, accentuated by poor visibility. The wind blows from the south-southeast, and the outside temperature is cool or cold with an increasing dew point. Finally, as the warm front approaches, the barometric pressure continues to fall until the front passes completely.

During the passage of a warm front, stratiform clouds are visible and drizzle may be falling. The visibility is generally poor, but improves with variable winds. The temperature rises steadily from the inflow of relatively warmer air. For the most part, the dew point remains steady and the pressure levels off. After the passage of a warm front, stratocumulus clouds predominate and rain showers are possible. The visibility eventually improves, but hazy conditions may exist for a short period after passage. The wind blows from the south-southwest. With warming temperatures, the dew point rises and then levels off. There is generally a slight rise in barometric pressure, followed by a decrease of barometric pressure.

Flight Toward an Approaching Warm Front

By studying a typical warm front, much can be learned about the general patterns and atmospheric conditions that exist when a warm front is encountered in flight. *Figure 12-25* depicts a warm front advancing eastward from St. Louis, Missouri, toward Pittsburgh, Pennsylvania during a flight from Pittsburgh to St. Louis.

At the time of departure from Pittsburgh, the weather is good VFR with a scattered layer of cirrus clouds at 15,000 feet. As the flight progresses westward to Columbus and closer to the oncoming warm front, the clouds deepen and become increasingly stratiform in appearance with a ceiling of 6,000 feet. The visibility decreases to six miles in haze with a falling

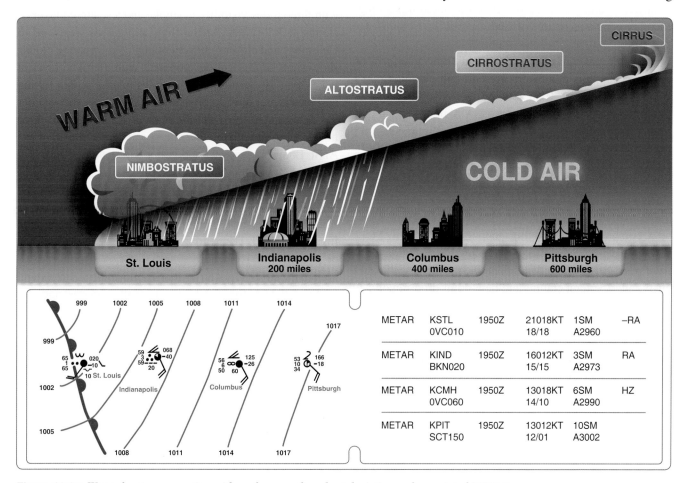

Figure 12-25. *Warm front cross-section with surface weather chart depiction and associated METAR.*

barometric pressure. Approaching Indianapolis, the weather deteriorates to broken clouds at 2,000 feet with three miles visibility and rain. With the temperature and dew point the same, fog is likely to develop. At St. Louis, the sky is overcast with low clouds and drizzle and the visibility is one mile. Beyond Indianapolis, the ceiling and visibility are too low to continue VFR. Therefore, it would be wise to remain in Indianapolis until the warm front passes, which may take up to two days.

Cold Front

A cold front occurs when a mass of cold, dense, and stable air advances and replaces a body of warmer air.

Cold fronts move more rapidly than warm fronts, progressing at a rate of 25 to 30 mph. However, extreme cold fronts have been recorded moving at speeds of up to 60 mph. A typical cold front moves in a manner opposite that of a warm front. It is so dense, it stays close to the ground and acts like a snowplow, sliding under the warmer air and forcing the less dense air aloft. The rapidly ascending air causes the temperature to decrease suddenly, forcing the creation of clouds. The type of clouds that form depends on the stability of the warmer air mass. A cold front in the Northern Hemisphere is normally oriented in a northeast to southwest manner and can be several hundred miles long, encompassing a large area of land.

Prior to the passage of a typical cold front, cirriform or towering cumulus clouds are present, and cumulonimbus clouds may develop. Rain showers may also develop due to the rapid development of clouds. A high dew point and falling barometric pressure are indicative of imminent cold front passage.

As the cold front passes, towering cumulus or cumulonimbus clouds continue to dominate the sky. Depending on the intensity of the cold front, heavy rain showers form and may be accompanied by lightning, thunder, and/or hail. More severe cold fronts can also produce tornadoes. During cold front passage, the visibility is poor with winds variable and gusty, and the temperature and dew point drop rapidly. A quickly falling barometric pressure bottoms out during frontal passage, then begins a gradual increase.

After frontal passage, the towering cumulus and cumulonimbus clouds begin to dissipate to cumulus clouds with a corresponding decrease in the precipitation. Good visibility eventually prevails with the winds from the west-northwest. Temperatures remain cooler and the barometric pressure continues to rise.

Fast-Moving Cold Front

Fast-moving cold fronts are pushed by intense pressure systems far behind the actual front. The friction between the ground and the cold front retards the movement of the front and creates a steeper frontal surface. This results in a very narrow band of weather, concentrated along the leading edge of the front. If the warm air being overtaken by the cold front is relatively stable, overcast skies and rain may occur for some distance behind the front. If the warm air is unstable, scattered thunderstorms and rain showers may form. A continuous line of thunderstorms, or squall line, may form along or ahead of the front. Squall lines present a serious hazard to pilots as squall-type thunderstorms are intense and move quickly. Behind a fast-moving cold front, the skies usually clear rapidly, and the front leaves behind gusty, turbulent winds and colder temperatures.

Flight Toward an Approaching Cold Front

Like warm fronts, not all cold fronts are the same. Examining a flight toward an approaching cold front, pilots can get a better understanding of the type of conditions that can be encountered in flight. *Figure 12-26* depicts a flight from Pittsburgh, Pennsylvania, toward St. Louis, Missouri.

At the time of departure from Pittsburgh, the weather is VFR with three miles visibility in smoke and a scattered layer of clouds at 3,500 feet. As the flight progresses westward to Columbus and closer to the oncoming cold front, the clouds show signs of vertical development with a broken layer at 2,500 feet. The visibility is six miles in haze with a falling barometric pressure. Approaching Indianapolis, the weather has deteriorated to overcast clouds at 1,000 feet and three miles visibility with thunderstorms and heavy rain showers. At St. Louis, the weather gets better with scattered clouds at 1,000 feet and a ten mile visibility.

A pilot using sound judgment based on the knowledge of frontal conditions will likely remain in Indianapolis until the front has passed. Trying to fly below a line of thunderstorms or a squall line is hazardous, and flight over the top of or around the storm is not an option. Thunderstorms can extend up to well over the capability of small airplanes and can extend in a line for 300 to 500 miles.

Comparison of Cold and Warm Fronts

Warm fronts and cold fronts are very different in nature as are the hazards associated with each front. They vary in speed, composition, weather phenomenon, and prediction. Cold fronts, which move at 20 to 35 mph, travel faster than warm fronts, which move at only 10 to 25 mph. Cold fronts also possess a

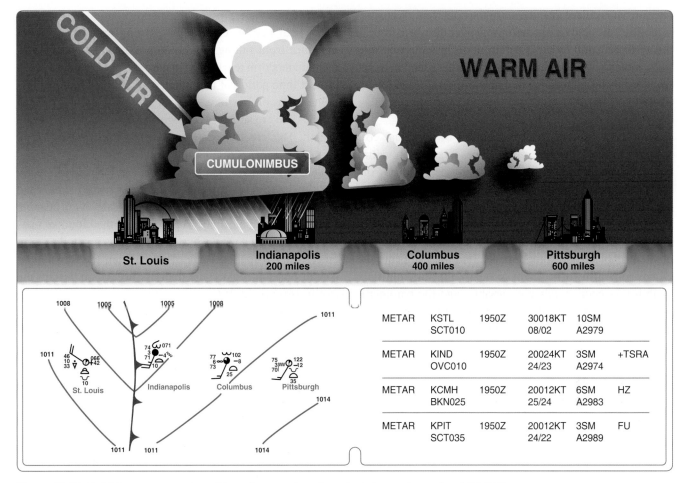

Figure 12-26. *Cold front cross-section with surface weather chart depiction and associated METAR.*

steeper frontal slope. Violent weather activity is associated with cold fronts, and the weather usually occurs along the frontal boundary, not in advance. However, squall lines can form during the summer months as far as 200 miles in advance of a strong cold front. Whereas warm fronts bring low ceilings, poor visibility, and rain, cold fronts bring sudden storms, gusty winds, turbulence, and sometimes hail or tornadoes.

Cold fronts are fast approaching with little or no warning, and they bring about a complete weather change in just a few hours. The weather clears rapidly after passage and drier air with unlimited visibilities prevail. Warm fronts, on the other hand, provide advance warning of their approach and can take days to pass through a region.

Wind Shifts

Wind around a high-pressure system rotates clockwise, while low-pressure winds rotate counter-clockwise. When two high pressure systems are adjacent, the winds are almost in direct opposition to each other at the point of contact. Fronts are the boundaries between two areas of high pressure, and therefore, wind shifts are continually occurring within a front. Shifting wind direction is most pronounced in conjunction with cold fronts.

Stationary Front

When the forces of two air masses are relatively equal, the boundary or front that separates them remains stationary and influences the local weather for days. This front is called a stationary front. The weather associated with a stationary front is typically a mixture that can be found in both warm and cold fronts.

Occluded Front

An occluded front occurs when a fast-moving cold front catches up with a slow-moving warm front. As the occluded front approaches, warm front weather prevails but is immediately followed by cold front weather. There are two types of occluded fronts that can occur, and the temperatures of the colliding frontal systems play a large part in defining the type of front and the resulting weather. A cold front occlusion occurs when a fast moving cold front is colder than the air ahead of the slow moving warm front. When this occurs, the cold air replaces the cool air and forces the warm front aloft into the atmosphere. Typically, the cold front occlusion creates a mixture of weather found in both warm and cold fronts, providing the air is relatively stable. A warm front occlusion occurs when the air ahead of the warm front is colder than the air of the cold front. When this is the case, the cold front rides up and over the warm front. If the air forced aloft by the warm front occlusion is unstable, the weather is more severe than the weather found in a cold front occlusion. Embedded thunderstorms, rain, and fog are likely to occur.

Figure 12-27 depicts a cross-section of a typical cold front occlusion. The warm front slopes over the prevailing cooler air and produces the warm front type weather. Prior to the passage of the typical occluded front, cirriform and stratiform clouds prevail, light to heavy precipitation falls, visibility is poor, dew point is steady, and barometric pressure drops. During the passage of the front, nimbostratus and cumulonimbus clouds predominate, and towering cumulus clouds may also form. Light to heavy precipitation falls, visibility is poor, winds are variable, and the barometric pressure levels off. After the passage of the front, nimbostratus and altostratus clouds are visible, precipitation decreases, and visibility improves.

Thunderstorms

A thunderstorm makes its way through three distinct stages before dissipating. It begins with the cumulus stage, in which lifting action of the air begins. If sufficient moisture and instability are present, the clouds continue to increase in vertical height. Continuous, strong updrafts prohibit moisture from falling. Within approximately 15 minutes, the thunderstorm reaches the mature stage, which is the most violent time period of the thunderstorm's life cycle. At this point, drops of moisture, whether rain or ice, are too heavy for the cloud to support and begin falling in the form of rain or hail. This creates a downward motion of the air. Warm, rising air; cool, precipitation-induced descending air; and violent turbulence all exist within and near the cloud. Below the cloud, the down-rushing air increases surface winds and decreases the temperature. Once the vertical motion near the top of the cloud slows down, the top of the cloud spreads out and takes on an anvil-like shape. At this point, the storm enters the dissipating stage. This is when the downdrafts spread out and replace the updrafts needed to sustain the storm. *[Figure 12-28]*

It is impossible to fly over thunderstorms in light aircraft. Severe thunderstorms can punch through the tropopause and reach staggering heights of 50,000 to 60,000 feet depending on latitude. Flying under thunderstorms can subject aircraft to rain, hail, damaging lightning, and violent turbulence. A good rule of thumb is to circumnavigate thunderstorms identified as severe or giving an extreme radar echo by at

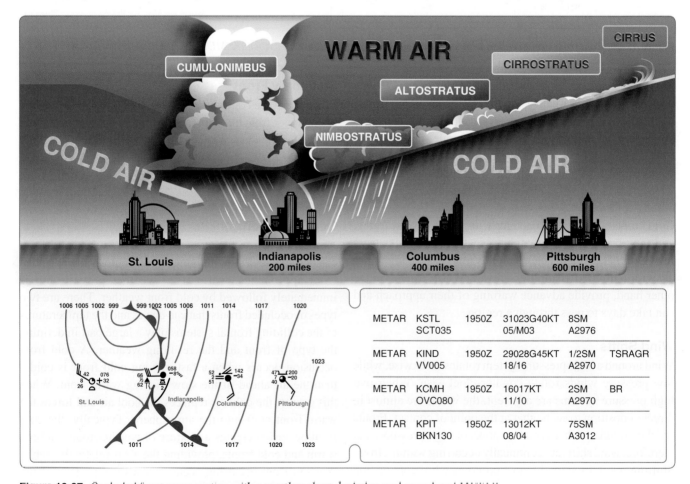

Figure 12-27. *Occluded front cross-section with a weather chart depiction and associated METAR.*

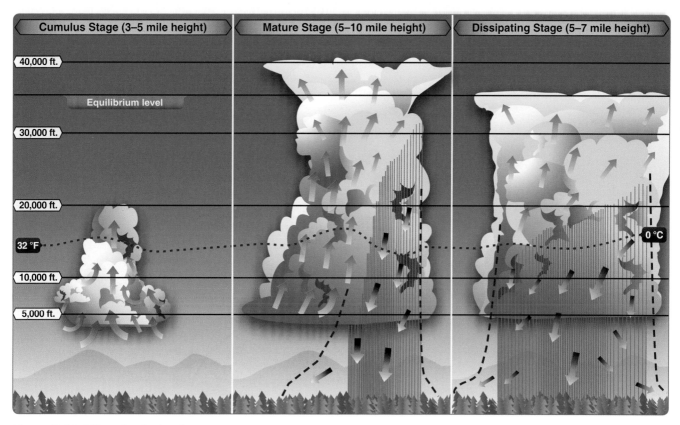

| Cumulus Stage (3–5 mile height) | Mature Stage (5–10 mile height) | Dissipating Stage (5–7 mile height) |

40,000 ft.

Equilibrium level

30,000 ft.

20,000 ft.

32 °F

10,000 ft.

5,000 ft.

0 °C

Figure 12-28. *Life cycle of a thunderstorm.*

least 20 nautical miles (NM) since hail may fall for miles outside of the clouds. If flying around a thunderstorm is not an option, stay on the ground until it passes.

For a thunderstorm to form, the air must have sufficient water vapor, an unstable lapse rate, and an initial lifting action to start the storm process. Some storms occur at random in unstable air, last for only an hour or two, and produce only moderate wind gusts and rainfall. These are known as air mass thunderstorms and are generally a result of surface heating. Steady-state thunderstorms are associated with weather systems. Fronts, converging winds, and troughs aloft force upward motion spawning these storms that often form into squall lines. In the mature stage, updrafts become stronger and last much longer than in air mass storms, hence the name steady state. *[Figure 12-29]*

Knowledge of thunderstorms and the hazards associated with them is critical to the safety of flight.

Hazards

All thunderstorms have conditions that are a hazard to aviation. These hazards occur in numerous combinations. While not every thunderstorm contains all hazards, it is not possible to visually determine which hazards a thunderstorm contains.

Squall Line

A squall line is a narrow band of active thunderstorms. Often it develops on or ahead of a cold front in moist, unstable air, but it may develop in unstable air far removed from any front. The line may be too long to detour easily and too wide and severe to penetrate. It often contains steady-state thunderstorms and presents the single most intense weather hazard to aircraft. It usually forms rapidly, generally reaching maximum intensity during the late afternoon and the first few hours of darkness.

Tornadoes

The most violent thunderstorms draw air into their cloud bases with great vigor. If the incoming air has any initial rotating motion, it often forms an extremely concentrated vortex from the surface well into the cloud. Meteorologists have estimated that wind in such a vortex can exceed 200 knots with pressure inside the vortex quite low. The strong winds gather dust and debris and the low pressure generates a funnel-shaped cloud extending downward from the cumulonimbus base. If the cloud does not reach the surface, it is a funnel cloud; if it touches a land surface, it is a tornado; and if it touches water, it is a "waterspout."

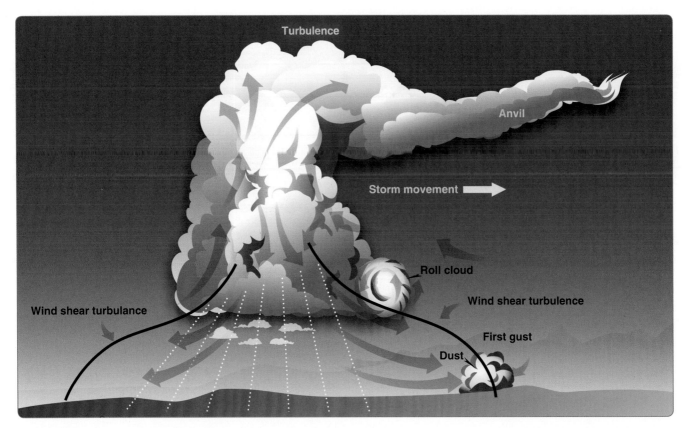

Figure 12-29. *Movement and turbulence of a maturing thunderstorm.*

Tornadoes occur with both isolated and squall line thunderstorms. Reports for forecasts of tornadoes indicate that atmospheric conditions are favorable for violent turbulence. An aircraft entering a tornado vortex is almost certain to suffer loss of control and structural damage. Since the vortex extends well into the cloud, any pilot inadvertently caught on instruments in a severe thunderstorm could encounter a hidden vortex.

Families of tornadoes have been observed as appendages of the main cloud extending several miles outward from the area of lightning and precipitation. Thus, any cloud connected to a severe thunderstorm carries a threat of violence.

Turbulence

Potentially hazardous turbulence is present in all thunderstorms, and a severe thunderstorm can destroy an aircraft. Strongest turbulence within the cloud occurs with shear between updrafts and downdrafts. Outside the cloud, shear turbulence has been encountered several thousand feet above and 20 miles laterally from a severe storm. A low-level turbulent area is the shear zone associated with the gust front. Often, a "roll cloud" on the leading edge of a storm marks the top of the eddies in this shear, and it signifies an extremely turbulent zone. Gust fronts often move far ahead (up to 15 miles) of associated precipitation. The gust front causes a rapid, and sometimes drastic, change in surface wind ahead

of an approaching storm. Advisory Circular (AC) 00-54, Pilot Windshear Guide, explains gust front hazards associated with thunderstorms. Figure 2 in the AC shows a cross section of a mature stage thunderstorm with a gust front area where very serious turbulence may be encountered.

Icing

Updrafts in a thunderstorm support abundant liquid water with relatively large droplet sizes. When carried above the freezing level, the water becomes supercooled. When temperature in the upward current cools to about –15 °C, much of the remaining water vapor sublimates as ice crystals. Above this level, at lower temperatures, the amount of supercooled water decreases.

Supercooled water freezes on impact with an aircraft. Clear icing can occur at any altitude above the freezing level, but at high levels, icing from smaller droplets may be rime or mixed rime and clear ice. The abundance of large, supercooled water droplets makes clear icing very rapid between 0 °C and –15 °C and encounters can be frequent in a cluster of cells. Thunderstorm icing can be extremely hazardous.

Thunderstorms are not the only area where pilots could encounter icing conditions. Pilots should be alert for icing anytime the temperature approaches 0 °C and visible moisture is present.

Hail

Hail competes with turbulence as the greatest thunderstorm hazard to aircraft. Supercooled drops above the freezing level begin to freeze. Once a drop has frozen, other drops latch on and freeze to it, so the hailstone grows—sometimes into a huge ice ball. Large hail occurs with severe thunderstorms with strong updrafts that have built to great heights. Eventually, the hailstones fall, possibly some distance from the storm core. Hail may be encountered in clear air several miles from thunderstorm clouds.

As hailstones fall through air whose temperature is above 0 °C, they begin to melt and precipitation may reach the ground as either hail or rain. Rain at the surface does not mean the absence of hail aloft. Possible hail should be anticipated with any thunderstorm, especially beneath the anvil of a large cumulonimbus. Hailstones larger than one-half inch in diameter can significantly damage an aircraft in a few seconds.

Ceiling and Visibility

Generally, visibility is near zero within a thunderstorm cloud. Ceiling and visibility also may be restricted in precipitation and dust between the cloud base and the ground. The restrictions create the same problem as all ceiling and visibility restrictions; but the hazards are multiplied when associated with the other thunderstorm hazards of turbulence, hail, and lightning.

Effect on Altimeters

Pressure usually falls rapidly with the approach of a thunderstorm, rises sharply with the onset of the first gust and arrival of the cold downdraft and heavy rain showers, and then falls back to normal as the storm moves on. This cycle of pressure change may occur in 15 minutes. If the pilot does not receive a corrected altimeter setting, the altimeter may be more than 100 feet in error.

Lightning

A lightning strike can puncture the skin of an aircraft and damage communications and electronic navigational equipment. Although lightning has been suspected of igniting fuel vapors and causing an explosion, serious accidents due to lightning strikes are rare. Nearby lightning can blind the pilot, rendering him or her momentarily unable to navigate either by instrument or by visual reference. Nearby lightning can also induce permanent errors in the magnetic compass. Lightning discharges, even distant ones, can disrupt radio communications on low and medium frequencies. Though lightning intensity and frequency have no simple relationship to other storm parameters, severe storms, as a rule, have a high frequency of lightning.

Engine Water Ingestion

Turbine engines have a limit on the amount of water they can ingest. Updrafts are present in many thunderstorms, particularly those in the developing stages. If the updraft velocity in the thunderstorm approaches or exceeds the terminal velocity of the falling raindrops, very high concentrations of water may occur. It is possible that these concentrations can be in excess of the quantity of water turbine engines are designed to ingest. Therefore, severe thunderstorms may contain areas of high water concentration, which could result in flameout and/or structural failure of one or more engines.

Chapter Summary

Knowledge of the atmosphere and the forces acting within it to create weather is essential to understand how weather affects a flight. By understanding basic weather theories, a pilot can make sound decisions during flight planning after receiving weather briefings. For additional information on the topics discussed in this chapter, see the following publications as amended: AC 00-6, Aviation Weather For Pilots and Flight Operations Personnel; AC 00-24, Thunderstorms; AC 00-45, Aviation Weather Services; AC 91-74, Pilot Guide: Flight in Icing Conditions; and chapter 7, section 2 of the Aeronautical Information Manual (AIM).

Chapter 13

Aviation Weather Services

Introduction

In aviation, weather service is a combined effort of the National Weather Service (NWS), Federal Aviation Administration (FAA), Department of Defense (DOD), other aviation groups, and individuals. Because of the increasing need for worldwide weather services, foreign weather organizations also provide vital input.

While weather forecasts are not 100 percent accurate, meteorologists, through careful scientific study and computer modeling, have the ability to predict weather patterns, trends, and characteristics with increasing accuracy. Through a complex system of weather services, government agencies, and independent weather observers, pilots and other aviation professionals receive the benefit of this vast knowledge base in the form of up-to-date weather reports and forecasts. These reports and forecasts enable pilots to make informed decisions regarding weather and flight safety before and during a flight.

	Intensity
(none)	Light
	Moderate
+	Heavy
++	Very Heavy
x	Intense
xx	Extreme

Contraction	Operational Status
PPINE	Radar is operating normally but there echoes being detected.
PPINA	Radar observation is not available.
PPIOM	Radar is inoperative or out of service.
AUTO	Automated radar report from WSR-88

Observations

The data gathered from surface and upper altitude observations form the basis of all weather forecasts, advisories, and briefings. There are four types of weather observations: surface, upper air, radar, and satellite.

Surface Aviation Weather Observations

Surface aviation weather observations (METARs) are a compilation of elements of the current weather at individual ground stations across the United States. The network is made up of government and privately contracted facilities that provide continuous up-to-date weather information. Automated weather sources, such as the Automated Weather Observing Systems (AWOS), Automated Surface Observing Systems (ASOS), as well as other automated facilities, also play a major role in the gathering of surface observations.

Surface observations provide local weather conditions and other relevant information for a specific airport. This information includes the type of report, station identifier, date and time, modifier (as required), wind, visibility, runway visual range (RVR), weather phenomena, sky condition, temperature/dew point, altimeter reading, and applicable remarks. The information gathered for the surface observation may be from a person, an automated station, or an automated station that is updated or enhanced by a weather observer. In any form, the surface observation provides valuable information about individual airports around the country. Although the reports cover only a small radius, the pilot can generate a good picture of the weather over a wide area when many reporting stations are viewed together.

Air Route Traffic Control Center (ARTCC)

The Air Route Traffic Control Center (ARTCC) facilities are responsible for maintaining separation between flights conducted under instrument flight rules (IFR) in the en route structure. Center radars (Air Route Surveillance Radar (ARSR)) acquire and track transponder returns using the same basic technology as terminal radars. Earlier center radars displayed weather as an area of slashes (light precipitation) and Hs (moderate rainfall). Because the controller could not detect higher levels of precipitation, pilots had to be wary of areas showing moderate rainfall. Newer radar displays show weather as three shades of blue. Controllers can select the level of weather to be displayed. Weather displays of higher levels of intensity make it difficult for controllers to see aircraft data blocks, so pilots should not expect air traffic control (ATC) to keep weather displayed continuously.

Upper Air Observations

Observations of upper air weather are more challenging than surface observations. There are several methods by which upper air weather phenomena can be observed: radiosonde observations, pilot weather reports (PIREPs), Aircraft Meteorological Data Relay (AMDAR) and the Meteorological Data Collection and Reporting System (MDCRS). A radiosonde is a small cubic instrumentation package that is suspended below a six foot hydrogen- or helium-filled balloon. Once released, the balloon rises at a rate of approximately 1,000 feet per minute (fpm). As it ascends, the instrumentation gathers various pieces of data, such as air temperature, moisture, and pressure, as well as wind speed and direction. Once the information is gathered, it is relayed to ground stations via a 300 milliwatt radio transmitter.

The balloon flight can last as long as 2 hours or more and can ascend to altitudes as high as 115,000 feet and drift as far as 125 miles. The temperatures and pressures experienced during the flight can be as low as -130 °F and pressures as low as a few thousandths of what is experienced at sea level.

Since the pressure decreases as the balloon rises in the atmosphere, the balloon expands until it reaches the limits of its elasticity. This point is reached when the diameter has increased to over 20 feet. At this point, the balloon pops and the radiosonde falls back to Earth. The descent is slowed by means of a parachute. The parachute aids in protecting people and objects on the ground. Each year over 75,000 balloons are launched. Of that number, 20 percent are recovered and returned for reconditioning. Return instructions are printed on the side of each radiosonde.

Pilots also provide vital information regarding upper air weather observations and remain the only real-time source of information regarding turbulence, icing, and cloud heights. This information is gathered and filed by pilots in flight. Together, PIREPs and radiosonde observations provide information on upper air conditions important for flight planning. Many domestic and international airlines have equipped their aircraft with instrumentation that automatically transmits in flight weather observations through the DataLink system.

The Aircraft Meteorological Data Relay (AMDAR) is an international program utilizing commercial aircraft to provide automated weather observations. The AMDAR program provides approximately 220,000-230,000 aircraft observations per day on a worldwide basis utilizing aircraft onboard sensors and probes that measure wind, temperature, humidity/water vapor, turbulence and icing data. AMDAR vertical profiles and en route observations provide significant benefits to the aviation community by enhancing aircraft safety and operating efficiency through improved weather analysis and forecasting. The AMDAR program also contributes to improved short and medium term numerical weather forecasts for a wide range of services including severe weather, defense, marine, public weather and environmental monitoring. The information is down linked either via Very High Frequency (VHF) communications through the Aircraft Communications Addressing and Reporting System (ACARS) or via satellite link through the Aircraft to Satellite Data Acquisition and Relay (ASDAR).

The Meteorological Data Collection and Reporting System (MDCRS) is an automated airborne weather observation program that is used in the U.S. This program collects and disseminates real-time upper-air weather observations from participating airlines. The weather elements are down linked via ACARS and are managed by Aeronautical Radio, Inc. (ARINC) who then forwards them in Binary Universal Form for the Representation of Meteorological Data (BUFR) format to the NWS and in raw data form to the Earth Science Research Laboratory (ESRL) and the participating airline. More than 1,500 aircraft report wind and temperature data with some of these same aircraft also providing turbulence and humidity/water vapor information. In conjunction with avionics manufacturers, each participating airline programs their equipment to provide certain levels of meteorological data. The monitoring and collection of climb, en route, and descent data is accomplished through the aircraft's Flight Data Acquisition and Monitoring System (FDAMS) and is then transmitted via ACARS. When aircraft are out of ACARS range, reports can be relayed through ASDAR. However, in most cases, the reports are buffered until the aircraft comes within ACARS range, at which point they are downloaded.

Radar Observations

There are four types of radars which provide information about precipitation and wind.

1. The WSR-88D NEXRAD radar, commonly called Doppler radar, provides in-depth observations that inform surrounding communities of impending weather. Doppler radar has two operational modes: clear air and precipitation. In clear air mode, the radar is in its most sensitive operational mode because a slow antenna rotation allows the radar to sample the atmosphere longer. Images are updated about every 10 minutes in this mode.

   Precipitation targets provide stronger return signals; therefore, the radar is operated in the Precipitation mode when precipitation is present. A faster antenna rotation in this mode allows images to update at a faster rate, approximately every 4 to 6 minutes. Intensity values in both modes are measured in dBZ (decibels of Z) and are depicted in color on the radar image. *[Figure 13-1]* Intensities are correlated to intensity terminology (phraseology) for ATC purposes. *[Figures 13-2* and *13-3]*

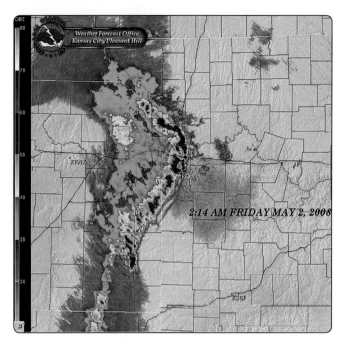

Figure 13-1. *Example of a weather radar scope.*

Clean Air Mode	Precipitation Mode
DBZ	DBZ
+28	75
+24	70
+20	65
+16	60
+12	55
+8	50
+4	45
0	40
-4	35
-8	30
-12	25
-16	20
-20	15
-24	10
-28	5
ND	ND

Figure 13-2. *WSR-88D Weather Radar Echo Intensity Legend.*

Reflectivity (dBZ) Ranges	Weather Radar Echo Intensity
<30 dBZ	Light
30–40 dBZ	Moderate
>40–50	Heavy
50+ dBZ	Extreme

Figure 13-3. *WSR-88D Weather Radar Precipitation Intensity Terminology.*

2. FAA terminal Doppler weather radar (TDWR), installed at some major airports around the country, also aids in providing severe weather alerts and warnings to ATC. Terminal radar ensures pilots are aware of wind shear, gust fronts, and heavy precipitation, all of which are dangerous to arriving and departing aircraft.

3. The third type of radar commonly used in the detection of precipitation is the FAA airport surveillance radar. This radar is used primarily to detect aircraft, but it also detects the location and intensity of precipitation, which is used to route aircraft traffic around severe weather in an airport environment.

4. Airborne radar is equipment carried by aircraft to locate weather disturbances. The airborne radars generally operate in the C or X bands (around 6 GHz or around 10 GHz, respectively) permitting both penetration of heavy precipitation, required for determining the extent of thunderstorms, and sufficient reflection from less intense precipitation.

Satellite

Advancement in satellite technologies has recently allowed for commercial use to include weather uplinks. Through the use of satellite subscription services, individuals are now able to receive satellite transmitted signals that provide near real-time weather information for the North American continent.

Service Outlets

Service outlets are government, government contract, or private facilities that provide aviation weather services. Several different government agencies, including the FAA, National Oceanic and Atmospheric Administration (NOAA), and the NWS work in conjunction with private aviation companies to provide different means of accessing weather information.

Flight Service Station (FSS)

The FSS is the primary source for preflight weather information. A preflight weather briefing from an FSS can be obtained 24 hours a day by calling 1-800-WX BRIEF from anywhere in the United States and Puerto Rico. Telephone numbers for FSS can be found in the Chart Supplement U.S. (formerly Airport/Facility Directory) or in the United States Government section of the telephone book.

The FSS also provides inflight weather briefing services and weather advisories to flights within the FSS area of responsibility.

Telephone Information Briefing Service (TIBS)

The Telephone Information Briefing Service (TIBS), provided by FSS, is a system of automated telephone recordings of meteorological and aeronautical information. TIBS provides area and route briefings, airspace procedures, and special announcements. The recordings are automatically updated as changes occur. It is designed to be a preliminary briefing tool and is not intended to replace a standard briefing from a FSS specialist. The TIBS service can only be accessed by a touchtone phone. The phone numbers for the TIBS service are listed in the Chart Supplement U.S. (formerly Airport/Facility Directory).

Hazardous Inflight Weather Advisory Service (HIWAS)

Hazardous Inflight Weather Advisory Service (HIWAS), available in the 48 conterminous states, is an automated continuous broadcast of hazardous weather information over selected VOR navigational aids (NAVAIDs). The broadcasts include advisories such as AIRMETS, SIGMETS, convective SIGMETS, and urgent PIREPs. The broadcasts are automatically updated as changes occur. Pilots should contact a FSS or EFAS for additional information. VORs that have HIWAS capability are depicted on aeronautical charts with an "H" in the upper right corner of the identification box. *[Figure 13-4]*

Transcribed Weather Broadcast (TWEB) (Alaska Only)

A continuous automated broadcast of meteorological and aeronautical data over selected low or medium frequency (L/MF) and very high frequency (VHF) omnidirectional range (VOR) NAVAID facilities. The broadcasts are automatically updated as changes occur. The broadcast contains adverse conditions, surface weather observations, PIREPS, and a density altitude statement (if applicable). Recordings may also include a synopsis, winds aloft forecast, en route and terminal forecast data, and radar reports. At selected locations, telephone access to the TWEB has been provided (TEL-TWEB). Telephone numbers for this service are found

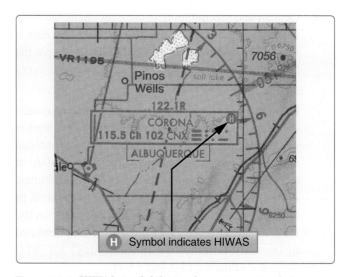

Figure 13-4. *HIWAS availability is shown on sectional chart.*

in the Alaska Chart Supplement U.S. (formerly Airport/ Facility Directory). These broadcasts are made available primarily for preflight and inflight planning, and as such, should not be considered as a substitute for specialist-provided preflight briefings.

Weather Briefings

Prior to every flight, pilots should gather all information vital to the nature of the flight. This includes an appropriate weather briefing obtained from a specialist at a FSS.

For weather specialists to provide an appropriate weather briefing, they need to know which of the three types of briefings is needed—standard, abbreviated, or outlook. Other helpful information is whether the flight is visual flight rules (VFR) or IFR, aircraft identification and type, departure point, estimated time of departure (ETD), flight altitude, route of flight, destination, and estimated time en route (ETE).

This information is recorded in the flight plan system and a note is made regarding the type of weather briefing provided. If necessary, it can be referenced later to file or amend a flight plan. It is also used when an aircraft is overdue or is reported missing.

Standard Briefing

A standard briefing provides the most complete information and a more complete weather picture. This type of briefing should be obtained prior to the departure of any flight and should be used during flight planning. A standard briefing provides the following information in sequential order if it is applicable to the route of flight.

1. Adverse conditions—this includes information about adverse conditions that may influence a decision to cancel or alter the route of flight. Adverse conditions include significant weather, such as thunderstorms or aircraft icing, or other important items such as airport closings.

2. VFR flight not recommended—if the weather for the route of flight is below VFR minimums, or if it is doubtful the flight could be made under VFR conditions due to the forecast weather, the briefer may state "VFR flight not recommended." It is the pilot's decision whether or not to continue the flight under VFR, but this advisory should be weighed carefully.

3. Synopsis—an overview of the larger weather picture. Fronts and major weather systems that affect the general area are provided.

4. Current conditions—the current ceilings, visibility, winds, and temperatures. If the departure time is more than 2 hours away, current conditions are not included in the briefing.

5. En route forecast—a summary of the weather forecast for the proposed route of flight.

6. Destination forecast—a summary of the expected weather for the destination airport at the estimated time of arrival (ETA).

7. Forecast winds and temperatures aloft—a forecast of the winds at specific altitudes for the route of flight. The forecast temperature information aloft is provided only upon request.

8. Notices to Airmen (NOTAM)—information pertinent to the route of flight that has not been published in the NOTAM publication. Published NOTAM information is provided during the briefing only when requested.

9. ATC delays—an advisory of any known ATC delays that may affect the flight.

10. Other information—at the end of the standard briefing, the FSS specialist provides the radio frequencies needed to open a flight plan and to contact EFAS. Any additional information requested is also provided at this time.

Abbreviated Briefing

An abbreviated briefing is a shortened version of the standard briefing. It should be requested when a departure has been delayed or when weather information is needed to update the previous briefing. When this is the case, the weather specialist needs to know the time and source of the previous briefing so the necessary weather information is not omitted inadvertently. It is always a good idea for the pilot to update the weather information whenever he/she has additional time.

Outlook Briefing

An outlook briefing should be requested when a planned departure is 6 hours or more away. It provides initial forecast information that is limited in scope due to the time frame of the planned flight. This type of briefing is a good source of flight planning information that can influence decisions regarding route of flight, altitude, and ultimately the go/no-go decision. A prudent pilot requests a follow-up briefing prior to departure since an outlook briefing generally only contains information based on weather trends and existing weather in geographical areas at or near the departure airport. A standard briefing near the time of departure ensures that the pilot has the latest information available prior to his/her flight.

Aviation Weather Reports

Aviation weather reports are designed to give accurate depictions of current weather conditions. Each report provides current information that is updated at different times. Some typical reports are METARs and PIREPs.

Aviation Routine Weather Report (METAR)

A METAR is an observation of current surface weather reported in a standard international format. While the METAR code has been adopted worldwide, each country is allowed to make modifications to the code. Normally, these differences are minor but necessary to accommodate local procedures or particular units of measure. This discussion of METAR covers elements used in the United States.

METARs are issued on a regularly scheduled basis unless significant weather changes have occurred. A special METAR (SPECI) can be issued at any time between routine METAR reports.

Example:
METAR KGGG 161753Z AUTO 14021G26KT 3/4SM +TSRA BR BKN008 OVC012CB 18/17 A2970 RMK PRESFR

A typical METAR report contains the following information in sequential order:

1. Type of report—there are two types of METAR reports. The first is the routine METAR report that is transmitted on a regular time interval. The second is the aviation selected SPECI. This is a special report that can be given at any time to update the METAR for rapidly changing weather conditions, aircraft mishaps, or other critical information.

2. Station identifier—a four-letter code as established by the International Civil Aviation Organization (ICAO). In the 48 contiguous states, a unique three-letter identifier is preceded by the letter "K." For example, Gregg County Airport in Longview, Texas, is identified by the letters "KGGG," K being the country designation and GGG being the airport identifier. In other regions of the world, including Alaska and Hawaii, the first two letters of the four-letter ICAO identifier indicate the region, country, or state. Alaska identifiers always begin with the letters "PA" and Hawaii identifiers always begin with the letters "PH." Station identifiers can be found by calling the FSS, a NWS office, or by searching various websites such as DUATS and NOAA's Aviation Weather Aviation Digital Data Services (ADDS).

3. Date and time of report—depicted in a six-digit group (161753Z). The first two digits are the date. The last four digits are the time of the METAR/SPECI, which is always given in coordinated universal time (UTC). A "Z" is appended to the end of the time to denote the time is given in Zulu time (UTC) as opposed to local time.

4. Modifier—denotes that the METAR/SPECI came from an automated source or that the report was corrected. If the notation "AUTO" is listed in the METAR/SPECI, the report came from an automated source. It also lists "AO1" (for no precipitation discriminator) or "AO2" (with precipitation discriminator) in the "Remarks" section to indicate the type of precipitation sensors employed at the automated station.

When the modifier "COR" is used, it identifies a corrected report sent out to replace an earlier report that contained an error (for example: METAR KGGG 161753Z COR).

5. Wind—reported with five digits (14021KT) unless the speed is greater than 99 knots, in which case the wind is reported with six digits. The first three digits indicate the direction the true wind is blowing from in tens of degrees. If the wind is variable, it is reported as "VRB." The last two digits indicate the speed of the wind in knots unless the wind is greater than 99 knots, in which case it is indicated by three digits. If the winds are gusting, the letter "G" follows the wind speed (G26KT). After the letter "G," the peak gust recorded is provided. If the wind direction varies more than 60° and the wind speed is greater than six knots, a separate group of numbers, separated by a "V," will indicate the extremes of the wind directions.

6. Visibility—the prevailing visibility (¾ SM) is reported in statute miles as denoted by the letters "SM." It is reported in both miles and fractions of miles. At times, runway visual range (RVR) is reported following the prevailing visibility. RVR is the distance a pilot can see down the runway in a moving aircraft. When RVR is reported, it is shown with an R, then the runway number followed by a slant, then the visual range in feet. For example, when the RVR is reported as R17L/1400FT, it translates to a visual range of 1,400 feet on runway 17 left.

7. Weather—can be broken down into two different categories: qualifiers and weather phenomenon (+TSRA BR). First, the qualifiers of intensity, proximity, and the descriptor of the weather are given. The intensity may be light (–), moderate ( ), or heavy (+). Proximity only depicts weather phenomena that are in the airport vicinity. The notation "VC" indicates a specific weather phenomenon is in the vicinity of five to ten miles from the airport. Descriptors are used to describe certain types of precipitation and obscurations. Weather phenomena may be reported as being precipitation, obscurations, and other phenomena, such as squalls or funnel clouds.

Descriptions of weather phenomena as they begin or end and hailstone size are also listed in thc "Remarks" sections of the report. *[Figure 13-5]*

8. Sky condition—always reported in the sequence of amount, height, and type or indefinite ceiling/height (vertical visibility) (BKN008 OVC012CB, VV003). The heights of the cloud bases are reported with a three-digit number in hundreds of feet AGL. Clouds above 12,000 feet are not detected or reported by an automated station. The types of clouds, specifically towering cumulus (TCU) or cumulonimbus (CB) clouds, are reported with their height. Contractions are used to describe the amount of cloud coverage and obscuring phenomena. The amount of sky coverage is reported in eighths of the sky from horizon to horizon. *[Figure 13-6]*
9. Temperature and dew point—the air temperature and dew point are always given in degrees Celsius (C) or (18/17). Temperatures below 0 °C are preceded by the letter "M" to indicate minus.
10. Altimeter setting—reported as inches of mercury ("Hg) in a four-digit number group (A2970). It is always preceded by the letter "A." Rising or falling pressure may also be denoted in the "Remarks" sections as "PRESRR" or "PRESFR," respectively.
11. Zulu time—a term used in aviation for UTC, which places the entire world on one time standard.

Sky Cover	Contraction
Less than ⅛ (Clear)	SKC, CLR, FEW
⅛–²⁄₈ (Few)	FEW
³⁄₈–⁴⁄₈ (Scattered)	SCT
⁵⁄₈–⅞ (Broken)	BKN
⁸⁄₈ or (Overcast)	OVC

Figure 13-6. *Reportable contractions for sky condition.*

12. Remarks—the remarks section always begins with the letters "RMK." Comments may or may not appear in this section of the METAR. The information contained in this section may include wind data, variable visibility, beginning and ending times of particular phenomenon, pressure information, and various other information deemed necessary. An example of a remark regarding weather phenomenon that does not fit in any other category would be: OCNL LTGICCG. This translates as occasional lightning in the clouds and from cloud to ground. Automated stations also use the remarks section to indicate the equipment needs maintenance.

Example:
METAR KGGG 161753Z AUTO 14021G26KT 3/4SM +TSRA BR BKN008 OVC012CB 18/17 A2970 RMK PRESFR

Qualifier		Weather Phenomena		
Intensity or Proximity 1	Descriptor 2	Precipitation 3	Obscuration 4	Other 5
– Light	**MI** Shallow	**DZ** Drizzle	**BR** Mist	**PO** Dust/sand whirls
Moderate (no qualifier)	**BC** Patches	**RA** Rain	**FG** Fog	**SQ** Squalls
+ Heavy	**DR** Low drifting	**SN** Snow	**FU** Smoke	**FC** Funnel cloud
VC in the vicinity	**BL** Blowing	**SG** Snow grains	**DU** Dust	**+FC** Tornado or waterspout
	SH Showers	**IC** Ice crystals (diamond dust)	**SA** Sand	**SS** Sandstorm
	TS Thunderstorms	**PL** Ice pellets	**HZ** Haze	**DS** Dust storm
	FZ Freezing	**GR** Hail	**PY** Spray	
	PR Partial	**GS** Small hail or snow pellets	**VA** Volcanic ash	
		UP *Unknown precipitation		

The weather groups are constructed by considering columns 1–5 in this table in sequence:
intensity, followed by descriptor, followed by weather phenomena (e.g., heavy rain showers(s) is coded as +SHRA).
* Automated stations only

Figure 13-5. *Descriptors and weather phenomena used in a typical METAR.*

Explanation:

Routine METAR for Gregg County Airport for the 16th day of the month at 1753Z automated source. Winds are 140 at 21 knots gusting to 26. Visibility is ¾ statute mile. Thunderstorms with heavy rain and mist. Ceiling is broken at 800 feet, overcast at 1,200 feet with cumulonimbus clouds. Temperature 18 °C and dew point 17 °C. Barometric pressure is 29.70 "Hg and falling rapidly.

Pilot Weather Reports (PIREPs)

PIREPs provide valuable information regarding the conditions as they actually exist in the air, which cannot be gathered from any other source. Pilots can confirm the height of bases and tops of clouds, locations of wind shear and turbulence, and the location of inflight icing. If the ceiling is below 5,000 feet, or visibility is at or below five miles, ATC facilities are required to solicit PIREPs from pilots in the area. When unexpected weather conditions are encountered, pilots are encouraged to make a report to a FSS or ATC. When a pilot weather report is filed, the ATC facility or FSS adds it to the distribution system to brief other pilots and provide inflight advisories.

PIREPs are easy to file and a standard reporting form outlines the manner in which they should be filed. *Figure 13-7* shows the elements of a PIREP form. Item numbers 1 through 5 are required information when making a report, as well as at least one weather phenomenon encountered. A PIREP is normally transmitted as an individual report but may be appended to a surface report. Pilot reports are easily decoded, and most contractions used in the reports are self-explanatory.

Example:

UA/OV GGG 090025/TM 1450/FL 060/TP C182/SK 080 OVC/WX FV04SM RA/TA 05/WV 270030KT/TB LGT/RM HVY RAIN

Explanation:

Type:	Routine pilot report
Location:	25 NM out on the 090° radial, Gregg County VOR
Time:	1450 Zulu
Altitude or Flight Level:	6,000 feet
Aircraft Type:	Cessna 182
Sky Cover:	8,000 overcast
Visibility/Weather:	4 miles in rain
Temperature:	5 °Celsius
Wind:	270° at 30 knots
Turbulence:	Light
Icing:	None reported
Remarks:	Rain is heavy

Encoding Pilot Weather Reports (PIREPS)			
1	XXX	3-letter station identifier	Nearest weather reporting location to the reported phenomenon
2	UA	Routine PIREP, UUA-Urgent PIREP.	

3	/OV	Location	Use 3-letter NAVAID idents only.
			  a. Fix: /OV ABC, /OV ABC 090025.
			  b. Fix: /OV ABC 045020-DEF, /OV ABC-DEF-GHI

4	/TM	Time	4 digits in UTC: /TM 0915.
5	/FL	Altitude/flight level	3 digits for hundreds of feet. If not known, use UNKN: /FL095, /FL310, /FLUNKN.
6	/TP	Type aircraft	4 digits maximum. If not known, use UNKN: /TP L329, /TP B727, /TP UNKN.

7	/SK	Sky cover/cloud layers	Describe as follows:
			  a. Height of cloud base in hundreds of feet. If unknown, use UNKN.
			  b. Cloud cover symbol.
			  c. Height of cloud tops in hundreds of feet.

8	/WX	Weather	Flight visibility reported first: Use standard weather symbols: /WX FV02SM RA HZ, /WX FV01SM TSRA.
9	/TA	Air temperature in celsius (C)	If below zero, prefix with a hyphen: /TA 15, /TA M06.
10	/WV	Wind	Direction in degrees magnetic north and speed in six digits: /WV270045KT, WV 280110KT.
11	/TB	Turbulence	Use standard contractions for intensity and type (use CAT or CHOP when appropriate). Include altitude only if different from /FL, /TB EXTRM, /TB LGT-MOD BLO 090.
12	/IC	Icing	Describe using standard intensity and type contractions. Include altitude only if different than /FL: /IC LGT-MOD RIME, /IC SEV CLR 028-045.
13	/RM	Remarks	Use free form to clarify the report and type hazardous elements first: /RM LLWS -15KT SFC-030 DURC RY22 JFK.

Figure 13-7. *PIREP encoding and decoding.*

Aviation Forecasts

Observed weather condition reports are often used in the creation of forecasts for the same area. A variety of different forecast products are produced and designed to be used in the preflight planning stage. The printed forecasts that pilots need to be familiar with are the terminal aerodrome forecast (TAF), aviation area forecast (FA), inflight weather advisories (SIGMET, AIRMET), and the winds and temperatures aloft forecast (FB).

Terminal Aerodrome Forecasts (TAF)

A TAF is a report established for the five statute mile radius around an airport. TAF reports are usually given for larger airports. Each TAF is valid for a 24 or 30-hour time period and is updated four times a day at 0000Z, 0600Z, 1200Z, and 1800Z. The TAF utilizes the same descriptors and abbreviations as used in the METAR report. The TAF includes the following information in sequential order:

1. Type of report—a TAF can be either a routine forecast (TAF) or an amended forecast (TAF AMD).

2. ICAO station identifier—the station identifier is the same as that used in a METAR.

3. Date and time of origin—time and date (081125Z) of TAF origination is given in the six-number code with the first two being the date, the last four being the time. Time is always given in UTC as denoted by the Z following the time block.

4. Valid period dates and times—The TAF valid period (0812/0912) follows the date/time of forecast origin group. Scheduled 24 and 30 hour TAFs are issued four times per day, at 0000, 0600, 1200, and 1800Z. The first two digits (08) are the day of the month for the start of the TAF. The next two digits (12) are the starting hour (UTC). 09 is the day of the month for the end of the TAF, and the last two digits (12) are the ending hour (UTC) of the valid period. A forecast period that begins at midnight UTC is annotated as 00. If the end time of a valid period is at midnight UTC, it is annotated as 24. For example, a 00Z TAF issued on the 9th of the month and valid for 24 hours would have a valid period of 0900/0924.

5. Forecast wind—the wind direction and speed forecast are coded in a five-digit number group. An example would be 15011KT. The first three digits indicate the direction of the wind in reference to true north. The last two digits state the windspeed in knots appended with "KT." Like the METAR, winds greater than 99 knots are given in three digits.

6. Forecast visibility—given in statute miles and may be in whole numbers or fractions. If the forecast is greater than six miles, it is coded as "P6SM."

7. Forecast significant weather—weather phenomena are coded in the TAF reports in the same format as the METAR.

8. Forecast sky condition—given in the same format as the METAR. Only cumulonimbus (CB) clouds are forecast in this portion of the TAF report as opposed to CBs and towering cumulus in the METAR.

9. Forecast change group—for any significant weather change forecast to occur during the TAF time period, the expected conditions and time period are included in this group. This information may be shown as from (FM), and temporary (TEMPO). "FM" is used when a rapid and significant change, usually within an hour, is expected. "TEMPO" is used for temporary fluctuations of weather, expected to last less than 1 hour.

10. PROB30—a given percentage that describes the probability of thunderstorms and precipitation occurring in the coming hours. This forecast is not used for the first 6 hours of the 24-hour forecast.

Example:
TAF
KPIR 111130Z 1112/1212
TEMPO 1112/1114 5SM BR
FM1500 16015G25KT P6SM SCT040 BKN250
FM120000 14012KT P6SM BKN080 OVC150 PROB30
1200/1204 3SM TSRA BKN030CB
FM120400 1408KT P6SM SCT040 OVC080
TEMPO 1204/1208 3SM TSRA OVC030CB

Explanation:
Routine TAF for Pierre, South Dakota…on the 11th day of the month, at 1130Z…valid for 24 hours from 1200Z on the 11th to 1200Z on the 12th…wind from 150° at 12 knots… visibility greater than 6 SM…broken clouds at 9,000 feet… temporarily, between 1200Z and 1400Z, visibility 5 SM in mist…from 1500Z winds from 160° at 15 knots, gusting to 25 knots visibility greater than 6 SM…clouds scattered at 4,000 feet and broken at 25,000 feet…from 0000Z wind from 140° at 12 knots…visibility greater than 6 SM…clouds broken at 8,000 feet, overcast at 15,000 feet…between 0000Z and 0400Z, there is 30 percent probability of visibility 3 SM…thunderstorm with moderate rain showers…clouds broken at 3,000 feet with cumulonimbus clouds…from 0400Z…winds from 140° at 8 knots…visibility greater than 6 miles…clouds at 4,000 scattered and overcast at 8,000… temporarily between 0400Z and 0800Z…visibility 3 miles… thunderstorms with moderate rain showers…clouds overcast at 3,000 feet with cumulonimbus clouds…end of report (=).

Area Forecasts (FA)

The FA gives a picture of clouds, general weather conditions, and visual meteorological conditions (VMC) expected over a large area encompassing several states. There are six areas for which area forecasts are published in the contiguous 48 states. Area forecasts are issued three times a day and are valid for 18 hours. This type of forecast gives information vital to en route operations, as well as forecast information for smaller airports that do not have terminal forecasts.

Area forecasts are typically disseminated in four sections and include the following information:

1. Header—gives the location identifier of the source of the FA, the date and time of issuance, the valid forecast time, and the area of coverage.

Example:
DFWC FA 120945
SYNOPSIS AND VFR CLDS/WX
SYNOPSIS VALID UNTIL 130400
CLDS/WX VALID UNTIL 122200...OTLK VALID
122200-130400
OK TX AR LA MS AL AND CSTL WTRS

Explanation:
The area forecast shows information given by Dallas Fort Worth, for the region of Oklahoma, Texas, Arkansas, Louisiana, Mississippi, and Alabama, as well as a portion of the Gulf coastal waters. It was issued on the 12th day of the month at 0945. The synopsis is valid from the time of issuance until 0400 hours on the 13th. VFR clouds and weather information on this area forecast are valid until 2200 hours on the 12th and the outlook is valid from 2200Z on the 12th to 0400Z on the 13th.

2. Precautionary statements—IFR conditions, mountain obscurations, and thunderstorm hazards are described in this section. Statements made here regarding height are given in MSL, and if given otherwise, AGL or ceiling (CIG) is noted.

Example:
SEE AIRMET SIERRA FOR IFR CONDS AND MTN OBSCN.
TS IMPLY SEV OR GTR TURB SEV ICE LLWS AND IFR CONDS.
NON MSL HGTS DENOTED BY AGL OR CIG.

Explanation:
The area forecast covers VFR clouds and weather, so the precautionary statement warns that AIRMET Sierra should be referenced for IFR conditions and mountain obscuration. The code TS indicates the possibility of thunderstorms and implies there may be occurrences of severe or greater turbulence, severe icing, low-level wind shear, and IFR conditions. The final line of the precautionary statement alerts the user that heights, for the most part, are MSL. Those that are not MSL will state AGL or CIG.

3. Synopsis—gives a brief summary identifying the location and movement of pressure systems, fronts, and circulation patterns.

Example:
SYNOPSIS...LOW PRES TROF 10Z OK/TX PNHDL AREA FCST MOV EWD INTO CNTRL-SWRN OK BY 04Z. WRMFNT 10Z CNTRL OK-SRN AR-NRN MS FCST LIFT NWD INTO NERN OK-NRN AR EXTRM NRN MS BY 04Z.

Explanation:
As of 1000Z, there is a low pressure trough over the Oklahoma and Texas panhandle area, which is forecast to move eastward into central to southwestern Oklahoma by 0400Z. A warm front located over central Oklahoma, southern Arkansas, and northern Mississippi at 1000Z is forecast to lift northwestward into northeastern Oklahoma, northern Arkansas, and extreme northern Mississippi by 0400Z.

4. VFR Clouds and Weather—This section lists expected sky conditions, visibility, and weather for the next 12 hours and an outlook for the following 6 hours.

Example:
S CNTRL AND SERN TX
AGL SCT-BKN010. TOPS 030. VIS 3-5SM BR. 14-16Z
BECMG AGL SCT030. 19Z AGL SCT050.
OTLK...VFR
OK
PNDLAND NW...AGL SCT030 SCT-BKN100.
TOPS FL200.
15Z AGL SCT040 SCT100. AFT 20Z SCT TSRA DVLPG..
FEW POSS SEV. CB TOPS FL450.
OTLK...VFR

Explanation:
In south central and southeastern Texas, there is a scattered to broken layer of clouds from 1,000 feet AGL with tops at 3,000 feet, visibility is 3 to 5 SM in mist. Between 1400Z and 1600Z, the cloud bases are expected to increase to 3,000 feet AGL. After 1900Z, the cloud bases are expected to continue to increase to 5,000 feet AGL and the outlook is VFR.

In northwestern Oklahoma and panhandle, the clouds are scattered at 3,000 feet with another scattered to broken layer at 10,000 feet AGL, with the tops at 20,000 feet. At 1500 Z, the lowest cloud base is expected to increase to 4,000 feet AGL with a scattered layer at 10,000 feet AGL. After

2000Z, the forecast calls for scattered thunderstorms with rain developing and a few becoming severe; the CB clouds have tops at flight level (FL) 450 or 45,000 feet MSL.

It should be noted that when information is given in the area forecast, locations may be given by states, regions, or specific geological features such as mountain ranges. *Figure 13-8* shows an area forecast chart with six regions of forecast, states, regional areas, and common geographical features.

Inflight Weather Advisories

Inflight weather advisories, which are provided to en route aircraft, are forecasts that detail potentially hazardous weather. These advisories are also available to pilots prior to departure for flight planning purposes. An inflight weather advisory is issued in the form of either an AIRMET, SIGMET, or convective SIGMET.

AIRMET

AIRMETs (WAs) are examples of inflight weather advisories that are issued every 6 hours with intermediate updates issued as needed for a particular area forecast region. The information contained in an AIRMET is of operational interest to all aircraft, but the weather section concerns phenomena considered potentially hazardous to light aircraft and aircraft with limited operational capabilities.

An AIRMET includes forecast of moderate icing, moderate turbulence, sustained surface winds of 30 knots or greater, widespread areas of ceilings less than 1,000 feet and/or visibilities less than three miles, and extensive mountain obscurement.

Each AIRMET bulletin has a fixed alphanumeric designator, numbered sequentially for easy identification, beginning with the first issuance of the day. Sierra is the AIRMET code used to denote IFR and mountain obscuration; Tango is used to denote turbulence, strong surface winds, and low-level wind shear; and Zulu is used to denote icing and freezing levels.

Example:
BOSS WA 211945
AIRMET SIERRA UPDT 3 FOR IFR AND MTN OBSCN
VALID UNTIL 220200
AIRMET IFT…ME NH VT MA CT RI NY NJ AND CSTL WTRS FROM CAR TO YSJ TO 150E ACK TO EWR TO YOW TO CAR OCNL CIG BLW 010/VIS BLW 3SM PCPN/BR. CONDS CONT BYD 02Z THRU 08Z

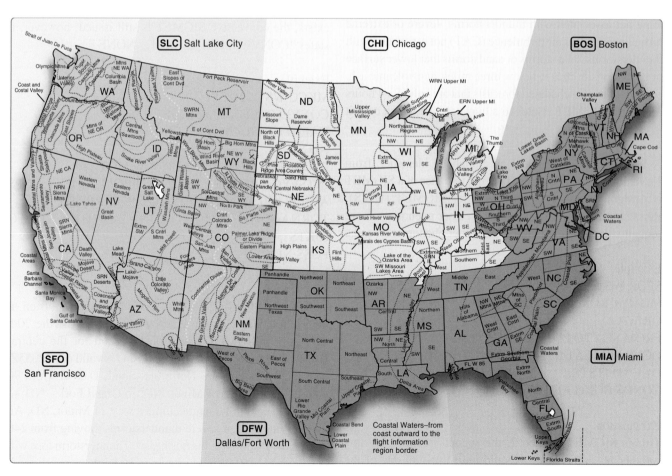

Figure 13-8. *Area forecast region map.*

AIRMET MTN OBSCN...ME NH VT MA NY PA
FROM CAR TO MLT TO CON TO SLT TO SYR TO CAR
MTNS OCNLY OBSCD BY CLDS/PCPN/BR. CONDS
CONT BYD 02Z THRU 08Z

Explanation:
AIRMET SIERRA was issued for the Boston area at 1945Z
on the 21st day of the month. SIERRA contains information
on IFR and/or mountain obscurations. This is the third
updated issuance of this Boston AIRMET series as indicated
by "SIERRA UPDT 3" and is valid until 0200Z on the
22nd. The affected states within the BOS area are: Maine,
New Hampshire, Vermont, Massachusetts, New York, and
Pennsylvania. Within an area bounded by: Caribou, ME;
to Saint Johns, New Brunswick; to 150 nautical miles east
of Nantucket, MA; to Newark, NJ; to Ottawa, Ontario;
to Caribou, ME. The effected states within Caribou, ME
to Millinocket, ME to Concord, NH to Slate Run, PA to
Syracuse, NY to Caribou, ME will experience ceilings
below 1,000 feet/visibility below 3 SM, precipitation/mist.
Conditions will continue beyond 0200Z through 0800Z.

SIGMET

SIGMETs (WSs) are inflight advisories concerning non-
convective weather that is potentially hazardous to all
aircraft. They report weather forecasts that include severe
icing not associated with thunderstorms, severe or extreme
turbulence or clear air turbulence (CAT) not associated with
thunderstorms, dust storms or sandstorms that lower surface
or inflight visibilities to below three miles, and volcanic ash.
SIGMETs are unscheduled forecasts that are valid for 4 hours
unless the SIGMET relates to a hurricane, in which case it
is valid for 6 hours.

A SIGMET is issued under an alphabetic identifier, from
November through Yankee. The first issuance of a SIGMET
is designated as an Urgent Weather SIGMET (UWS).
Reissued SIGMETs for the same weather phenomenon are
sequentially numbered until the weather phenomenon ends.

Example:
SFOR WS 100130
SIGMET ROME02 VALID UNTIL 100530
OR WA
FROM SEA TO PDT TO EUG TO SEA
OCNL SEV CAT BTN FL280 AND FL350 EXPCD
DUE TO JTSTR.
CONDS BGNG AFT 0200Z CONTG BYD 0530Z .

Explanation:
This is SIGMET Romeo 2, the second issuance for this
weather phenomenon. It is valid until the 10th day of the
month at 0530Z time. This SIGMET is for Oregon and

Washington, for a defined area from Seattle to Portland to
Eugene to Seattle. It calls for occasional severe clear air
turbulence between FL280 and FL350 due to the location of
the jet stream. These conditions will begin after 0200Z and
continue beyond the forecast scope of this SIGMET of 0530Z.

Convective Significant Meteorological Information (WST)

A Convective SIGMET (WST) is an inflight weather
advisory issued for hazardous convective weather that affects
the safety of every flight. Convective SIGMETs are issued
for severe thunderstorms with surface winds greater than 50
knots, hail at the surface greater than or equal to ¾ inch in
diameter, or tornadoes. They are also issued to advise pilots
of embedded thunderstorms, lines of thunderstorms, or
thunderstorms with heavy or greater precipitation that affect
40 percent or more of a 3,000 square mile or greater region.

Convective SIGMETs are issued for each area of the
contiguous 48 states but not Alaska or Hawaii. Convective
SIGMETs are issued for the eastern (E), western (W), and
central (C) United States. Each report is issued at 55 minutes
past the hour, but special Convective SIGMETs can be
issued during the interim for any reason. Each forecast is
valid for 2 hours. They are numbered sequentially each day
from 1–99, beginning at 00Z time. If no hazardous weather
exists, the convective SIGMET is still issued; however, it
states "CONVECTIVE SIGMET...NONE."

Example:
MKCC WST 221855
CONVECTIVE SIGMET 20C
VALID UNTIL 2055Z
ND SD
FROM 90W MOT-GFK-ABR-90W MOT
INTSFYG AREA SEV TS MOVG FROM 24045KT. TOPS
ABV FL450. WIND GUSTS TO 60KTS RPRTD.
TORNADOES...HAIL TO 2 IN... WIND GUSTS TO
65KTS
POSS ND PTN

Explanation:
Convective SIGMET was issued for the central portion of
the United States on the 22nd at 1855Z. This is the 20th
Convective SIGMET issued on the 22nd for the central
United States as indicated by "20C" and is valid until 2055Z.
The affected states are North and South Dakota, from 90
nautical miles west of Minot, ND; to Grand Forks, ND; to
Aberdeen, SD; to 90 nautical miles west of Minot, ND. An
intensifying area of severe thunderstorms moving from 240
degrees at 45 knots (to the northeast). Thunderstorm tops will
be above FL 450. Wind gusts up to 60 knots were reported.

Also reported were tornadoes, hail to 2 inches in diameter, and wind gusts to 65 knots possible in the North Dakota portion.

Winds and Temperature Aloft Forecast (FB)

Winds and temperatures aloft forecasts (FB) provide wind and temperature forecasts for specific locations throughout the United States, including network locations in Hawaii and Alaska. The forecasts are made twice a day based on the radiosonde upper air observations taken at 0000Z and 1200Z.

Altitudes through 12,000 feet are classified as true altitudes, while altitudes 18,000 feet and above are classified as altitudes and are termed flight levels. Wind direction is always in reference to true north, and wind speed is given in knots. The temperature is given in degrees Celsius. No winds are forecast when a given level is within 1,500 feet of the station elevation. Similarly, temperatures are not forecast for any station within 2,500 feet of the station elevation.

If the wind speed is forecast to be greater than 99 knots but less than 199 knots, the computer adds 50 to the direction and subtracts 100 from the speed. To decode this type of data group, the reverse must be accomplished. For example, when the data appears as "731960," subtract 50 from the 73 and add 100 to the 19, and the wind would be 230° at 119 knots with a temperature of –60 °C. If the wind speed is forecast to be 200 knots or greater, the wind group is coded as 99 knots. For example, when the data appears as "7799," subtract 50 from 77 and add 100 to 99, and the wind is 270° at 199 knots or greater. When the forecast wind speed is calm, or less than 5 knots, the data group is coded "9900," which means light and variable. *[Figure 13-9]*

Explanation of *Figure 13-9:*
The heading indicates that this FB was transmitted on the 15th of the month at 1640Z and is based on the 1200Z upper air data. The valid time is 1800Z on the same day and should be used for the period between 1400Z and 2100Z. The heading also indicates that the temperatures above FL240 are negative. Therefore, the minus sign will be omitted for all forecast temperatures above FL240.

A four-digit data group shows the wind direction in reference to true north and the wind speed in knots. The elevation at Amarillo, Texas (AMA) is 3,605 feet, so the lowest reportable

altitude is 6,000 feet for the forecast winds. In this case, "2714" means the wind is forecast to be from 270° at a speed of 14 knots.

A six-digit group includes the forecast temperature aloft. The elevation at Denver (DEN) is 5,431 feet, so the lowest reportable altitude is 9,000 feet for the winds and temperature forecast. In this case, "2321-04" indicates the wind is forecast to be from 230° at a speed of 21 knots with a temperature of –4 °C.

Weather Charts

Weather charts are graphic charts that depict current or forecast weather. They provide an overall picture of the United States and should be used in the beginning stages of flight planning. Typically, weather charts show the movement of major weather systems and fronts. Surface analysis, weather depiction, and significant weather prognostic charts are sources of current weather information. Significant weather prognostic charts provide an overall forecast weather picture.

Surface Analysis Chart

The surface analysis chart depicts an analysis of the current surface weather. *[Figure 13-10]* This chart is transmitted every 3 hours and covers the contiguous 48 states and adjacent areas. A surface analysis chart shows the areas of high and low pressure, fronts, temperatures, dew points, wind directions and speeds, local weather, and visual obstructions. Surface weather observations for reporting points across the United States are also depicted on this chart. Each of these reporting points is illustrated by a station model. *[Figure 13-11]* A station model includes:

- Sky cover—the station model depicts total sky cover and is shown as clear, scattered, broken, overcast, or obscured/partially obscured.

- Sea level pressure—given in three digits to the nearest tenth of a millibar (mb). For 1,000 mbs or greater, prefix a 10 to the three digits. For less than 1,000 mbs, prefix a 9 to the three digits.

- Pressure change/tendency—pressure change in tenths of mb over the past 3 hours. This is depicted directly below the sea level pressure.

- Dew point—given in degrees Fahrenheit.

- Present weather—over 100 different standard weather symbols are used to describe the current weather.

- Temperature—given in degrees Fahrenheit.

- Wind—true direction of wind is given by the wind pointer line, indicating the direction from which the wind is blowing. A short barb is equal to 5 knots of wind, a long barb is equal to 10 knots of wind, and a pennant is equal to 50 knots.

```
FB KWBC 151640
DATA BASED ON 151200Z
VALID 151800Z FOR USE 1400-2100Z
TEMPS NEGATIVE ABV 24000
```

FB	3000	6000	9000	12000	18000	24000	30000
AMA		2714	2725+00	2625-04	2531-15	2542-27	265842
DEN			2321-04	2532-08	2434-19	2441-31	235347

Figure 13-9. *Winds and temperature aloft forecast.*

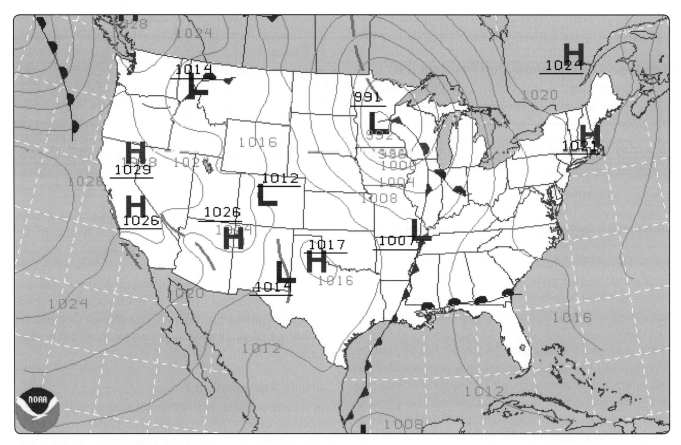

Figure 13-10. *Surface analysis chart.*

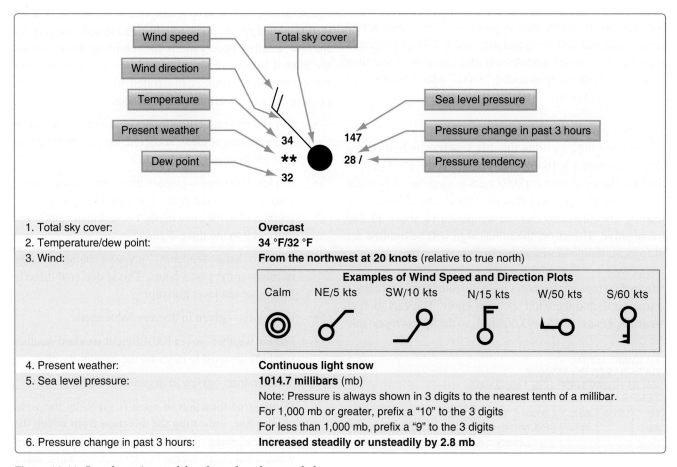

Figure 13-11. *Sample station model and weather chart symbols.*

Weather Depiction Chart

A weather depiction chart details surface conditions as derived from METAR and other surface observations. The weather depiction chart is prepared and transmitted by computer every 3 hours beginning at 0100Z time and is valid data for the forecast period. It is designed to be used for flight planning by giving an overall picture of the weather across the United States. *[Figure 13-12]*

The weather depiction chart also provides a graphic display of IFR, VFR, and marginal VFR (MVFR) weather. Areas of IFR conditions (ceilings less than 1,000 feet and visibility less than three miles) are shown by a hatched area outlined by a smooth line. MVFR regions (ceilings 1,000 to 3,000 feet, visibility 3 to 5 miles) are shown by a nonhatched area outlined by a smooth line. Areas of VFR (no ceiling or ceiling greater than 3,000 feet and visibility greater than five miles) are not outlined. Also plotted are fronts, troughs, and squall lines from the previous hours surface analysis chart.

Weather depiction charts show a modified station model that provides sky conditions in the form of total sky cover, ceiling height, weather, and obstructions to visibility, but does not include winds or pressure readings like the surface analysis chart. A bracket ( ]) symbol to the right of the station indicates the observation was made by an automated station.

Significant Weather Prognostic Charts

Significant weather prognostic charts are available for low-level significant weather from the surface to FL 240 (24,000 feet), also referred to as the 400 mb level and high-level significant weather from FL 250 to FL 630 (25,000 to 63,000 feet). The primary concern of this discussion is the low-level significant weather prognostic chart.

The low-level chart is is a forecast of aviation weather hazards, primarily intended to be used as a guidance product for briefing the VFR pilot. The forecast domain covers the 48 contiguous states, southern Canada and the coastal waters for altitudes below 24,000 ft. Low altitude Significant Weather charts are issued four times daily and are valid at fixed times: 0000, 0600, 1200, and 1800 UTC. Each chart is divided on the left and right into 12 and 24 hour forecast intervals (based on the current NAM model available).

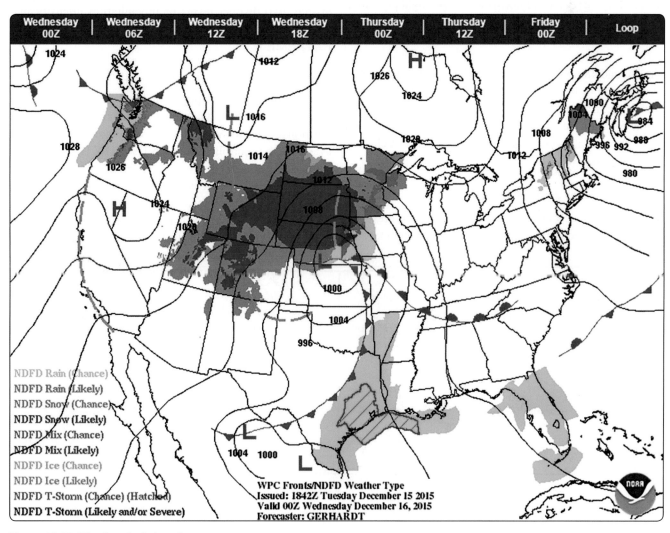

Figure 13-12. *Weather depiction chart.*

Effective September 1, 2015, the four-panel Low Level SFC-240 chart was replaced with a two-panel chart. The new two-panel chart will be the same as the top two panels in the former four-panel chart, depicting the freezing level and areas of IFR, MVFR, and moderate or greater turbulence. The bottom two panels of the chart have been removed. In lieu of these bottom two panels, an enhanced surface chart that includes fronts, pressure, precipitation type, precipitation intensity, and weather type, is displayed. The green precipitation polygons will be replaced by shaded precipitation areas using the National Digital Forecast Database (NDFD) weather grid.

Figure 13-13 depicts the new two-panel significant weather prognostic chart, as well as the symbols typically used to depict precipitation. The two panels depict freezing levels, turbulence, and low cloud ceilings and/or restrictions to visibility (shown as contoured areas of MVFR and IFR conditions). These charts enable the pilot to pictorially evaluate existing and potential weather hazards they may encounter. Pilots can balance weather phenomena with their aircraft capability and skill set resulting in aeronautical decision-making appropriate to the flight. Prognostic charts are an excellent source of information for preflight planning; however, this chart should be viewed in light of current conditions and specific local area forecasts.

The 36- and 48-hour significant weather prognostic chart is an extension of the 12- and 24-hour forecast. This chart is issued twice a day. It typically contains forecast positions and characteristics of pressure patterns, fronts, and precipitation. An example of a 36- and 48-hour surface prognostic chart is shown in *Figure 13-14.*

ATC Radar Weather Displays

Although ATC systems cannot always detect the presence or absence of clouds, they can often determine the intensity of a precipitation area, but the specific character of that area (snow, rain, hail, VIRGA, etc.) cannot be determined. For this reason, ATC refers to all weather areas displayed on ATC radar scopes as "precipitation."

ARTCC facilities normally use a Weather and Radar Processor (WARP) to display a mosaic of data obtained from multiple NEXRAD sites. There is a time delay between actual conditions and those displayed to the controller. The precipitation data on the ARTCC controller's display could be up to 6 minutes old. The WARP processor is only used in ARTCC facilities. All ATC facilities using radar weather processors with the ability to determine precipitation intensity, describe the intensity to pilots as:

- Light
- Moderate
- Heavy
- Extreme

When the WARP is not available, a second system, the narrowband Air Route Surveillance Radar (ARSR) can display two distinct levels of precipitation intensity that is described to pilots as "MODERATE and "HEAVY TO EXTREME."

ATC facilities that cannot display the intensity levels of precipitation due to equipment limitations describe the location of the precipitation area by geographic position or position relative to the aircraft. Since the intensity level is not available, the controller states "INTENSITY UNKNOWN."

ATC radar is not able to detect turbulence. Generally, turbulence can be expected to occur as the rate of rainfall or intensity of precipitation increases. Turbulence associated with greater rates of precipitation is normally more severe than

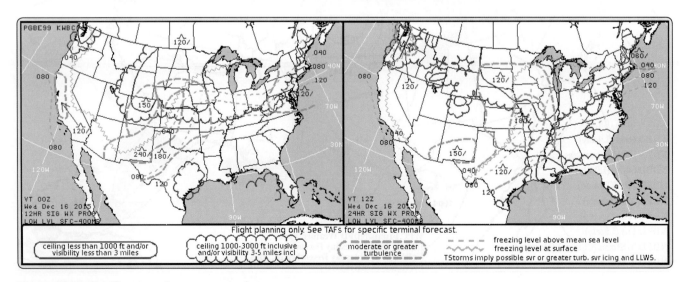

Figure 13-13. *Significant weather prognostic chart.*

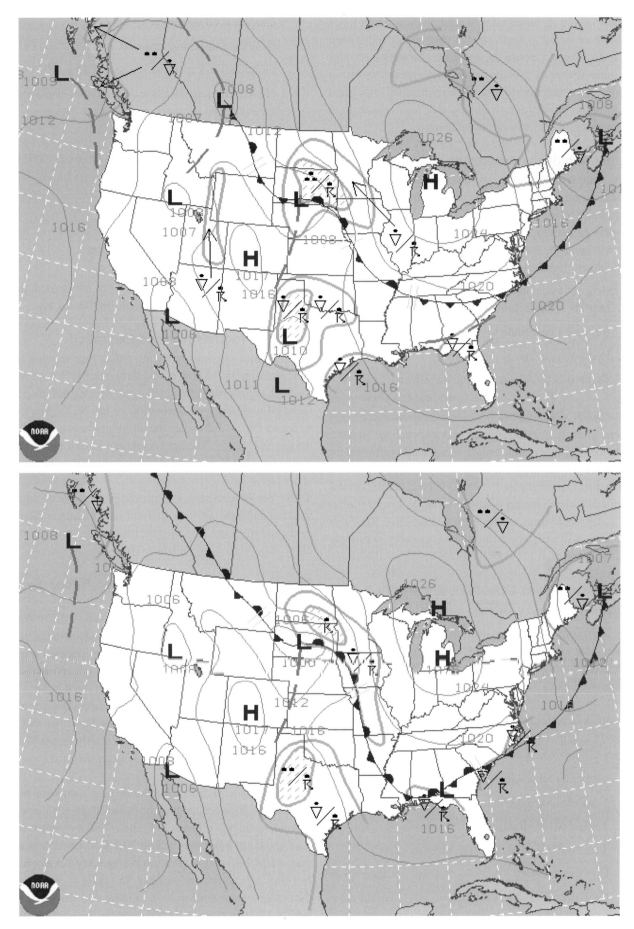

Figure 13-14. *36- (top) and 48-hour (bottom) surface prognostic chart.*

any associated with lesser rates of precipitation. Turbulence should be expected to occur near convective activity, even in clear air. Thunderstorms are a form of convective activity that imply severe or greater turbulence. Operation within 20 miles of thunderstorms should be approached with great caution, as the severity of turbulence can be much greater than the precipitation intensity might indicate.

Weather Avoidance Assistance

To the extent possible, controllers will issue pertinent information on weather and assist pilots in avoiding such areas when requested. Pilots should respond to a weather advisory by either acknowledging the advisory or by acknowledging the advisory and requesting an alternative course of action as follows:

- Request to deviate off course by stating the number of miles and the direction of the requested deviation.

- Request a new route to avoid the affected area.

- Request a change of altitude.

- Request radar vectors around the affected areas.

The controller's primary function is to provide safe separation between aircraft. Any additional service, such as weather avoidance assistance, can only be provided to the extent that it does not detract from the primary function. It's also worth noting that the separation workload is generally greater than normal when weather disrupts the usual flow of traffic. ATC radar limitations and frequency congestion may also be a factor in limiting the controller's capability to provide additional service.

Electronic Flight Displays (EFD) /Multi-Function Display (MFD) Weather

Many aircraft manufacturers now include data link weather services with new electronic flight display (EFD) systems. EFDs give a pilot access to many of the data link weather services available.

Products available to a pilot on the display pictured in *Figure 13-15* are listed as follows. The letters in parentheses indicate the soft key to press in order to access the data.

- Graphical NEXRAD data (NEXRAD)

- Graphical METAR data (METAR)

- Textual METAR data

- Textual terminal aerodrome forecasts (TAF)

- City forecast data

- Graphical wind data (WIND)

- Graphical echo tops (ECHO T,,,OPS)

- Graphical cloud tops (CLD TOPS)

- Graphical lightning strikes (LTNG)

- Graphical storm cell movement (CELL MOV)

- NEXRAD radar coverage (information displayed with the NEXRAD data)

- SIGMETs/AIRMETs (SIG/AIR)

- Surface analysis to include city forecasts (SFC)

- County warnings (COUNTY)

- Freezing levels (FRZ LVL)

- Hurricane track (CYCLONE)

- Temporary flight restrictions (TFR)

Pilots must be familiar with any EFD or MFD used and the data link weather products available on the display.

Weather Products Age and Expiration

The information displayed using a data link weather link is near real time but should not be thought of as instantaneous, up-to-date information. Each type of weather display is stamped with the age information on the MFD. The time is referenced from Zulu when the information was assembled at the ground station. The age should not be assumed to be the time when the FIS received the information from the data link.

Two types of weather are displayed on the screen: "current" weather and forecast data. Current information is displayed by an age while the forecast data has a data stamp in the form of "__ / __ __ : __." *[Figure 13-16]*

The Next Generation Weather Radar System (NEXRAD)

The NEXRAD system is comprised of a series of 159 Weather Surveillance Radar–1988 Doppler (WSR-88D) sites situated throughout the United States, as well as selected overseas sites. The NEXRAD system is a joint venture between the United States Department of Commerce (DOC), the United States DOD, as well as the United States Department of Transportation (DOT). The individual agencies that have control over the system are the NWS, Air Force Weather Agency (AFWA) and the FAA. *[Figure 13-17]*

NEXRAD data for up to a 2,000 mile range can be displayed. It is important to realize that the radar image is not real time and can be up to 5 minutes old. The NTSB has reported on 2 fatal accidents where in-cockpit NEXRAD mosaic imagery was available to pilots operating near quickly-developing and fast-moving convective weather. In one of these accidents, the images were from 6 to 8 minutes old. In some cases, NEXRAD data can age significantly by the time the mosaic image is created. In some extreme latency cases, the actual

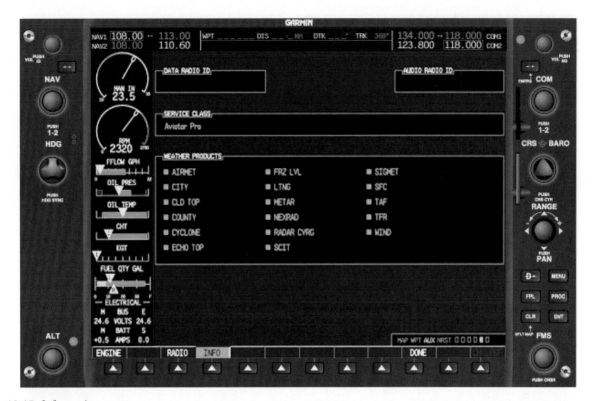

Figure 13-15. *Information page.*

age of the oldest NEXRAD data in the mosaic can exceed the age indication in the cockpit by 15 to 20 minutes. Even small-time differences between the age indicator and actual conditions can be important for safety of flight, especially when considering fast-moving weather hazards, quickly developing weather scenarios, and/or fast-moving aircraft. At no time should the images be used as storm penetrating radar nor to navigate through a line of storms. The images display should only be used as a reference.

NEXRAD radar is mutually exclusive of Topographic (TOPO), TERRAIN and STORMSCOPE. When NEXRAD is turned on, TOPO, TERRAIN, and STORMSCOPE are turned off because the colors used to display intensities are very similar.

Lightning information is available to assist when NEXRAD is enabled. This presents a more comprehensive picture of the weather in the surrounding area.

In addition to utilizing the soft keys to activate the NEXRAD display, the pilot also has the option of setting the desired range. It is possible to zoom in on a specific area of the display in order to gain a more detailed picture of the radar display. *[Figure 13-18]*

What Can Pilots Do?

Remember that the in-cockpit NEXRAD display depicts where the weather WAS, not where it IS. The age indicator does not show the age of the actual weather conditions, but rather the age of the mosaic image. The actual weather conditions could be up to 15 to 20 minutes OLDER than the age indicated on the display. You should consider this

Figure 13-16. *List of weather products and the expiration times of each.*

Figure 13-17. *NEXRAD radar display.*

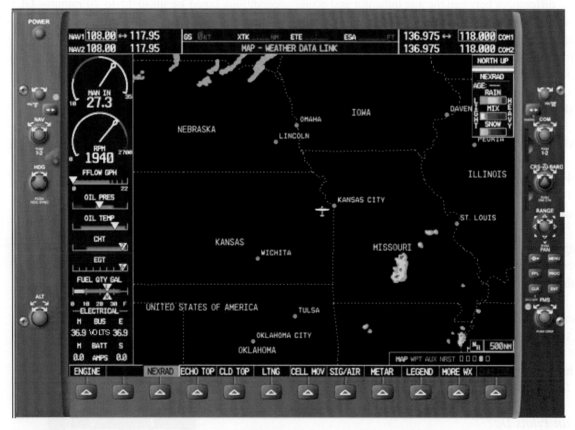

Figure 13-18. *NEXRAD radar display (500 mile range). The individual color gradients can be easily discerned and interpreted via the legend in the upper right corner of the screen. Additional information can be gained by pressing the LEGEND soft key, which displays the legend page.*

potential delay when using in-cockpit NEXRAD capabilities, as the movement and/or intensification of weather could adversely affect safety of flight.

- Understand that the common perception of a "5-minute latency" with radar data is not always correct.

- Get your preflight weather briefing! Having in-cockpit weather capabilities does not circumvent the need for a complete weather briefing before takeoff.

- Use all appropriate sources of weather information to make in-flight decisions.

- Let your fellow pilots know about the limitations of in-cockpit NEXRAD.

NEXRAD Abnormalities

Although NEXRAD is a compilation of stations across the country, there can be abnormalities associated with the system. Some of the abnormalities are listed below.

- Ground clutter

- Strobes and spurious radar data

- Sun strobes, when the radar antenna points directly at the sun

- Interference from buildings or mountains that may cause shadows

- Military aircraft that deploy metallic dust and may reflect the radar signature

NEXRAD Limitations

In addition to the abnormalities listed, the NEXRAD system does have some specific limitations.

Base Reflectivity

The NEXRAD base reflectivity does not provide adequate information from which to determine cloud layers or type of precipitation with respect to hail versus rain. Therefore, a pilot may mistake rain for hail.

In addition, the base reflectivity is sampled at the minimum antenna elevation angle. With this minimum angle, an individual site cannot depict high altitude storms directly over the station. This leaves an area of null coverage if an adjacent site does not also cover the affected area.

Resolution Display

The resolution of the displayed data poses additional concerns when the range is decreased. The minimum resolution for NEXRAD returns is 1.24 miles. This means that when the display range is zoomed in to approximately ten miles, the individual square return boxes are more prevalent. Each square indicates the strongest display return within that 1.24 mile square area.

AIRMET/SIGMET Display

AIRMET/SIGMET information is available for the displayed viewing range on the MFD. Some displays are capable of displaying weather information for a 2,000 mile range. AIRMETS/SIGMETS are displayed by dashed lines on the map. [Figure 13-19]

The legend box denotes the various colors used to depict the AIRMETs, such as icing, turbulence, IFR weather, mountain obscuration, and surface winds. [Figure 13-20] The great advantage of the graphically displayed AIRMET/SIGMET boundary box is the pilot can see the extent of the area that the advisory covers. The pilot does not need to manually plot the points to determine the full extent of the coverage area.

Graphical METARs

METARs can be displayed on the MFD. Each reporting station that has a METAR/TAF available is depicted by a flag from the center of the airport symbol. Each flag is color coded to depict the type of weather that is currently reported at that station. A legend is available to assist users in determining what each flag color represents. [Figure 13-21]

The graphical METAR display shows all available reporting stations within the set viewing range. By setting the range knob up to a 2,000 mile range, pilots can pan around the display map to check the current conditions of various airports along the route of flight.

By understanding what each colored flag indicates, a pilot can quickly determine where weather patterns display marginal weather, IFR, or areas of VFR. These flags make it easy to determine weather at a specific airport should the need arise to divert from the intended airport of landing.

Data Link Weather

Pilots now have the capability of receiving continuously updated weather across the entire country at any altitude. No longer are pilots restricted by radio range or geographic isolations, such as mountains or valleys.

In addition, pilots no longer have to request specific information from weather briefing personnel directly. When the weather becomes questionable, radio congestion often increases, delaying the timely exchange of valuable inflight weather updates for a pilot's specific route of flight. Flight Service Station (FSS) personnel can communicate with only one pilot at a time, which leaves other pilots waiting and flying in uncertain weather conditions. Data link weather

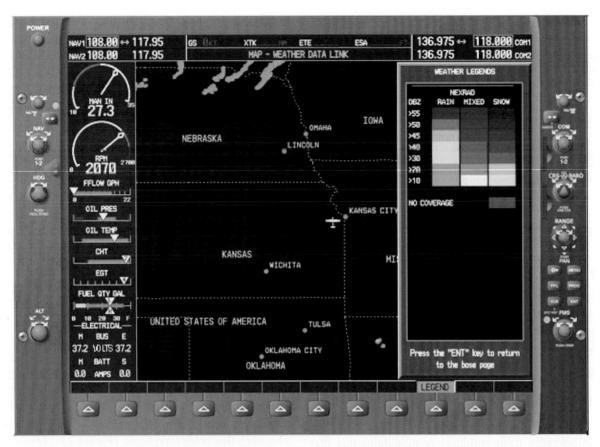

Figure 13-19. *The AIRMET information box instructs the pilot to press the ENTER button soft key (ENT) to gain additional information on the selected area of weather. Once the ENTER soft key (ENT) is depressed, the specific textual information is displayed on the right side of the screen.*

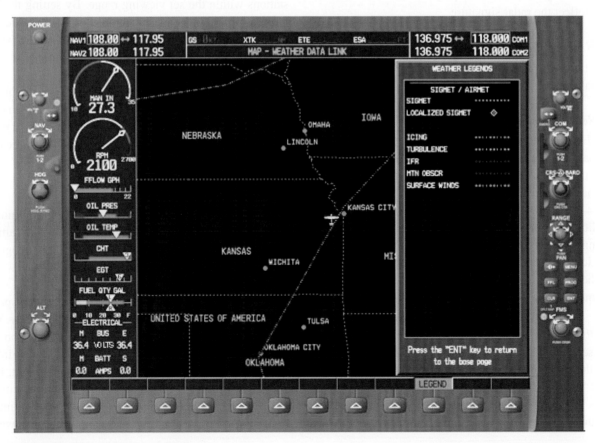

Figure 13-20. *SIGMET/AIRMET legend display.*

Figure 13-21. *Graphical METAR legend display.*

provides the pilot with a powerful resource for enhanced situational awareness at any time. Due to continuous data link broadcasts, pilots can obtain a weather briefing by looking at a display screen. Pilots have a choice between FAA-certified devices or portable receivers as a source of weather data.

Data Link Weather Products
Flight Information Service- Broadcast (FIS-B)

Flight Information Service–Broadcast (FIS-B) is a ground broadcast service provided through the Automatic Dependent Surveillance–Broadcast (ADS-B) Services network over the 978 MHz UAT data link. The FAA FIS-B system provides pilots and flight crews of properly-equipped aircraft with a flightdeck display of certain aviation weather and aeronautical information which are listed below.

- Aviation Routine Weather Reports (METARs)

- Special Aviation Reports (SPECIs)

- Terminal Area Forecasts (TAFs) and their amendments

- NEXRAD (regional and CONUS) precipitation maps

- Notice to Airmen (NOTAM) Distant and Flight Data Center

- Airmen's Meteorological Conditions (AIRMET)

- Significant Meteorological Conditions (SIGMET) and Convective SIGMET

- Status of Special Use Airspace (SUA)

- Temporary Flight Restrictions (TFRs)

- Winds and Temperatures Aloft.

- Pilot Reports (PIREPS)

- TIS-B service status

The weather products provided by FIS-B are for information only. Therefore, these products do not meet the safety and regulatory requirements of official weather products. The weather products displayed on FIS-B should not be used as primary weather products (i.e., aviation weather to meet operational and safety requirements). Each aircraft system is different and some of the data that is rendered can be up to 20 or 30 minutes old and not current. Pilots should consult the individual equipment manuals for specific delay times.

Pilot Responsibility

It is important for pilots to understand the realization that the derived safety benefits of data link depends heavily upon the pilot's understanding of the specific system's capabilities and limitations which are listed below.

- Product latency—be aware of the time stamp or "valid until" time on the particular data link information displayed in the flightdeck. For example, since initial processing and transmission of NEXRAD data can take several minutes, pilots must assume that data link weather information will always be a minimum of seven to eight minutes older than shown on the time stamp and only use data link weather radar images for broad strategic avoidance of adverse weather.

- Product update cycles—be aware of when and how often a product is updated as well as the Data Link Service Providers (DLSP) update rate for particular products.

- Indication of system failure—be aware of partial or total system failure indications.

- Coverage areas/service volume—coverage limitations are associated with the type of data link network being used. For example, ground-based systems that require a line-of-sight may have relatively limited coverage below 5,000 feet AGL. Satellite-based data link weather systems can have limitations stemming from whether the network is in geosynchronous orbit or low earth orbit. Also, NWS NEXRAD coverage has gaps, especially in the western states.

- Content/format—since service providers often refine or enhance data link products for flightdeck display, pilots must be familiar with the content, format, and meaning of symbols and displays (i.e., the legend) in the specific system.

- Data integrity/limitations to use—reliability of information depicted. Be aware of any applicable disclaimer provided by the service provider.

- Use of equipment/avionics display—pilots remain responsible for the proper use of an electronic flight bag (EFB) or installed avionics. Pilots should be cognizant that, per the FAA Practical Test Standards, they may be evaluated on the use and interpretation of an EFB or installed avionics on the aircraft.

- Overload of Information—most DLSPs offer numerous products with information that can be layered on top of each other. Pilots need to be aware that too much information can have a negative effect on their cognitive work load. Pilots need to manage the amount of information to a level that offers the most pertinent information to that specific flight without creating a distraction. Pilots may need to adjust the amount of information based on numerous factors including, but not limited to, the phase of flight, single pilot operation, autopilot availability, class of airspace, and the weather conditions encountered.

Chapter Summary

While no weather forecast is guaranteed to be 100 percent accurate, pilots have access to a myriad of weather information on which to base flight decisions. Weather products available for preflight planning to en route information received over the radio or via data link provide the pilot with the most accurate and up-to-date information available. Each report provides a piece of the weather puzzle. Pilots must use several reports to get an overall picture and gain an understanding of the weather that affects the safe completion of a flight.

Chapter 14
Airport Operations

Introduction

Each time a pilot operates an aircraft, the flight normally begins and ends at an airport. An airport may be a small sod field or a large complex utilized by air carriers. This chapter examines airport operations, identifies features of an airport complex, and provides information on operating on or in the vicinity of an airport.

Airport Categories

The definition for airports refers to any area of land or water used or intended for landing or takeoff of aircraft. This includes, within the five categories of airports listed below, special types of facilities including seaplane bases, heliports, and facilities to accommodate tilt rotor aircraft. An airport includes an area used or intended for airport buildings, facilities, as well as rights of way together with the buildings and facilities.

LEGEND

Recommended standard left-hand traffic pattern (depicted) ▓▓▓ (standard right-hand traffic pattern would be mirror image)

Application of traffic pattern indicators

Entry

Downwind

Segmented circle

Crosswind

Departure

Base

Final

RUNWAY

Departure

Depart

Tetrahedron

Wind Tee

Wind sock or cone

Color and Type of Signal	Movement of Vehicles, Equipment and Personnel	Aircraft on the Ground	Aircraft in Flight
Steady green	Cleared to cross, proceed or go	Cleared for takeoff	Cleared to land
Flashing green	Not applicable	Cleared for taxi	Return for landing (to be followed by steady green at the proper time
Steady red	Stop	Stop	Give way to other aircraft and continue circling
Flashing red	Clear the taxiway/runway	Taxi clear of the runway in use	Airport unsafe, do not land
Flashing white	Return to starting point on airport	Return to starting point on airport	Not applicable
Alternating red and green	Exercise extreme caution!!!!	Exercise extreme caution!!!!	

The law defines airports by categories of airport activities, including commercial service, primary, cargo service, reliever, and general aviation airports, as shown below:

- Commercial Service Airports—publicly owned airports that have at least 2,500 passenger boardings each calendar year and receive scheduled passenger service. Passenger boardings refer to revenue passenger boardings on an aircraft in service in air commerce whether or not in scheduled service. The definition also includes passengers who continue on an aircraft in international flight that stops at an airport in any of the 50 States for a non-traffic purpose, such as refueling or aircraft maintenance rather than passenger activity. Passenger boardings at airports that receive scheduled passenger service are also referred to as Enplanements.

- Cargo Service Airports—airports that, in addition to any other air transportation services that may be available, are served by aircraft providing air transportation of only cargo with a total annual landed weight of more than 100 million pounds. "Landed weight" means the weight of aircraft transporting only cargo in intrastate, interstate, and foreign air transportation. An airport may be both a commercial service and a cargo service airport.

- Reliever Airports—airports designated by the FAA to relieve congestion at Commercial Service Airports and to provide improved general aviation access to the overall community. These may be publicly or privately-owned.

- General Aviation Airports — the remaining airports are commonly described as General Aviation Airports. This airport type is the largest single group of airports in the U.S. system. The category also includes privately owned, public use airports that enplane 2500 or more passengers annually and receive scheduled airline service.

Types of Airports

There are two types of airports—towered and nontowered. These types can be further subdivided to:

- Civil Airports—airports that are open to the general public.

- Military/Federal Government airports—airports operated by the military, National Aeronautics and Space Administration (NASA), or other agencies of the Federal Government.

- Private Airports—airports designated for private or restricted use only, not open to the general public.

Towered Airport

A towered airport has an operating control tower. Air traffic control (ATC) is responsible for providing the safe, orderly, and expeditious flow of air traffic at airports where the type of operations and/or volume of traffic requires such a service. Pilots operating from a towered airport are required to maintain two-way radio communication with ATC and to acknowledge and comply with their instructions. Pilots must advise ATC if they cannot comply with the instructions issued and request amended instructions. A pilot may deviate from an air traffic instruction in an emergency, but must advise ATC of the deviation as soon as possible.

Nontowered Airport

A nontowered airport does not have an operating control tower. Two-way radio communications are not required, although it is a good operating practice for pilots to transmit their intentions on the specified frequency for the benefit of other traffic in the area. The key to communicating at an airport without an operating control tower is selection of the correct common frequency. The acronym CTAF, which stands for Common Traffic Advisory Frequency, is synonymous with this program. A CTAF is a frequency designated for the purpose of carrying out airport advisory practices while operating to or from an airport without an operating control tower. The CTAF may be a Universal Integrated Community (UNICOM), MULTICOM, Flight Service Station (FSS), or tower frequency and is identified in appropriate aeronautical publications. UNICOM is a nongovernment air/ground radio communication station that may provide airport information at public use airports where there is no tower or FSS. On pilot request, UNICOM stations may provide pilots with weather information, wind direction, the recommended runway, or other necessary information. If the UNICOM frequency is designated as the CTAF, it is identified in appropriate aeronautical publications. *Figure 14-1* lists recommended communication procedures. More information regarding radio communications is provided later in this chapter.

Nontowered airport traffic patterns are always entered at pattern altitude. How you enter the pattern depends upon the direction of arrival. The preferred method for entering from the downwind side of the pattern is to approach the pattern on a course 45 degrees to the downwind leg and join the pattern at midfield.

There are several ways to enter the pattern if you're coming from the upwind leg side of the airport. One method of entry from the opposite side of the pattern is to announce your intentions and cross over midfield at least 500 feet above

Facility at Airport	Frequency Use	Communication/Broadcast Procedures		
		Outbound	Inbound	Practice Instrument Approach
UNICOM (no tower or FSS)	Communicate with UNICOM station on published CTAF frequency (122.7, 122.8, 122.725, 122.975, or 123.0). If unable to contact UNICOM station, use self-announce procedures on CTAF.	Before taxiing and before taxiing on the runway for departure.	10 miles out. Entering downwind, base, and final. Leaving the runway.	
No tower, FSS, or UNICOM	Self-announce on MULTICOM frequency 122.9.	Before taxiing and before taxiing on the runway for departure.	10 miles out. Entering downwind, base, and final. Leaving the runway.	Departing final approach fix (name) or on final approach segment inbound.
No tower in operation, FSS open	Communicate with FSS on CTAF frequency.	Before taxiing and before taxiing on the runway for departure.	10 miles out. Entering downwind, base, and final. Leaving the runway.	Approach completed/terminated.
FSS closed (no tower)	Self-announce on CTAF.	Before taxiing and before taxiing on the runway for departure.	10 miles out. Entering downwind, base, and final. Leaving the runway.	
Tower or FSS not in operation	Self-announce on CTAF.	Before taxiing and before taxiing on the runway for departure.	10 miles out. Entering downwind, base, and final. Leaving the runway.	

Figure 14-1. *Recommended communication procedures.*

pattern altitude (normally 1,500 feet AGL.) However, if large or turbine aircraft operate at your airport, it is best to remain 2,000 feet AGL so you are not in conflict with their traffic pattern. When well clear of the pattern—approximately 2 miles–scan carefully for traffic, descend to pattern altitude, then turn right to enter at 45° to the downwind leg at midfield. *[Figure 14-2]*

An alternate method is to enter on a midfield crosswind at pattern altitude, carefully scan for traffic, announce your intentions, and then turn downwind. *[Figure 14-3]* This technique should not be used if the pattern is busy. Always remember to give way to aircraft on the preferred 45° entry and to aircraft already established on downwind.

In either case, it is vital to announce your intentions, and remember to scan outside. Before joining the downwind leg, adjust your course or speed to blend into the traffic. Adjust power on the downwind leg, or sooner, to fit into the flow of traffic. Avoid flying too fast or too slow. Speeds recommended by the airplane manufacturer should be used. They will generally fall between 70 to 80 knots for fixed-gear singles and 80 to 90 knots for high-performance retractable.

Sources for Airport Data

When a pilot flies into a different airport, it is important to review the current data for that airport. This data provides the pilot with information, such as communication frequencies, services available, closed runways, or airport construction. Three common sources of information are:

- Aeronautical Charts

- Chart Supplement U.S. (formerly Airport/Facility Directory)

- Notices to Airmen (NOTAMs)

- Automated Terminal Information Service (ATIS)

Aeronautical Charts

Aeronautical charts provide specific information on airports. Chapter 16, "Navigation," contains an excerpt from an aeronautical chart and an aeronautical chart legend, which provides guidance on interpreting the information on the chart.

Chart Supplement U.S. (formerly Airport/Facility Directory)

The Chart Supplement U.S. (formerly Airport/Facility Directory) provides the most comprehensive information on a given airport. It contains information on airports, heliports, and seaplane bases that are open to the public. The Chart Supplement U.S. is published in seven books, which are organized by regions and are revised every 56 days. The Chart Supplement U.S. is also available digitally at www.faa.

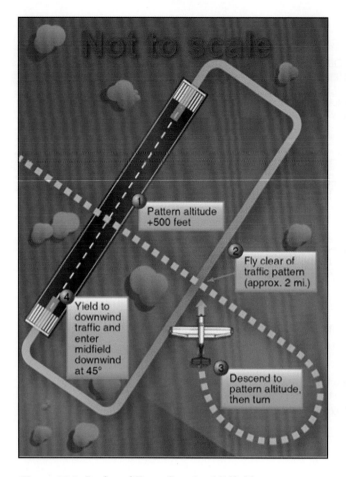

Figure 14-2. *Preferred Entry-Crossing Midfield.*

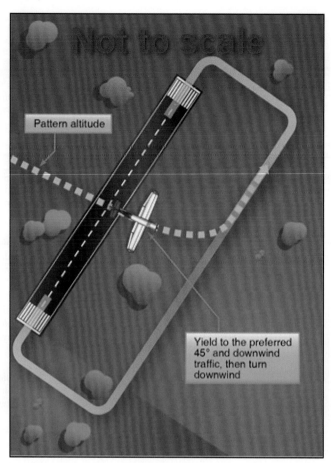

Figure 14-3. *Alternate Midfield Entry.*

gov/air_traffic/flight_info/aeronav. *Figure 14-4* contains an excerpt from a directory. For a complete listing of information provided in a Chart Supplement U.S. and how the information may be decoded, refer to the "Legend Sample" located in the front of each Chart Supplement U.S.

In addition to airport information, each Chart Supplement U.S. contains information such as special notices, Federal Aviation Administration (FAA) and National Weather Service (NWS) telephone numbers, preferred instrument flight rules (IFR) routing, visual flight rules (VFR) waypoints, a listing of very high frequency (VHF) omnidirectional range (VOR) receiver checkpoints, aeronautical chart bulletins, land and hold short operations (LAHSO) for selected airports, airport diagrams for selected towered airports, en route flight advisory service (EFAS) outlets, parachute jumping areas, and facility telephone numbers. It is beneficial to review a Chart Supplement U.S. to become familiar with the information it contains.

Notices to Airmen (NOTAM)

Time-critical aeronautical information, which is of a temporary nature or not sufficiently known in advance to permit publication, on aeronautical charts or in other operational

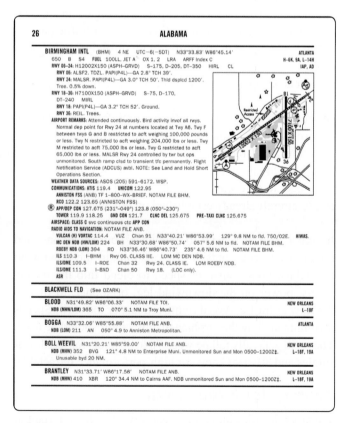

Figure 14-4. *Chart Supplement U.S. (formerly Airport/Facility Directory excerpt.*

publications receives immediate dissemination by the NOTAM system. The NOTAM information could affect your decision to make the flight. It includes such information as taxiway and runway closures, construction, communications, changes in status of navigational aids, and other information essential to planned en route, terminal, or landing operations. Exercise good judgment and common sense by carefully regarding the information readily available in NOTAMs.

Prior to any flight, pilots should check for any NOTAMs that could affect their intended flight. For more information on NOTAMs, refer back to Chapter 1, "Pilot and Aeronautical Information" section.

Automated Terminal Information Service (ATIS)

The Automated Terminal Information Service (ATIS) is a recording of the local weather conditions and other pertinent non-control information broadcast on a local frequency in a looped format. It is normally updated once per hour but is updated more often when changing local conditions warrant. Important information is broadcast on ATIS including weather, runways in use, specific ATC procedures, and any airport construction activity that could affect taxi planning.

When the ATIS is recorded, it is given a code. This code is changed with every ATIS update. For example, ATIS Alpha is replaced by ATIS Bravo. The next hour, ATIS Charlie is recorded, followed by ATIS Delta and progresses down the alphabet.

Prior to calling ATC, tune to the ATIS frequency and listen to the recorded broadcast. The broadcast ends with a statement containing the ATIS code. For example, "Advise on initial contact, you have information Bravo." Upon contacting the tower controller, state information Bravo was received. This allows the tower controller to verify the pilot has the current local weather and airport information without having to state it all to each pilot who calls. This also clears the tower frequency from being overtaken by the constant relay of the same information, which would result without an ATIS broadcast. The use of ATIS broadcasts at departure and arrival airports is not only a sound practice but a wise decision.

Airport Markings and Signs

There are markings and signs used at airports that provide directions and assist pilots in airport operations. It is important for you to know the meanings of the signs, markings, and lights that are used on airports as surface navigational aids. All airport markings are painted on the surface, whereas some signs are vertical and some are painted on the surface. An overview of the most common signs and markings are described on the following pages. Additional information may be found in Chapter 2,

"Aeronautical Lighting and Other Airport Visual Aids," of the Aeronautical Information Manual (AIM).

Runway Markings and Signs

Runway markings vary depending on the type of operations conducted at the airport. A basic VFR runway may only have centerline markings and runway numbers. Refer to Appendix C of this publication for an example of the most common runway markings that are found at airports.

Since aircraft are affected by the wind during takeoffs and landings, runways are laid out according to the local prevailing winds. Runway numbers are in reference to magnetic north. Certain airports have two or even three runways laid out in the same direction. These are referred to as parallel runways and are distinguished by a letter added to the runway number (e.g., runway 36L (left), 36C (center), and 36R (right)).

Relocated Runway Threshold

It is sometimes necessary, due to construction or runway maintenance, to close only a portion of a runway. When a portion of a runway is closed, the runway threshold is relocated as necessary. It is referred to as a relocated threshold and methods for identifying the relocated threshold vary. A common way for the relocated threshold to be marked is a ten foot wide white bar across the width of the runway. *[Figure 14-5A and B]*

When the threshold is relocated, the closed portion of the runway is not available for use by aircraft for takeoff or landing, but it is available for taxi. When a threshold is relocated, it closes not only a set portion of the approach end of a runway, but also shortens the length of the opposite direction runway. Yellow arrow heads are placed across the width of the runway just prior to the threshold bar.

Displaced Threshold

A displaced threshold is a threshold located at a point on the runway other than the designated beginning of the runway. Displacement of a threshold reduces the length of runway available for landings. The portion of runway behind a displaced threshold is available for takeoffs in either direction, or landings from the opposite direction. A ten feet wide white threshold bar is located across the width of the runway at the displaced threshold, and white arrows are located along the centerline in the area between the beginning of the runway and displaced threshold. White arrow heads are located across the width of the runway just prior to the threshold bar. *[Figure 14-6A and B]*

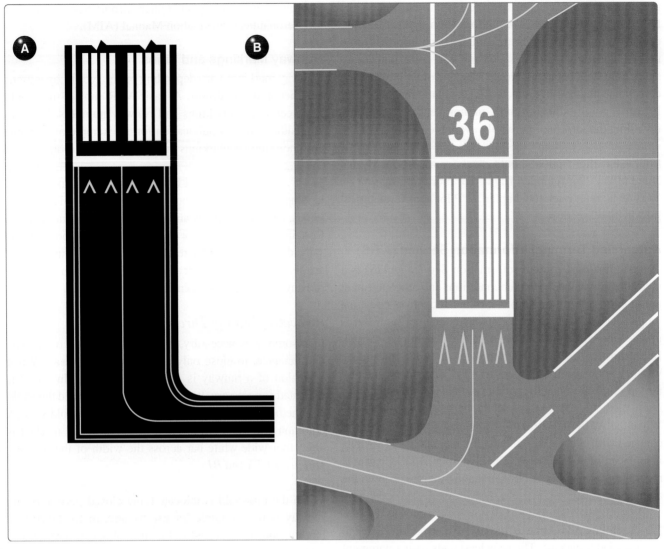

Figure 14-5. *(A) Relocated runway threshold drawing. (B) Relocated threshold for Runway 36 at Joplin Regional Airport (JLN).*

Runway Safety Area

The runway safety area (RSA) is a defined surface surrounding the runway prepared, or suitable, for reducing the risk of damage to airplanes in the event of an undershoot, overshoot, or excursion from the runway. The dimensions of the RSA vary and can be determined by using the criteria contained within AC 150/5300-13, Airport Design, Chapter 3. Figure 3-1 in AC 150/5300-13 depicts the RSA. Additionally, it provides greater accessibility for firefighting and rescue equipment in emergency situations.

The RSA is typically graded and mowed. The lateral boundaries are usually identified by the presence of the runway holding position signs and markings on the adjoining taxiway stubs. Aircraft should not enter the RSA without making sure of adequate separation from other aircraft during operations at uncontrolled airports. *[Figure 14-7]*

Runway Safety Area Boundary Sign

Some taxiway stubs also have a runway safety area boundary sign that faces the runway and is visible to you only when exiting the runway. This sign has a yellow background with black markings and is typically used at towered airports where a controller commonly requests you to report clear of a runway. This sign is intended to provide you with another visual cue that is used as a guide to determine when you are clear of the runway safety boundary area. The sign shown in *Figure 14-8* is what you would see when exiting the runway at Taxiway Kilo. You are out of the runway safety area boundary when the entire aircraft passes the sign and the accompanying surface painted marking.

Runway Holding Position Sign

Noncompliance with a runway holding position sign may result in the FAA filing a Pilot Deviation against you. A

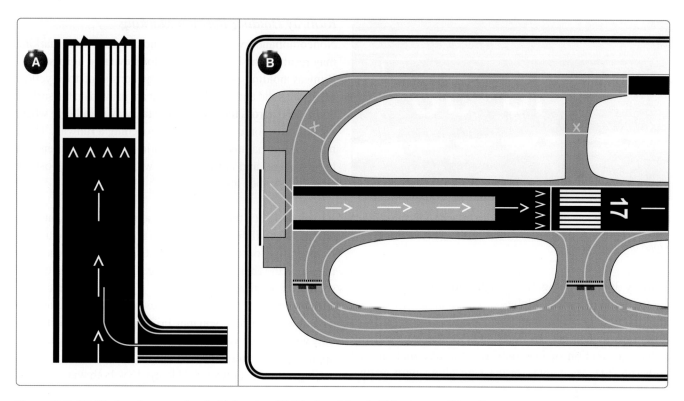

Figure 14-6. *(A) Displaced runway threshold drawing. (B) Displaced threshold for Runway 17 at Albuquerque International Airport (ABQ).*

runway holding position sign is an airport version of a stop sign. *[Figure 14-9]* It may be seen as a sign and/or its characters painted on the airport pavement. The sign has white characters outlined in black on a red background. It is always collocated with the surface painted holding position markings and is located where taxiways intersect runways. On taxiways that intersect the threshold of the takeoff runway, only the designation of the runway may appear on the sign.

If a taxiway intersects a runway somewhere other than at the threshold, the sign has the designation of the intersecting runway. The runway numbers on the sign are arranged to correspond to the relative location of the respective runway thresholds. *Figure 14-10* shows "18-36" to indicate the threshold for Runway 18 is to the left and the threshold for

Figure 14-8. *Runway safety area boundary sign and marking located on Taxiway Kilo.*

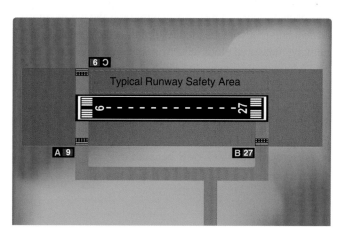

Figure 14-7. *Runway Safety Area.*

Figure 14-9. *Runway holding position sign at takeoff end of Runway 14 with collocated Taxiway Alpha location sign.*

Figure 14-10. *Runway holding position sign at a location other than the takeoff end of Runway 18-36 with collocated Taxiway Alpha location sign.*

Runway 36 is to the right. The sign also indicates that you are located on Taxiway Alpha.

If the runway holding position sign is located on a taxiway at the intersection of two runways, the designations for both runways are shown on the sign along with arrows showing the approximate alignment of each runway. *[Figure 14-11A and B]* In addition to showing the approximate runway alignment, the arrows indicate the direction(s) to the threshold of the runway whose designation is immediately next to each corresponding arrow.

This type of taxiway and runway/runway intersection geometry can be very confusing and create navigational challenges. Extreme caution must be exercised when taxiing onto or crossing this type of intersection. *Figure 14-11A and B* shows a depiction of a taxiway, runway/runway intersection and is also designated as a **"hot spot"** on the airport diagram. In the example, Taxiway Bravo intersects with two runways, 31-13 and 35-17, which cross each other.

Surface painted runway holding position signs may also be used to aid you in determining the holding position. These markings consist of white characters on a red background and are painted on the left side of the taxiway centerline. *Figure 14-12* shows a surface painted runway holding position sign that is the holding point for Runway 32R-14L.

You should never allow any part of your aircraft to cross the runway holding position sign (either a vertical or surface painted sign) without a clearance from ATC. Doing so poses a hazard to yourself and others.

When the tower is closed or you are operating at a nontowered airport, you may taxi past a runway holding position sign only when the runway is clear of aircraft, and there are no aircraft on final approach. You may then proceed with extreme caution.

Runway Holding Position Marking

Noncompliance with a runway holding position marking may result in the FAA filing a Pilot Deviation against you. Runway holding position markings consist of four yellow lines, two solid and two dashed, that are painted on the surface and extend across the width of the taxiway to indicate where the aircraft should stop when approaching a runway. These markings are painted across the entire taxiway pavement, are in alignment, and are collocated with the holding position sign as described above.

As you approach the runway, two solid yellow lines and two dashed lines will be visible. Prior to reaching the solid lines, it is imperative to stop and ensure that no portion of the aircraft intersects the first solid yellow line. Do not cross the double solid lines until a clearance from ATC has been received. *[Figure 14-13]* When the tower is closed or when operating at a nontowered airport, you may taxi onto or across the runway only when the runway is clear and there are no aircraft on final approach. You should use extreme caution when crossing or taxiing onto the runway and always look both ways.

When exiting the runway, the same markings will be seen except the aircraft will be approaching the double dashed lines. *[Figure 14-14]* In order to be clear of the runway, the entire aircraft must cross both the dashed and solid lines. An ATC clearance is not needed to cross this marking when exiting the runway.

Runway Distance Remaining Signs

Runway distance remaining signs have a black background with a white number and may be installed along one or both sides of the runway. *[Figure 14-15]* The number on the signs indicates the distance, in thousands of feet, of landing runway remaining. The last sign, which has the numeral "1," is located at least 950 feet from the runway end.

Runway Designation Marking

Runway numbers and letters are determined from the approach direction. The runway number is the whole number nearest one-tenth the magnetic azimuth of the centerline of the runway, measured clockwise from the magnetic north. In the case where there are parallel runways, the letters differentiate between left (L), right (R), or center (C). *[Figure 14-16]* For example, if there are two parallel runways, they would show the designation number and then either L or R beneath it. For three parallel runways, the designation number would be presented with L, C, or R beneath it.

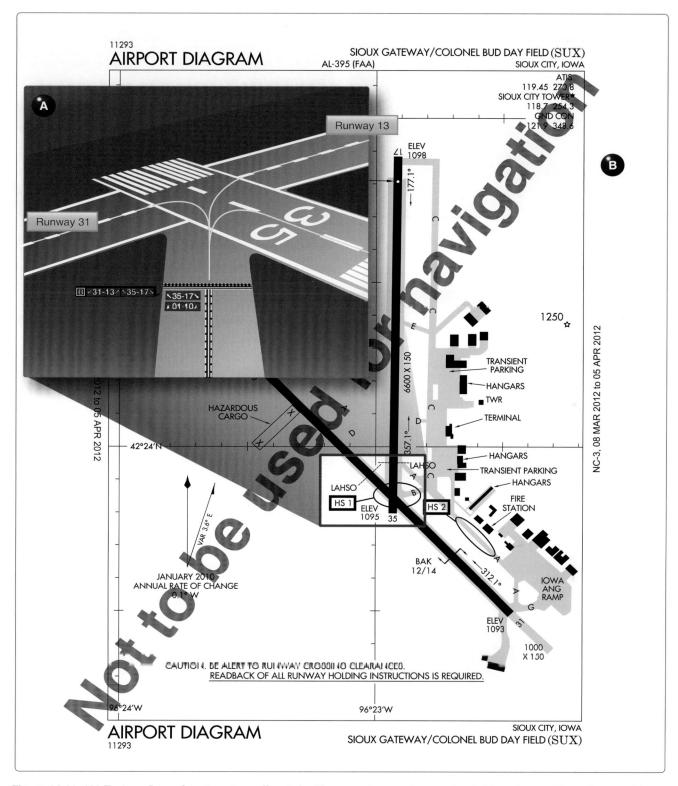

Figure 14-11. *(A) Taxiway Bravo location sign collocated with runway/runway intersection holding signs at Sioux Gateway Airport (SUX) (B) Airport diagram of Sioux Gateway Airport (SUX), Sioux City, Iowa. The area outlined in red is a designated "**hot spot**" (HS1).*

Figure 14-12. *Surface painted runway holding position signs for Runway 32R-14L along with the enhanced taxiway centerline marking.*

Figure 14-13. *Surface painted holding position marking along with enhanced taxiway centerline.*

Figure 14-14. *Runway holding position markings as seen when exiting the runway. When exiting the runway, no ATC clearance is required to cross.*

Land and Hold Short Operations (LAHSO)

When simultaneous operations (takeoffs and landings) are being conducted on intersecting runways, Land and Hold Short Operations (LAHSO) may also be in effect. LAHSO is an ATC procedure that may require your participation and

Figure 14-15. *Runway distance remaining sign indicating that there is 2,000 feet of runway remaining.*

compliance. As pilot in command (PIC), you have the final authority to accept or decline any LAHSO clearance.

If issued a land and hold short clearance, you must be aware of the reduced runway distances and whether or not you can comply before accepting the clearance. You do not have to accept a LAHSO clearance. Pilots should only receive a LAHSO clearance when there is a minimum ceiling of 1,000 feet and 3 statute miles of visibility.

Runway holding position signs and markings are installed on those runways used for LAHSO. The signs and markings are placed at the LAHSO point to aid you in determining where to stop and hold the aircraft and are located prior to the runway/runway intersection. *[Figure 14-17]*

The holding position sign has a white inscription with black border around the numbers on a red background and is installed adjacent to the holding position markings. If you accept a land and hold short clearance, you must comply so that no portion of the aircraft extends beyond these hold markings.

If receiving "cleared to land" instructions from ATC, you are authorized to use the entire landing length of the runway and should disregard any LAHSO holding position markings located on the runway. If you receive and accept LAHSO instructions, you must stop short of the intersecting runway prior to the LAHSO signs and markings.

Below is a list of items which, if thoroughly understood and complied with, will ensure that LAHSO operations are conducted properly.

- Know landing distance available.

- Be advised by ATC as to why LAHSO are being conducted.

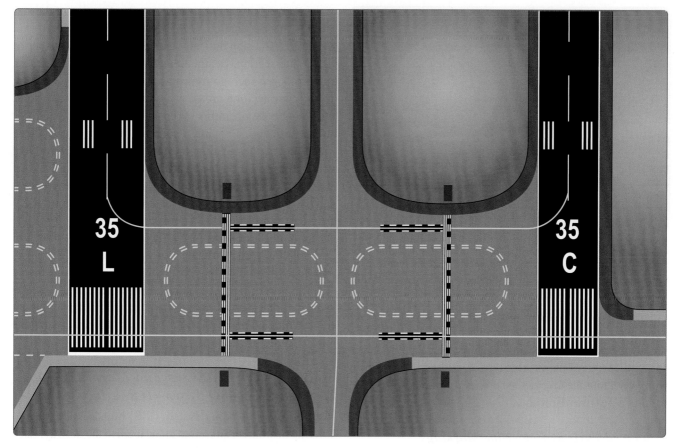

Figure 14-16. *Two of three parallel runways.*

- Advise ATC if you cannot comply with LAHSO.

- Know what signs and markings are at the LAHSO point.

- LAHSO are not authorized for student pilots who are performing a solo flight.

- At many airports air carrier aircraft are not authorized to participate in LAHSO if the other aircraft is a general aviation aircraft.

- Generally, LAHSO are not authorized at night.

- LAHSO are not authorized on wet runways.

Figure 14-17. *Runway holding position sign and marking for LAHSO.*

If you accept the following clearance from ATC: "Cleared to land Runway 36 hold short of Runway 23," you must either exit Runway 36 or stop at the holding position prior to Runway 23.

Taxiway Markings and Signs

Taxiway direction signs have a yellow background and black characters, which identifies the designation or intersecting taxiways. Arrows indicate the direction of turn that would place the aircraft on the designated taxiway. [*Figure 14-18*] Direction signs are normally located on the left side of the taxiway and prior to the intersection. These signs and markings (with a yellow background and black characters) indicate the direction toward a different taxiway, leading off a runway, or out of an intersection. *Figure 14-18* shows Taxiway Delta and how Taxiway Bravo intersects ahead at 90° both left and right.

Taxiway direction signs can also be displayed as surface painted markings. *Figure 14-19* shows Taxiway Bravo as proceeding straight ahead while Taxiway Alpha turns to the right at approximately 45°.

Figure 14-18. *Taxiway Bravo direction sign with a collocated Taxiway Delta location sign. When the arrow on the direction sign indicates a turn, the sign is located prior to the intersection.*

Figure 14-20A and *B* shows an example of a direction sign at a complex taxiway intersection. *Figure 14-20A* and *B* shows Taxiway Bravo intersects with Taxiway Sierra at 90°, but at 45° with Taxiway Foxtrot. This type of array can be displayed with or without the taxiway location sign, which in this case would be Taxiway Bravo.

Enhanced Taxiway Centerline Markings

At most towered airports, the enhanced taxiway centerline marking is used to warn you of an upcoming runway. It consists of yellow dashed lines on either side of the normal solid taxiway centerline and the dashes extend up to 150 feet prior to a runway holding position marking. *[Figure 14-21A and B]* They are used to aid you in maintaining awareness during surface movement to reduce runway incursions.

Destination Signs

Destination signs have black characters on a yellow background indicating a destination at the airport. These

Figure 14-19. *Surface painted taxiway direction signs.*

signs always have an arrow showing the direction of the taxi route to that destination. *[Figure 14-22]* When the arrow on the destination sign indicates a turn, the sign is located prior to the intersection. Destinations commonly shown on these types of signs include runways, aprons, terminals, military areas, civil aviation areas, cargo areas, international areas, and fixed-base operators. When the inscription for two or more destinations having a common taxi route are placed on a sign, the destinations are separated by a "dot" (•) and one arrow would be used as shown in *Figure 14-22*. When the inscription on a sign contains two or more destinations having different taxi routes, each destination is accompanied by an arrow and separated from the other destination(s) on the sign with a vertical black message divider as shown in *Figure 14-23*. The example shown in *Figure 14-23* shows two signs. The sign in the foreground explains that Runway 20 threshold is to the left, and Runways 32, 2, and 14 are to the right. The sign in the background indicates that you are located on Taxiway Bravo and Taxiway November will take you to those runways.

Holding Position Signs and Markings for an Instrument Landing System (ILS) Critical Area

The instrument landing system (ILS) broadcasts signals to arriving instrument aircraft to guide them to the runway. Each of these ILSs have critical areas that must be kept clear of all obstacles in order to ensure quality of the broadcast signal. At many airports, taxiways extend into the ILS critical area. Most of the time, this is of no concern; however, during times of poor weather, an aircraft on approach may depend on a good signal quality. When necessary, ATC will protect the ILS critical area for arrival instrument traffic by instructing taxiing aircraft to "**hold short**" of Runway (XX) ILS critical area.

The ILS critical area hold sign has white characters, outlined in black, on a red background and is installed adjacent to the ILS holding position markings. *[Figure 14-24]* The holding position markings for the ILS critical area appear on the pavement as a horizontal yellow ladder extending across the width of the taxiway.

When instructed to "hold short of Runway (XX) ILS critical area," you must ensure no portion of the aircraft extends beyond these markings. *[Figure 14-25]* If ATC does not instruct you to hold at this point, then you may bypass the ILS critical area hold position markings and continue with your taxi. *Figure 14-24* shows that the ILS hold sign is located on Taxiway Golf and the ILS ladder hold position marking is adjacent to the hold sign.

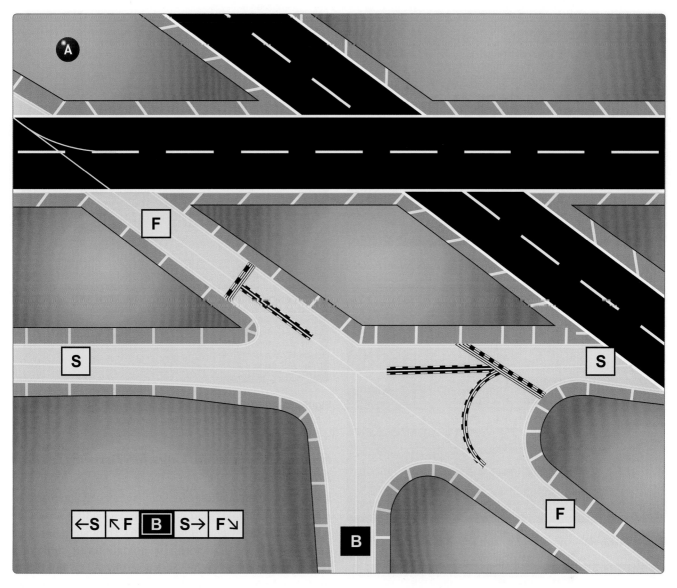

Figure 14-20. *Orientation of signs is from left to right in a clockwise manner. Left turn signs are on the left and right turn on the right. In this view, the pilot is on Taxiway Bravo.*

Enhanced taxiway centerline marking extends 150 feet prior to a runway holding position marking. **Prepare to STOP.**

Prepare to STOP unless you have been cleared onto or across the runway by ATC.

Figure 14-21. *(A) Enhanced taxiway centerline marking. (B) Enhanced taxiway centerline marking and runway holding position marking.*

Figure 14-22. *Destination sign to the fixed-base operator (FBO).*

Figure 14-23. *Runway destination sign with different taxi routes.*

Figure 14-24. *Instrument landing system (ILS) holding position sign and marking on Taxiway Golf.*

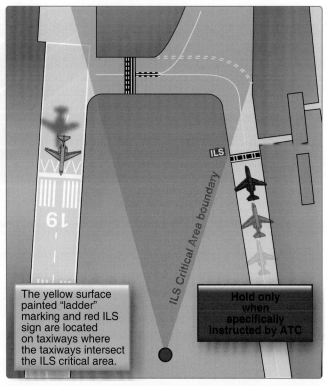

The yellow surface painted "ladder" marking and red ILS sign are located on taxiways where the taxiways intersect the ILS critical area.

ILS Critical Area boundary

Hold only when specifically instructed by ATC

Figure 14-25. *Holding position sign and marking for instrument landing system (ILS) critical area boundary.*

Holding Position Markings for Taxiway/Taxiway Intersections

Holding position markings for taxiway/taxiway intersections consist of a single dashed yellow line extending across the width of the taxiway. *[Figure 14-26]* They are painted on taxiways where ATC normally holds aircraft short of a taxiway intersection. When instructed by ATC "hold short of Taxiway X," you should stop so that no part of your aircraft extends beyond the holding position marking. When the marking is not present, you should stop your aircraft at a point that provides adequate clearance from an aircraft on the intersecting taxiway.

Marking and Lighting of Permanently Closed Runways and Taxiways

For runways and taxiways that are permanently closed, the lighting circuits are disconnected. The runway threshold, runway designation, and touchdown markings are obliterated and yellow "Xs" are placed at each end of the runway and at 1,000-foot intervals.

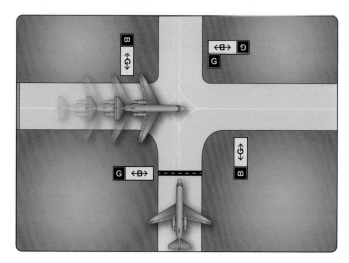

Figure 14-26. *Holding position marking on a taxiway.*

Temporarily Closed Runways and Taxiways

For temporarily closed runways and taxiways, a visual indication is often provided with yellow "Xs" or raised lighted yellow "Xs" placed at each end of the runway. Depending on the reason for the closure, duration of closure, airfield configuration, and the existence and the hours of operation of an ATC tower, a visual indication may not be present. As discussed previously in the chapter, you must always check NOTAMs and ATIS for runway and taxiway closure information.

Figure 14-27A shows an example of a yellow "X" laid flat with an adequate number of heavy sand bags to keep the wind from getting under and displacing the vinyl material.

A very effective and preferable visual aid to depict temporary closure is the lighted "X" placed on or near the runway designation numbers. *[Figure 14-27B and C]* This device is much more discernible to approaching aircraft than the other materials described above.

Other Markings

Some other markings found on the airport include vehicle roadway markings, VOR receiver checkpoint markings, and non-movement area boundary markings.

Airport Signs

There are six types of signs that may be found at airports. The more complex the layout of an airport, the more important the signs become to pilots. Appendix C of this publication shows examples of some signs that are found at most airports, their purpose, and appropriate pilot action. The six types of signs are:

- Mandatory instruction signs—red background with white inscription. These signs denote an entrance to a runway, critical area, or prohibited area.

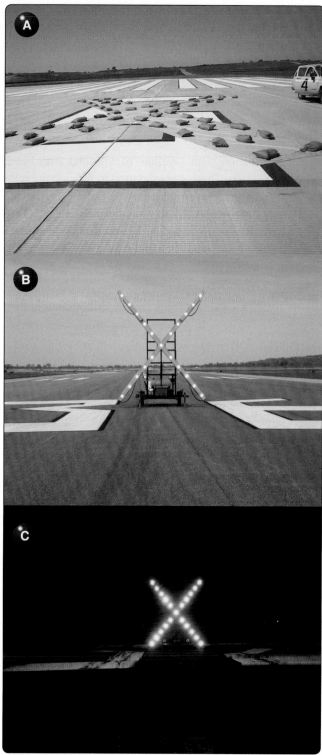

Figure 14-27. *(A) Yellow "X" placed on surface of temporarily closed runways. (B) Lighted "X" placed on temporarily closed runways. (C) Lighted "X" at night showing a temporarily closed runway.*

- Location signs—black with yellow inscription and a yellow border, no arrows. They are used to identify a taxiway or runway location, to identify the boundary of the runway, or identify an instrument landing system (ILS) critical area.

- Direction signs—yellow background with black inscription. The inscription identifies the designation of the intersecting taxiway(s) leading out of an intersection.

- Destination signs—yellow background with black inscription and arrows. These signs provide information on locating areas, such as runways, terminals, cargo areas, and civil aviation areas.

- Information signs—yellow background with black inscription. These signs are used to provide the pilot with information on areas that cannot be seen from the control tower, applicable radio frequencies, and noise abatement procedures. The airport operator determines the need, size, and location of these signs.

- Runway distance remaining signs—black background with white numbers. The numbers indicate the distance of the remaining runway in thousands of feet.

Airport Lighting

The majority of airports have some type of lighting for night operations. The variety and type of lighting systems depends on the volume and complexity of operations at a given airport. Airport lighting is standardized so that airports use the same light colors for runways and taxiways.

Airport Beacon

Airport beacons help a pilot identify an airport at night. The beacons are normally operated from dusk until dawn. Sometimes they are turned on if the ceiling is less than 1,000 feet and/or the ground visibility is less than 3 statute miles (VFR minimums). However, there is no requirement for this, so a pilot has the responsibility of determining if the weather meets VFR requirements. The beacon has a vertical light distribution to make it most effective from 1–10° above the horizon, although it can be seen well above or below this spread. The beacon may be an omnidirectional capacitor-discharge device, or it may rotate at a constant speed, that produces the visual effect of flashes at regular intervals. The combination of light colors from an airport beacon indicates the type of airport. *[Figure 14-28]* Some of the most common beacons are:

- Flashing white and green for civilian land airports

- Flashing white and yellow for a water airport

- Flashing white, yellow, and green for a heliport

- Two quick white flashes alternating with a green flash identifying a military airport

Approach Light Systems

Approach light systems are primarily intended to provide a means to transition from instrument flight to visual flight for landing. The system configuration depends on whether the

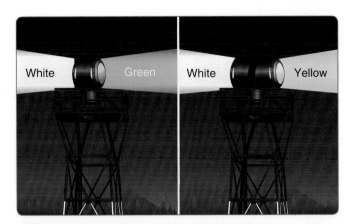

Figure 14-28. *Airport rotating beacons.*

runway is a precision or nonprecision instrument runway. Some systems include sequenced flashing lights that appear to the pilot as a ball of light traveling toward the runway at high speed. Approach lights can also aid pilots operating under VFR at night.

Visual Glideslope Indicators

Visual glideslope indicators provide the pilot with glidepath information that can be used for day or night approaches. By maintaining the proper glidepath as provided by the system, a pilot should have adequate obstacle clearance and should touch down within a specified portion of the runway.

Visual Approach Slope Indicator (VASI)

VASI installations are the most common visual glidepath systems in use. The VASI provides obstruction clearance within 10° of the extended runway centerline and up to four nautical miles (NM) from the runway threshold.

The VASI consists of light units arranged in bars. There are 2-bar and 3-bar VASIs. The 2-bar VASI has near and far light bars and the 3-bar VASI has near, middle, and far light bars. Two-bar VASI installations provide one visual glidepath that is normally set at 3°. The 3-bar system provides two glidepaths, the lower glidepath normally set at 3° and the upper glidepath ¼ degree above the lower glidepath.

The basic principle of the VASI is that of color differentiation between red and white. Each light unit projects a beam of light, a white segment in the upper part of the beam and a red segment in the lower part of the beam. The lights are arranged so the pilot sees the combination of lights shown in *Figure 14-29* to indicate below, on, or above the glidepath.

Other Glidepath Systems

A precision approach path indicator (PAPI) uses lights similar to the VASI system, except they are installed in a single row, normally on the left side of the runway. *[Figure 14-30]*

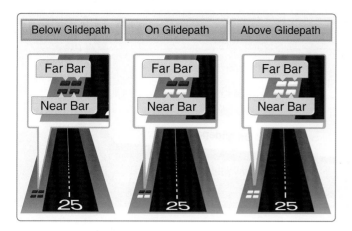

Figure 14-29. *Two-bar VASI system.*

A tri-color system consists of a single-light unit projecting a three-color visual approach path. Below the glidepath is indicated by red, on the glidepath is indicated by green, and above the glidepath is indicated by amber. When descending below the glidepath, there is a small area of dark amber. Pilots should not mistake this area for an "above the glidepath" indication. *[Figure 14-31]*

Pulsating VASIs normally consist of a single-light unit projecting a two-color visual approach path into the final approach area of the runway upon which the indicator is installed. The "on glidepath" indication is a steady white light. The "slightly below glidepath" indication is a steady red light. If the aircraft descends further below the glidepath, the red light starts to pulsate. The "above glidepath" indication is a pulsating white light. The pulsating rate increases as the aircraft gets further above or below the desired glideslope. The useful range of the system is about four miles during the day and up to ten miles at night. *[Figure 14-32]*

Runway Lighting
There are various lights that identify parts of the runway complex. These assist a pilot in safely making a takeoff or landing during night operations.

Runway End Identifier Lights (REIL)
Runway end identifier lights (REIL) are installed at many airfields to provide rapid and positive identification of the approach end of a particular runway. The system consists of a pair of synchronized flashing lights located laterally on each side of the runway threshold. REILs may be either omnidirectional or unidirectional facing the approach area.

Runway Edge Lights
Runway edge lights are used to outline the edges of runways at night or during low visibility conditions. *[Figure 14-33]* These lights are classified according to the intensity they are capable of producing: high intensity runway lights (HIRL), medium intensity runway lights (MIRL), and

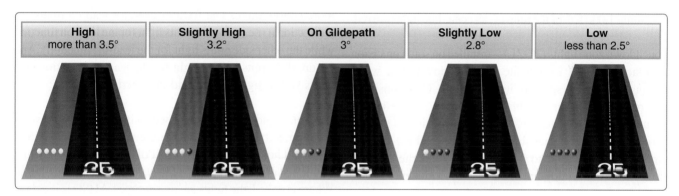

Figure 14-30. *Precision approach path indicator for a typical 3° glide slope.*

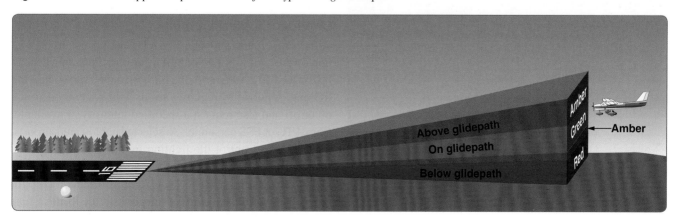

Figure 14-31. *Tri-color visual approach slope indicator.*

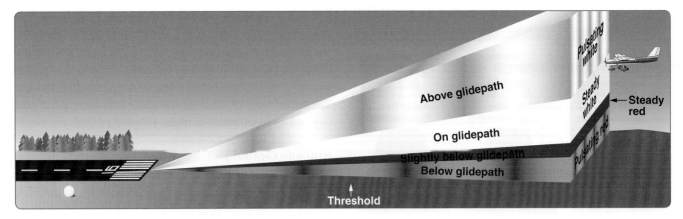

Figure 14-32. *Pulsating visual approach slope indicator.*

low intensity runway lights (LIRL). The HIRL and MIRL have variable intensity settings. These lights are white, except on instrument runways where amber lights are used on the last 2,000 feet or half the length of the runway, whichever is less. The lights marking the end of the runway are red.

In-Runway Lighting

Runway centerline lighting system (RCLS)—installed on some precision approach runways to facilitate landing under adverse visibility conditions. They are located along the runway centerline and are spaced at 50-foot intervals. When viewed from the landing threshold, the runway centerline lights are white until the last 3,000 feet of the runway. The white lights begin to alternate with red for the next 2,000 feet. For the remaining 1,000 feet of the runway, all centerline lights are red.

Touchdown zone lights (TDZL)—installed on some precision approach runways to indicate the touchdown zone when landing under adverse visibility conditions. They consist of two rows of transverse light bars disposed symmetrically about the runway centerline. The system consists of steady-burning white lights that start 100 feet beyond the landing threshold and extend to 3,000 feet beyond the landing threshold or to the midpoint of the runway, whichever is less.

Figure 14-33. *Runway lights.*

Taxiway centerline lead-off lights—provide visual guidance to persons exiting the runway. They are color-coded to warn pilots and vehicle drivers that they are within the runway environment or ILS critical area, whichever is more restrictive. Alternate green and yellow lights are installed, beginning with green, from the runway centerline to one centerline light position beyond the runway holding position or ILS critical area holding position.

Taxiway centerline lead-on lights—provide visual guidance to persons entering the runway. These "lead-on" lights are also color-coded with the same color pattern as lead-off lights to warn pilots and vehicle drivers that they are within the runway environment or ILS critical area, whichever is more conservative. The fixtures used for lead-on lights are bidirectional (i.e., one side emits light for the lead-on function while the other side emits light for the lead-off function). Any fixture that emits yellow light for the lead-off function also emits yellow light for the lead-on function.

Land and hold short lights—used to indicate the hold short point on certain runways which are approved for LAHSO. Land and hold short lights consist of a row of pulsing white lights installed across the runway at the hold short point. Where installed, the lights are on anytime LAHSO is in effect. These lights are off when LAHSO is not in effect.

Control of Airport Lighting

Airport lighting is controlled by ATC at towered airports. At nontowered airports, the lights may be on a timer, or where an FSS is located at an airport, the FSS personnel may control the lighting. A pilot may request various light systems be turned on or off and also request a specified intensity, if available, from ATC or FSS personnel. At selected nontowered airports, the pilot may control the lighting by using the radio. This is done by selecting a specified frequency and clicking the radio microphone. *[Figure 14-34]* For information on pilot controlled lighting at various airports, refer to the Chart Supplement U.S. (formerly Airport/Facility Directory).

Key Mike	Function
7 times within 5 seconds	Highest intensity available
5 times within 5 seconds	Medium or lower intensity (Lower REIL or REIL off)
3 times within 5 seconds	Lowest intensity available (Lower REIL or REIL off)

Figure 14-34. *Radio controlled runway lighting.*

Taxiway Lights

Similar to runway lighting, taxiways also have various lights which help pilots identify areas of the taxiway and any surrounding runways.

Omnidirectional

Omnidirectional taxiway lights outline the edges of the taxiway and are blue in color. At many airports, these edge lights may have variable intensity settings that may be adjusted by an ATC when deemed necessary or when requested by the pilot. Some airports also have taxiway centerline lights that are green in color.

Clearance Bar Lights

Clearance bar lights are installed at holding positions on taxiways in order to increase the conspicuity of the holding position in low visibility conditions. They may also be installed to indicate the location of an intersecting taxiway during periods of darkness. Clearance bars consist of three in-pavement steady-burning yellow lights.

Runway Guard Lights

Runway guard lights are installed at taxiway/runway intersections. They are primarily used to enhance the conspicuity of taxiway/runway intersections during low visibility conditions, but may be used in all weather conditions. Runway guard lights consist of either a pair of elevated flashing yellow lights installed on either side of the taxiway, or a row of in-pavement yellow lights installed across the entire taxiway, at the runway holding position marking.

Note: Some airports may have a row of three or five in-pavement yellow lights installed at taxiway/runway intersections. They should not be confused with clearance bar lights described previously in this section.

Stop Bar Lights

Stop bar lights, when installed, are used to confirm the ATC clearance to enter or cross the active runway in low visibility conditions (below 1,200 ft Runway Visual Range (RVR)). A stop bar consists of a row of red, unidirectional, steady-burning in-pavement lights installed across the entire taxiway at the runway holding position, and elevated steady-burning

red lights on each side. A controlled stop bar is operated in conjunction with the taxiway centerline lead-on lights which extend from the stop bar toward the runway. Following the ATC clearance to proceed, the stop bar is turned off and the lead-on lights are turned on. The stop bar and lead-on lights are automatically reset by a sensor or backup timer.

Obstruction Lights

Obstructions are marked or lighted to warn pilots of their presence during daytime and nighttime conditions. Obstruction lighting can be found both on and off an airport to identify obstructions. They may be marked or lighted in any of the following conditions.

- Red obstruction lights—flash or emit a steady red color during nighttime operations, and the obstructions are painted orange and white for daytime operations.

- High intensity white obstruction lights—flash high intensity white lights during the daytime with the intensity reduced for nighttime.

- Dual lighting—a combination of flashing red beacons and steady red lights for nighttime operation and high intensity white lights for daytime operations.

New Lighting Technologies

A top priority of the FAA is to continue to enhance airport safety while maintaining airport capacity. Reducing runway incursions is a major component of this effort. Runway incursions develop quickly and without warning during routine traffic situations on the airport surface, leaving little time for corrective action. The Runway Status Lights (RWSL) System is designed to provide a direct indication to you that it is unsafe to enter a runway, cross a runway, or takeoff from or land on a runway when the system is activated.

Runway status lights are red in color and indicate runway status only; they do not indicate clearance to enter a runway or clearance to takeoff. The RWSL system provides warning lights on runways and taxiways, illuminating when it is unsafe to enter, cross, or begin takeoff on a runway. Currently, there are two types: Runway Entrance Lights (REL) and Takeoff Hold Lights (THL). *[Figures 14-35 and 14-36]*

REL provide a warning to aircraft crossing or entering a runway from intersecting taxiways that there is conflicting traffic on the runway. THL provide a warning signal to aircraft in position for takeoff that the runway is occupied and it is unsafe to take off. As of 2016, the RWSL system is operational at 14 of the nation's busiest airports with 3 more airports scheduled to receive the system by 2017.

Figure 14-35. *Runway Entrance Lights (REL).*

Figure 14-36. *Takeoff Hold Lights (THL).*

Wind Direction Indicators

It is important for a pilot to know the direction of the wind. At facilities with an operating control tower, this information is provided by ATC. Information may also be provided by FSS personnel either located at a particular airport or remotely available through a remote communication outlet (RCO), or by requesting information on a CTAF at airports that have the capacity to receive and broadcast on this frequency.

When none of these services is available, it is possible to determine wind direction and runway in use by visual wind indicators. A pilot should check these wind indicators even when information is provided on the CTAF at a given airport because there is no assurance that the information provided is accurate.

The wind direction indicator can be a wind cone, wind sock, tetrahedron, or wind tee. These are usually located in a central location near the runway and may be placed in the center of a segmented circle, which identifies the traffic pattern direction if it is other than the standard left-hand pattern. *[Figures 14-37 and 14-38]*

The wind sock is a good source of information since it not only indicates wind direction but allows the pilot to estimate the wind velocity and/or gust factor. The wind sock extends

out straighter in strong winds and tends to move back and forth when the wind is gusting. Wind tees and tetrahedrons can swing freely and align themselves with the wind direction. Since a wind tee or tetrahedron can also be manually set to align with the runway in use, a pilot should also look at the wind sock for wind information, if one is available.

Traffic Patterns

At airports without an operating control tower, a segmented circle visual indicator system, if installed, is designed to provide traffic pattern information. *[Figure 14-38]* Usually located in a position affording maximum visibility to pilots in the air and on the ground and providing a centralized location for other elements of the system, the segmented circle consists of the following components: wind direction indicators, landing direction indicators, landing strip indicators, and traffic pattern indicators.

A tetrahedron is installed to indicate the direction of landings and takeoffs when conditions at the airport warrant its use. It may be located at the center of a segmented circle and may be lighted for night operations. The small end of the tetrahedron points in the direction of landing. Pilots are cautioned against using a tetrahedron for any purpose other than as an indicator of landing direction. At airports with control towers, the tetrahedron should only be referenced when the control tower is not in operation. Tower instructions supersede tetrahedron indications.

Landing strip indicators are installed in pairs and are used to show the alignment of landing strips. *[Figure 14-38]* Traffic pattern indicators are arranged in pairs in conjunction with landing strip indicators and used to indicate the direction of turns when there is a variation from the normal left traffic pattern. (If there is no segmented circle installed at the airport, traffic pattern indicators may be installed on or near the end of the runway.)

At most airports and military air bases, traffic pattern altitudes for propeller-driven aircraft generally extend from 600 feet to as high as 1,500 feet above ground level (AGL). Pilots can obtain the traffic pattern altitude for an airport from the Chart Supplement U.S. (formerly Airport/Facility Directory). Also, traffic pattern altitudes for military turbojet aircraft sometimes extend up to 2,500 feet AGL. Therefore, pilots of en route aircraft should be constantly on alert for other aircraft in traffic patterns and avoid these areas whenever possible. When operating at an airport, traffic pattern altitudes should be maintained unless otherwise required by the applicable distance from cloud criteria according to Title 14 of the Code of Federal Regulations (14 CFR) part 91, section 91.155. Additional information on airport traffic pattern operations

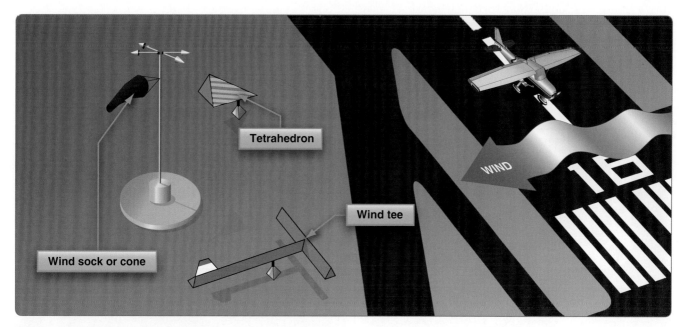

Figure 14-37. *Wind direction indicators.*

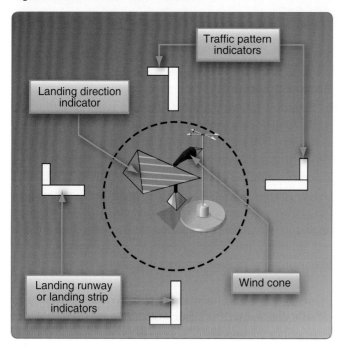

Figure 14-38. *Segmented circle.*

can be found in Chapter 4, "Air Traffic Control," of the AIM. Pilots can find traffic pattern information and restrictions, such as noise abatement in the Chart Supplement U.S. (formerly Airport/Facility Directory).

Example: Key to Traffic Pattern Operations— Single Runway

1. Enter pattern in level flight, abeam the midpoint of the runway, at pattern altitude. (1,000' AGL is recommended pattern altitude unless otherwise established.) *[Figure 14-39]*

2. Maintain pattern altitude until abeam approach end of the landing runway on downwind leg. *[Figure 14-39]*

3. Complete turn to final at least ¼ mile from the runway. *[Figure 14-39]*

4. After takeoff or go-around, continue straight ahead until beyond departure end of runway. *[Figure 14-39]*

5. If remaining in the traffic pattern, commence turn to crosswind leg beyond the departure end of the runway within 300 feet of pattern altitude. *[Figure 14-39]*

6. If departing the traffic pattern, continue straight out, or exit with a 45° turn (to the left when in a left-hand traffic pattern; to the right when in a right-hand traffic pattern) beyond the departure end of the runway, after reaching pattern altitude. *[Figure 14-39]*

Example: Key to Traffic Pattern Operations— Parallel Runways

1. Enter pattern in level flight, abeam the midpoint of the runway, at pattern altitude. (1,000' AGL is recommended pattern altitude unless otherwise established.) *[Figure 14-40]*

2. Maintain pattern altitude until abeam approach end of the landing runway on downwind leg. *[Figure 14-40]*

3. Complete turn to final at least ¼ mile from the runway. *[Figure 14-40]*

4. Do not overshoot final or continue on a track that penetrates the final approach of the parallel runway

5. After takeoff or go-around, continue straight ahead until beyond departure end of runway. *[Figure 14-40]*

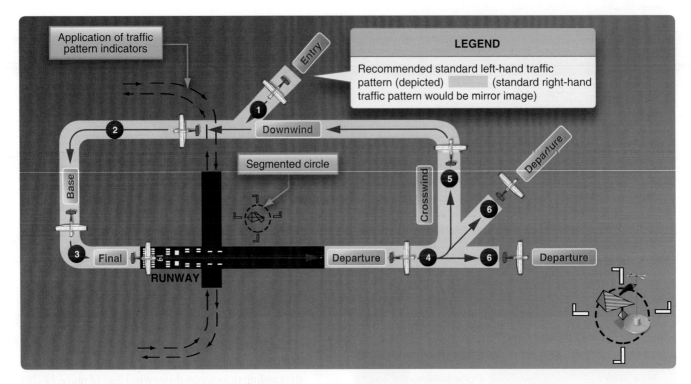

Figure 14-39. *Traffic pattern operations—single runway.*

6. If remaining in the traffic pattern, commence turn to crosswind leg beyond the departure end of the runway within 300 feet of pattern altitude. *[Figure 14-40]*

7. If departing the traffic pattern, continue straight out, or exit with a 45° turn (to the left when in a left-hand traffic pattern; to the right when in a right-hand traffic pattern) beyond the departure end of the runway, after reaching pattern altitude. *[Figure 14-40]*

8. Do not continue on a track that penetrates the departure path of the parallel runway. *[Figure 14-40]*

Radio Communications

Operating in and out of a towered airport, as well as in a good portion of the airspace system, requires that an aircraft have two-way radio communication capability. For this reason, a pilot should be knowledgeable of radio station license requirements and radio communications equipment and procedures.

Radio License

There is no license requirement for a pilot operating in the United States; however, a pilot who operates internationally is required to hold a restricted radiotelephone permit issued by the Federal Communications Commission (FCC). There is also no station license requirement for most general aviation aircraft operating in the United States. A station license is required, however, for an aircraft that is operating internationally, that uses other than a VHF radio, and that meets other criteria.

Radio Equipment

In general aviation, the most common types of radios are VHF. A VHF radio operates on frequencies between 118.0 megahertz (MHz) and 136.975 MHz and is classified as 720 or 760 depending on the number of channels it can accommodate. The 720 and 760 use .025 MHz (25 kilohertz (KHz) spacing (118.025, 118.050) with the 720 having a frequency range up to 135.975 MHz and the 760 reaching up to 136.975 MHz. VHF radios are limited to line of sight transmissions; therefore, aircraft at higher altitudes are able to transmit and receive at greater distances.

In March of 1997, the International Civil Aviation Organization (ICAO) amended its International Standards and Recommended Practices to incorporate a channel plan specifying 8.33 kHz channel spacings in the Aeronautical Mobile Service. The 8.33 kHz channel plan was adopted to alleviate the shortage of VHF ATC channels experienced in western Europe and in the United Kingdom. Seven western European countries and the United Kingdom implemented the 8.33 kHz channel plan on January 1, 1999. Accordingly, aircraft operating in the airspace of these countries must have the capability of transmitting and receiving on the 8.33 kHz spaced channels.

Using Proper Radio Procedures

Using proper radio phraseology and procedures contribute to a pilot's ability to operate safely and efficiently in the airspace system. A review of the Pilot/Controller Glossary contained in the AIM assists a pilot in the use and understanding of

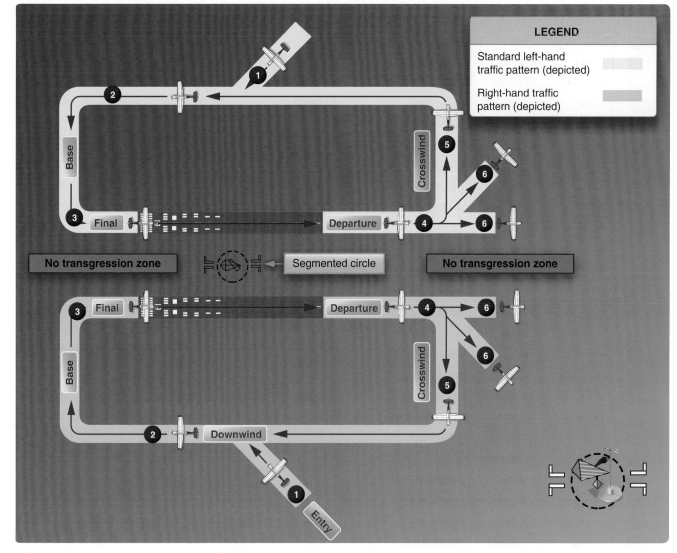

Figure 14-40. *Traffic pattern operation—parallel runways.*

standard terminology. The AIM also contains many examples of radio communications.

ICAO has adopted a phonetic alphabet that should be used in radio communications. When communicating with ATC, pilots should use this alphabet to identify their aircraft. *[Figure 14-41]*

Lost Communication Procedures

It is possible that a pilot might experience a malfunction of the radio. This might cause the transmitter, receiver, or both to become inoperative. If a receiver becomes inoperative and a pilot needs to land at a towered airport, it is advisable to remain outside or above Class D airspace until the direction and flow of traffic is determined. A pilot should then advise the tower of the aircraft type, position, altitude, and intention to land. The pilot should continue, enter the pattern, report a position as appropriate, and watch for light signals from the tower. Light signal colors and their meanings are contained in *Figure 14-42.*

If the transmitter becomes inoperative, a pilot should follow the previously stated procedures and also monitor the appropriate ATC frequency. During daylight hours, ATC transmissions may be acknowledged by rocking the wings and at night by blinking the landing light.

When both receiver and transmitter are inoperative, the pilot should remain outside of Class D airspace until the flow of traffic has been determined and then enter the pattern and watch for light signals.

Radio malfunctions should be repaired before further flight. If this is not possible, ATC may be contacted by telephone requesting a VFR departure without two-way radio communications. No radio (NORDO) procedure arrivals are not accepted at busy airports. If authorization is given to depart, the pilot is advised to monitor the appropriate frequency and/or watch for light signals as appropriate.

Character	Morse Code	Telephony	Phonic Pronunciation
A	•—	Alfa	(AL-FAH)
B	—•••	Bravo	(BRAH-VOH)
C	—•—•	Charlie	(CHAR-LEE) or (SHAR-LEE)
D	—••	Delta	(DELL-TAH)
E	•	Echo	(ECK-OH)
F	••—•	Foxtrot	(FOKS-TROT)
G	——•	Golf	(GOLF)
H	••••	Hotel	(HOH-TEL)
I	••	India	(IN-DEE-AH)
J	•———	Juliett	(JEW-LEE-ETT)
K	—•—	Kilo	(KEY-LOH)
L	•—••	Lima	(LEE-MAH)
M	——	Mike	(MIKE)
N	—•	November	(NO-VEM-BER)
O	———	Oscar	(OSS-CAH)
P	•——•	Papa	(PAH-PAH)
Q	——•—	Quebec	(KEH-BECK)
R	•—•	Romeo	(ROW-ME-OH)
S	•••	Sierra	(SEE-AIR-RAH)
T	—	Tango	(TANG-GO)
U	••—	Uniform	(YOU-NEE-FORM) or (OO-NEE-FORM)
V	•••—	Victor	(VIK-TAH)
W	•——	Whiskey	(WISS-KEY)
X	—••—	Xray	(ECKS-RAY)
Y	—•——	Yankee	(YANG-KEY)
Z	——••	Zulu	(ZOO-LOO)
1	•————	One	(WUN)
2	••———	Two	(TOO)
3	•••——	Three	(TREE)
4	••••—	Four	(FOW-ER)
5	•••••	Five	(FIFE)
6	—••••	Six	(SIX)
7	——•••	Seven	(SEV-EN)
8	———••	Eight	(AIT)
9	————•	Nine	(NIN-ER)
0	—————	Zero	(ZEE-RO)

Figure 14-41. *Phonetic alphabet.*

If radio communication is lost, it may be a prudent decision to land at a non-towered airport with lower traffic volume, if practical. When operating at a non-towered airport, no radio communication is necessary. However, pilots should be extra vigilant when not using the radio. Other traffic may not as

easily be aware of your presence when they are expecting the standard radio calls.

Air Traffic Control (ATC) Services

Besides the services provided by an FSS as discussed in Chapter 12, "Aviation Weather Services," numerous other services are provided by ATC. In many instances a pilot is required to have contact with ATC, but even when not required, a pilot may find their services helpful.

Primary Radar

Radar is a device that provides information on range, azimuth, and/or elevation of objects in the path of the transmitted pulses. It measures the time interval between transmission and reception of radio pulses and correlates the angular orientation of the radiated antenna beam or beams in azimuth and/or elevation. Range is determined by measuring the time it takes for the radio wave to go out to the object and then return to the receiving antenna. The direction of a detected object from a radar site is determined by the position of the rotating antenna when the reflected portion of the radio wave is received.

Modern radar is very reliable and there are seldom outages. This is due to reliable maintenance and improved equipment. There are, however, some limitations that may affect ATC services and prevent a controller from issuing advisories concerning aircraft that are not under his or her control and cannot be seen on radar.

The characteristics of radio waves are such that they normally travel in a continuous straight line unless they are "bent" by atmospheric phenomena, such as temperature inversions, reflected or attenuated by dense objects such as heavy clouds and precipitation, or screened by high terrain features. Radar signals degrade over distance, cannot penetrate through solid objects such as mountains, and the fastest radar updates every 4.7 seconds. By contrast, the satellite signals used with Automatic Dependent Surveillance–Broadcast (ADS–B) do not degrade over distance, provide better visibility around mountainous terrain and allows equipped aircraft to update their own position once a second with better accuracy.

ATC Radar Beacon System (ATCRBS)

The ATC radar beacon system (ATCRBS) is often referred to as "secondary surveillance radar." This system consists of three components and helps in alleviating some of the limitations associated with primary radar. The three components are an interrogator, transponder, and radarscope. The advantages of ATCRBS are the reinforcement of radar targets, rapid target identification, and a unique display of selected codes.

Growing air traffic in the National Airspace System (NAS) will be addressed through the use of ADS-B, which not only

Color and Type of Signal	Movement of Vehicles, Equipment and Personnel	Aircraft on the Ground	Aircraft In Flight
Steady green	Cleared to cross, proceed or go	Cleared for takeoff	Cleared to land
Flashing green	Not applicable	Cleared for taxi	Return for landing (to be followed by steady green at the proper time)
Steady red	Stop	Stop	Give way to other aircraft and continue circling
Flashing red	Clear the taxiway/runway	Taxi clear of the runway in use	Airport unsafe, do not land
Flashing white	Return to starting point on airport	Return to starting point on airport	Not applicable
Alternating red and green	Exercise extreme caution!!!!	Exercise extreme caution!!!!	Exercise extreme caution!!!!

Figure 14-42. *Light gun signals.*

provides all the same information the ATCRBS, but will do so more rapidly and with significantly more accuracy. By broadcasting aircraft position information to a ground station, ADS–B can also provide coverage in areas that do not have radar coverage. In addition, ADS–B provides trajectory information that includes speed and direction of motion.

Transponder

The transponder is the airborne portion of the secondary surveillance radar system and a system with which a pilot should be familiar. The ATCRBS cannot display the secondary information unless an aircraft is equipped with a transponder. A transponder is also required to operate in certain controlled airspace as discussed in Chapter 15, "Airspace."

A transponder code consists of four numbers from 0 to 7 (4,096 possible codes). There are some standard codes or ATC may issue a four-digit code to an aircraft. When a controller requests a code or function on the transponder, the word "squawk" may be used. *Figure 14-43* lists some standard transponder phraseology. Additional information concerning transponder operation can be found in the AIM, Chapter 4.

Radar Beacon Phraseology	
SQUAWK (number)	Operate radar beacon transponder on designated code in MODE A/3.
IDENT	Engage the "IDENT" feature (military I/P) of the transponder.
SQUAWK (number) and IDENT	Operate transponder on specified code in MODE A/3 and engage the "IDENT" (military I/P) feature.
SQUAWK Standby	Switch transponder to standby position.
SQUAWK Low/Normal	Operate transponder on low or normal sensitivity as specified. Transponder is operated in "NORMAL" position unless ATC specifies "LOW" ("ON" is used instead of "NORMAL" as a master control label on some types of transponders).
SQUAWK Altitude	Activate MODE C with automatic altitude reporting.
STOP Altitude SQUAWK	Turn off altitude reporting switch and continue transmitting MODE C framing pulses. If your equipment does not have this capability, turn off MODE C.
STOP SQUAWK (mode in use)	Switch off specified mode. (Used for military aircraft when the controller is unaware of military service requirements for the aircraft to continue operation on another MODE.)
STOP SQUAWK	Switch off transponder.
SQUAWK Mayday	Operate transponder in the emergency position (MODE A Code 7700 for civil transponder, MODE 3 Code 7700 and emergency feature for military transponder).
SQUAWK VFR	Operate radar beacon transponder on Code 1200 in MODE A/3, or other appropriate VFR code.

Figure 14-43. *Transponder phraseology.*

Automatic Dependent Surveillance–Broadcast (ADS-B)

Automatic Dependent Surveillance–Broadcast (ADS–B) is a surveillance technology being deployed throughout the NAS to facilitate improvements needed to increase the capacity and efficiency of the NAS, while maintaining safety. ADS-B supports these improvements by providing a higher update rate and enhanced accuracy of surveillance information over the current radar-based surveillance system. In addition, ADS-B enables the expansion of air traffic control (ATC) surveillance services into areas where none existed previously. The ADS-B ground system also provides Traffic Information Services-Broadcast (TIS-B) and Flight Information Services-Broadcast (FIS-B) for use on appropriately equipped aircraft, enhancing the user's situational awareness (SA) and improving the overall safety of the NAS.

The ADS–B system is composed of aircraft avionics and a ground infrastructure. Onboard avionics determine the position of the aircraft by using the GPS and transmit its position, along with additional information about the aircraft, to ground stations for use by ATC and nearby ADS-B equipped aircraft.

In the United States, ADS–B equipped aircraft exchange information on one of two frequencies: 978 or 1090 MHz. The 1090 MHz frequency is associated with Mode A, C, and S transponder operations. 1090 MHz transponders with integrated ADS–B functionality extend the transponder message sets with additional ADS–B information. This additional information is known as an "extended squitter" message and referred to as 1090ES. ADS–B equipment operating on 978 MHz is known as the Universal Access Transceiver (UAT).

Radar Traffic Advisories

Radar equipped ATC facilities provide radar assistance to aircraft on instrument flight plans and VFR aircraft provided the aircraft can communicate with the facility and are within radar coverage. This basic service includes safety alerts, traffic advisories, limited vectoring when requested, and sequencing at locations where this procedure has been established. ATC issues traffic advisories based on observed radar targets. The traffic is referenced by azimuth from the aircraft in terms of the 12-hour clock. Also, distance in nautical miles, direction in which the target is moving, and type and altitude of the aircraft, if known, are given.

An example would be: "Traffic 10 o'clock 5 miles east bound, Cessna 152, 3,000 feet." The pilot should note that traffic position is based on the aircraft track and that wind correction can affect the clock position at which a pilot locates traffic. This service is not intended to relieve the pilot of the responsibility to see and avoid other aircraft. *[Figure 14-44]* In addition to basic radar service, terminal radar service

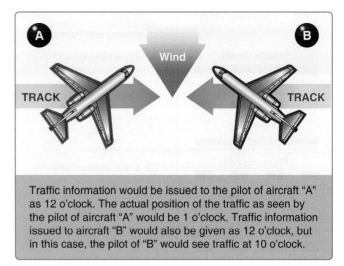

Traffic information would be issued to the pilot of aircraft "A" as 12 o'clock. The actual position of the traffic as seen by the pilot of aircraft "A" would be 1 o'clock. Traffic information issued to aircraft "B" would also be given as 12 o'clock, but in this case, the pilot of "B" would see traffic at 10 o'clock.

Figure 14-44. *Traffic advisories.*

area (TRSA) has been implemented at certain terminal locations. TRSAs are depicted on sectional aeronautical charts and listed in the Chart Supplement U.S. (formerly Airport/Facility Directory). The purpose of this service is to provide separation between all participating VFR aircraft and all IFR aircraft operating within the TRSA. Class C service provides approved separation between IFR and VFR aircraft and sequencing of VFR aircraft to the primary airport. Class B service provides approved separation of aircraft based on IFR, VFR, and/or weight and sequencing of VFR arrivals to the primary airport(s).

Wake Turbulence

All aircraft generate wake turbulence during flight. This disturbance is caused by a pair of counter-rotating vortices trailing from the wingtips. The vortices from larger aircraft pose problems to encountering aircraft. The wake of these aircraft can impose rolling moments exceeding the roll-control authority of the encountering aircraft. Also, the turbulence generated within the vortices can damage aircraft components and equipment if encountered at close range. For this reason, a pilot must envision the location of the vortex wake and adjust the flight path accordingly.

Vortex Generation

Lift is generated by the creation of a pressure differential over the wing surface. The lowest pressure occurs over the upper wing surface and the highest pressure under the wing. This pressure differential triggers the rollup of the airflow aft of the wing resulting in swirling air masses trailing downstream of the wingtips. After the rollup is completed, the wake consists of two counter rotating cylindrical vortices. Most of the energy lies within a few feet of the center of each vortex. *[Figure 14-45]*

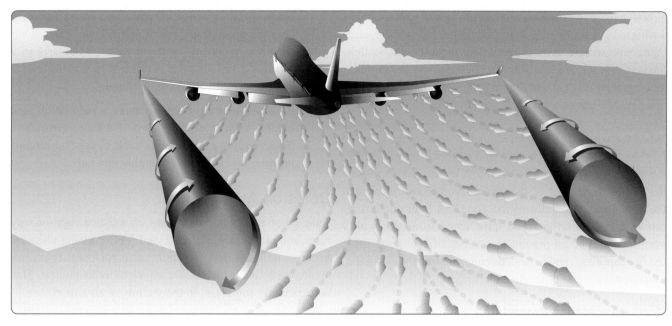

Figure 14-45. *Vortex generation.*

Vortex Strength

Terminal Area

Wake turbulence has historically been thought of as only a function of aircraft weight, but recent research considers additional parameters, such as speed, aspects of the wing, wake decay rates, and aircraft resistance to wake, just to name a few. The vortex characteristics of any aircraft will be changed with the extension of flaps or other wing configuration devices, as well as changing speed. However, as the basic factors are weight and speed, the vortex strength increases proportionately with an increase in aircraft operating weight or decrease in aircraft speed. The greatest vortex strength occurs when the generating aircraft is heavy, slow, and clean, since the turbulence from a "dirty" aircraft configuration hastens wake decay.

En Route

En route wake turbulence events have been influenced by changes to the aircraft fleet mix that have more "Super" (A380) and "Heavy" (B-747, B-777, A340, etc.) aircraft operating in the NAS. There have been wake turbulence events in excess of 30NM and 2000 feet lower than the wake generating aircraft. Air density is also a factor in wake strength. Even though the speeds are higher in cruise at high altitude, the reduced air density may result in wake strength comparable to that in the terminal area. In addition, for a given separation distance, the higher speeds in cruise result in less time for the wake to decay before being encountered by a trailing aircraft.

Vortex Behavior

Trailing vortices have certain behavioral characteristics that can help a pilot visualize the wake location and take avoidance precautions.

Vortices are generated from the moment an aircraft leaves the ground (until it touches down), since trailing vortices are the byproduct of wing lift. *[Figure 14-46]* The vortex circulation is outward, upward, and around the wingtips when viewed from either ahead or behind the aircraft. Tests with large

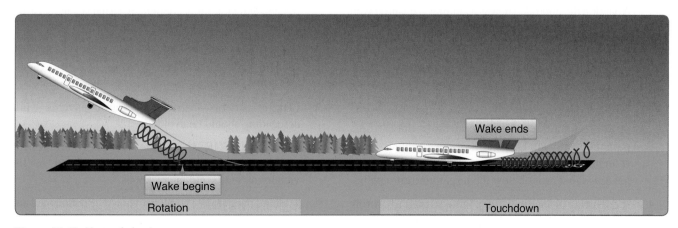

Figure 14-46. *Vortex behavior.*

aircraft have shown that vortices remain spaced a bit less than a wingspan apart, drifting with the wind, at altitudes greater than a wingspan from the ground. Tests have also shown that the vortices sink at a rate of several hundred feet per minute, slowing their descent and diminishing in strength with time and distance behind the generating aircraft.

When the vortices of larger aircraft sink close to the ground (within 100 to 200 feet), they tend to move laterally over the ground at a speed of 2–3 knots. A crosswind decreases the lateral movement of the upwind vortex and increases the movement of the downwind vortex. A light quartering tailwind presents the worst case scenario as the wake vortices could be all present along a significant portion of the final approach and extended centerline and not just in the touchdown zone as typically expected.

Vortex Avoidance Procedures

The following procedures are in place to assist pilots in vortex avoidance in the given scenario.

- Landing behind a larger aircraft on the same runway—stay at or above the larger aircraft's approach flight path and land beyond its touchdown point. *[Figure 14-47A]*

- Landing behind a larger aircraft on a parallel runway closer than 2,500 feet—consider the possibility of drift and stay at or above the larger aircraft's final approach flight path and note its touchdown point. *[Figure 14-47B]*

- Landing behind a larger aircraft on crossing runway—cross above the larger aircraft's flight path.

- Landing behind a departing aircraft on the same runway—land prior to the departing aircraft's rotating point.

- Landing behind a larger aircraft on a crossing runway—note the aircraft's rotation point and, if that point is past the intersection, continue and land prior to the intersection. If the larger aircraft rotates prior to the intersection, avoid flight below its flight path. Abandon the approach unless a landing is ensured well before reaching the intersection. *[Figure 14-47C]*

- Departing behind a large aircraft—rotate prior to the large aircraft's rotation point and climb above its climb path until turning clear of the wake.

- For intersection takeoffs on the same runway—be alert to adjacent larger aircraft operations, particularly upwind of the runway of intended use. If an intersection takeoff clearance is received, avoid headings that cross below the larger aircraft's path.

- If departing or landing after a large aircraft executing a low approach, missed approach, or touch-and-go

landing (since vortices settle and move laterally near the ground, the vortex hazard may exist along the runway and in the flight path, particularly in a quartering tailwind), it is prudent to wait at least 2 minutes prior to a takeoff or landing.

- En route, it is advisable to avoid a path below and behind a large aircraft, and if a large aircraft is observed above on the same track, change the aircraft position laterally and preferably upwind.

Collision Avoidance

Title 14 of the CFR part 91 has established right-of-way rules, minimum safe altitudes, and VFR cruising altitudes to enhance flight safety. The pilot can contribute to collision avoidance by being alert and scanning for other aircraft. This is particularly important in the vicinity of an airport.

Effective scanning is accomplished with a series of short, regularly spaced eye movements that bring successive areas of the sky into the central visual field. Each movement should not exceed 10°, and each should be observed for at least 1 second to enable detection. Although back and forth eye movements seem preferred by most pilots, each pilot should develop a scanning pattern that is most comfortable and then adhere to it to assure optimum scanning. Even if entitled to the right-of-way, a pilot should yield if another aircraft seems too close.

Clearing Procedures

The following procedures and considerations are in place to assist pilots in collision avoidance under various situations:

- Before takeoff—prior to taxiing onto a runway or landing area in preparation for takeoff, pilots should scan the approach area for possible landing traffic, executing appropriate maneuvers to provide a clear view of the approach areas.

- Climbs and descents—during climbs and descents in flight conditions that permit visual detection of other traffic, pilots should execute gentle banks left and right at a frequency that permits continuous visual scanning of the airspace.

- Straight and level—during sustained periods of straight-and-level flight, a pilot should execute appropriate clearing procedures at periodic intervals.

- Traffic patterns—entries into traffic patterns while descending should be avoided.

- Traffic at VOR sites—due to converging traffic, sustained vigilance should be maintained in the vicinity of VORs and intersections.

- Training operations—vigilance should be maintained and clearing turns should be made prior to a practice maneuver. During instruction, the pilot should be asked to verbalize the clearing procedures (call out "clear left, right, above, and below").

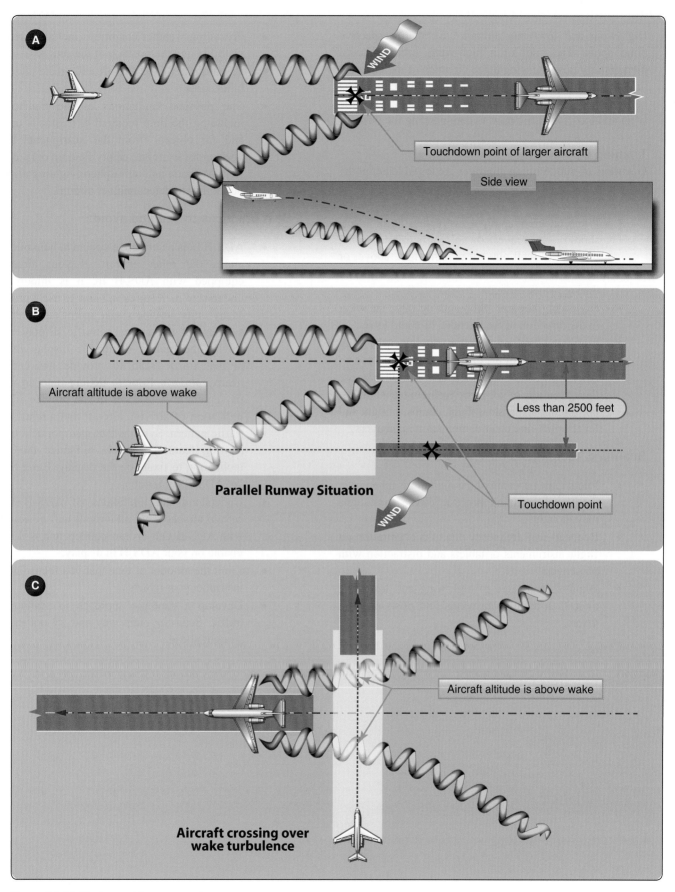

Figure 14-47. *Vortex avoidance procedures.*

High-wing and low-wing aircraft have their respective blind spots. The pilot of a high-wing aircraft should momentarily raise the wing in the direction of the intended turn and look for traffic prior to commencing the turn. The pilot of a low- wing aircraft should momentarily lower the wing and look for traffic prior to commencing the turn.

Training Operations

Operators of pilot training programs are urged to adopt the following practices:

- Pilots undergoing flight instruction at all levels should be requested to verbalize clearing procedures (call out "clear" left, right, above, or below) to instill and sustain the habit of vigilance during maneuvering.
- High-wing airplane. Momentarily raise the wing in the direction of the intended turn and look.
- Low-wing airplane. Momentarily lower the wing in the direction of the intended turn and look.
- Appropriate clearing procedures should precede the execution of all turns including chandelles, lazy eights, stalls, slow flight, climbs, straight and level, spins, and other combination maneuvers.

Scanning Techniques for Traffic Avoidance

- Pilots must be aware of the limitations inherent in the visual scanning process. These limitations may include:
- Reduced scan frequency due to concentration on flight instruments or tablets and distraction with passengers.
- Blind spots related to high-wing and low-wing aircraft in addition to windshield posts and sun visors.

- Prevailing weather conditions including reduced visibility and the position of the sun.
- The attitude of the aircraft will create additional blind spots.
- The physical limitations of the human eye, including the time required to (re)focus on near and far objects, from the instruments to the horizon for example; empty field myopia, narrow field of vision and atmospheric lighting all affect our ability to detect another aircraft.

Best practices to see and avoid:

- ADS-B In is an effective system to help pilots see and avoid other aircraft. If your aircraft is equipped with ADS-B In, it is important to understand its features and how to use it properly. Many units provide visual and/or audio alerts to supplement the system's traffic display. Pilots should incorporate the traffic display in their normal traffic scan to provide awareness of nearby aircraft. Prior to taxiing onto an airport movement area, ADS-B In can provide advance indication of arriving aircraft and aircraft in the traffic pattern. Systems that incorporate a traffic-alerting feature can help minimize the pilot's inclination to fixate on the display. Refer to 4-5-7e, ADS-B Limitations.
- Understand the limitations of ADS-B In. In certain airspace, not all aircraft will be equipped with ADS-B Out or transponders and will not be visible on your ADS-B In display.
- Limit the amount of time that you focus on flight instruments or tablets.
- Develop a strategic approach to scanning for traffic. Scan the entire sky and try not to focus straight ahead.

Pilot Deviations (PDs)

A pilot deviation (PD) is an action of a pilot that violates any Federal Aviation Regulation. While PDs should be avoided, the regulations do authorize deviations from a clearance in response to a traffic alert and collision avoidance system resolution advisory. You must notify ATC as soon as possible following a deviation.

Pilot deviations can occur in several different ways. Airborne deviations result when a pilot strays from an assigned heading or altitude or from an instrument procedure, or if the pilot penetrates controlled or restricted airspace without ATC clearance.

To prevent airborne deviations, follow these steps:

- Plan each flight—you may have flown the flight many times before but conditions and situations can change rapidly, such as in the case of a pop-up temporary flight restriction (TFR). Take a few minutes prior to each flight to plan accordingly.

- Talk and squawk—Proper communication with ATC has its benefits. Flight following often makes the controller's job easier because they can better integrate VFR and IFR traffic.

- Give yourself some room—GPS is usually more precise than ATC radar. Using your GPS to fly up to and along the line of the airspace you are trying to avoid could result in a pilot deviation because ATC radar may show you within the restricted airspace.

Ground deviations (also called surface deviations) include taxiing, taking off, or landing without clearance, deviating from an assigned taxi route, or failing to hold short of an assigned clearance limit. To prevent ground deviations, stay alert during ground operations. Pilot deviations can and frequently do occur on the ground. Many strategies and tactics pilots use to avoid airborne deviations also work on the ground.

Pilots should also remain vigilant about vehicle/pedestrian deviations (V/PDs). A vehicle or pedestrian deviation includes pedestrians, vehicles or other objects interfering with aircraft operations by entering or moving on the runway movement area without authorization from air traffic control. In serious instances, any ground deviation (PD or VPD) can result in a runway incursion. Best practices in preventing ground deviations can be found in the following section under runway incursion avoidance.

Runway Incursion Avoidance

A runway incursion is "any occurrence in the airport runway environment involving an aircraft, vehicle, person, or object on the ground that creates a collision hazard or results in a loss of required separation with an aircraft taking off, intending to take off, landing, or intending to land." It is important to give the same attention to operating on the surface as in other phases of flights. Proper planning can prevent runway incursions and the possibility of a ground collision. A pilot should always be aware of the aircraft's position on the surface at all times and be aware of other aircraft and vehicle operations on the airport. At times, towered airports can be busy and taxi instructions complex. In this situation, it may be advisable to write down taxi instructions. The following are some practices to help prevent a runway incursion:

- Read back all runway crossing and/or hold instructions.

- Review airport layouts as part of preflight planning, before descending to land and while taxiing, as needed.

- Know airport signage.

- Review NOTAM for information on runway/taxiway closures and construction areas.

- Request progressive taxi instructions from ATC when unsure of the taxi route.

- Check for traffic before crossing any runway hold line and before entering a taxiway.

- Turn on aircraft lights and the rotating beacon or strobe lights while taxing.

- When landing, clear the active runway as soon as possible, then wait for taxi instructions before further movement.

- Study and use proper phraseology in order to understand and respond to ground control instructions.

- Write down complex taxi instructions at unfamiliar airports.

Approximately three runway incursions occur each day at towered airports within the United States. The potential that these numbers present for a catastrophic accident is unacceptable. The following are examples of pilot deviations, operational incidents (OI), and vehicle (driver) deviations that may lead to runway incursions.

Pilot Deviations:

- Crossing a runway hold marking without clearance from ATC
- Taking off without clearance
- Landing without clearance

Operational Incidents (OI):

- Clearing an aircraft onto a runway while another aircraft is landing on the same runway
- Issuing a takeoff clearance while the runway is occupied by another aircraft or vehicle

Vehicle (Driver) Deviations:

- Crossing a runway hold marking without ATC clearance

According to FAA data, approximately 65 percent of all runway incursions are caused by pilots. Of the pilot runway incursions, FAA data shows almost half of those incursions are caused by GA pilots.

Causal Factors of Runway Incursions

Detailed investigations of runway incursions over the past 10 years have identified three major areas contributing to these events:

- Failure to comply with ATC instructions
- Lack of airport familiarity
- Nonconformance with standard operating procedures

Clear, concise, and effective pilot/controller communication is paramount to safe airport surface operations. You must fully understand and comply with all ATC instructions. It is mandatory to read back all runway "**hold short**" instructions verbatim.

Taxiing on an unfamiliar airport can be very challenging, especially during hours of darkness or low visibility. A request may be made for progressive taxi instructions which include step by step taxi routing instructions. Ensure you have a current airport diagram, remain "heads-up" with eyes outside, and devote your entire attention to surface navigation per ATC clearance. All checklists should be completed while the aircraft is stopped. There is no place for non-essential chatter or other activities while maintaining vigilance during taxi. *[Figure 14-48]*

Runway Confusion

Runway confusion is a subset of runway incursions and often results in you unintentionally taking off or landing on a taxiway or wrong runway. Generally, you are unaware of the mistake until after it has occurred.

Figure 14-48. *Heads-up, eyes outside.*

In August 2006, the flight crew of a commercial regional jet was cleared for takeoff on Runway 22 but mistakenly lined up and departed on Runway 26, a much shorter runway. As a result, the aircraft crashed off the end of the runway.

Causal Factors of Runway Confusion

There are three major factors that increase the risk of runway confusion and can lead to a wrong runway departure:

- Airport complexity
- Close proximity of runway thresholds
- Joint use of a runway as a taxiway

Not only can airport complexity contribute to a runway incursion; it can also play a significant role in runway confusion. If you are operating at an unfamiliar airport and need assistance in executing the taxi clearance, do not hesitate to ask ATC for help. Always carry a current airport diagram and trace or highlight your taxi route to the departure runway prior to leaving the ramp.

If you are operating from an airport with runway thresholds in close proximity to one another, exercise extreme caution when taxiing onto the runway. *Figure 14-49* shows a perfect example of a taxiway leading to multiple runways that may cause confusion. If departing on Runway 36, ensure that you set your aircraft heading "bug" to 360°, and align your aircraft to the runway heading to avoid departing from the wrong runway. Before adding power, make one last instrument scan to ensure the aircraft heading and runway heading are aligned. Under certain circumstances, it may be necessary to use a runway as a taxiway. For example, during airport construction some taxiways may be closed requiring re-routing of traffic onto runways. In other cases, departing traffic may be required to back taxi on the runway in order to utilize the full runway length.

Figure 14-49. *Confusing runway/runway intersection.*

Since inattention and confusion often are factors contributing to runway incursion, it is important to remain extremely cautious and maintain situational awareness (SA). When instructed to use a runway as a taxiway, do not become confused and take off on the runway you are using as a taxiway.

ATC Instructions

Title 14 of the Code of Federal Regulations (14 CFR) part 91, section 91.123 requires you to follow all ATC clearances and instructions. Request clarification if you are unsure of the clearance or instruction to be followed. If you are unfamiliar with the airport or unsure of a taxi route, ask ATC for a "progressive taxi." Progressive taxi requires the controller to provide step-by-step taxi instructions.

The final decision to act on ATC's instruction rests with you. If you cannot safely comply with any of ATC's instructions, inform them immediately by using the word "UNABLE." There is nothing wrong with telling a controller that you are unable to safely comply with the clearance.

Another way to mitigate the risk of runway incursions is to write down all taxi instructions as soon as they are received from ATC. *[Figure 14-50]* It is also helpful to monitor ATC clearances and instructions that are issued to other aircraft. You should be especially vigilant if another aircraft has a similar sounding call sign so there is no mistake about who ATC is contacting or to whom they are giving instructions and clearances.

Read back your complete ATC clearance with your aircraft call sign. This gives ATC the opportunity to clarify any misunderstandings and ensure that instructions were given to the correct aircraft. If, at any time, there is uncertainty about any ATC instructions or clearances, ask ATC to "say again" or ask for progressive taxi instructions.

ATC Instructions—"Hold Short"

The most important sign and marking on the airport is the hold sign and hold marking. These are located on a stub taxiway leading directly to a runway. They depict the holding position or the location where the aircraft is to stop so as not to enter the runway environment. *[Figure 14-51]* For example, *Figure 14-52* shows the holding position sign and marking for Runway 13 and Runway 31.

When ATC issues a "**hold short**" clearance, you are expected to taxi up to, but not cross any part of the runway holding marking. At a towered airport, runway hold markings should never be crossed without explicit ATC instructions. Do not enter a runway at a towered airport unless instructions are given from ATC to cross, takeoff from, or "line up and wait" on that specific runway.

ATC is required to obtain a read-back from the pilot of all runway "**hold short**" instructions. Therefore, you must read back the entire clearance and "**hold short**" instruction, to include runway identifier and your call sign.

Figure 14-50. *A sound practice is to write down taxi instructions from ATC.*

Figure 14-51. *Do NOT cross a runway holding position marking without ATC clearance. If the tower is closed or you are operating from a non-towered airport, check both directions for conflicting traffic before crossing the hold position marking.*

Figure 14-53 shows an example of a controller's taxi and "**hold short**" instructions and the reply from the pilot.

ATC Instructions—Explicit Runway Crossing

As of June 30, 2010, ATC is required to issue explicit instructions to "**cross**" or "**hold short**" of each runway. Instructions to "**cross**" a runway are normally issued one at a time, and an aircraft must have crossed the previous runway before another runway crossing is issued. Exceptions may apply for closely spaced runways that have less than 1,000 feet between centerlines. This applies to all runways to include active, inactive, or closed. *Figure 14-54* shows communication between ATC and a pilot who is requesting a taxi clearance. Extra caution should be used when directed by ATC to taxi onto or across a runway, especially at night and during reduced visibility conditions. Always comply with "**hold**

Figure 14-52. *Runway 13-31 holding position sign and marking located on Taxiway Charlie.*

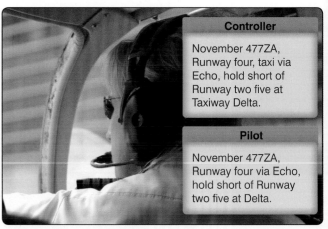

Figure 14-53. *Example of taxi and "**hold short**" instructions from ATC to a pilot.*

short" or crossing instructions when approaching an entrance to a runway. Scan the full length of the runway and the final approaches before entering or crossing any runway, even if ATC has issued a clearance.

ATC Instructions—"Line Up and Wait" (LUAW)

ATC now uses the "line up and wait" (LUAW) instruction when a takeoff clearance cannot be issued immediately due to traffic or other reasons. The words "line up and wait" have replaced "position and hold" in directing you to taxi onto a runway and await takeoff clearance.

An ATC instruction to "line up and wait" is not a clearance for takeoff. It is only a clearance to enter the runway and hold in position for takeoff. Under LUAW phraseology, the controller states the aircraft call sign, departure runway, and "line up and wait." Be aware that "traffic holding in position" will continue to be used to advise other aircraft that traffic has been authorized to line up and wait on an active runway. Pay close attention when instructed to "line up and wait," especially at night or during periods of low visibility. Before

Figure 14-54. *Communication between ATC and a pilot who is requesting taxi procedures.*

entering the runway, remember to scan the full length of the runway and its approach end for other aircraft.

There have been collisions and incidents involving aircraft instructed to "line up and wait" while ATC waits for the necessary conditions to issue a takeoff clearance. An OI caused a 737 to land on a runway occupied by a twin-engine turboprop. The turboprop was holding in position awaiting takeoff clearance. Upon landing, the 737 collided with the twin-engine turboprop.

When ATC instructs you to "line up and wait," they should advise you of any anticipated delay in receiving your takeoff clearance. Possible reasons for ATC takeoff clearance delays may include other aircraft landing and/or departing, wake turbulence, or traffic crossing an intersecting runway.

- If advised of a reason for the delay, or the reason is clearly visible, expect an imminent takeoff clearance once the reason is no longer an issue.

- If a takeoff clearance is not received within 90 seconds after receiving the "line up and wait" instruction, contact ATC immediately.

- When ATC issues "line up and wait" instructions and takeoff clearances from taxiway intersection, the taxiway designator is included.

    Example – "N123AG Runway One-Eight, at Charlie Three, line up and wait."

    Example – "N123AG Runway One-Eight, at Charlie Three, cleared for takeoff."

If LUAW procedures are being used and landing traffic is a factor, ATC is required to:

- Inform the aircraft in the LUAW position of the closest aircraft that is requesting a full-stop, touch-and-go, stop-and-go, option, or unrestricted low approach.

    Example – "N123AG, Runway One-Eight, line up and wait, traffic a Cessna 210 on a six-mile final."

- In some cases, where safety logic is being used, ATC is permitted to issue landing clearances with traffic in the LUAW position. Traffic information is issued to the landing traffic.

    Example – "N456HK, Runway One-Eight, cleared to land, traffic a DeHavilland Otter holding in position."

    NOTE: ATC will/must issue a takeoff clearance to the traffic holding in position in sufficient time to ensure no conflict exists with landing aircraft. Prescribed runway separation must exist no later than when the landing aircraft crosses the threshold.

- In cases where ATC is not permitted to issue landing clearances with traffic in the LUAW position, traffic information is issued to the closest aircraft that is requesting a full-stop, touch-and-go, stop-and-go, option, or unrestricted low approach.

    Example – "N456HK, Runway One-Eight, continue, traffic holding in position."

ATC Instructions—"Runway Shortened"

You should review NOTAMs in your preflight planning to determine any airport changes that will affect your departure or arrival. When the available runway length has been temporarily or permanently shortened due to construction, the ATIS includes the words "warning" and "shortened" in the text of the message. For the duration of the construction when the runway is temporarily shortened, ATC will include the word "shortened" in their clearance instructions. Furthermore, the use of the term "full length" will not be used by ATC during this period of the construction.

Some examples of ATC instructions are:

- "Runway three six shortened, line up and wait."
- "Runway three six shortened, cleared for takeoff."
- "Runway three six shortened, cleared to land."

When an intersection departure is requested on a temporarily or permanently shortened runway during the construction, the remaining length of runway is included in the clearance. For example, "Runway three six at Echo, intersection departure, 5,600 feet available." If following the construction, the runway is permanently shortened, ATC will include the word "shortened" until the Chart Supplement U.S. (formerly Airport/Facility Directory) is updated to include the permanent changes to the runway length.

Pre-Landing, Landing, and After-Landing

While en route and after receiving the destination airport ATIS/landing information, review the airport diagram and brief yourself as to your exit taxiway. Determine the following:

- Are there any runway hold markings in close proximity to the exit taxiway?

- **Do not cross any hold markings or exit onto any runways without ATC clearance.**

After landing, use the utmost caution where the exit taxiways intersect another runway, and do not exit onto another runway without ATC authorization. Do not accept last minute turnoff instructions from the control tower unless you clearly understand the instructions and are at a speed that ensures you can safely comply. Finally, after landing and upon exiting the runway, ensure your aircraft has completely crossed over the runway hold markings. Once all parts of the aircraft have crossed the runway holding position markings, you must hold unless further instructions have been issued by ATC. Do not initiate non-essential communications or actions until the aircraft has stopped and the brakes set.

Engineered Materials Arresting Systems (EMAS)

Aircraft can and do overrun the ends of runways and sometimes with devastating results. An overrun occurs when an aircraft passes beyond the end of a runway during an aborted takeoff or on landing rollout. To minimize the hazards of overruns, the FAA incorporated the concept of a runway safety area (RSA) beyond the runway end into airport design standards. At most commercial airports, the RSA is 500 feet wide and extends 1,000 feet beyond each end of the runway. The FAA implemented this requirement in the event that an aircraft overruns, undershoots, or veers off the side of the runway.

The most dangerous of these incidents are overruns, but since many airports were built before the 1,000-foot RSA length was adopted some 20 years ago, the area beyond the end of the runway is where many airports cannot achieve the full standard RSA. This is due to obstacles, such as bodies of water, highways, railroads, populated areas, or severe drop-off of terrain. Under these specific circumstances, the installation of an Engineered Materials Arresting System (EMAS) is an acceptable alternative to a RSA beyond the runway end. It provides a level of safety that is generally equivalent to a full RSA. *[Figure 14-55]*

An EMAS uses materials of closely controlled strength and density placed at the end of a runway to stop or greatly slow an aircraft that overruns the runway. The best material found to date is a lightweight, crushable concrete. When an aircraft rolls into an EMAS arrestor bed, the tires of the aircraft sink into the lightweight concrete and the aircraft is decelerated by having to roll through the material. *[Figure 14-56]*

Incidents

To date, there have been several incidents listed below where the EMAS technology has worked successfully to arrest aircraft that overrun the runway. All cases have resulted in minimal to do damage to the aircraft. The only known injury was an ankle injury to a passenger during egress following the arrestment. *[Figure 14-57]*

- May 1999—A Saab 340 commuter aircraft overran the runway at John F. Kennedy International Airport (JFK).

Figure 14-55. *Engineered material arresting system (EMAS) located at Yeager Airport, Charleston, West Virginia.*

- May 2003—A Cargo McDonnell Douglas (MD)-11 overran the runway at JFK.

- January 2005—A Boeing 747 overran the runway at JFK.

- July 2006—A Mystere Falcon 900 overran the runway at Greenville Downtown Airport (KGMU) in Greenville, South Carolina.

- July 2008—An Airbus A320 overran the runway at O'Hare International Airport (ORD).

- January 2010—A Bombardier CRJ-200 regional jet overran the runway at Yeager Airport (KCRW) in Charleston, West Virginia (WV). *[Figure 14-58]*

- October 2010—A G-4 Gulfstream overran the runway at Teterboro Airport (KTEB) in Teterboro, New Jersey (NJ).

- November 2011—A Cessna Citation 550 overran the runway at Key West International Airport (KEYW) in Key West, Florida.

EMAS Installations and Information

Currently, EMAS is installed at 63 runway ends at 42 airports in the United States with plans to install more throughout the next few years.

EMAS information is available in the Chart Supplement U.S. (formerly Airport/Facility Directory) under the specific airport information. *Figure 14-59* shows airport information for Boston Logan International Airport. At the bottom of the page, it shows which runways are equipped with arresting systems and the type that they have. It is important for pilots to study airport information, become familiar with the details and limitations of the arresting system, and the runways that are equipped with them. *[Figure 14-60]*

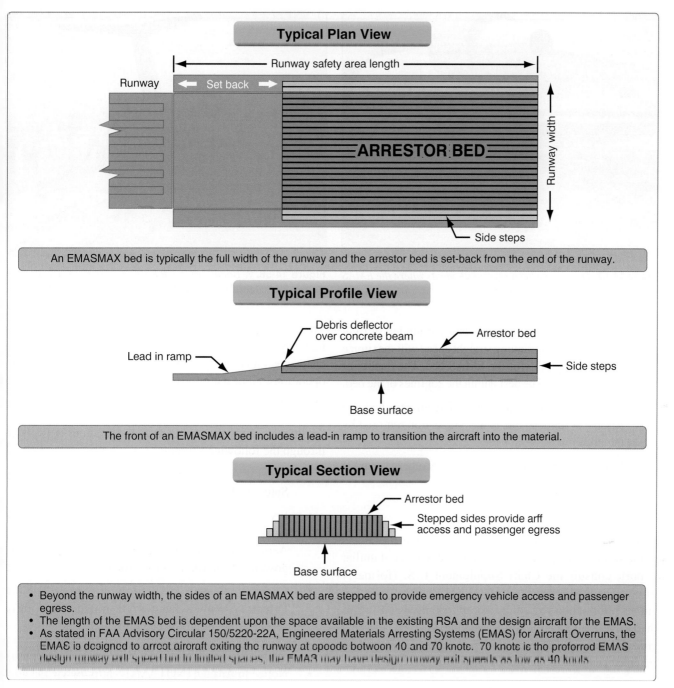

Typical Plan View

Runway safety area length

Runway

Set back

ARRESTOR BED

Runway width

Side steps

An EMASMAX bed is typically the full width of the runway and the arrestor bed is set-back from the end of the runway.

Typical Profile View

Debris deflector
over concrete beam

Arrestor bed

Lead in ramp

Side steps

Base surface

The front of an EMASMAX bed includes a lead-in ramp to transition the aircraft into the material.

Typical Section View

Arrestor bed

Stepped sides provide arff
access and passenger egress

Base surface

- Beyond the runway width, the sides of an EMASMAX bed are stepped to provide emergency vehicle access and passenger egress.
- The length of the EMAS bed is dependent upon the space available in the existing RSA and the design aircraft for the EMAS.
- As stated in FAA Advisory Circular 150/5220-22A, Engineered Materials Arresting Systems (EMAS) for Aircraft Overruns, the EMAS is designed to arrest aircraft exiting the runway at speeds between 40 and 70 knots. 70 knots is the preferred EMAS design runway exit speed but in limited spaces, the EMAS may have design runway exit speeds as low as 40 knots.

Figure 14-56. *Diagram of an EMASMAX system.*

Pilot Considerations

Although engaging an EMAS should not be a desired outcome for the end of a flight, pilots need to know what EMAS is, how to identify it on the airfield diagram and on the airfield, as well as knowing what to do should they find themselves approaching an installation in an overrun situation. *[Figure 14-59* and *Figure 14-60]* Pilots also need to know that an EMAS may not stop lightweight general aviation aircraft that are not heavy enough to sink into the crushable concrete. The time to discuss whether or not a runway has an EMAS at the end is during the pre-departure briefing prior to takeoff or during the approach briefing prior to commencing the approach. Following the guidance below ensures that the aircraft engages the EMAS according to the design entry parameters.

During the takeoff or landing phase, if a pilot determines that the aircraft will exit the runway end and enter the EMAS, the following guidance should be adhered to:

1. Continue deceleration - Regardless of aircraft speed upon exiting the runway, continue to follow Rejected/ Aborted Takeoff procedures, or if landing, Maximum Braking procedures outlined in the Flight Manual.

Figure 14-57. *There have been several incidents where the EMAS has successfully arrested the aircraft.*

Figure 14-58. *A Bombardier CRJ-200 regional jet overran the runway at Yeager Airport (KCRW) in Charleston, West Virginia.*

2. Maintain runway centerline - Not veering left or right of the bed and continuing straight ahead will maximize stopping capability of the EMAS bed. The quality of deceleration will be best within the confines of the bed.

3. Maintain deceleration efforts - The arrestor bed is a passive system, so this is the only action required by the pilot.

4. Once stopped, do not attempt to taxi or otherwise move the aircraft.

Chapter Summary

This chapter focused on airport operations both in the air and on the surface. For specific information about an unfamiliar airport, consult the Chart Supplement U.S. (formerly Airport/Facility Directory) and NOTAMS before flying. For further information regarding procedures discussed in this chapter, refer to 14 CFR part 91 and the AIM. By adhering to established procedures, both airport operations and safety are enhanced.

This chapter is also designed to help you attain an understanding of the risks associated with surface navigation and is intended to provide you with basic information regarding the safe operation of aircraft at towered and nontowered airports. This chapter focuses on the following major areas:

- Runway incursion overview
- Taxi route planning
- Taxi procedures
- Communications
- Airport signs, markings and lighting

The chapter identifies best practices to help you avoid errors that may potentially lead to runway incursions. Although the chapter pertains mostly to surface movements for single-pilot operations, all of the information is relevant for flight crew operations as well.

Additional information about surface operations is available through the following sources:

- Federal Aviation Administration (FAA) Runway Safety website—www.faa.gov/go/runwaysafety

- FAA National Aeronautical Navigation Services (AeroNav), formerly known as the National Aeronautical Charting Office (NACO)—www.faa.gov/air_traffic/flight_info/aeronav

- Chart Supplement U.S. (formerly Airport/Facility Directory)—www.faa.gov/air_traffic/flight_info/aeronav/digital_products/dafd/search/

- Automatic Terminal Information Service (ATIS)

- Notice to Airmen (NOTAMs)—http://www.faa.gov/pilots/flt_plan/notams

- Advisory Circular (AC) 91-73, part 91 and part 135, Single-Pilot and Flight School Procedures During Taxi Operations

- Aeronautical Information Manual (AIM)—www.faa.gov/air_traffic/publications/atpubs/aim/

- AC 120-74, parts 91, 121, 125, and 135, Flight Crew Procedures During Taxi Operations

BOSTON

GENERAL EDWARD LAWRENCE LOGAN INTL (BOS) 1 E UTC−5(−4DT)

N42°21.78′ W71°00.39′ NEW YORK
COPTER
20 B S4 **FUEL** 100LL, JET A OX 1, 2, 3, 4 LRA Class I, ARFF Index E H−10J, 11D, 12K, L−33D, 34J
NOTAM FILE BOS IAP, AD

RWY 15R−33L: H10083X150 (ASPH−GRVD) S−200, D−200, 2S−175,
2D−400, 2D/2D2−800 HIRL CL

RWY 15R: MALSR. TDZL. PAPI(P4L)—GA 3.0° TCH 60′. Thld dsplcd
880′. Trees.

RWY 33L: MALSR. TDZL. PAPI(P4R)—GA 3.0° TCH 57′. Boat.

RWY 04R−22L: H10005X150 (ASPH−GRVD) S−200, D−200, 2S−175,
2D−400, 2D/2D2−800 HIRL CL

RWY 04R: ALSF2. TDZL. PAPI(P4L)—GA 3.0° TCH 67′.Thld dsplcd
1154′. Boat.

RWY 22L: MALSF. PAPI(P4R)—GA 3.0° TCH 55′. Thld dsplcd 1199′.
Boat.

RWY 04L−22R: H7861X150 (ASPH−GRVD) S−200, D−200, 2S−175,
2D 400, 2D/2D2 800 HIRL

RWY 04L: REIL. PAPI(P4L)—GA 3.0°TCH 50′. Boat.

RWY 22R: PAPI(P4L)—GA 3.0° TCH 50′. Thld dsplcd 815′. Boat.

RWY 09−27: H7000X150 (ASPH−GRVD) S−200, D−200, 2S−175,
2D−400, 2D/2D2−800 HIRL CL

RWY 09: Boat.

RWY 27: REIL. PAPI(P4L)—GA 3.0° TCH 71′. Boat.

RWY 14−32: H5000X100 (ASPH−GRVD) S−75, D−200, 2S−175,
2D−400, 2D/2D2−875 HIRL

RWY 14: Bldg. RWY 32: REIL. PAPI (P4L)—GA 3.0° TCH 45′.

RWY 15L−33R: H2557X100 (ASPH) S−200, D−200, 2S−175, 2D−400, 2D/2D2−800 MIRL

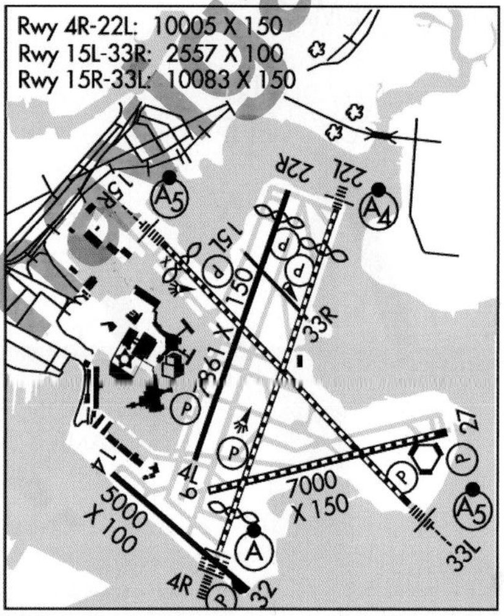

Rwy 4R-22L: 10005 X 150
Rwy 15L-33R: 2557 X 100
Rwy 15R-33L: 10083 X 150

LAND AND HOLD SHORT OPERATIONS

LANDING	HOLD SHORT POINT	DIST AVBL
RWY 04L	15L−33R	5250
RWY 15R	09−27	6800
RWY 22L	09−27	6400
RWY 27	04R−22L	5650

RUNWAY DECLARED DISTANCE INFORMATION

RWY 04L:	TORA−7861	TODA−7861	ASDA−7861	LDA−7861
RWY 04R:	TORA−10005	TODA−10005	ASDA−10005	LDA−8851
RWY 09:	TORA−7000	TODA−7000	ASDA−7000	LDA−7000
RWY 14:	TORA−5000	TODA−5000	ASDA−5000	LDA−5000
RWY 15L:	TORA−2557	TODA−2557	ASDA−2557	LDA−2557
RWY 15R:	TORA−10083	TODA−10083	ASDA−10083	LDA−9203
RWY 22L:	TORA−10005	TODA−10005	ASDA−10005	LDA−8806
RWY 22R:	TORA−7861	TODA−7861	ASDA−7861	LDA−7046
RWY 27:	TORA−7000	TODA−7000	ASDA−7000	LDA−7000
RWY 32:	TORA−5000	TODA−5000	ASDA−5000	LDA−5000
RWY 33L:	TORA−10083	TODA−10083	ASDA−10083	LDA−10083
RWY 33R:	TORA−2557	TODA−2557	ASDA−2557	LDA−2557

ARRESTING GEAR/SYSTEM

RWY 04L: EMAS

RWY 15R: EMAS

Figure 14-59. *EMAS information for Boston Logan International Airport located in the Chart Supplement U.S. (formerly Airport/ Facility Directory).*

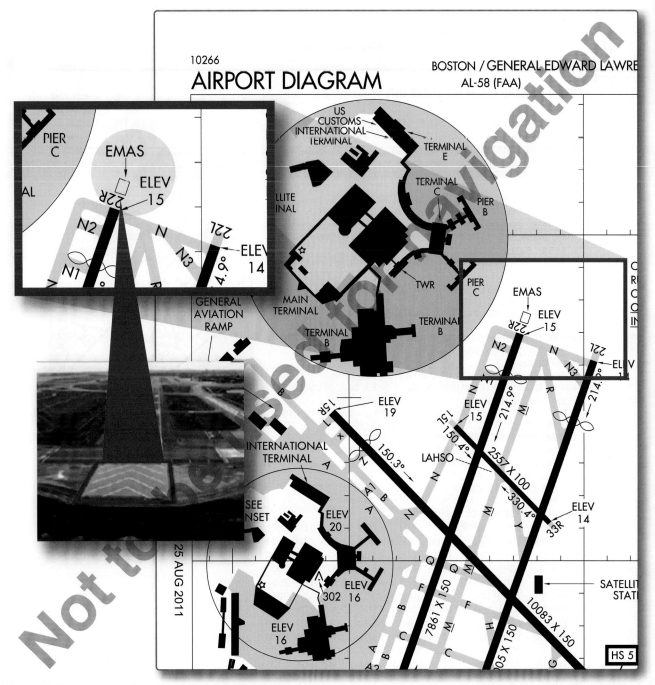

Figure 14-60. *An airport diagram with EMAS information.*

Chapter 15

Airspace

Introduction

The two categories of airspace are: regulatory and nonregulatory. Within these two categories, there are four types: controlled, uncontrolled, special use, and other airspace. The categories and types of airspace are dictated by the complexity or density of aircraft movements, nature of the operations conducted within the airspace, the level of safety required, and national and public interest. *Figure 15-1* presents a profile view of the dimensions of various classes of airspace. Also, there are excerpts from sectional charts that are discussed in Chapter 16, Navigation, that are used to illustrate how airspace is depicted.

Basic VFR Weather Minimums		Flight Visibility	Distance from Clouds
Airspace			
Class A		Not applicable	Not applicable
Class B		3 statute miles	Clear of clouds
Class C		3 statute miles	1,000 feet above 500 feet below 2,000 feet horizontal
Class D		3 statute miles	1,000 feet above 500 feet below 2,000 feet horizontal
Class E	At or above 10,000 feet MSL	5 statute miles	1,000 feet above 1,000 feet below 1 statute mile horizontal
	Less than 10,000 feet MSL	3 statute miles	1,000 feet above 500 feet below 2,000 feet horizontal
Class G	1,200 feet or less above the surface (regardless of MSL altitude).	Day, except as provided in section 91.155(b) — 1 statute mile	Clear of clouds
		Night, except as provided in section 91.155(b) — 3 statute miles	1,000 feet above 500 feet below 2,000 feet horizontal
	More than 1,200 feet above the surface but less than 10,000 feet MSL.	Day — 1 statute mile	1,000 feet above 500 feet below 2,000 feet horizontal
		Night — 3 statute miles	1,000 feet above 500 feet below 2,000 feet horizontal
	More than 1,200 feet above the surface and at or above 10,000 feet MSL.	5 statute miles	1,000 feet above 1,000 feet below 1 statute mile horizontal

Class Airspace	Entry Requirements	Equipment	Minimum Pilot
Class A	ATC clearance	IFR equipped	Instrument rating
Class B	ATC clearance	Two-way radio, transponder with altitude reporting capability	Private—(However, recreational pilot other than those seeking private if regulator...)
Class C	Two-way radio communications prior to entry	Two-way radio, transponder with altitude reporting capability	No spe...
Class D	Two-way radio communications prior to entry	Two-way radio	No
Class E	None for VFR	No specific requirement	
Class G	None	No specific requirement	

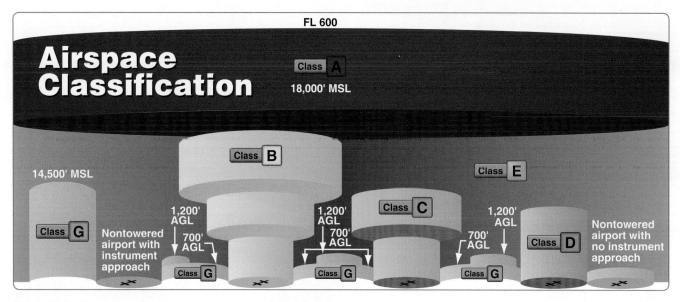

Figure 15-1. *Airspace profile.*

Controlled Airspace

Controlled airspace is a generic term that covers the different classifications of airspace and defined dimensions within which air traffic control (ATC) service is provided in accordance with the airspace classification. Controlled airspace consists of:

- Class A
- Class B
- Class C
- Class D
- Class E

Class A Airspace

Class A airspace is generally the airspace from 18,000 feet mean sea level (MSL) up to and including flight level (FL) 600, including the airspace overlying the waters within 12 nautical miles (NM) of the coast of the 48 contiguous states and Alaska. Unless otherwise authorized, all operation in Class A airspace is conducted under instrument flight rules (IFR).

Class B Airspace

Class B airspace is generally airspace from the surface to 10,000 feet MSL surrounding the nation's busiest airports in terms of airport operations or passenger enplanements. The configuration of each Class B airspace area is individually tailored, consists of a surface area and two or more layers (some Class B airspace areas resemble upside-down wedding cakes), and is designed to contain all published instrument procedures once an aircraft enters the airspace. ATC clearance is required for all aircraft to operate in the area, and all aircraft that are so cleared receive separation services within the airspace.

Class C Airspace

Class C airspace is generally airspace from the surface to 4,000 feet above the airport elevation (charted in MSL) surrounding those airports that have an operational control tower, are serviced by a radar approach control, and have a certain number of IFR operations or passenger enplanements. Although the configuration of each Class C area is individually tailored, the airspace usually consists of a surface area with a five NM radius, an outer circle with a ten NM radius that extends from 1,200 feet to 4,000 feet above the airport elevation. Each aircraft must establish two-way radio communications with the ATC facility providing air traffic services prior to entering the airspace and thereafter must maintain those communications while within the airspace.

Class D Airspace

Class D airspace is generally airspace from the surface to 2,500 feet above the airport elevation (charted in MSL) surrounding those airports that have an operational control tower. The configuration of each Class D airspace area is individually tailored and, when instrument procedures are published, the airspace is normally designed to contain the procedures. Arrival extensions for instrument approach procedures (IAPs) may be Class D or Class E airspace. Unless otherwise authorized, each aircraft must establish two-way radio communications with the ATC facility providing air traffic services prior to entering the airspace and thereafter maintain those communications while in the airspace.

Class E Airspace

Class E airspace is the controlled airspace not classified as Class A, B, C, or D airspace. A large amount of the airspace over the United States is designated as Class E airspace.

This provides sufficient airspace for the safe control and separation of aircraft during IFR operations. Chapter 3 of the Aeronautical Information Manual (AIM) explains the various types of Class E airspace.

Sectional and other charts depict all locations of Class E airspace with bases below 14,500 feet MSL. In areas where charts do not depict a class E base, class E begins at 14,500 feet MSL.

In most areas, the Class E airspace base is 1,200 feet AGL. In many other areas, the Class E airspace base is either the surface or 700 feet AGL. Some Class E airspace begins at an MSL altitude depicted on the charts, instead of an AGL altitude.

Class E airspace typically extends up to, but not including, 18,000 feet MSL (the lower limit of Class A airspace). All airspace above FL 600 is Class E airspace.

Uncontrolled Airspace

Class G Airspace

Uncontrolled airspace or Class G airspace is the portion of the airspace that has not been designated as Class A, B, C, D, or E. It is therefore designated uncontrolled airspace. Class G airspace extends from the surface to the base of the overlying Class E airspace. Although ATC has no authority or responsibility to control air traffic, pilots should remember there are visual flight rules (VFR) minimums that apply to Class G airspace.

Special Use Airspace

Special use airspace or special area of operation (SAO) is the designation for airspace in which certain activities must be confined, or where limitations may be imposed on aircraft operations that are not part of those activities. Certain special use airspace areas can create limitations on the mixed use of airspace. The special use airspace depicted on instrument charts includes the area name or number, effective altitude, time and weather conditions of operation, the controlling agency, and the chart panel location. On National Aeronautical Charting Group (NACG) en route charts, this information is available on one of the end panels. Special use airspace usually consists of:

- Prohibited areas
- Restricted areas
- Warning areas
- Military operation areas (MOAs)
- Alert areas
- Controlled firing areas (CFAs)

Prohibited Areas

Prohibited areas contain airspace of defined dimensions within which the flight of aircraft is prohibited. Such areas are established for security or other reasons associated with the national welfare. These areas are published in the Federal Register and are depicted on aeronautical charts. The area is charted as a "P" followed by a number (e.g., P-40). Examples of prohibited areas include Camp David and the National Mall in Washington, D.C., where the White House and the Congressional buildings are located. *[Figure 15-2]*

Restricted Areas

Restricted areas are areas where operations are hazardous to nonparticipating aircraft and contain airspace within which the flight of aircraft, while not wholly prohibited, is subject to restrictions. Activities within these areas must be confined because of their nature, or limitations may be imposed upon aircraft operations that are not a part of those activities, or both. Restricted areas denote the existence of unusual, often invisible, hazards to aircraft (e.g., artillery firing, aerial gunnery, or guided missiles). IFR flights may be authorized to transit the airspace and are routed accordingly. Penetration of restricted areas without authorization from the using or controlling agency may be extremely hazardous to the aircraft and its occupants. ATC facilities apply the following procedures when aircraft are operating on an IFR clearance (including those cleared by ATC to maintain VFR on top) via a route that lies within joint-use restricted airspace:

1. If the restricted area is not active and has been released to the Federal Aviation Administration (FAA), the ATC facility allows the aircraft to operate in the restricted airspace without issuing specific clearance for it to do so.

2. If the restricted area is active and has not been released to the FAA, the ATC facility issues a clearance that ensures the aircraft avoids the restricted airspace.

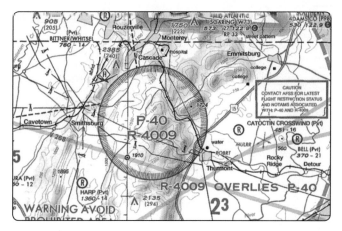

Figure 15-2. *An example of a prohibited area, P-40 around Camp David.*

Restricted areas are charted with an "R" followed by a number (e.g., R-4401) and are depicted on the en route chart appropriate for use at the altitude or FL being flown. *[Figure 15-3]* Restricted area information can be obtained on the back of the chart.

Warning Areas

Warning areas are similar in nature to restricted areas; however, the United States government does not have sole jurisdiction over the airspace. A warning area is airspace of defined dimensions, extending from 3 NM outward from the coast of the United States, containing activity that may be hazardous to nonparticipating aircraft. The purpose of such areas is to warn nonparticipating pilots of the potential danger. A warning area may be located over domestic or international waters or both. The airspace is designated with a "W" followed by a number (e.g., W-237). *[Figure 15-4]*

Military Operation Areas (MOAs)

MOAs consist of airspace with defined vertical and lateral limits established for the purpose of separating certain military training activities from IFR traffic. Whenever an MOA is being used, nonparticipating IFR traffic may be cleared through an MOA if IFR separation can be provided by ATC. Otherwise, ATC reroutes or restricts nonparticipating IFR traffic. MOAs are depicted on sectional, VFR terminal area, and en route low altitude charts and are not numbered (e.g., "Camden Ridge MOA"). *[Figure 15-5]* However, the MOA is also further defined on the back of the sectional charts with times of operation, altitudes affected, and the controlling agency.

Alert Areas

Alert areas are depicted on aeronautical charts with an "A" followed by a number (e.g., A-211) to inform nonparticipating

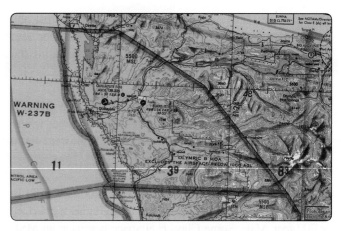

Figure 15-4. *Requirements for airspace operations.*

pilots of areas that may contain a high volume of pilot training or an unusual type of aerial activity. Pilots should exercise caution in alert areas. All activity within an alert area shall be conducted in accordance with regulations, without waiver, and pilots of participating aircraft, as well as pilots transiting the area, shall be equally responsible for collision avoidance. *[Figure 15-6]*

Controlled Firing Areas (CFAs)

CFAs contain activities that, if not conducted in a controlled environment, could be hazardous to nonparticipating aircraft. The difference between CFAs and other special use airspace is that activities must be suspended when a spotter aircraft, radar, or ground lookout position indicates an aircraft might be approaching the area. There is no need to chart CFAs since they do not cause a nonparticipating aircraft to change its flight path.

Other Airspace Areas

"Other airspace areas" is a general term referring to the majority of the remaining airspace. It includes:

- Local airport advisory (LAA)
- Military training route (MTR)
- Temporary flight restriction (TFR)
- Parachute jump aircraft operations
- Published VFR routes
- Terminal radar service area (TRSA)
- National security area (NSA)
- Air Defense Identification Zones (ADIZ) land and water based and need for Defense VFR (DVFR) flight plan to operate VFR in this airspace
- Intercept Procedures and use of 121.5 for communication if not on ATC already

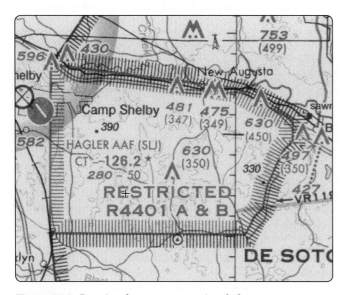

Figure 15-3. *Restricted areas on a sectional chart.*

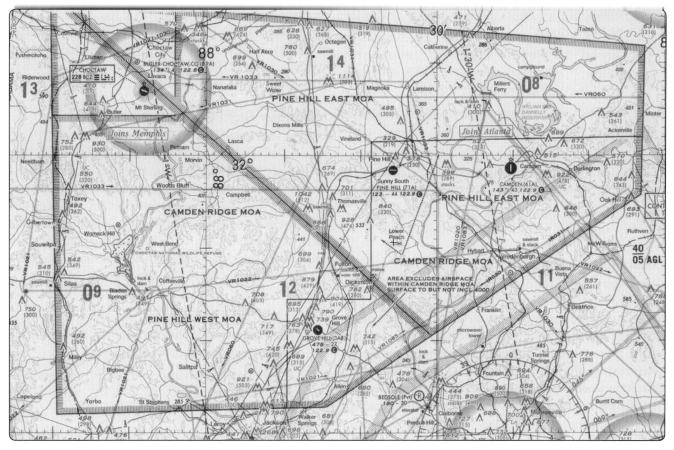

Figure 15-5. *Camden Ridge MOA is an example of a military operations area.*

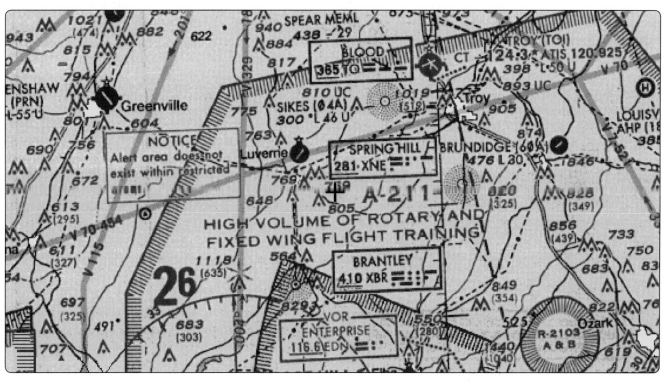

Figure 15-6. *Alert area (A-211).*

- Flight Restricted Zones (FRZ) in vicinity of Capitol and White House

- Special Awareness Training required by 14 CFR 91.161 for pilots to operate VFR within 60 NM of the Washington, DC VOR/DME

- Wildlife Areas/Wilderness Areas/National Parks and request to operate above 2,000 AGL

- National Oceanic and Atmospheric Administration (NOAA) Marine Areas off the coast with requirement to operate above 2,000 AGL

- Tethered Balloons for observation and weather recordings that extend on cables up to 60,000

Local Airport Advisory (LAA)

An advisory service provided by Flight Service Station (FSS) facilities, which are located on the landing airport, using a discrete ground-to-air frequency or the tower frequency when the tower is closed. LAA services include local airport advisories, automated weather reporting with voice broadcasting, and a continuous Automated Surface Observing System (ASOS)/Automated Weather Observing Station (AWOS) data display, other continuous direct reading instruments, or manual observations available to the specialist.

Military Training Routes (MTRs)

MTRs are routes used by military aircraft to maintain proficiency in tactical flying. These routes are usually established below 10,000 feet MSL for operations at speeds in excess of 250 knots. Some route segments may be defined at higher altitudes for purposes of route continuity. Routes are identified as IFR (IR), and VFR (VR), followed by a number. *[Figure 15-7]* MTRs with no segment above 1,500 feet AGL are identified by four number characters (e.g., IR1206, VR1207). MTRs that include one or more segments above 1,500 feet AGL are identified by three number characters (e.g., IR206, VR207). IFR low altitude en route charts depict all IR routes and all VR routes that

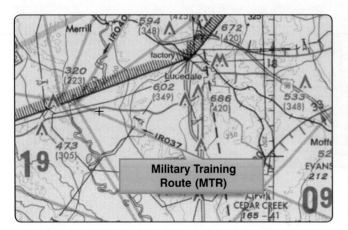

Figure 15-7. *Military training route (MTR) chart symbols.*

accommodate operations above 1,500 feet AGL. IR routes are conducted in accordance with IFR regardless of weather conditions. VFR sectional charts depict military training activities, such as IR, VR, MOA, restricted area, warning area, and alert area information.

Temporary Flight Restrictions (TFR)

A flight data center (FDC) Notice to Airmen (NOTAM) is issued to designate a TFR. The NOTAM begins with the phrase "FLIGHT RESTRICTIONS" followed by the location of the temporary restriction, effective time period, area defined in statute miles, and altitudes affected. The NOTAM also contains the FAA coordination facility and telephone number, the reason for the restriction, and any other information deemed appropriate. The pilot should check the NOTAMs as part of flight planning.

Some of the purposes for establishing a TFR are:

- Protect persons and property in the air or on the surface from an existing or imminent hazard.

- Provide a safe environment for the operation of disaster relief aircraft.

- Prevent an unsafe congestion of sightseeing aircraft above an incident or event, that may generate a high degree of public interest.

- Protect declared national disasters for humanitarian reasons in the State of Hawaii.

- Protect the President, Vice President, or other public figures.

- Provide a safe environment for space agency operations.

Since the events of September 11, 2001, the use of TFRs has become much more common. There have been a number of incidents of aircraft incursions into TFRs that have resulted in pilots undergoing security investigations and certificate suspensions. It is a pilot's responsibility to be aware of TFRs in their proposed area of flight. One way to check is to visit the FAA website, www.tfr.faa.gov, and verify that there is not a TFR in the area.

Parachute Jump Aircraft Operations

Parachute jump aircraft operations are published in the Chart Supplement U.S. (formerly Airport/Facility Directory). Sites that are used frequently are depicted on sectional charts.

Published VFR Routes

Published VFR routes are for transitioning around, under, or through some complex airspace. Terms such as VFR flyway, VFR corridor, Class B airspace VFR transition route, and terminal area VFR route have been applied to such routes.

These routes are generally found on VFR terminal area planning charts.

Terminal Radar Service Areas (TRSAs)

TRSAs are areas where participating pilots can receive additional radar services. The purpose of the service is to provide separation between all IFR operations and participating VFR aircraft.

The primary airport(s) within the TRSA become(s) Class D airspace. The remaining portion of the TRSA overlies other controlled airspace, which is normally Class E airspace beginning at 700 or 1,200 feet and established to transition to/from the en route/terminal environment. TRSAs are depicted on VFR sectional charts and terminal area charts with a solid black line and altitudes for each segment. The Class D portion is charted with a blue segmented line. Participation in TRSA services is voluntary; however, pilots operating under VFR are encouraged to contact the radar approach control and take advantage of TRSA service.

National Security Areas (NSAs)

NSAs consist of airspace of defined vertical and lateral dimensions established at locations where there is a requirement for increased security and safety of ground facilities. Flight in NSAs may be temporarily prohibited by regulation under the provisions of Title 14 of the Code of Federal Regulations (14 CFR) part 99, and prohibitions are disseminated via NOTAM. Pilots are requested to voluntarily avoid flying through these depicted areas.

Air Traffic Control and the National Airspace System

The primary purpose of the ATC system is to prevent a collision between aircraft operating in the system and to organize and expedite the flow of traffic. In addition to its primary function, the ATC system has the capability to provide (with certain limitations) additional services. The ability to provide additional services is limited by many factors, such as the volume of traffic, frequency congestion, quality of radar, controller workload, higher priority duties, and the pure physical inability to scan and detect those situations that fall in this category. It is recognized that these services cannot be provided in cases in which the provision of services is precluded by the above factors.

Consistent with the aforementioned conditions, controllers shall provide additional service procedures to the extent permitted by higher priority duties and other circumstances. The provision of additional services is not optional on the part of the controller, but rather is required when the work situation permits. Provide ATC service in accordance with the procedures and minima in this order except when:

1. A deviation is necessary to conform to ICAO Documents, National Rules of the Air, or special agreements where the United States provides ATC service in airspace outside the country and its possessions

2. Other procedures/minima are prescribed in a letter of agreement, FAA directive, or a military document

3. A deviation is necessary to assist an aircraft when an emergency has been declared

Coordinating the Use of Airspace

ATC is responsible for ensuring that the necessary coordination has been accomplished before allowing an aircraft under their control to enter another controller's area of jurisdiction.

Before issuing control instructions directly or relaying through another source to an aircraft that is within another controller's area of jurisdiction that will change that aircraft's heading, route, speed, or altitude, ATC ensures that coordination has been accomplished with each of the controllers listed below whose area of jurisdiction is affected by those instructions unless otherwise specified by a letter of agreement or a facility directive:

1. The controller within whose area of jurisdiction the control instructions are issued

2. The controller receiving the transfer of control

3. Any intervening controller(s) through whose area of jurisdiction the aircraft will pass

If ATC issues control instructions to an aircraft through a source other than another controller (e.g., Aeronautical Radio, Incorporated (ARINC), FSS, another pilot), they ensure that the necessary coordination has been accomplished with any controllers listed above, whose area of jurisdiction is affected by those instructions unless otherwise specified by a letter of agreement or a facility directive.

Operating in the Various Types of Airspace

It is important that pilots be familiar with the operational requirements for each of the various types or classes of airspace. Subsequent sections cover each class in sufficient detail to facilitate understanding regarding weather, type of pilot certificate held, and equipment required.

Basic VFR Weather Minimums

No pilot may operate an aircraft under basic VFR when the flight visibility is less, or at a distance from clouds that is less, than that prescribed for the corresponding altitude and class of airspace. [Figure 15-8] Except as provided in 14 CFR part 91, section 91.157, "Special VFR Weather Minimums,"

Basic VFR Weather Minimums				
Airspace			Flight Visibility	Distance from Clouds
Class A			Not applicable	Not applicable
Class B			3 statute miles	Clear of clouds
Class C			3 statute miles	1,000 feet above 500 feet below 2,000 feet horizontal
Class D			3 statute miles	1,000 feet above 500 feet below 2,000 feet horizontal
Class E	At or above 10,000 feet MSL		5 statute miles	1,000 feet above 1,000 feet below 1 statute mile horizontal
	Less than 10,000 feet MSL		3 statute miles	1,000 feet above 500 feet below 2,000 feet horizontal
Class G	1,200 feet or less above the surface (regardless of MSL altitude).	Day, except as provided in section 91.155(b)	1 statute mile	Clear of clouds
		Night, except as provided in section 91.155(b)	3 statute miles	1,000 feet above 500 feet below 2,000 feet horizontal
	More than 1,200 feet above the surface but less than 10,000 feet MSL.	Day	1 statute mile	1,000 feet above 500 feet below 2,000 feet horizontal
		Night	3 statute miles	1,000 feet above 500 feet below 2,000 feet horizontal
	More than 1,200 feet above the surface and at or above 10,000 feet MSL.		5 statute miles	1,000 feet above 1,000 feet below 1 statute mile horizontal

Figure 15-8. *Visual flight rule weather minimums.*

no person may operate an aircraft beneath the ceiling under VFR within the lateral boundaries of controlled airspace designated to the surface for an airport when the ceiling is less than 1,000 feet. Additional information can be found in 14 CFR part 91, section 91.155(c).

Operating Rules and Pilot/Equipment Requirements
The safety of flight is a top priority of all pilots and the responsibilities associated with operating an aircraft should always be taken seriously. The air traffic system maintains a high degree of safety and efficiency with strict regulatory oversight of the FAA. Pilots fly in accordance with regulations that have served the United States well, as evidenced by the fact that the country has the safest aviation system in the world.

All aircraft operating in today's National Airspace System (NAS) has complied with the CFR governing its certification and maintenance; all pilots operating today have completed rigorous pilot certification training and testing. Of equal importance is the proper execution of preflight planning, aeronautical decision-making (ADM) and risk management. ADM involves a systematic approach to risk assessment and stress management in aviation, illustrates how personal attitudes can influence decision-making, and how those attitudes can be modified to enhance safety in the flight deck. More detailed information regarding ADM and risk mitigation can be found in Chapter 2, "Aeronautical Decision-Making."

Pilots also comply with very strict FAA general operating and flight rules as outlined in the CFR, including the FAA's important "see and avoid" mandate. These regulations provide the historical foundation of the FAA regulations governing the aviation system and the individual classes of airspace. *Figure 15-9* lists the operational and equipment requirements for these various classes of airspace. It is helpful to refer to this figure as the specific classes are discussed in greater detail.

Class A

Pilots operating an aircraft in Class A airspace must conduct that operation under IFR and only under an ATC clearance received prior to entering the airspace. Unless otherwise authorized by ATC, each aircraft operating in Class A airspace must be equipped with a two-way radio capable of communicating with ATC on a frequency assigned by ATC. Unless otherwise authorized by ATC, all aircraft within Class A airspace must be equipped with the appropriate transponder equipment meeting all applicable specifications found in 14 CFR part 91, section 91.215. Additionally, beginning January 1, 2020, aircraft operating in the Class A airspace described in 14 CFR part 91, section 91.225, must have ADS-B Out equipment installed, which meets the performance requirements of 14 CFR part 91, section 91.227.

Class B

All pilots operating an aircraft within a Class B airspace area must receive an ATC clearance from the ATC facility having jurisdiction for that area. The pilot in command (PIC) may not take off or land an aircraft at an airport within a Class B airspace unless he or she has met one of the following requirements:

1. A private pilot certificate

2. A recreational pilot certificate and all requirements contained within 14 CFR part 61, section 61.101(d), or the requirements for a student pilot seeking a recreational pilot certificate in 14 CFR part 61, section 61.94.

3. A sport pilot certificate and all requirements contained within 14 CFR part 61, section 61.325, or the requirements for a student pilot seeking a recreational pilot certificate in 14 CFR part 61, section 61.94, or the aircraft is operated by a student pilot who has met the requirements of 14 CFR part 61, sections 61.94 and 61.95, as applicable.

Unless otherwise authorized by ATC, all aircraft within Class B airspace must be equipped with the applicable operating transponder and automatic altitude reporting equipment specified in 14 CFR part 91, section 91.215(a) and an operable two-way radio capable of communications with ATC on appropriate frequencies for that Class B airspace area. Additionally, beginning January 1, 2020, aircraft operating in the Class B airspace described in 14 CFR part 91, section 91.225, must have ADS-B Out equipment installed, which meets the performance requirements of 14 CFR part 91, section 91.227.

Class C

For the purpose of this section, the primary airport is the airport for which the Class C airspace area is designated. A satellite airport is any other airport within the Class C airspace area. No pilot may take off or land an aircraft at a satellite airport within a Class C airspace area except in compliance with FAA arrival and departure traffic patterns.

Two-way radio communications must be established and maintained with the ATC facility providing air traffic services

Class Airspace	Entry Requirements	Equipment*	Minimum Pilot Certificate
Class A	ATC clearance	IFR equipped	Instrument rating
Class B	ATC clearance	Two-way radio, transponder with altitude reporting capability	Private—(However, a student or recreational pilot may operate at other than the primary airport if seeking private pilot certification and if regulatory requirements are met.)
Class C	Two-way radio communications prior to entry	Two-way radio, transponder with altitude reporting capability	No specific requirement
Class D	Two-way radio communications prior to entry	Two-way radio	No specific requirement
Class E	None for VFR	No specific requirement	No specific requirement
Class G	None	No specific requirement	No specific requirement
*Beginning January 1, 2020, ADS-B Out equipment may be required in accordance with 14 CFR part 91, section 91.225.			

Figure 15-9. *Requirements for airspace operations.*

prior to entering the airspace and thereafter maintained while within the airspace.

A pilot departing from the primary airport or satellite airport with an operating control tower must establish and maintain two-way radio communications with the control tower, and thereafter as instructed by ATC while operating in the Class C airspace area. If departing from a satellite airport without an operating control tower, the pilot must establish and maintain two-way radio communications with the ATC facility having jurisdiction over the Class C airspace area as soon as practicable after departing.

Unless otherwise authorized by the ATC having jurisdiction over the Class C airspace area, all aircraft within Class C airspace must be equipped with the appropriate transponder equipment meeting all applicable specifications found in 14 CFR part 91, section 91.215. Additionally, beginning January 1, 2020, aircraft operating in the Class C airspace described in 14 CFR part 91, section 91.225, must have ADS-B Out equipment installed, which meets the performance requirements of 14 CFR part 91, section 91.227.

Class D

No pilot may take off or land an aircraft at a satellite airport within a Class D airspace area except in compliance with FAA arrival and departure traffic patterns. A pilot departing from the primary airport or satellite airport with an operating control tower must establish and maintain two-way radio communications with the control tower, and thereafter as instructed by ATC while operating in the Class D airspace area. If departing from a satellite airport without an operating control tower, the pilot must establish and maintain two-way radio communications with the ATC facility having jurisdiction over the Class D airspace area as soon as practicable after departing.

Two-way radio communications must be established and maintained with the ATC facility providing air traffic services prior to entering the airspace and thereafter maintained while within the airspace.

If the aircraft radio fails in flight under IFR, the pilot should continue the flight by the route assigned in the last ATC clearance received; or, if being radar vectored, by the direct route from the point of radio failure to the fix, route, or airway specified in the vector clearance. In the absence of an assigned route, the pilot should continue by the route that ATC advised may be expected in a further clearance; or, if a route had not been advised, by the route filed in the flight plan.

If the aircraft radio fails in flight under VFR, the PIC may operate that aircraft and land if weather conditions are at or above basic VFR weather minimums, visual contact with the tower is maintained, and a clearance to land is received.

Class E

Unless otherwise required by 14 CFR part 93 or unless otherwise authorized or required by the ATC facility having jurisdiction over the Class E airspace area, each pilot operating an aircraft on or in the vicinity of an airport in a Class E airspace area must comply with the requirements of Class G airspace. Each pilot must also comply with any traffic patterns established for that airport in 14 CFR part 93.

Unless otherwise authorized or required by ATC, no person may operate an aircraft to, from, through, or on an airport having an operational control tower unless two-way radio communications are maintained between that aircraft and the control tower. Communications must be established within four nautical miles from the airport, up to and including 2,500 feet AGL. However, if the aircraft radio fails in flight, the PIC may operate that aircraft and land if weather conditions are at or above basic VFR weather minimums, visual contact with the tower is maintained, and a clearance to land is received.

If the aircraft radio fails in flight under IFR, the pilot should continue the flight by the route assigned in the last ATC clearance received; or, if being radar vectored, by the direct route from the point of radio failure to the fix, route, or airway specified in the vector clearance. In the absence of an assigned route, the pilot should continue by the route that ATC advised may be expected in a further clearance; or, if a route had not been advised, by the route filed in the flight plan. Additionally, beginning January 1, 2020, aircraft operating in the Class E airspace described in 14 CFR part 91, section 91.225, must have ADS-B Out equipment installed, which meets the performance requirements of 14 CFR part 91, section 91.227.

Class G

When approaching to land at an airport without an operating control tower in Class G airspace:

1. Each pilot of an airplane must make all turns of that airplane to the left unless the airport displays approved light signals or visual markings indicating that turns should be made to the right, in which case the pilot must make all turns to the right.

2. Each pilot of a helicopter or a powered parachute must avoid the flow of fixed-wing aircraft.

Unless otherwise authorized or required by ATC, no person may operate an aircraft to, from, through, or on an airport having an operational control tower unless two-way radio communications are maintained between that aircraft and the control tower. Communications must be established within four nautical miles from the airport, up to and including 2,500 feet AGL. However, if the aircraft radio fails in flight, the PIC may operate that aircraft and land if weather conditions are at or above basic VFR weather minimums, visual contact with the tower is maintained, and a clearance to land is received.

If the aircraft radio fails in flight under IFR, the pilot should continue the flight by the route assigned in the last ATC clearance received; or, if being radar vectored, by the direct route from the point of radio failure to the fix, route, or airway specified in the vector clearance. In the absence of an assigned route, the pilot should continue by the route that ATC advised may be expected in a further clearance; or, if a route had not been advised, by the route filed in the flight plan.

Uncontrolled Airspace

It is possible for some airports within Class G airspace to have a control tower (Lake City, FL, for example). Be sure to check the Chart Supplement U.S. (formerly Airport/Facility Directory) to be familiar with the airport and associated airspace prior to flight.

Ultralight Vehicles

No person may operate an ultralight vehicle within Class A, Class B, Class C, or Class D airspace or within the lateral boundaries of the surface area of Class E airspace designated for an airport unless that person has prior authorization from the ATC facility having jurisdiction over that airspace. (See 14 CFR part 103.)

Unmanned Free Balloons

Unless otherwise authorized by ATC, no person may operate an unmanned free balloon below 2,000 feet above the surface within the lateral boundaries of Class B, Class C, Class D, or Class E airspace designated for an airport. (See 14 CFR part 101.)

Unmanned Aircraft Systems

Regulations regarding unmanned aircraft systems (UAS) are currently being developed and are expected to be published by summer 2016 as 14 CFR part 107.

Parachute Jumps

No person may make a parachute jump, and no PIC may allow a parachute jump to be made from an aircraft, in or into Class A, Class B, Class C, or Class D airspace without, or in violation of, the terms of an ATC authorization issued by the ATC facility having jurisdiction over the airspace. (See 14 CFR part 105.)

Chapter Summary

This chapter introduces the various classifications of airspace and provides information on the requirements to operate in such airspace. For further information, consult the AIM and 14 CFR parts 71, 73, and 91.

Chapter 16
Navigation

Introduction

This chapter provides an introduction to cross-country flying under visual flight rules (VFR). It contains practical information for planning and executing cross-country flights for the beginning pilot.

Air navigation is the process of piloting an aircraft from one geographic position to another while monitoring one's position as the flight progresses. It introduces the need for planning, which includes plotting the course on an aeronautical chart, selecting checkpoints, measuring distances, obtaining pertinent weather information, and computing flight time, headings, and fuel requirements. The methods used in this chapter include pilotage—navigating by reference to visible landmarks, dead reckoning—computations of direction and distance from a known position, and radio navigation—by use of radio aids.

Aeronautical Charts

An aeronautical chart is the road map for a pilot flying under VFR. The chart provides information that allows pilots to track their position and provides available information that enhances safety. The three aeronautical charts used by VFR pilots are:

- Sectional
- VFR Terminal Area
- World Aeronautical

A free catalog listing aeronautical charts and related publications including prices and instructions for ordering is available at the Aeronautical Navigation Products website: www.aeronav.faa.gov.

Sectional Charts

Sectional charts are the most common charts used by pilots today. The charts have a scale of 1:500,000 (1 inch = 6.86 nautical miles (NM) or approximately 8 statute miles (SM)), which allows for more detailed information to be included on the chart.

The charts provide an abundance of information, including airport data, navigational aids, airspace, and topography. *Figure 16-1* is an excerpt from the legend of a sectional chart. By referring to the chart legend, a pilot can interpret most of the information on the chart. A pilot should also check the chart for other legend information, which includes air traffic control (ATC) frequencies and information on airspace. These charts are revised semiannually except for some areas outside the conterminous United States where they are revised annually.

VFR Terminal Area Charts

VFR terminal area charts are helpful when flying in or near Class B airspace. They have a scale of 1:250,000 (1 inch = 3.43 NM or approximately 4 SM). These charts provide a more detailed display of topographical information and are revised semiannually, except for several Alaskan and Caribbean charts. *[Figure 16-2]*

World Aeronautical Charts

World aeronautical charts are designed to provide a standard series of aeronautical charts, covering land areas of the world,

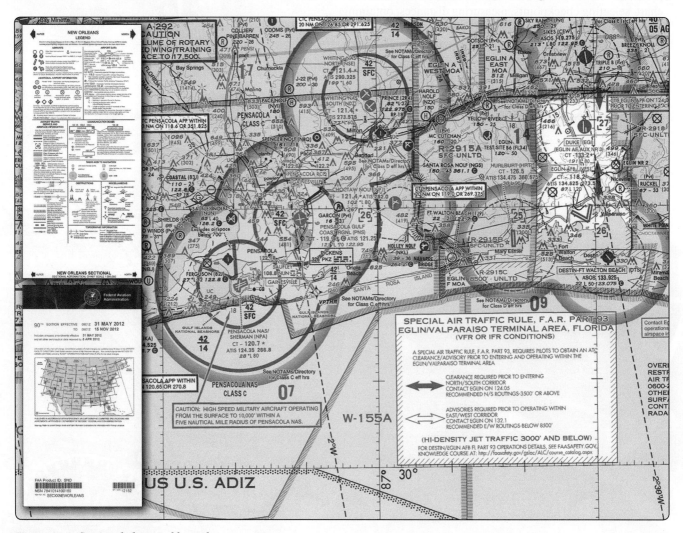

Figure 16-1. *Sectional chart and legend.*

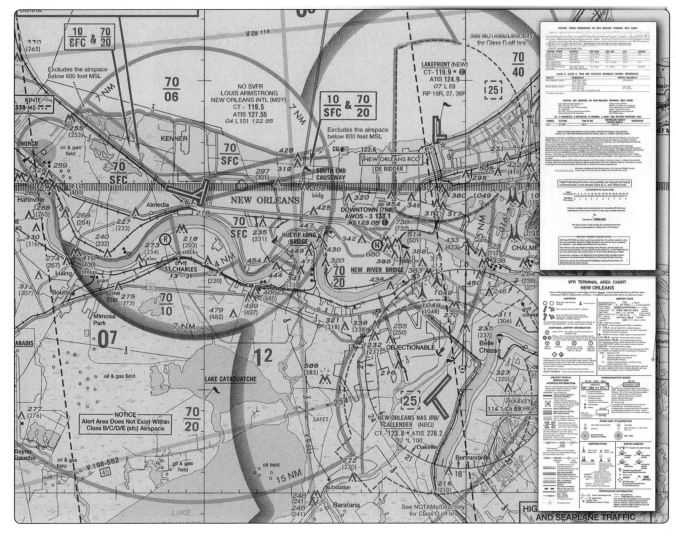

Figure 16-2. *VFR Terminal Area Chart and legend.*

at a size and scale convenient for navigation by moderate speed aircraft. They are produced at a scale of 1:1,000,000 (1 inch = 13.7 NM or approximately 16 SM). These charts are similar to sectional charts, and the symbols are the same except there is less detail due to the smaller scale. *[Figure 16-3]* These charts are revised annually except several Alaskan charts and the Mexican/Caribbean charts, which are revised every 2 years.

Latitude and Longitude (Meridians and Parallels)

The equator is an imaginary circle equidistant from the poles of the Earth. Circles parallel to the equator (lines running east and west) are parallels of latitude. They are used to measure degrees of latitude north (N) or south (S) of the equator. The angular distance from the equator to the pole is one-fourth of a circle or 90°. The 48 conterminous states of the United States are located between 25° and 49° N latitude. The arrows in *Figure 16-4* labeled "Latitude" point to lines of latitude. Meridians of longitude are drawn from the North Pole to the South Pole and are at right angles to the Equator. The "Prime

Meridian," which passes through Greenwich, England, is used as the zero line from which measurements are made in degrees east (E) and west (W) to 180°. The 48 conterminous states of the United States are between 67° and 125° W longitude. The arrows in *Figure 16-4* labeled "Longitude" point to lines of longitude.

Any specific geographical point can be located by reference to its longitude and latitude. Washington, D.C., for example, is approximately 39° N latitude, 77° W longitude. Chicago is approximately 42° N latitude, 88° W longitude.

Time Zones

The meridians are also useful for designating time zones. A day is defined as the time required for the Earth to make one complete rotation of 360°. Since the day is divided into 24 hours, the Earth revolves at the rate of 15° an hour. Noon is the time when the sun is directly above a meridian; to the west of that meridian is morning, to the east is afternoon.

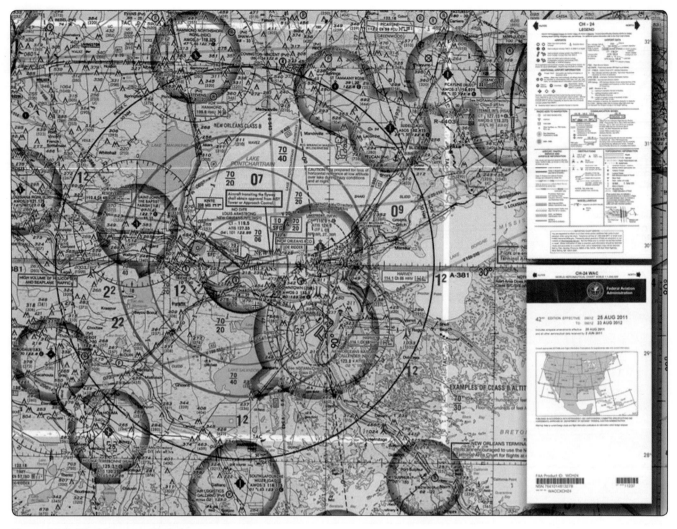

Figure 16-3. *World aeronautical chart.*

Figure 16-4. *Meridians and parallels—the basis of measuring time, distance, and direction.*

The standard practice is to establish a time zone for each 15° of longitude. This makes a difference of exactly 1 hour between each zone. In the conterminous United States, there are four time zones. The time zones are Eastern (75°), Central (90°), Mountain (105°), and Pacific (120°). The dividing lines are somewhat irregular because communities near the boundaries often find it more convenient to use time designations of neighboring communities or trade centers.

Figure 16-5 shows the time zones in the conterminous United States. When the sun is directly above the 90th meridian, it is noon Central Standard Time. At the same time, it is 1 p.m. Eastern Standard Time, 11 a.m. Mountain Standard Time, and 10 a.m. Pacific Standard Time. When Daylight Saving Time is in effect, generally between the second Sunday in March and the first Sunday in November, the sun is directly above the 75th meridian at noon, Central Daylight Time.

These time zone differences must be taken into account during long flights eastward—especially if the flight must be completed before dark. Remember, an hour is lost when

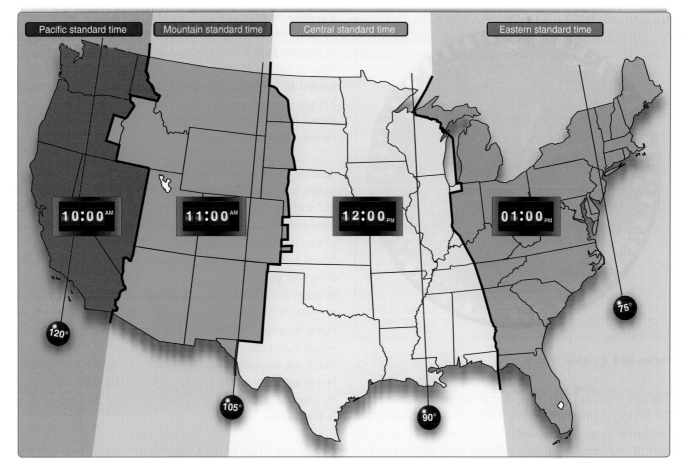

Figure 16-5. *Time zones in the conterminous United States.*

flying eastward from one time zone to another, or perhaps even when flying from the western edge to the eastern edge of the same time zone. Determine the time of sunset at the destination by consulting the flight service station (FSS) and take this into account when planning an eastbound flight.

In most aviation operations, time is expressed in terms of the 24-hour clock. ATC instructions, weather reports and broadcasts, and estimated times of arrival are all based on this system. For example: 9 a.m. is expressed as 0900, 1 p.m. is 1300, and 10 p.m. is 2200.

Because a pilot may cross several time zones during a flight, a standard time system has been adopted. It is called Universal Coordinated Time (UTC) and is often referred to as Zulu time. UTC is the time at the 0° line of longitude which passes through Greenwich, England. All of the time zones around the world are based on this reference. To convert to this time, a pilot should do the following:

      Eastern Standard Time ....Add 5 hours

      Central Standard Time ....Add 6 hours

      Mountain Standard Time.......Add 7 hours

      Pacific Standard Time ....Add 8 hours

For Daylight Saving Time, 1 hour should be subtracted from the calculated times.

Measurement of Direction

By using the meridians, direction from one point to another can be measured in degrees, in a clockwise direction from true north. To indicate a course to be followed in flight, draw a line on the chart from the point of departure to the destination and measure the angle that this line forms with a meridian. Direction is expressed in degrees, as shown by the compass rose in *Figure 16-6.*

Because meridians converge toward the poles, course measurement should be taken at a meridian near the midpoint of the course rather than at the point of departure. The course measured on the chart is known as the true course (TC). This is the direction measured by reference to a meridian or true north (TN). It is the direction of intended flight as measured in degrees clockwise from TN.

As shown in *Figure 16-7,* the direction from A to B would be a TC of 065°, whereas the return trip (called the reciprocal) would be a TC of 245°.

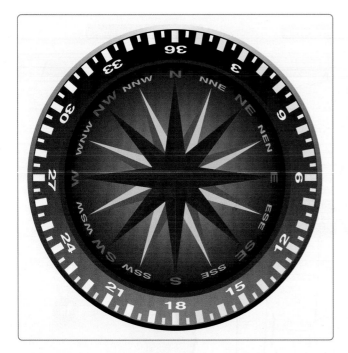

Figure 16-6. *Compass rose.*

The true heading (TH) is the direction in which the nose of the aircraft points during a flight when measured in degrees clockwise from TN. Usually, it is necessary to head the aircraft in a direction slightly different from the TC to offset the effect of wind. Consequently, numerical value of the TH may not correspond with that of the TC. This is discussed more fully in subsequent sections in this chapter. For the purpose of this discussion, assume a no-wind condition exists under which heading and course would coincide. Thus, for a TC of 065°, the TH would be 065°. To use the compass accurately, however, corrections must be made for magnetic variation and compass deviation.

Variation

Variation is the angle between TN and magnetic north (MN). It is expressed as east variation or west variation depending upon whether MN is to the east or west of TN.

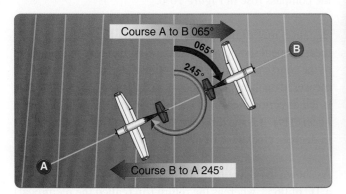

Figure 16-7. *Courses are determined by reference to meridians on aeronautical charts.*

The north magnetic pole is located close to 71° N latitude, 96° W longitude and is about 1,300 miles from the geographic or true north pole, as indicated in *Figure 16-8*. If the Earth were uniformly magnetized, the compass needle would point toward the magnetic pole, in which case the variation between TN (as shown by the geographical meridians) and MN (as shown by the magnetic meridians) could be measured at any intersection of the meridians.

Actually, the Earth is not uniformly magnetized. In the United States, the needle usually points in the general direction of the magnetic pole, but it may vary in certain geographical localities by many degrees. Consequently, the exact amount of variation at thousands of selected locations in the United States has been carefully determined. The amount and the direction of variation, which change slightly from time to time, are shown on most aeronautical charts as broken magenta lines called isogonic lines that connect points of equal magnetic variation. (The line connecting points at which there is no variation between TN and MN is the agonic line.) An isogonic chart is shown in *Figure 16-9*. Minor bends and turns in the isogonic and agonic lines are caused by unusual geological conditions affecting magnetic forces in these areas.

On the west coast of the United States, the compass needle points to the east of TN; on the east coast, the compass needle points to the west of TN.

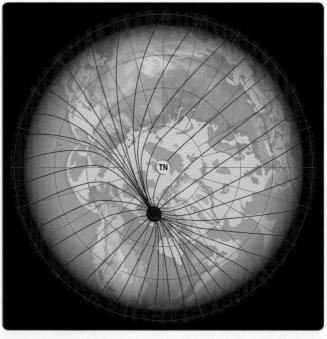

Figure 16-8. *Magnetic meridians are in red while the lines of longitude and latitude are in blue. From these lines of variation (magnetic meridians), one can determine the effect of local magnetic variations on a magnetic compass.*

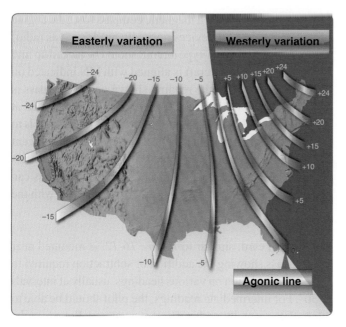

Figure 16-9. *Note the agonic line where magnetic variation is zero.*

Zero degree variation exists on the agonic line where MN and TN coincide. This line runs roughly west of the Great Lakes, south through Wisconsin, Illinois, western Tennessee, and along the border of Mississippi and Alabama. Compare *Figures 16-9* and *16-10*.

Because courses are measured in reference to geographical meridians that point toward TN, and these courses are maintained by reference to the compass that points along a magnetic meridian in the general direction of MN, the true direction must be converted into magnetic direction for the purpose of flight. This conversion is made by adding or subtracting the variation indicated by the nearest isogonic line on the chart.

For example, a line drawn between two points on a chart is called a TC as it is measured from TN. However, flying this course off the magnetic compass would not provide an accurate course between the two points due to three elements that must be considered. The first is magnetic variation, the second is compass deviation, and the third is wind correction. All three must be considered for accurate navigation.

Magnetic Variation

As mentioned in the paragraph discussing variation, the appropriate variation for the geographical location of the flight must be considered and added or subtracted as appropriate. If flying across an area where the variation changes, then the values must be applied along the route of flight appropriately. Once applied, this new course is called the magnetic course.

Magnetic Deviation

Because each aircraft has its own internal effect upon the onboard compass systems from its own localized magnetic influencers, the pilot must add or subtract these influencers based upon the direction he or she is flying. The application of deviation (taken from a compass deviation card) compensates the magnetic course unique to that aircraft's compass system (as affected by localized magnetic influencers) and it now becomes the compass course. Therefore, the compass course, when followed (in a no wind condition), takes the aircraft from point A to point B even though the aircraft heading may not match the original course line drawn on the chart.

If the variation is shown as "9° E," this means that MN is 9° east of TN. If a TC of 360° is to be flown, 9° must be subtracted from 360°, which results in a magnetic heading of 351°. To fly east, a magnetic course of 081° (090° − 9°) would be flown. To fly south, the magnetic course would be 171° (180° − 9°). To fly west, it would be 261° (270° − 9°). To fly a TH of 060°, a magnetic course of 051° (060° − 9°) would be flown.

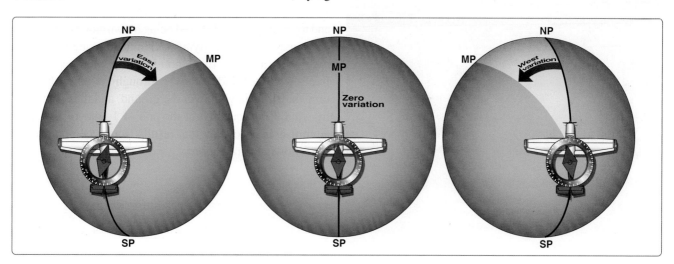

Figure 16-10. *Effect of variation on the compass.*

Remember, if variation is west, add; if east, subtract. One method for remembering whether to add or subtract variation is the phrase "east is least (subtract) and west is best (add)."

Deviation

Determining the magnetic heading is an intermediate step necessary to obtain the correct compass heading for the flight. To determine compass heading, a correction for deviation must be made. Because of magnetic influences within an aircraft, such as electrical circuits, radio, lights, tools, engine, and magnetized metal parts, the compass needle is frequently deflected from its normal reading. This deflection is called deviation. The deviation is different for each aircraft, and it also may vary for different headings in the same aircraft. For instance, if magnetism in the engine attracts the north end of the compass, there would be no effect when the plane is on a heading of MN. On easterly or westerly headings, however, the compass indications would be in error, as shown in *Figure 16-11*. Magnetic attraction can come from many other parts of the aircraft; the assumption of attraction in the engine is merely used for the purpose of illustration.

Some adjustment of the compass, referred to as compensation, can be made to reduce this error, but the remaining correction must be applied by the pilot.

Proper compensation of the compass is best performed by a competent technician. Since the magnetic forces within the aircraft change because of landing shocks, vibration, mechanical work, or changes in equipment, the pilot should occasionally have the deviation of the compass checked. The procedure used to check the deviation is called "swinging the compass" and is briefly outlined as follows.

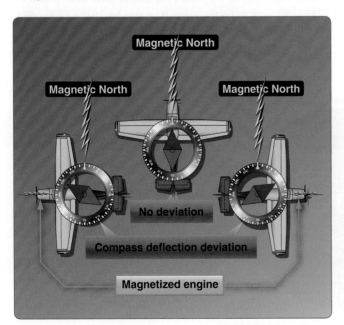

Figure 16-11. *Magnetized portions of the airplane cause the compass to deviate from its normal indications.*

The aircraft is placed on a magnetic compass rose, the engine started, and electrical devices normally used (such as radio) are turned on. Tailwheel-type aircraft should be jacked up into flying position. The aircraft is aligned with MN indicated on the compass rose and the reading shown on the compass is recorded on a deviation card. The aircraft is then aligned at 30° intervals and each reading is recorded. If the aircraft is to be flown at night, the lights are turned on and any significant changes in the readings are noted. If so, additional entries are made for use at night. The accuracy of the compass can also be checked by comparing the compass reading with the known runway headings.

A deviation card, similar to *Figure 16-12,* is mounted near the compass showing the addition or subtraction required to correct for deviation on various headings, usually at intervals of 30°. For intermediate readings, the pilot should be able to interpolate mentally with sufficient accuracy. For example, if the pilot needed the correction for 195° and noted the correction for 180° to be 0° and for 210° to be +2°, it could be assumed that the correction for 195° would be +1°. The magnetic heading, when corrected for deviation, is known as compass heading.

Effect of Wind

The preceding discussion explained how to measure a TC on the aeronautical chart and how to make corrections for variation and deviation, but one important factor has not been considered—wind. As discussed in the study of the atmosphere, wind is a mass of air moving over the surface of the Earth in a definite direction. When the wind is blowing from the north at 25 knots, it simply means that air is moving southward over the Earth's surface at the rate of 25 NM in 1 hour.

Under these conditions, any inert object free from contact with the Earth is carried 25 NM southward in 1 hour. This effect becomes apparent when such things as clouds, dust, and toy balloons are observed being blown along by the wind. Obviously, an aircraft flying within the moving mass of air is similarly affected. Even though the aircraft does not float freely with the wind, it moves through the air at the same time the air is moving over the ground, and thus is affected by wind. Consequently, at the end of 1 hour of flight, the aircraft is in a position that results from a combination of the following two motions:

For (Magnetic)	N	30	60	E	120	150
Steer (Compass)	0	28	57	86	117	148
For (Magnetic)	S	210	240	W	300	330
Steer (Compass)	180	212	243	274	303	332

Figure 16-12. *Compass deviation card.*

- Movement of the air mass in reference to the ground
- Forward movement of the aircraft through the air mass

Actually, these two motions are independent. It makes no difference whether the mass of air through which the aircraft is flying is moving or is stationary. A pilot flying in a 70-knot gale would be totally unaware of any wind (except for possible turbulence) unless the ground were observed. In reference to the ground, however, the aircraft would appear to fly faster with a tailwind or slower with a headwind, or to drift right or left with a crosswind.

As shown in *Figure 16-13,* an aircraft flying eastward at an airspeed of 120 knots in still air has a groundspeed (GS) exactly the same—120 knots. If the mass of air is moving eastward at 20 knots, the airspeed of the aircraft is not affected, but the progress of the aircraft over the ground is 120 plus 20 or a GS of 140 knots. On the other hand, if the mass of air is moving westward at 20 knots, the airspeed of the aircraft remains the same, but GS becomes 120 minus 20 or 100 knots.

Assuming no correction is made for wind effect, if an aircraft is heading eastward at 120 knots and the air mass moving southward at 20 knots, the aircraft at the end of 1 hour is almost 120 miles east of its point of departure because of its progress through the air. It is 20 miles south because of the motion of the air. Under these circumstances, the airspeed remains 120 knots, but the GS is determined by combining the movement of the aircraft with that of the air mass. GS can be measured as the distance from the point of departure to the position of the aircraft at the end of 1 hour. The GS can be computed by the time required to fly between two points a known distance apart. It also can be determined before flight by constructing a wind triangle, which is explained later in this chapter. *[Figure 16-14]*

The direction in which the aircraft is pointing as it flies is called heading. Its actual path over the ground, which is a combination of the motion of the aircraft and the motion of the air, is called track. The angle between the heading and the track is called drift angle. If the aircraft heading coincides with the TC and the wind is blowing from the left, the track does not coincide with the TC. The wind causes the aircraft to drift to the right, so the track falls to the right of the desired course or TC. *[Figure 16-15]*

The following method is used by many pilots to determine compass heading: after the TC is measured, and wind correction applied resulting in a TH, the sequence TH ± variation (V) = magnetic heading (MH) ± deviation (D) = compass heading (CH) is followed to arrive at compass heading. *[Figure 16-16]*

By determining the amount of drift, the pilot can counteract the effect of the wind and make the track of the aircraft coincide with the desired course. If the mass of air is moving across the course from the left, the aircraft drifts to the right, and a correction must be made by heading the aircraft sufficiently to the left to offset this drift. In other words, if the wind is from the left, the correction is made by pointing the aircraft to the left a certain number of degrees, therefore correcting for wind drift. This is the wind correction angle (WCA) and is expressed in terms of degrees right or left of the TC. *[Figure 16-17]*

Figure 16-13. *Motion of the air affects the speed with which aircraft move over the Earth's surface. Airspeed, the rate at which an aircraft moves through the air, is not affected by air motion.*

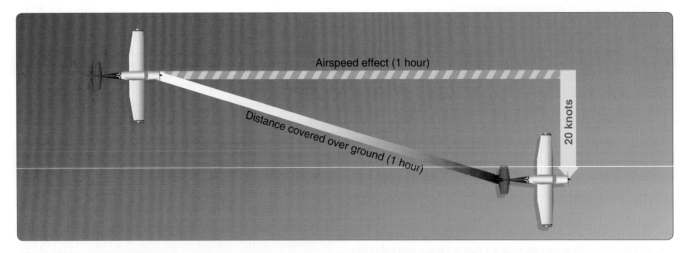

Figure 16-14. *Aircraft flight path resulting from its airspeed and direction and the wind speed and direction.*

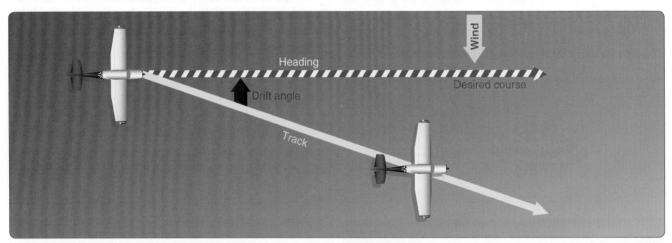

Figure 16-15. *Effects of wind drift on maintaining desired course.*

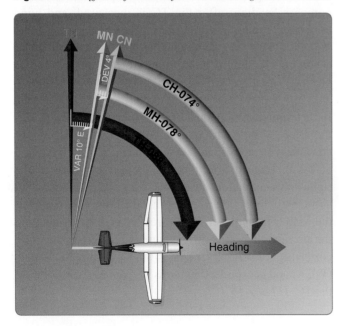

Figure 16-16. *Relationship between true, magnetic, and compass headings for a particular instance.*

To summarize:

- Course—intended path of an aircraft over the ground or the direction of a line drawn on a chart representing the intended aircraft path, expressed as the angle measured from a specific reference datum clockwise from 0° through 360° to the line.

- Heading—direction in which the nose of the aircraft points during flight.

- Track—actual path made over the ground in flight. (If proper correction has been made for the wind, track and course are identical.)

- Drift angle—angle between heading and track.

- WCA—correction applied to the course to establish a heading so that track coincides with course.

- Airspeed—rate of the aircraft's progress through the air.

- GS—rate of the aircraft's inflight progress over the ground.

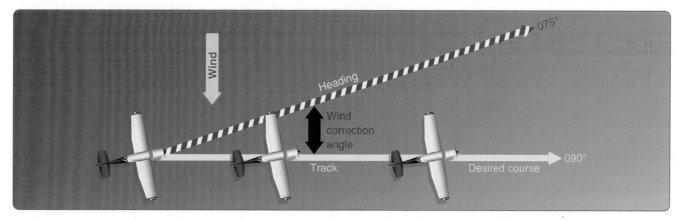

Figure 16-17. *Establishing a wind correction angle that counteracts wind drift and maintains the desired course.*

Basic Calculations

Before a cross-country flight, a pilot should make common calculations for time, speed, and distance, and the amount of fuel required.

Converting Minutes to Equivalent Hours

Frequently, it is necessary to convert minutes into equivalent hours when solving speed, time, and distance problems. To convert minutes to hours, divide by 60 (60 minutes = 1 hour). Thus, 30 minutes is 30/60 = 0.5 hour. To convert hours to minutes, multiply by 60. Thus, 0.75 hour equals 0.75 × 60 = 45 minutes.

Time T = D/GS

To find the time (T) in flight, divide the distance (D) by the GS. The time to fly 210 NM at a GS of 140 knots is 210 ÷ 140 or 1.5 hours. (The 0.5 hour multiplied by 60 minutes equals 30 minutes.) Answer: 1:30.

Distance D = GS X T

To find the distance flown in a given time, multiply GS by time. The distance flown in 1 hour 45 minutes at a GS of 120 knots is 120 × 1.75 or 210 NM.

GS GS = D/T

To find the GS, divide the distance flown by the time required. If an aircraft flies 270 NM in 3 hours, the GS is 270 ÷ 3 = 90 knots.

Converting Knots to Miles Per Hour

Another conversion is that of changing knots to miles per hour (mph). The aviation industry is using knots more frequently than mph, but is important to understand the conversion for those that use mph when working with speed problems. The NWS reports both surface winds and winds aloft in knots. However, airspeed indicators in some aircraft are calibrated in mph (although many are now calibrated in both mph and

knots). Pilots, therefore, should learn to convert wind speeds that are reported in knots to mph.

A knot is 1 nautical mile per hour (NMPH). Because there are 6,076.1 feet in 1 NM and 5,280 feet in 1 SM, the conversion factor is 1.15. To convert knots to mph, multiply speed in knots by 1.15. For example: a wind speed of 20 knots is equivalent to 23 mph.

Most flight computers or electronic calculators have a means of making this conversion. Another quick method of conversion is to use the scales of NM and SM at the bottom of aeronautical charts.

Fuel Consumption

To ensure that sufficient fuel is available for your intended flight, you must be able to accurately compute aircraft fuel consumption during preflight planning. Typically, fuel consumption in gasoline-fueled aircraft is measured in gallons per hour. Since turbine engines consume much more fuel than reciprocating engines, turbine-powered aircraft require much more fuel, and thus much larger fuel tanks. When determining these large fuel quantities, using a volume measurement such as gallons presents a problem because the volume of fuel varies greatly in relation to temperature. In contrast, density (weight) is less affected by temperature and therefore, provides a more uniform and repeatable measurement. For this reason, jet fuel is generally quantified by its density and volume.

This standard industry convention yields a pounds-of-fuel-per-hour value which, when divided into the nautical miles (NM) per hour of travel (TAS ± winds) value, results in a specific range value. The typical label for specific range is NM per pound of fuel, or often NM per 1,000 pounds of fuel. Preflight planning should be supported by proper monitoring of past fuel consumption as well as use of specified fuel management and mixture adjustment procedures in flight.

For simple aircraft with reciprocating engines, the Aircraft Flight Manual/Pilot's Operating Handbook (AFM/POH) supplied by the aircraft manufacturer provides gallons-per-hour values to assist with preflight planning.

When planning a flight, you must determine how much fuel is needed to reach your destination by calculating the distance the aircraft can travel (with winds considered) at a known rate of fuel consumption (gal/hr or lbs/hr) for the expected groundspeed (GS) and ensure this amount, plus an adequate reserve, is available on board. GS determines the time the flight will take. The amount of fuel needed for a given flight can be calculated by multiplying the estimated flight time by the rate of consumption. For example, a flight of 400 NM at 100 knots GS takes 4 hours to complete. If an aircraft consumes 5 gallons of fuel per hour, the total fuel consumption is 20 gallons (4 hours times 5 gallons). In this example, there is no wind; therefore, true airspeed (TAS) is also 100 knots, the same as GS. Since the rate of fuel consumption remains relatively constant at a given TAS, you must use GS to calculate fuel consumption when wind is present. Specific range (NM/lb or NM/gal) is also useful in calculating fuel consumption when wind is a factor.

You should always plan to be on the surface before any of the following occur:

- Your flight time exceeds the amount of flight time you calculated for the consumption of your preflight fuel amount

- Your fuel gauge indicates low fuel level

The rate of fuel consumption depends on many factors: condition of the engine, propeller/rotor pitch, propeller/rotor revolutions per minute (rpm), richness of the mixture, and the percentage of horsepower used for flight at cruising speed. The pilot should know the approximate consumption rate from cruise performance charts or from experience. In addition to the amount of fuel required for the flight, there should be sufficient fuel for reserve. When estimating consumption you must plan for cruise flight as well as startup and taxi, and higher fuel burn during climb. Remember that ground speed during climb is less than during cruise flight at the same airspeed. Additional fuel for adequate reserve should also be added as a safety measure.

Flight Computers

Up to this point, only mathematical formulas have been used to determine such items as time, distance, speed, and fuel consumption. In reality, most pilots use a mechanical flight computer called an E6B or electronic flight calculator. These devices can compute numerous problems associated with flight planning and navigation. The mechanical or electronic computer has an instruction book that probably includes sample problems so the pilot can become familiar with its functions and operation. *[Figure 16-18]*

Plotter

Another aid in flight planning is a plotter, which is a protractor and ruler. The pilot can use this when determining TC and measuring distance. Most plotters have a ruler that measures in both NM and SM and has a scale for a sectional chart on one side and a world aeronautical chart on the other. *[Figure 16-18]*

Pilotage

Pilotage is navigation by reference to landmarks or checkpoints. It is a method of navigation that can be used on any course that has adequate checkpoints, but it is more commonly used in conjunction with dead reckoning and VFR radio navigation.

The checkpoints selected should be prominent features common to the area of the flight. Choose checkpoints that can be readily identified by other features, such as roads, rivers, railroad tracks, lakes, and power lines. If possible, select features that make useful boundaries or brackets on each side of the course, such as highways, rivers, railroads, and mountains. A pilot can keep from drifting too far off course by referring to and not crossing the selected brackets. Never place complete reliance on any single checkpoint. Choose ample checkpoints. If one is missed, look for the next one while maintaining the heading. When determining position from checkpoints, remember that the scale of a sectional chart is 1 inch = 8 SM or 6.86 NM. For example, if a checkpoint selected was approximately one-half inch from the course line on the chart, it is 4 SM or 3.43 NM from the course on the ground. In the more congested areas, some of the smaller features are not included on the chart. If confused, hold the heading. If a turn is made away from the heading, it is easy to become lost.

Roads shown on the chart are primarily the well-traveled roads or those most apparent when viewed from the air. New roads and structures are constantly being built and may not be shown on the chart until the next chart is issued. Some structures, such as antennas, may be difficult to see. Sometimes TV antennas are grouped together in an area near a town. They are supported by almost invisible guy wires. Never approach an area of antennas less than 500 feet above the tallest one. Most of the taller structures are marked with strobe lights to make them more visible to pilots. However, some weather conditions or background lighting may make them difficult to see. Aeronautical charts display the best information available at the time of printing, but a pilot should be cautious for new structures or changes that have occurred since the chart was printed.

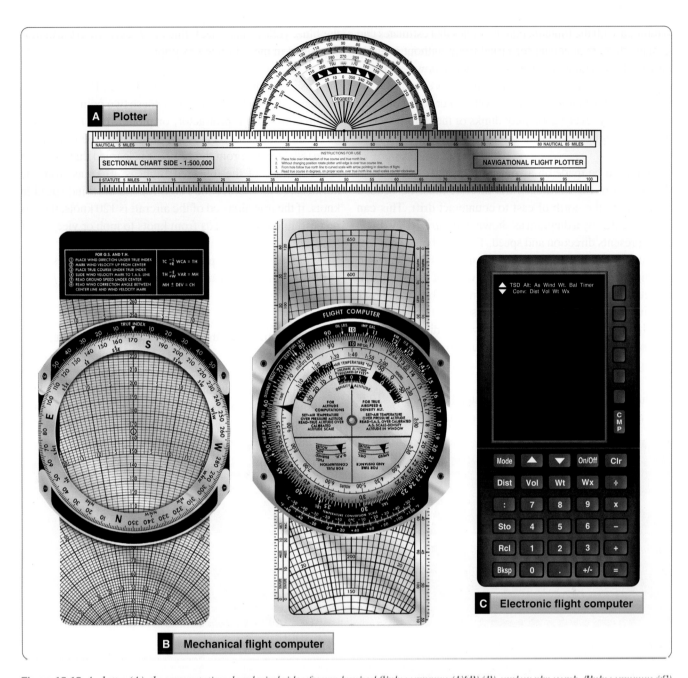

Figure 10-10. A plotter (A), the computational and wind side of a mechanical flight computer (E6B) (B), and an electronic flight computer (C).

Dead Reckoning

Dead reckoning is navigation solely by means of computations based on time, airspeed, distance, and direction. The products derived from these variables, when adjusted by wind speed and velocity, are heading and GS. The predicted heading takes the aircraft along the intended path and the GS establishes the time to arrive at each checkpoint and the destination. Except for flights over water, dead reckoning is usually used with pilotage for cross-country flying. The heading and GS, as calculated, is constantly monitored and corrected by pilotage as observed from checkpoints.

Wind Triangle or Vector Analysis

If there is no wind, the aircraft's ground track is the same as the heading and the GS is the same as the true airspeed. This condition rarely exists. A wind triangle, the pilot's version of vector analysis, is the basis of dead reckoning.

The wind triangle is a graphic explanation of the effect of wind upon flight. GS, heading, and time for any flight can be determined by using the wind triangle. It can be applied to the simplest kind of cross-country flight, as well as the most complicated instrument flight. The experienced pilot becomes so familiar with the fundamental principles that estimates can be made that are adequate for visual flight without actually drawing the diagrams. The beginning student, however, needs to develop skill in constructing these diagrams as an aid to the complete understanding of wind effect. Either consciously or unconsciously, every good pilot thinks of the flight in terms of wind triangle.

If flight is to be made on a course to the east, with a wind blowing from the northeast, the aircraft must be headed somewhat to the north of east to counteract drift. This can be represented by a diagram as shown in *Figure 16-19*. Each line represents direction and speed. The long blue and white hashed line shows the direction the aircraft is heading, and its length represents the distance traveled at the indicated airspeed for 1 hour. The short blue arrow at the right shows the wind direction, and its length represents the wind velocity for 1 hour. The solid yellow line shows the direction of the track or the path of the aircraft as measured over the earth, and its length represents the distance traveled in 1 hour or the GS.

In actual practice, the triangle illustrated in *Figure 16-19* is not drawn; instead, construct a similar triangle as shown by the blue, yellow, and black lines in *Figure 16-20,* which is explained in the following example.

Suppose a flight is to be flown from E to P. Draw a line on the aeronautical chart connecting these two points; measure its direction with a protractor, or plotter, in reference to a meridian. This is the TC, which in this example is assumed to be 090° (east). From the NWS, it is learned that the wind at the altitude of the intended flight is 40 knots from the northeast (045°). Since the NWS reports the wind speed in knots, if the true airspeed of the aircraft is 120 knots, there is no need to convert speeds from knots to mph or vice versa.

Now, on a plain sheet of paper draw a vertical line representing north to south. (The various steps are shown in *Figure 16-21*.)

Step 1

Place the protractor with the base resting on the vertical line and the curved edge facing east. At the center point of the base, make a dot labeled "E" (point of departure) and at the curved edge, make a dot at 90° (indicating the direction of the true course) and another at 45° (indicating wind direction).

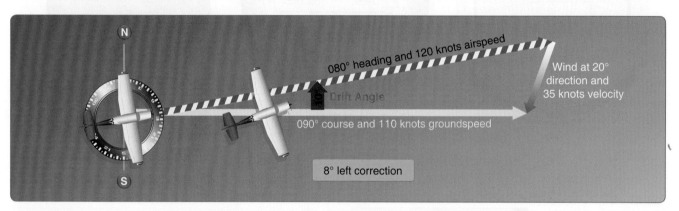

Figure 16-19. *Principle of the wind triangle.*

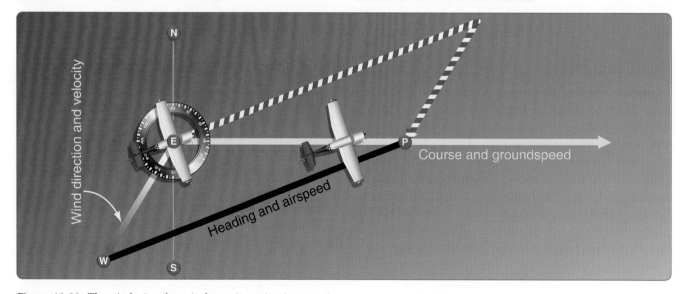

Figure 16-20. *The wind triangle as is drawn in navigation practice.*

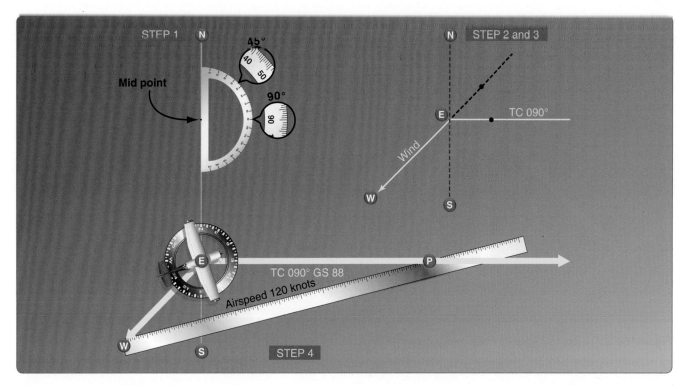

Figure 16-21. *Steps in drawing the wind triangle.*

Step 2

With the ruler, draw the true course line from E, extending it somewhat beyond the dot by 90°, and labeling it "TC 090°."

Step 3

Next, align the ruler with E and the dot at 45°, and draw the wind arrow from E, not toward 045°, but downwind in the direction the wind is blowing making it 40 units long to correspond with the wind velocity of 40 knots. Identify this line as the wind line by placing the letter "W" at the end to show the wind direction.

Step 4

Finally, measure 120 units on the ruler to represent the airspeed, making a dot on the ruler at this point. The units used may be of any convenient scale or value (such as ¼ inch = 10 knots), but once selected, the same scale must be used for each of the linear movements involved. Then place the ruler so that the end is on the arrowhead (W) and the 120-knot dot intercepts the TC line. Draw the line and label it "AS 120." The point "P" placed at the intersection represents the position of the aircraft at the end of 1 hour. The diagram is now complete.

The distance flown in 1 hour (GS) is measured as the numbers of units on the TC line (88 NMPH or 88 knots). The TH necessary to offset drift is indicated by the direction of the airspeed line, which can be determined in one of two ways:

- By placing the straight side of the protractor along the north-south line, with its center point at the intersection of the airspeed line and north-south line, read the TH directly in degrees (076°). *[Figure 16-22]*

- By placing the straight side of the protractor along the TC line, with its center at P, read the angle between the TC and the airspeed line. This is the WCA, which must be applied to the TC to obtain the TH. If the wind blows from the right of TC, the angle is added; if from the left, it is subtracted. In the example given, the WCA is 14° and the wind is from the left; therefore, subtract 14° from TC of 090°, making the TH 076°. *[Figure 16-23]*

After obtaining the TH, apply the correction for magnetic variation to obtain magnetic heading and the correction for compass deviation to obtain a compass heading. The compass heading can be used to fly to the destination by dead reckoning.

To determine the time and fuel required for the flight, first find the distance to your destination by measuring the length of the course line drawn on the aeronautical chart (using the appropriate scale at the bottom of the chart). If the distance measures 220 NM, divide by the GS of 88 knots, which gives 2.5 hours, or 2:30, as the time required. If fuel consumption is 8 gallons an hour, 8 × 2.5 or about 20 gallons is used.

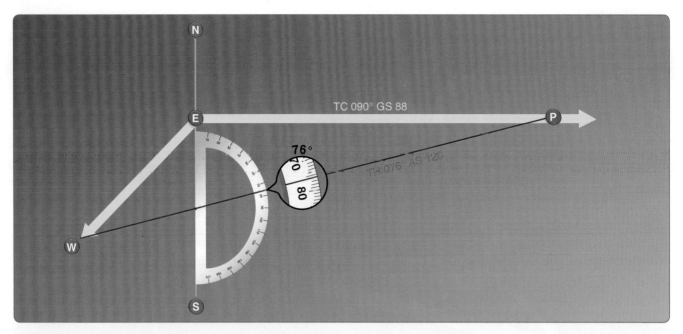

Figure 16-22. *Finding true heading by the wind correction angle.*

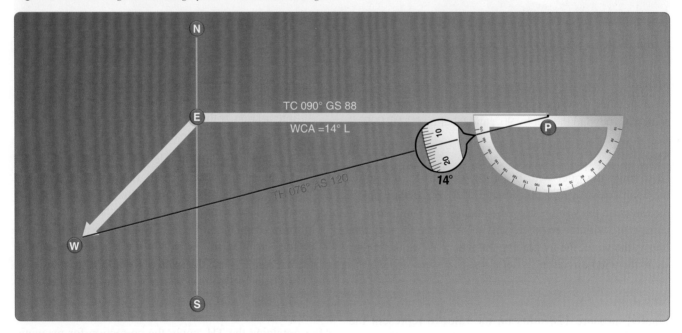

Figure 16-23. *Finding true heading by direct measurement.*

Briefly summarized, the steps in obtaining flight information are as follows:

- TC—direction of the line connecting two desired points, drawn on the chart and measured clockwise in degrees from TN on the mid-meridian

- WCA—determined from the wind triangle. (Added to TC if the wind is from the right; subtracted if wind is from the left)

- TH—direction measured in degrees clockwise from TN, in which the nose of the plane should point to remain on the desired course

- Variation—obtained from the isogonic line on the chart (added to TH if west; subtracted if east)

- MH—an intermediate step in the conversion (obtained by applying variation to TH)

- Deviation—obtained from the deviation card on the aircraft (added to or subtracted from MH, as indicated)

- Compass heading—reading on the compass (found by applying deviation to MH) that is followed to remain on the desired course

- Total distance—obtained by measuring the length of the TC line on the chart (using the scale at the bottom of the chart)

- GS—obtained by measuring the length of the TC line on the wind triangle (using the scale employed for drawing the diagram)

- Estimated time en route (ETE)—total distance divided by GS

- Fuel rate—predetermined gallons per hour used at cruising speed

NOTE: Additional fuel for adequate reserve should be added as a safety measure.

Flight Planning

Title 14 of the Code of Federal Regulations (14 CFR) part 91 states, in part, that before beginning a flight, the pilot in command (PIC) of an aircraft shall become familiar with all available information concerning that flight. For flights not in the vicinity of an airport, this must include information on available current weather reports and forecasts, fuel requirements, alternatives available if the planned flight cannot be completed, and any known traffic delays of which the PIC has been advised by ATC.

Assembling Necessary Material

The pilot should collect the necessary material well before beginning the flight. An appropriate current sectional chart and charts for areas adjoining the flight route should be among this material if the route of flight is near the border of a chart.

Additional equipment should include a flight computer or electronic calculator, plotter, and any other item appropriate to the particular flight. For example, if a night flight is to be undertaken, carry a flashlight; if a flight is over desert country, carry a supply of water and other necessities.

Weather Check

It is wise to check the weather before continuing with other aspects of flight planning to see, first of all, if the flight is feasible and, if it is, which route is best. Chapter 12, "Aviation Weather Services," discusses obtaining a weather briefing.

Use of Chart Supplement U.S. (formerly Airport/Facility Directory)

Study available information about each airport at which a landing is intended. This should include a study of the Notices to Airmen (NOTAMs) and the Chart Supplement U.S. (formerly Airport/Facility Directory). [Figure 16-24] This includes location, elevation, runway and lighting facilities, available services, availability of aeronautical advisory station frequency (UNICOM), types of fuel available (use to

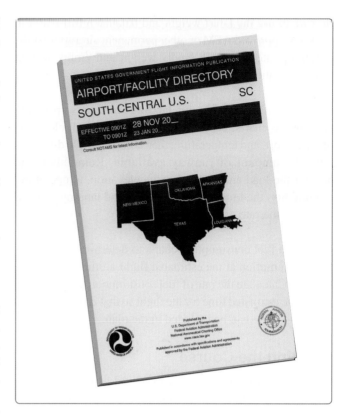

Figure 16-24. *Chart Supplement U.S. (formerly Airport/Facility Directory).*

decide on refueling stops), FSS located on the airport, control tower and ground control frequencies, traffic information, remarks, and other pertinent information. The NOTAMs, issued every 28 days, should be checked for additional information on hazardous conditions or changes that have been made since issuance of the Chart Supplement U.S.

The sectional chart bulletin subsection should be checked for major changes that have occurred since the last publication date of each sectional chart being used. Remember, the chart may be up to 6 months old. The effective date of the chart appears at the top of the front of the chart. The Chart Supplement U.S. generally has the latest information pertaining to such matters and should be used in preference to the information on the back of the chart, if there are differences.

Airplane Flight Manual or Pilot's Operating Handbook (AFM/POH)

The Aircraft Flight Manual or Pilot's Operating Handbook (AFM/POH) should be checked to determine the proper loading of the aircraft (weight and balance data). The weight of the usable fuel and drainable oil aboard must be known. Also, check the weight of the passengers, the weight of all baggage to be carried, and the empty weight of the aircraft to be sure that the total weight does not exceed the maximum allowable weight. The distribution of the load must be known to tell if the resulting center of gravity (CG) is within limits.

Be sure to use the latest weight and balance information in the FAA-approved AFM or other permanent aircraft records, as appropriate, to obtain empty weight and empty weight CG information.

Determine the takeoff and landing distances from the appropriate charts, based on the calculated load, elevation of the airport, and temperature; then compare these distances with the amount of runway available. Remember, the heavier the load and the higher the elevation, temperature, or humidity, the longer the takeoff roll and landing roll and the lower the rate of climb.

Check the fuel consumption charts to determine the rate of fuel consumption at the estimated flight altitude and power settings. Calculate the rate of fuel consumption, and compare it with the estimated time for the flight so that refueling points along the route can be included in the plan.

Charting the Course

Once the weather has been checked and some preliminary planning completed, it is time to chart the course and determine the data needed to accomplish the flight. The following sections provide a logical sequence to follow in charting the course, complete a flight log, and filing a flight plan. In the following example, a trip is planned based on the following data and the sectional chart excerpt in *Figure 16-25*.

Route of flight: Chickasha Airport direct to Guthrie Airport

True airspeed (TAS)	115 knots
Winds aloft	360° at 10 knots
Usable fuel	38 gallons
Fuel rate	8 GPH
Deviation	+2°

Steps in Charting the Course

The following is a suggested sequence for arriving at the pertinent information for the trip. As information is determined, it may be noted as illustrated in the example of a flight log in *Figure 16-26*. Where calculations are required, the pilot may use a mathematical formula or a manual or electronic flight computer. If unfamiliar with the use of a manual or electronic computer, it would be advantageous to read the operation manual and work several practice problems at this point.

First, draw a line from Chickasha Airport (point A) directly to Guthrie Airport (point F). The course line should begin at the center of the airport of departure and end at the center of the destination airport. If the route is direct, the course line consists of a single straight line. If the route is not direct, it consists of two or more straight line segments. For example, a

VOR station that is off the direct route, but makes navigating easier, may be chosen (radio navigation is discussed later in this chapter).

Appropriate checkpoints should be selected along the route and noted in some way. These should be easy-to-locate points, such as large towns, large lakes and rivers, or combinations of recognizable points, such as towns with an airport, towns with a network of highways, and railroads entering and departing.

Normally, choose only towns indicated by splashes of yellow on the chart. Do not choose towns represented by a small circle—these may turn out to be only a half-dozen houses. (In isolated areas, however, towns represented by a small circle can be prominent checkpoints.) For this trip, four checkpoints have been selected. Checkpoint 1 consists of a tower located east of the course and can be further identified by the highway and railroad track, which almost parallels the course at this point. Checkpoint 2 is the obstruction just to the west of the course and can be further identified by Will Rogers World Airport, which is directly to the east. Checkpoint 3 is Wiley Post Airport, which the aircraft should fly directly over. Checkpoint 4 is a private, non-surfaced airport to the west of the course and can be further identified by the railroad track and highway to the east of the course.

The course and areas on either side of the planned route should be checked to determine if there is any type of airspace with which the pilot should be concerned or which has special operational requirements. For this trip, it should be noted that the course passes through a segment of the Class C airspace surrounding Will Rogers World Airport where the floor of the airspace is 2,500 feet mean sea level (MSL) and the ceiling is 5,300 feet MSL (point B). Also, there is Class D airspace from the surface to 3,800 feet MSL surrounding Wiley Post Airport (point C) during the time the control tower is in operation.

Study the terrain and obstructions along the route. This is necessary to determine the highest and lowest elevations, as well as the highest obstruction to be encountered so an appropriate altitude that conforms to 14 CFR part 91 regulations can be selected. If the flight is to be flown at an altitude of more than 3,000 feet above the terrain, conformance to the cruising altitude appropriate to the direction of flight is required. Check the route for particularly rugged terrain so it can be avoided. Areas where a takeoff or landing is made should be carefully checked for tall obstructions. Television transmitting towers may extend to altitudes over 1,500 feet above the surrounding terrain. It is essential that pilots be aware of their presence and location. For this trip, it should be noted that the tallest obstruction is

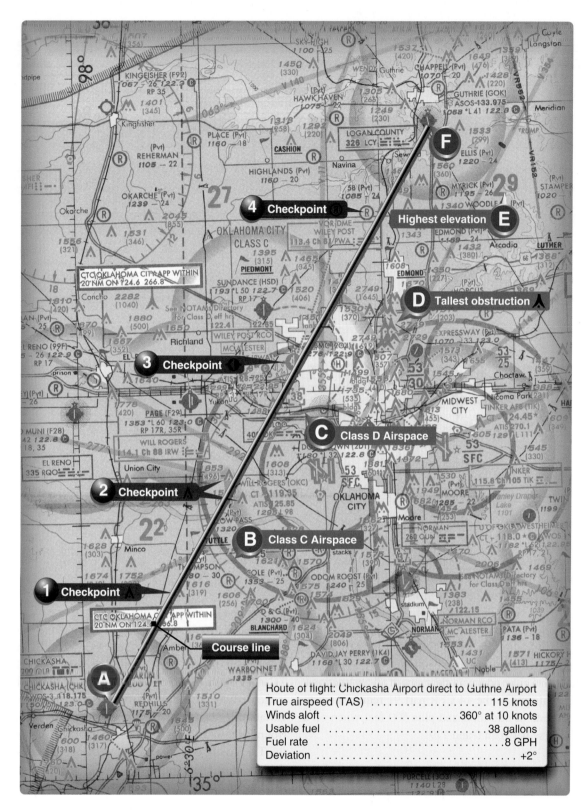

Figure 16-25. *Sectional chart excerpt.*

PILOT'S PLANNING SHEET

PLANE IDENTIFICATION N123DB   **DATE**

COURSE	TC	WIND Knots	WIND From	ALTITUDE	WCA R+ L-	TH	MAG VAR W+ E-	MH	DEV	CH	TOTAL MILES	GS	TOTAL TIME	FUEL RATE	TOTAL FUEL
From Chickasha **To** Guthrie	031°	10	360°	8000	3° L	28	7° E	21°	+2°	23	53	106 kts	35 min	8 GPH	38 gal
From **To**															

VISUAL FLIGHT LOG

TIME OF DEPARTURE	NAVIGATION AIDS	COURSE	ALTITUDE	DISTANCE	ELAPSED TIME	GS	CH	REMARKS
POINT OF DEPARTURE Chickasha Airport	NAVAID IDENT. FREQ.	TO / FROM	TO / FROM	POINT TO POINT / CUMULATIVE	ESTIMATED / ACTUAL	ESTIMATED / ACTUAL	ESTIMATED / ACTUAL	WEATHER AIRSPACE ETC.
CHECKPOINT #1			8000 / 10000	11 NM	6 min / +5	106 kts	023°	
CHECKPOINT #2			8000 / 10000	10 NM / 21 NM	6 min	106 kts	023°	
CHECKPOINT #3			8000 / 10000	10.5 NM / 31.5 NM	6 min	106 kts	023°	
CHECKPOINT #4			8000 / 10000	13 NM / 44.5 NM	7 min	106 kts	023°	
DESTINATION Guthrie Airport				8.5 NM / 53 NM	5 min			

Figure 16-26. *Pilot's planning sheet and visual flight log.*

part of a series of antennas with a height of 2,749 feet MSL (point D). The highest elevation should be located in the northeast quadrant and is 2,900 feet MSL (point E).

Since the wind is no factor and it is desirable and within the aircraft's capability to fly above the Class C and D airspace to be encountered, an altitude of 5,500 feet MSL is chosen. This altitude also gives adequate clearance of all obstructions, as well as conforms to the 14 CFR part 91 requirement to fly at an altitude of odd thousand plus 500 feet when on a magnetic course between 0 and 179°.

Next, the pilot should measure the total distance of the course, as well as the distance between checkpoints. The total distance is 53 NM, and the distance between checkpoints is as noted on the flight log in *Figure 16-26*.

After determining the distance, the TC should be measured. If using a plotter, follow the directions on the plotter. The TC is 031°. Once the TH is established, the pilot can determine the compass heading. This is done by following the formula given earlier in this chapter.

The formula is:

$$TC \pm WCA = TH \pm V = MH \pm D = CH$$

The WCA can be determined by using a manual or electronic flight computer. Using a wind of 360° at 10 knots, it is determined the WCA is 3° left. This is subtracted from the TC making the TH 28°. Next, the pilot should locate the isogonic line closest to the route of the flight to determine variation. *Figure 16-25* shows the variation to be 6.30° E (rounded to 7° E), which means it should be subtracted from the TH, giving an MH of 21°. Next, add 2° to the MH for the deviation correction. This gives the pilot the compass heading of 23°.

Now, the GS can be determined. This is done using a manual or electronic calculator. The GS is determined to be 106 knots. Based on this information, the total trip time, as well as time between checkpoints, and the fuel burned can be determined. These numbers can be calculated by using a manual or electronic calculator.

For this trip, the GS is 106 knots and the total time is 35 minutes (30 minutes plus 5 minutes for climb) with a fuel burn of 4.7 gallons. Refer to the flight log in *Figure 16-26* for the time between checkpoints.

As the trip progresses, the pilot can note headings and time and make adjustments in heading, GS, and time.

Filing a VFR Flight Plan

Filing a flight plan is not required by regulations; however, it is a good operating practice since the information contained in the flight plan can be used in search and rescue in the event of an emergency.

Flight plans can be filed in the air by radio, but it is best to file a flight plan by phone just before departing. After takeoff, contact the FSS by radio and give them the takeoff time so the flight plan can be activated.

When a VFR flight plan is filed, it is held by the FSS until 1 hour after the proposed departure time and then canceled unless: the actual departure time is received; a revised proposed departure time is received; or at the time of filing, the FSS is informed that the proposed departure time is met, but actual time cannot be given because of inadequate communication. The FSS specialist who accepts the flight plan does not inform the pilot of this procedure, however.

Figure 16-27 shows the flight plan form a pilot files with the FSS. When filing a flight plan by telephone or radio, give the information in the order of the numbered spaces. This enables the FSS specialist to copy the information more efficiently. Most of the fields are either self-explanatory or non-applicable to the VFR flight plan (such as item 13). However, some fields may need explanation.

- Item 3 is the aircraft type and special equipment. An example would be C-150/X, which means the aircraft has no transponder. A listing of special equipment codes is found in the Aeronautical Information Manual (AIM).
- Item 6 is the proposed departure time in UTC (indicated by the "Z").
- Item 7 is the cruising altitude. Normally, "VFR" can be entered in this block since the pilot chooses a cruising altitude to conform to FAA regulations.

Form Approved OMB No. 2120-0026

U.S. DEPARTMENT OF TRANSPORTATION
FEDERAL AVIATION ADMINISTRATION

FLIGHT PLAN

(FAA USE ONLY) ☐ PILOT BRIEFING ☐ VNR ☐ STOPOVER | TIME STARTED | SPECIALIST INITIALS

1. TYPE	2. AIRCRAFT IDENTIFICATION	3. AIRCRAFT TYPE/ SPECIAL EQUIPMENT	4. TRUE AIRSPEED	5. DEPARTURE POINT	6. DEPARTURE TIME PROPOSED (Z)	6. DEPARTURE TIME ACTUAL (Z)	7. CRUISING ALTITUDE
X VFR / IFR / DVFR	N123DB	C150/X	115 KTS	CHK, CHICKASHA AIRPORT	1400		5500

8. ROUTE OF FLIGHT: **Chickasha direct Guthrie**

9. DESTINATION (Name of airport and city)	10. EST. TIME ENROUTE HOURS	10. EST. TIME ENROUTE MINUTES	11. REMARKS
GOK, Guthrie Airport Guthrie, OK		35	

12. FUEL ON BOARD HOURS	12. FUEL ON BOARD MINUTES	13. ALTERNATE AIRPORT(S)	14. PILOT'S NAME, ADDRESS & TELEPHONE NUMBER & AIRCRAFT HOME BASE	15. NUMBER ABOARD
4	45		Jane Smith Aero Air, Oklahoma City, OK (405) 555-4149 / 17. DESTINATION CONTACT/TELEPHONE (OPTIONAL)	1

16. COLOR OF AIRCRAFT: **Red/White**

CIVIL AIRCRAFT PILOTS. 14 CFR Part 91 requires you file an IFR flight plan to operate under instrument flight rules in controlled airspace. Failure to file could result in a civil penalty not to exceed $1,000 for each violation (Section 901 of the Federal Aviation Act of 1958, as amended). Filing of a VFR flight plan is recommended as a good operating practice. See also Part 99 for requirements concerning DVFR flight plans.

FAA Form 7233-1 (8-82) CLOSE VFR FLIGHT PLAN WITH McAlester FSS ON ARRIVAL

Figure 16-27. *Domestic flight plan form.*

- Item 8 is the route of flight. If the flight is to be direct, enter the word "direct;" if not, enter the actual route to be followed, such as via certain towns or navigation aids.

- Item 10 is the estimated time en route. In the sample flight plan, 5 minutes was added to the total time to allow for the climb.

- Item 12 is the fuel on board in hours and minutes. This is determined by dividing the total usable fuel aboard in gallons by the estimated rate of fuel consumption in gallons.

Remember, there is every advantage in filing a flight plan; but do not forget to close the flight plan upon arrival. This should be done via telephone to avoid radio congestion.

Ground-Based Navigation

Advances in navigational radio receivers installed in aircraft, the development of aeronautical charts that show the exact location of ground transmitting stations and their frequencies, along with refined flight deck instrumentation make it possible for pilots to navigate with precision to almost any point desired. Although precision in navigation is obtainable through the proper use of this equipment, beginning pilots should use this equipment to supplement navigation by visual reference to the ground (pilotage). This method provides the pilot with an effective safeguard against disorientation in the event of radio malfunction.

There are three radio navigation systems available for use for VFR navigation. These are:

- VHF Omnidirectional Range (VOR)

- Nondirectional Radio Beacon (NDB)

- Global Positioning System (GPS)

Very High Frequency (VHF) Omnidirectional Range (VOR)

The VOR system is present in three slightly different navigation aids (NAVAIDs): VOR, VOR/distance measuring equipment (DME)(discussed in a later section), and VORTAC. By itself it is known as a VOR, and it provides magnetic bearing information to and from the station. When DME is also installed with a VOR, the NAVAID is referred to as a VOR/DME. When military tactical air navigation (TACAN) equipment is installed with a VOR, the NAVAID is known as a VORTAC. DME is always an integral part of a VORTAC. Regardless of the type of NAVAID utilized (VOR, VOR/DME, or VORTAC), the VOR indicator behaves the same. Unless otherwise noted in this section, VOR, VOR/DME, and VORTAC NAVAIDs are all referred to hereafter as VORs.

The prefix "omni-" means all, and an omnidirectional range is a VHF radio transmitting ground station that projects straight line courses (radials) from the station in all directions. From a top view, it can be visualized as being similar to the spokes from the hub of a wheel. The distance VOR radials are projected depends upon the power output of the transmitter.

The course or radials projected from the station are referenced to MN. Therefore, a radial is defined as a line of magnetic bearing extending outward from the VOR station. Radials are identified by numbers beginning with 001, which is 1° east of MN and progress in sequence through all the degrees of a circle until reaching 360. To aid in orientation, a compass rose reference to magnetic north is superimposed on aeronautical charts at the station location.

VOR ground stations transmit within a VHF frequency band of 108.0–117.95 MHz. Because the equipment is VHF, the signals transmitted are subject to line-of-sight restrictions. Therefore, its range varies in direct proportion to the altitude of receiving equipment. Generally, the reception range of the signals at an altitude of 1,000 feet above ground level (AGL) is about 40 to 45 miles. This distance increases with altitude. *[Figure 16-28]*

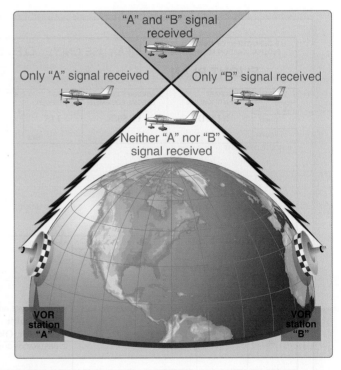

Figure 16-28. *VHF transmissions follow a line-of-sight course.*

VORs and VORTACs are classed according to operational use. There are three classes:

- T (Terminal)
- L (Low altitude)
- H (High altitude)

The normal useful range for the various classes is shown in the following table:

VOR/VORTAC NAVAIDS
Normal Usable Altitudes and Radius Distances

Class	Altitudes	Distance (Miles)
T	12,000' and below	25
L	Below 18,000'	40
H	Below 14,500'	40
H	Within the conterminous 48 states only, between 14,500 and 17,999'	100
H	18,000'—FL 450	130
H	FL 450—60,000'	100

The useful range of certain facilities may be less than 50 miles. For further information concerning these restrictions, refer to the Communication/NAVAID Remarks in the Chart Supplement U.S.

The accuracy of course alignment of VOR radials is considered to be excellent. It is generally within plus or minus 1°. However, certain parts of the VOR receiver equipment deteriorate, affecting its accuracy. This is particularly true at great distances from the VOR station. The best assurance of maintaining an accurate VOR receiver is periodic checks and calibrations. VOR accuracy checks are not a regulatory requirement for VFR flight. However, to assure accuracy of the equipment, these checks should be accomplished quite frequently and a complete calibration should be performed each year. The following means are provided for pilots to check VOR accuracy:

- FAA VOR test facility (VOT)
- Certified airborne checkpoints
- Certified ground checkpoints located on airport surfaces

If an aircraft has two VOR receivers installed, a dual VOR receiver check can be made. To accomplish the dual receiver check, a pilot must tune both VOR receivers to the same VOR ground facility. The maximum permissible variation between the two indicated bearings is 4°. A list of the airborne and ground checkpoints is published in the Chart Supplement U.S.

Basically, these checks consist of verifying that the VOR radials the aircraft equipment receives are aligned with the radials the station transmits. There are not specific tolerances in VOR checks required for VFR flight. But as a guide to assure acceptable accuracy, the required IFR tolerances can be used—±4° for ground checks and ±6° for airborne checks. These checks can be performed by the pilot.

The VOR transmitting station can be positively identified by its Morse code identification or by a recorded voice identification that states the name of the station followed by "VOR." Many FSSs transmit voice messages on the same frequency that the VOR operates. Voice transmissions should not be relied upon to identify stations because many FSSs remotely transmit over several omniranges that have names different from that of the transmitting FSS. If the VOR is out of service for maintenance, the coded identification is removed and not transmitted. This serves to alert pilots that this station should not be used for navigation. VOR receivers are designed with an alarm flag to indicate when signal strength is inadequate to operate the navigational equipment. This happens if the aircraft is too far from the VOR or the aircraft is too low and, therefore, is out of the line of sight of the transmitting signals.

Using the VOR

In review, for VOR radio navigation, there are two components required: ground transmitter and aircraft receiving equipment. The ground transmitter is located at a specific position on the ground and transmits on an assigned frequency. The aircraft equipment includes a receiver with a tuning device and a VOR or omninavigation instrument. The navigation instrument could be a course deviation indicator (CDI), horizontal situation indicator (HSI), or a radio magnetic indicator (RMI). Each of these instruments indicates the course to the tuned VOR.

Course Deviation Indicator (CDI)

The CDI is found in most training aircraft. It consists of an omnibearing selector (OBS) sometimes referred to as the course selector, a CDI needle (left-right needle), and a TO/FROM indicator.

The course selector is an azimuth dial that can be rotated to select a desired radial or to determine the radial over which the aircraft is flying. In addition, the magnetic course "TO" or "FROM" the station can be determined.

When the course selector is rotated, it moves the CDI or needle to indicate the position of the radial relative to the aircraft. If the course selector is rotated until the deviation needle is centered, the radial (magnetic course "FROM" the station) or its reciprocal (magnetic course "TO" the station) can be determined. The course deviation needle also moves to the right or left if the aircraft is flown or drifting away from the radial which is set in the course selector.

By centering the needle, the course selector indicates either the course "FROM" the station or the course "TO" the station. If the flag displays a "TO," the course shown on the course selector must be flown to the station. [Figure 16-29] If "FROM" is displayed and the course shown is followed, the aircraft is flown away from the station.

Horizontal Situation Indicator

The HSI is a direction indicator that uses the output from a flux valve to drive the compass card. The HSI [Figure 16-30] combines the magnetic compass with navigation signals and a glideslope. The HSI gives the pilot an indication of the location of the aircraft in relation to the chosen course or radial.

In Figure 16-30, the aircraft magnetic heading displayed on the compass card under the lubber line is 184°. The course select pointer shown is set to 295°; the tail of the pointer indicates the reciprocal, 115°. The course deviation bar operates with a VOR/Localizer (VOR/LOC) or GPS navigation receiver to indicate left or right deviations from the course selected with the course select pointer; operating in the same manner, the angular movement of a conventional VOR/LOC needle indicates deviation from course.

The desired course is selected by rotating the course select pointer, in relation to the compass card, by means of the

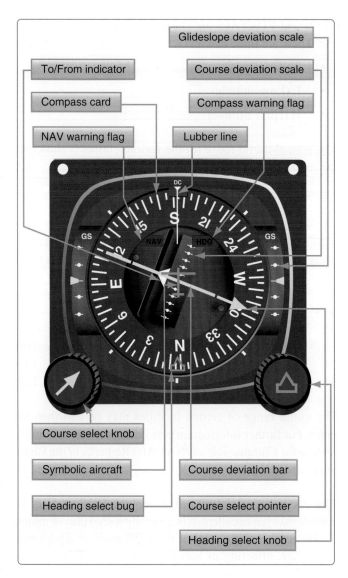

Figure 16-30. Horizontal situation indicator.

course select knob. The HSI has a fixed aircraft symbol and the course deviation bar displays the aircraft's position relative to the selected course. The TO/FROM indicator is a triangular pointer. When the indicator points to the head of the course select pointer, the arrow shows the course selected. If properly intercepted and flown, the course takes the aircraft to the chosen facility. When the indicator points to the tail of the course, the arrow shows that the course selected, if properly intercepted and flown, takes the aircraft directly away from the chosen facility.

When the NAV warning flag appears, it indicates no reliable signal is being received. The appearance of the HDG flag indicates the compass card is not functioning properly.

Radio Magnetic Indicator (RMI)

The RMI is a navigational aid providing aircraft magnetic or directional gyro heading and very high frequency omnidirectional range (VOR), GPS, and automatic direction

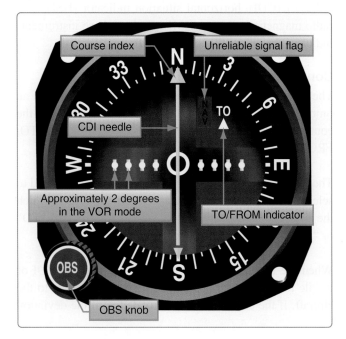

Figure 16-29. VOR indicator.

Figure 16-31. *Radio magnetic indicator.*

finder (ADF) bearing information. *[Figure 16-31]* Remote indicating compasses were developed to compensate for errors in and limitations of older types of heading indicators. The remote compass transmitter is a separate unit usually mounted in a wingtip to eliminate the possibility of magnetic interference. The RMI consists of a compass card, a heading index, two bearing pointers, and pointer function switches. The two pointers are driven by any two combinations of a GPS, an ADF, and/or a VOR. The pilot has the ability to select the navigation aid to be indicated. The pointer indicates the course to the selected NAVAID or waypoint. In *Figure 16-31,* the green pointer is indicating the station tuned on the ADF. The yellow pointer is indicating the course to a VOR or GPS waypoint. Note that there is no requirement for a pilot to select a course with the RMI. Only the selected navigation source is pointed to by the needle(s).

Tracking With VOR

The following describes a step-by-step procedure for tracking to and from a VOR station using a CDI. *Figure 16-32* illustrates the procedure.

First, tune the VOR receiver to the frequency of the selected VOR station. For example, 115.0 to receive Bravo VOR. Next, check the identifiers to verify that the desired VOR is being received. As soon as the VOR is properly tuned, the course deviation needle deflects either left or right. Then, rotate the azimuth dial to the course selector until the course deviation needle centers and the TO-FROM indicator indicates "TO." If the needle centers with a "FROM" indication, the azimuth should be rotated 180° because, in this case, it is desired to fly "TO" the station. Now, turn the aircraft to the heading indicated on the VOR azimuth dial or course selector, 350° in this example.

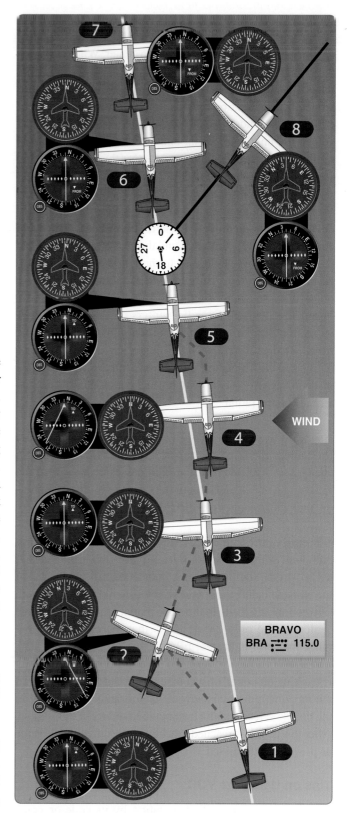

Figure 16-32. *Tracking a radial in a crosswind.*

If a heading of 350° is maintained with a wind from the right as shown, the aircraft drifts to the left of the intended track. As the aircraft drifts off course, the VOR course deviation needle gradually moves to the right of center or indicates the direction of the desired radial or track.

To return to the desired radial, the aircraft heading must be altered to the right. As the aircraft returns to the desired track, the deviation needle slowly returns to center. When centered, the aircraft is on the desired radial and a left turn must be made toward, but not to the original heading of 350° because a wind drift correction must be established. The amount of correction depends upon the strength of the wind. If the wind velocity is unknown, a trial-and-error method can be used to find the correct heading. Assume, for this example, a 10° correction for a heading of 360° is maintained.

While maintaining a heading of 360°, assume that the course deviation begins to move to the left. This means that the wind correction of 10° is too great and the aircraft is flying to the right of course. A slight turn to the left should be made to permit the aircraft to return to the desired radial.

When the deviation needle centers, a small wind drift correction of 5° or a heading correction of 355° should be flown. If this correction is adequate, the aircraft remains on the radial. If not, small variations in heading should be made to keep the needle centered and consequently keep the aircraft on the radial.

As the VOR station is passed, the course deviation needle fluctuates, then settles down, and the "TO" indication changes to "FROM." If the aircraft passes to one side of the station, the needle deflects in the direction of the station as the indicator changes to "FROM."

Generally, the same techniques apply when tracking outbound as those used for tracking inbound. If the intent is to fly over the station and track outbound on the reciprocal of the inbound radial, the course selector should not be changed. Corrections are made in the same manner to keep the needle centered. The only difference is that the omnidirectional range indicator indicates "FROM."

If tracking outbound on a course other than the reciprocal of the inbound radial, this new course or radial must be set in the course selector and a turn made to intercept this course. After this course is reached, tracking procedures are the same as previously discussed.

Tips on Using the VOR

- Positively identify the station by its code or voice identification.

- Remember that VOR signals are "line-of-sight." A weak signal or no signal at all is received if the aircraft is too low or too far from the station.

- When navigating to a station, determine the inbound radial and use this radial. Fly a heading that will maintain the course. If the aircraft drifts, fly a heading to re-intercept the course then apply a correction to compensate for wind drift.

- If minor needle fluctuations occur, avoid changing headings immediately. Wait a moment to see if the needle recenters; if it does not, then you must correctly recenter the course to the needle.

- When flying "TO" a station, always fly the selected course with a "TO" indication. When flying "FROM" a station, always fly the selected course with a "FROM" indication. If this is not done, the action of the course deviation needle is reversed. To further explain this reverse action, if the aircraft is flown toward a station with a "FROM" indication or away from a station with a "TO" indication, the course deviation needle indicates in a direction opposite to that which it should indicate. For example, if the aircraft drifts to the right of a radial being flown, the needle moves to the right or points away from the radial. If the aircraft drifts to the left of the radial being flown, the needle moves left or in the direction opposite of the radial.

- When navigating using the VOR, it is important to fly headings that maintain or re-intercept the course. Just turning toward the needle will cause overshooting the radial and flying an S turn to the left and right of course.

Time and Distance Check From a Station Using a RMI

To compute time and distance from a station, first turn the aircraft to place the RMI bearing pointer on the nearest 90° index. Note the time and maintain the heading. When the RMI bearing pointer has moved 10°, note the elapsed time in seconds and apply the formulas in the following example to determine the approximate time and distance from a given station. [Figure 16-33]

The time from station may also be calculated by using a short method based on the above formula, if a 10° bearing change is flown. If the elapsed time for the bearing change is noted

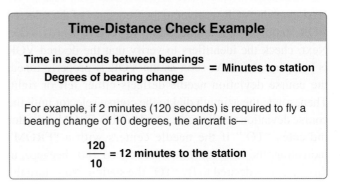

Time-Distance Check Example

$$\frac{\text{Time in seconds between bearings}}{\text{Degrees of bearing change}} = \text{Minutes to station}$$

For example, if 2 minutes (120 seconds) is required to fly a bearing change of 10 degrees, the aircraft is—

$$\frac{120}{10} = 12 \text{ minutes to the station}$$

Figure 16-33. *Time-distance check example.*

in seconds and a 10° bearing change is made, the time from the station, in minutes, is determined by counting off one decimal point. Thus, if 75 seconds are required to fly a 10° bearing change, the aircraft is 7.5 minutes from the station. When the RMI bearing pointer is moving rapidly or when several corrections are required to place the pointer on the wingtip position, the aircraft is at station passage.

The distance from the station is computed by multiplying TAS or GS (in miles per minute) by the previously determined time in minutes. For example, if the aircraft is 7.5 minutes from station, flying at a TAS of 120 knots or 2 NM per minute, the distance from station is 15 NM (7.5 × 2 = 15).

The accuracy of time and distance checks is governed by existing wind, degree of bearing change, and accuracy of timing. The number of variables involved causes the result to be only an approximation. However, by flying an accurate heading and checking the time and bearing closely, the pilot can make a reasonable estimate of time and distance from the station.

Time and Distance Check From a Station Using a CDI

To compute time and distance from a station using a CDI, first tune and identify the VOR station and determine the radial on which you are located. Then turn inbound and re-center the needle if necessary. Turn 90° left or right, of the inbound course, rotating the OBS to the nearest 10° increment opposite the direction of turn. Maintain heading and when the CDI centers, note the time. Maintaining the same heading, rotate the OBS 10° in the same direction as was done previously and note the elapsed time when the CDI again centers. Time and distance from the station is determined from the formula shown in *Figure 16-34*.

Course Intercept

Course interceptions are performed in most phases of instrument navigation. The equipment used varies, but an intercept heading must be flown that results in an angle or rate of intercept sufficient for solving a particular problem.

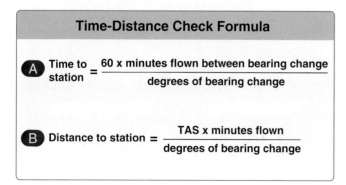

Time-Distance Check Formula

A Time to station = $\dfrac{60 \times \text{minutes flown between bearing change}}{\text{degrees of bearing change}}$

B Distance to station = $\dfrac{\text{TAS} \times \text{minutes flown}}{\text{degrees of bearing change}}$

Figure 16-34. *Time-distance check formula using a CDI.*

Rate of Intercept

Rate of intercept, seen by the aviator as bearing pointer or HSI movement, is a result of the following factors:

- The angle at which the aircraft is flown toward a desired course (angle of intercept)
- True airspeed and wind (GS)
- Distance from the station

Angle of Intercept

The angle of intercept is the angle between the heading of the aircraft (intercept heading) and the desired course. Controlling this angle by selection/adjustment of the intercept heading is the easiest and most effective way to control course interceptions. Angle of intercept must be greater than the degrees from course, but should not exceed 90°. Within this limit, make adjustments as needed, to achieve the most desirable rate of intercept.

When selecting an intercept heading, the key factor is the relationship between distance from the station and degrees from the course. Each degree, or radial, is 1 NM wide at a distance of 60 NM from the station. Width increases or decreases in proportion to the 60 NM distance. For example, 1 degree is 2 NM wide at 120 NM—and ½ NM wide at 30 NM. For a given GS and angle of intercept, the resultant rate of intercept varies according to the distance from the station. When selecting an intercept heading to form an angle of intercept, consider the following factors:

- Degrees from course
- Distance from the station
- True airspeed and wind (GS)

Distance Measuring Equipment (DME)

Distance measuring equipment (DME) consists of an ultra high frequency (UHF) navigational aid with VOR/DMEs and VORTACs. It measures, in NM, the slant range distance of an aircraft from a VOR/DME or VORTAC (both hereafter referred to as a VORTAC). Although DME equipment is very popular, not all aircraft are DME equipped.

To utilize DME, the pilot should select, tune, and identify a VORTAC, as previously described. The DME receiver, utilizing what is called a "paired frequency" concept, automatically selects and tunes the UHF DME frequency associated with the VHF VORTAC frequency selected by the pilot. This process is entirely transparent to the pilot. After a brief pause, the DME display shows the slant range distance to or from the VORTAC. Slant range distance is the direct distance between the aircraft and the VORTAC and is therefore affected by aircraft altitude. (Station passage directly over a VORTAC from an altitude of 6,076 feet AGL would show approximately 1.0 NM on the DME.) DME is a very useful adjunct to VOR navigation. A VOR radial alone merely gives line of position information. With DME, a pilot may precisely locate the aircraft on a given line (radial).

Most DME receivers also provide GS and time-to-station modes of operation. The GS is displayed in knots (NMPH). The time-to-station mode displays the minutes remaining to VORTAC station passage, predicated upon the present GS. GS and time-to-station information is only accurate when tracking directly to or from a VORTAC. DME receivers typically need a minute or two of stabilized flight directly to or from a VORTAC before displaying accurate GS or time-to-station information.

Some DME installations have a hold feature that permits a DME signal to be retained from one VORTAC while the course indicator displays course deviation information from an ILS or another VORTAC.

VOR/DME RNAV

Area navigation (RNAV) permits electronic course guidance on any direct route between points established by the pilot. While RNAV is a generic term that applies to a variety of NAVAIDS, such as GPS and others, this section deals with VOR/DME-based RNAV. VOR/DME RNAV is not a separate ground-based NAVAID, but a method of navigation using VOR/DME and VORTAC signals specially processed by the aircraft's RNAV computer. *[Figure 16-35]*

NOTE: In this section, the term "VORTAC" also includes VOR/DME NAVAIDs.

In its simplest form, VOR/DME RNAV allows the pilot to electronically move VORTACs around to more convenient locations. Once electronically relocated, they are referred to as waypoints. These waypoints are described as a combination of a selected radial and distance within the service volume of the VORTAC to be used. These waypoints allow a straight course to be flown between almost any origin and destination, without regard to the orientation of VORTACs or the existence of airways.

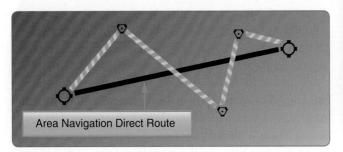

Area Navigation Direct Route

Figure 16-35. *Flying an RNAV course.*

While the capabilities and methods of operation of VOR/DME RNAV units differ, there are basic principles of operation that are common to all. Pilots are urged to study the manufacturer's operating guide and receive instruction prior to the use of VOR/DME RNAV or any unfamiliar navigational system. Operational information and limitations should also be sought from placards and the supplement section of the AFM/POH.

VOR/DME-based RNAV units operate in at least three modes: VOR, en route, and approach. A fourth mode, VOR Parallel, may also be found on some models. The units need both VOR and DME signals to operate in any RNAV mode. If the NAVAID selected is a VOR without DME, RNAV mode will not function.

In the VOR (or non-RNAV) mode, the unit simply functions as a VOR receiver with DME capability. *[Figure 16-36]* The unit's display on the VOR indicator is conventional in all respects. For operation on established airways or any other ordinary VOR navigation, the VOR mode is used.

To utilize the unit's RNAV capability, the pilot selects and establishes a waypoint or a series of waypoints to define a course. A VORTAC (or VOR/DME) needs to be selected as a NAVAID, since both radial and distance signals are available from these stations. To establish a waypoint, a point somewhere within the service range of a VORTAC is defined on the basis of radial and distance. Once the waypoint is entered into the unit and the RNAV en route mode is selected, the CDI displays course guidance to the waypoint, not the original VORTAC. DME also displays distance to the waypoint. Many units have the capability to store several waypoints, allowing them to be programmed prior to flight, if desired, and called up in flight.

RNAV waypoints are entered into the unit in magnetic bearings (radials) of degrees and tenths (i.e., 275.5°) and distances in NM and tenths (i.e., 25.2 NM). When plotting RNAV waypoints on an aeronautical chart, pilots find it

Figure 16-36. *RNAV controls.*

difficult to measure to that level of accuracy, and in practical application, it is rarely necessary. A number of flight planning publications publish airport coordinates and waypoints with this precision and the unit accepts those figures. There is a subtle but important difference in CDI operation and display in the RNAV modes.

In the RNAV modes, course deviation is displayed in terms of linear deviation. In the RNAV en route mode, maximum deflection of the CDI typically represents 5 NM on either side of the selected course without regard to distance from the waypoint. In the RNAV approach mode, maximum deflection of the CDI typically represents 1¼ NM on either side of the selected course. There is no increase in CDI sensitivity as the aircraft approaches a waypoint in RNAV mode.

The RNAV approach mode is used for instrument approaches. Its narrow scale width (¼ of the en route mode) permits very precise tracking to or from the selected waypoint. In VFR cross-country navigation, tracking a course in the approach mode is not desirable because it requires a great deal of attention and soon becomes tedious.

A fourth, lesser-used mode on some units is the VOR Parallel mode. This permits the CDI to display linear (not angular) deviation as the aircraft tracks to and from VORTACs. It derives its name from permitting the pilot to offset (or parallel) a selected course or airway at a fixed distance of the pilot's choosing, if desired. The VOR parallel mode has the same effect as placing a waypoint directly over an existing VORTAC. Some pilots select the VOR parallel mode when utilizing the navigation (NAV) tracking function of their autopilot for smoother course following near the VORTAC.

Navigating an aircraft with VOR/DME-based RNAV can be confusing, and it is essential that the pilot become familiar with the equipment installed. It is not unknown for pilots to operate inadvertently in one of the RNAV modes when the operation was not intended, by overlooking switch positions or annunciators. The reverse has also occurred with a pilot neglecting to place the unit into one of the RNAV modes by overlooking switch positions or annunciators. As always, the prudent pilot is not only familiar with the equipment used, but never places complete reliance in just one method of navigation when others are available for cross-check.

Automatic Direction Finder (ADF)

Many general aviation-type aircraft are equipped with ADF radio receiving equipment. To navigate using the ADF, the pilot tunes the receiving equipment to a ground station known as a nondirectional radio beacon (NDB). The NDB stations normally operate in a low or medium frequency band

of 200 to 415 kHz. The frequencies are readily available on aeronautical charts or in the Chart Supplement U.S.

All radio beacons, except compass locators, transmit a continuous three-letter identification in code, except during voice transmissions. A compass locator, which is associated with an instrument landing system, transmits a two-letter identification.

Standard broadcast stations can also be used in conjunction with ADF. Positive identification of all radio stations is extremely important and this is particularly true when using standard broadcast stations for navigation.

NDBs have one advantage over the VOR in that low or medium frequencies are not affected by line-of-sight. The signals follow the curvature of the Earth; therefore, if the aircraft is within the range of the station, the signals can be received regardless of altitude.

The following table gives the class of NDB stations, their power, and their usable range:

NONDIRECTIONAL RADIO BEACON (NDB)
(Usable radius distances for all altitudes)

Class	Power (Watts)	Distance (Miles)
Compass Locator	Under 25	15
MH	Under 50	25
H	50–1999	*50
HH	2000 or more	75

*Service range of individual facilities may be less than 50 miles.

One of the disadvantages that should be considered when using low frequency (LF) for navigation is that LF signals are very susceptible to electrical disturbances, such as lightning. These disturbances create excessive static, needle deviations, and signal fades. There may be interference from distant stations. Pilots should know the conditions under which these disturbances can occur so they can be more alert to possible interference when using the ADF.

Basically, the ADF aircraft equipment consists of a tuner, which is used to set the desired station frequency, and the navigational display.

The navigational display consists of a dial upon which the azimuth is printed and a needle which rotates around the dial and points to the station to which the receiver is tuned.

Some of the ADF dials can be rotated to align the azimuth with the aircraft heading; others are fixed with 0° representing the nose of the aircraft and 180° representing the tail. Only the fixed azimuth dial is discussed in this handbook. [Figure 16-37]

Figure 16-38 illustrates terms that are used with the ADF and should be understood by the pilot.

To determine the magnetic bearing "FROM" the station, 180° is added to or subtracted from the magnetic bearing to the station. This is the reciprocal bearing and is used when plotting position fixes.

Figure 16-37. *ADF with fixed azimuth and magnetic compass.*

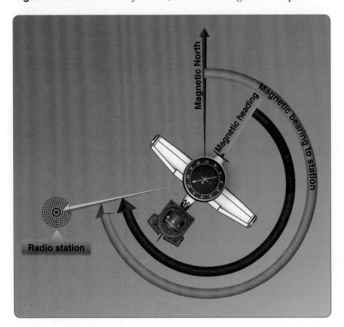

Figure 16-38. *ADF terms.*

Keep in mind that the needle of fixed azimuth points to the station in relation to the nose of the aircraft. If the needle is deflected 30° to the left for a relative bearing of 330°, this means that the station is located 30° left. If the aircraft is turned left 30°, the needle moves to the right 30° and indicates a relative bearing of 0° meaning that the aircraft is pointing toward the station. If the pilot continues flight toward the station keeping the needle on 0°, the procedure is called homing to the station. If a crosswind exists, the ADF needle continues to drift away from zero. To keep the needle on zero, the aircraft must be turned slightly resulting in a curved flight path to the station. Homing to the station is a common procedure but may result in drifting downwind, thus lengthening the distance to the station.

Tracking to the station requires correcting for wind drift and results in maintaining flight along a straight track or bearing to the station. When the wind drift correction is established, the ADF needle indicates the amount of correction to the right or left. For instance, if the magnetic bearing to the station is 340°, a correction for a left crosswind would result in a magnetic heading of 330°, and the ADF needle would indicate 10° to the right or a relative bearing of 010°. [Figure 16-39]

When tracking away from the station, wind corrections are made similar to tracking to the station, but the ADF needle points toward the tail of the aircraft or the 180° position on the azimuth dial. Attempting to keep the ADF needle on the 180° position during winds results in the aircraft flying a curved flight leading further and further from the desired track. When tracking outbound, corrections for wind should be made in the direction opposite of that in which the needle is pointing.

Although the ADF is not as popular as the VOR for radio navigation, with proper precautions and intelligent use, the ADF can be a valuable aid to navigation.

Global Positioning System

The GPS is a satellite-based radio navigation system. Its RNAV guidance is worldwide in scope. There are no symbols for GPS on aeronautical charts as it is a space-based system with global coverage. Development of the system is underway so that GPS is capable of providing the primary means of electronic navigation. Portable and yoke-mounted units are proving to be very popular in addition to those permanently installed in the aircraft. Extensive navigation databases are common features in aircraft GPS receivers.

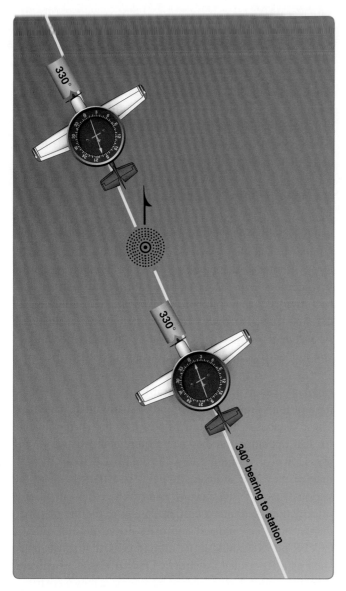

Figure 16-39. *ADF tracking.*

The GPS is a satellite radio navigation and time dissemination system developed and operated by the U.S. Department of Defense (DOD). Civilian interface and GPS system status is available from the U.S. Coast Guard.

It is not necessary to understand the technical aspects of GPS operation to use it in VFR/IFR navigation. It does differ significantly from conventional, ground-based electronic navigation and awareness of those differences is important. Awareness of equipment approvals and limitations is critical to the safety of flight.

The GPS navigation system broadcasts a signal that is used by receivers to determine precise position anywhere in the world. The receiver tracks multiple satellites and determines a pseudorange measurement to determine the user location. A minimum of four satellites is necessary to establish an accurate three-dimensional position. The Department of Defense (DOD) is responsible for operating the GPS satellite constellation and monitors the GPS satellites to ensure proper operation.

The status of a GPS satellite is broadcast as part of the data message transmitted by the satellite. GPS status information is also available from the U.S. Coast Guard navigation information service at (703) 313-5907 or online at www. navcen.uscg.gov. Additionally, satellite status is available through the NOTAM system.

The GPS receiver verifies the integrity (usability) of the signals received from the GPS constellation through receiver autonomous integrity monitoring (RAIM) to determine if a satellite is providing corrupted information. At least one satellite, in addition to those required for navigation, must be in view for the receiver to perform the RAIM function; thus, RAIM needs a minimum of five satellites in view or four satellites and a barometric altimeter (baro-aiding) to detect an integrity anomaly. For receivers capable of doing so, RAIM needs six satellites in view (or five satellites with baro-aiding) to isolate the corrupt satellite signal and remove it from the navigation solution. Baro-aiding is a method of augmenting the GPS integrity solution by using a nonsatellite input source. GPS derived altitude should not be relied upon to determine aircraft altitude since the vertical error can be quite large and no integrity is provided. To ensure that baro-aiding is available, the current altimeter setting must be entered into the receiver as described in the operating manual.

RAIM messages vary somewhat between receivers; however, generally there are two types. One type indicates that there are not enough satellites available to provide RAIM integrity monitoring and another type indicates that the RAIM integrity monitor has detected a potential error that exceeds the limit for the current phase of flight. Without RAIM capability, the pilot has no assurance of the accuracy of the GPS position.

Selective Availability

Selective Availability (SA) is a method by which the accuracy of GPS is intentionally degraded. This feature is designed to deny hostile use of precise GPS positioning data. SA was discontinued on May 1, 2000, but many GPS receivers are designed to assume that SA is still active.

The baseline GPS satellite constellation consists of 24 satellites positioned in six earth-centered orbital planes with four operation satellites and a spare satellite slot in each orbital plane. The system can support a constellation of up to thirty satellites in orbit. The orbital period of a GPS satellite is one-half of a sidereal day or 11 hours 58 minutes. The orbits are nearly circular and equally spaced about the equator at a 60-degree separation with an inclination of

55 degrees relative to the equator. The orbital radius (i.e. distance from the center of mass of the earth to the satellite) is approximately 26,600 km.

With the baseline satellite constellation, users with a clear view of the sky have a minimum of four satellites in view. It is more likely that a user would see six to eight satellites. The satellites broadcast ranging signals and navigation data allowing users to measure their pseudoranges in order to estimate their position, velocity and time, in a passive, listen-only mode. The receiver uses data from a minimum of four satellites above the mask angle (the lowest angle above the horizon at which a receiver can use a satellite). The exact number of satellites operating at any one particular time varies depending on the number of satellite outages and operational spares in orbit. For current status of the GPS constellation, please visit http://tycho.usno.navy.mil/gpscurr.html. *[Figure 16-40]*

VFR Use of GPS

GPS navigation has become a great asset to VFR pilots providing increased navigation capability and enhanced situational awareness while reducing operating costs due to greater ease in flying direct routes. While GPS has many benefits to the VFR pilot, care must be exercised to ensure that system capabilities are not exceeded.

Types of receivers used for GPS navigation under VFR are varied from a full IFR installation being used to support a VFR flight to a VFR only installation (in either a VFR or IFR capable aircraft) to a hand-held receiver. The limitations of

Figure 16-40. *Satellite constellation.*

each type of receiver installation or use must be understood by the pilot to avoid misusing navigation information. In all cases, VFR pilots should never rely solely on one system of navigation. GPS navigation must be integrated with other forms of electronic navigation, as well as pilotage and dead reckoning. Only through the integration of these techniques can the VFR pilot ensure accuracy in navigation. Some critical concerns in VFR use of GPS include RAIM capability, database currency, and antenna location.

RAIM Capability

Many VFR GPS receivers and all hand-held units are not equipped with RAIM alerting capability. Loss of the required number of satellites in view, or the detection of a position error, cannot be displayed to the pilot by such receivers. In receivers with no RAIM capability, no alert would be provided to the pilot that the navigation solution had deteriorated and an undetected navigation error could occur. A systematic cross-check with other navigation techniques would identify this failure and prevent a serious deviation.

In many receivers, an updatable database is used for navigation fixes, airports, and instrument procedures. These databases must be maintained to the current update for IFR operation, but no such requirement exists for VFR use. However, in many cases, the database drives a moving map display that indicates Special Use Airspace and the various classes of airspace in addition to other operational information. Without a current database, the moving map display may be outdated and offer erroneous information to VFR pilots wishing to fly around critical airspace areas, such as a Restricted Area or a Class B airspace segment. Numerous pilots have ventured into airspace they were trying to avoid by using an outdated database. If there is not a current database in the receiver, disregard the moving map display when making critical navigation decisions.

In addition, waypoints are added, removed, relocated, or re-named as required to meet operational needs. When using GPS to navigate relative to a named fix, a current database must be used to properly locate a named waypoint. Without the update, it is the pilot's responsibility to verify the waypoint location referencing to an official current source, such as the Chart Supplement U.S., sectional chart, or en route chart.

In many VFR installations of GPS receivers, antenna location is more a matter of convenience than performance. In IFR installations, care is exercised to ensure that an adequate clear view is provided for the antenna to communicate with satellites. If an alternate location is used, some portion of the aircraft may block the view of the antenna increasing the possibility of losing navigation signal.

This is especially true in the case of hand-held receivers. The use of hand-held receivers for VFR operations is a growing trend, especially among rental pilots. Typically, suction cups are used to place the GPS antennas on the inside of aircraft windows. While this method has great utility, the antenna location is limited by aircraft structure for optimal reception of available satellites. Consequently, signal loss may occur in certain situations where aircraft-satellite geometry causes a loss of navigation signal. These losses, coupled with a lack of RAIM capability, could present erroneous position and navigation information with no warning to the pilot.

While the use of hand-held GPS receivers for VFR operations is not limited by regulation, modification of the aircraft, such as installing a panel- or yoke-mounted holder, is governed by 14 CFR part 43. Pilots should consult a mechanic to ensure compliance with the regulation and a safe installation.

Tips for Using GPS for VFR Operations

Always check to see if the unit has RAIM capability. If no RAIM capability exists, be suspicious of a GPS displayed position when any disagreement exists with the position derived from other radio navigation systems, pilotage, or dead reckoning.

Check the currency of the database, if any. If expired, update the database using the current revision. If an update of an expired database is not possible, disregard any moving map display of airspace for critical navigation decisions. Be aware that named waypoints may no longer exist or may have been relocated since the database expired. At a minimum, the waypoints to be used should be verified against a current official source, such as the Chart Supplement U.S. or a Sectional Aeronautical Chart.

While a hand-held GPS receiver can provide excellent navigation capability to VFR pilots, be prepared for intermittent loss of navigation signal, possibly with no RAIM warning to the pilot. If mounting the receiver in the aircraft, be sure to comply with 14 CFR part 43.

Plan flights carefully before taking off. If navigating to user-defined waypoints, enter them prior to flight, not on the fly. Verify the planned flight against a current source, such as a current sectional chart. There have been cases in which one pilot used waypoints created by another pilot that were not where the pilot flying was expecting. This generally resulted in a navigation error. Minimize head-down time in the aircraft and maintain a sharp lookout for traffic, terrain, and obstacles. Just a few minutes of preparation and planning on the ground makes a great difference in the air.

Another way to minimize head-down time is to become very familiar with the receiver's operation. Most receivers are not intuitive. The pilot must take the time to learn the various keystrokes, knob functions, and displays that are used in the operation of the receiver. Some manufacturers provide computer-based tutorials or simulations of their receivers. Take the time to learn about the particular unit before using it in flight.

In summary, be careful not to rely on GPS to solve all VFR navigational problems. Unless an IFR receiver is installed in accordance with IFR requirements, no standard of accuracy or integrity can be assured. While the practicality of GPS is compelling, the fact remains that only the pilot can navigate the aircraft, and GPS is just one of the pilot's tools to do the job.

VFR Waypoints

VFR waypoints provide VFR pilots with a supplementary tool to assist with position awareness while navigating visually in aircraft equipped with area navigation receivers. VFR waypoints should be used as a tool to supplement current navigation procedures. The use of VFR waypoints include providing navigational aids for pilots unfamiliar with an area, waypoint definition of existing reporting points, enhanced navigation in and around Class B and Class C airspace, and enhanced navigation around Special Use Airspace. VFR pilots should rely on appropriate and current aeronautical charts published specifically for visual navigation. If operating in a terminal area, pilots should take advantage of the Terminal Area Chart available for the area, if published. The use of VFR waypoints does not relieve the pilot of any responsibility to comply with the operational requirements of 14 CFR part 91.

VFR waypoint names (for computer entry and flight plans) consist of five letters beginning with the letters "VP" and are retrievable from navigation databases. The VFR waypoint names are not intended to be pronounceable, and they are not for use in ATC communications. On VFR charts, a stand-alone VFR waypoint is portrayed using the same four-point star symbol used for IFR waypoints. VFR waypoint collocated with a visual checkpoint on the chart is identified by a small magenta flag symbol. A VFR waypoint collocated with a visual checkpoint is pronounceable based on the name of the visual checkpoint and may be used for ATC communications. Each VFR waypoint name appears in parentheses adjacent to the geographic location on the chart. Latitude/longitude data for all established VFR waypoints may be found in the appropriate regional Chart Supplement U.S.

When filing VFR flight plans, use the five-letter identifier as a waypoint in the route of flight section if there is an intended course change at that point or if used to describe the planned route of flight. This VFR filing would be similar to VOR use in a route of flight. Pilots must use the VFR waypoints only when operating under VFR conditions.

Any VFR waypoints intended for use during a flight should be loaded into the receiver while on the ground and prior to departure. Once airborne, pilots should avoid programming routes or VFR waypoint chains into their receivers.

Pilots should be especially vigilant for other traffic while operating near VFR waypoints. The same effort to see and avoid other aircraft near VFR waypoints is necessary, as is the case when operating near VORs and NDBs. In fact, the increased accuracy of navigation through the use of GPS demands even greater vigilance as there are fewer off-course deviations among different pilots and receivers. When operating near a VFR waypoint, use all available ATC services, even if outside a class of airspace where communications are required. Regardless of the class of airspace, monitor the available ATC frequency closely for information on other aircraft operating in the vicinity. It is also a good idea to turn on landing light(s) when operating near a VFR waypoint to make the aircraft more conspicuous to other pilots, especially when visibility is reduced.

Lost Procedures

Getting lost in flight is a potentially dangerous situation, especially when low on fuel. If a pilot becomes lost, there are some good common sense procedures to follow. If a town or city cannot be seen, the first thing to do is climb, being mindful of traffic and weather conditions. An increase in altitude increases radio and navigation reception range and also increases radar coverage. If flying near a town or city, it may be possible to read the name of the town on a water tower.

If the aircraft has a navigational radio, such as a VOR or ADF receiver, it can be possible to determine position by plotting an azimuth from two or more navigational facilities. If GPS is installed, or a pilot has a portable aviation GPS on board, it can be used to determine the position and the location of the nearest airport.

Communicate with any available facility using frequencies shown on the sectional chart. If contact is made with a controller, radar vectors may be offered. Other facilities may offer direction finding (DF) assistance. To use this procedure, the controller requests the pilot to hold down the transmit button for a few seconds and then release it. The controller may ask the pilot to change directions a few times and repeat the transmit procedure. This gives the controller enough

information to plot the aircraft position and then give vectors to a suitable landing site. If the situation becomes threatening, transmit the situation on the emergency frequency 121.5 MHz and set the transponder to 7700. Most facilities, and even airliners, monitor the emergency frequency.

Flight Diversion

There may come a time when a pilot is not able to make it to the planned destination. This can be the result of unpredicted weather conditions, a system malfunction, or poor preflight planning. In any case, the pilot needs to be able to safely and efficiently divert to an alternate destination. Risk management procedures become a priority during any type of flight diversion and should be used the pilot. For example, the hazards of inadvertent VFR into IMC involve a risk that the pilot can identify and assess and then mitigate through a pre-planned or in-flight diversion around hazardous weather. Before any cross-country flight, check the charts for airports or suitable landing areas along or near the route of flight. Also, check for navigational aids that can be used during a diversion. Risk management is explained in greater detail in Chapter 2, Aeronautical Decision-making.

Computing course, time, speed, and distance information in flight requires the same computations used during preflight planning. However, because of the limited flight deck space and because attention must be divided between flying the aircraft, making calculations, and scanning for other aircraft, take advantage of all possible shortcuts and rule-of-thumb computations.

When in flight, it is rarely practical to actually plot a course on a sectional chart and mark checkpoints and distances. Furthermore, because an alternate airport is usually not very far from your original course, actual plotting is seldom necessary.

The course to an alternate destination can be measured accurately with a protractor or plotter but can also be measured with reasonable accuracy using a straightedge and the compass rose depicted around VOR stations. This approximation can be made on the basis of a radial from a nearby VOR or an airway that closely parallels the course to your alternate destination. However, remember that the magnetic heading associated with a VOR radial or printed airway is outbound from the station. To find the course to the station, it may be necessary to determine the reciprocal of that heading. It is typically easier to navigate to an alternate airport that has a VOR or NDB facility on the field.

After selecting the most appropriate alternate destination, approximate the magnetic course to the alternate using a compass rose or airway on the sectional chart. If time permits, try to start the diversion over a prominent ground feature.

However, in an emergency, divert promptly toward your alternate destination. Attempting to complete all plotting, measuring, and computations involved before diverting to the alternate destination may only aggravate an actual emergency.

Once established on course, note the time, and then use the winds aloft nearest to your diversion point to calculate a heading and GS. Once a GS has been calculated, determine a new arrival time and fuel consumption. Give priority to flying the aircraft while dividing attention between navigation and planning. When determining an altitude to use while diverting, consider cloud heights, winds, terrain, and radio reception.

Chapter Summary

This chapter has discussed the fundamentals of VFR navigation. Beginning with an introduction to the charts that can be used for navigation to the more technically advanced concept of GPS, there is one aspect of navigation that remains the same—the pilot is responsible for proper planning and the execution of that planning to ensure a safe flight.

Chapter 17

Aeromedical Factors

Introduction

It is important for a pilot to be aware of the mental and physical standards required for the type of flying performed. This chapter provides information on medical certification and on a variety of aeromedical factors related to flight activities.

Fovea centralis

Optic disk (blind spot)

Rod concentration

Lens

Iris

Optic nerve

Retina

PUPIL

The **pupil** (aperture) is the opening at the center of the iris. The size of the pupil is adjusted to control the amount of light entering the eye.

CORNEA

Light passes through the **cornea** (the transparent window on the front of the eye) and then through the **lens** to focus on the retina.

YAW

PITCH

ROLL

ROLL

PITCH

YAW

Ampulla of se

Sem

The semicircular tubes are arranged at approximately right angles to each other in the roll, pitch, and yaw axes.

Endolymph fluid

Vestibular nerve

Cup

Hair cells

Obtaining a Medical Certificate

Most pilots must have a valid medical certificate to exercise the privileges of their airman certificates. Glider and free balloon pilots are not required to hold a medical certificate. Sport pilots may hold either a medical certificate or a valid state driver's license. Regardless of whether a medical certificate or drivers license is required, 14 CFR 61.53 requires every pilot not to act as a crewmember if they know, or have reason to know, of any medical condition that would make them unable to operate the aircraft in a safe manner.

Acquisition of a medical certificate requires an examination by an aviation medical examiner (AME), a physician with training in aviation medicine designated by the Civil Aerospace Medical Institute (CAMI). There are three classes of medical certificates. The class of certificate needed depends on the type of flying the pilot plans to perform.

A third-class medical certificate is required for a private or recreational pilot certificate. It is valid for 5 years for those individuals who have not reached the age of 40; otherwise it is valid for 2 years. A commercial pilot certificate requires at least a second-class medical certificate, which is valid for 1 year. First-class medical certificates are required for airline transport pilots and are valid for one year if the airman is 40 or younger; 40 and older it is valid for 6 months.

The standards are more rigorous for the higher classes of certificates. A pilot with a higher class medical certificate has met the requirements for the lower classes as well. Since the required medical class applies only when exercising the privileges of the pilot certificate for which it is required, a first-class medical certificate would be valid for 1 year if exercising the privileges of a commercial certificate and 2 or 5 years, as appropriate, for exercising the privileges of a private or recreational certificate. The same applies for a second-class medical certificate. The standards for medical certification are contained in Title 14 of the Code of Federal Regulations (14 CFR) part 67 and the requirements for obtaining medical certificates can be found in 14 CFR part 61.

Students who have physical limitations, such as impaired vision, loss of a limb, or hearing impairment may be issued a medical certificate valid for "student pilot privileges only" while learning to fly. Pilots with disabilities may require special equipment to be installed in the aircraft, such as hand controls for pilots with paraplegia. Some disabilities necessitate a limitation on the individual's certificate; for example, impaired hearing would require the limitation "not valid for flight requiring the use of radio." When all the knowledge, experience, and proficiency requirements have been met and a student can demonstrate the ability to operate the aircraft with the normal level of safety, a "statement of demonstrated ability" (SODA) can be issued. This waiver, or SODA, is valid as long as the physical impairment does not worsen. Contact the local Flight Standards District Office (FSDO) for more information on this subject.

The FAA medical standards, 14 CFR part 67, specify fifteen medical conditions that are considered disqualifying by "history or clinical diagnosis." Regardless of when one of these conditions was diagnosed and treated, an airman may not be issued a medical certificate except through a process called a "Special Issuance Authorization," as explained in 14 CFR part 67, section 67.401. A special issuance is a discretionary issuance by the FAA Federal Air Surgeon and requires satisfactory completion of special testing determined by the FAA to demonstrate that an airman is safe to fly for the duration of the medical certificate issued. The specific disqualifying conditions include:

- Diabetes mellitus requiring oral hypoglycemic medication or insulin
- Angina pectoris
- Coronary heart disease that has been treated or, if untreated, that has been symptomatic or clinically significant
- Myocardial infarction
- Cardiac valve replacement
- Permanent cardiac pacemaker
- Heart replacement
- Psychosis
- Bipolar disorder
- Personality disorder that is severe enough to have repeatedly manifested itself by overt acts
- Substance dependence (including alcohol)
- Substance abuse
- Epilepsy
- Disturbance of consciousness and without satisfactory explanation of cause
- Transient loss of control of nervous system function(s) without satisfactory explanation of cause

However, this list includes only the mandatory disqualifying conditions. There are many other medical conditions that fall into the General Medical Condition section of the regulations that are considered by the FAA to be disqualifying even though they are not stated in the regulations. Conditions such as cancer, kidney stones, neurologic and neuromuscular conditions including Parkinson's disease and multiple sclerosis, certain blood disorders, and other conditions that may progress over time require review by the FAA before a medical certificate may be issued.

The important thing to remember is that with very few exceptions, all disqualifying medical conditions may be considered for special issuance. If you can present satisfactory medical documentation to the FAA that your condition is stable, the chances are good that you will be able to qualify for an Authorization.

Health and Physiological Factors Affecting Pilot Performance

A number of health factors and physiological effects can be linked to flying. Some are minor, while others are important enough to require special attention to ensure safety of flight. In some cases, physiological factors can lead to inflight emergencies. Some important medical factors that a pilot should be aware of include hypoxia, hyperventilation, middle ear and sinus problems, spatial disorientation, motion sickness, carbon monoxide (CO) poisoning, stress and fatigue, dehydration, and heatstroke. Other subjects include the effects of alcohol and drugs, anxiety, and excess nitrogen in the blood after scuba diving.

Hypoxia

Hypoxia means "reduced oxygen" or "not enough oxygen." Although any tissue will die if deprived of oxygen long enough, the greatest concern regarding hypoxia during flight is lack of oxygen to the brain, since it is particularly vulnerable to oxygen deprivation. Any reduction in mental function while flying can result in life-threatening errors. Hypoxia can be caused by several factors, including an insufficient supply of oxygen, inadequate transportation of oxygen, or the inability of the body tissues to use oxygen. The forms of hypoxia are based on their causes:

- Hypoxic hypoxia
- Hypemic hypoxia
- Stagnant hypoxia
- Histotoxic hypoxia

Hypoxic Hypoxia

Hypoxic hypoxia is a result of insufficient oxygen available to the body as a whole. A blocked airway and drowning are obvious examples of how the lungs can be deprived of oxygen, but the reduction in partial pressure of oxygen at high altitude is an appropriate example for pilots. Although the percentage of oxygen in the atmosphere is constant, its partial pressure decreases proportionately as atmospheric pressure decreases. As an aircraft ascends during flight, the percentage of each gas in the atmosphere remains the same, but there are fewer molecules available at the pressure required for them to pass between the membranes in the respiratory system. This decrease in number of oxygen molecules at sufficient pressure can lead to hypoxic hypoxia.

Dangers of Transporting Dry Ice

Sublimation is a process in which a substance transitions from a solid to a gaseous state without passing through an intermediate liquid state. Dry ice sublimates into large quantities of CO_2 gas, which can rapidly displace oxygen-containing air and potentially cause hypoxia via carbon dioxide intoxication. Case studies have shown that both illness and death can be caused by occupational and/or unintentional exposure when transporting dry ice in small, confined spaces such as a flightdeck or airplane. Exposure to high concentration of CO_2 gas may lead to increased respiration, tachycardia, cardiac arrhythmia, and unconsciousness. Exposure to concentration of CO_2 gas in excess of 10 percent may cause convulsions, coma, and/or death.

The tendency of dry ice to rapidly sublimate also means that without proper ventilation, it can rapidly pressurize. For this reason, dry ice should never be placed inside a sealed transport container (i.e., leak-proof secondary container) and must be placed within an outer shipping container or storage container that allows adequate ventilation to release the CO_2 gas and avoid pressurization. Sealing dry ice within a leak-proof container may result in explosion of the container potentially leading to serious physical injury or death.

Hypemic Hypoxia

Hypemic hypoxia occurs when the blood is not able to take up and transport a sufficient amount of oxygen to the cells in the body. Hypemic means "not enough blood." This type of hypoxia is a result of oxygen deficiency in the blood, rather than a lack of inhaled oxygen, and can be caused by a variety of factors. It may be due to reduced blood volume (from severe bleeding), or it may result from certain blood diseases, such as anemia. More often, hypemic hypoxia occurs because hemoglobin, the actual blood molecule that transports oxygen, is chemically unable to bind oxygen molecules. The most common form of hypemic hypoxia is CO poisoning. This is explained in greater detail later in this chapter. Hypemic hypoxia can also be caused by the loss of blood due to blood donation. Blood volume can require several weeks to return to normal following a donation. Although the effects of the blood loss are slight at ground level, there are risks when flying during this time.

Stagnant Hypoxia

Stagnant means "not flowing," and stagnant hypoxia or ischemia results when the oxygen-rich blood in the lungs is not moving, for one reason or another, to the tissues that need it. An arm or leg "going to sleep" because the blood flow has accidentally been shut off is one form of stagnant hypoxia. This kind of hypoxia can also result from shock, the heart failing to pump blood effectively, or a constricted artery. During flight, stagnant hypoxia can occur with excessive acceleration of gravity (Gs). Cold temperatures can also reduce circulation and decrease the blood supplied to extremities.

Histotoxic Hypoxia

The inability of the cells to effectively use oxygen is defined as histotoxic hypoxia. "Histo" refers to tissues or cells, and "toxic" means poisonous. In this case, enough oxygen is being transported to the cells that need it, but they are unable to make use of it. This impairment of cellular respiration can be caused by alcohol and other drugs, such as narcotics and poisons. Research has shown that drinking one ounce of alcohol can equate to an additional 2,000 feet of physiological altitude.

Symptoms of Hypoxia

High-altitude flying can place a pilot in danger of becoming hypoxic. Oxygen starvation causes the brain and other vital organs to become impaired. The first symptoms of hypoxia can include euphoria and a carefree feeling. With increased oxygen starvation, the extremities become less responsive and flying becomes less coordinated. The symptoms of hypoxia vary with the individual, but common symptoms include:

- Cyanosis (blue fingernails and lips)
- Headache
- Decreased response to stimuli and increased reaction time
- Impaired judgment
- Euphoria
- Visual impairment
- Drowsiness
- Lightheaded or dizzy sensation
- Tingling in fingers and toes
- Numbness

As hypoxia worsens, the field of vision begins to narrow and instrument interpretation can become difficult. Even with all these symptoms, the effects of hypoxia can cause a pilot to have a false sense of security and be deceived into believing everything is normal.

Treatment of Hypoxia

Treatment for hypoxia includes flying at lower altitudes and/or using supplemental oxygen. All pilots are susceptible to the effects of oxygen starvation, regardless of physical endurance or acclimatization. When flying at high altitudes, it is paramount that oxygen be used to avoid the effects of hypoxia. The term "time of useful consciousness" describes the maximum time the pilot has to make rational, life-saving decisions and carry them out at a given altitude without supplemental oxygen. As altitude increases above 10,000 feet, the symptoms of hypoxia increase in severity, and the time of useful consciousness rapidly decreases. *[Figure 17-1]* Since symptoms of hypoxia can be different for each individual, the ability to recognize hypoxia can be greatly improved by experiencing and witnessing the effects of it during an altitude chamber "flight." The Federal Aviation Administration (FAA) provides this opportunity through aviation physiology training, which is conducted at the FAA CAMI in Oklahoma City, Oklahoma, and at many military facilities across the United States. For information about the FAA's one-day physiological training course with altitude chamber and vertigo demonstrations, visit the FAA website at www.faa.gov.

Hyperventilation

Hyperventilation is the excessive rate and depth of respiration leading to abnormal loss of carbon dioxide from the blood. This condition occurs more often among pilots than is generally recognized. It seldom incapacitates completely, but it causes disturbing symptoms that can alarm the uninformed pilot. In such cases, increased breathing rate and anxiety further aggravate the problem. Hyperventilation can lead to unconsciousness due to the respiratory system's overriding mechanism to regain control of breathing.

Pilots encountering an unexpected stressful situation may subconsciously increase their breathing rate. If flying at higher altitudes, either with or without oxygen, a pilot may have a tendency to breathe more rapidly than normal, which often leads to hyperventilation.

Since many of the symptoms of hyperventilation are similar to those of hypoxia, it is important to correctly diagnose and treat the proper condition. If using supplemental oxygen, check the equipment and flow rate to ensure the symptoms are

Altitude	Time of useful consciousness
45,000 feet MSL	9 to 15 seconds
40,000 feet MSL	15 to 20 seconds
35,000 feet MSL	30 to 60 seconds
30,000 feet MSL	1 to 2 minutes
28,000 feet MSL	2½ to 3 minutes
25,000 feet MSL	3 to 5 minutes
22,000 feet MSL	5 to 10 minutes
20,000 feet MSL	30 minutes or more

Figure 17-1. *Time of useful consciousness.*

not hypoxia related. Common symptoms of hyperventilation include:

- Visual impairment
- Unconsciousness
- Lightheaded or dizzy sensation
- Tingling sensations
- Hot and cold sensations
- Muscle spasms

The treatment for hyperventilation involves restoring the proper carbon dioxide level in the body. Breathing normally is both the best prevention and the best cure for hyperventilation. In addition to slowing the breathing rate, breathing into a paper bag or talking aloud helps to overcome hyperventilation. Recovery is usually rapid once the breathing rate is returned to normal.

Middle Ear and Sinus Problems

During climbs and descents, the free gas formerly present in various body cavities expands due to a difference between the pressure of the air outside the body and that of the air inside the body. If the escape of the expanded gas is impeded, pressure builds up within the cavity and pain is experienced. Trapped gas expansion accounts for ear pain and sinus pain, as well as a temporary reduction in the ability to hear.

The middle ear is a small cavity located in the bone of the skull. It is closed off from the external ear canal by the eardrum. Normally, pressure differences between the middle ear and the outside world are equalized by a tube leading from inside each ear to the back of the throat on each side called the Eustachian tube. These tubes are usually closed but open during chewing, yawning, or swallowing to equalize pressure. Even a slight difference between external pressure and middle ear pressure can cause discomfort. *[Figure 17-2]*

During a climb, middle ear air pressure may exceed the pressure of the air in the external ear canal causing the eardrum to bulge outward. Pilots become aware of this pressure change when they experience alternate sensations of "fullness" and "clearing." During descent, the reverse happens. While the pressure of the air in the external ear canal increases, the middle ear cavity, which equalized with the lower pressure at altitude, is at lower pressure than the external ear canal. This results in the higher outside pressure causing the eardrum to bulge inward.

This condition can be more difficult to relieve due to the fact that the partial vacuum tends to constrict the walls of the Eustachian tube. To remedy this often painful condition, which also causes a temporary reduction in hearing

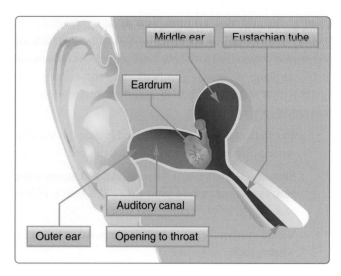

Figure 17-2. *The Eustachian tube allows air pressure to equalize in the middle ear.*

sensitivity, pinch the nostrils shut, close the mouth and lips, and blow slowly and gently into the mouth and nose.

This procedure forces air through the Eustachian tube into the middle ear. It may not be possible to equalize the pressure in the ears if a pilot has a cold, an ear infection, or sore throat. A flight in this condition can be extremely painful, as well as damaging to the eardrums. If experiencing minor congestion, nose drops or nasal sprays may reduce the risk of a painful ear blockage. Before using any medication, check with an AME to ensure that it will not affect the ability to fly.

In a similar way, air pressure in the sinuses equalizes with the pressure in the flight deck through small openings that connect the sinuses to the nasal passages. An upper respiratory infection, such as a cold or sinusitis, or a nasal allergic condition can produce enough congestion around an opening to slow equalization. As the difference in pressure between the sinuses and the flight deck increases, congestion may plug the opening. This "sinus block" occurs most frequently during descent. Slow descent rates can reduce the associated pain. A sinus block can occur in the frontal sinuses, located above each eyebrow, or in the maxillary sinuses, located in each upper cheek. It usually produces excruciating pain over the sinus area. A maxillary sinus block can also make the upper teeth ache. Bloody mucus may discharge from the nasal passages.

Sinus block can be avoided by not flying with an upper respiratory infection or nasal allergic condition. Adequate protection is usually not provided by decongestant sprays or drops to reduce congestion around the sinus openings. Oral decongestants have side effects that can impair pilot performance. If a sinus block does not clear shortly after landing, a physician should be consulted.

Spatial Disorientation and Illusions

Spatial disorientation specifically refers to the lack of orientation with regard to the position, attitude, or movement of the airplane in space. The body uses three integrated systems that work together to ascertain orientation and movement in space.

- Vestibular system—organs found in the inner ear that sense position by the way we are balanced

- Somatosensory system—nerves in the skin, muscles, and joints that, along with hearing, sense position based on gravity, feeling, and sound

- Visual system—eyes, which sense position based on what is seen

All this information comes together in the brain and, most of the time, the three streams of information agree, giving a clear idea of where and how the body is moving. Flying can sometimes cause these systems to supply conflicting information to the brain, which can lead to disorientation. During flight in visual meteorological conditions (VMC), the eyes are the major orientation source and usually prevail over false sensations from other sensory systems. When these visual cues are removed, as they are in instrument meteorological conditions (IMC), false sensations can cause a pilot to quickly become disoriented.

The vestibular system in the inner ear allows the pilot to sense movement and determine orientation in the surrounding environment. In both the left and right inner ear, three semicircular canals are positioned at approximate right angles to each other. *[Figure 17-3]* Each canal is filled with fluid and has a section full of fine hairs. Acceleration of the inner ear in any direction causes the tiny hairs to deflect, which in turn stimulates nerve impulses, sending messages to the brain. The vestibular nerve transmits the impulses from the utricle, saccule, and semicircular canals to the brain to interpret motion.

The somatosensory system sends signals from the skin, joints, and muscles to the brain that are interpreted in relation to the Earth's gravitational pull. These signals determine posture. Inputs from each movement update the body's position to the brain on a constant basis. "Seat of the pants" flying is largely dependent upon these signals. Used in conjunction with visual and vestibular clues, these sensations can be fairly reliable. However, the body cannot distinguish between acceleration forces due to gravity and those resulting from maneuvering the aircraft, which can lead to sensory illusions and false impressions of an aircraft's orientation and movement.

Under normal flight conditions, when there is a visual reference to the horizon and ground, the sensory system in the inner ear helps to identify the pitch, roll, and yaw movements of the aircraft. When visual contact with the horizon is lost, the vestibular system becomes unreliable. Without visual references outside the aircraft, there are many situations in which combinations of normal motions and forces create convincing illusions that are difficult to overcome.

Prevention is usually the best remedy for spatial disorientation. Unless a pilot has many hours of training in instrument flight, flight should be avoided in reduced visibility or at night when the horizon is not visible. A pilot can reduce susceptibility to disorienting illusions through training and awareness and learning to rely totally on flight instruments.

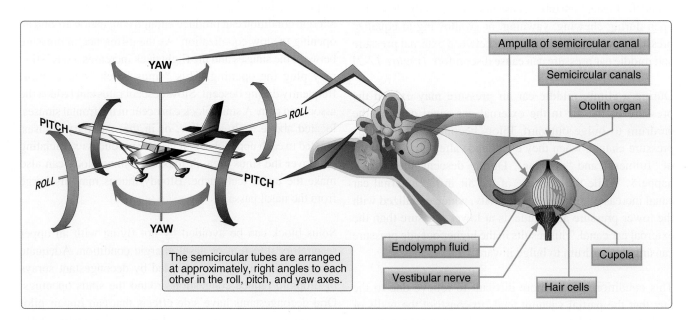

Figure 17-3. *The semicircular canals lie in three planes and sense motions of roll, pitch, and yaw.*

Vestibular Illusions

The Leans

A condition called the leans, is the most common illusion during flight and is caused by a sudden return to level flight following a gradual and prolonged turn that went unnoticed by the pilot. The reason a pilot can be unaware of such a gradual turn is that human exposure to a rotational acceleration of 2 degrees per second or lower is below the detection threshold of the semicircular canals. [Figure 17-4] Leveling the wings after such a turn may cause an illusion that the aircraft is banking in the opposite direction. In response to such an illusion, a pilot may lean in the direction of the original turn in a corrective attempt to regain the perception of a correct vertical posture.

Coriolis Illusion

The "coriolis illusion" occurs when a pilot has been in a turn long enough for the fluid in the ear canal to move at the same speed as the canal. A movement of the head in a different plane, such as looking at something in a different part of the flight deck, may set the fluid moving, creating the illusion of turning or accelerating on an entirely different axis. This action causes the pilot to think the aircraft is performing a maneuver it is not. The disoriented pilot may maneuver the aircraft into a dangerous attitude in an attempt to correct the aircraft's perceived attitude.

For this reason, it is important that pilots develop an instrument cross-check or scan that involves minimal head movement. Take care when retrieving charts and other objects in the flight deck—if something is dropped, retrieve it with minimal head movement and be alert for the coriolis illusion.

Graveyard Spiral

As in other illusions, a pilot in a prolonged coordinated, constant-rate turn may experience the illusion of not turning. During the recovery to level flight, the pilot will then experience the sensation of turning in the opposite direction causing the disoriented pilot to return the aircraft to its original turn. Because an aircraft tends to lose altitude in turns unless the pilot compensates for the loss in lift, the pilot may notice a loss of altitude. The absence of any sensation of turning creates the illusion of being in a level descent. The pilot may pull back on the controls in an attempt to climb or stop the descent. This action tightens the spiral and increases the loss of altitude; this illusion is referred to as a "graveyard spiral." [Figure 17-5] This may lead to a loss of aircraft control.

Somatogravic Illusion

A rapid acceleration, such as experienced during takeoff, stimulates the otolith organs in the same way as tilting the head backwards. This action may create what is known as the "somatogravic illusion" of being in a nose-up attitude, especially in conditions with poor visual references. The disoriented pilot may push the aircraft into a nose-low or dive attitude. A rapid deceleration by quick reduction of the throttle(s) can have the opposite effect, with the disoriented pilot pulling the aircraft into a nose-up or stall attitude.

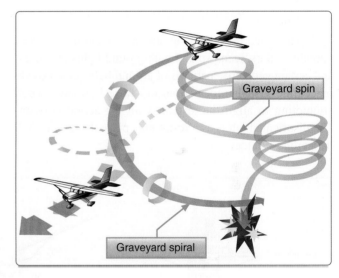

Figure 17-5. *Graveyard spiral.*

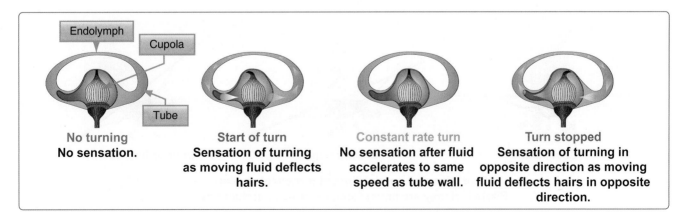

Figure 17-4. *Human sensation of angular acceleration.*

Inversion Illusion

An abrupt change from climb to straight-and-level flight can stimulate the otolith organs enough to create the illusion of tumbling backwards, known as "inversion illusion." The disoriented pilot may push the aircraft abruptly into a nose-low attitude, which may intensify this illusion.

Elevator Illusion

An abrupt upward vertical acceleration, as can occur in an updraft, can stimulate the otolith organs to create the illusion of being in a climb. This is known as "elevator illusion." The disoriented pilot may push the aircraft into a nose-low attitude. An abrupt downward vertical acceleration, usually in a downdraft, has the opposite effect with the disoriented pilot pulling the aircraft into a nose-up attitude.

Visual Illusions

Visual illusions are especially hazardous because pilots rely on their eyes for correct information. Two illusions that lead to spatial disorientation, false horizon and autokinesis, affect the visual system only.

False Horizon

A sloping cloud formation, an obscured horizon, an aurora borealis, a dark scene spread with ground lights and stars, and certain geometric patterns of ground lights can provide inaccurate visual information, or "false horizon," when attempting to align the aircraft with the actual horizon. The disoriented pilots as a result may place the aircraft in a dangerous attitude.

Autokinesis

When flying in the dark, a stationary light may appear to move if it is stared at for a prolonged period of time. As a result, a pilot may attempt to align the aircraft with the perceived moving light potentially causing him/her to lose control of the aircraft. This illusion is known as "autokinesis."

Postural Considerations

The postural system sends signals from the skin, joints, and muscles to the brain that are interpreted in relation to the Earth's gravitational pull. These signals determine posture. Inputs from each movement update the body's position to the brain on a constant basis. "Seat of the pants" flying is largely dependent upon these signals. Used in conjunction with visual and vestibular clues, these sensations can be fairly reliable. However, because of the forces acting upon the body in certain flight situations, many false sensations can occur due to acceleration forces overpowering gravity. [Figure 17-6] These situations include uncoordinated turns, climbing turns, and turbulence.

Demonstration of Spatial Disorientation

There are a number of controlled aircraft maneuvers a pilot can perform to experiment with spatial disorientation. While each maneuver normally creates a specific illusion, any false sensation is an effective demonstration of disorientation. Thus, even if there is no sensation during any of these maneuvers, the absence of sensation is still an effective demonstration because it illustrates the inability to detect bank or roll.

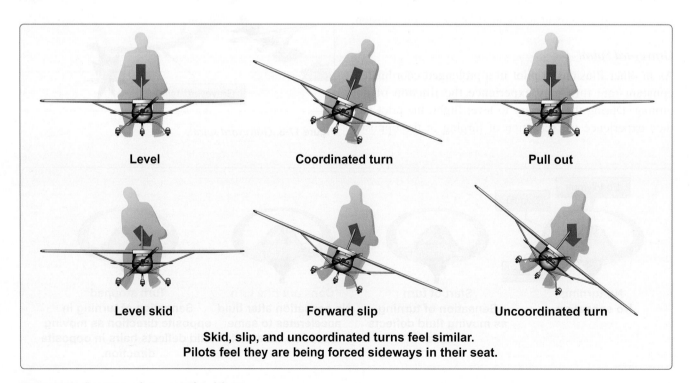

Level   **Coordinated turn**   **Pull out**

Level skid   **Forward slip**   **Uncoordinated turn**

Skid, slip, and uncoordinated turns feel similar.
Pilots feel they are being forced sideways in their seat.

Figure 17-6. *Sensations from centrifugal force.*

There are several objectives in demonstrating these various maneuvers.

1. They teach pilots to understand the susceptibility of the human system to spatial disorientation.

2. They demonstrate that judgments of aircraft attitude based on bodily sensations are frequently false.

3. They help decrease the occurrence and degree of disorientation through a better understanding of the relationship between aircraft motion, head movements, and resulting disorientation.

4. They help instill a greater confidence in relying on flight instruments for assessing true aircraft attitude.

A pilot should not attempt any of these maneuvers at low altitudes or in the absence of an instructor pilot or an appropriate safety pilot.

Climbing While Accelerating

With the pilot's eyes closed, the instructor pilot maintains approach airspeed in a straight-and-level attitude for several seconds, then accelerates while maintaining straight-and-level attitude. The usual illusion during this maneuver, without visual references, is that the aircraft is climbing.

Climbing While Turning

With the pilot's eyes still closed and the aircraft in a straight-and-level attitude, the instructor pilot now executes, with a relatively slow entry, a well coordinated turn of about 1.5 positive G (approximately 50° bank) for 90°. While in the turn, without outside visual references and under the effect of the slight positive G, the usual illusion produced is that of a climb. Upon sensing the climb, the pilot should immediately open the eyes to see that a slowly established, coordinated turn produces the same sensation as a climb.

Diving While Turning

Repeating the previous procedure, but with the pilot's eyes should be kept closed until recovery from the turn is approximately one-half completed, can create the illusion of diving while turning.

Tilting to Right or Left

While in a straight-and-level attitude, with the pilot's eyes closed, the instructor pilot executes a moderate or slight skid to the left with wings level. This creates the illusion of the body being tilted to the right.

Reversal of Motion

This illusion can be demonstrated in any of the three planes of motion. While straight and level, with the pilot's eyes closed, the instructor pilot smoothly and positively rolls the aircraft to approximately 45° bank attitude while maintaining heading and pitch attitude. This creates the illusion of a strong sense of rotation in the opposite direction. After this illusion is noted, the pilot should open his or her eyes and observe that the aircraft is in a banked attitude.

Diving or Rolling Beyond the Vertical Plane

This maneuver may produce extreme disorientation. While in straight-and-level flight, the pilot should sit normally, either with eyes closed or gaze lowered to the floor. The instructor pilot starts a positive, coordinated roll toward a 30° or 40° angle of bank. As this is in progress, the pilot tilts his or her head forward, looks to the right or left, then immediately returns his or her head to an upright position. The instructor pilot should time the maneuver so the roll is stopped as the pilot returns his or her head upright. An intense disorientation is usually produced by this maneuver, and the pilot experiences the sensation of falling downward into the direction of the roll.

In the descriptions of these maneuvers, the instructor pilot is doing the flying, but having the pilot do the flying can also be a very effective demonstration. The pilot should close his or her eyes and tilt the head to one side. The instructor pilot tells the pilot what control inputs to perform. The pilot then attempts to establish the correct attitude or control input with eyes closed and head tilted. While it is clear the pilot has no idea of the actual attitude, he or she will react to what the senses are saying. After a short time, the pilot will become disoriented and the instructor pilot will tell the pilot to look up and recover. This exercise allows the pilot to experience the disorientation while flying the aircraft.

Coping with Spatial Disorientation

To prevent illusions and their potentially disastrous consequences, pilots can:

1. Understand the causes of these illusions and remain constantly alert for them. Take the opportunity to experience spatial disorientation illusions in a device, such as a Barany chair, a Vertigon, or a Virtual Reality Spatial Disorientation Demonstrator.

2. Always obtain and understand preflight weather briefings.

3. Before flying in marginal visibility (less than 3 miles) or where a visible horizon is not evident, such as flight over open water during the night, obtain training and maintain proficiency in aircraft control by reference to instruments.

4. Do not fly into adverse weather conditions or into dusk or darkness unless proficient in the use of flight instruments. If intending to fly at night, maintain

night-flight currency and proficiency. Include cross-country and local operations at various airfields.

5. Ensure that when outside visual references are used, they are reliable, fixed points on the Earth's surface.

6. Avoid sudden head movement, particularly during takeoffs, turns, and approaches to landing.

7. Be physically tuned for flight into reduced visibility. Ensure proper rest, adequate diet, and, if flying at night, allow for night adaptation. Remember that illness, medication, alcohol, fatigue, sleep loss, and mild hypoxia are likely to increase susceptibility to spatial disorientation.

8. Most importantly, become proficient in the use of flight instruments and rely upon them. Trust the instruments and disregard your sensory perceptions.

The sensations that lead to illusions during instrument flight conditions are normal perceptions experienced by pilots. These undesirable sensations cannot be completely prevented, but through training and awareness, pilots can ignore or suppress them by developing absolute reliance on the flight instruments. As pilots gain proficiency in instrument flying, they become less susceptible to these illusions and their effects.

Optical Illusions

Of the senses, vision is the most important for safe flight. However, various terrain features and atmospheric conditions can create optical illusions. These illusions are primarily associated with landing. Since pilots must transition from reliance on instruments to visual cues outside the flight deck for landing at the end of an instrument approach, it is imperative that they be aware of the potential problems associated with these illusions and take appropriate corrective action. The major illusions leading to landing errors are described below.

Runway Width Illusion

A narrower-than-usual runway can create an illusion that the aircraft is at a higher altitude than it actually is, especially when runway length-to-width relationships are comparable. *[Figure 17-7]* The pilot who does not recognize this illusion will fly a lower approach, with the risk of striking objects along the approach path or landing short. A wider-than-usual runway can have the opposite effect with the risk of the pilot leveling out the aircraft high and landing hard or overshooting the runway.

Runway and Terrain Slopes Illusion

An upsloping runway, upsloping terrain, or both can create an illusion that the aircraft is at a higher altitude than it actually is. *[Figure 17-7]* The pilot who does not recognize this illusion will fly a lower approach. Downsloping runways and downsloping approach terrain can have the opposite effect.

Featureless Terrain Illusion

An absence of surrounding ground features, as in an overwater approach over darkened areas or terrain made featureless by snow, can create an illusion that the aircraft is at a higher altitude than it actually is. This illusion, sometimes referred to as the "black hole approach," causes pilots to fly a lower approach than is desired.

Water Refraction

Rain on the windscreen can create an illusion of being at a higher altitude due to the horizon appearing lower than it is. This can result in the pilot flying a lower approach.

Haze

Atmospheric haze can create an illusion of being at a greater distance and height from the runway. As a result, the pilot has a tendency to be low on the approach. Conversely, extremely clear air (clear bright conditions of a high attitude airport) can give the pilot the illusion of being closer than he or she actually is, resulting in a high approach that may result in an overshoot or go around. The diffusion of light due to water particles on the windshield can adversely affect depth perception. The lights and terrain features normally used to gauge height during landing become less effective for the pilot.

Fog

Flying into fog can create an illusion of pitching up. Pilots who do not recognize this illusion often steepen the approach abruptly.

Ground Lighting Illusions

Lights along a straight path, such as a road or lights on moving trains, can be mistaken for runway and approach lights. Bright runway and approach lighting systems, especially where few lights illuminate the surrounding terrain, may create the illusion of less distance to the runway. The pilot who does not recognize this illusion will often fly a higher approach.

How To Prevent Landing Errors Due to Optical Illusions

To prevent these illusions and their potentially hazardous consequences, pilots can:

1. Anticipate the possibility of visual illusions during approaches to unfamiliar airports, particularly at night or in adverse weather conditions. Consult airport

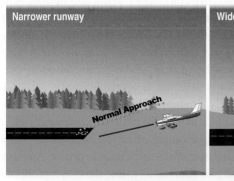
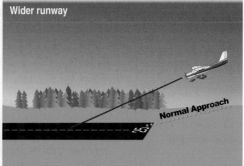

Runway width illusion

- A narrower-than-usual runway can create an illusion that the aircraft is higher than it actually is, leading to a lower approach.

- A wider-than-usual runway can create an illusion that the aircraft is lower than it actually is, leading to a higher approach.

Runway slope illusion

- A downsloping runway can create the illusion that the aircraft is lower than it actually is, leading to a higher approach.

- An upsloping runway can create the illusion that the aircraft is higher than it actually is, leading to a lower approach.

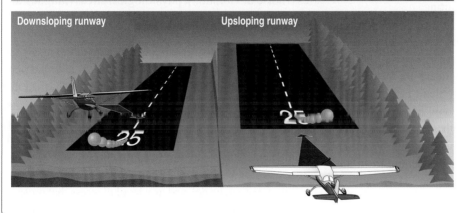

···· Normal approach
◄ Approach due to illusion

Figure 17-7. *Runway illusions.*

diagrams and the Chart Supplement U.S. (formerly Airport/Facility Directory) for information on runway slope, terrain, and lighting.

2. Make frequent reference to the altimeter, especially during all approaches, day and night.

3. If possible, conduct an aerial visual inspection of unfamiliar airports before landing.

4. Use Visual Approach Slope Indicator (VASI) or Precision Approach Path Indicator (PAPI) systems for a visual reference, or an electronic glideslope, whenever they are available.

5. Utilize the visual descent point (VDP) found on many nonprecision instrument approach procedure charts.

6. Recognize that the chances of being involved in an approach accident increase when an emergency or other activity distracts from usual procedures.

7. Maintain optimum proficiency in landing procedures.

In addition to the sensory illusions due to misleading inputs to the vestibular system, a pilot may also encounter various visual illusions during flight. Illusions rank among the most common factors cited as contributing to fatal aviation accidents.

Sloping cloud formations, an obscured horizon, a dark scene spread with ground lights and stars, and certain geometric patterns of ground light can create illusions of not being aligned correctly with the actual horizon. Various surface features and atmospheric conditions encountered in landing can create illusions of being on the wrong approach path. Landing errors due to these illusions can be prevented by anticipating them during approaches, inspecting unfamiliar airports before landing, using electronic glideslope or VASI systems when available, and maintaining proficiency in landing procedures.

Motion Sickness

Motion sickness, or airsickness, is caused by the brain receiving conflicting messages about the state of the body. A pilot may experience motion sickness during initial flights, but it generally goes away within the first few lessons. Anxiety and stress, which may be experienced at the beginning of flight training, can contribute to motion sickness. Symptoms of motion sickness include general discomfort, nausea, dizziness, paleness, sweating, and vomiting.

It is important to remember that experiencing airsickness is no reflection on one's ability as a pilot. If prone to motion sickness, let the flight instructor know, there are techniques that can be used to overcome this problem. For example, avoid lessons in turbulent conditions until becoming more comfortable in the aircraft or start with shorter flights and graduate to longer instruction periods. If symptoms of motion sickness are experienced during a lesson, opening fresh air vents, focusing on objects outside the airplane, and avoiding unnecessary head movements may help alleviate some of the discomfort. Although medications like Dramamine can prevent airsickness in passengers, they are not recommended while flying since they can cause drowsiness and other problems.

Carbon Monoxide (CO) Poisoning

CO is a colorless and odorless gas produced by all internal combustion engines. Attaching itself to the hemoglobin in the blood about 200 times more easily than oxygen, CO prevents the hemoglobin from carrying oxygen to the cells, resulting in hypemic hypoxia. The body requires up to 48 hours to dispose of CO. If severe enough, the CO poisoning can result in death. Aircraft heater vents and defrost vents may provide CO a passageway into the cabin, particularly if the engine exhaust system has a leak or is damaged. If a strong odor of exhaust gases is detected, assume that CO is present. However, CO may be present in dangerous amounts even if no exhaust odor is detected. Disposable, inexpensive CO detectors are widely available. In the presence of CO, these detectors change color to alert the pilot of the presence of CO. Some effects of CO poisoning are headache, blurred vision, dizziness, drowsiness, and/or loss of muscle power. Any time a pilot smells exhaust odor, or any time these symptoms are experienced, immediate corrective action should be taken including turning off the heater, opening fresh air vents and windows, and using supplemental oxygen, if available.

Tobacco smoke also causes CO poisoning. Smoking at sea level can raise the CO concentration in the blood and result in physiological effects similar to flying at 8,000 feet. Besides hypoxia, tobacco causes diseases and physiological debilitation that can be medically disqualifying for pilots.

Stress

Stress is the body's response to physical and psychological demands placed upon it. The body's reaction to stress includes releasing chemical hormones (such as adrenaline) into the blood and increasing metabolism to provide more energy to the muscles. Blood sugar, heart rate, respiration, blood pressure, and perspiration all increase. The term "stressor" is used to describe an element that causes an individual to experience stress. Examples of stressors include physical stress (noise or vibration), physiological stress (fatigue), and psychological stress (difficult work or personal situations).

Stress falls into two broad categories: acute (short term) and chronic (long term). Acute stress involves an immediate threat that is perceived as danger. This is the type of stress that triggers a "fight or flight" response in an individual, whether the threat is real or imagined. Normally, a healthy person can cope with acute stress and prevent stress overload. However, ongoing acute stress can develop into chronic stress.

Chronic stress can be defined as a level of stress that presents an intolerable burden, exceeds the ability of an individual to cope, and causes individual performance to fall sharply. Unrelenting psychological pressures, such as loneliness, financial worries, and relationship or work problems can produce a cumulative level of stress that exceeds a person's ability to cope with the situation. When stress reaches these levels, performance falls off rapidly. Pilots experiencing this level of stress are not safe and should not exercise their airman privileges. Pilots who suspect they are suffering from chronic stress should consult a physician.

Fatigue

Fatigue is frequently associated with pilot error. Some of the effects of fatigue include degradation of attention and concentration, impaired coordination, and decreased ability to communicate. These factors seriously influence the ability to make effective decisions. Physical fatigue results from sleep loss, exercise, or physical work. Factors such as stress and prolonged performance of cognitive work result in mental fatigue.

Like stress, fatigue falls into two broad categories: acute and chronic. Acute fatigue is short term and is a normal occurrence in everyday living. It is the kind of tiredness people feel after a period of strenuous effort, excitement, or lack of sleep. Rest after exertion and 8 hours of sound sleep ordinarily cures this condition.

A special type of acute fatigue is skill fatigue. This type of fatigue has two main effects on performance:

- Timing disruption—appearing to perform a task as usual, but the timing of each component is slightly off. This makes the pattern of the operation less smooth because the pilot performs each component as though it were separate, instead of part of an integrated activity.

- Disruption of the perceptual field—concentrating attention upon movements or objects in the center of vision and neglecting those in the periphery. This is accompanied by loss of accuracy and smoothness in control movements.

Acute fatigue has many causes, but the following are among the most important for the pilot:

- Mild hypoxia (oxygen deficiency)
- Physical stress
- Psychological stress
- Depletion of physical energy resulting from psychological stress
- Sustained psychological stress

Sustained psychological stress accelerates the glandular secretions that prepare the body for quick reactions during an emergency. These secretions make the circulatory and respiratory systems work harder, and the liver releases energy to provide the extra fuel needed for brain and muscle work. When this reserve energy supply is depleted, the body lapses into generalized and severe fatigue.

Acute fatigue can be prevented by proper diet and adequate rest and sleep. A well-balanced diet prevents the body from needing to consume its own tissues as an energy source. Adequate rest maintains the body's store of vital energy.

Chronic fatigue, extending over a long period of time, usually has psychological roots, although an underlying disease is sometimes responsible. Continuous high-stress levels produce chronic fatigue. Chronic fatigue is not relieved by proper diet and adequate rest and sleep and usually requires treatment by a physician. An individual may experience this condition in the form of weakness, tiredness, palpitations of the heart, breathlessness, headaches, or irritability. Sometimes chronic fatigue even creates stomach or intestinal problems and generalized aches and pains throughout the body. When the condition becomes serious enough, it leads to emotional illness.

If suffering from acute fatigue, stay on the ground. If fatigue occurs in the flight deck, no amount of training or experience can overcome the detrimental effects. Getting adequate rest is the only way to prevent fatigue from occurring. Avoid flying without a full night's rest, after working excessive hours, or after an especially exhausting or stressful day. Pilots who suspect they are suffering from chronic fatigue should consult a physician.

Exposure to Chemicals

When conducting preflight and post-flight inspections, pilots must verify that the fluid levels in their aircraft meet the levels specified for safe operations as stated in the Pilot's Operating Handbook. These fluids include, but are not limited to hydraulic fluid, engine oil, and fuel.

It is important that every pilot recognize the potential hazards of working with these fluids as well as the recommended first aid measures to follow should any of these fluids come in contact with their eyes, skin, and/or respiratory system. As the specific first aid measures for dealing with exposure to these chemicals can vary by chemical type, it is important that every pilot be familiar with the location and use of the Material Safety Data Sheet (MSDS) for each chemical they encounter.

The procedures described in the following sections are minimum guideline for first aid for each of the indicated scenarios. Ultimately, the pilot should consult the MSDS for first aid procedures specific to the type of chemical and exposure scenario.

Hydraulic Fluid

- Eye Contact—immediately flush the eyes with clean water and seek medical attention if irritation occurs.

- Skin Contact—remove all contaminated clothing and thoroughly cleanse the affected areas with mild soap and water or a waterless hand cleaner. If irritation or redness develops and persists, seek medical attention. Should the hydraulic fluid get into or under the skin, or into any other part of the body, regardless of the appearance of the wound or its size, seek medical attention immediately.

- Inhalation—if respiratory symptoms develop, move away from the source of exposure and into fresh air in a position comfortable for breathing. If symptoms persist, seek medical attention.

- Ingestion—first aid is not normally required; however, if swallowed and symptoms develop, seek medical attention.

Engine Oil

- Eye Contact—immediately flush the eyes with clean water and seek medical attention if irritation occurs.

- Skin Contact—remove all contaminated clothing and thoroughly cleanse the affected areas with soap and water. Launder contaminated clothing before reuse.

- Inhalation—move away from the source of exposure and into fresh air. If respiratory irritation, dizziness, nausea, or unconsciousness occurs, seek immediate medical attention. If breathing stops, assisted ventilation is required via a bag-valve-mask or cardiopulmonary resuscitation (CPR).

- Ingestion—seek immediate medical attention. If immediate medical attention is not available, contact a regional poison control center or emergency medical professional regarding the induction of vomiting or use of activated charcoal. Vomiting should never be induced to a person who is groggy or unconscious.

Fuel

- Eye Contact—immediately flush the eyes with clean water for at least 15 minutes and seek medical attention immediately.

- Skin Contact—remove all contaminated clothing and thoroughly cleanse the affected areas with mild soap and water or a waterless hand cleaner. If skin surface is damaged, apply a clean dressing and seek medical attention. If irritation or redness develops, seek medical attention. Launder contaminated clothing before reuse.

- Inhalation—move away from the source of exposure and into fresh air. If breathing stops, assisted ventilation is required via a bag-valve-mask or cardiopulmonary resuscitation (CPR). Once breathing is restored, the use of additional oxygen may be necessary. Seek medical attention immediately.

- Ingestion—seek immediate medical attention. Do not induce vomiting or take anything by mouth as this may cause the material to enter the lungs and cause severe lung damage. Should vomiting occur, keep head below the hips to reduce the risks of aspiration. Monitor for breathing difficulties. Rinse out any material which enters the mouth until the taste is dissipated.

Dehydration and Heatstroke

Dehydration is the term given to a critical loss of water from the body. Causes of dehydration are hot flight decks and flight lines, wind, humidity, and diuretic drinks—coffee, tea, alcohol, and caffeinated soft drinks. Some common signs of dehydration are headache, fatigue, cramps, sleepiness, and dizziness.

The first noticeable effect of dehydration is fatigue, which in turn makes top physical and mental performance difficult, if not impossible. Flying for long periods in hot summer temperatures or at high altitudes increases the susceptibility to dehydration because these conditions tend to increase the rate of water loss from the body.

To help prevent dehydration, drink two to four quarts of water every 24 hours. Since each person is physiologically different, this is only a guide. Most people are aware of the eight-glasses-a-day guide: If each glass of water is eight ounces, this equates to 64 ounces, which is two quarts. If this fluid is not replaced, fatigue progresses to dizziness, weakness, nausea, tingling of hands and feet, abdominal cramps, and extreme thirst.

The key for pilots is to be continually aware of their condition. Most people become thirsty with a 1.5 quart deficit or a loss of 2 percent of total body weight. This level of dehydration triggers the "thirst mechanism." The problem is that the thirst mechanism arrives too late and is turned off too easily. A small amount of fluid in the mouth turns this mechanism off and the replacement of needed body fluid is delayed.

Other steps to prevent dehydration include:

- Carrying a container in order to measure daily water intake.

- Staying ahead—not relying on the thirst sensation as an alarm. If plain water is not preferred, add some sport drink flavoring to make it more acceptable.

- Limiting daily intake of caffeine and alcohol (both are diuretics and stimulate increased production of urine).

Heatstroke is a condition caused by any inability of the body to control its temperature. Onset of this condition may be recognized by the symptoms of dehydration, but also has been known to be recognized only upon complete collapse.

To prevent these symptoms, it is recommended that an ample supply of water be carried and used at frequent intervals on any long flight, whether thirsty or not. The body normally absorbs water at a rate of 1.2 to 1.5 quarts per hour. Individuals should drink one quart per hour for severe heat stress conditions or one pint per hour for moderate stress conditions. If the aircraft has a canopy or roof window, wearing light-colored, porous clothing and a hat will help provide protection from the sun. Keeping the flight deck well ventilated aids in dissipating excess heat.

Alcohol

Alcohol impairs the efficiency of the human body. *[Figure 17-8]* Studies have shown that consuming alcohol is closely linked to performance deterioration. Pilots must make hundreds of decisions, some of them time-critical, during the course of a flight. The safe outcome of any flight depends on the ability to make the correct decisions and take the appropriate actions during routine occurrences, as well as abnormal situations. The influence of alcohol drastically reduces the chances of completing a flight without incident. Even in small amounts, alcohol can impair judgment, decrease sense of responsibility, affect coordination, constrict visual field, diminish memory, reduce reasoning ability, and lower attention span. As little as one ounce of alcohol can decrease the speed and strength of muscular reflexes, lessen the efficiency of eye movements while reading, and increase the frequency at which errors are committed. Impairments in vision and hearing can occur from consuming as little as one drink.

The alcohol consumed in beer and mixed drinks is ethyl alcohol, a central nervous system depressant. From a medical point of view, it acts on the body much like a general anesthetic. The "dose" is generally much lower and more slowly consumed in the case of alcohol, but the basic effects on the human body are similar. Alcohol is easily and quickly absorbed by the digestive tract. The bloodstream absorbs about 80 to 90 percent of the alcohol in a drink within 30 minutes when ingested on an empty stomach. The body requires about 3 hours to rid itself of all the alcohol contained in one mixed drink or one beer.

While experiencing a hangover, a pilot is still under the influence of alcohol. Although a pilot may think he or she is functioning normally, motor and mental response impairment is still present. Considerable amounts of alcohol can remain in the body for over 16 hours, so pilots should be cautious about flying too soon after drinking.

Altitude multiplies the effects of alcohol on the brain. When combined with altitude, the alcohol from two drinks may have the same effect as three or four drinks. Alcohol interferes with the brain's ability to utilize oxygen, producing a form of histotoxic hypoxia. The effects are rapid because alcohol passes quickly into the bloodstream. In addition, the brain is a highly vascular organ that is immediately sensitive to changes in the blood's composition. For a pilot, the lower oxygen availability at altitude and the lower capability of the brain to use the oxygen that is available can add up to a deadly combination.

Intoxication is determined by the amount of alcohol in the bloodstream. This is usually measured as a percentage by weight in the blood. 14 CFR part 91 requires that blood alcohol level be less than .04 percent and that 8 hours pass between drinking alcohol and piloting an aircraft. A pilot with a blood alcohol level of .04 percent or greater after 8 hours cannot fly until the blood alcohol falls below that amount. Even though blood alcohol may be well below .04 percent, a pilot cannot fly sooner than 8 hours after drinking alcohol.

Type Beverage	Typical Serving (oz)	Pure Alcohol Content (oz)
Table wine	4.0	.48
Light beer	12.0	.48
Aperitif liquor	1.5	.38
Champagne	4.0	.48
Vodka	1.0	.50
Whiskey	1.25	.50
0.01–0.05% (10–50 mg)	average individual appears normal	
0.03–0.12%* (30–120 mg)	mild euphoria, talkativeness, decreased inhibitions, decreased attention, impaired judgment, increased reaction time	
0.09–0.25% (90–250 mg)	emotional instability, loss of critical judgment, impairment of memory and comprehension, decreased sensory response, mild muscular incoordination	
0.18–0.30% (180–300 mg)	confusion, dizziness, exaggerated emotions (anger, fear, grief), impaired visual perception, decreased pain sensation, impaired balance, staggering gait, slurred speech, moderate muscular incoordination	
0.27–0.40% (270–400 mg)	apathy, impaired consciousness, stupor, significantly decreased response to stimulation, severe muscular incoordination, inability to stand or walk, vomiting, incontinence of urine and feces	
0.35–0.50% (350–500 mg)	unconsciousness, depressed or abolished reflexes, abnormal body temperature, coma, possible death from respiratory paralysis (450 mg or above)	

* Legal limit for motor vehicle operation in most states is 0.08 or 0.10% (80–100 mg of alcohol per dL of blood).

Figure 17-8. *Impairment scale with alcohol use.*

Although the regulations are quite specific, it is a good idea to be more conservative than the regulations.

Drugs

The Federal Aviation Regulations include no specific references to medication usage. Two regulations, though, are important to keep in mind. Title 14 of the CFR part 61, section 61.53 prohibits acting as pilot-in-command or in any other capacity as a required pilot flight crewmember, while that person:

1. Knows or has reason to know of any medical condition that would make the person unable to meet the requirement for the medical certificate necessary for the pilot operation, or

2. Is taking medication or receiving other treatment for a medical condition that results in the person being unable to meet the requirements for the medical certificate necessary for the pilot operation.

Further, 14 CFR part 91, section 91.17 prohibits the use of any drug that affects the person's faculties in any way contrary to safety.

There are several thousand medications currently approved by the U.S. Food and Drug Administration (FDA), not including OTC (over the counter) drugs. Virtually all medications have the potential for adverse side effects in some people. Additionally, herbal and dietary supplements, sport and energy boosters, and some other "natural" products are derived from substances often found in medications that could also have adverse side effects. While some individuals experience no side effects with a particular drug or product, others may be noticeably affected. The FAA regularly reviews FDA and other data to assure that medications found acceptable for aviation duties do not pose an adverse safety risk. Drugs that cause no apparent side effects on the ground can create serious problems at even relatively low altitudes. Even at typical general aviation altitudes, the changes in concentrations of atmospheric gases in the blood can enhance the effects of seemingly innocuous drugs that can result in impaired judgment, decision-making, and performance. In addition, fatigue, stress, dehydration, and inadequate nutrition can increase an airman's susceptibility to adverse effects from various drugs, even if they appeared to tolerate them in the past. If multiple medications are being taken at the same time, the adverse effects can be even more pronounced.

Another important consideration is that the medical condition for which a medication is prescribed may itself be disqualifying. The FAA will consider the condition in the context of risk for medical incapacitation, and the medication as well for cognitive impairment, and either or both could be found unacceptable for medical certification.

Some of the most commonly used OTC drugs, antihistamines and decongestants, have the potential to cause noticeable adverse side effects, including drowsiness and cognitive deficits. The symptoms associated with common upper respiratory infections, including the common cold, often suppress a pilot's desire to fly, and treating symptoms with a drug that causes adverse side effects only compounds the problem. Particularly, medications containing diphenhydramine (e.g., Benadryl) are known to cause drowsiness and have a prolonged half life, meaning the drugs stay in one's system for an extended time, which lengthens the time that side effects are present.

Many medications, such as tranquilizers, sedatives, strong pain relievers, and cough suppressants, have primary effects that may impair judgment, memory, alertness, coordination, vision, and the ability to make calculations. [Figure 17-9] Others, such as antihistamines, blood pressure drugs, muscle relaxants, and agents to control diarrhea and motion sickness, have side effects that may impair the same critical functions. Any medication that depresses the nervous system, such as a sedative, tranquilizer, or antihistamine, can make a pilot more susceptible to hypoxia.

Painkillers are grouped into two broad categories: analgesics and anesthetics. Analgesics are drugs that reduce pain, while anesthetics are drugs that deaden pain or cause loss of consciousness.

Over-the-counter analgesics, such as acetylsalicylic acid (aspirin), acetaminophen (Tylenol), and ibuprofen (Advil), have few side effects when taken in the correct dosage. Although some people are allergic to certain analgesics or may suffer from stomach irritation, flying usually is not restricted when taking these drugs. However, flying is almost always precluded while using prescription analgesics, such as drugs containing propoxyphene (e.g., Darvon), oxycodone (e.g., Percodan), meperidine (e.g., Demerol), and codeine, since these drugs are known to cause side effects, such as mental confusion, dizziness, headaches, nausea, and vision problems.

Anesthetic drugs are commonly used for dental and surgical procedures. Most local anesthetics used for minor dental and outpatient procedures wear off within a relatively short period of time. The anesthetic itself may not limit flying as much as the actual procedure and subsequent pain.

There is evidence taking illicit drugs significantly elevates the risk of having an aviation accident. Even though the Drug Enforcement Administration (DEA) defines marijuana as a Schedule I drug on its controlled substances list, states have taken steps to allow the possession, sale, and use of marijuana within their border. The FAA has stated, "Marijuana is an illicit drug per federal law and its use by airmen is prohibited."

Stimulants are drugs that excite the central nervous system and produce an increase in alertness and activity. Amphetamines, caffeine, and nicotine are all forms of stimulants. Common uses of these drugs include appetite suppression, fatigue reduction, and mood elevation. Some of these drugs may cause a stimulant reaction, even though this reaction is not their primary function. In some cases, stimulants can produce anxiety and mood swings, both of which are dangerous when flying.

Depressants are drugs that reduce the body's functioning in many areas. These drugs lower blood pressure, reduce mental processing, and slow motor and reaction responses. There are several types of drugs that can cause a depressing effect on the body, including tranquilizers, motion sickness medication, some types of stomach medication, decongestants, and antihistamines. The most common depressant is alcohol.

Substance	Generic Or Brand Name	Treatment for	Possible Side Effects
Alcohol	Beer Liquor Wine	N/A	Impaired judgment and perception Impaired coordination and motor control Reduced reaction time Impaired sensory perception Reduced intellectual functions Reduced tolerance to G-forces Inner-ear disturbance and spatial disorientation (up to 48 hours) Central nervous system depression
Nicotine	Cigars Cigarettes Pipe tobacco Chewing tobacco Snuff	N/A	Sinus and respiratory system infection and irritation Impaired night vision Hypertension Carbon monoxide poisoning (from smoking)
Amphetamines	Ritalin Obetrol Eskatrol	Obesity (diet pills) Tiredness	Prolonged wakefulness Nervousness Impaired vision Suppressed appetite Shakiness Excessive sweating Rapid heart rate Sleep disturbance Seriously impaired judgment
Caffeine	Coffee Tea Chocolate No-Doz	N/A	Impaired judgment Reduced reaction time Sleep disturbance Increased motor activity and tremors Hypertension Irregular heart rate Rapid heart rate Body dehydration (through increased urine output) Headaches
Antacid	Alka-2 Di-Gel Maalox	Stomach acids	Liberations of carbon dioxide at altitude (distension may cause acute abdominal pain and may mask other medical problems)
Antihistamines	Coricidin Contac Dristan Dimetapp Omade Chlor-Trimeton Diphenhydramine	Allergies Colds	Drowsiness and dizziness (sometimes recurring) Visual disturbances (when medications also contain antispasmodic drugs)
Aspirin	Bayer Bufferin Alka-Seltzer	Headaches Fevers Aches Pains	Irregular body temperature Variation in rate and depth of respiration Hypoxia and hyperventilation (two aspirin can contribute to) Nausea, ringing in ears, deafness, diarrhea, and hallucinations when taken in excessive dosages Corrosive action on the stomach lining Gastrointestinal problems Decreased clotting ability of the blood (clotting ability could be the difference between life and death in a survival situation)

Figure 17-9. *Adverse affects of various drugs.*

Some drugs that are classified as neither stimulants nor depressants have adverse effects on flying. For example, some antibiotics can produce dangerous side effects, such as balance disorders, hearing loss, nausea, and vomiting. While many antibiotics are safe for use while flying, the infection requiring the antibiotic may prohibit flying. In addition, unless specifically prescribed by a physician, do not take more than one drug at a time, and never mix drugs with alcohol because the effects are often unpredictable.

The dangers of illegal drugs also are well documented. Certain illegal drugs can have hallucinatory effects that occur days or weeks after the drug is taken. Obviously, these drugs have no place in the aviation community.

14 CFR prohibits pilots from performing crewmember duties while using any medication that affects the body in any way contrary to safety. The safest rule is not to fly as a crewmember while taking any medication, unless approved to do so by the FAA. If there is any doubt regarding the effects of any medication, consult an AME before flying.

Prior to each and every flight, all pilots must do a proper physical self-assessment to ensure safety. A great mnemonic, covered in Chapter 2 on Aeronautical Decision-Making, is IMSAFE, which stands for Illness, Medication, Stress, Alcohol, Fatigue, and Emotion.

For the medication component of IMSAFE, pilots need to ask themselves, "Am I taking any medicines that might affect my judgment or make me drowsy? For any new medication, OTC or prescribed, you should wait at least 48 hours after the first dose before flying to determine you do not have any adverse side effects that would make it unsafe to operate an aircraft. In addition to medication questions, pilots should also consider the following –

- Do not take any unnecessary or elective medications;

- Make sure you eat regular balanced meals;

- Bring a snack for both you and your passengers for the flight;

- Maintain good hydration - bring plenty of water;

- Ensure adequate sleep the night prior to the flight; and

- Stay physically fit.

Additionally, you should wait at least five maximal dosing intervals, the time between recommended or prescribed dosing, (e.g., a dosing interval of 5 to 6 hours would require you to wait 30 hours) before flying after taking any medication that has potentially adverse side effects (e.g., sedating or dizziness). Observing the recommended dosing interval doesn't eliminate the risk for adverse side effects because everyone metabolizes medications differently. However, five times the dosing interval is a reasonable rule of thumb.

Altitude-Induced Decompression Sickness (DCS)

Decompression sickness (DCS) describes a condition characterized by a variety of symptoms resulting from exposure to low barometric pressures that cause inert gases (mainly nitrogen), normally dissolved in body fluids and tissues, to come out of physical solution and form bubbles. Nitrogen is an inert gas normally stored throughout the human body (tissues and fluids) in physical solution. When the body is exposed to decreased barometric pressures (as in flying an unpressurized aircraft to altitude or during a rapid decompression), the nitrogen dissolved in the body comes out of solution. If the nitrogen is forced to leave the solution too rapidly, bubbles form in different areas of the body causing a variety of signs and symptoms. The most common symptom is joint pain, which is known as "the bends." [Figure 17-10]

What to do when altitude-induced DCS occurs:

- Put on oxygen mask immediately and switch the regulator to 100 percent oxygen.

- Begin an emergency descent and land as soon as possible. Even if the symptoms disappear during descent, land and seek medical evaluation while continuing to breathe oxygen.

- If one of the symptoms is joint pain, keep the affected area still; do not try to work pain out by moving the joint around.

- Upon landing, seek medical assistance from an FAA medical officer, AME, military flight surgeon, or a hyperbaric medicine specialist. Be aware that a physician not specialized in aviation or hypobaric medicine may not be familiar with this type of medical problem.

- Definitive medical treatment may involve the use of a hyperbaric chamber operated by specially-trained personnel.

- Delayed signs and symptoms of altitude-induced DCS can occur after return to ground level regardless of presence during flight.

DCS After Scuba Diving

Scuba diving subjects the body to increased pressure, which allows more nitrogen to dissolve in body tissues and fluids. [Figure 17-11] The reduction of atmospheric pressure that accompanies flying can produce physical problems for scuba divers. A pilot or passenger who intends to fly after scuba diving should allow the body sufficient time to rid itself of excess nitrogen absorbed during diving. If not, DCS due to

DCS Type	Bubble Location	Signs and Symptoms (Clinical Manifestations)
BENDS	Mostly large joints of the body (elbows, shoulders, hip, wrists, knees, ankles)	• Localized deep pain, ranging from mild (a "niggle") to excruciating–sometimes a dull ache, but rarely a sharp pain • Active and passive motion of the joint aggravating the pain • Pain occurring at altitude, during the descent, or many hours later
NEUROLOGIC Manifestations	Brain	• Confusion or memory loss • Headache • Spots in visual field (scotoma), tunnel vision, double vision (diplopia), or blurry vision • Unexplained extreme fatigue or behavior changes • Seizures, dizziness, vertigo, nausea, vomiting, and unconsciousness
	Spinal cord	• Abnormal sensations, such as burning, stinging, and tingling, around the lower chest and back • Symptoms spreading from the feet up and possibly accompanied by ascending weakness or paralysis • Girdling abdominal or chest pain
	Peripheral nerves	• Urinary and rectal incontinence • Abnormal sensations, such as numbness, burning, stinging and tingling (paresthesia) • Muscle weakness or twitching
CHOKES	Lungs	• Burning deep chest pain (under the sternum) • Pain aggravated by breathing • Shortness of breath (dyspnea) • Dry constant cough
SKIN BENDS	Skin	• Itching usually around the ears, face, neck, arms, and upper torso • Sensation of tiny insects crawling over the skin • Mottled or marbled skin usually around the shoulders, upper chest, and abdomen accompanied by itching • Swelling of the skin, accompanied by tiny scar-like skin depressions (pitting edema)

Figure 17-10. *Signs and symptoms of altitude decompression sickness.*

evolved gas can occur during exposure to low altitude and create a serious inflight emergency.

The recommended waiting time before going to flight altitudes of up to 8,000 feet is at least 12 hours after diving that does not require controlled ascent (nondecompression stop diving), and at least 24 hours after diving that does require controlled ascent (decompression stop diving). The waiting time before going to flight altitudes above 8,000 feet should be at least 24 hours after any scuba dive. These recommended altitudes are actual flight altitudes above mean sea level (MSL) and not pressurized cabin altitudes. This takes into consideration the risk of decompression of the aircraft during flight.

Vision in Flight

Of all the senses, vision is the most important for safe flight. Most of the things perceived while flying are visual or heavily supplemented by vision. As remarkable and vital as it is, vision is subject to limitations, such as illusions and blind spots. The more a pilot understands about the eyes and how they function, the easier it is to use vision effectively and compensate for potential problems.

The eye functions much like a camera. Its structure includes an aperture, a lens, a mechanism for focusing, and a surface for registering images. Light enters through the cornea at the front of the eyeball, travels through the lens, and falls on the retina. The retina contains light sensitive cells that convert

Figure 17-11. *To avoid the bends, scuba divers must not fly for specific time periods following dives.*

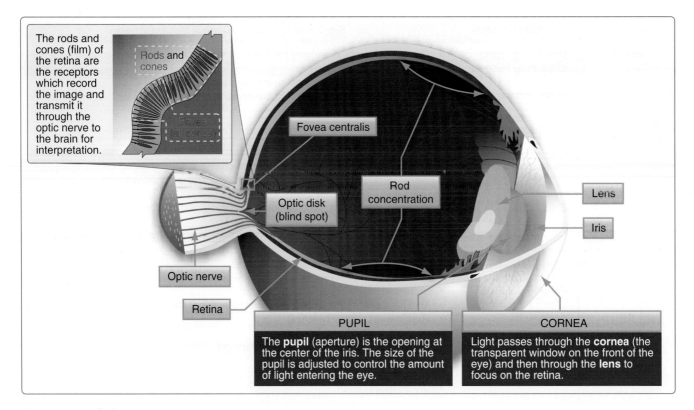

The rods and cones (film) of the retina are the receptors which record the image and transmit it through the optic nerve to the brain for interpretation.

Rods and cones

Fovea (all cones)

Fovea centralis

Rod concentration

Optic disk (blind spot)

Lens

Iris

Optic nerve

Retina

PUPIL
The **pupil** (aperture) is the opening at the center of the iris. The size of the pupil is adjusted to control the amount of light entering the eye.

CORNEA
Light passes through the **cornea** (the transparent window on the front of the eye) and then through the **lens** to focus on the retina.

Figure 17-12. *The human eye.*

light energy into electrical impulses that travel through nerves to the brain. The brain interprets the electrical signals to form images. There are two kinds of light-sensitive cells in the eyes: rods and cones. *[Figure 17-12]*

The cones are responsible for all color vision, from appreciating a glorious sunset to discerning the subtle shades in a fine painting. Cones are present throughout the retina, but are concentrated toward the center of the field of vision at the back of the retina. There is a small pit called the fovea where almost all the light sensing cells are cones. This is the area where most "looking" occurs (the center of the visual field where detail, color sensitivity, and resolution are highest).

While the cones and their associated nerves are well suited to detecting fine detail and color in high light levels, the rods are better able to detect movement and provide vision in dim light. The rods are unable to discern color but are very sensitive at low-light levels. The trouble with rods is

that a large amount of light overwhelms them, and they take longer to "reset" and adapt to the dark again. There are so many cones in the fovea that are at the very center of the visual field but virtually has no rods at all. So in low light, the middle of the visual field is not very sensitive, but farther from the fovea, the rods are more numerous and provide the major portion of night vision.

Vision Types

There are three types of vision: photopic, mesopic, and scotopic. Each type functions under different sensory stimuli or ambient light conditions. *[Figure 17-13]*

Photopic Vision

Photopic vision provides the capability for seeing color and resolving fine detail (20/20 or better), but it functions only in good illumination. Photopic vision is experienced during daylight or when a high level of artificial illumination exists.

Types of Vision						
Types of vision used	Light level	Technique of viewing	Color perception	Receptors used	Acuity best	Blind spot
Photopic	High	Central	Good	Cones	20/20	Day
Mesopic	Medium/Low	Both	Some	Cones/Rods	Varies	Day/Night
Scotopic	Low	Scanning	None	Rods	20/200	Day/Night

Figure 17-13. *Types of vision.*

The cones concentrated in the fovea centralis of the eye are primarily responsible for vision in bright light. [*Figure 17-12*] Because of the high light level, rhodopsin, which is a biological pigment of the retina that is responsible for both the formation of the photoreceptor cells and the first events in the perception of light, is bleached out causing the rod cells to become less effective.

Mesopic Vision

Mesopic vision is achieved by a combination of rods and cones and is experienced at dawn, dusk, and during full moonlight. Visual acuity steadily decreases as available light decreases and color perception changes because the cones become less effective. Mesopic viewing period is considered the most dangerous period for viewing. As cone sensitivity decreases, pilots should use off-center vision and proper scanning techniques to detect objects during low-light levels.

Scotopic Vision

Scotopic vision is experienced under low-light levels and the cones become ineffective, resulting in poor resolution of detail. Visual acuity decreases to 20/200 or less and enables a person to see only objects the size of or larger than the big "E" on visual acuity testing charts from 20 feet away. In other words, a person must stand at 20 feet to see what can normally be seen at 200 feet under daylight conditions. When using scotopic vision, color perception is lost and a night blind spot in the central field of view appears at low light levels when the cone-cell sensitivity is lost.

Central Blind Spot

The area where the optic nerve connects to the retina in the back of each eye is known as the optic disk. There is a total absence of cones and rods in this area, and consequently, each eye is completely blind in this spot. [*Figure 17-14*] As a result, it is referred to as the blind spot that everyone

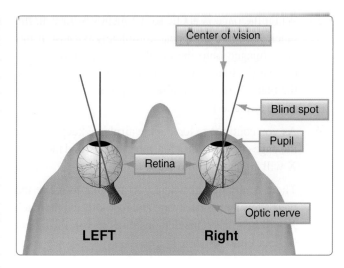

Figure 17-14. *Central blind spot.*

has in each eye. Under normal binocular vision conditions (both eyes are used together), this is not a problem because an object cannot be in the blind spot of both eyes at the same time. On the other hand, where the field of vision of one eye is obstructed by an object (windshield divider or another aircraft), a visual target could fall in the blind spot of the other eye and remain undetected.

Figure 17-15 provides a dramatic example of the eye's blind spot.

1. Hold this page at an arm's length.

2. Completely cover your left eye (without closing or pressing on it) using your hand or other flat object.

3. With your right eye, stare directly at the airplane on the left side of the picture page. In your periphery, you will notice the black X on the right side of the picture.

4. Slowly move the page closer to you while continuing to stare at the airplane.

Figure 17-15. *The eye's blind spot.*

5. When the page is about 16–18 inches from you, the black X should disappear completely because it has been imaged onto the blind spot of your right eye. (Resist the temptation to move your right eye while the black X is gone or else it reappears. Keep staring at the airplane.)

6. As you continue to look at the airplane, keep moving the page closer to you a few more inches, and the black X will come back into view.

7. There is an interval where you are able to move the page a few inches backward and forward, and the black X will be gone. This demonstrates to you the extent of your blind spot.

8. You can try the same thing again, except this time with your right eye covered stare at the black X with your left eye. Move the page in closer and the airplane will disappear.

Another way to check your blind spot is to do a similar test outside at night when there is a full moon. Cover your left eye, looking at the full moon with your right eye. Gradually move your right eye to the left (and maybe slightly up or down). Before long, all you will be able to see is the large halo around the full moon; the entire moon itself will seem to have disappeared.

Empty-Field Myopia

Empty-field myopia is a condition that usually occurs when flying above the clouds or in a haze layer that provides nothing specific to focus on outside the aircraft. This causes the eyes to relax and seek a comfortable focal distance that may range from 10 to 30 feet. For the pilot, this means looking without seeing, which is dangerous. Searching out and focusing on distant light sources, no matter how dim, helps prevent the onset of empty-field myopia.

Night Vision

There are many good reasons to fly at night, but pilots must keep in mind that the risks of night flying are different than during the day and often times higher. *[Figure 17-16]* Pilots who are cautious and educated on night-flying techniques can mitigate those risks and become very comfortable and proficient in the task.

Night Blind Spot

It is estimated that once fully adapted to darkness, the rods are 10,000 times more sensitive to light than the cones, making them the primary receptors for night vision. Since the cones are concentrated near the fovea, the rods are also responsible for much of the peripheral vision. The concentration of cones in the fovea can make a night blind spot in the center of the field of vision. To see an object clearly at night, the pilot must

Figure 17-16. *Night vision.*

expose the rods to the image. This can be done by looking 5° to 10° off center of the object to be seen. This can be tried in a dim light in a darkened room. When looking directly at the light, it dims or disappears altogether. When looking slightly off center, it becomes clearer and brighter.

When looking directly at an object, the image is focused mainly on the fovea, where detail is best seen. At night, the ability to see an object in the center of the visual field is reduced as the cones lose much of their sensitivity and the rods become more sensitive. Looking off center can help compensate for this night blind spot. Along with the loss of

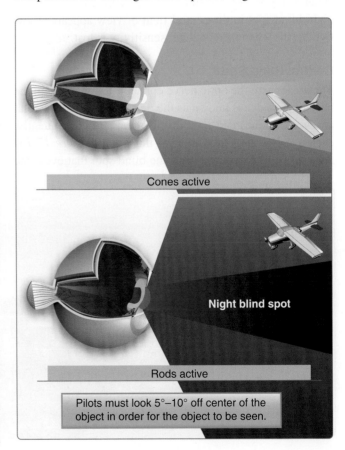

Cones active

Night blind spot

Rods active

Pilots must look 5°–10° off center of the object in order for the object to be seen.

Figure 17-17. *Night blind spot.*

sharpness (acuity) and color at night, depth perception and judgment of size may be lost. [Figure 17-17]

Dark Adaptation

Dark adaptation is the adjustment of the human eye to a dark environment. That adjustment takes longer depending on the amount of light in the environment that a person has just left. Moving from a bright room into a dark one takes longer than moving from a dim room and going into a dark one.

While the cones adapt rapidly to changes in light intensities, the rods take much longer. Walking from bright sunlight into a dark movie theater is an example of this dark adaptation period experience. The rods can take approximately 30 minutes to fully adapt to darkness. A bright light, however, can completely destroy night adaptation, leaving night vision severely compromised while the adaptation process is repeated.

Scanning Techniques

Scanning techniques are very important in identifying objects at night. To scan effectively, pilots must look from right to left or left to right. They should begin scanning at the greatest distance an object can be perceived (top) and move inward toward the position of the aircraft (bottom). For each stop, an area approximately 30° wide should be scanned. The duration of each stop is based on the degree of detail that is required, but no stop should last longer than 2 to 3 seconds. When moving from one viewing point to the next, pilots should overlap the previous field of view by 10°. [Figure 17-18]

Off-center viewing is another type of scan that pilots can use during night flying. It is a technique that requires an object be viewed by looking 10° above, below, or to either side of the object. [Figure 17-19] In this manner, the peripheral vision can maintain contact with an object.

With off-center vision, the images of an object viewed longer than 2 to 3 seconds will disappear. This occurs because the rods reach a photochemical equilibrium that prevents any further response until the scene changes. This produces a potentially unsafe operating condition. To overcome this night vision limitation, pilots must be aware of the phenomenon and avoid viewing an object for longer than 2 or 3 seconds. The peripheral field of vision will continue to pick up the object when the eyes are shifted from one off-center point to another.

Night Vision Protection

Several things can be done to help with the dark adaptation process and to keep the eyes adapted to darkness. Some of the steps pilots and flight crews can take to protect their night vision are described in the following paragraphs.

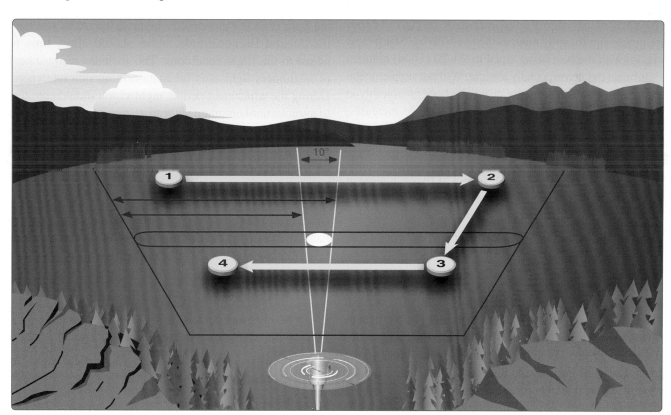

Figure 17-18. *Scanning techniques.*

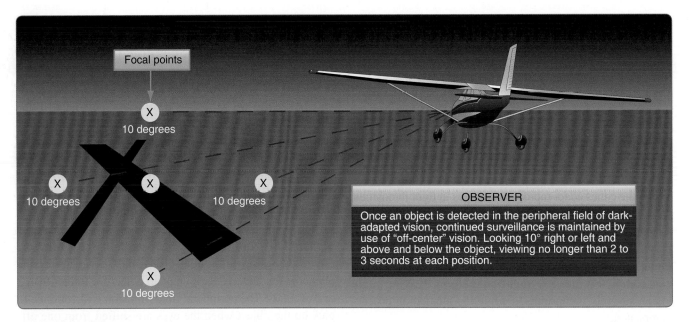

Figure 17-19. *Off-center viewing.*

Sunglasses

If a night flight is scheduled, pilots and crew members should wear neutral density (N-15) sunglasses or equivalent filter lenses when exposed to bright sunlight. This precaution increases the rate of dark adaptation at night and improves night visual sensitivity.

Oxygen Supply

Unaided night vision depends on optimum function and sensitivity of the rods of the retina. Lack of oxygen to the rods (hypoxia) significantly reduces their sensitivity. Sharp clear vision (with the best being equal to 20–20 vision) requires significant oxygen especially at night. Without supplemental oxygen, an individual's night vision declines measurably at pressure altitudes above 4,000 feet. As altitude increases, the available oxygen decreases, degrading night vision. Compounding the problem is fatigue, which minimizes physiological well being. Adding fatigue to high altitude exposure is a recipe for disaster. In fact, if flying at night at an altitude of 12,000 feet, the pilot may actually see elements of his or her normal vision missing or not in focus. Missing visual elements resemble the missing pixels in a digital image while unfocused vision is dim and washed out.

For the pilot suffering the effects of hypoxic hypoxia, a simple descent to a lower altitude may not be sufficient to reestablish vision. For example, a climb from 8,000 feet to 12,000 feet for 30 minutes does not mean a descent to 8,000 feet will rectify the problem. Visual acuity may not be regained for over an hour. Thus, it is important to remember, altitude and fatigue have a profound effect on a pilot's ability to see.

High Intensity Lighting

If, during the flight, any high intensity lighting areas are encountered, attempt to turn the aircraft away and fly in the periphery of the lighted area. This will not expose the eyes to such a large amount of light all at once. If possible, plan your route to avoid direct over flight of built-up, brightly lit areas.

Flightdeck Lighting

Flightdeck lighting should be kept as low as possible so that the light does not monopolize night vision. After reaching the desired flight altitude, pilots should allow time to adjust to the flight conditions. This includes readjustment of instrument lights and orientation to outside references. During the adjustment period, night vision should continue to improve until optimum night adaptation is achieved. When it is necessary to read maps, charts, and checklists, use a dim white light flashlight and avoid shining it in your or any other crewmember's eyes.

Airfield Precautions

Often time, pilots have no say in how airfield operations are handled, but listed below are some precautions that can be taken to make night flying safer and help protect night vision.

- Airfield lighting should be reduced to the lowest usable intensity.

- Maintenance personnel should practice light discipline with headlights and flashlights.

- Position the aircraft at a part of the airfield where the least amount of lighting exists.

- **Select** approach and departure routes that avoid highways and residential areas where illumination can impair night vision.

Self-Imposed Stress

Night flight can be more fatiguing and stressful than day flight, and many self-imposed stressors can limit night vision. Pilots can control this type of stress by knowing the factors that can cause self-imposed stressors. Some of these factors are listed in the following paragraphs. *[Figure 17-20]*

Drugs

Drugs can seriously degrade visual acuity during the day and especially at night. Pilots who become ill should consult an aviation medical examiner (AME) or flight surgeon as to which drugs are appropriate to take while flying.

Exhaustion

Pilots who become fatigued during a night flight will not be mentally alert and will respond more slowly to situations requiring immediate action. Exhausted pilots tend to concentrate on one aspect of a situation without considering the total requirement. Their performance may become a safety hazard depending on the degree of fatigue and instead of using proper scanning techniques may get fixated on the instruments or stare off rather than multitask.

Poor Physical Conditioning

To overcome poor physical conditioning, pilots should participate in regular exercise programs. People who are physically fit become less fatigued during flight and have better night scanning efficiency. However, too much exercise in a given day may leave crew members too fatigued for night flying.

Alcohol

Alcohol is a sedative and its use impairs both coordination and judgment. As a result, pilots who are impaired by alcohol fail to apply the proper techniques of night vision. They are likely to stare at objects and to neglect scanning techniques. The amount of alcohol consumed determines the degree to which night vision is affected. The effects of alcohol are long lasting and the residual effects of alcohol can also impair visual scanning efficiency.

Tobacco

Of all the self-imposed stressors, cigarette smoking most decreases visual sensitivity at night. Smoking significantly increases the amount of carbon monoxide carried by the hemoglobin in red blood cells. This reduces the blood's capacity to combine with oxygen, so less oxygen is carried in the blood. Hypoxia caused by carbon monoxide poisoning

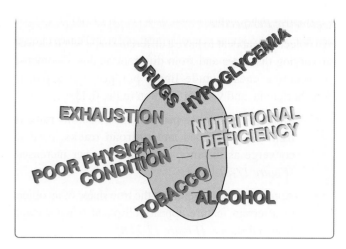

Figure 17-20. *Self-imposed stress.*

affects peripheral vision and dark adaptation. The results are the same as those for hypoxia caused by high altitude. Smoking 3 cigarettes in rapid succession or 20 to 30 cigarettes within a 24-hour period may saturate from 8 to 10 percent of the capacity of hemoglobin. Smokers lose 20 percent of their night vision capability at sea level, which is equal to a physiological altitude of 5,000 feet.

Hypoglycemia and Nutritional Deficiency

Missing or postponing meals can cause low blood sugar, which impairs night flight performance. Low blood sugar levels may result in stomach contractions, distraction, breakdown in habit pattern, and a shortened attention span. Likewise, an insufficient consumption of vitamin A may also impair night vision. Foods high in vitamin A include eggs, butter, cheese, liver, apricots, peaches, carrots, squash, spinach, peas, and most types of greens. High quantities of vitamin A do not increase night vision but a lack of vitamin A certainly impairs it.

Distance Estimation and Depth Perception

Knowledge of the mechanisms and cues affecting distance estimation and depth perception assist pilots in judging distances at night. These cues may be monocular or binocular. The monocular cues that aid in distance estimation and depth perception include motion parallax, geometric perspective, retinal image size, and aerial perspective.

Motion Parallax

Motion parallax refers to the apparent motion of stationary objects as viewed by an observer moving across the landscape. When the pilot or crewmember looks outside the aircraft perpendicular to the direction of travel, near objects appear to move backward, past, or opposite the path of motion; far objects seem to move in the direction of motion or remain fixed. The rate of apparent movement depends on the distance the observer is from the object.

Geometric Perspective

An object may appear to have a different shape when viewed at varying distances and from different angles. Geometric perspective cues include linear perspective, apparent foreshortening, and vertical position in the field.

- Linear perspective—parallel lines, such as runway lights, power lines and railroad tracks, tend to converge as distance from the observer increases. *[Figure 17-21A]*

- Apparent foreshortening—the true shape of an object or a terrain feature appears elliptical when viewed from a distance. *[Figure 17-21B]*

- Vertical position in the field—objects or terrain features farther away from the observer appear higher on the horizon than those closer to the observer. *[Figure 17-21C]*

Aerial Perspective

The clarity of an object and the shadow cast by it are perceived by the brain and are cues for estimating distance. Subtle variations in color or shade are clearer the closer the observer is to an object. However, as distance increases, these distinctions may become blurry. The same applies to an object detail or texture. As a person gets farther from an object, its discrete details become less apparent. Another important fact to remember while flying at night is that every object casts a shadow from a light source. The direction in which the shadow is cast depends on the position of the light source. If the shadow of an object is cast toward the observer, the object is closer than the light source is to the observer.

Binocular Cues

Binocular cues of an object are dependent upon the slightly different viewing angle of each eye of an object. Binocular perception is useful only when the object is close enough to make an obvious difference in the viewing angle of both eyes. In the flight environment, most distances outside the cockpit are so great that binocular cues are of little, if any, value. In addition, binocular cues operate on a more subconscious level than monocular cues and are performed automatically.

Night Vision Illusions

There are many different types of visual illusions that commonly occur at night. Anticipating and maintaining awareness of them is usually the best way to avoid them.

Autokinesis

Autokinesis is caused by staring at a single point of light against a dark background for more than a few seconds. After a few moments, the light appears to move on its own. Apparent movement of the light source will begin in about 8 to 10 seconds. To prevent this illusion, focus the eyes on objects at varying distances and avoid fixating on one source of light. This illusion can be eliminated or reduced by visual scanning, by increasing the number of lights, or by varying the light intensity. The most important of the three solutions is visual scanning. A light or lights should not be stared at for more than 10 seconds.

False Horizon

A false horizon can occur when the natural horizon is obscured or not readily apparent. It can be generated by confusing bright stars and city lights. It can also occur while flying toward the shore of an ocean or a large lake. Because of the relative darkness of the water, the lights along the shoreline can be mistaken for stars in the sky. *[Figure 17-22]*

Reversible Perspective Illusion

At night, an aircraft may appear to be moving away from a second aircraft when it is, in fact, approaching a second aircraft. This illusion often occurs when an aircraft is flying

Figure 17-21. *Geometric perspective.*

Figure 17-22. *At night, the horizon may be hard to discern due to dark terrain and misleading light patterns on the ground.*

parallel to another's course. To determine the direction of flight, pilots should observe aircraft lights and their relative position to the horizon. If the intensity of the lights increases, the aircraft is approaching; if the lights dim, the aircraft is moving away.

Size-Distance Illusion

This illusion results from viewing a source of light that is increasing or decreasing in luminance (brightness). Pilots may interpret the light as approaching or retreating.

Fascination (Fixation)

This illusion occurs when pilots ignore orientation cues and fix their attention on a goal or an object. Student pilots tend to have this happen when they are concentrating on the aircraft instruments or attempting to land. They become fixated on one task and forget to look at what is going on around them. At night, this can be especially dangerous because aircraft ground-closure rates are difficult to determine, and there may be minimal time to correct the situation.

Flicker Vertigo

A light flickering at a rate between 4 and 20 cycles per second can produce unpleasant and dangerous reactions. Such conditions as nausea, vomiting, and vertigo may occur. On rare occasions, convulsions and unconsciousness may also occur. Proper scanning techniques at night can prevent pilots from getting flicker vertigo.

Night Landing Illusions

Landing illusions occur in many forms. Above featureless terrain at night, there is a natural tendency to fly a lower-than-normal approach. Elements that cause any type of visual obscurities, such as rain, haze, or a dark runway environment, can also cause low approaches. Bright lights, steep surrounding terrain, and a wide runway can produce the illusion of being too low with a tendency to fly a higher-than-normal approach. A set of regularly spaced lights along a road or highway can appear to be runway lights. Pilots have even mistaken the lights on moving trains as runway or approach lights. Bright runway or approach lighting systems can create the illusion that the aircraft is closer to the runway, especially where few lights illuminate the surrounding terrain.

Prior to flying at night, it is best to learn and know the challenges of the area in which you are flying in. Study the area and know how to navigate your way through areas that may pose a problem at night. For example, many areas near water may be obscured by low lying clouds or fog. To help deal with this type of situation, it is important to have a plan before you leave the ground. In the daytime, fly the routes and passes that you will be flying at night and determine the minimum altitude you are willing to use at night. If weather prevents you from maintaining the altitude that you planned, make a decision early to turn 180° and land at an alternate airport with better weather conditions. Always consider safer alternatives rather than hope things will work out by taking a chance.

Pilots who fly at night should strongly consider oxygen supplementation at altitudes and times not required by the FAA, especially at night when critical judgment and hand-eye coordination is necessary (e.g., IFR) or if he/she is a smoker or not perfectly healthy.

Enhanced Night Vision Systems

Synthetic Vision Systems (SVS) and Enhanced Flight Vision Systems (EFVS) are two systems that can improve the safety of flight at night. The technology of both is evolving rapidly and being used more and more. *[Figure 17-23]*

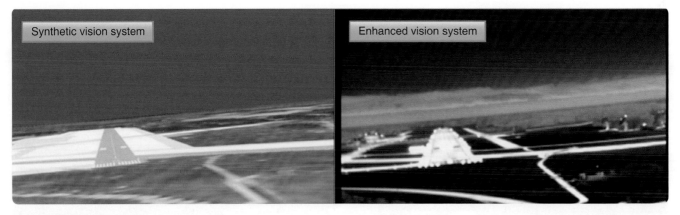

Figure 17-23. *Synthetic and enhanced vision systems.*

Synthetic Vision System

A Synthetic Vision System (SVS) is an electronic means to display a synthetic vision image of the external scene topography to the flight crew. *[Figure 17-24]* It is not a real-time image like that produced by an EFVS. Unlike EFVS, SVS requires a terrain and obstacle database, a precise navigation solution, and a display. The terrain image is based on the use of data from a Digital Elevation Model (DEM) that is stored within the SVS. With SVS, the synthetic terrain/vision image is intended to enhance pilot awareness of spatial position relative to important features in all visibility conditions. This is particularly useful during critical phases of flight, such as takeoff, approach, and landing, where important features, such as terrain, obstacles, runways, and landmarks, may be depicted on the SVS display. *[Figure 17-25]* During approach operations, the

obvious advantages of SVS are that the digital terrain image remains on the pilot's display regardless of how poor the visibility is outside.

An SVS image can be displayed on either a head-down display or head-up display (HUD); however, to date, SVS has only been certified on head-down displays. Development efforts to display a synthetic image on a HUD are currently underway as are efforts that would combine SVS with a real-time sensor image produced by an EFVS. These systems are known as Combined Vision Systems. While SVS is currently certified as an aid to situation awareness only, the FAA and aviation industry are working on defining operational concepts and airworthiness criteria that would enable SVS to be used for operational credit in certain low visibility conditions. Other future enhancements to SVS displays could include integrating ADS-B to display traffic information.

Enhanced Flight Vision System

Enhanced Vision (EV) or Enhanced Flight Vision System (EFVS) is an electronic means to provide a display of

Figure 17-24. *SVS system.*

Figure 17-25. *Night time SVS system.*

the external scene by use of an imaging sensor, such as a Forward Looking InfraRed (FLIR) or millimeter wave radar (MMWR). In 2004, 14 CFR part 91, section 91.175 was amended to reflect that operators conducting straight-in instrument approach procedures (in other than Category II or Category III operations) may now operate below the published decision height (DH) or minimum descent altitude (MDA) when using an approved EFVS shown on the pilot's HUD. This rule change provides "operational credit" for EV equipage. No such credit exists for SV.

Chapter Summary

This chapter provides an introduction to aeromedical factors relating to flight activities. More detailed information on the subjects discussed in this chapter is available in the Aeronautical Information Manual (AIM) and online at www.faa.gov.

Performance Data for Cessna Model 172R and Challenger 605

Short Field Takeoff Distance at 2,450 Pounds for a Cessna Model 172R

CONDITIONS:

Flaps 10°
Full Throttle Prior to Brake Release
Paved, level, dry runway
Zero Wind
Lift Off: 51 KIAS
Speed at 50 Ft: 57 KIAS

Press Alt In Feet	0°C		10°C		20°C		30°C		40°C	
	Grnd Roll Ft	Total Ft To Clear 50 Ft Obst	Grnd Roll Ft	Total Ft To Clear 50 Ft Obst	Grnd Roll Ft	Total Ft To Clear 50 Ft Obst	Grnd Roll Ft	Total Ft To Clear 50 Ft Obst	Grnd Roll Ft	Total Ft To Clear 50 Ft Obst
S. L.	845	1510	910	1625	980	1745	1055	1875	1135	2015
1000	925	1660	1000	1790	1075	1925	1160	2070	1245	2220
2000	1015	1830	1095	1970	1185	2125	1275	2290	1365	2455
3000	1115	2020	1205	2185	1305	2360	1400	2540	1505	2730
4000	1230	2245	1330	2430	1435	2630	1545	2830	1655	3045
5000	1355	2500	1470	2715	1585	2945	1705	3175	1830	3430
6000	1500	2805	1625	3060	1750	3315	1880	3590	2020	3895
7000	1660	3170	1795	3470	1935	3770	2085	4105	2240	4485
8000	1840	3620	1995	3975	2150	4345	2315	4775	---	---

NOTES:

1. Short field technique as specified in Section 4.
2. Prior to takeoff from fields above 3000 feet elevation, the mixture should be leaned to give maximum RPM in a full throttle, static runup.
3. Decrease distances 10% for each 9 knots headwind. For operation with tail winds up to 10 knots, increase distances by 10% for each 2 knots.
4. For operation on dry, grass runway, increase distances by 15% of the "ground roll" figure.
5. Where distance value has been deleted, climb performance is minimal.

Time, Fuel, and Distance to Climb at 2,450 Pounds for a Cessna Model 172R

CONDITIONS:

Flaps Up
Full Throttle
Standard Temperature

PRESS ALT FT	TEMP °C	CLIMB SPEED KIAS	RATE OF CLIMB FPM	FROM SEA LEVEL		
				TIME IN MIN	FUEL USED GAL	DIST NM
S.L.	15	79	720	0	0.0	0
1000	13	78	670	1	0.4	2
2000	11	77	625	3	0.7	4
3000	9	76	575	5	1.2	6
4000	7	76	560	6	1.5	8
5000	5	75	515	8	1.8	11
6000	3	74	465	10	2.1	14
7000	1	73	415	13	2.5	17
8000	-1	72	365	15	3.0	21
9000	-3	72	315	18	3.4	25
10,000	-5	71	270	22	4.0	29
11,000	-7	70	220	26	4.6	35
12,000	-9	69	170	31	5.4	43

NOTES:

1. Add 1.1 gallons of fuel for engine start, taxi and takeoff allowance.
2. Mixture leaned above 3000 feet for maximum RPM.
3. Increase time, fuel and distance by 10% for each 10°C above standard temperature.
4. Distances shown are based on zero wind.

Cruise Performance for a Cessna Model 172R

CONDITIONS:
2450 Pounds
Recommended Lean Mixture At All Altitudes (Refer to Section 4, Cruise)

PRESS ALT FT	RPM	20°C BELOW STANDARD TEMP			STANDARD TEMPERATURE			20°C ABOVE STANDARD TEMP		
		% BHP	KTAS	GPH	% BHP	KTAS	GPH	% BHP	KTAS	GPH
2000	2250	---	---	---	79	115	9.0	74	114	8.5
	2200	79	112	9.1	74	112	8.5	70	111	8.0
	2100	69	107	7.9	65	106	7.5	62	105	7.1
	2000	61	101	7.0	58	99	6.6	55	97	6.4
	1900	54	94	6.2	51	91	5.9	50	89	5.8
4000	2300	--	---	---	79	117	9.1	75	117	8.6
	2250	80	115	9.2	75	114	8.6	70	114	8.1
	2200	75	112	8.6	70	111	8.1	66	110	7.6
	2100	66	106	7.6	62	105	7.1	59	103	6.8
	2000	58	100	6.7	55	98	6.4	53	95	6.2
	1900	52	92	6.0	50	90	5.8	49	87	5.6
6000	2350	--	---	---	80	120	9.2	75	119	8.6
	2300	80	117	9.2	75	117	8.6	71	116	8.1
	2250	76	115	8.7	71	114	8.1	67	113	7.7
	2200	71	112	8.1	67	111	7.7	64	109	7.3
	2100	63	105	7.2	60	104	6.9	57	101	6.6
	2000	56	98	6.4	53	96	6.2	52	93	6.0

NOTE:

1. Cruise speeds are shown for an airplane equipped with speed fairings. Without speed fairings, decrease speeds shown by 2 knots.

Short Field Landing Distance at 2,450 Pounds for a Cessna Model 172R

CONDITIONS:

Flaps 30°
Power Off
Maximum Braking
Paved, level, dry runway
Zero Wind
Speed at 50 Ft: 62 KIAS

Press Alt In Feet	0°C		10°C		20°C		30°C		40°C	
	Grnd Roll Ft	Total Ft To Clear 50 Ft Obst	Grnd Roll Ft	Total Ft To Clear 50 Ft Obst	Grnd Roll Ft	Total Ft To Clear 50 Ft Obst	Grnd Roll Ft	Total Ft To Clear 50 Ft Obst	Grnd Roll Ft	Total Ft To Clear 50 Ft Obst
S. L.	525	1250	540	1280	560	1310	580	1340	600	1370
1000	545	1280	560	1310	580	1345	600	1375	620	1405
2000	565	1310	585	1345	605	1375	625	1410	645	1440
3000	585	1345	605	1380	625	1415	650	1445	670	1480
4000	605	1380	630	1415	650	1450	670	1485	695	1520
5000	630	1415	650	1455	675	1490	700	1525	720	1560
6000	655	1455	675	1490	700	1530	725	1565	750	1605
7000	680	1495	705	1535	730	1570	755	1610	775	1650
8000	705	1535	730	1575	755	1615	780	1655	810	1695

NOTES:

1. Short field technique as specified in Section 4.
2. Decrease distances 10% for each 9 knots headwind. For operation with tail winds up to 10 knots, increase distances by 10% for each 2 knots.
3. For operation on dry, grass runway, increase distances by 45% of the "ground roll" figure.
4. If landing with flaps up, increase the approach speed by 7 KIAS and allow for 35% longer distances.

Challenger 605 Range/Payload Profile

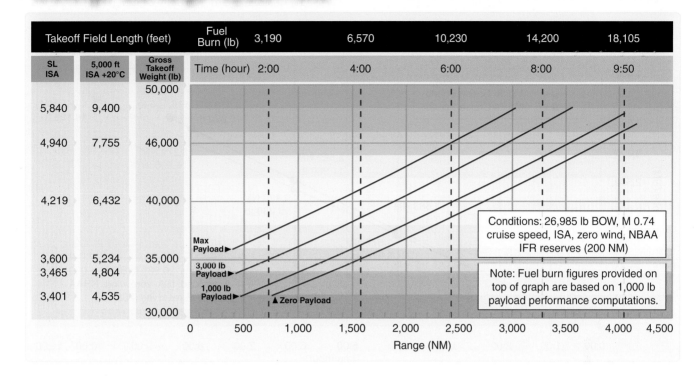

Takeoff Field Length (feet)			Fuel Burn (lb)	3,190	6,570	10,230	14,200	18,105
SL ISA	5,000 ft ISA +20°C	Gross Takeoff Weight (lb)	Time (hour)	2:00	4:00	6:00	8:00	9:50
		50,000						
5,840	9,400							
4,940	7,755	46,000						
4,219	6,432	40,000						
			Max Payload ▶					
3,600	5,234	35,000						
3,465	4,804		3,000 lb Payload ▶					
3,401	4,535		1,000 lb Payload ▶					
		30,000	▲ Zero Payload					

Conditions: 26,985 lb BOW, M 0.74 cruise speed, ISA, zero wind, NBAA IFR reserves (200 NM)

Note: Fuel burn figures provided on top of graph are based on 1,000 lb payload performance computations.

Range (NM)
0   500   1,000   1,500   2,000   2,500   3,000   3,500   4,000   4,500

Challenger 605 Time and Fuel Versus Distance

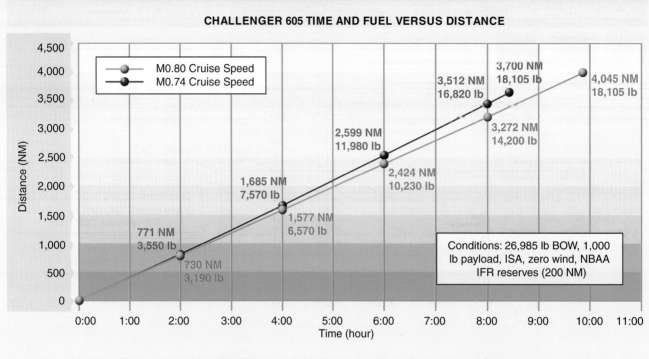

CHALLENGER 605 TIME AND FUEL VERSUS DISTANCE

	M0.80 Cruise Speed					
M0.80 Cruise Speed	Time	0:00	2:00	4:00	6:00	8:00	8:25
	Distance (NM)	0	771	1,685	2,599	3,512	3,701
	Fuel (lb)	0	3,550	7,570	11,980	16,820	18,105
M0.74 Cruise Speed	Time	0:00	2:00	4:00	6:00	8:00	9:50
	Distance (NM)	0	730	1,577	2,424	3,272	4,045
	Fuel (lb)	0	3,190	6,570	10,230	14,200	18,105

Conditions: 1,000 lb payload, ISA, zero wind, NBAA IFR reserves (200 NM alternate), 26,985 lb BOW

Note: All Challenger 605 performance data are for discussion purposes only. By this document, Bombardier Inc., does not intend to make, and is not making, any offer, commitment, representation or warranty of any kind whatsoever. All data are subject to change without prior notice.

CHALLENGER 605 SPECIFIC RANGE

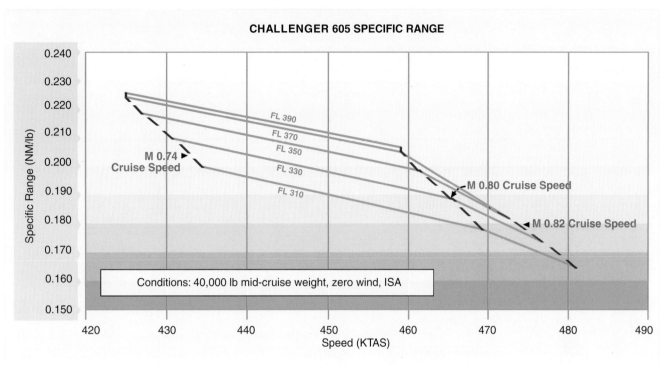

Plotting of constant FL lines Flight Level		M0.82	M0.80	M0.74
290	Speed Spc Range			
310	Speed Spc Range	481 0.165	469 0.178	434 0.199
330	Speed Spc Range	477 0.174	465 0.188	430 0.208
350	Speed Spc Range	473 0.181	461 0.197	427 0.216
370	Speed Spc Range	470 0.185	459 0.204	424 0.222
390	Speed Spc Range		459 0.205	424 0.223

Plotting of Long Range Cruise and High Speed Cruise lines

	FL290	FL310	FL330	FL350	FL370	FL390
M0.74 "X"		404	400	427	424	424
M0.74 "Y"		0.199	0.208	0.216	0.222	0.223
M0.80 "X"		469	465	461	459	459
M0.80 "Y"		0.178	0.188	0.197	0.204	0.205
M0.82 "X"		481	477	473	470	
M0.82 "Y"		0.165	0.174	0.181	0.185	

Note: Based on 40,000 lb mid-cruise weight, ISA Conditions, zero wind

Acronyms, Abbreviations, and NOTAM Contractions

This is a list of common acronyms and abbreviations used in the aviation industry as well as NOTAM contractions. For a more complete list of contractions used in aviation, see FAA Order JO 7340.2 (as amended). Additional information regarding NOTAMs can be found at pilotweb.nas.faa.gov/PilotWeb/.

A

A/C—aircraft
A/FD—airport/facility directory
A/G—air to ground
A/HA—altitude/height
AAF—Army Air Field
AAI —arrival aircraft interval
AAP—advanced automation program
AAR—airport acceptance rate
ABDIS—Automated Data Interchange System Service B
ABN—aerodrome beacon
ABV—above
ACAIS—air carrier activity information system
ACAS—aircraft collision avoidance system
ACC—area control center; Airports Consultants Council
ACCT—accounting records
ACCUM—accumulate
ACD—Automatic Call Distributor
ACDO—Air Carrier District Office
ACF—Area Control Facility
ACFO—Aircraft Certification Field Office
ACFT—aircraft
ACID—aircraft identification
ACI-NA—Airports Council International-North America
ACIP—airport capital improvement plan
ACLS—automatic carrier landing system
ACLT—actual landing time calculated
ACO—Office of Airports Compliance and Field Operations; Aircraft Certification Office
ACR—air carrier
ACRP—Airport Cooperative Research Program
ACS—Airman Certification Standard
ACT—active, activated, or activity
ADA—air defense area
ADAP—Airport Development Aid Program
ADAS—AWOS data acquisition system
ADCCP—advanced data communications control procedure
ADDA—administrative data
ADF—automatic direction finding
ADI—automatic de-ice and inhibitor

ADIN—AUTODIN service
ADIZ—air defense identification zone
ADJ—adjacent
ADL—aeronautical data-link
ADLY—arrival delay
ADO—airline dispatch office
ADP—automated data processing
ADS—automatic dependent surveillance
ADSIM —airfield delay simulation model
ADSY—administrative equipment systems
ADTN—Administrative Data Transmission Network
ADTN2000—Administrative Data Transmission Network 2000
ADVO—administrative voice
ADZD—advised
AEG—Aircraft Evaluation Group
AERA—automated en route air traffic control
AEX—automated execution
AF—airway facilities
AFB—Air Force Base
AFIS—automated flight inspection system
AFP—area flight plan
AFRES—Air Force Reserve Station
AFS—airways facilities sector
AFSFO—AFS field office
AFSFU—AFS field unit
AFSOU—AFS field office unit (standard is AFSFOU)
AFSS—automated flight service station
AFTN—Automated Fixed Telecommunications Network
AGIS—airports geographic information system
AGL—above ground level
AID—airport information desk
AIG—Airbus Industries Group
AIM—Airman's Information Manual
AIP—airport improvement plan
AIRMET—Airmen's Meteorological Information
AIRNET—Airport Network Simulation Model
AIS—aeronautical Information service
AIT—automated information transfer
ALP—airport layout plan

ALS—approach light system
ALSFl—ALS with sequenced flashers I
ALSF2—ALS with sequenced flashers II
ALSIP—Approach Lighting System Improvement Plan
ALSTG—altimeter setting
ALT—altitude
ALTM—altimeter
ALTN—alternate
ALTNLY—alternately
ALTRV—altitude reservation
AMASS—airport movement area safety system
AMCC—ADF/ARTCC Maintenance Control Center
AMDT—amendment
AMGR—Airport Manager
AMOS—Automatic meteorological observing system
AMP—ARINC Message Processor; Airport Master Plan
AMVER—automated mutual assistance vessel rescue system
ANC—alternate network connectivity
ANCA—Airport Noise and Capacity Act
ANG—Air National Guard
ANGB—Air National Guard Base
ANMS—automated network monitoring system
ANSI—American National Standards Group
AOA—air operations area
AP—airport; acquisition plan
APCH—approach
APL—airport lights
APP—approach; approach control; Approach Control Office
APS—airport planning standard
AQAFO—Aeronautical Quality Assurance Field Office
ARAC—Army Radar Approach Control (AAF); Aviation Rulemaking Advisory Committee
ARCTR—FAA Aeronautical Center or Academy
ARF—airport reservation function
ARFF—aircraft rescue and fire fighting
ARINC—Aeronautical Radio, Inc.
ARLNO—Airline Office
ARO—Airport Reservation Office
ARP—airport reference point
ARR—arrive; arrival
ARRA—American Recovery and Reinvestment Act of 2009
ARSA—airport service radar area
ARSR—air route surveillance radar
ARTCC—air route traffic control center
ARTS—automated radar terminal system
ASAS—aviation safety analysis system
ASC—AUTODIN switching center
ASCP—Aviation System Capacity Plan
ASD—aircraft situation display
ASDA—accelerate-stop distance available
ASLAR—aircraft surge launch and recovery
ASM—available seat mile
ASOS—automated surface observing system

ASP—arrival sequencing program
ASPH—asphalt
ASQP—airline service quality performance
ASR—airport surveillance radar
ASTA—airport surface traffic automation
ASV—airline schedule vendor
AT—air traffic
ATA—Air Transport Association of America
ATAS—airspace and traffic advisory service
ATC—air traffic control
ATCAA—air traffic control assigned airspace
ATCBI—air traffic control beacon indicator
ATCCC—Air Traffic Control Command Center
ATCO—Air Taxi Commercial Operator
ATCRB—air traffic control radar beacon
ATCRBS—air traffic control radar beacon system
ATCSCC—Air Traffic Control System Command Center
ATCT—airport traffic control tower
ATIS—automatic terminal information service
ATISR—ATIS recorder
ATM—air traffic management; asynchronous transfer mode
ATMS—advanced traffic management system
ATN—Aeronautical Telecommunications Network
ATODN—AUTODIN terminal (FUS)
ATOMS—air traffic operations management system
ATOVN—AUOTVON (facility)
ATS—air traffic service
ATSCCP—ATS contingency command post
AUTH—authority
AUTOB—automatic weather reporting system
AUTODIN—DoD Automatic Digital Network
AUTOVON—DoD Automatic Voice Network
AVBL—available
AVN—Aviation Standards National Field Office, Oklahoma City
AVON—AUTOVON service
AWIS—airport weather information
AWOS—automatic weather; observing/reporting system
AWP—Aviation Weather Processor
AWPG—aviation weather products generator
AWS—air weather station
AWY—airway
AZM—azimuth

B

BA FAIR—braking action fair
BA NIL—braking action nil
BA POOR—braking action poor
BANS—BRITE alphanumeric system
BART—billing analysis reporting tool (GSA software tool)
BASIC—basic contract observing station
BASOP—military base operations

BC—back course
BCA—benefit/cost analysis
BCN—beacon
BCR—benefit/cost ratio
BDAT—digitized beacon data
BERM—snowbank(s) containing earth/gravel
BLW—below
BMP—best management practices
BND—bound
BOC—Bell Operating Company
bps—bits per second
BRG—bearing
BRI—basic rate interface
BRITE—bright radar indicator terminal equipment
BRL—building restriction line
BUEC—back-up emergency communications
BUECE—back-up emergency communications equipment
BYD—beyond

C

C/S/S/N—capacity/safety/security/noise
CAA—civil aviation authority; Clean Air Act
CAAS—Class A Airspace
CAB—civil aeronautics board
CARF—Central Altitude Reservation Facility
CASFO—Civil Aviation Security Office
CAT—category; clear-air turbulence
CAU—Crypto Ancillary Unit
CBAS—Class B airspace
CBI—computer based instruction
CBSA—Class B surface area
CC&O—customer cost and obligation
CCAS—Class C Airspace
CCC—Communications Command Center
CCCC—staff communications
CCCH—central computer complex host
CCLKWS—counterclockwise
CCS7-NI—Communication Channel Signal-7-Network Interconnect
CCSA—Class C surface area
CCSD—Command Communications Service Designator
CCU—Central Control Unit
CD—clearance delivery; common digitizer
CDAS—Class D Airspace
CDR—cost detail report
CDSA—Class D surface area
CDT—controlled departure time
CDTI—cockpit display of traffic information
CEAS—Class E Airspace
CENTX—central telephone exchange
CEP—capacity enhancement program
CEQ—council on environmental quality

CERAP—center radar approach control; combined center radar approach control
CESA—Class E surface area
CFC—central flow control
CFCF—Central Flow Control Facility
CFCS—central flow control service
CFR—Code of Federal Regulations
CFWP—central flow weather processor
CFWU—central flow weather unit
CGAS—Class G Airspace; Coast Guard Air Station
CHG—change
CIG—ceiling
CK—check
CL—centerline
CLC—course line computer
CLIN—contract line item
CLKWS—clockwise
CLR—clearance, clear(s), cleared to
CLSD—closed
CLT—calculated landing time
CM—commercial service airport
CMB—climb
CMSND—commissioned
CNL—cancel
CNMPS—Canadian Minimum Navigation Performance Specification Airspace
CNS—consolidated NOTAM system
CNSP—consolidated NOTAM system processor
CO—central office
COE—U.S. Army Corps of Engineers
COM—communications
COMCO—command communications outlet
CONC—concrete
CONUS—Continental United States
CORP—private corporation other than ARINC or MITRE
CPD—coupled
CPE—customer premise equipment
CPMIS—consolidated personnel management information system
CRA—conflict resolution advisory
CRDA—converging runway display aid
CRS—course
CRT—cathode ray tube
CSA—communications service authorization
CSIS—centralized storm information system
CSO—customer service office
CSR—communications service request
CSS—central site system
CTA—controlled time of arrival; control area
CTA/FIR—control area/flight information region
CTAF—common traffic advisory frequency
CTAS—center-TRACON automation system

CTC—contact
CTL—control
CTMA—Center Traffic Management Advisor
CUPS—consolidated uniform payroll system
CVFR—controlled visual flight rules
CVTS—compressed video transmission service
CW—continuous wave
CWSU—Central Weather Service Unit
CWY—clearway

D

DA—direct access; decision altitude/decision height; Descent Advisor
DABBS—DITCO automated bulletin board system
DAIR—direct altitude and identity readout
DALGT—daylight
DAR—Designated Agency Representative
DARC—direct access radar channel
dBA—decibels A-weighted
DBCRC—Defense Base Closure and Realignment Commission
DBE—disadvantaged business enterprise
DBMS—database management system
DBRITE—digital bright radar indicator tower equipment
DCA—Defense Communications Agency
DCAA—dual call, automatic answer device
DCCU—Data Communications Control Unit
DCE—data communications equipment
DCMSND—decommissioned
DCT—direct
DDA—dedicated digital access
DDD—direct distance dialing
DDM—difference in depth of modulation
DDS—Digital Data Service
DEA—Drug Enforcement Agency
DEDS—data entry and display system
DEGS—degrees
DEIS—Draft Environmental Impact Statement
DEP—depart/departure
DEPPROC—departure procedures
DEWIZ—distance early warning identification zone
DF—direction finder
DFAX—digital facsimile
DFI—direction finding indicator
DGPS—Differential Global Positioning Satellite (System)
DH—decision height
DID—direct inward dial
DIP—drop and insert point
DIRF—direction finding
DISABLD—disabled
DIST—distance
DITCO—Defense Information Technology Contracting Office Agency

DLA—delay or delayed
DLT—delete
DLY—daily
DME—distance measuring equipment
DME/P—precision distance measuring equipment
DMN—Data Multiplexing Network
DMSTN—demonstration
DNL—day-night equivalent sound level (also called Ldn)
DOD—direct outward dial
DoD—Department of Defense
DOI—Department of Interior
DOS—Department of State
DOT—Department of Transportation
DOTCC—Department of Transportation Computer Center
DOTS—dynamic ocean tracking system
DP—dew point temperature
DRFT—snowbank(s) caused by wind action
DSCS—digital satellite compression service
DSPLCD—displaced
DSUA—dynamic special use airspace
DTS—dedicated transmission service
DUAT—direct user access terminal
DVFR—defense visual flight rules; day visual flight rules
DVOR—doppler very high frequency omni-directional range
DYSIM—dynamic simulator

E

E—east
EA—environmental assessment
EARTS—en route automated radar tracking system
EB—eastbound
ECOM—en route communications
ECVFP—expanded charted visual flight procedures
EDCT—expedite departure path
EFC—expect further clearance
EFIS—electronic flight information systems
EIAF—expanded inward access features
EIS—environmental impact statement
ELEV—elevation
ELT—emergency locator transmitter
ELWRT—electrowriter
EMAS—engineered materials arresting system
EMPS—en route maintenance processor system
EMS—environmental management system
E-MSAW—en route automated minimum safe altitude warning
ENAV—en route navigational aids
ENG—engine
ENRT—en route
ENTR—entire
EOF—emergency Operating Facility
EPA—Environmental Protection Agency
EPS—Engineered Performance Standards

EPSS—enhanced packet switched service
ERAD—en route broadband radar
ESEC—en route broadband secondary radar
ESF—extended superframe format
ESP—en route spacing program
ESYS—en route equipment systems
ETA—estimated time of arrival
ETE—estimated time en route
ETG—enhanced target generator
ETMS—enhanced traffic management system
ETN—Electronic Telecommunications Network
EVAS—enhanced vortex advisory system
EVCS—emergency voice communications system
EXC—except

F

F&E—facility and equipment
FAA—Federal Aviation Administration
FAAAC—FAA aeronautical center
FAACIS—FAA communications information system
FAATC—FAA technical center
FAATSAT—FAA telecommunications satellite
FAC—facility/facilities
FAF—final approach fix
FAN—MKR fan marker
FAP—final approach point
FAPM—FTS2000 associate program manager
FAR—Federal Aviation Regulation
FAST—final approach spacing tool
FAX—facsimile equipment
FBO—fixed base operator
FBS—fall back switch
FCC—Federal Communications Commission
FCLT—freeze calculated landing time
FCOM—FSS radio voice communications
FCPU—Facility Central Processing Unit
FDAT—flight data entry and printout (FDEP) and flight data service
FDC—flight data center
FDE—flight data entry
FDEP—flight data entry and printout
FDIO—flight data input/output
FDIOC—flight data input/output center
FDIOR—flight data input/output remote
FDM—frequency division multiplexing
FDP—flight data processing
FED—federal
FEIS—Final Environmental Impact Statement
FEP—front end processor
FFAC—from facility
FI/P—flight inspection permanent
FI/T—flight inspection temporary
FIFO—Flight Inspection Field Office

FIG—flight inspection group
FINO—Flight Inspection National Field Office
FIPS—federal information publication standard
FIR—flight information region
FIRE—fire station
FIRMR—Federal Information Resource Management Regulation
FL—flight level
FLOWSIM—traffic flow planning simulation
FM—from
FMA—final monitor aid
FMF—facility master file
FMIS—FTS2000 management information system
FMS—flight management system
FNA—final approach
FNMS—FTS2000 network management system
FOIA—Freedom Of Information Act
FONSI—finding of no significant impact
FP—flight plan
FPM—feet per minute
FRC—request full route clearance
FREQ—frequency
FRH—fly runway heading
FRI—Friday
FRZN—frozen
FSAS—flight service automation system
FSDO—Flight Standards District Office
FSDPS—flight service data processing system
FSEP—facility/service/equipment profile
FSP—flight strip printer
FSPD—freeze speed parameter
FSS—flight service station
FSSA—flight service station automated service
FSTS—federal secure telephone service
FSYS—flight service station equipment systems
FTS—federal telecommunications system
FT—feet/foot
FTS2000—Federal Telecommunications System 2000
FUS—functional units or systems
FWCS—flight watch control station

G

GA—general aviation
GAA—general aviation activity
GAAA—general aviation activity and avionics
GADO—General Aviation District Office
GC—ground control
GCA—ground control approach
GIS—geographic information system
GNAS—general national airspace system
GNSS—global navigation satellite system
GOES—Geostationary Operational Environmental Satellite
GOESF—GOES feed point

GOEST—GOES terminal equipment
GOVT—government
GP—glide path
GPRA—Government Performance Results Act
GPS—global positioning system
GPWS—ground proximity warning system
GRADE—graphical airspace design environment
GRVL—gravel
GS—glide slope indicator
GSA—General Services Administration
GSE—ground support equipment

H

H—non-directional radio homing beacon (NDB)
HAA—height above airport
HAL—height above landing
HARS—high altitude route system
HAT—height above touchdown
HAZMAT—hazardous materials
HCAP—high capacity carriers
HDG—heading
HDME—NDB with distance measuring equipment
HDQ—FAA headquarters
HEL—helicopter
HELI—heliport
HF—high frequency
HH—NDB, 2kw or more
HI-EFAS—high altitude EFAS
HIRL—high intensity runway lights
HIWAS—Hazardous Inflight Weather Advisory Service
HLDC—high level data link control
HLDG—holding
HOL—holiday
HOV—high occupancy vehicle
HP—holding pattern
HR—hour
HSI—horizontal situation indicators
HUD—housing and urban development
HWAS—hazardous in-flight weather advisory
Hz—Hertz

I

I/AFSS—international AFSS
IA—indirect access
IAF—initial approach fix
IAP—instrument approach procedures
IAPA—instrument approach procedures automation
IBM—International Business Machines
IBP—international boundary point
IBR—intermediate bit rate
ICAO—International Civil Aviation Organization
ICSS—international communications switching systems

ID—identification
IDAT—interfacility data
IDENT—identify/identifier/identification
IF—intermediate fix
IFCP—interfacility communications processor
IFDS—interfacility data system
IFEA—in-flight emergency assistance
IFO—International Field Office
IFR—instrument flight rules
IFSS—international flight service station
ILS—instrument landing system
IM—inner marker
IMC—instrument meteorological conditions
IN—inch/inches
INBD—inbound
INDEFLY—indefinitely
INFO—information
INM—integrated noise model
INOP—inoperative
INS—inertial navigation system
INSTR—instrument
INT—intersection
INTL—international
INTST—intensity
IR—ice on runway(s)
IRMP—information resources management plan
ISDN—integrated services digital network
ISMLS—interim standard microwave landing system
ITI—interactive terminal interface
IVRS—interim voice response system
IW—inside wiring

K

Kbps—Kilobits per second
Khz—Kilohertz
KT—knots
KVDT—keyboard video display terminal

L

L—left
LAA—local airport advisory
LAAS—low altitude alert system
LABS—leased A B service
LABSC—LABS GS-200 computer
LABSR—LABS remote equipment
LABSW—LABS switch system
LAHSO—land and hold short operation
LAN—local area network
LAT—latitude
LATA—local access and transport area
LAWRS—limited aviation weather reporting station
LB—pound/pounds

LC—local control
LCF—local control facility
LCN—local communications network
LCTD—located
LDA—localizer-type directional aid; landing directional aid
LDG—landing
LDIN—lead-in lights
LEC—local exchange carrier
LF—low frequency
LGT—light or lighting
LGTD—lighted
LINCS—leased interfacility NAS C
LIRL—low intensity runway lights
LIS—logistics and inventory system
LLWAS—low level wind shear alert system
LLZ—localizer
LM—compass locator at ILS middle marker
LM/MS—low/medium frequency
LMM—locator middle marker
LO—compass locator at ILS outer marker
LOC—local; locally; location; localizer
LOCID—location identifier
LOI—letter of intent
LOM—compass locator at outer marker
LONG—longitude
LPV—lateral precision performance with vertical guidance
LRCO—limited remote communications outlet
LRNAV—long range navigation
LRR—long range radar
LSR—loose snow on runway(s)
LT—left turn

M

MAA—maximum authorized altitude
MAG—magnetic
MAINT—maintain, maintenance
MALS—medium intensity approach light system
MALSF—medium intensity approach light system with sequenced flashers
MALSR—medium intensity approach light system with runway alignment indicator lights
MAP—maintenance automation program; military airport program; missed approach point; modified access pricing
MAPT—missed approach point
Mbps—megabits per second
MCA—minimum crossing altitude
MCAS—Marine Corps air station
MCC—maintenance control center
MCL—middle compass locater
MCS—maintenance and control system
MDA—minimum descent altitude
MDT—maintenance data terminal
MEA—minimum en route altitude

MED—medium
METI—meteorological information
MF—middle frequency
MFJ—modified final judgment
MFT—meter fix crossing time/slot time
MHA—minimum holding altitude
Mhg—Meghertz
MIA—minimum IFR altitudes
MIDO—Manufacturing Inspection District Office
MIN—minute
MIRL—medium intensity runway lights
MIS—Meteorological Impact Statement
MISC—miscellaneous
MISO—Manufacturing Inspection Satellite Office
MIT—miles in trail
MITRE—Mitre Corporation
MLS—microwave landing system
MM—middle marker
MMAC—Mike Monroney Aeronautical Center
MMC—maintenance monitoring console
MMS—maintenance monitoring system
MNM—minimum
MNPS—minimum navigation performance specification
MNPSA—minimum navigation performance specifications airspace
MNT—monitor; monitoring; monitored
MOA—memorandum of agreement; military operations area
MOC—minimum obstruction clearance
MOCA—minimum obstruction clearance altitude
MODE C—altitude-encoded beacon reply; altitude reporting mode of secondary radar
MODE S—mode select beacon system
MON—Monday
MOU—memorandum of understanding
MPO—Metropolitan Planning Organization
MPS—maintenance processor subsystem or master plan supplement
MRA—minimum reception altitude
MRC—monthly recurring charge
MSA—minimum safe altitude; minimum sector altitude
MSAW—minimum safe altitude warning
MSG—message
MSL—mean sea level
MSN—message switching network
MTCS—modular terminal communications system
MTI—moving target indicator
MU—mu meters
MUD—mud
MUNI—municipal
MUX—multiplexor
MVA—minimum vectoring altitude
MVFR—marginal visual flight rules

N

N—north
NA—not authorized
NAAQS—national ambient air quality standards
NADA—ADIN concentrator
NADIN—National Airspace Data Interchange Network
NADSW—NADIN switches
NAILS—National Airspace Integrated Logistics Support
NAMS—NADIN IA
NAPRS—National Airspace Performance Reporting System
NAS—National Airspace System or Naval Air Station
NASDC—National Aviation Safety Data
NASP—National Airspace System Plan
NASPAC—National Airspace System Performance Analysis Capability
NATCO—National Communications Switching Center
NAV—navigation
NAVAID—navigation aid
NAVMN—navigation monitor and control
NAWAU—National Aviation Weather Advisory Unit
NAWPF—National Aviation Weather Processing Facility
NB—northbound
NCAR—National Center for Atmospheric Research, Boulder, CO
NCF—National Control Facility
NCIU—NEXRAD Communications Interface Unit
NCP—noise compatibility program
NCS—national communications system
NDB—non-directional radio beacon
NDNB—NADIN II
NE—northeast
NEM—noise exposure map
NEPA—National Environmental Policy Act
NEXRAD—next generation weather radar
NFAX—National Facsimile Service
NFDC—National Flight Data Center
NFIS—NAS Facilities Information System
NGT—night
NI—network interface
NICS—national interfacility communications system
NM—nautical mile(s)
NMAC—near mid-air collision
NMC—National Meteorological Center
NMCE—network monitoring and control equipment
NMCS—network monitoring and control system
NMR—nautical mile radius
NOAA—National Oceanic and Atmospheric Administration
NOC—notice of completion
NONSTD—nonstandard
NOPT—no procedure turn required
NOTAM—notice to airmen
NPDES—National pollutant discharge elimination system
NPE—non-primary airport entitlement

NPIAS—national plan of integrated airport systems
NR—number
NRC—non-recurring charge
NRCS—national radio communications systems
NSAP—National Service Assurance Plan
NSRCATN—National Strategy to Reduce Congestion on America's Transportation Network
NSSFC—National Severe Storms Forecast Center
NSSL—National Severe Storms Laboratory, Norman, OK
NSWRH—NWS Regional Headquarters
NTAP—Notices To Airmen Publication
NTP—National Transportation Policy
NTSB—National Transportation Safety Board
NTZ—no transgression zone
NW—northwest
NWS—National Weather Service
NWSR—NWS weather excluding NXRD
NXRD—advanced weather radar system

O

OAG—official airline guide
OALT—operational acceptable level of traffic
OAW—off-airway weather station
OBSC—obscured
OBST—obstruction
ODAL—omnidirectional approach lighting system
ODAPS—oceanic display and processing station
OEP—operational evolution plan/partnership
OFA—object free area
OFDPS—offshore flight data processing system
OFT—outer fix time
OFZ—obstacle free zone
OM—outer marker
OMB—Office Of Management and Budget
ONER—Oceanic Navigational Error Report
OPLT—operational acceptable level of traffic
OPR—operate
OPS—operation
OPSW—operational switch
OPX—off premises exchange
ORD—operational readiness demonstration
ORIG—original
OTR—oceanic transition route
OTS—out of service; organized track system
OVR—over

P

PABX—private automated branch exchange
PAD—packet assembler/disassembler
PAEW—personnel and equipment working
PAM—peripheral adapter module
PAPI—precision approach path indicator
PAR—precision approach radar; preferential arrival route

PARL—parallel

PAT—pattern

PATWAS—Pilots Automatic Telephone Weather Answering Service

PAX—passenger

PBCT—proposed boundary crossing time

PBRF—pilot briefing

PBX—private branch exchange

PCA—positive control airspace

PCL—pilot controlled lighting

PCM—pulse code modulation

PD—Pilot Deviation

PDAR—preferential arrival and departure route

PDC—pre-departure clearance; program designator code

PDN—Public Data Network

PDR—preferential departure route

PERM—permanent/permanently

PFC—passenger facility charge

PGP—planning grant program

PIC—principal interexchange carrier

PIDP—programmable indicator data processor

PIREP—pilot weather report

PJE—parachute jumping exercise

PLA—practice low approach

PLW—plow/plowed

PMS—program management system

PNR—prior notice required

POLIC—police station

POP—point of presence

POT—point of termination

PPIMS—personal property information management system

PPR—prior permission required

PR—primary commercial service airport

PREV—previous

PRI—primary rate interface

PRM—precision runway monitor

PRN—pseudo random noise

PROC—procedure

PROP—propeller

PSDN—public switched data network

PSN—packet switched network

PSR—packed snow on runway(s)

PSS—packet switched service

PSTN—public switched telephone network

PTC—presumed-to-conform

PTCHY—patchy

PTN—procedure turn

PUB—publication

PUP—principal user processor

PVC—permanent virtual circuit

PVD—plan view display

PVT—private

R

RAIL—runway alignment indicator lights

RAMOS—remote automatic meteorological observing system

RAPCO—radar approach control (USAF)

RAPCON—radar approach control (FAA)

RATCC—Radar Air Traffic Control Center

RATCF—Radar Air Traffic Control Facility (USN)

RBC—rotating beam ceilometer

RBDPE—radar beacon data processing equipment

RBSS—Radar Bomb Scoring Squadron

RCAG—remote communications air/ground facility

RCC—Rescue Coordination Center

RCCC—Regional Communications Control Centers

RCF—Remote Communication Facility

RCIU— Remote Control Interface Unit

RCL— runway centerline; radio communications link

RCLL—runway centerline light system

RCLR—RCL repeater

RCLT—RCL terminal

RCO—remote communications outlet

RCU—remote control unit

RDAT—digitized radar data

RDP—radar data processing

RDSIM—runway delay simulation model

REC—receive/receiver

REIL—runway end identifier lights

RELCTD—relocated

REP—report

RF—radio frequency

RL—General Aviation Reliever Airport

RLLS—runway lead-in lights system

RMCC—Remote Monitor Control Center

RMCF—Remote Monitor Control Facility

RML—radio microwave link

RMLR—RML repeater

RMLT—RML terminal

RMM—remote maintenance monitoring

RMMS—remote maintenance monitoring system

RMNDR—remainder

RMS—remote monitoring subsystem

RMSC—remote monitoring subsystem concentrator

RNAV—area navigation

RNP—required navigation performance

ROD—record of decision

ROSA—report of service activity

ROT—runway occupancy time

RP—restoration priority

RPC—restoration priority code

RPG—radar processing group

RPLC—replace

RPZ—runway protection zone
RQRD—required
RRH—remote reading hygrothermometer
RRHS—remote reading hydrometer
RRL—runway remaining lights
RRWDS—remote radar weather display
RRWSS—RWDS sensor site
RSA—runway safety area
RSAT—runway safety action team
RSR—en route surveillance radar
RSS—remote speaking system
RSVN—reservation
RT—right turn; remote transmitter
RT & BTL—radar tracking and beacon tracking level
RTAD—remote tower alphanumerics display
RTCA—Radio Technical Commission for Aeronautics
RTE—route
RTP—regional transportation plan
RTR—remote transmitter/receiver
RTRD—remote tower radar display
RTS—return to service
RUF—rough
RVR—runway visual range
RVRM—runway visual range midpoint
RVRR—runway visual range rollout
RVRT—runway visual range touchdown
RW—runway
RWDS—same as RRWDS
RWP—real-time weather processor
RWY—runway

S

S—south
S/S—sector suite
SA—sand, sanded
SAC—Strategic Air Command
SAFI—semi-automatic flight inspection
SALS—short approach lighting system
SAT—Saturday
SATCOM—satellite communications
SAWR—Supplementary Aviation Weather Reporting Station
SAWRS—Supplementary Aviation Weather Reporting System
SB—southbound
SBGP—state block grant program
SCC—System Command Center
SCVTS—Switched Compressed Video Telecommunications Service
SDF—simplified directional facility; simplified direction finding; software defined network
SDIS—switched digital integrated service
SDP—service delivery point

SD-ROB—radar weather report
SDS—switched data service
SE—southeast
SEL—single event level
SELF—simplified short approach lighting system with sequenced flashing lights
SFAR-38—Special Federal Aviation Regulation 38
SFL—sequence flashing lights
SHPO—State Historic Preservation Officer
SIC—service initiation charge
SID— standard instrument departure; station identifier
SIGMET—significant meteorological information
SIMMOD—airport and airspace simulation model
SIMUL—simultaneous
SIP—state implementation plan
SIR—packed or compacted snow and ice on runway(s)
SKED—scheduled
SLR—slush on runway(s)
SM—statute miles
SMGC—surface movement guidance and control
SMPS—sector maintenance processor subsystem
SMS—safety management system; simulation modeling system
SN—snow
SNBNK—snowbank(s) caused by plowing
SNGL—single
SNR—signal-to-noise ratio, also: S/N
SOAR—system of airports reporting
SOC—service oversight center
SOIR—simultaneous operations on intersecting runways
SOIWR—simultaneous operations on intersecting wet runways
SPD—speed
SRAP—sensor receiver and processor
SSALF—simplified short approach lighting system with sequenced flashers
SSALR—simplified short approach lighting system with runway alignment indicator lights
SSALS—simplified short approach lighting system
SSB—single side band
SSR—secondary surveillance radar
STA—straight-in approach
STAR—standard terminal arrival route
STD—standard
STMUX—statistical data multiplexer
STOL—short takeoff and landing
SUN—Sunday
SURPIC—surface picture
SVC—service
SVCA—service A
SVCB—service B
SVCC—service C

SVCU—service U
SVFB—interphone service F (B)
SVFC—interphone service F (C)
SVFD—interphone service F (D)
SVFO—interphone service F (A)
SVFR—special visual flight rules
SW—southwest
SWEPT—swept or broom/broomed

T

T—temperature
T1MUX—T1 multiplexer
TAA—terminal arrival area
TAAS—terminal advance automation system
TACAN—tactical air navigation
TACR—TACAN at VOR, TACAN only
TAF—terminal area forecast
TAR—terminal area surveillance radar
TARS—terminal automated radar service
TAS—true air speed
TATCA—terminal air traffic control automation
TAVT—terminal airspace visualization tool
TCA—traffic control airport or tower control airport; terminal control area
TCACCIS—Transportation Coordinator Automated Command And Control Information System
TCAS—Traffic Alert and Collision Avoidance System
TCC—DOT Transportation Computer Center
TCCC—Tower Control Computer Complex
TCE—tone control equipment
TCLT—tentative calculated landing time
TCO—Telecommunications Certification Officer
TCOM—Terminal Communications
TCS—tower communications system
TDLS—Tower Data-Link Services
TDMUX—time division data multiplexer
TDWR—terminal doppler weather radar
TDZ—touchdown zone
TDZ LG—touchdown zone lights
TELCO—telephone company
TELMS—telecommunications management system
TEMPO—temporary
TERPS—terminal instrument procedures
TFAC—to facility
TFC—traffic
TFR—temporary flight restriction
TGL—touch-and-go landings
TH—threshold
THN—thin
THR—threshold
THRU—through
THU—Thursday
TIL—until

TIMS—telecommunications information management system
TIPS—terminal information processing system
TKOF—takeoff
TL—taxilane
TM—traffic management
TM&O—telecommunications management and operations
TMA—Traffic Management Advisor
TMC—Traffic Management Coordinator
TMC/MC—Traffic Management Coordinator/Military Coordinator
TMCC—terminal information processing system; Traffic Management Computer Complex
TMF—Traffic Management Facility
TML—television microwave link
TMLI—television microwave link indicator
TMLR—television microwave link repeater
TMLT—television microwave link terminal
TMP—Traffic Management Processor
TMPA—traffic management program alert
TMS—traffic management system
TMSPS—traffic management specialists
TMU—traffic management unit
TNAV—terminal navigational aids
TODA—takeoff distance available
TOF—time of flight
TOFMS—time of flight mass spectrometer
TOPS—Telecommunications Ordering And Pricing System (GSA software tool)
TORA—take-off run available
TR—telecommunications request
TRACAB—terminal radar approach control in tower cab
TRACON—Terminal Radar Approach Control Facility
TRAD—terminal radar service
TRB—Transportation Research Board
TRML—terminal
TRNG—training
TRSN—transition
TSA—taxiway safety area; Transportation Security Administration
TSEC—terminal secondary radar service
TSNT—transient
TSP—telecommunications service priority
TSR—telecommunications service request
TSYS—terminal equipment systems
TTMA—TRACON Traffic Management Advisor
TTY—teletype
TUE—Tuesday
TVOR—terminal VHF omnidirectional range
TW—taxiway
TWEB—transcribed weather broadcast
TWR—tower
TWY—taxiway
TY—type (FAACIS)

U

UAS—unmanned aircraft systems
UFN—until further notice
UHF—ultra high frequency
UNAVBL—unavailable
UNLGTD—unlighted
UNMKD—unmarked
UNMNT—unmonitored
UNREL—unreliable
UNUSBL—unusable
URA—Uniform Relocation Assistance and Real Property Acquisition Policies Act of 1970
USAF—United States Air Force
USC—United States Code
USOC—Uniform Service Order Code

V

V/PD—Vehicle/pedestrian deviation
VALE—voluntary airport low emission
VASI—visual approach slope indicator
VDME—VOR with distance measuring equipment
VDP—visual descent point
VF—voice frequency
VFR—visual flight rules
VGSI—visual glide slope indicator
VHF—very high frequency
VIA—by way of
VICE—instead/versus
VIS—visibility
VLF—very low frequency
VMC—visual meteorological conditions
VNAV—visual navigational aids
VNTSC—Volpe National Transportation System Center
VOL—volume
VON—virtual on-net
VOR—VHF omnidirectional range
VOR/DME—VHF omnidirectional range/distance measuring equipment
VORTAC—VOR and TACAN (collocated)
VOT—VOR Test Facility
VP/D—vehicle/pedestrian deviation
VRS—voice recording system
VSCS—voice switching and control system
VTA—vertex time of arrival
VTAC—VOR and TACAN (collocated)
VTOL—vertical takeoff and landing
VTS—voice telecommunications system

W

W—west
WAAS—Wide Area Augmentation System
WAN—wide area network
WB—westbound
WC—work center
WCP—Weather Communications Processor
WECO—Western Electric Company
WED—Wednesday
WEF—with effect from; effective from
WESCOM—Western Electric Satellite Communications
WI—within
WIE—with immediate effect, or effective immediately
WKDAYS—Monday through Friday
WKEND—Saturday and Sunday
WMSC—Weather Message Switching Center
WMSCR—Weather Message Switching Center Replacement
WND—wind
WPT—waypoint
WSCMO—Weather Service Contract Meteorological Observatory
WSFO—Weather Service Forecast Office
WSMO—Weather Service Meteorological Observatory
WSO—Weather Service Office
WSR—wet snow on runway(s)
WTHR—weather
WTR—water on runway(s)
WX—weather

Appendix C

Airport Signs and Markings

Airport Signs			
Type of Sign	**Action or Purpose**	**Type of Sign**	**Action or Purpose**
A 4-22	**Taxiway/Runway Hold Position:** Holding position for RWY 4-22 on TWY A.	☰	**Runway Safety Area Boundary:** Identifies exit boundary of runway safety area.
26-8	**Runway/Runway Intersection:** Identifies intersecting runways or holding position for LAHSO operations.	⊞	**ILS Critical Area Boundary:** Identifies exit boundary of ILS critical area.
B 8-APCH	**Runway Approach Hold Position:** Runway approach holding position for RWY 8 on TWY B.	↙J↗	**Taxiway Direction:** Defines direction and designation of intersecting taxiway(s).
C ILS	**ILS Critical Area Hold Position:** Holding position for the ILS critical area on TWY C.	←K	**Runway Exit:** Defines direction and designation of exit taxiway from runway.
⊖	**No Entry:** Identifies paved areas where aircraft entry is prohibited.	22↑	**Outbound Destination:** Defines directions to takeoff runway(s).
B	**Taxiway Location:** Identifies taxiway on which aircraft is located.	↖MIL	**Inbound Destination:** Defines directions to destination for arriving aircraft.
22	**Runway Location:** Identifies runway on which aircraft is located.	⬛	**Taxiway Ending Marker:** Indicates taxiway does not continue.
4	**Runway Distance Remaining:** Provides remaining runway length in 1,000-foot increments.	↙A G L→	**Direction Sign Array:** Identifies location in conjunction with multiple intersecting taxiways.

Figure C-1. *Samples and explanations of standard airport signs.*

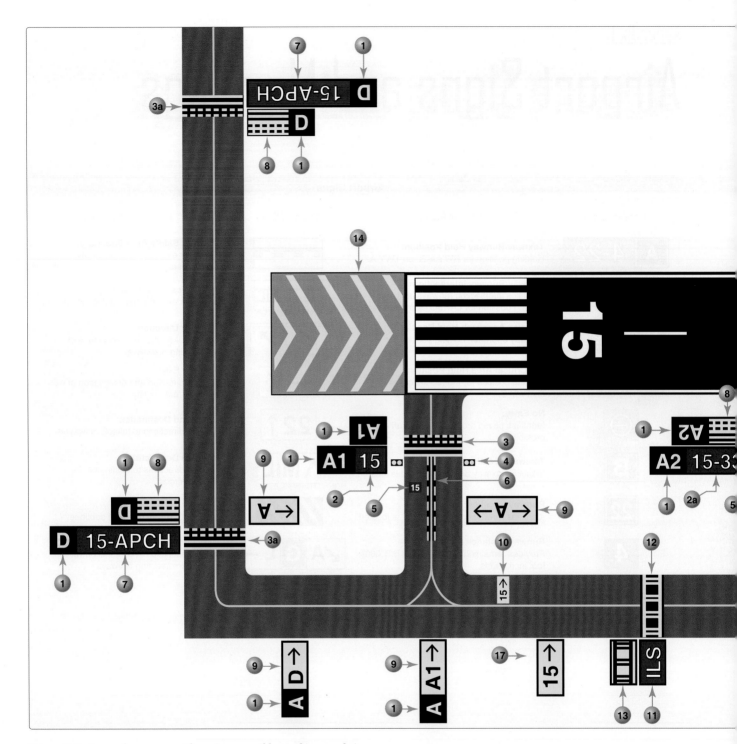

Figure C-2. *A sample runway with various possible markings and signs.*

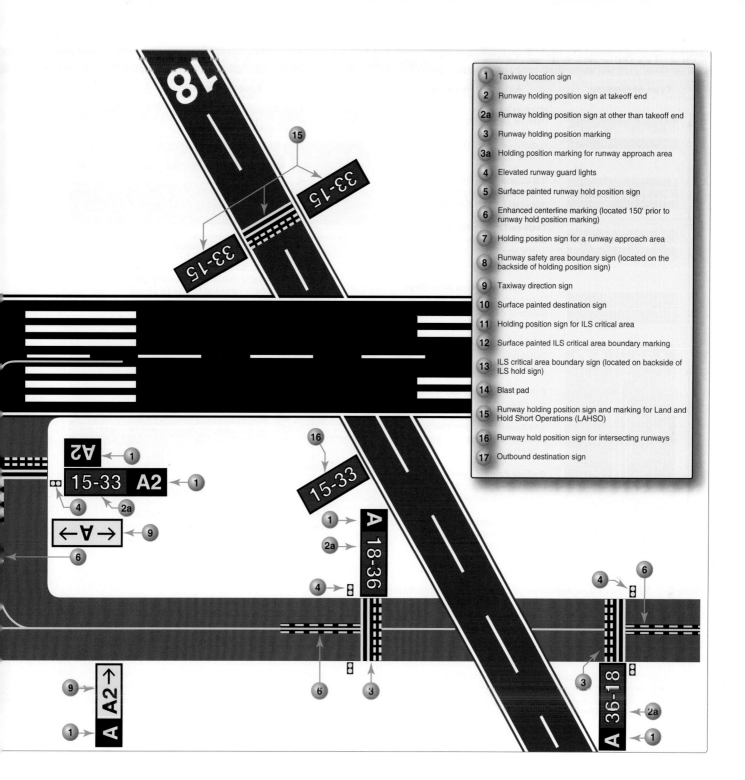

1 Taxiway location sign

2 Runway holding position sign at takeoff end

2a Runway holding position sign at other than takeoff end

3 Runway holding position marking

3a Holding position marking for runway approach area

4 Elevated runway guard lights

5 Surface painted runway hold position sign

6 Enhanced centerline marking (located 150' prior to runway hold position marking)

7 Holding position sign for a runway approach area

8 Runway safety area boundary sign (located on the backside of holding position sign)

9 Taxiway direction sign

10 Surface painted destination sign

11 Holding position sign for ILS critical area

12 Surface painted ILS critical area boundary marking

13 ILS critical area boundary sign (located on backside of ILS hold sign)

14 Blast pad

15 Runway holding position sign and marking for Land and Hold Short Operations (LAHSO)

16 Runway hold position sign for intersecting runways

17 Outbound destination sign

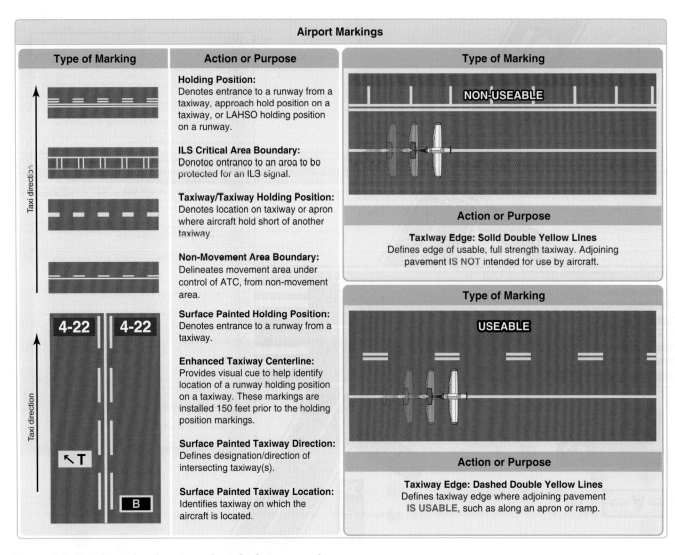

Figure C-3. *Samples and explanations of standard airport markings.*

Glossary

A

14 CFR. See Title 14 of the Code of Federal Regulations.

100-hour inspection. An inspection identical in scope to an annual inspection. Conducted every 100 hours of flight on aircraft of under 12,500 pounds that are used to carry passengers for hire.

Absolute accuracy. The ability to determine present position in space independently, and is most often used by pilots.

Absolute altitude. The actual distance between an aircraft and the terrain over which it is flying.

Absolute pressure. Pressure measured from the reference of zero pressure, or a vacuum.

A.C. Alternating current.

Acceleration. Force involved in overcoming inertia, and which may be defined as a change in velocity per unit of time.

Acceleration error. A magnetic compass error apparent when the aircraft accelerates while flying on an easterly or westerly heading, causing the compass card to rotate toward North.

Accelerate-go distance. The distance required to accelerate to V_1 with all engines at takeoff power, experience an engine failure at V_1, and continue the takeoff on the remaining engine(s). The runway required includes the distance required to climb to 35 feet by which time V_2 speed must be attained.

Accelerate-stop distance. The distance required to accelerate to V_1 with all engines at takeoff power, experience an engine failure at V_1, and abort the takeoff and bring the airplane to a stop using braking action only (use of thrust reversing is not considered).

Accelerometer. A part of an inertial navigation system (INS) that accurately measures the force of acceleration in one direction.

ADC. See air data computer.

ADF. See automatic direction finder.

ADI. See attitude director indicator.

Adiabatic cooling. A process of cooling the air through expansion. For example, as air moves up slope it expands with the reduction of atmospheric pressure and cools as it expands.

Adiabatic heating. A process of heating dry air through compression. For example, as air moves down a slope it is compressed, which results in an increase in temperature.

Adjustable-pitch propeller. A propeller with blades whose pitch can be adjusted on the ground with the engine not running, but which cannot be adjusted in flight. Also referred to as a ground adjustable propeller. Sometimes also used to refer to constant-speed propellers that are adjustable in flight.

Adjustable stabilizer. A stabilizer that can be adjusted in flight to trim the airplane, thereby allowing the airplane to fly hands-off at any given airspeed.

ADM. See aeronautical decision-making.

ADS-B. See automatic dependent surveillance-broadcast.

Advection fog. Fog resulting from the movement of warm, humid air over a cold surface.

Adverse yaw. A condition of flight in which the nose of an airplane tends to yaw toward the outside of the turn. This is caused by the higher induced drag on the outside wing, which is also producing more lift. Induced drag is a by-product of the lift associated with the outside wing.

Aerodynamics. The science of the action of air on an object, and with the motion of air on other gases. Aerodynamics deals with the production of lift by the aircraft, the relative wind, and the atmosphere.

Aeronautical chart. A map used in air navigation containing all or part of the following: topographic features, hazards and obstructions, navigation aids, navigation routes, designated airspace, and airports.

Aeronautical decision-making (ADM). A systematic approach to the mental process used by pilots to consistently determine the best course of action in response to a given set of circumstances.

Agonic line. An irregular imaginary line across the surface of the Earth along which the magnetic and geographic poles are in alignment, and along which there is no magnetic variation.

Ailerons. Primary flight control surfaces mounted on the trailing edge of an airplane wing, near the tip. Ailerons control roll about the longitudinal axis.

Aircraft. A device that is used, or intended to be used, for flight.

Aircraft altitude. The actual height above sea level at which the aircraft is flying.

Aircraft approach category. A performance grouping of aircraft based on a speed of 1.3 times the stall speed in the landing configuration at maximum gross landing weight.

Air data computer (ADC). An aircraft computer that receives and processes pitot pressure, static pressure, and temperature to calculate very precise altitude, indicated airspeed, true airspeed, and air temperature.

Airfoil. Any surface, such as a wing, propeller, rudder, or even a trim tab, which provides aerodynamic force when it interacts with a moving stream of air.

Air mass. An extensive body of air having fairly uniform properties of temperature and moisture.

AIRMET. Inflight weather advisory issued as an amendment to the area forecast, concerning weather phenomena of operational interest to all aircraft and that is potentially hazardous to aircraft with limited capability due to lack of equipment, instrumentation, or pilot qualifications.

Airplane. An engine-driven, fixed-wing aircraft heavier than air that is supported in flight by the dynamic reaction of air against its wings.

Airplane Flight Manual (AFM). A document developed by the airplane manufacturer and approved by the Federal Aviation Administration (FAA). It is specific to a particular make and model airplane by serial number and it contains operating procedures and limitations.

Airplane Owner/Information Manual. A document developed by the airplane manufacturer containing general information about the make and model of an airplane. The airplane owner's manual is not FAA approved and is not specific to a particular serial numbered airplane. This manual is not kept current, and therefore cannot be substituted for the AFM/POH.

Airport diagram. The section of an instrument approach procedure chart that shows a detailed diagram of the airport. This diagram includes surface features and airport configuration information.

Airport/Facility Directory (A/FD). See Chart Supplement U.S.

Airport surface detection equipment (ASDE). Radar equipment specifically designed to detect all principal features and traffic on the surface of an airport, presenting the entire image on the control tower console; used to augment visual observation by tower personnel of aircraft and/or vehicular movements on runways and taxiways.

Airport surveillance radar (ASR). Approach control radar used to detect and display an aircraft's position in the terminal area.

Airport surveillance radar approach. An instrument approach in which ATC issues instructions for pilot compliance based on aircraft position in relation to the final approach course and the distance from the end of the runway as displayed on the controller's radar scope.

Air route surveillance radar (ARSR). Air route traffic control center (ARTCC) radar used primarily to detect and display an aircraft's position while en route between terminal areas.

Air route traffic control center (ARTCC). Provides ATC service to aircraft operating on IFR flight plans within controlled airspace and principally during the en route phase of flight.

Airspeed. Rate of the aircraft's progress through the air.

Airspeed indicator. A differential pressure gauge that measures the dynamic pressure of the air through which the aircraft is flying. Displays the craft's airspeed, typically in knots, to the pilot.

Air traffic control radar beacon system (ATCRBS). Sometimes called secondary surveillance radar (SSR), which utilizes a transponder in the aircraft. The ground equipment is an interrogating unit, in which the beacon antenna is mounted so it rotates with the surveillance antenna. The interrogating unit transmits a coded pulse sequence that actuates the aircraft transponder. The transponder answers the coded sequence by transmitting a preselected coded sequence back to the ground equipment, providing a strong return signal and positive aircraft identification, as well as other special data.

Airway. An airway is based on a centerline that extends from one navigation aid or intersection to another navigation aid (or through several navigation aids or intersections); used to establish a known route for en route procedures between terminal areas.

Airworthiness Certificate. A certificate issued by the FAA to all aircraft that have been proven to meet the minimum standards set down by the Code of Federal Regulations.

Airworthiness Directive. A regulatory notice sent out by the FAA to the registered owner of an aircraft informing the owner of a condition that prevents the aircraft from continuing to meet its conditions for airworthiness. Airworthiness Directives (AD notes) are to be complied with within the required time limit, and the fact of compliance, the date of compliance, and the method of compliance are recorded in the aircraft's maintenance records.

Alert area. An area in which there is a high volume of pilot training or an unusual type of aeronautical activity.

Almanac data. Information the global positioning system (GPS) receiver can obtain from one satellite which describes the approximate orbital positioning of all satellites in the constellation. This information is necessary for the GPS receiver to know what satellites to look for in the sky at a given time.

ALS. See approach lighting system.

Alternate airport. An airport designated in an IFR flight plan, providing a suitable destination if a landing at the intended airport becomes inadvisable.

Alternate static source valve. A valve in the instrument static air system that supplies reference air pressure to the altimeter, airspeed indicator, and vertical speed indicator if the normal static pickup should become clogged or iced over.

Altimeter. A flight instrument that indicates altitude by sensing pressure changes.

Altimeter setting. Station pressure (the barometric pressure at the location the reading is taken) which has been corrected for the height of the station above sea level.

Altitude engine. A reciprocating aircraft engine having a rated takeoff power that is producible from sea level to an established higher altitude.

Ambient pressure. The pressure in the area immediately surrounding the aircraft.

Ambient temperature. The temperature in the area immediately surrounding the aircraft.

AME. See aviation medical examiner.

Amendment status. The circulation date and revision number of an instrument approach procedure, printed above the procedure identification.

Ammeter. An instrument installed in series with an electrical load used to measure the amount of current flowing through the load.

Aneroid. The sensitive component in an altimeter or barometer that measures the absolute pressure of the air. It is a sealed, flat capsule made of thin disks of corrugated metal soldered together and evacuated by pumping all of the air out of it.

Aneroid barometer. An instrument that measures the absolute pressure of the atmosphere by balancing the weight of the air above it against the spring action of the aneroid.

Angle of attack. The angle of attack is the angle at which relative wind meets an airfoil. It is the angle that is formed by the chord of the airfoil and the direction of the relative wind or between the chord line and the flight path. The angle of attack changes during a flight as the pilot changes the direction of the aircraft and is related to the amount of lift being produced.

Angle of incidence. The acute angle formed between the chord line of an airfoil and the longitudinal axis of the aircraft on which it is mounted.

Anhedral. A downward slant from root to tip of an aircraft's wing or horizontal tail surface.

Annual inspection. A complete inspection of an aircraft and engine, required by the Code of Federal Regulations, to be accomplished every 12 calendar months on all certificated aircraft. Only an A&P technician holding an Inspection Authorization can conduct an annual inspection.

Anti-ice. Preventing the accumulation of ice on an aircraft structure via a system designed for that purpose.

Antiservo tab. An adjustable tab attached to the trailing edge of a stabilator that moves in the same direction as the primary control. It is used to make the stabilator less sensitive.

Approach lighting system (ALS). Provides lights that will penetrate the atmosphere far enough from touchdown to give directional, distance, and glidepath information for safe transition from instrument to visual flight.

Area chart. Part of the low-altitude en route chart series, this chart furnishes terminal data at a larger scale for congested areas.

Area forecast (FA). A report that gives a picture of clouds, general weather conditions, and visual meteorological conditions (VMC) expected over a large area encompassing several states.

Area navigation (RNAV). Allows a pilot to fly a selected course to a predetermined point without the need to overfly ground-based navigation facilities, by using waypoints.

Arm. See moment arm.

ARSR. See air route surveillance radar.

ARTCC. See air route traffic control center.

ASDE. See airport surface detection equipment.

ASOS. See Automated Surface Observing System.

Aspect ratio. Span of a wing divided by its average chord.

ASR. See airport surveillance radar.

Asymmetric thrust. Also known as P-factor. A tendency for an aircraft to yaw to the left due to the descending propeller blade on the right producing more thrust than the ascending blade on the left. This occurs when the aircraft's longitudinal axis is in a climbing attitude in relation to the relative wind. The P-factor would be to the right if the aircraft had a counterclockwise rotating propeller.

ATC. Air Traffic Control.

ATCRBS. See air traffic control radar beacon system.

ATIS. See automatic terminal information service.

Atmospheric propagation delay. A bending of the electromagnetic (EM) wave from the satellite that creates an error in the GPS system.

Attitude. A personal motivational predisposition to respond to persons, situations, or events in a given manner that can, nevertheless, be changed or modified through training as sort of a mental shortcut to decision-making.

Attitude and heading reference system (AHRS). A system composed of three-axis sensors that provide heading, attitude, and yaw information for aircraft. AHRS are designed to replace traditional mechanical gyroscopic flight instruments and provide superior reliability and accuracy.

Attitude director indicator (ADI). An aircraft attitude indicator that incorporates flight command bars to provide pitch and roll commands.

Attitude indicator. The foundation for all instrument flight, this instrument reflects the airplane's attitude in relation to the horizon.

Attitude instrument flying. Controlling the aircraft by reference to the instruments rather than by outside visual cues.

Attitude management. The ability to recognize hazardous attitudes in oneself and the willingness to modify them as necessary through the application of an appropriate antidote thought.

Autokinesis. Nighttime visual illusion that a stationary light is moving, which becomes apparent after several seconds of staring at the light.

Automated Surface Observing System (ASOS). Weather reporting system which provides surface observations every minute via digitized voice broadcasts and printed reports.

Automated Weather Observing System (AWOS). Automated weather reporting system consisting of various sensors, a processor, a computer-generated voice subsystem, and a transmitter to broadcast weather data.

Automatic dependent surveillance—broadcast (ADS–B). A function on an aircraft or vehicle that periodically broadcasts its state vector (i.e., horizontal and vertical position, horizontal and vertical velocity) and other information.

Automatic direction finder (ADF). Electronic navigation equipment that operates in the low- and medium-frequency bands. Used in conjunction with the ground-based nondirectional beacon (NDB), the instrument displays the number of degrees clockwise from the nose of the aircraft to the station being received.

Automatic terminal information service (ATIS). The continuous broadcast of recorded non-control information in selected terminal areas. Its purpose is to improve controller effectiveness and relieve frequency congestion by automating repetitive transmission of essential but routine information.

Autopilot. An automatic flight control system which keeps an aircraft in level flight or on a set course. Automatic pilots can be directed by the pilot, or they may be coupled to a radio navigation signal.

Aviation medical examiner (AME). A physician with training in aviation medicine designated by the Civil Aerospace Medical Institute (CAMI).

Aviation Routine Weather Report (METAR). Observation of current surface weather reported in a standard international format.

AWOS. See Automated Weather Observing System.

Axes of an aircraft. Three imaginary lines that pass through an aircraft's center of gravity. The axes can be considered as imaginary axles around which the aircraft rotates. The three axes pass through the center of gravity at 90° angles to each other. The axis from nose to tail is the longitudinal axis (pitch), the axis that passes from wingtip to wingtip is the lateral axis (roll), and the axis that passes vertically through the center of gravity is the vertical axis (yaw).

Axial flow compressor. A type of compressor used in a turbine engine in which the airflow through the compressor is essentially linear. An axial-flow compressor is made up of several stages of alternate rotors and stators. The compressor ratio is determined by the decrease in area of the succeeding stages.

Azimuth card. A card that may be set, gyroscopically controlled, or driven by a remote compass.

B

Back course (BC). The reciprocal of the localizer course for an ILS. When flying a back-course approach, an aircraft approaches the instrument runway from the end at which the localizer antennas are installed.

Balance tab. An auxiliary control mounted on a primary control surface, which automatically moves in the direction opposite the primary control to provide an aerodynamic assist in the movement of the control.

Baro-aiding. A method of augmenting the GPS integrity solution by using a nonsatellite input source. To ensure that baro-aiding is available, the current altimeter setting must be entered as described in the operating manual.

Barometric scale. A scale on the dial of an altimeter to which the pilot sets the barometric pressure level from which the altitude shown by the pointers is measured.

Basic empty weight (GAMA). Basic empty weight includes the standard empty weight plus optional and special equipment that has been installed.

BC. See back course.

Bernoulli's Principle. A principle that explains how the pressure of a moving fluid varies with its speed of motion. An increase in the speed of movement causes a decrease in the fluid's pressure.

Biplanes. Airplanes with two sets of wings.

Block altitude. A block of altitudes assigned by ATC to allow altitude deviations; for example, "Maintain block altitude 9 to 11 thousand."

Bypass ratio. The ratio of the mass airflow in pounds per second through the fan section of a turbofan engine to the mass airflow that passes through the gas generator portion of the engine.

C

Cabin altitude. Cabin pressure in terms of equivalent altitude above sea level.

Cage. The black markings on the ball instrument indicating its neutral position.

Calibrated. The instrument indication compared with a standard value to determine the accuracy of the instrument.

Calibrated orifice. A hole of specific diameter used to delay the pressure change in the case of a vertical speed indicator.

Calibrated airspeed. The speed at which the aircraft is moving through the air, found by correcting IAS for instrument and position errors.

Camber. The camber of an airfoil is the characteristic curve of its upper and lower surfaces. The upper camber is more pronounced, while the lower camber is comparatively flat. This causes the velocity of the airflow immediately above the wing to be much higher than that below the wing.

Canard. A horizontal surface mounted ahead of the main wing to provide longitudinal stability and control. It may be a fixed, movable, or variable geometry surface, with or without control surfaces.

Canard configuration. A configuration in which the span of the forward wings is substantially less than that of the main wing.

Cantilever. A wing designed to carry loads without external struts.

CAS. Calibrated airspeed.

CDI. Course deviation indicator.

Ceiling. The height above the earth's surface of the lowest layer of clouds, which is reported as broken or overcast, or the vertical visibility into an obscuration.

Center of gravity (CG). The point at which an airplane would balance if it were possible to suspend it at that point. It is the mass center of the airplane, or the theoretical point at which the entire weight of the airplane is assumed to be concentrated. It may be expressed in inches from the reference datum, or in percentage of mean aerodynamic chord (MAC). The location depends on the distribution of weight in the airplane.

Center of gravity limits. The specified forward and aft points within which the CG must be located during flight. These limits are indicated on pertinent airplane specifications.

Center of gravity range. The distance between the forward and aft CG limits indicated on pertinent airplane specifications.

Center of pressure. A point along the wing chord line where lift is considered to be concentrated. For this reason, the center of pressure is commonly referred to as the center of lift.

Centrifugal flow compressor. An impeller-shaped device that receives air at its center and slings the air outward at high velocity into a diffuser for increased pressure. Also referred to as a radial outflow compressor.

Centrifugal force. An outward force that opposes centripetal force, resulting from the effect of inertia during a turn.

Centripetal force. A center-seeking force directed inward toward the center of rotation created by the horizontal component of lift in turning flight.

CG. See center of gravity.

Changeover point (COP). A point along the route or airway segment between two adjacent navigation facilities or waypoints where changeover in navigation guidance should occur.

Chart Supplement U.S. (formerly Airport/Facility Directory). An FAA publication containing information on all airports, communications, and NAVAIDs.

Checklist. A tool that is used as a human factors aid in aviation safety. It is a systematic and sequential list of all operations that must be performed to properly accomplish a task.

Chord line. An imaginary straight line drawn through an airfoil from the leading edge to the trailing edge.

Circling approach. A maneuver initiated by the pilot to align the aircraft with a runway for landing when a straight-in landing from an instrument approach is not possible or is not desirable.

Class A airspace. Airspace from 18,000 feet MSL up to and including FL 600, including the airspace overlying the waters within 12 NM of the coast of the 48 contiguous states and Alaska; and designated international airspace beyond 12 NM of the coast of the 48 contiguous states and Alaska within areas of domestic radio navigational signal or ATC radar coverage, and within which domestic procedures are applied.

Class B airspace. Airspace from the surface to 10,000 feet MSL surrounding the nation's busiest airports in terms of IFR operations or passenger numbers. The configuration of each Class B airspace is individually tailored and consists of a surface area and two or more layers, and is designed to contain all published instrument procedures once an aircraft enters the airspace. For all aircraft, an ATC clearance is required to operate in the area, and aircraft so cleared receive separation services within the airspace.

Class C airspace. Airspace from the surface to 4,000 feet above the airport elevation (charted in MSL) surrounding those airports having an operational control tower, serviced by radar approach control, and having a certain number of IFR operations or passenger numbers. Although the configuration of each Class C airspace area is individually tailored, the airspace usually consists of a 5 NM radius core surface area that extends from the surface up to 4,000 feet above the airport elevation, and a 10 NM radius shelf area that extends from 1,200 feet to 4,000 feet above the airport elevation.

Class D airspace. Airspace from the surface to 2,500 feet above the airport elevation (charted in MSL) surrounding those airports that have an operational control tower. The configuration of each Class D airspace area is individually tailored, and when instrument procedures are published, the airspace is normally designed to contain the procedures.

Class E airspace. Airspace that is not Class A, Class B, Class C, or Class D, and is controlled airspace.

Class G airspace. Airspace that is uncontrolled, except when associated with a temporary control tower, and has not been designated as Class A, Class B, Class C, Class D, or Class E airspace.

Clean configuration. A configuration in which all flight control surfaces have been placed to create minimum drag. In most aircraft this means flaps and gear retracted.

Clearance. ATC permission for an aircraft to proceed under specified traffic conditions within controlled airspace, for the purpose of providing separation between known aircraft.

Clearance delivery. Control tower position responsible for transmitting departure clearances to IFR flights.

Clearance limit. The fix, point, or location to which an aircraft is cleared when issued an air traffic clearance.

Clearance on request. An IFR clearance not yet received after filing a flight plan.

Clearance void time. Used by ATC, the time at which the departure clearance is automatically canceled if takeoff has not been made. The pilot must obtain a new clearance or cancel the IFR flight plan if not off by the specified time.

Clear ice. Glossy, clear, or translucent ice formed by the relatively slow freezing of large, supercooled water droplets.

Coefficient of lift (C_L). The ratio between lift pressure and dynamic pressure.

Cold front. The boundary between two air masses where cold air is replacing warm air.

Compass course. A true course corrected for variation and deviation errors.

Compass locator. A low-power, low- or medium-frequency (L/MF) radio beacon installed at the site of the outer or middle marker of an ILS.

Compass rose. A small circle graduated in 360° increments, to show direction expressed in degrees.

Complex aircraft. An aircraft with retractable landing gear, flaps, and a controllable-pitch propeller.

Compressor pressure ratio. The ratio of compressor discharge pressure to compressor inlet pressure.

Compressor stall. In gas turbine engines, a condition in an axial-flow compressor in which one or more stages of rotor blades fail to pass air smoothly to the succeeding stages. A stall condition is caused by a pressure ratio that is incompatible with the engine rpm. Compressor stall will be indicated by a rise in exhaust temperature or rpm fluctuation, and if allowed to continue, may result in flameout and physical damage to the engine.

Computer navigation fix. A point used to define a navigation track for an airborne computer system such as GPS or FMS.

Concentric rings. Dashed-line circles depicted in the plan view of IAP charts, outside of the reference circle, that show en route and feeder facilities.

Condensation. A change of state of water from a gas (water vapor) to a liquid.

Condensation nuclei. Small particles of solid matter in the air on which water vapor condenses.

Cone of confusion. A cone-shaped volume of airspace directly above a VOR station where no signal is received, causing the CDI to fluctuate.

Configuration. This is a general term, which normally refers to the position of the landing gear and flaps.

Constant-speed propeller. A controllable-pitch propeller whose pitch is automatically varied in flight by a governor to maintain a constant rpm in spite of varying air loads.

Continuous flow oxygen system. System that supplies a constant supply of pure oxygen to a rebreather bag that dilutes the pure oxygen with exhaled gases and thus supplies a healthy mix of oxygen and ambient air to the mask. Primarily used in passenger cabins of commercial airliners.

Control and performance. A method of attitude instrument flying in which one instrument is used for making attitude changes, and the other instruments are used to monitor the progress of the change.

Control display unit. A display interfaced with the master computer, providing the pilot with a single control point for all navigations systems, thereby reducing the number of required flight deck panels.

Controllability. A measure of the response of an aircraft relative to the pilot's flight control inputs.

Controllable-pitch propeller (CPP). A type of propeller with blades that can be rotated around their long axis to change their pitch. If the pitch can be set to negative values, the reversible propeller can also create reverse thrust for braking or reversing without the need of changing the direction of shaft revolutions.

Controlled airspace. An airspace of defined dimensions within which ATC service is provided to IFR and VFR flights in accordance with the airspace classification. It includes Class A, Class B, Class C, Class D, and Class E airspace.

Control pressures. The amount of physical exertion on the control column necessary to achieve the desired attitude.

Convective weather. Unstable, rising air found in cumiliform clouds.

Convective SIGMET. Weather advisory concerning convective weather significant to the safety of all aircraft, including thunderstorms, hail, and tornadoes.

Conventional landing gear. Landing gear employing a third rear-mounted wheel. These airplanes are also sometimes referred to as tailwheel airplanes.

Coordinated flight. Flight with a minimum disturbance of the forces maintaining equilibrium, established via effective control use.

COP. See changeover point.

Coriolis illusion. The illusion of rotation or movement in an entirely different axis, caused by an abrupt head movement, while in a prolonged constant-rate turn that has ceased to stimulate the brain's motion sensing system.

Coupled ailerons and rudder. Rudder and ailerons are connected with interconnected springs in order to counteract adverse yaw. Can be overridden if it becomes necessary to slip the aircraft.

Course. The intended direction of flight in the horizontal plane measured in degrees from north.

Cowl flaps. Shutter-like devices arranged around certain air-cooled engine cowlings, which may be opened or closed to regulate the flow of air around the engine.

Crew resource management (CRM). The application of team management concepts in the flight deck environment. It was initially known as cockpit resource management, but as CRM programs evolved to include cabin crews, maintenance personnel, and others, the phrase "crew resource management" was adopted. This includes single pilots, as in most general aviation aircraft. Pilots of small aircraft, as well as crews of larger aircraft, must make effective use of all available resources; human resources, hardware, and information. A current definition includes all groups routinely working with the flight crew who are involved in decisions required to operate a flight safely. These groups include, but are not limited to pilots, dispatchers, cabin crewmembers, maintenance personnel, and air traffic controllers. CRM is one way of addressing the challenge of optimizing the human/machine interface and accompanying interpersonal activities.

Critical altitude. The maximum altitude under standard atmospheric conditions at which a turbocharged engine can produce its rated horsepower.

Critical angle of attack. The angle of attack at which a wing stalls regardless of airspeed, flight attitude, or weight.

Critical areas. Areas where disturbances to the ILS localizer and glideslope courses may occur when surface vehicles or aircraft operate near the localizer or glideslope antennas.

CRM. See crew resource management.

Cross-check. The first fundamental skill of instrument flight, also known as "scan," the continuous and logical observation of instruments for attitude and performance information.

Cruise clearance. An ATC clearance issued to allow a pilot to conduct flight at any altitude from the minimum IFR altitude up to and including the altitude specified in the clearance. Also authorizes a pilot to proceed to and make an approach at the destination airport.

Current induction. An electrical current being induced into, or generated in, any conductor that is crossed by lines of flux from any magnet.

D

DA. See decision altitude.

Datum (Reference Datum). An imaginary vertical plane or line from which all measurements of arm are taken. The datum is established by the manufacturer. Once the datum has been selected, all moment arms and the location of CG range are measured from this point.

D.C. Direct current.

Dark adaptation. Physical and chemical adjustments of the eye that make vision possible in relative darkness.

Dead reckoning. Navigation of an airplane solely by means of computations based on airspeed, course, heading, wind direction and speed, groundspeed, and elapsed time.

Deceleration error. A magnetic compass error that occurs when the aircraft decelerates while flying on an easterly or westerly heading, causing the compass card to rotate toward South.

Decision altitude (DA). A specified altitude in the precision approach, charted in feet MSL, at which a missed approach must be initiated if the required visual reference to continue the approach has not been established.

Decision height (DH). A specified altitude in the precision approach, charted in height above threshold elevation, at which a decision must be made either to continue the approach or to execute a missed approach.

Deice. The act of removing ice accumulation from an aircraft structure.

Delta. A Greek letter expressed by the symbol Δ to indicate a change of values. As an example, ΔCG indicates a change (or movement) of the CG.

Density altitude. Pressure altitude corrected for nonstandard temperature. Density altitude is used in computing the performance of an aircraft and its engines.

Departure procedure (DP). Preplanned IFR ATC departure, published for pilot use, in textual and graphic format.

Deposition. The direct transformation of a gas to a solid state, in which the liquid state is bypassed. Some sources use sublimation to describe this process instead of deposition.

Detonation. The sudden release of heat energy from fuel in an aircraft engine caused by the fuel-air mixture reaching its critical pressure and temperature. Detonation occurs as a violent explosion rather than a smooth burning process.

Deviation. A magnetic compass error caused by local magnetic fields within the aircraft. Deviation error is different on each heading.

Dew. Moisture that has condensed from water vapor. Usually found on cooler objects near the ground, such as grass, as the near-surface layer of air cools faster than the layers of air above it.

Dewpoint. The temperature at which air reaches a state where it can hold no more water.

DGPS. Differential global positioning system.

DH. See decision height.

Differential ailerons. Control surface rigged such that the aileron moving up moves a greater distance than the aileron moving down. The up aileron produces extra parasite drag to compensate for the additional induced drag caused by the down aileron. This balancing of the drag forces helps minimize adverse yaw.

Differential Global Positioning System (DGPS). A system that improves the accuracy of Global Navigation Satellite Systems (GNSS) by measuring changes in variables to provide satellite positioning corrections.

Differential pressure. A difference between two pressures. The measurement of airspeed is an example of the use of differential pressure.

Dihedral. The positive acute angle between the lateral axis of an airplane and a line through the center of a wing or horizontal stabilizer. Dihedral contributes to the lateral stability of an airplane.

Diluter-demand oxygen system. An oxygen system that delivers oxygen mixed or diluted with air in order to maintain a constant oxygen partial pressure as the altitude changes.

Direct indication. The true and instantaneous reflection of aircraft pitch-and-bank attitude by the miniature aircraft, relative to the horizon bar of the attitude indicator.

Direct User Access Terminal System (DUATS). A system that provides current FAA weather and flight plan filing services to certified civil pilots, via personal computer, modem, or telephone access to the system. Pilots can request specific types of weather briefings and other pertinent data for planned flights.

Directional stability. Stability about the vertical axis of an aircraft, whereby an aircraft tends to return, on its own, to flight aligned with the relative wind when disturbed from that equilibrium state. The vertical tail is the primary contributor to directional stability, causing an airplane in flight to align with the relative wind.

Distance circle. See reference circle.

Distance measuring equipment (DME). A pulse-type electronic navigation system that shows the pilot, by an instrument-panel indication, the number of nautical miles between the aircraft and a ground station or waypoint.

DME. See distance measuring equipment.

DME arc. A flight track that is a constant distance from the station or waypoint.

DOD. Department of Defense.

Doghouse. A turn-and-slip indicator dial mark in the shape of a doghouse.

Domestic Reduced Vertical Separation Minimum (DRVSM). Additional flight levels between FL 290 and FL 410 to provide operational, traffic, and airspace efficiency.

Double gimbal. A type of mount used for the gyro in an attitude instrument. The axes of the two gimbals are at right angles to the spin axis of the gyro, allowing free motion in two planes around the gyro.

DP. See departure procedure.

Drag. The net aerodynamic force parallel to the relative wind, usually the sum of two components: induced drag and parasite drag.

Drag curve. The curve created when plotting induced drag and parasite drag.

Drift angle. Angle between heading and track.

DRVSM. See Domestic Reduced Vertical Separation Minimum.

DUATS. See direct user access terminal system.

Duplex. Transmitting on one frequency and receiving on a separate frequency.

Dutch roll. A combination of rolling and yawing oscillations that normally occurs when the dihedral effects of an aircraft are more powerful than the directional stability. Usually dynamically stable but objectionable in an airplane because of the oscillatory nature.

Dynamic hydroplaning. A condition that exists when landing on a surface with standing water deeper than the tread depth of the tires. When the brakes are applied, there is a possibility that the brake will lock up and the tire will ride on the surface of the water, much like a water ski. When the tires are hydroplaning, directional control and braking action are virtually impossible. An effective anti-skid system can minimize the effects of hydroplaning.

Dynamic stability. The property of an aircraft that causes it, when disturbed from straight-and-level flight, to develop forces or moments that restore the original condition of straight and level.

E

Eddy currents. Current induced in a metal cup or disc when it is crossed by lines of flux from a moving magnet.

Eddy current damping. The decreased amplitude of oscillations by the interaction of magnetic fields. In the case of a vertical card magnetic compass, flux from the oscillating permanent magnet produces eddy currents in a damping disk or cup. The magnetic flux produced by the eddy currents opposes the flux from the permanent magnet and decreases the oscillations.

EFC. See expect-further-clearance.

EFD. See electronic flight display.

EGT. See exhaust gas temperature.

Electronic flight display (EFD). For the purpose of standardization, any flight instrument display that uses LCD or other image-producing system (cathode ray tube (CRT), etc.)

Elevator. The horizontal, movable primary control surface in the tail section, or empennage, of an airplane. The elevator is hinged to the trailing edge of the fixed horizontal stabilizer.

Elevator illusion. The sensation of being in a climb or descent, caused by the kind of abrupt vertical accelerations that result from up- or downdrafts.

Emergency. A distress or urgent condition.

Empennage. The section of the airplane that consists of the vertical stabilizer, the horizontal stabilizer, and the associated control surfaces.

Emphasis error. The result of giving too much attention to a particular instrument during the cross-check, instead of relying on a combination of instruments necessary for attitude and performance information.

Empty-field myopia. Induced nearsightedness that is associated with flying at night, in instrument meteorological conditions and/or reduced visibility. With nothing to focus on, the eyes automatically focus on a point just slightly ahead of the airplane.

EM wave. Electromagnetic wave.

Encoding altimeter. A special type of pressure altimeter used to send a signal to the air traffic controller on the ground, showing the pressure altitude the aircraft is flying.

Engine pressure ratio (EPR). The ratio of turbine discharge pressure divided by compressor inlet pressure, which is used as an indication of the amount of thrust being developed by a turbine engine.

En route facilities ring. Depicted in the plan view of IAP charts, a circle which designates NAVAIDs, fixes, and intersections that are part of the en route low altitude airway structure.

En route high-altitude charts. Aeronautical charts for en route instrument navigation at or above 18,000 feet MSL.

En route low-altitude charts. Aeronautical charts for en route IFR navigation below 18,000 feet MSL.

EPR. See engine pressure ratio.

Equilibrium. A condition that exists within a body when the sum of the moments of all of the forces acting on the body is equal to zero. In aerodynamics, equilibrium is when all opposing forces acting on an aircraft are balanced (steady, unaccelerated flight conditions).

Equivalent airspeed. Airspeed equivalent to CAS in standard atmosphere at sea level. As the airspeed and pressure altitude increase, the CAS becomes higher than it should be, and a correction for compression must be subtracted from the CAS.

Evaporation. The transformation of a liquid to a gaseous state, such as the change of water to water vapor.

Exhaust gas temperature (EGT). The temperature of the exhaust gases as they leave the cylinders of a reciprocating engine or the turbine section of a turbine engine.

Expect-further-clearance (EFC). The time a pilot can expect to receive clearance beyond a clearance limit.

Explosive decompression. A change in cabin pressure faster than the lungs can decompress. Lung damage is possible.

F

FA. See area forecast.

FAA. Federal Aviation Administration.

FAF. See final approach fix.

False horizon. Inaccurate visual information for aligning the aircraft, caused by various natural and geometric formations that disorient the pilot from the actual horizon.

FDI. See flight director indicator.

Federal airways. Class E airspace areas that extend upward from 1,200 feet to, but not including, 18,000 feet MSL, unless otherwise specified.

Feeder facilities. Used by ATC to direct aircraft to intervening fixes between the en route structure and the initial approach fix.

Final approach. Part of an instrument approach procedure in which alignment and descent for landing are accomplished.

Final approach fix (FAF). The fix from which the IFR final approach to an airport is executed, and which identifies the beginning of the final approach segment. An FAF is designated on government charts by a Maltese cross symbol for nonprecision approaches, and a lightning bolt symbol for precision approaches.

Fixating. Staring at a single instrument, thereby interrupting the cross-check process.

Fixed-pitch propellers. Propellers with fixed blade angles. Fixed-pitch propellers are designed as climb propellers, cruise propellers, or standard propellers.

Fixed slot. A fixed, nozzle shaped opening near the leading edge of a wing that ducts air onto the top surface of the wing. Its purpose is to increase lift at higher angles of attack.

FL. See flight level.

Flameout. A condition in the operation of a gas turbine engine in which the fire in the engine goes out due to either too much or too little fuel sprayed into the combustors.

Flaps. Hinged portion of the trailing edge between the ailerons and fuselage. In some aircraft ailerons and flaps are interconnected to produce full-span "flaperons." In either case, flaps change the lift and drag on the wing.

Floor load limit. The maximum weight the floor can sustain per square inch/foot as provided by the manufacturer.

Flight configurations. Adjusting the aircraft control surfaces (including flaps and landing gear) in a manner that will achieve a specified attitude.

Flight director indicator (FDI). One of the major components of a flight director system, it provides steering commands that the pilot (or the autopilot, if coupled) follows.

Flight level (FL). A measure of altitude (in hundreds of feet) used by aircraft flying above 18,000 feet with the altimeter set at 29.92 "Hg.

Flight management system (FMS). Provides pilot and crew with highly accurate and automatic long-range navigation capability, blending available inputs from long- and short-range sensors.

Flight path. The line, course, or track along which an aircraft is flying or is intended to be flown.

Flight patterns. Basic maneuvers, flown by reference to the instruments rather than outside visual cues, for the purpose of practicing basic attitude flying. The patterns simulate maneuvers encountered on instrument flights such as holding patterns, procedure turns, and approaches.

Flight strips. Paper strips containing instrument flight information, used by ATC when processing flight plans.

FMS. See flight management system.

FOD. See foreign object damage.

Fog. Cloud consisting of numerous minute water droplets and based at the surface; droplets are small enough to be suspended in the earth's atmosphere indefinitely. (Unlike drizzle, it does not fall to the surface. Fog differs from a cloud only in that a cloud is not based at the surface, and is distinguished from haze by its wetness and gray color.)

Force (F). The energy applied to an object that attempts to cause the object to change its direction, speed, or motion. In aerodynamics, it is expressed as F, T (thrust), L (lift), W (weight), or D (drag), usually in pounds.

Foreign object damage (FOD). Damage to a gas turbine engine caused by some object being sucked into the engine while it is running. Debris from runways or taxiways can cause foreign object damage during ground operations, and the ingestion of ice and birds can cause FOD in flight.

Form drag. The drag created because of the shape of a component or the aircraft.

Frise-type aileron. Aileron having the nose portion projecting ahead of the hinge line. When the trailing edge of the aileron moves up, the nose projects below the wing's lower surface and produces some parasite drag, decreasing the amount of adverse yaw.

Front. The boundary between two different air masses.

Frost. Ice crystal deposits formed by sublimation when temperature and dewpoint are below freezing.

Fuel load. The expendable part of the load of the airplane. It includes only usable fuel, not fuel required to fill the lines or that which remains trapped in the tank sumps.

Fundamental skills. Pilot skills of instrument cross-check, instrument interpretation, and aircraft control.

Fuselage. The section of the airplane that consists of the cabin and/or cockpit, containing seats for the occupants and the controls for the airplane.

G

GAMA. General Aviation Manufacturers Association.

Gimbal ring. A type of support that allows an object, such as a gyroscope, to remain in an upright condition when its base is tilted.

Glideslope (GS). Part of the ILS that projects a radio beam upward at an angle of approximately 3° from the approach end of an instrument runway. The glideslope provides vertical guidance to aircraft on the final approach course for the aircraft to follow when making an ILS approach along the localizer path.

Glideslope intercept altitude. The minimum altitude of an intermediate approach segment prescribed for a precision approach that ensures obstacle clearance.

Global landing system (GLS). An instrument approach with lateral and vertical guidance with integrity limits (similar to barometric vertical navigation (BARO VNAV).

Global navigation satellite system (GNSS). Satellite navigation system that provides autonomous geospatial positioning with global coverage. It allows small electronic receivers to determine their location (longitude, latitude, and altitude) to within a few meters using time signals transmitted along a line of sight by radio from satellites.

Global positioning system (GPS). Navigation system that uses satellite rather than ground-based transmitters for location information.

GLS. See global landing system.

GNSS. See global navigation satellite system.

Goniometer. As used in radio frequency (RF) antenna systems, a direction-sensing device consisting of two fixed loops of wire oriented 90° from each other, which separately sense received signal strength and send those signals to two rotors (also oriented 90°) in the sealed direction-indicating instrument. The rotors are attached to the direction-indicating needle of the instrument and rotated by a small motor until minimum magnetic field is sensed near the rotors.

GPS. See global positioning system.

GPS Approach Overlay Program. An authorization for pilots to use GPS avionics under IFR for flying designated existing nonprecision instrument approach procedures, with the exception of LOC, LDA, and SDF procedures.

GPWS. See ground proximity warning system.

Graveyard spiral. The illusion of the cessation of a turn while still in a prolonged, coordinated, constant rate turn, which can lead a disoriented pilot to a loss of control of the aircraft.

Great circle route. The shortest distance across the surface of a sphere (the Earth) between two points on the surface.

Ground adjustable trim tab. Non-movable metal trim tab on a control surface. Bent in one direction or another while on the ground to apply trim forces to the control surface.

Ground effect. The condition of slightly increased air pressure below an airplane wing or helicopter rotor system that increases the amount of lift produced. It exists within approximately one wing span or one rotor diameter from the ground. It results from a reduction in upwash, downwash, and wingtip vortices, and provides a corresponding decrease in induced drag.

Ground proximity warning system (GPWS). A system designed to determine an aircraft's clearance above the Earth and provides limited predictability about aircraft position relative to rising terrain.

Groundspeed. Speed over the ground, either closing speed to the station or waypoint, or speed over the ground in whatever direction the aircraft is going at the moment, depending upon the navigation system used.

GS. See glideslope.

GWPS. See ground proximity warning system.

Gyroscopic precession. An inherent quality of rotating bodies, which causes an applied force to be manifested 90° in the direction of rotation from the point where the force is applied.

H

HAA. See height above airport.

HAL. See height above landing.

HAT. See height above touchdown elevation.

Hazardous attitudes. Five aeronautical decision-making attitudes that may contribute to poor pilot judgment: anti-authority, impulsivity, invulnerability, machismo, and resignation.

Hazardous Inflight Weather Advisory Service (HIWAS). An en route FSS service providing continuously updated automated of hazardous weather within 150 nautical miles of selected VORs, available only in the conterminous 48 states.

Head-up display (HUD). A special type of flight viewing screen that allows the pilot to watch the flight instruments and other data while looking through the windshield of the aircraft for other traffic, the approach lights, or the runway.

Heading. The direction in which the nose of the aircraft is pointing during flight.

Heading indicator. An instrument which senses airplane movement and displays heading based on a 360° azimuth, with the final zero omitted. The heading indicator, also called a directional gyro (DG), is fundamentally a mechanical instrument designed to facilitate the use of the magnetic compass. The heading indicator is not affected by the forces that make the magnetic compass difficult to interpret.

Headwork. Required to accomplish a conscious, rational thought process when making decisions. Good decision-making involves risk identification and assessment, information processing, and problem solving.

Height above airport (HAA). The height of the MDA above the published airport elevation.

Height above landing (HAL). The height above a designated helicopter landing area used for helicopter instrument approach procedures.

Height above touchdown elevation (HAT). The DA/DH or MDA above the highest runway elevation in the touchdown zone (first 3,000 feet of the runway).

HF. High frequency.

Hg. Abbreviation for mercury, from the Latin hydrargyrum.

High performance aircraft. An aircraft with an engine of more than 200 horsepower.

Histotoxic hypoxia. The inability of cells to effectively use oxygen. Plenty of oxygen is being transported to the cells that need it, but they are unable to use it.

HIWAS. See Hazardous Inflight Weather Advisory Service.

Holding. A predetermined maneuver that keeps aircraft within a specified airspace while awaiting further clearance from ATC.

Holding pattern. A racetrack pattern, involving two turns and two legs, used to keep an aircraft within a prescribed airspace with respect to a geographic fix. A standard pattern uses right turns; nonstandard patterns use left turns.

Homing. Flying the aircraft on any heading required to keep the needle pointing to the 0° relative bearing position.

Horizontal situation indicator (HSI). A flight navigation instrument that combines the heading indicator with a CDI, in order to provide the pilot with better situational awareness of location with respect to the courseline.

Horsepower. The term, originated by inventor James Watt, means the amount of work a horse could do in one second. One horsepower equals 550 foot-pounds per second, or 33,000 foot-pounds per minute.

Hot start. In gas turbine engines, a start which occurs with normal engine rotation, but exhaust temperature exceeds prescribed limits. This is usually caused by an excessively rich mixture in the combustor. The fuel to the engine must be terminated immediately to prevent engine damage.

HSI. See horizontal situation indicator.

HUD. See head-up display.

Human factors. A multidisciplinary field encompassing the behavioral and social sciences, engineering, and physiology, to consider the variables that influence individual and crew performance for the purpose of optimizing human performance and reducing errors.

Hung start. In gas turbine engines, a condition of normal light off but with rpm remaining at some low value rather than increasing to the normal idle rpm. This is often the result of insufficient power to the engine from the starter. In the event of a hung start, the engine should be shut down.

Hydroplaning. A condition that exists when landing on a surface with standing water deeper than the tread depth of the tires. When the brakes are applied, there is a possibility that the brake will lock up and the tire will ride on the surface of the water, much like a water ski. When the tires are hydroplaning, directional control and braking action are virtually impossible. An effective anti-skid system can minimize the effects of hydroplaning.

Hypemic hypoxia. A type of hypoxia that is a result of oxygen deficiency in the blood, rather than a lack of inhaled oxygen. It can be caused by a variety of factors. Hypemic means "not enough blood."

Hyperventilation. Occurs when an individual is experiencing emotional stress, fright, or pain, and the breathing rate and depth increase, although the carbon dioxide level in the blood is already at a reduced level. The result is an excessive loss of carbon dioxide from the body, which can lead to unconsciousness due to the respiratory system's overriding mechanism to regain control of breathing.

Hypoxia. A state of oxygen deficiency in the body sufficient to impair functions of the brain and other organs.

Hypoxic hypoxia. This type of hypoxia is a result of insufficient oxygen available to the lungs. A decrease of oxygen molecules at sufficient pressure can lead to hypoxic hypoxia.

I

IAF. See initial approach fix.

IAP. See instrument approach procedures.

IAS. See indicated airspeed.

ICAO. See International Civil Aviation Organization.

Ident. Air Traffic Control request for a pilot to push the button on the transponder to identify return on the controller's scope.

IFR. See instrument flight rules.

ILS. See instrument landing system.

ILS categories. Categories of instrument approach procedures allowed at airports equipped with the following types of instrument landing systems:

> **ILS Category I:** Provides for approach to a height above touchdown of not less than 200 feet, and with runway visual range of not less than 1,800 feet.

> **ILS Category II:** Provides for approach to a height above touchdown of not less than 100 feet and with runway visual range of not less than 1,200 feet.

> **ILS Category IIIA:** Provides for approach without a decision height minimum and with runway visual range of not less than 700 feet.

> **ILS Category IIIB:** Provides for approach without a decision height minimum and with runway visual range of not less than 150 feet.

> **ILS Category IIIC:** Provides for approach without a decision height minimum and without runway visual range minimum.

IMC. See instrument meteorological conditions.

Inclinometer. An instrument consisting of a curved glass tube, housing a glass ball, and damped with a fluid similar to kerosene. It may be used to indicate inclination, as a level, or, as used in the turn indicators, to show the relationship between gravity and centrifugal force in a turn.

Indicated airspeed (IAS). Shown on the dial of the instrument airspeed indicator on an aircraft. Indicated airspeed (IAS) is the airspeed indicator reading uncorrected for instrument, position, and other errors. Indicated airspeed means the speed of an aircraft as shown on its pitot static airspeed indicator calibrated to reflect standard atmosphere adiabatic compressible flow at sea level uncorrected for airspeed system errors. Calibrated airspeed (CAS) is IAS corrected for instrument errors, position error (due to incorrect pressure at the static port) and installation errors.

Indicated altitude. The altitude read directly from the altimeter (uncorrected) when it is set to the current altimeter setting.

Indirect indication. A reflection of aircraft pitch-and-bank attitude by instruments other than the attitude indicator.

Induced drag. Drag caused by the same factors that produce lift; its amount varies inversely with airspeed. As airspeed decreases, the angle of attack must increase, in turn increasing induced drag.

Induction icing. A type of ice in the induction system that reduces the amount of air available for combustion. The most commonly found induction icing is carburetor icing.

Inertial navigation system (INS). A computer-based navigation system that tracks the movement of an aircraft via signals produced by onboard accelerometers. The initial location of the aircraft is entered into the computer, and all subsequent movement of the aircraft is sensed and used to keep the position updated. An INS does not require any inputs from outside signals.

Initial approach fix (IAF). The fix depicted on IAP charts where the instrument approach procedure (IAP) begins unless otherwise authorized by ATC.

Inoperative components. Higher minimums are prescribed when the specified visual aids are not functioning; this information is listed in the Inoperative Components Table found in the United States Terminal Procedures Publications.

INS. See inertial navigation system.

Instantaneous vertical speed indicator (IVSI). Assists in interpretation by instantaneously indicating the rate of climb or descent at a given moment with little or no lag as displayed in a vertical speed indicator (VSI).

Instrument approach procedures (IAP). A series of predetermined maneuvers for the orderly transfer of an aircraft under IFR from the beginning of the initial approach to a landing or to a point from which a landing may be made visually.

Instrument flight rules (IFR). Rules and regulations established by the Federal Aviation Administration to govern flight under conditions in which flight by outside visual reference is not safe. IFR flight depends upon flying by reference to instruments in the flight deck, and navigation is accomplished by reference to electronic signals.

Instrument landing system (ILS). An electronic system that provides both horizontal and vertical guidance to a specific runway, used to execute a precision instrument approach procedure.

Instrument meteorological conditions (IMC). Meteorological conditions expressed in terms of visibility, distance from clouds, and ceiling less than the minimums specified for visual meteorological conditions, requiring operations to be conducted under IFR.

Instrument takeoff. Using the instruments rather than outside visual cues to maintain runway heading and execute a safe takeoff.

Intercooler. A device used to reduce the temperatures of the compressed air before it enters the fuel metering device. The resulting cooler air has a higher density, which permits the engine to be operated with a higher power setting.

Interference drag. Drag generated by the collision of airstreams creating eddy currents, turbulence, or restrictions to smooth flow.

International Civil Aviation Organization (ICAO). The United Nations agency for developing the principles and techniques of international air navigation, and fostering planning and development of international civil air transport.

International standard atmosphere (IAS). A model of standard variation of pressure and temperature.

Interpolation. The estimation of an intermediate value of a quantity that falls between marked values in a series. Example: In a measurement of length, with a rule that is marked in eighths of an inch, the value falls between 3/8 inch and 1/2 inch. The estimated (interpolated) value might then be said to be 7/16 inch.

Inversion. An increase in temperature with altitude.

Inversion illusion. The feeling that the aircraft is tumbling backwards, caused by an abrupt change from climb to straight-and-level flight while in situations lacking visual reference.

Inverter. A solid-state electronic device that converts D.C. into A.C. current of the proper voltage and frequency to operate A.C. gyro instruments.

Isobars. Lines which connect points of equal barometric pressure.

Isogonic lines. Lines drawn across aeronautical charts to connect points having the same magnetic variation.

IVSI. See instantaneous vertical speed indicator.

J

Jet route. A route designated to serve flight operations from 18,000 feet MSL up to and including FL 450.

Jet stream. A high-velocity narrow stream of winds, usually found near the upper limit of the troposphere, which flows generally from west to east.

Judgment. The mental process of recognizing and analyzing all pertinent information in a particular situation, a rational evaluation of alternative actions in response to it, and a timely decision on which action to take.

K

KIAS. Knots indicated airspeed.

Knot. The knot is a unit of speed equal to one nautical mile (1.852 km) per hour, approximately 1.151 mph.

Kollsman window. A barometric scale window of a sensitive altimeter used to adjust the altitude for the altimeter setting.

L

LAAS. See local area augmentation system.

Lag. The delay that occurs before an instrument needle attains a stable indication.

Land breeze. A coastal breeze flowing from land to sea caused by temperature differences when the sea surface is warmer than the adjacent land. The land breeze usually occurs at night and alternates with the sea breeze that blows in the opposite direction by day.

Land as soon as possible. Land without delay at the nearest suitable area, such as an open field, at which a safe approach and landing is assured.

Land as soon as practical. The landing site and duration of flight are at the discretion of the pilot. Extended flight beyond the nearest approved landing area is not recommended.

Land immediately. The urgency of the landing is paramount. The primary consideration is to ensure the survival of the occupants. Landing in trees, water, or other unsafe areas should be considered only as a last resort.

Lateral axis. An imaginary line passing through the center of gravity of an airplane and extending across the airplane from wingtip to wingtip.

Lateral stability (rolling). The stability about the longitudinal axis of an aircraft. Rolling stability or the ability of an airplane to return to level flight due to a disturbance that causes one of the wings to drop.

Latitude. Measurement north or south of the equator in degrees, minutes, and seconds. Lines of latitude are also referred to as parallels.

LDA. See localizer-type directional aid.

Lead radial. The radial at which the turn from the DME arc to the inbound course is started.

Leading edge. The part of an airfoil that meets the airflow first.

Leading edge devices. High lift devices which are found on the leading edge of the airfoil. The most common types are fixed slots, movable slats, and leading edge flaps.

Leading-edge flap. A portion of the leading edge of an airplane wing that folds downward to increase the camber, lift, and drag of the wing. The leading-edge flaps are extended for takeoffs and landings to increase the amount of aerodynamic lift that is produced at any given airspeed.

Leans, the. A physical sensation caused by an abrupt correction of a banked attitude entered too slowly to stimulate the motion sensing system in the inner ear. The abrupt correction can create the illusion of banking in the opposite direction.

Licensed empty weight. The empty weight that consists of the airframe, engine(s), unusable fuel, and undrainable oil plus standard and optional equipment as specified in the equipment list. Some manufacturers used this term prior to GAMA standardization.

Lift. A component of the total aerodynamic force on an airfoil and acts perpendicular to the relative wind.

Limit load factor. Amount of stress, or load factor, that an aircraft can withstand before structural damage or failure occurs.

Lines of flux. Invisible lines of magnetic force passing between the poles of a magnet.

L/MF. See low or medium frequency.

LMM. See locator middle marker.

Load factor. The ratio of a specified load to the total weight of the aircraft. The specified load is expressed in terms of any of the following: aerodynamic forces, inertial forces, or ground or water reactions.

Loadmeter. A type of ammeter installed between the generator output and the main bus in an aircraft electrical system.

LOC. See localizer.

Local area augmentation system (LAAS). A differential global positioning system (DGPS) that improves the accuracy of the system by determining position error from the GPS satellites, then transmitting the error, or corrective factors, to the airborne GPS receiver.

Localizer (LOC). The portion of an ILS that gives left/right guidance information down the centerline of the instrument runway for final approach.

Localizer-type directional aid (LDA). A NAVAID used for nonprecision instrument approaches with utility and accuracy comparable to a localizer but which is not a part of a complete ILS and is not aligned with the runway. Some LDAs are equipped with a glideslope.

Locator middle marker (LMM). Nondirectional radio beacon (NDB) compass locator, collocated with a middle marker (MM).

Locator outer marker (LOM). NDB compass locator, collocated with an outer marker (OM).

LOM. See locator outer marker.

Longitude. Measurement east or west of the Prime Meridian in degrees, minutes, and seconds. The Prime Meridian is 0° longitude and runs through Greenwich, England. Lines of longitude are also referred to as meridians.

Longitudinal axis. An imaginary line through an aircraft from nose to tail, passing through its center of gravity. The longitudinal axis is also called the roll axis of the aircraft. Movement of the ailerons rotates an airplane about its longitudinal axis.

Longitudinal stability (pitching). Stability about the lateral axis. A desirable characteristic of an airplane whereby it tends to return to its trimmed angle of attack after displacement.

Low or medium frequency. A frequency range between 190 and 535 kHz with the medium frequency above 300 kHz. Generally associated with nondirectional beacons transmitting a continuous carrier with either a 400 or 1,020 Hz modulation.

Lubber line. The reference line used in a magnetic compass or heading indicator.

M

MAA. See maximum authorized altitude.

MAC. See mean aerodynamic chord.

Mach number. The ratio of the true airspeed of the aircraft to the speed of sound in the same atmospheric conditions, named in honor of Ernst Mach, late 19th century physicist.

Mach meter. The instrument that displays the ratio of the speed of sound to the true airspeed an aircraft is flying.

Magnetic bearing (MB). The direction to or from a radio transmitting station measured relative to magnetic north.

Magnetic compass. A device for determining direction measured from magnetic north.

Magnetic dip. A vertical attraction between a compass needle and the magnetic poles. The closer the aircraft is to a pole, the more severe the effect.

Magnetic heading (MH). The direction an aircraft is pointed with respect to magnetic north.

Magneto. A self-contained, engine-driven unit that supplies electrical current to the spark plugs; completely independent of the airplane's electrical system. Normally there are two magnetos per engine.

Magnus effect. Lifting force produced when a rotating cylinder produces a pressure differential. This is the same effect that makes a baseball curve or a golf ball slice.

Mandatory altitude. An altitude depicted on an instrument approach chart with the altitude value both underscored and overscored. Aircraft are required to maintain altitude at the depicted value.

Mandatory block altitude. An altitude depicted on an instrument approach chart with two underscored and overscored altitude values between which aircraft are required to maintain altitude.

Maneuverability. Ability of an aircraft to change directions along a flight path and withstand the stresses imposed upon it.

Maneuvering speed (V$_A$). The design maneuvering speed. Operating at or below design maneuvering speed does not provide structural protection against multiple full control inputs in one axis or full control inputs in more than one axis at the same time.

Manifold absolute pressure. The absolute pressure of the fuel/air mixture within the intake manifold, usually indicated in inches of mercury.

MAP. See missed approach point.

Margin identification. The top and bottom areas on an instrument approach chart that depict information about the procedure, including airport location and procedure identification.

Marker beacon. A low-powered transmitter that directs its signal upward in a small, fan-shaped pattern. Used along the flight path when approaching an airport for landing, marker beacons indicate both aurally and visually when the aircraft is directly over the facility.

Mass. The amount of matter in a body.

Maximum altitude. An altitude depicted on an instrument approach chart with overscored altitude value at which or below aircraft are required to maintain altitude.

Maximum authorized altitude (MAA). A published altitude representing the maximum usable altitude or flight level for an airspace structure or route segment.

Maximum landing weight. The greatest weight that an airplane normally is allowed to have at landing.

Maximum ramp weight. The total weight of a loaded aircraft, including all fuel. It is greater than the takeoff weight due to the fuel that will be burned during the taxi and runup operations. Ramp weight may also be referred to as taxi weight.

Maximum takeoff weight. The maximum allowable weight for takeoff.

Maximum weight. The maximum authorized weight of the aircraft and all of its equipment as specified in the Type Certificate Data Sheets (TCDS) for the aircraft.

Maximum zero fuel weight (GAMA). The maximum weight, exclusive of usable fuel.

MB. See magnetic bearing.

MCA. See minimum crossing altitude.

MDA. See minimum descent altitude.

MEA. See minimum en route altitude.

Mean aerodynamic chord (MAC). The average distance from the leading edge to the trailing edge of the wing.

Mean sea level. The average height of the surface of the sea at a particular location for all stages of the tide over a 19-year period.

MEL. See minimum equipment list.

Meridians. Lines of longitude.

Mesosphere. A layer of the atmosphere directly above the stratosphere.

METAR. See Aviation Routine Weather Report.

MFD. See multi-function display.

MH. See magnetic heading.

MHz. Megahertz.

Microbursts. A strong downdraft which normally occurs over horizontal distances of 1 NM or less and vertical distances of less than 1,000 feet. In spite of its small horizontal scale, an intense microburst could induce windspeeds greater than 100 knots and downdrafts as strong as 6,000 feet per minute.

Microwave landing system (MLS). A precision instrument approach system operating in the microwave spectrum which normally consists of an azimuth station, elevation station, and precision distance measuring equipment.

Mileage breakdown. A fix indicating a course change that appears on the chart as an "x" at a break between two segments of a federal airway.

Military operations area (MOA). Airspace established for the purpose of separating certain military training activities from IFR traffic.

Military training route (MTR). Airspace of defined vertical and lateral dimensions established for the conduct of military training at airspeeds in excess of 250 knots indicated airspeed (KIAS).

Minimum altitude. An altitude depicted on an instrument approach chart with the altitude value underscored. Aircraft are required to maintain altitude at or above the depicted value.

Minimum crossing altitude (MCA). The lowest allowed altitude at certain fixes an aircraft must cross when proceeding in the direction of a higher minimum en route altitude (MEA).

Minimum descent altitude (MDA). The lowest altitude (in feet MSL) to which descent is authorized on final approach, or during circle-to-land maneuvering in execution of a nonprecision approach.

Minimum drag. The point on the total drag curve where the lift-to-drag ratio is the greatest. At this speed, total drag is minimized.

Minimum en route altitude (MEA). The lowest published altitude between radio fixes that ensures acceptable navigational signal coverage and meets obstacle clearance requirements between those fixes.

Minimum equipment list (MEL). A list developed for larger aircraft that outlines equipment that can be inoperative for various types of flight including IFR and icing conditions. This list is based on the master minimum equipment list (MMEL) developed by the FAA and must be approved by the FAA for use. It is specific to an individual aircraft make and model.

Minimum obstruction clearance altitude (MOCA). The lowest published altitude in effect between radio fixes on VOR airways, off-airway routes, or route segments, which meets obstacle clearance requirements for the entire route segment and which ensures acceptable navigational signal coverage only within 25 statute (22 nautical) miles of a VOR.

Minimum reception altitude (MRA). The lowest altitude at which an airway intersection can be determined.

Minimum safe altitude (MSA). The minimum altitude depicted on approach charts which provides at least 1,000 feet of obstacle clearance for emergency use within a specified distance from the listed navigation facility.

Minimum vectoring altitude (MVA). An IFR altitude lower than the minimum en route altitude (MEA) that provides terrain and obstacle clearance.

Minimums section. The area on an IAP chart that displays the lowest altitude and visibility requirements for the approach.

Missed approach. A maneuver conducted by a pilot when an instrument approach cannot be completed to a landing.

Missed approach point (MAP). A point prescribed in each instrument approach at which a missed approach procedure shall be executed if the required visual reference has not been established.

Mixed ice. A mixture of clear ice and rime ice.

MLS. See microwave landing system.

MM. Middle marker.

MOA. See military operations area.

MOCA. See minimum obstruction clearance altitude.

Mode C. Altitude reporting transponder mode.

Moment. The product of the weight of an item multiplied by its arm. Moments are expressed in pound-inches (lb-in). Total moment is the weight of the airplane multiplied by the distance between the datum and the CG.

Moment arm. The distance from a datum to the applied force.

Moment index (or index). A moment divided by a constant such as 100, 1,000, or 10,000. The purpose of using a moment index is to simplify weight and balance computations of airplanes where heavy items and long arms result in large, unmanageable numbers.

Monocoque. A shell-like fuselage design in which the stressed outer skin is used to support the majority of imposed stresses. Monocoque fuselage design may include bulkheads but not stringers.

Monoplanes. Airplanes with a single set of wings.

Movable slat. A movable auxiliary airfoil on the leading edge of a wing. It is closed in normal flight but extends at high angles of attack. This allows air to continue flowing over the top of the wing and delays airflow separation.

MRA. See minimum reception altitude.

MSA. See minimum safe altitude.

MSL. See mean sea level.

MTR. See military training route.

Multi-function display (MFD). Small screen (CRT or LCD) in an aircraft that can be used to display information to the pilot in numerous configurable ways. Often an MFD will be used in concert with a primary flight display.

MVA. See minimum vectoring altitude.

N

N₁. Rotational speed of the low pressure compressor in a turbine engine.

N₂. Rotational speed of the high pressure compressor in a turbine engine.

Nacelle. A streamlined enclosure on an aircraft in which an engine is mounted. On multiengine propeller-driven airplanes, the nacelle is normally mounted on the leading edge of the wing.

NACG. See National Aeronautical Charting Group.

NAS. See National Airspace System.

National Airspace System (NAS). The common network of United States airspace—air navigation facilities, equipment and services, airports or landing areas; aeronautical charts, information and services; rules, regulations and procedures, technical information; and manpower and material.

National Aeronautical Charting Group (NACG). A Federal agency operating under the FAA, responsible for publishing charts such as the terminal procedures and en route charts.

National Route Program (NRP). A set of rules and procedures designed to increase the flexibility of user flight planning within published guidelines.

National Security Area (NSA). Areas consisting of airspace of defined vertical and lateral dimensions established at locations where there is a requirement for increased security and safety of ground facilities. Pilots are requested to voluntarily avoid flying through the depicted NSA. When it is necessary to provide a greater level of security and safety, flight in NSAs may be temporarily prohibited. Regulatory prohibitions are disseminated via NOTAMs.

National Transportation Safety Board (NTSB). A United States Government independent organization responsible for investigations of accidents involving aviation, highways, waterways, pipelines, and railroads in the United States. NTSB is charged by congress to investigate every civil aviation accident in the United States.

NAVAID. Navigational aid.

NAV/COM. Navigation and communication radio.

NDB. See nondirectional radio beacon.

Negative static stability. The initial tendency of an aircraft to continue away from the original state of equilibrium after being disturbed.

Neutral static stability. The initial tendency of an aircraft to remain in a new condition after its equilibrium has been disturbed.

NM. Nautical mile.

NOAA. National Oceanic and Atmospheric Administration.

No-gyro approach. A radar approach that may be used in case of a malfunctioning gyro-compass or directional gyro. Instead of providing the pilot with headings to be flown, the controller observes the radar track and issues control instructions "turn right/left" or "stop turn," as appropriate.

Nondirectional radio beacon (NDB). A ground-based radio transmitter that transmits radio energy in all directions.

Nonprecision approach. A standard instrument approach procedure in which only horizontal guidance is provided.

No procedure turn (NoPT). Term used with the appropriate course and altitude to denote that the procedure turn is not required.

NoPT. See no procedure turn.

NOTAM. See Notice to Airmen.

Notice to Airmen (NOTAM). A notice filed with an aviation authority to alert aircraft pilots of any hazards en route or at a specific location. The authority in turn provides means of disseminating relevant NOTAMs to pilots.

NRP. See National Route Program.

NSA. See National Security Area.

NTSB. See National Transportation Safety Board.

NWS. National Weather Service.

O

Obstacle departure procedures (ODP). A preplanned instrument flight rule (IFR) departure procedure printed for pilot use in textual or graphic form to provide obstruction clearance via the least onerous route from the terminal area to the appropriate en route structure. ODPs are recommended for obstruction clearance and may be flown without ATC clearance unless an alternate departure procedure (SID or radar vector) has been specifically assigned by ATC.

Obstruction lights. Lights that can be found both on and off an airport to identify obstructions.

Occluded front. A frontal occlusion occurs when a fast-moving cold front catches up with a slow moving warm front. The difference in temperature within each frontal system is a major factor in determining whether a cold or warm front occlusion occurs.

ODP. See obstacle departure procedures.

OM. Outer marker.

Omission error. The failure to anticipate significant instrument indications following attitude changes; for example, concentrating on pitch control while forgetting about heading or roll information, resulting in erratic control of heading and bank.

Optical illusion. A misleading visual image. For the purpose of this handbook, the term refers to the brain's misinterpretation of features on the ground associated with landing, which causes a pilot to misread the spatial relationships between the aircraft and the runway.

Orientation. Awareness of the position of the aircraft and of oneself in relation to a specific reference point.

Otolith organ. An inner ear organ that detects linear acceleration and gravity orientation.

Outer marker. A marker beacon at or near the glideslope intercept altitude of an ILS approach. It is normally located four to seven miles from the runway threshold on the extended centerline of the runway.

Outside air temperature (OAT). The measured or indicated air temperature (IAT) corrected for compression and friction heating. Also referred to as true air temperature.

Overcontrolling. Using more movement in the control column than is necessary to achieve the desired pitch-and-bank condition.

Overboost. A condition in which a reciprocating engine has exceeded the maximum manifold pressure allowed by the manufacturer. Can cause damage to engine components.

Overpower. To use more power than required for the purpose of achieving a faster rate of airspeed change.

P

P-static. See precipitation static.

PAPI. See precision approach path indicator.

PAR. See precision approach radar.

Parallels. Lines of latitude.

Parasite drag. Drag caused by the friction of air moving over the aircraft structure; its amount varies directly with the airspeed.

Payload (GAMA). The weight of occupants, cargo, and baggage.

Personality. The embodiment of personal traits and characteristics of an individual that are set at a very early age and extremely resistant to change.

P-factor. A tendency for an aircraft to yaw to the left due to the descending propeller blade on the right producing more thrust than the ascending blade on the left. This occurs when the aircraft's longitudinal axis is in a climbing attitude in relation to the relative wind. The P-factor would be to the right if the aircraft had a counterclockwise rotating propeller.

PFD. See primary flight display.

Phugoid oscillations. Long-period oscillations of an aircraft around its lateral axis. It is a slow change in pitch accompanied by equally slow changes in airspeed. Angle of attack remains constant, and the pilot often corrects for phugoid oscillations without even being aware of them.

PIC. See pilot in command.

Pilotage. Navigation by visual reference to landmarks.

Pilot in command (PIC). The pilot responsible for the operation and safety of an aircraft.

Pilot report (PIREP). Report of meteorological phenomena encountered by aircraft.

Pilot's Operating Handbook/Airplane Flight Manual (POH/AFM). FAA-approved documents published by the airframe manufacturer that list the operating conditions for a particular model of aircraft.

PIREP. See pilot report.

Pitot pressure. Ram air pressure used to measure airspeed.

Pitot-static head. A combination pickup used to sample pitot pressure and static air pressure.

Plan view. The overhead view of an approach procedure on an instrument approach chart. The plan view depicts the routes that guide the pilot from the en route segments to the IAF.

Planform. The shape or form of a wing as viewed from above. It may be long and tapered, short and rectangular, or various other shapes.

Pneumatic. Operation by the use of compressed air.

POH/AFM. See Pilot's Operating Handbook/Airplane Flight Manual.

Point-in-space approach. A type of helicopter instrument approach procedure to a missed approach point more than 2,600 feet from an associated helicopter landing area.

Poor judgment chain. A series of mistakes that may lead to an accident or incident. Two basic principles generally associated with the creation of a poor judgment chain are: (1) one bad decision often leads to another; and (2) as a string of bad decisions grows, it reduces the number of subsequent alternatives for continued safe flight. ADM is intended to break the poor judgment chain before it can cause an accident or incident.

Position error. Error in the indication of the altimeter, ASI, and VSI caused by the air at the static system entrance not being absolutely still.

Position report. A report over a known location as transmitted by an aircraft to ATC.

Positive static stability. The initial tendency to return to a state of equilibrium when disturbed from that state.

Power. Implies work rate or units of work per unit of time, and as such, it is a function of the speed at which the force is developed. The term "power required" is generally associated with reciprocating engines.

Powerplant. A complete engine and propeller combination with accessories.

Precession. The characteristic of a gyroscope that causes an applied force to be felt, not at the point of application, but 90° from that point in the direction of rotation.

Precipitation. Any or all forms of water particles (rain, sleet, hail, or snow) that fall from the atmosphere and reach the surface.

Precipitation static (P-static). A form of radio interference caused by rain, snow, or dust particles hitting the antenna and inducing a small radio-frequency voltage into it.

Precision approach. A standard instrument approach procedure in which both vertical and horizontal guidance is provided.

Precision approach path indicator (PAPI). A system of lights similar to the VASI, but consisting of one row of lights in two- or four-light systems. A pilot on the correct glideslope will see two white lights and two red lights. See VASI.

Precision approach radar (PAR). A type of radar used at an airport to guide an aircraft through the final stages of landing, providing horizontal and vertical guidance. The radar operator directs the pilot to change heading or adjust the descent rate to keep the aircraft on a path that allows it to touch down at the correct spot on the runway.

Precision runway monitor (PRM). System allows simultaneous, independent instrument flight rules (IFR) approaches at airports with closely spaced parallel runways.

Preferred IFR routes. Routes established in the major terminal and en route environments to increase system efficiency and capacity. IFR clearances are issued based on these routes, listed in the Chart Supplement U.S. except when severe weather avoidance procedures or other factors dictate otherwise.

Preignition. Ignition occurring in the cylinder before the time of normal ignition. Preignition is often caused by a local hot spot in the combustion chamber igniting the fuel-air mixture.

Pressure altitude. Altitude above the standard 29.92 "Hg plane.

Pressure demand oxygen system. A demand oxygen system that supplies 100 percent oxygen at sufficient pressure above the altitude where normal breathing is adequate. Also referred to as a pressure breathing system.

Prevailing visibility. The greatest horizontal visibility equaled or exceeded throughout at least half the horizon circle (which is not necessarily continuous).

Preventive maintenance. Simple or minor preservative operations and the replacement of small standard parts not involving complex assembly operation as listed in 14 CFR part 43, appendix A. Certificated pilots may perform preventive maintenance on any aircraft that is owned or operated by them provided that the aircraft is not used in air carrier service.

Primary and supporting. A method of attitude instrument flying using the instrument that provides the most direct indication of attitude and performance.

Primary flight display (PFD). A display that provides increased situational awareness to the pilot by replacing the traditional six instruments used for instrument flight with an easy-to-scan display that provides the horizon, airspeed, altitude, vertical speed, trend, trim, and rate of turn among other key relevant indications.

PRM. See precision runway monitor.

Procedure turn. A maneuver prescribed when it is necessary to reverse direction to establish an aircraft on the intermediate approach segment or final approach course.

Profile view. Side view of an IAP chart illustrating the vertical approach path altitudes, headings, distances, and fixes.

Prohibited area. Designated airspace within which flight of aircraft is prohibited.

Propeller. A device for propelling an aircraft that, when rotated, produces by its action on the air, a thrust approximately perpendicular to its plane of rotation. It includes the control components normally supplied by its manufacturer.

Propeller/rotor modulation error. Certain propeller rpm settings or helicopter rotor speeds can cause the VOR course deviation indicator (CDI) to fluctuate as much as ±6°. Slight changes to the rpm setting will normally smooth out this roughness.

R

Rabbit, the. High-intensity flasher system installed at many large airports. The flashers consist of a series of brilliant blue-white bursts of light flashing in sequence along the approach lights, giving the effect of a ball of light traveling toward the runway.

Radar. A system that uses electromagnetic waves to identify the range, altitude, direction, or speed of both moving and fixed objects such as aircraft, weather formations, and terrain. The term RADAR was coined in 1941 as an acronym for Radio Detection and Ranging. The term has since entered the English language as a standard word, radar, losing the capitalization in the process.

Radar approach. The controller provides vectors while monitoring the progress of the flight with radar, guiding the pilot through the descent to the airport/heliport or to a specific runway.

Radar services. Radar is a method whereby radio waves are transmitted into the air and are then received when they have been reflected by an object in the path of the beam. Range is determined by measuring the time it takes (at the speed of light) for the radio wave to go out to the object and then return to the receiving antenna. The direction of a detected object from a radar site is determined by the position of the rotating antenna when the reflected portion of the radio wave is received.

Radar summary chart. A weather product derived from the national radar network that graphically displays a summary of radar weather reports.

Radar weather report (SD). A report issued by radar stations at 35 minutes after the hour, and special reports as needed. Provides information on the type, intensity, and location of the echo tops of the precipitation.

Radials. The courses oriented from a station.

Radio or radar altimeter. An electronic altimeter that determines the height of an aircraft above the terrain by measuring the time needed for a pulse of radio-frequency energy to travel from the aircraft to the ground and return.

Radio frequency (RF). A term that refers to alternating current (AC) having characteristics such that, if the current is input to antenna, an electromagnetic (EM) field is generated suitable for wireless broadcasting and/or communications.

Radio magnetic indicator (RMI). An electronic navigation instrument that combines a magnetic compass with an ADF or VOR. The card of the RMI acts as a gyro-stabilized magnetic compass, and shows the magnetic heading the aircraft is flying.

Radiosonde. A weather instrument that observes and reports meteorological conditions from the upper atmosphere. This instrument is typically carried into the atmosphere by some form of weather balloon.

Radio wave. An electromagnetic (EM) wave with frequency characteristics useful for radio transmission.

RAIM. See receiver autonomous integrity monitoring.

RAM recovery. The increase in thrust as a result of ram air pressures and density on the front of the engine caused by air velocity.

Random RNAV routes. Direct routes, based on area navigation capability, between waypoints defined in terms of latitude/longitude coordinates, degree-distance fixes, or offsets from established routes/airways at a specified distance and direction.

Ranging signals. Transmitted from the GPS satellite, signals allowing the aircraft's receiver to determine range (distance) from each satellite.

Rapid decompression. The almost instantaneous loss of cabin pressure in aircraft with a pressurized cockpit or cabin.

RB. See relative bearing.

RBI. See relative bearing indicator.

RCO. See remote communications outlet.

Receiver autonomous integrity monitoring (RAIM). A system used to verify the usability of the received GPS signals and warns the pilot of any malfunction in the navigation system. This system is required for IFR-certified GPS units.

Recommended altitude. An altitude depicted on an instrument approach chart with the altitude value neither underscored nor overscored. The depicted value is an advisory value.

Receiver-transmitter (RT). A system that receives and transmits a signal and an indicator.

Reduced vertical separation minimum (RVSM). Reduces the vertical separation between flight levels (FL) 290 and 410 from 2,000 feet to 1,000 feet, and makes six additional FLs available for operation. Also see DRVSM.

Reference circle (also, distance circle). The circle depicted in the plan view of an IAP chart that typically has a 10 NM radius, within which chart the elements are drawn to scale.

Regions of command. The "regions of normal and reversed command" refers to the relationship between speed and the power required to maintain or change that speed in flight.

Region of reverse command. Flight regime in which flight at a higher airspeed requires a lower power setting and a lower airspeed requires a higher power setting in order to maintain altitude.

REIL. See runway end identifier lights.

Relative bearing (RB). The angular difference between the aircraft heading and the direction to the station, measured clockwise from the nose of the aircraft.

Relative bearing indicator (RBI). Also known as the fixed-card ADF, zero is always indicated at the top of the instrument and the needle indicates the relative bearing to the station.

Relative humidity. The ratio of the existing amount of water vapor in the air at a given temperature to the maximum amount that could exist at that temperature; usually expressed in percent.

Relative wind. Direction of the airflow produced by an object moving through the air. The relative wind for an airplane in flight flows in a direction parallel with and opposite to the direction of flight; therefore, the actual flight path of the airplane determines the direction of the relative wind.

Remote communications outlet (RCO). An unmanned communications facility that is remotely controlled by air traffic personnel.

Required navigation performance (RNP). A specified level of accuracy defined by a lateral area of confined airspace in which an RNP-certified aircraft operates.

Restricted area. Airspace designated under 14 CFR part 73 within which the flight of aircraft, while not wholly prohibited, is subject to restriction.

Reverse sensing. The VOR needle appearing to indicate the reverse of normal operation.

RF. Radio frequency.

Rhodopsin. The photosensitive pigments that initiate the visual response in the rods of the eye.

Rigging. The final adjustment and alignment of an aircraft and its flight control system that provides the proper aerodynamic characteristics.

Rigidity. The characteristic of a gyroscope that prevents its axis of rotation tilting as the Earth rotates.

Rigidity in space. The principle that a wheel with a heavily weighted rim spinning rapidly will remain in a fixed position in the plane in which it is spinning.

Rime ice. Rough, milky, opaque ice formed by the instantaneous freezing of small supercooled water droplets.

Risk. The future impact of a hazard that is not eliminated or controlled.

Risk elements. There are four fundamental risk elements in aviation: the pilot, the aircraft, the environment, and the type of operation that comprise any given aviation situation.

Risk management. The part of the decision-making process which relies on situational awareness, problem recognition, and good judgment to reduce risks associated with each flight.

RMI. See radio magnetic indicator.

RNAV. See area navigation.

RNP. See required navigation performance.

RT. See receiver-transmitter.

Rudder. The movable primary control surface mounted on the trailing edge of the vertical fin of an airplane. Movement of the rudder rotates the airplane about its vertical axis.

Ruddervator. A pair of control surfaces on the tail of an aircraft arranged in the form of a V. These surfaces, when moved together by the control wheel, serve as elevators, and when moved differentially by the rudder pedals, serve as a rudder.

Runway centerline lights. Runway lighting which consists of flush centerline lights spaced at 50-foot intervals beginning 75 feet from the landing threshold.

Runway edge lights. A component of the runway lighting system that is used to outline the edges of runways at night or during low visibility conditions. These lights are classified according to the intensity they are capable of producing.

Runway end identifier lights (REIL). A pair of synchronized flashing lights, located laterally on each side of the runway threshold, providing rapid and positive identification of the approach end of a runway.

Runway visibility value (RVV). The visibility determined for a particular runway by a transmissometer.

Runway visual range (RVR). The instrumentally derived horizontal distance a pilot should be able to see down the runway from the approach end, based on either the sighting of high-intensity runway lights, or the visual contrast of other objects.

RVR. See runway visual range.

RVV. See runway visibility value.

S

SA. See selective availability.

St. Elmo's Fire. A corona discharge which lights up the aircraft surface areas where maximum static discharge occurs.

Satellite ephemeris data. Data broadcast by the GPS satellite containing very accurate orbital data for that satellite, atmospheric propagation data, and satellite clock error data.

Sea breeze. A coastal breeze blowing from sea to land caused by the temperature difference when the land surface is warmer than the sea surface. The sea breeze usually occurs during the day and alternates with the land breeze that blows in the opposite direction at night.

Sea level engine. A reciprocating aircraft engine having a rated takeoff power that is producible only at sea level.

Scan. The first fundamental skill of instrument flight, also known as "cross-check;" the continuous and logical observation of instruments for attitude and performance information.

Sectional aeronautical charts. Designed for visual navigation of slow- or medium-speed aircraft. Topographic information on these charts features the portrayal of relief, and a judicious selection of visual check points for VFR flight. Aeronautical information includes visual and radio aids to navigation, airports, controlled airspace, restricted areas, obstructions and related data.

Selective availability (SA). A satellite technology permitting the Department of Defense (DOD) to create, in the interest of national security, a significant clock and ephemeris error in the satellites, resulting in a navigation error.

Semicircular canal. An inner ear organ that detects angular acceleration of the body.

Semimonocoque. A fuselage design that includes a substructure of bulkheads and/or formers, along with stringers, to support flight loads and stresses imposed on the fuselage.

Sensitive altimeter. A form of multipointer pneumatic altimeter with an adjustable barometric scale that allows the reference pressure to be set to any desired level.

Service ceiling. The maximum density altitude where the best rate-of-climb airspeed will produce a 100-feet-per-minute climb at maximum weight while in a clean configuration with maximum continuous power.

Servo. A motor or other form of actuator which receives a small signal from the control device and exerts a large force to accomplish the desired work.

Servo tab. An auxiliary control mounted on a primary control surface, which automatically moves in the direction opposite the primary control to provide an aerodynamic assist in the movement of the control.

SIDS. See standard instrument departure procedures.

SIGMET. The acronym for Significant Meteorological information. A weather advisory in abbreviated plain language concerning the occurrence or expected occurrence of potentially hazardous en route weather phenomena that may affect the safety of aircraft operations. SIGMET is warning information, hence it is of highest priority among other types of meteorological information provided to the aviation users.

Signal-to-noise ratio. An indication of signal strength received compared to background noise, which is a measure of the adequacy of the received signal.

Significant weather prognostic. Presents four panels showing forecast significant weather.

Simplex. Transmission and reception on the same frequency.

Simplified directional facility (SDF). A NAVAID used for nonprecision instrument approaches. The final approach course is similar to that of an ILS localizer; however, the SDF course may be offset from the runway, generally not more than 3°, and the course may be wider than the localizer, resulting in a lower degree of accuracy.

Single-pilot resource management (SRM). The ability for a pilot to manage all resources effectively to ensure the outcome of the flight is successful.

Situational awareness. Pilot knowledge of where the aircraft is in regard to location, air traffic control, weather, regulations, aircraft status, and other factors that may affect flight.

Skidding turn. An uncoordinated turn in which the rate of turn is too great for the angle of bank, pulling the aircraft to the outside of the turn.

Skills and procedures. The procedural, psychomotor, and perceptual skills used to control a specific aircraft or its systems. They are the airmanship abilities that are gained through conventional training, are perfected, and become almost automatic through experience.

Skin friction drag. Drag generated between air molecules and the solid surface of the aircraft.

Slant range. The horizontal distance from the aircraft antenna to the ground station, due to line-of-sight transmission of the DME signal.

Slaved compass. A system whereby the heading gyro is "slaved to," or continuously corrected to bring its direction readings into agreement with a remotely located magnetic direction sensing device (usually a flux valve or flux gate compass).

Slipping turn. An uncoordinated turn in which the aircraft is banked too much for the rate of turn, so the horizontal lift component is greater than the centrifugal force, pulling the aircraft toward the inside of the turn.

Small airplane. An airplane of 12,500 pounds or less maximum certificated takeoff weight.

Somatogravic illusion. The misperception of being in a nose-up or nose-down attitude, caused by a rapid acceleration or deceleration while in flight situations that lack visual reference.

Spatial disorientation. The state of confusion due to misleading information being sent to the brain from various sensory organs, resulting in a lack of awareness of the aircraft position in relation to a specific reference point.

Special flight permit. A flight permit issued to an aircraft that does not meet airworthiness requirements but is capable of safe flight. A special flight permit can be issued to move an aircraft for the purposes of maintenance or repair, buyer delivery, manufacturer flight tests, evacuation from danger, or customer demonstration. Also referred to as a ferry permit.

Special use airspace. Airspace in which flight activities are subject to restrictions that can create limitations on the mixed use of airspace. Consists of prohibited, restricted, warning, military operations, and alert areas.

Special fuel consumption. The amount of fuel in pounds per hour consumed or required by an engine per brake horsepower or per pound of thrust.

Speed. The distance traveled in a given time.

Spin. An aggravated stall that results in an airplane descending in a helical, or corkscrew path.

Spiral instability. A condition that exists when the static directional stability of the airplane is very strong as compared to the effect of its dihedral in maintaining lateral equilibrium.

Spiraling slipstream. The slipstream of a propeller-driven airplane rotates around the airplane. This slipstream strikes the left side of the vertical fin, causing the aircraft to yaw slightly. Rudder offset is sometimes used by aircraft designers to counteract this tendency.

Spoilers. High-drag devices that can be raised into the air flowing over an airfoil, reducing lift and increasing drag. Spoilers are used for roll control on some aircraft. Deploying spoilers on both wings at the same time allows the aircraft to descend without gaining speed. Spoilers are also used to shorten the ground roll after landing.

SRM. See single-pilot resource management.

SSR. See secondary surveillance radar.

SSV. See standard service volume.

Stabilator. A single-piece horizontal tail surface on an airplane that pivots around a central hinge point. A stabilator serves the purposes of both the horizontal stabilizer and the elevators.

Stability. The inherent quality of an airplane to correct for conditions that may disturb its equilibrium, and to return or to continue on the original flight path. It is primarily an airplane design characteristic.

Stagnant hypoxia. A type of hypoxia that results when the oxygen-rich blood in the lungs is not moving to the tissues that need it.

Stall. A rapid decrease in lift caused by the separation of airflow from the wing's surface, brought on by exceeding the critical angle of attack. A stall can occur at any pitch attitude or airspeed.

Standard atmosphere. At sea level, the standard atmosphere consists of a barometric pressure of 29.92 inches of mercury ("Hg) or 1013.2 millibars, and a temperature of 15 °C (59 °F). Pressure and temperature normally decrease as altitude increases. The standard lapse rate in the lower atmosphere for each 1,000 feet of altitude is approximately 1 "Hg and 2 °C (3.5 °F). For example, the standard pressure and temperature at 3,000 feet mean sea level (MSL) are 26.92 "Hg (29.92 "Hg – 3 "Hg) and 9 °C (15 °C – 6 °C).

Standard empty weight (GAMA). This weight consists of the airframe, engines, and all items of operating equipment that have fixed locations and are permanently installed in the airplane including fixed ballast, hydraulic fluid, unusable fuel, and full engine oil.

Standard holding pattern. A holding pattern in which all turns are made to the right.

Standard instrument departure procedures (SIDS). Published procedures to expedite clearance delivery and to facilitate transition between takeoff and en route operations.

Standard rate turn. A turn in which an aircraft changes its direction at a rate of 3° per second (360° in 2 minutes) for low- or medium-speed aircraft. For high-speed aircraft, the standard rate turn is 1½° per second (360° in 4 minutes).

Standard service volume (SSV). Defines the limits of the volume of airspace which the VOR serves.

Standard terminal arrival route (STAR). A preplanned IFR ATC arrival procedure published for pilot use in graphic and/or textual form.

Standard weights. Weights established for numerous items involved in weight and balance computations. These weights should not be used if actual weights are available.

Static longitudinal stability. The aerodynamic pitching moments required to return the aircraft to the equilibrium angle of attack.

Static pressure. Pressure of air that is still or not moving, measured perpendicular to the surface of the aircraft.

Static stability. The initial tendency an aircraft displays when disturbed from a state of equilibrium.

Station. A location in the airplane that is identified by a number designating its distance in inches from the datum. The datum is, therefore, identified as station zero. An item located at station +50 would have an arm of 50 inches.

Stationary front. A front that is moving at a speed of less than 5 knots.

Steep turns. In instrument flight, any turn greater than standard rate; in visual flight, anything greater than a 45° bank.

Stepdown fix. The point after which additional descent is permitted within a segment of an IAP.

Strapdown system. An INS in which the accelerometers and gyros are permanently "strapped down" or aligned with the three axes of the aircraft.

Stratoshere. A layer of the atmosphere above the tropopause extending to a height of approximately 160,000 feet.

Stress. The body's response to demands placed upon it.

Stress management. The personal analysis of the kinds of stress experienced while flying, the application of appropriate stress assessment tools, and other coping mechanisms.

Structural icing. The accumulation of ice on the exterior of the aircraft.

Sublimation. Process by which a solid is changed to a gas without going through the liquid state.

Suction relief valve. A relief valve in an instrument vacuum system required to maintain the correct low pressure inside the instrument case for the proper operation of the gyros.

Supercharger. An engine- or exhaust-driven air compressor used to provide additional pressure to the induction air so the engine can produce additional power.

Supercooled water droplets. Water droplets that have been cooled below the freezing point, but are still in a liquid state.

Surface analysis chart. A report that depicts an analysis of the current surface weather. Shows the areas of high and low pressure, fronts, temperatures, dewpoints, wind directions and speeds, local weather, and visual obstructions.

Synchro. A device used to transmit indications of angular movement or position from one location to another.

Synthetic vision. A realistic display depiction of the aircraft in relation to terrain and flight path.

T

TAA. See terminal arrival area.

TACAN. See tactical air navigation.

Tactical air navigation (TACAN). An electronic navigation system used by military aircraft, providing both distance and direction information.

Takeoff decision speed (V_1). Per 14 CFR section 23.51: "the calibrated airspeed on the ground at which, as a result of engine failure or other reasons, the pilot assumed to have made a decision to continue or discontinue the takeoff."

Takeoff distance. The distance required to complete an all-engines operative takeoff to the 35-foot height. It must be at least 15 percent less than the distance required for a one-engine inoperative engine takeoff. This distance is not normally a limiting factor as it is usually less than the one-engine inoperative takeoff distance.

Takeoff safety speed (V_2). Per 14 CFR part 1: "A referenced airspeed obtained after lift-off at which the required one-engine-inoperative climb performance can be achieved."

TAWS. See terrain awareness and warning system.

Taxiway lights. Omnidirectional lights that outline the edges of the taxiway and are blue in color.

Taxiway turnoff lights. Lights that are flush with the runway which emit a steady green color.

TCAS. See traffic alert collision avoidance system.

TCH. See threshold crossing height.

TDZE. See touchdown zone elevation.

TEC. See Tower En Route Control.

Technique. The manner in which procedures are executed.

Telephone information briefing service (TIBS). An FSS service providing continuously updated automated telephone recordings of area and/or route weather, airspace procedures, and special aviation-oriented announcements.

Temporary flight restriction (TFR). Restriction to flight imposed in order to:

1. Protect persons and property in the air or on the surface from an existing or imminent flight associated hazard;

2. Provide a safe environment for the operation of disaster relief aircraft;

3. Prevent an unsafe congestion of sightseeing aircraft above an incident;

4. Protect the President, Vice President, or other public figures; and,

5. Provide a safe environment for space agency operations.

Pilots are expected to check appropriate NOTAMs during flight planning when conducting flight in an area where a temporary flight restriction is in effect.

Tension. Maintaining an excessively strong grip on the control column, usually resulting in an overcontrolled situation.

Terminal aerodrome forecast (TAF). A report established for the 5 statute mile radius around an airport. Utilizes the same descriptors and abbreviations as the METAR report.

Terminal arrival area (TAA). A procedure to provide a new transition method for arriving aircraft equipped with FMS and/or GPS navigational equipment. The TAA contains a "T" structure that normally provides a NoPT for aircraft using the approach.

Terminal instrument approach procedure (TERP). Prescribes standardized methods for use in designing instrument flight procedures.

TERP. See terminal instrument approach procedure.

Terminal radar service areas (TRSA). Areas where participating pilots can receive additional radar services. The purpose of the service is to provide separation between all IFR operations and participating VFR aircraft.

Terrain awareness and warning system (TAWS). A timed-based system that provides information concerning potential hazards with fixed objects by using GPS positioning and a database of terrain and obstructions to provide true predictability of the upcoming terrain and obstacles.

TFR. See temporary flight restriction.

Thermosphere. The last layer of the atmosphere that begins above the mesosphere and gradually fades away into space.

Threshold crossing height (TCH). The theoretical height above the runway threshold at which the aircraft's glideslope antenna would be if the aircraft maintained the trajectory established by the mean ILS glideslope or MLS glidepath.

Thrust. The force which imparts a change in the velocity of a mass. This force is measured in pounds but has no element of time or rate. The term "thrust required" is generally associated with jet engines. A forward force which propels the airplane through the air.

Thrust (aerodynamic force). The forward aerodynamic force produced by a propeller, fan, or turbojet engine as it forces a mass of air to the rear, behind the aircraft.

Thrust line. An imaginary line passing through the center of the propeller hub, perpendicular to the plane of the propeller rotation.

Time and speed table. A table depicted on an instrument approach procedure chart that identifies the distance from the FAF to the MAP, and provides the time required to transit that distance based on various groundspeeds.

Timed turn. A turn in which the clock and the turn coordinator are used to change heading a definite number of degrees in a given time.

TIS. See traffic information service.

Title 14 of the Code of Federal Regulations (14 CFR). Includes the federal aviation regulations governing the operation of aircraft, airways, and airmen.

Torque. (1) A resistance to turning or twisting. (2) Forces that produce a twisting or rotating motion. (3) In an airplane, the tendency of the aircraft to turn (roll) in the opposite direction of rotation of the engine and propeller. (4) In helicopters with a single, main rotor system, the tendency of the helicopter to turn in the opposite direction of the main rotor rotation.

Torquemeter. An instrument used with some of the larger reciprocating engines and turboprop or turboshaft engines to measure the reaction between the propeller reduction gears and the engine case.

Total drag. The sum of the parasite drag and induced drag.

Touchdown zone elevation (TDZE). The highest elevation in the first 3,000 feet of the landing surface, TDZE is indicated on the instrument approach procedure chart when straight-in landing minimums are authorized.

Touchdown zone lights. Two rows of transverse light bars disposed symmetrically about the runway centerline in the runway touchdown zone.

Tower En Route Control (TEC). The control of IFR en route traffic within delegated airspace between two or more adjacent approach control facilities, designed to expedite traffic and reduce control and pilot communication requirements.

TPP. See United States Terminal Procedures Publication.

Track. The actual path made over the ground in flight.

Tracking. Flying a heading that will maintain the desired track to or from the station regardless of crosswind conditions.

Traffic Alert Collision Avoidance System (TCAS). An airborne system developed by the FAA that operates independently from the ground-based Air Traffic Control system. Designed to increase flight deck awareness of proximate aircraft and to serve as a "last line of defense" for the prevention of midair collisions.

Traffic information service (TIS). A ground-based service providing information to the flight deck via data link using the S-mode transponder and altitude encoder to improve the safety and efficiency of "see and avoid" flight through an automatic display that informs the pilot of nearby traffic.

Trailing edge. The portion of the airfoil where the airflow over the upper surface rejoins the lower surface airflow.

Transcribed Weather Broadcast (TWEB). An FSS service, available in Alaska only, providing continuously updated automated broadcast of meteorological and aeronautical data over selected L/MF and VOR NAVAIDs.

Transponder. The airborne portion of the ATC radar beacon system.

Transponder code. One of 4,096 four-digit discrete codes ATC assigns to distinguish between aircraft.

Trend. Immediate indication of the direction of aircraft movement, as shown on instruments.

Tricycle gear. Landing gear employing a third wheel located on the nose of the aircraft.

Trim. To adjust the aerodynamic forces on the control surfaces so that the aircraft maintains the set attitude without any control input.

Trim tab. A small auxiliary hinged portion of a movable control surface that can be adjusted during flight to a position resulting in a balance of control forces.

Tropopause. The boundary layer between the troposphere and the stratosphere which acts as a lid to confine most of the water vapor, and the associated weather, to the troposphere.

Troposphere. The layer of the atmosphere extending from the surface to a height of 20,000 to 60,000 feet, depending on latitude.

True airspeed. Actual airspeed, determined by applying a correction for pressure altitude and temperature to the CAS.

True altitude. The vertical distance of the airplane above sea level—the actual altitude. It is often expressed as feet above mean sea level (MSL). Airport, terrain, and obstacle elevations on aeronautical charts are true altitudes.

Truss. A fuselage design made up of supporting structural members that resist deformation by applied loads. The truss-type fuselage is constructed of steel or aluminum tubing. Strength and rigidity is achieved by welding the tubing together into a series of triangular shapes, called trusses.

T-tail. An aircraft with the horizontal stabilizer mounted on the top of the vertical stabilizer, forming a T.

Turbine discharge pressure. The total pressure at the discharge of the low-pressure turbine in a dual-turbine axial-flow engine.

Turbine engine. An aircraft engine which consists of an air compressor, a combustion section, and a turbine. Thrust is produced by increasing the velocity of the air flowing through the engine.

Turbocharger. An air compressor driven by exhaust gases, which increases the pressure of the air going into the engine through the carburetor or fuel injection system.

Turbofan engine. A fanlike turbojet engine designed to create additional thrust by diverting a secondary airflow around the combustion chamber.

Turbojet engine. A turbine engine which produces its thrust entirely by accelerating the air through the engine.

Turboprop engine. A turbine engine which drives a propeller through a reduction gearing arrangement. Most of the energy in the exhaust gases is converted into torque, rather than using its acceleration to drive the aircraft.

Turboshaft engine. A gas turbine engine that delivers power through a shaft to operate something other than a propeller.

Turn-and-slip indicator. A flight instrument consisting of a rate gyro to indicate the rate of yaw and a curved glass inclinometer to indicate the relationship between gravity and centrifugal force. The turn-and-slip indicator indicates the relationship between angle of bank and rate of yaw. Also called a turn-and-bank indicator.

Turn coordinator. A rate gyro that senses both roll and yaw due to the gimbal being canted. Has largely replaced the turn-and-slip indicator in modern aircraft.

TWEB. See Transcribed Weather Broadcast.

U

UHF. See ultra-high frequency.

Ultra-high frequency (UHF). The range of electromagnetic frequencies between 300 MHz and 3,000 MHz.

Ulitimate load factor. In stress analysis, the load that causes physical breakdown in an aircraft or aircraft component during a strength test, or the load that according to computations, should cause such a breakdown.

Uncaging. Unlocking the gimbals of a gyroscopic instrument, making it susceptible to damage by abrupt flight maneuvers or rough handling.

Uncontrolled airspace. Class G airspace that has not been designated as Class A, B, C, D, or E. It is airspace in which air traffic control has no authority or responsibility to control air traffic; however, pilots should remember there are VFR minimums which apply to this airspace.

Underpower. Using less power than required for the purpose of achieving a faster rate of airspeed change.

United States Terminal Procedures Publication (TPP). Booklets published in regional format by FAA Aeronautical Navigation Products (AeroNav Products) that include DPs, STARs, IAPs, and other information pertinent to IFR flight.

Unusual attitude. An unintentional, unanticipated, or extreme aircraft attitude.

Useful load. The weight of the pilot, copilot, passengers, baggage, usable fuel, and drainable oil. It is the basic empty weight subtracted from the maximum allowable gross weight. This term applies to general aviation aircraft only.

User-defined waypoints. Waypoint location and other data which may be input by the user, this is the only GPS database information that may be altered (edited) by the user.

V

V_1. See takeoff decision speed.

V_2. See takeoff safety speed.

V_A. See maneuvering speed.

Vapor lock. A problem that mostly affects gasoline-fuelled internal combustion engines. It occurs when the liquid fuel changes state from liquid to gas while still in the fuel delivery system. This disrupts the operation of the fuel pump, causing loss of feed pressure to the carburetor or fuel injection system, resulting in transient loss of power or complete stalling. Restarting the engine from this state may be difficult. The fuel can vaporize due to being heated by the engine, by the local climate or due to a lower boiling point at high altitude.

Variation. Compass error caused by the difference in the physical locations of the magnetic north pole and the geographic north pole.

VASI. See visual approach slope indicator.

VDP. See visual descent point.

Vector. A force vector is a graphic representation of a force and shows both the magnitude and direction of the force.

Vectoring. Navigational guidance by assigning headings.

VEF. Calibrated airspeed at which the critical engine of a multi-engine aircraft is assumed to fail.

Velocity. The speed or rate of movement in a certain direction.

Venturi tube. A specially shaped tube attached to the outside of an aircraft to produce suction to allow proper operation of gyro instruments.

Vertical axis. An imaginary line passing vertically through the center of gravity of an aircraft. The vertical axis is called the z-axis or the yaw axis.

Vertical card compass. A magnetic compass that consists of an azimuth on a vertical card, resembling a heading indicator with a fixed miniature airplane to accurately present the heading of the aircraft. The design uses eddy current damping to minimize lead and lag during turns.

Vertical speed indicator (VSI). A rate-of-pressure change instrument that gives an indication of any deviation from a constant pressure level.

Vertical stability. Stability about an aircraft's vertical axis. Also called yawing or directional stability.

Very-high frequency (VHF). A band of radio frequencies falling between 30 and 300 MHz.

Very-high frequency omnidirectional range (VOR). Electronic navigation equipment in which the flight deck instrument identifies the radial or line from the VOR station, measured in degrees clockwise from magnetic north, along which the aircraft is located.

Vestibule. The central cavity of the bony labyrinth of the ear, or the parts of the membranous labyrinth that it contains.

V_{FE}. The maximum speed with the flaps extended. The upper limit of the white arc.

VFR. See visual flight rules.

VFR on top. ATC authorization for an IFR aircraft to operate in VFR conditions at any appropriate VFR altitude.

VFR over the top. A VFR operation in which an aircraft operates in VFR conditions on top of an undercast.

VFR terminal area chart. At a scale of 1:250,000, a chart that depicts Class B airspace, which provides for the control or segregation of all the aircraft within the Class B airspace. The chart depicts topographic information and aeronautical information including visual and radio aids to navigation, airports, controlled airspace, restricted areas, obstructions, and related data.

V-G diagram. A chart that relates velocity to load factor. It is valid only for a specific weight, configuration and altitude and shows the maximum amount of positive or negative lift the airplane is capable of generating at a given speed. Also shows the safe load factor limits and the load factor that the aircraft can sustain at various speeds.

Victor airways. Airways based on a centerline that extends from one VOR or VORTAC navigation aid or intersection, to another navigation aid (or through several navigation aids or intersections); used to establish a known route for en route procedures between terminal areas.

Visual approach slope indicator (VASI). A visual aid of lights arranged to provide descent guidance information during the approach to the runway. A pilot on the correct glideslope will see red lights over white lights.

Visual descent point (VDP). A defined point on the final approach course of a nonprecision straight-in approach procedure from which normal descent from the MDA to the runway touchdown point may be commenced, provided the runway environment is clearly visible to the pilot.

Visual flight rules (VFR). Flight rules adopted by the FAA governing aircraft flight using visual references. VFR operations specify the amount of ceiling and the visibility the pilot must have in order to operate according to these rules. When the weather conditions are such that the pilot cannot operate according to VFR, he or she must use instrument flight rules (IFR).

Visual meteorological conditions (VMC). Meteorological conditions expressed in terms of visibility, distance from cloud, and ceiling meeting or exceeding the minimums specified for VFR.

V_{LE}. Landing gear extended speed. The maximum speed at which an airplane can be safely flown with the landing gear extended.

V_{LO}. Landing gear operating speed. The maximum speed for extending or retracting the landing gear if using an airplane equipped with retractable landing gear.

V_{MC}. Minimum control airspeed. This is the minimum flight speed at which a light, twin-engine airplane can be satisfactorily controlled when an engine suddenly becomes inoperative and the remaining engine is at takeoff power.

VMC. See visual meteorological conditions.

V_{NE}. The never-exceed speed. Operating above this speed is prohibited since it may result in damage or structural failure. The red line on the airspeed indicator.

V_{NO}. The maximum structural cruising speed. Do not exceed this speed except in smooth air. The upper limit of the green arc.

VOR. See very-high frequency omnidirectional range.

VORTAC. A facility consisting of two components, VOR and TACAN, which provides three individual services: VOR azimuth, TACAN azimuth, and TACAN distance (DME) at one site.

VOR test facility (VOT). A ground facility which emits a test signal to check VOR receiver accuracy. Some VOTs are available to the user while airborne, while others are limited to ground use only.

VOT. See VOR test facility.

VSI. See vertical speed indicator.

V_{S0}. The stalling speed or the minimum steady flight speed in the landing configuration. In small airplanes, this is the power-off stall speed at the maximum landing weight in the landing configuration (gear and flaps down). The lower limit of the white arc.

V_{S1}. The stalling speed or the minimum steady flight speed obtained in a specified configuration. For most airplanes, this is the power-off stall speed at the maximum takeoff weight in the clean configuration (gear up, if retractable, and flaps up). The lower limit of the green arc.

V-tail. A design which utilizes two slanted tail surfaces to perform the same functions as the surfaces of a conventional elevator and rudder configuration. The fixed surfaces act as both horizontal and vertical stabilizers.

V_X. Best angle-of-climb speed. The airspeed at which an airplane gains the greatest amount of altitude in a given distance. It is used during a short-field takeoff to clear an obstacle.

V_Y. Best rate-of-climb speed. This airspeed provides the most altitude gain in a given period of time.

V_{YSE}. Best rate-of-climb speed with one engine inoperative. This airspeed provides the most altitude gain in a given period of time in a light, twin-engine airplane following an engine failure.

W

WAAS. See wide area augmentation system.

Wake turbulence. Wingtip vortices that are created when an airplane generates lift. When an airplane generates lift, air spills over the wingtips from the high pressure areas below the wings to the low pressure areas above them. This flow causes rapidly rotating whirlpools of air called wingtip vortices or wake turbulence.

Warm front. The boundary area formed when a warm air mass contacts and flows over a colder air mass. Warm fronts cause low ceilings and rain.

Warning area. An area containing hazards to any aircraft not participating in the activities being conducted in the area. Warning areas may contain intensive military training, gunnery exercises, or special weapons testing.

WARP. See weather and radar processing.

Waste gate. A controllable valve in the tailpipe of an aircraft reciprocating engine equipped with a turbocharger. The valve is controlled to vary the amount of exhaust gases forced through the turbocharger turbine.

Waypoint. A designated geographical location used for route definition or progress-reporting purposes and is defined in terms of latitude/longitude coordinates.

WCA. See wind correction angle.

Weather and radar processor (WARP). A device that provides real-time, accurate, predictive, and strategic weather information presented in an integrated manner in the National Airspace System (NAS).

Weather depiction chart. Details surface conditions as derived from METAR and other surface observations.

Weight. The force exerted by an aircraft from the pull of gravity.

Wide area augmentation system (WAAS). A differential global positioning system (DGPS) that improves the accuracy of the system by determining position error from the GPS satellites, then transmitting the error, or corrective factors, to the airborne GPS receiver.

Wind correction angle (WCA). The angle between the desired track and the heading of the aircraft necessary to keep the aircraft tracking over the desired track.

Wind direction indicators. Indicators that include a wind sock, wind tee, or tetrahedron. Visual reference will determine wind direction and runway in use.

Wind shear. A sudden, drastic shift in windspeed, direction, or both that may occur in the horizontal or vertical plane.

Winds and temperature aloft forecast (FB). A twice daily forecast that provides wind and temperature forecasts for specific locations in the contiguous United States.

Wing area. The total surface of the wing (in square feet), which includes control surfaces and may include wing area covered by the fuselage (main body of the airplane), and engine nacelles.

Wings. Airfoils attached to each side of the fuselage and are the main lifting surfaces that support the airplane in flight.

Wing root. The wing root is the part of the wing on a fixed-wing aircraft that is closest to the fuselage. Wing roots usually bear the highest bending forces in flight and during landing, and they often have fairings to reduce interference drag between the wing and the fuselage. The opposite end of a wing from the wing root is the wing tip.

Wing span. The maximum distance from wingtip to wingtip.

Wingtip vortices. The rapidly rotating air that spills over an airplane's wings during flight. The intensity of the turbulence depends on the airplane's weight, speed, and configuration. Also referred to as wake turbulence. Vortices from heavy aircraft may be extremely hazardous to small aircraft.

Wing twist. A design feature incorporated into some wings to improve aileron control effectiveness at high angles of attack during an approach to a stall.

Work. A measurement of force used to produce movement.

World Aeronautical Charts (WAC). A standard series of aeronautical charts covering land areas of the world at a size and scale convenient for navigation (1:1,000,000) by moderate speed aircraft. Topographic information includes cities and towns, principal roads, railroads, distinctive landmarks, drainage, and relief. Aeronautical information includes visual and radio aids to navigation, airports, airways, restricted areas, obstructions and other pertinent data.

Z

Zone of confusion. Volume of space above the station where a lack of adequate navigation signal directly above the VOR station causes the needle to deviate.

Zulu time. A term used in aviation for coordinated universal time (UTC) which places the entire world on one time standard.

Index

NOTES

NOTES

NOTES

NOTES